MACHINE DESIGN

MACHINE DESIGN
A CAD Approach

ANDREW D. DIMAROGONAS
W. Palm Professor of Mechanical Design
Washington University, St. Louis, Missouri, USA

A Wiley-Interscience Publication
JOHN WILEY & SONS, INC.
New York / Chichester / Weinheim / Brisbane / Singapore / Toronto

Library of Congress Cataloging-in-Publication Data:

Dimarogonas, Andrew D., 1938–
   Machine design : a CAD approach / by Andrew D. Dimarogonas.
      p.   cm.
   ISBN 0-471-31528-1 (cloth : alk. paper)
      1. Machine design—Computer-aided design.   2. CAD/CAM systems.   I. Title.

TJ233 .D53 2000
621.8′15′0285—dc21

                                                                                00-039932

Printed in the United States of America

10 9 8 7 6 5 4 3 2 1

To the memory of
George N. Sandor (1920–1996)
great engineer, researcher, educator, and gentleman

CONTENTS

PREFACE

30 years ago this author, having spent some years with a major machine maker where all design calculations were done already in the computer, interviewed for a mid-level faculty position in mechanical design and was asked by the senior professor of design, "What do you think of computer aided design?" The answer was, "Is there any other kind of design?" The interview went well.

Fifteen years later, the same author was interviewed by a dean for a senior faculty position. The dean said, "Our vision is to develop a model manufacturing facility where any machine or product will be designed in a computer and with a push of a button the manufacturing facility will make a fully functional prototype. Thus, we can make several machines and choose the best." This author could not resist a joke and remarked, "Excellent! Then we can feed the manufacturing system a pig and ask for several sausage outputs. If we do not like the taste, we feed the sausage backwards and get the pig back!" The humor was apparently trivial, at any rate not appreciated and this interview did not go well.

The moral of these two stories is that 30 years ago our system of university instruction was experimenting with hesitation with the idea of computer aided design, while 15 years later it had gone overboard with it. The aim of this book is to find a sensible balance between the intellectual activity that goes into a machine design effort and the computer assistance that a machine designer can have. *Computer aided design* here means computer "aided" design, not computer "made" design.

Computer aided design (CAD) emerged in the 1960s out of the general acceptance of fast digital computers as tools to aid the design of complex systems. At that time, the most progressive industries used computer methods to aid their machine design efforts, using modern optimization techniques with the aid of computer graphics and computerized structural analysis.

Since then, the capacity of computers has increased by orders of magnitude while their price has decreased by orders of magnitude, making them a generally available tool to the student and the designer. The matching of computer techniques with the most traditional mechanical engineering courses such as Machine Design (or Design of Machine Elements) came rather late. The purpose of this textbook is exactly the introduction of computer methodology in the design of machines and machine elements.

The need to rewrite such a classical subject from the new point of view created the opportunity for a fresh look at the presentation of the subject.

All over the world, machine design textbooks have presented the material with component orientation, such as bolts, shafts, bearings, and gears, following the structure of the classical Reuleaux text of 1854. However, with the vast number of different elements of machines now available, such presentations are in any case incomplete. This has led the author to try a unifying approach, dividing the material from the point of view of design methodology rather than element function, to the extent possible. Therefore, the material is presented as, for example, design for strength. The student has the opportunity to observe the common features of designing different elements with the same methods and with the same considerations. The aim is to guide the student through the design methodology rather than merely to teach how to design particular machine elements.

Emphasis is given to subjects more particular to computer application. It was natural also to emphasize design methods that are particularly suitable for computer implementation. However, since learning in depth is achieved through full student involvement in the design process, the book does not rely exclusively on computers but first presents the methodology for longhand calculated solutions of machine design problems in the tradition of Reuleaux and the newer texts by Niemann and Shigley.

The text is supplemented by appropriate software and spreadsheet solutions and templates. The spreadsheets listed are not intended to substitute for design calculations nor to be a complete design system, although they cover a wide range of machine design applications. They have a dual purpose:

1. To serve as a guide and a base for the development of design applications by the student. Most of them can be modified and extended for particular applications.
2. To form the basis for design automation, since most solid modelers, such as SOLID EDGE, UNIGRAPHICS, and PRO ENGINEER, accept spreadsheet inputs for basic dimensioning. This can lead towards complete design automation. The latter is left to the individual instructor because it depends on the software available.

Most of the chapters of this book are self-contained. Therefore, the text can be used flexibly manner to meet specific requirements.

Every chapter concludes with a selected list of references, not as a general bibliography on the subject but for referencing purposes and as a guide for further reading.

This text was intended to be timely and teachable and to lead the student to become acquainted with modern machine design and computer methods in machine design and, furthermore, to develop skills for applying contemporary CAMD methodology in practice.

The first seven chapters deal with the machine as a system. In particular:

Chapter 1 deals with the machine design methodology. The several stages of machine design are presented together with the general machine design principles. Computer methods are not yet introduced at this point, to emphasize the fact that no matter how much work is delegated to the computer, the designer has to make the decisions based on judgment, experience, and ingenuity.

Chapter 2 is a review of basic machine kinematic design that can be used as a refresher for those readers who have had a course on or exposure to mechanisms but also an introduction for those who do not have such knowledge. With many universities having eliminated a required course in this area, this task becomes essential for a machine design textbook. Moreover, it introduces the idea of kinematic abstraction in the Ampère sense, treating the machine as a kinematically compatible system.

Chapter 3 is a review of basic machine statics that can be used as a refresher for those readers who have had a course in statics and strength of materials and want to connect that information to methods and applications related to machine design.

Chapter 4 is an introduction to design for manufacturing, a subject that cannot be overemphasized in a competitive world.

Chapter 5 presents basic methodologies for stress analysis and sizing of machine members on the basis of their strength and reliability. Probablistic design, reliability, and the statistical character of the material properties and the safety factor are also introduced. Material properties for machine design applications are discussed.

Chapter 6 presents some concepts of computer methods in machine design and graphics that useful not only useful for actual construction of machine drawings but to an even greater degree for an adequate presentation of computer solutions in the chapters to follow. Stress analysis methods in machine design are presented with extension to computer methods, in particular transfer matrix and finite element methods. The introduction of transfer matrices first leads by a natural extension to finite elements. The basic principles of the latter are presented with certain applications that are very common in machine design. This presentation is intended to familiarize the reader with the subject to the extent that he or she will be able to develop simple finite element programs to gain some insight into the commonly used commercial codes, if a course in finite elements is not in the curriculum. It also provides the student and the practicing engineer (who was educated with the strength of materials methods) with a practical introduction, without mystique to computer methods in structural analysis, a topic covered many times with unnecessary complication. After all, designers were using finite element methods many decades before researchers formally discovered them.

In Chapter 7, the basic mathematical approach is presented for machine modeling and design optimization together with its computer implementation. Emphasis is given to numerical methods which are suitable for machine design applications. A spreadsheet is used later on with applications. The Solver module of EXCEL used here has several optimization methods incorporated, so that the reader can study their particular features.

The next seven chapters are devoted to machine element design methodology:

In Chapter 8, design methods for strength are applied for the design of fasteners and welded joints. Analysis of several types of joints is unified and optimization methods are used.

Chapter 9 presents design methods for flexible elements, such as springs and machine mountings.

Friction and wear considerations are the main features in many applications, such as clutches, brakes, and belts. The relevant desgin methods are presented in Chapter 10.

Elements associated with fluid flow, such as fluid bearings, together with basic hydrodynamic and hydrostatic lubrication theory, are presented in Chapter 11.

Surface strength, associated mainly with high contact stresses and boundary lubrication, is a common design problem in many machine elements. The associated design methods are presented in Chapter 12 with emphasis on antifriction bearings.

Chapter 13 deals with gearing transmission systems, where almost all the previously presented design considerations appear together with problems of efficiency and heat flow. Such systems consist in general of several types of gear elements. The student can learn a great deal about the involute gears and their limitations by experimenting with software accompanying the text and building up gear design spreadsheets.

Certain machine members, such as rotating shafts and couplings, have inherent dynamic loads. The associated design approach is presented in Chapter 14.

Every section is concluded with formalized design procedures and worked out examples and problems. Examples are worked out with longhand calculations, in general, and some of them are further solved with computer methods, most in a spreadsheet.

Many case studies and problems are from the author's industrial experience. They have two aims:

1. To introduce the student to real engineering problems that are simple enough to be understood and worked out.
2. To help the student achieve computer literacy in an applied subject such as machine design.

The problems are designed to form at least five complete sets, so that the course can be repeated five times without repetition of the problems. Furthermore, there is a sufficient number of problems that the instructor can emphasize analysis, design, or CAD, depending on the needs of the course.

Today it is very difficult for a designer to follow the fast pace of development of computer languages. For this reason, for those instructors and readers who want to use computer methods, we have included spreadsheet solutions and executable code so that no programming skills are necessary.

The book can be used in many ways, including:

- As an introductory machine design book (Chapters 1–5, 8–13)
- As a computer aided design book (Chapters 1–7)
- As a companion for a capstone design course (study Chapters 1, 4, 6, 7, reference Chapters 2, 3, 5, 8–14)

A number of projects are included in Chapter 1 for such a course and for project assignments for the machine design course. Suggested syllabi are posted at the book's homepage.

ANDREW DIMAROGONAS, *St. Louis*

2000

Note from the Publisher

Wiley wishes to express its sincere regrets that the author, Professor Andrew Dimarogonas, passed away during the production of this book. The efforts of his son, James Dimarogonas, and his colleague, Professor Sophia Panteliou, to ensure the book's timely publication are appreciated.

NOMENCLATURE

A = area
a = acceleration
b = width
C = antifriction bearing dynamic capacity, fatigue derating factor
C_0 = antifriction bearing static capacity
c = bearing radial clearance
c_p = specific heat
D = diameter, outer diameter
d = diameter, inner diameter
E = elasticity modulus, energy
e = eccentricity
F = force
f = friction coefficient
g = acceleration of gravity, equality constraint
H = height
h = height, inequality constraint
H_B = Brinell hardness number
I = area moment of inertia
J = mass moment of inertia
K = stress concentration factor, fatigue derating factor, penalty function constant
k = leg of the weld length, rigidity, spring constant
L = length
M = moment
m = mass, distributed moment per unit length of a beam
N = safety factor
N = speed of rotation, rps
N_a = active turns

N_f = normal force
n = speed of rotation, rpm
p_m = bearing mean pressure = W/LD
P force = power
p = pressure
q = fluid flow rate, mass of belt per unit length, beam load per unit length
Q = fluid flow rate, heat flow rate
R gear ratio = ω_1/ω_2, radius
r = radius, length of a link, radius of gyration, radial coordinate
rpm = revolutions per minute
rps = revolutions per second
S = material strength, Sommerfeld number, standard deviation (estimate)
T = temperature, torque
t = time, thickness, normalized statistical distribution parameter
U = peripheral force
u = peripheral velocity
V = velocity, volume, shear force
v = velocity
x = Cartesian coordinate
y = Cartesian coordinate
W = load
w = load per unit width, Cartesian coordinate
Δp = difference in the pressure causing the flow
α = angular acceleration, thermal expansion coefficient, angle
β, γ = angles
δ = elongation, displacement
θ = slope, angle
ε = strain, bearing eccentricity ratio
λ = Lagrange multiplier
μ = absolute viscosity, column end factor
v = kinematic viscosity, Poisson ratio, weld derating factor
ρ = density
σ = normal stress, standard deviation
τ = shear stress
ϕ angle = input angle of a linkage
ψ angle = output angle of a linkage
ω = angular velocity

Subscripts

1 = cylindrical elements: smaller of two diameters. Linkages: base link
2 = cylindrical elements: larger of two diameters. Linkages: crank
3 = Linkages: coupler
4 = Linkages: follower
a = axial

b = bearing
cr = critical
d = direct
el = elastic
f = frictional, fatigue
ind = indirect
m = mean
n = normal
pl = plastic
r = radial, range
s = shear
T = torsional
t = tangential, theoretical
u = ultimate
y = yield

Prefixes

μ = micro ($\times 10^{-6}$)
m = milli ($\times 10^{-3}$)
k = kilo ($\times 10^{3}$)
M = mega ($\times 10^{6}$)
Δ = small increment

THE MACHINE:
A HISTORICAL INTRODUCTION

1. WHAT IS A MACHINE?

The term *engineering* has been used, especially in literature on the history of engineering, as synonymous with technology and, in many instances, with craft (Kirby et al. 1956, De Camp 1960, Landels 1978, Hill 1984). What is engineering? The famous jazz musician Louis Armstrong was once asked to define Jazz. He answered, "If you got to ask, you'll never know." Schopenhauer (1911) called engineering the "battle between weight and rigidity." Aristotle gave engineering a sense of wonder (Aristotle 1936):

> Nature works against man's needs, because it always takes its own course. Thus, when it is necessary to do something that goes beyond Nature, the difficulties can be overcome with the assistance of Engineering. Mechanics is the name of the engineering discipline that helps us over those difficulties; as the poet Antiphon put it, "engineering brings the victory that Nature impedes."

Technology encompasses craft, invention, and engineering. There is no clear dividing line—each of the three includes some part of the other two, as shown in Figure I.1—but there are many attributes that *causa sine qua non* distinguish engineering from craft and invention.

Some of the attributes of engineering are:

1. *Size.* Engineers deal with systems of substantial size, such that there is no substantial number of systems of the same size that would allow the craftsman to make them by mere repetition.

2. *Complexity.* Engineers deal with systems of substantial complexity, such that it is difficult for the craftsman to make them merely by building them up from smaller parts.

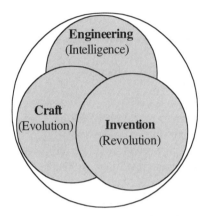

Figure I.1. Technology: the interactions of engineering, craft, and invention.

3. *Innovation.* If substantial innovation is required to design a new system, an engineer is needed who will put together knowledge from different components and other systems, together with knowledge from engineering science, and synthesize a substantially novel system.

4. *Change.* An existing system needs an engineer to redesign it if any of the design parameters, such as the material available, change substantially.

5. *Rational use of natural resources.* An engineer is supposed to accomplish a given task with minimum utilization of resources.

Craft is characterized by repetition and extrapolation. The apprentice learns the trade by long and repeated observation of the master craftsman at work. He practices again and again during his apprenticeship whatever his master craftsman does. When he becomes a craftsman, he has little potential for deviation from what he has learned. He can extrapolate to a certain extent; that is, use the same techniques to build a larger system. By such evolution, craftsmen have built systems of substantial size, such as the Köln Dome, which is higher than 150 m. However, a craftsman who builds small churches, say up to 30 m high, cannot be entrusted with building a 150 m dome.

Invention is attributed to the impulse of the moment. It has been observed that specialization reduces creativity, and thus one should not expect the engineer or the craftsman to be inventors. Crossing the lines, however, is not unusual.

Reuleaux (1876) suggested that the earliest machine was the twirling stick used for starting fire and discussed further other early machinery, such as water mills. The lever and the wedge also stem from our heritage of technology from the Paleolithic era. By the beginning of the historical age, substantial technological developments had been arrived at by craftsmen on the basis of repetition and improvement (evolution) or by inventors (revolution). Engineering is distinguished from these processes, as discussed above, in that it arrives at a technological solution to a stated problem in a systematic manner (intelligence) (see Figure I.1). Overlap exists, and the distinguishing line cannot always be defined precisely. Numerous examples illustrating these points will be discussed in the following.

A particular field of engineering is machine design, which is the subject of this book. The *Encyclopedia Britannica* defines *machine* as an object, having a unique purpose, that augments or replaces human or animal effort for the accomplishment of physical tasks. This broad definition covers such simple devices as the lever, wedge, wheel and axle, pulley, and screw (the five so-called simple machines) as well as such complex mechanical systems as the modern aircraft.

Design of an object is the formulation of a plan for its structure, operation, and appearance and then the planning of these to fit efficient production, distribution, selling, servicing, and disposal procedures. Design refers to a multitude of objects, such building interiors, clothing, machines, industrial products, organizations, systems, and procedures.

In turn, *machine design* is the formulation of a plan for the structure, manufacturing, operation, and appearance of a machine and the planning of these attributes to fit efficient production, distribution, selling, servicing, and disposal procedures.

2. THE EMERGENCE OF CRAFTS

The earliest ancestors of the human race inhabited northeast Africa some 4 million years ago and probably used unmodified natural objects, *naturefacts,* for food collection and defense purposes. *Australopithecus afarensis* was probably the first hominid that modified natural objects, beginning a process that led gradually to tools and weapons of increasing amounts of modification and complexity, *artifacts.* About 2 million years ago stone artifacts were used by *Homo habilis* (domesticated man), and general use of man-made tools and weapons occurred with the emergence of *Homo sapiens* (thinking man—that is, modern humans), about 50 thousand years ago. By that time, speech and language had been developed. Verbal communication and language are essential ingredients of cultural development because only with them can any improvement achieved by someone be the basis on which others can build. From this period on, there is increasing evidence in surviving drawings of humans thinking beyond items of everyday utility to terrestrial objects and the heavens. From such drawings we obtain information about emerging technologies, such as hunting techniques.

The development of man-made tools undoubtedly followed the observation that certain shapes of pebbles were better for cutting meat or killing animals. Broken pieces of rock could have the attributes of such pebbles. Friction between two stones could be used to shape the stone to the form observed to have the desired qualities. Such observations accumulated for thousands of years and through verbal transmission they led to the development of *experience.*

During the Paleolithic (Old Stone) Age, 1 million to 10,000 BC, humans learned first to maintain the natural fire, then to produce fire technically. No doubt, observation of the heat produced by rubbing stone and wood to shape them led Paleolithic humans to the artificial creation of fire. The sparks associated with friction of certain type of stones, flint or pyrites, gave humans an invaluable experience. Fire could be created at will. Percussion was found to be more effective in producing sparks.

Fire helped create a more habitable environment but led to the observation that food exposed to fire could be preserved longer. Preservation of dried food and possibility of sustaining life were two of the very early victories of man against nature.

Preserved food required storage and this led to gradual accumulation of experience in container-making. Use of natural cavities was followed by carving in stone or earth, then by shaping clay. The last, in conjunction with the cooking of food, led to the observation that clay baked in fire would become hard, durable, and nondissolvable in water, and thus *pottery* was born. The first known pottery artifacts, found in the Jericho settlement (6000 BC), have the shape of gourds, the natural form having been adopted in the attempt to replace a natural function (Forbes 1967).

With simple one-piece tools, humans could reach as far as their arms would go. A spear could extend an arm's range, but its sharp end, made of wood, was soft. Attachment of a stone edge was the work of an inventor who combined different discoveries (the spear and the stone knife) and observation into a new tool.

3. THE FIRST DESIGNERS

One can find technological elements in the most ancient and primitive societies, such as in the finding and storing drinkable water, in irrigation, and in tool-making. Between 6000 and 4000 BC, cities were built and agriculture, pottery, textiles, and metal production and processing were introduced. Thus, in the great empires of Mesopotamia and Egypt and the great societies of the East, India and China of the Potamic (great river) age, parallel development of crafts and technology occurred without the concurrent development of sciences, which is the *sine qua non* of engineering. Technical advances were arrived at by long evolution or invention and not by a conscious search for the solution of a societal problem. Moreover, the political and social systems did not allow for the liberal thinking that is essential to the development of scientific thought, and knowledge was confined to the clergy or to the ruling cast. This did not prevent the construction of monumental works in these societies, such as the Great Pyramids and other monuments of Egypt. These were gigantic in size and complexity and must have demanded a substantial planning effort. It seems, however, that technology was developing by trial and error rather than by rational engineering in the above sense. King Zoser Neter-khet's pyramid at Saqqara (ca. 2800 BC), for example, the first one to be built of stone, was originally built as a tomb (see Figure I.2a). Imhotep, the first master builder whose name is known, later increased the dimensions of the structure in several stages (Figure I.2b to I.2e) (Strandh 1989). Strandh calls Imhotep "the first engineer whose name is known," but there is no evidence that Imhotep fits the description of an engineer as proposed above.

In the Bible, on the other hand, the literary and historical value of which has been greatly underestimated, King Solomon (I Kings 6:7) built a magnificent house for his Lord, ". . . of stone made ready before it was brought thither: so that there was neither hammer nor axe nor any tool of iron heard in the house, while it was in building." Such an undertaking would have required detailed engineering planning and drawings. Even if only the symbolic character of the biblical scripts is accepted, one has to accept that in King Solomon's time detailed engineering planning was taking place.

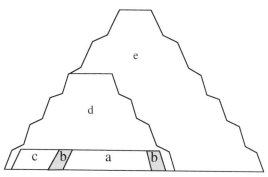

Figure I.2. Construction of King Zoser Neter-khet's pyramid at Saqqara by Imhotep (ca. 2800 BC)

At a later time, Uzziah, King of Judah (reigned 779–740 BC) constructed engines of war ("invented by cunning men") and other military works (2 Chronicles 26: 15). The careful planning in the biblical description and the fact that Uzziah built original engines might qualify him to be the first machine designer on record and the predecessor of Archimedes.

The scientific method of dealing with nature started with the Ionian school of natural philosophy, whose distinguished leader was Thales of Miletus (640–546 BC), the first of the Seven Wise Men of antiquity. He is perhaps best known for his legendary discovery of the electrical properties of yellow amber (*electron*) and for introducing the term *electricity,* the phenomenon of static charge that results when electron is rubbed against a wool cloth. More important, Thales introduced the concept of the logical proof for abstract propositions. Thales, who was also a very successful entrepreneur, traveled extensively in Mesopotamia and Egypt and became familiar with the knowledge of geometry (land surveying) and astronomy in those regions.

The search for reason led to the development of a generalized science as distinct from a set of unrelated empirical rules. The Pythagoreans, for example, sought the principles of geometry in ultimate ideas and investigated its theorems in an abstract and rigorous way (Proclus Diadochus, 410–485 AD). The rigorous proof, based in deductive logic and mathematical symbolism, was introduced. Experimentation was established as a method to support and lead scientific reasoning. Notable examples are the experimental investigations of sound and vibration by Pythagoras in the 6th century BC (Theon of Smyrna ca. 150 AD) and the experimental proof of the corporeality of air by Strato of Lampsacus (d. ca. 270 BC), known to Anaxagoras since the 5th Century BC.

Developments in natural philosophy and the scientific method in the 6th and 5th Centuries BC led to rapid development of engineering design in the 4th to 1st Centuries BC in the Greek and Hellenistic world, reaching maturity in the Roman Empire between the 1st Century BC and the 2nd Century AD.

4. THE ORIGINS OF SYSTEMATIC DESIGN

The first machine designers were the master builders of the Potamic Civilizations (Mesopotamia, India, China, Egypt). Those designers rose to the level of engineering in the Thalassic (great seas) societies of ancient Greece and Rome. They did not know much about marketing, but they did know the users of their products. They had continuous feedback concerning their acceptance and performance. During the manufacturing phase they kept in touch with the prospective users, and they were always in line with the contemporary ideas about acceptable appearance and utility. Moreover, conception, design, and manufacture were the work of a single person. Consequently, these products were simple and of human proportions. Much later, mass production caused the breaking of this process into distinct smaller ones and led to the separation of design from manufacturing. However, the principles underlying design activity were investigated very early in history. The first design theory was part of aesthetics, where *aesthetic* (beautiful) included also *functional* (useful) and *ethical* (the good) attributes. Function and ethics were inseparable from form. This society simply could not afford spending resources only for aesthetic pleasure. It was able, however, to afford a pleasing appearance for the useful goods of everyday life and to pay attention to more general societal needs.

Rapid advancement in natural sciences was followed by systematic attempts to organize knowledge in engineering and, in particular, in *machine design,* developing this body of knowledge beyond the level of a mere craft.

Kinematics and machine design have a distinct place in the history of engineering because they comprised the very first of its divisions to receive a mathematical foundation. Heraclitus of Ephesus (ca. 550–475 BC) appears to have been the first to separate the study of motion itself from dynamics, the forces causing the motion, and introduced the principle of *retribution,* or *change,* in the motion of celestial bodies.

The first known written record of the word "machine" appears in Homer and Herodotus to describe political manipulation (Dimarogonas 1999). The word was not used with its modern meaning until Aeschylus used it to describe the theatrical device used to bring the gods or the heroes of the drama on stage; whence the Latin term *deus ex machina. Mechanema* (mechanism), in turn, as used by Aristophanes, means "an assemblage of machines." None of these theatrical machines, made of perishable materials, is extant. However, there are numerous references to such machines in extant Greek plays and also in vase paintings, from which they can be reconstructed. They were large mechanisms consisting of booms, wheels, and ropes that could raise weights perhaps as great as one ton and, in some cases, move them back and forth violently to depict traveling through space, when the play demanded it.

The designers and builders of these mechanisms were called by Aristophanes *mechanopoioi* (machine-makers), meaning "machine designers" in modern terminology and identical with the corresponding German term for mechanical engineers, *Maschinenbauingenieure.* They designed and built the *mechane* not by evolution or invention but to the order and specifications of the playwright, Aeschylus in this case, and they truly deserve to be called the first mechanical designers.

Very few details are known about the *mechane,* but it is certain that they were substantial mechanisms for path and motion generation. They were probably several meters high and operated above the stage and roof of the theater. In a fully retracted position a *mechane* was hidden behind the stage, so it would have to have more than one moving link and it could carry substantial load. In Aeschylus's *Prometheus*

Bound, at the entrances of the Oceanids and Oceanus, a fifteen member chorus plus chariot and horses are lifted in the air, a total estimated weight of one ton. The simultaneous entrance of Oceanus must have necessitated the use of a second *mechane* or else a very complicated single one. *Mechane* could also be used to create very violent motions. In the *Peace* of Aristophanes, Trygaeus straddles a huge dung beetle and flies upward to the abode of Zeus in a swinging, violent motion to convey a sense of flying through space while he chants eighteen lines of anapestic dimeters; then he is lowered back to the terrace from which he started. As he flies upwards, he asks to the stage engineer to be careful so that he will not become food for the beetle.

The *mechane* must have been well balanced. Antiphanes in *Athenaeus* complains that when tragic poets have nothing left to say, they "raise the *Mechane* like a finger." Such undertakings would require mechanisms enormous in both size and complexity. The technical challenge is similar to that in certain scenes in opera, such as the flying maidens in *Valkyrie,* the swans in *Lohengrin,* and the dragon in *Siegfried.*

Aristotle (384–322 BC), Greek philosopher and scientist, was born at Stageira, Macedonia, the son of a physician to the royal court. A student at Plato's Academy, he became the tutor of the king's young son Alexander, later known as Alexander the Great. In 335, when Alexander became king, Aristotle returned to Athens and established his own school, the Lyceum. Upon the death of Alexander in 323, strong anti-Macedonian feeling developed in Athens, and Aristotle retired to a family estate in Euboea. He died there the following year. Aristotle's lecture notes for carefully outlined courses treating almost every branch of knowledge and art were collected and arranged by later editors. He wrote a treatise on logic, called *Organon* (instrument) because it provides the means by which positive knowledge is to be attained. His works on natural science include *Physics,* which gives a vast amount of information on astronomy, meteorology, plants, and animals. He wrote *Mechanical Problems,* the first treatise on the design of machines.

Aristotle, marble bust with restored nose, Roman copy of a Greek original, last quarter of the 4th Century BC. In the Kunsthistorisches Museum, Vienna.

The principles of statics and dynamics were discussed by Aristotle in *Problems of Machines* (*Mechanica*), the first extant treatise of this kind, probably written by one of Aristotle's students in Lyceum. *Mechanica* starts with the definition of *machine,* which in that era was synonymous with *mechanism.* In fact, mechanisms were the only machines known. To quote Aristotle, "Machine is a means of altering nature's course." The author further discusses several purely kinematic aspects of mechanisms, such as:

• The vectorial character of velocity, the superposition of velocities, and the parallelogram law for velocity addition

- The concepts of absolute and relative velocity of points along a link of a machine
- Circular motion: the velocity of every point on a wheel rotating about its geometric center is directed along a tangent to its circular path and is proportional to the radius of the circle and the angular velocity
- The motion of two wheels rolling against one another without slipping; they rotate in opposite directions
- The rhomboid four-bar linkage and the relative velocities of the opposing joints

For all of these aspects, Aristotle developed rational geometric methods and proofs. Moreover, *Mechanica* contains remarkable discussions of the mechanics of the lever, the balance, the wedge, rolling friction, the strength of beams, impact, mechanical advantage, and the difference between static and kinematic friction. The second successor of Aristotle in the Lyceum was Strato of Lampsacus (d. ca. 270 BC) who introduced an important kinematic criterion of equilibrium, the *principle of virtual velocities.*

The decline of Greece proper coincided with the rise of Alexandria, founded in honor of Alexander the Great (356–323 BC) in the Nile Delta. This city became the most important scientific center in the world. In its University, the *Museum* (meaning the house of the Muses, who were the protectresses of the arts and sciences), flourished a number of great mathematicians and engineers. One of the most important engineering professors of the Museum was Ctesibius (ca. 283–247 BC), the designer of the precision water clock. He left many writings, which were subsequently lost, and only references to them by his students, notably Philo and Hero, are extant.

Purely mechanical treatises on machinery go back to the 4th century BC. One of Plato's contemporaries and friends and a student of Pythagoras, Archytas of Tarentum (ca. 400–365 BC), is said to have written the first systematic treatise on machines based on mathematical principles. This, too, is lost. It is known, however, that Archytas built an air-propelled flying wooden dove (Aulus Gellius, ca. 150 AD). Details about Archytas's dove are not known but it seems to be the first flying machine.

Some fragments of and references to an extensive treatise, *Machine Syntaxis* (synonymous with *synthesis*), by Philo of Byzantium (ca. 250 BC). He was a student of Ctesibius at the Museum, and his treatise dealt among other things with the idea of *machine elements,* a small number of simple elements that constitute every machine. Different machines are constituted from different syntheses of these basic machine elements. Most of this treatise is also lost, and only parts of it as well as references to it are extant in other works. However, a comprehensive mechanical treatise by Hero of Alexandria (1st Century AD)[1] is extant in an Arab translation. Its title, *Mechanica* (the same as that of Aristotle's treatise) has been mistranslated and misunderstood. It is generally known as *Mechanics* (*Mechanike*), but the correct meaning of the title is *Problems of Machines.* The Greek word *Mechane* means

---

[1] Hero was the most prolific writer of his time, but we know very little about his life. From his writings, he seems to predate Vitruvius, but an observation of a solar eclipse of the year 62 AD that he describes in *Dioptra* places him in the 1st Century AD.

"Machine." It should be noted that *Mechanica* is a not a guide for the design of mechanical devices, because Hero wrote such texts for particular categories of devices, such as pneumatic machines, automata, optical instruments, balances, and artillery machines. *Problems of Machines* is really a study of design methodology. Hero separated the study of particular machines and the general concepts of machines from the study of standardized elements that, in different combinations, constitute any machine. He introduced five simple mechanical elements for the solution of the general problem of moving a weight with a given force: wheel and axle, lever, windlass, wedge, and screw. In fact, he asserted that all five solutions are physical devices embodying the lever principle, a simple function module in terms of contemporary literature. Hero's *Problems of Machines* gave the first systematic development of design solutions to a given mechanism problem. It is to be noted that a substantial part of Hero's work is probably based on earlier (lost) works of Philo and Ctesibius.

Hero was aware of the difference between the craftsman and the engineer-designer of mechanical devices. Pappus of Alexandria (4th Century AD) expounds on this:

> The mechanicians (more precisely: the mechanical engineers) of Hero's school told us that the study of machines consists of theoretical and practical part. The theoretical part includes the natural sciences while the practical part consists of the engineering disciplines. They postulate that a necessary condition for an able designer[2] of mechanical devices is a solid background in both natural sciences and practical skills.

In modern terminology, this could be a reasonable account of what should be required of a mechanical engineer today. "*The Hero School,*" Pappus continues, "recognized the need for specialization, for it is not possible for someone to have competence in all the related fields. Further, they classified the mechanical engineering specialties in five categories: handling of weights, military engineering, hydraulics, marvelous device making, sphere making."

Hero's works *Automata Making, Pneumatics,* and *On Water Clocks* (the last is not extant) contain descriptions of a number of devices that most modern commentators consider an exercise in toy-making. A careful observer can see instead that *Automata Making* and *Pneumatics* are design studies wherein Hero develops alternative solutions to some basic design problems. Hero was explicit about his intention to develop original designs: "Furthermore, one must avoid the predecessor's designs, so that the device will appear as something new."

Archimedes (ca. 287–212 BC), who was one of the Museum students but lived in Syracuse (in Sicily), mastered all the disciplines of mathematics and engineering and wrote separate treatises in those areas. Archimedes was both a great engineer and a great inventor, but his books concentrated on applied mathematics and mechanics and rigorous mathematical proofs. There is a clear distinction between his works and the later works of Hero. Even their methodologies were different: where

---

[2] Most scholars translate this word erroneously as "inventor." Hero uses the word $\varepsilon \upsilon \rho \varepsilon \tau \acute{\eta} \varsigma$ = discoverer, not by chance but by intelligence (Linddell and Scott 1843), and not $\varepsilon \phi \varepsilon \upsilon \rho \acute{\varepsilon} \tau \eta \varsigma$ = inventor.

mathematical rigor was used by Archimedes, Hero applied numerical methods. Hero reported on these in his *Metrica,* a treatise on numerical methods for area computations and the approximation of roots.

Archimedes (ca. 287–212 BC) was born in Syracuse, Sicily, a relative of its ruler Hieron. He studied mathematics at the Museum in Alexandria, and returned to Syracuse to devote his life to the study of mathematics and the design of machines. He introduced the windlass, gears, and the screw and wrote treatises on mathematics and the design of several types of machinery. He is the founder of statics and of hydrostatics, and his machine designs fascinated subsequent writers, who attributed to Archimedes, for example, a machine by which he could lift the ships of the besieging Roman fleet and destroy them by dropping them from a high level, which probably is an exaggeration. He did design and build, however, *Syracusia* ("The Lady of Syracuse"), the largest ship of his times, 80 m long, 4,000 ton displacement, with three decks. The ship made only its maiden trip to Alexandria because it was too slow and there were no harbor facilities anywhere to handle her.

Archimedes was killed during the capture of the city by the Roman general Marcellus while trying to solve a mathematical problem. To the soldier who ordered him to stand up, he answered, "Do not disturb my circles." The soldier, not knowing who he was (and against the orders of Marcellus), killed him.

Archimedes systematized the design of simple machines and the study of their functions. He was probably the inventor of the compound pulley and developed a rigorous theory of levers and the kinematics of the screw. Plutarch and Polybius describe giant mechanisms for lifting ships from the sea, ship-burning mirrors, and a steam gun designed and built by Archimedes. The last allegedly fascinated Leonardo da Vinci. However, the validity of these stories is questionable.

Philo of Byzantium wrote a treatise on artillery (*Technology of Arrow Making*) in which he described an analytical method for the design of *ballistae* (heavy stone throwers). Such machines were used by the peoples in the East. In the 2 Chronicles 26:15 for example, as mentioned above, Uzziah, the King of Judah, talks about such machines. Philo stands out because he devised a design equation obtained empirically on the basis of certain assumptions. In the case of the ballistae, Philo's equation was $d = (11/10)\sqrt[3]{100m}$, where d is the diameter of the twisted gut skeins providing the potential energy, in dactyls, and m is the mass of the stone, in hundreds of drachmas. In conjunction with this formula, Philo introduced the idea of *similitude,* stating that larger machines can be designed on the basis of the dimensions of smaller ones with the formula $d_1/d_2 = (m_1/m_2)^{1/3}$. Moreover, Philo's formula for the balistae was the first developed design equation in the contemporary sense that directly related the design objective (the payload) to the design parameter (the torsional spring diameter). In Philo's formula, the cube root relationship was derived empirically while the numerical constant 100 was derived from the units used and the constant 11/10 was used as a safety margin to account for the error involved in the approximation for the cube root, thus introducing the safety factor. Cotterell and Kamminga (1990) prove Philo's formula in the form $d = c\sqrt[3]{m}$, where c is a constant. Marcus Vitruvius Pollio (*On Architecture*) proposed a number of pairs (m, d), apparently obtained experimentally, that fit closely the equation $d = 3.3\sqrt[3]{m} + (1/60)(60 - m)$, in Roman units (Cohen and Drabkin 1958).

This method of systematic design was further exemplified by Hero in his *Arrow Making* (artillery), where he also introduced the idea of the sensitivity of the design to variations in the design parameters. It is not strange that Hero, driven by abstract reasoning and not by practical needs, designed devices using very advanced principles, such as those that can be conceived as very early models for the reaction machine (turbine) and the piston engine. Moreover, his design of machines was based on mathematical analysis and engineering science and is thus distinguished from mere empiricism.

In his *Problems of Machines,* Hero made extensive use of analytical methods for the study of machines and mechanisms. His static and kinematic analysis was

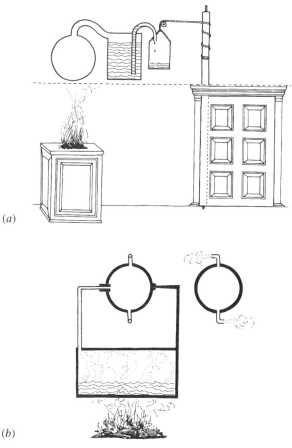

(a)

(b)

Figure I.3. Hero's thermal engines: (a) a piston engine to perform the miracle of opening the temple door; (b) a reaction turbine.

based on the conservation of energy ($F_1 s_1 = F_2 s_2$), although he did not recognize the product Fs as being energy (or work). On this basis, he studied the flow of motion and force in a machine. He also developed a method of number synthesis for finding the number of pulleys needed to raise a weight, given the weight and the pulling force, and a methodology for gear sizing on the basis of the desired mechanical advantage and the design of the screw threads.

The end of the Alexandrian era marked the eclipse of the ancient Greek science, and the systematic study of the design of machines became stagnant for a long period of time. During that time, the enormous emphasis of the Romans on practicability led to the separation of functional considerations from those of form and aesthetics.

5. ROMAN MACHINE DESIGN

As the slavery-based society of Rome reached maturity, productivity fell and once again utility prevailed over form in design considerations. The Romans were not great philosophers or artists, but they were great engineers and designers. They recorded and adopted the knowledge produced in the earlier dynamic societies of Greece, Egypt, and Mesopotamia to accomplish great feats of civil and mechanical engineering works needed by the Empire. Aesthetic and ethical dimensions were not important. Romans were mostly unaware of Greek mathematics until the 2nd Century AD, when Greek mathematical works started being translated into Latin. A substantial number of treatises in architectural and mechanical design exists, mainly encyclopedic in nature; the one by Vitruvius is the most notable. The Romans further gave the world sophisticated legal and administrative systems and separated the professions of civil and mechanical engineering. Moreover, they were careful recorders of the design rules of thumb. Analytical reasoning gave way to pure empiricism and improvement by repetition. Progress, sometimes very substantial (such as the invention of roller bearings, which according to some findings might have been used by Romans and Chinese in the 2nd and 1st Centuries BC (Hartenberg and Denavit 1964)), seems to have been the result of chance discoveries followed by trial and error experiments rather than the sustained use of intellect combined with analysis and experiments designed to explain the physical phenomena involved. In this respect, the design of fairly advanced mechanisms such as cranes and linkages must be classified as purely empirical.

Commentators on the classics flourished in Rome. They not only preserved most of the classical culture but made substantial advances of their own. Most notable was the architect Marcus Vitruvius Pollio, whose ten books *De Archetecture* (*On Architecture*) contained important material on the history of technology and on the design of machinery. Vitruvius defined a machine as "a combination of timbers fastened together, chiefly efficacious in moving great weights," and his book is imbued (Benvenuto 1991) with the words *firmitas, utilitas, venustas* (strength, utility, beauty).

At a very early stage, designers found that standardization is very important in design for both interchangeability and reducing manufacturing cost. Consider, for example, the standardization of the plumbing used to supply water in the city of

Figure I.4. Roman relief showing a Roman construction crane. It is nearly 60 ft tall.

Rome. The lead pipes had standard sizes that followed very closely the Renard Series (introduced in Napoleonic times) $\sqrt[m]{10}$, $m = 1, 2, 3, \ldots$, where m is the series index. In this series, for example, $R10$ means $m = 10$. Such series are used to standardize a small number of dimensions within a range of desirable dimensions in order to optimize the inventory of manufacturing tools and fixtures. The lead pipe diameters set by the Roman standards had sizes 5, 8, 10, 15, 20, 30, 50, 60, 100 digits (1 digit = 4.3 mm). This series is remarkably close to the $R7$ series (Figure I.5).

Evidence exists that there was concern for environmental issues in very early times. Hippocrates (5th Century BC) wrote a treatise, *Airs, Waters and Places,* in which he discussed the problem of water pollution (Landels 1978). In the design of the Roman aqueducts, elaborate measures were taken to guard against water pollution. Vitruvius (*De Architectura*) advised against the use of lead pipes, to protect the public from lead poisoning, which was apparent in the workers in lead-processing shops. He recognized that lead oxide (*cerussa*) was the toxic substance

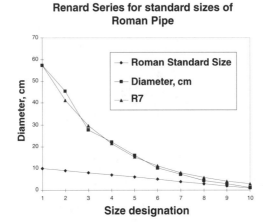

Figure I.5. Roman lead pipe standard sizes.

responsible for lead poisoning. It is also known that noise pollution was a problem in ancient Rome and vehicles were not allowed to operate in the center of the city during certain hours.

6. MACHINE DESIGN IN THE ARAB WORLD

The Arabs played an important role in the preservation of the Greek science and engineering and made substantial contributions of their own. They devised ingenious mechanisms with a high degree of automation and control. One of the Arabic writers of this era was Ibn al-Razzaz Jazari, who in his extensive and beautifully illustrated treatise on machines, *Book of Knowledge of Ingenious Mechanical Devices* (1206), described a great number of ingenious mechanisms and automata. One aspect of Jazari's book which has gone largely unnoticed is the systematic development of design concepts for particular design tasks. He divided his book into different *Categories,* each one describing a great variety of solutions to a specific problem. Category I describes six different clocks based on the principle of water flow and four of them employing the burning of a candle. Category II considers 10 solutions to the problem of facilitating the drinking of water. Categories III, IV, and V deal with the problems of hand washing and phlebotomy (10 solutions), fountains and perpetual flutes (10 solutions), and the raising of water (5 solutions), respectively. Later writers who commented on Jazari's work considered his designs as inventions of devices intended mostly for the entertainment of Muslim feudal lords. Many even dismissed Jazari's work altogether as exercises in toy-making, but the multiplicity of solutions given in the various categories (5 or 10) clearly indicates to the knowledgeable reader that his work contains studies of design alternatives embodying a certain design principle. Thus, all Jazari's work is really a study in systematic machine design!

Another, earlier book, *The Book of Ingenious Devices,* written by the brothers Ahmad and Hasen bin Musá Ibn Shakir (ca. 850 AD), also contains a wealth of

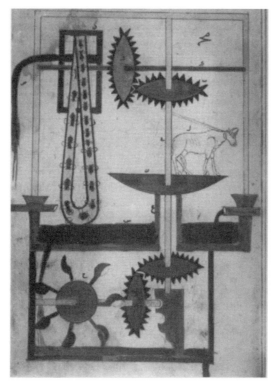

Figure I.6. Jazari's device for pumping water. Note the dual power source: a cow and an undershot wheel may drive the pump.

systematic design studies. It presents conceptual designs for 9 pitchers, 40 jars, 7 flasks, and 10 basins.

7. MACHINE DESIGN IN THE MIDDLE AGES

In the first half of the 2nd Millennium AD there was massive emigration of Greek scholars from Constantinople to Western Europe. This resulted in a substantial dissemination of the works of Aristotle and a renewed interest in mechanics and kinematics, in particular. Jean Buridan and Albert of Rückmersdorf (both ca. 1350), the latter known also as Albert of Saxony, were both rectors of the University of Paris at different times and were instrumental in reviving the writings of Aristotle by publishing translations of his works and writing commentaries on his theories (Dugas 1955).

Nicole Oresme (1330–1382), a Buridan disciple and Grand Master of the College of Navarre, was given by Charles V the task of translating into French some of the works of Aristotle. Oresme wrote treatises of his own on mechanics, introduced the use of the Cartesian coordinate system, and studied uniformly accelerated motion. He derived the equation $s = gt^2/2$ for the distance traveled, in contemporary terminology. Moreover, he introduced the basic ideas of continuously varying quantities and thus laid the foundations of continuum mechanics.

The diffusion of Roman culture into highly religious medieval Europe shifted the emphasis to the practical needs of worshipping God. In the medieval monasteries, important mechanical devices were conceived. Again, aesthetics became secondary to pragmatic and ethical dimensions. Intellect was reserved for religious investigations, and technological advancements had come as they did in Roman times. Design methodology returned to the level of a craft, according to the testimony of several authors, and no noticeable advancements were recorded until the time of Leonardo da Vinci. Although Leonardo did not explicitly enunciate a specific design methodology, in his notebooks one can see a systematic development of design concepts to arrive at specific solutions. This accomplishment had little impact on his contemporaries, however, because of Leonardo's secretive nature; his notebooks were published only centuries later. Leonardo's designs were in the style of the 16th Century, in pictorial books with a variety of drawings of machines. They were easy to draw and much easier for craftsmen to understand than text and mathematics. The building of machinery thus was just a craft then.

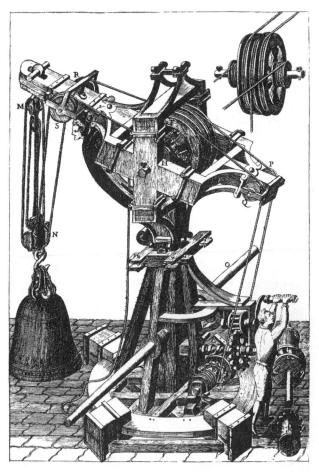

Figure I.7. A crane in Agostino Ramelli's (1531–1600) *Diverse et artificiose machine* (*Various and Ingenious Machines*), 1588.

The need for more accurate timekeeping in the monasteries brought about the development of the pendulum clock, a remarkable mechanism that had a paramount effect in the development of modern science and civilization, starting with Galileo Galilei (1564–1642). The predecessors of the pendulum clock were the water clock, known to the ancient Assyrians and Egyptians and developed to a precision instrument by the Ctesibius school of Alexandria, and the mechanical clock, which can rightly be called "the mother of all machines." The latter was gravity-driven with an escapement control. The origin of the mechanical clock has been placed as early as 1335, but it is probable that the clock mentioned by Dante (1235–1321) in his *Paradiso* was a mechanical clock. The mechanical escapement is usually attributed to the French architect Villard de Honnecourt (ca. 1225–1250), shown in his *Album,* a book of drawings. The mechanical clock played a fundamental role in the revival of mechanical design.

8. THE RENAISSANCE AND THE INDUSTRIAL REVOLUTION

The early modern era is highlighted by the works of Galileo and Newton and included the early stages of mechanization and the Industrial Revolution. The utilization of the high-density chemical energy of the fossil fuel and the consequent high power per unit of machinery volume spawned numerous problems in machine dynamics. The contemporary development of calculus and continuum mechanics led to the rapid development of mechanics by the mid-19th Century. By the time of Galileo and Newton, classical physics and mechanics were much further developed than is generally assumed. The fundamental contribution of Galileo and Newton is the revival and redefinition of these sciences just as greater progress was being demanded from natural science. There had been several previous attempts at the revival of physics and mechanics during the first half of the 2nd Millennium AD. The time, however, was not yet propitious.

Leonardo da Vinci (1452–1519). Leonardo was born in 1452 in Vinci, the illegitimate son of Master Piero, a public notary, and his companion Caterina. At age 17, Leonardo moved with his father to Florence, where Leonardo apprenticed to Verrocchio. In 1472, Leonardo became a member of the painter's guild of Florence. He subsequently offered his services to different feudal lords of Italy as artist, consultant for architectural matters, and military engineer. Having painted the *Mona Lisa,* he devoted much of his time to scientific studies and major engineering projects. In his notebooks, which were published centuries later, he developed many ingenious machines and mechanisms, such as flying machines, submarines, water turbines, cranes, a mechanical saw, a screw-cutting machine, and roller bearings. He also wrote treatises on architecture and mechanics. Leonardo died on May 2, 1519.

Leonardo da Vinci (self-portrait).

About the 17th Century, as societal development moved from the Mediterranean region to western Europe, political power was concentrated in the hands of centralized, monarchical governments. The concomitant generation and concentration of capital allowed ample patronage of arts and crafts. The aim of such patronage was to elevate the grandeur of life for the nobility. Craftsmen became organized in guilds that functioned mainly to preserve their design secrets. However, at the same time the guilds prevented the crafts from developing beyond mere empiricism.

By the second half of the 18th Century, machine design had begun to influence life beyond the royal court and the nobility. After the Revolution, the royal manufactories of France entered into commercial competition and the products made for the general public began to share the high aesthetic quality previously shared by the royalty.

Watt, James (1736–1819) Scottish instrument maker and inventor whose steam engine contributed substantially to the Industrial Revolution. Watt's father ran a successful ship- and house-building business. At age 17 he decided to be a mathematical-instrument maker. He was trained in Glasgow and London and opened a shop in Glasgow in 1757 at the university. While repairing a model Newcomen steam engine in 1764, Watt was impressed by its waste of steam. In May 1765, he suddenly came upon a solution—the separate condenser, his first and greatest invention. Shortly afterward, he met John Roebuck, the founder of the Carron Works, who urged him to make an engine. He entered into partnership with him in 1768. The following year Watt took out the famous patent for "A New Invented Method of Lessening the Consumption of Steam and Fuel in Fire Engines." After Roebuck went bankrupt in 1772, Matthew Boulton, the manufacturer of the Soho Works in Birmingham, took over a share in Watt's patent. This began a partnership that lasted 25 years. In 1781, Boulton urged Watt to invent a rotary motion for the steam engine, to replace the reciprocating action of the original. He did this in 1781 with his so-called sun-and-planet gear, by means of which a shaft produced two revolutions for each cycle of the engine. In 1782, he patented the double-acting engine, in which the piston pushed as well as pulled. The engine required a new method of rigidly connecting the piston to the beam. He solved this problem in 1784 with his invention of the parallel motion—an arrangement of connected rods that guided the piston rod in a perpendicular motion—which he described as "one of the most ingenious, simple pieces of mechanism I have contrived." Four years later his application of the centrifugal governor for automatic control of the speed of the engine, at Boulton's suggestion, and in 1790 his invention of a pressure gauge virtually completed the Watt engine. In 1785, he and Boulton were elected fellows of the Royal Society of London. Watt was made Doctor of Laws of the University of Glasgow in 1806 and a foreign associate of the French Academy of Sciences in 1814.

Developments in machines and mechanisms during the 17th and 18th Centuries were stimulated mainly by the development of the steam engine, just as the development of mechanical clocks had provided the stimulus in the 15th and 16th Centuries. Many of the advances in mechanical design in this era can be traced to some genius inventor, but many were conceived by the systematic reasoning of the engineer. Watt (1736–1819), in a letter to his son in November 1808, described the methodology he utilized to design a linkage to trace an approximately straight path. He was thus the first to consider coupler points and the systematic synthesis of linkages by means of kinematic considerations. Euler (1707–1783), a near contemporary of Watt, laid the foundations of kinematic analysis by treating planar motion as a superposition of translation of a point and rotation about it (Euler 1736–1742) and made a clear distinction between the kinematics and dynamics of a mechanism, the former having to do solely with the motion without reference to the forces causing it (Euler 1790). An important addition to the plane kinematics of Euler was the discovery by Johann Bernoulli of the instantaneous center of rotation in 1742.

Since earlier times, authors such as Valturius (1472), Agricola (1550), Ramelli (1580), and Branca (1629) had dealt either with whole machines or with their separate elements. Leupold (1724) seems to have been the first to have considered a mechanism, that is, a system consisting of different elements that are in some functional relationship with one another and classified by the transformations of motion that they effected. He described mechanisms that convert one type of motion to another.

9. DESIGN OF MACHINES AND MECHANISMS IN MODERN TIMES

The École Polytechnique, established in Paris in 1794, was the first school to separate the study of kinematics from the study of machinery, in general as pro-

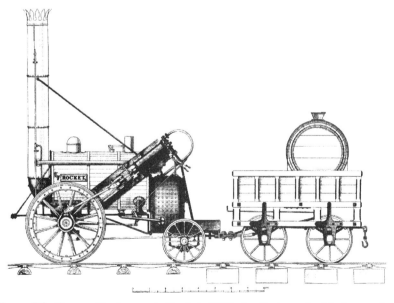

Figure I.8. George Stephenson's Rocket, 1829, the first practicable locomotive.

posed by Monge (1746–1818), the founder of descriptive geometry and one of the founders of the École Polytechnique, and L. N. M Carnot (1753–1823) and elaborated by his colleague Hachette in his book *Traité élémentaire des machines* (1811). The latter presented a systematic classification of mechanisms by capability and devised a classification chart and formed the basis for the kinematics course at the École Polytechnique and later at West Point. Hachette proposed the classification of the machine elements into six *orders: recepteurs,* receiving the motion from prime movers; *communicateurs,* passing on the motion; *modificateurs,* changing the type of the motion; *supports; regulateurs,* transmission ratio regulators; and *operateurs,* performing the machine function.

The separation of kinematics into a distinct discipline was concluded by Lanz and Betancourt (1808) in their *Essai sur le composition des machines.* This work and Hachette's book are the first textbooks in kinematics. Lanz and Betancourt separated the study of mechanisms on the basis of their motion: rectilinear, rotary, continuous, reciprocating in ten classes.

Borgnis (1818) in his *Traité complet de mechanique* further classified mechanisms into classes based on function, such as receiver, transmitter, or regulator. This received further development in the works of Coriolis and Poncelet. Reuleaux disputed this approach (*Amicus Plato, sed magis amica veritas*), although he used some of these ideas in his work.

In 1830, Ampère published his *Essai sur la philosophie des sciences* defined the machine as an instrument by the help of which the direction and velocity of a given motion can be altered. He excluded forces from kinematic analysis and gave to the study of such machines the name *kinematics,* from the Greek word *kinema* (motion), and the science of kinematics was born (kinetics would be a more accurate expression, but it was already used with another meaning). In the new science, the movements are to be studied as movements, independent of the power creating them, in the same way that we observe the static bodies around us, especially in such combinations as we call machines. This science ought to include all that can be said with respect to motion in its different kinds, independently of the forces by which it is produced.

In 1831, Coriolis investigated relative motions and posed the fundamental problem of kinematic analysis: *to find the motion of any machine in which certain parts are moved in a given way.* He introduced the ideas of relative velocity (known already to Aristotle) and relative acceleration.

Robert Willis, a professor of natural and experimental philosophy, taught kinematics at Cambridge. His lectures were published in a book, *Principles of Mechanisms* (1841), which dealt with applied kinematics in Ampère's sense, classified mechanisms on the basis on change in directional relation and velocity ratio, and identified different ways in achieving each function. Willis, who was also an accomplished historian of medieval architecture, further applied trigonometry and other mathematical methods to the kinematic analysis of specific mechanisms.

Among those who took up Willis's ideas was Ferdinand Redtenbacher, professor at Karlsruhe Technical University, a great teacher who played an important role in the dissemination of those ideas. He inspired one of his students, Franz Reuleaux, to become fascinated with kinematics.

Reuleaux was born to a family of master builders; his father was a designer and builder of steam engines. He studied engineering in Karlsruhe, natural sciences in

Berlin, and arts in Bonn. He taught a few years at the ETH in Zurich and moved to Berlin Technical University, where he compiled a "mechanical alphabet" of mechanical devices of his own. Advancing mechanization had created a great number of different machines and mechanisms, and Reuleaux embarked on an attempt to systematize and classify them. His aim was to achieve the same uncompromisingly logical order in engineering that Linnaeus had achieved in biology. Known already for his 1854 book *Machine Elements* and the 1861 handbook *The Designer,* Reuleaux in 1875 published *Theoretische Kinematik,* was translated immediately in English (1874–1876) by Kennedy under the title *Kinematics of Machinery,* in which a complete treatment of kinematics of mechanism was presented. The book was intended to be part of a trilogy, but only one more book was published, in 1900. Reuleaux, who was equally well known in cultural circles for his lectures and articles on "Culture and Technology," made extensive studies in the application of kinematics to human body mechanics, translated poetry, wrote essays on art and books on a variety of subjects, and had a great effect on the development of kinematics.

Rankine devoted a good part of his book *Machinery and Millworks* (1869) to machine kinematics. Rankine considered a machine to be made up of a *frame* and *moving pieces.* The frame is the fixed link, in the language of Reuleaux. Rankine also made extensive use of the instantaneous centers of rotation. Although he introduced the *elementary combinations* of elements, his treatment was less general than that of Reuleaux. On the other hand, he presented more specific results for particular forms of mechanisms.

Euler's ideas for planar motion found extensive application in Smith's (1825–1916) *Graphics* (1889), where extensive use was made of velocity and acceleration diagrams in mechanism analysis. In parallel, Burmester's (1840–1927) *Lehrbuch der Kinematic* (1888) adopted most of Reuleaux's methodology and also applied geometric methods for mechanism displacement, velocity, and acceleration analysis. Aronhold in Germany and Kennedy in England extended the use of instantaneous centers of velocity while introducing the theorem of three centers. Gruebler (1851–1935) developed criteria for the movability of a linkage; this can be considered an important step towards *number synthesis.* Finally, *dimensional synthesis* received attention in Burmester's work on geometric methods and was further extended by Tschebytschev (1821–1894) and S. Roberts (1827–1913), who introduced triple generation of the coupler curves, which was formalized much later by Freudenstein and Sandor.

Kinematicians were always fascinated by mechanisms for rectilinear motion. Watt considered his rectilinear motion mechanism his greatest achievement. Several mechanisms approximating linear motion were invented by Cartwright, White, Evans, Roberts, and others. In 1864, Nicholas Peaucellier, a French military engineer, wrote a letter to the editor of a mathematical journal, proposing another rectilinear motion mechanism, called *compas composé.* There was immediate reaction from kinematicians, including Pafnuti Tschebytschev (1821–1894), who had improved Watt's design. He tried to disprove mathematically Peaucellier's claim that his mechanism was producing an exact straight line. Proving that a kinematic chain, no matter how complex, could not produce an exact straight line became an obsession to him for the rest of his life, but he was unable to find the proof.

The 19th Century, in the words of Hartenberg and Denavit (1964),

. . . is the end of the formative period, of an era in which many principles had been uncovered, analytical methods established, and the road to synthesis opened. Subsequent growth of these ideas, no longer nurtured only in Europe, has been of such scope that no short overview is possible. . . . The full story of kinematics of mechanisms, doing justice to the many who practiced the art of mechanisms and contributed to the science of kinematics, is yet to be written. . . .

In the 19th Century, as the Industrial Revolution in England flourished, the mechanization of production was widely adopted. This did not always bring about changes in design methodology or production technology. The rising middle class demanded that their newly acquired wealth and social status be displayed in decorative, often ostentatious ways. While academics and intellectuals were debating aesthetic matters in industrial production, industry itself was striving for novelty by adopting the stylistic views of past cultures. These conflicting trends seriously impeded the natural and evolutionary development of form, in many cases in sharp contrast to the development of product function.

A growing concern with the imbalance between function and form can be discerned in an editorial of *Cole's Journal of Design* in 1850:

. . . Design has a twofold relation, having in the first place a strict reference to utility in the thing designed; and secondarily, to beautifying or ornamenting that utility. The word design, however, with the many has become identified rather with its secondary than with its whole signification with ornament, as apart from and often as opposed to, utility. This was the reason for many of those great errors in taste which are observed in the works of many designers. . . .

This view was further stressed by Colburn, a civil engineer, who in his 1871 book on locomotives wrote, ". . . none who aspire to become an engineer should encourage any play of imagination involving the forms or proportions of a mere mechanism. . . ."

10. SYMBOLIC REPRESENTATION—GROUP TECHNOLOGY

Group technology is usually thought to be, a contemporary issue but it really was developed some 200 years ago. Automation in the production of pulley blocks, introduced by Marc Brunel (1769–1849), a French-born engineer, generated a need for standardization of machine components. Christopher Pohlem (1661–1751), a Swedish engineer, introduced the *mechanical alphabet,* a collection of different components labeled with different letters of the alphabet. Every mechanism could be synthesized from elements of the mechanical alphabet just as a poem is made up of simple letters. One of Pohlem's student-assistants was Carl Cronstedt (1709–1779), who described in a sketchbook about one third of the elements of Pohlem's mechanical alphabet. He concentrated on the movements that can be generated, in particular by the *mighty five* simple elements, in contrast to Archimedes and the Alexandrians, who dealt mainly with their force-amplifying features.

During the 17th and 18th centuries the idea of symbolic representation of a mechanism consisting of basic elements was used by clock-makers for the wheel-

$$.....2.......b,a.........1..........2$$

$$C^+...\mid C^+_z, C^+_z ... \mid ...(C)... \parallel ...C^-_=$$

$$C^+...\mid C^+_z, C^+_z ... \mid ...(C)... \parallel ...C^-_=$$

$$...2.......c,d..........3..........2$$

$$-\ -\ -\ -\ -\ -\ -\ -\ -\ -\ -\ -\ -\ -$$

$$C^+...\mid ... \begin{Bmatrix} C^+_z, C^+_z ... \mid ...(C)... \\ C^+_z, C^+_z ... \mid ...(C)... \end{Bmatrix} ... \parallel ...C^-_=$$

$$...............c,d............3$$

Figure I.9. Symbolic representation of a gear train (Reuleaux 1876).

work of clocks, as for example by, Oughtred (1677), Derham (1696), and Alexandre (1735). Babbage (1826) proposed the first complete symbolic description for the function and geometry of a linkage: an *Alphabet of Form* and an *Alphabet of Motion*. Willis (1841) attempted to improve Babbage's kinematic notation to make it applicable to mechanism design. It was Reuleaux, however, who presented in his *Kinematics of Machinery* (1876) an extensive description and symbolic representation of mechanisms used in engineering. He defined the machine as "a combination of solid bodies, so arranged as to compel the mechanical forces of nature to perform work as a result of certain determinative movements." Moreover, he dealt with very important aspects of kinematics, such as pairing and inversion of mechanisms. Reuleaux stated at the outset that he was concerned primarily with the theoretical aspects of machine kinematics, not applications. His aim was not to deal with the kinematic behavior of any particular machine but ". . . to determine the conditions which are common to all machines. . . ." He considered it important to analyze not, for example, the straight path-generating linkages of Watt, Evans, or Reichenbach, but rather *how* their inventors arrived at them with no knowledge of the underlying processes of thought. In this respect Reuleaux's work was the first to be devoted to modern kinematic synthesis and further to a symbolic representation of knowledge. Synthesis in the Reuleaux sense was limited to *type* synthesis. Systematic study of *type* and *dimensional* synthesis, although identified long before, were developed later. However, it is fair to consider Reuleaux the father of *group technology*.

11. MACHINE DESIGN METHODOLOGY

During the first half of the 19th Century, following the Napoleonic wars, there was a systematic attempt to apply engineering science to the design analysis of machines. This movement matured about this time, culminating in Poncelet's *Introduction à la Mécanique Industrielle*. The focal point of the efforts to apply engineering science to mechanical design was the École Polytechnique, then as now a military school.

Jean Victor Poncelet (1788–1867), French general, mathematician and engineer. He participated in the Napoleonic wars and was captured in Russia in 1812. In captivity, he developed the foundations of projective geometry. After his return to France he published his work in his book *Manual for the Projective Properties of Solids*. In 1848, he was promoted to general and was appointed director of École Polytechnique, the famous French military school of engineering that Napoleon Bonaparte created. Poncelet wrote the first treatise on applications of mechanics in machine design, entitled *Introduction à la Méchanique Industrielle*.

J. V. Poncelet (Reprinted from S. P. Timoshenko, *History of the Strength of Materials,* by permission of the McGraw-Hill Book Co.)

The need to control the steam engine generated a wealth of investigations in kinematics and linkage designs (Hartenberg and Denavit 1964). The synthesis of linkages can be viewed as a formal design method. However, by its nature it cannot be considered as general mechanical design theory because it is limited to linkages.

F. Redtenbacher (1809–1863) He graduated in 1829 from the Vienna Polytechnic Institute and in 1833 became a teacher of mathematics at the Federal Institute of Technology in Zurich. He combined his teaching with a designer's job at Escher & Wyss Manufacturing Co. He later moved to the Technical University of Karlsruhe as a professor of applied mechanics and machine design. His books *Principien der Mechanik und des Maschinenbaues* (1852) and *Konstruktion ü den Maschinenbau* (1862) not only included mechanics and strength of materials considerations in the design analysis of machines, but also introduced the concept of *synthesis (Konstruktion)*, a systematic methodology for finding design solutions.

F. Redtenbacher (portrait at the Technical University of Karlsruhe by permission of the McGraw-Hill Book Co.)

In Germany, the von Humboldt model of universities entailed strong ties with industry, in contrast to the École Polytechnique, which was supported mainly by the French government. Consequently, in Germany, there was a rapid dissemination of engineering science in mechanical engineering practice. J. Weisbach wrote *Mechanics of Machinery and Engineering,* which had a great impact on design analysis. F. Redtenbacher introduced the general mechanical design principles, such as:

- Sufficient strength
- Sufficient resistance to wear
- Sufficiently low friction
- Optimum use of materials
- Ease of manufacturing
- Ease of maintenance
- Simplicity

Redtenbacher's principles are too contradictory and overlapping to be taken as a formal system of design axioms. A more abstract set of design principles was developed by his student Reuleaux (1854), who separately addressed considerations of function and form in the following two ground rules (*Grundsätze*) of machine design:

1. *Function:* The design must provide a uniform satisfaction of the design requirements. For example, in the case of loading by pressure, all parts should equally contribute to the support of the pressure.
2. *Form:* The form of the design embodiment must have the highest possible symmetry.

The first principle can be regarded as an optimization principle. Equally loaded parts mean maximum utilization of the material, as Redtenbacher's principle (*d*). Stated in another form, when the design fails, it fails everywhere.

The second principle is essentially a minimum information principle. To an extent dependent on the particular design, it is not totally independent of the first principle, since symmetry results, in general, in lower manufacturing costs. It is not clear to what extent Reuleaux was driven to this principle by considerations of mere aesthetics or optimum design.

Reuleaux's design principles led to a balance between form and function considerations, which had been separated since Alexandrian times and the decline of ancient Greek philosophy and science. Reuleaux saw another dimension in machine design:

> Just as the poet contrasts the gentle and lovable Odyssean wanderers with the untamable Cyclops, the "lawless-thoughted monsters," so appear to us the unrestrained power of natural forces, acting and reacting in limitless freedom, bringing forth from the struggle of all against all their inevitable but unknown results, compared with the action of forces in the machine, carefully constrained and guided so as to produce the single desired result.

Bach (Heinrich 1992) and Riedler (1913) realized that the choice of materials and manufacturing methods and the design for adequate strength are of equal importance and that they influence one another. Roetscher (1927) considered the following as the essential characteristics of design: specified purpose, effective load paths, and efficient production and assembly (Pahl and Beitz 1996). Load transmission should be along the shortest paths, and if possible through axial forces rather than by bending moments. Longer load paths not only waste materials and increase costs but also require unnecessarily complex shapes.

Laudien (1931) suggested for the load paths in machine parts:

- For a rigid connection, join the parts in the direction of the load.
- If flexibility is required, join the parts along indirect load paths.
- Do not make unnecessary provisions.
- Do not overspecify.
- Do not fulfill more demands than are required.
- Save by simplification and economical construction.

Figure I.10. Ford's Model A car (1927) built most probably with little theory, had a paramount influence on mechanical design.

Modern ideas of systematic machine design were pioneered by Erkens (1928). He introduced the *step-by-step approach,* based on systematic testing and evaluation and also on the balancing of conflicting demands during an iterative design process.

A more comprehensive account of the machine design methodology was introduced by Wögerbauer (1943), who introduced the idea of *task decomposition,* the separation of the design task into subsidiary tasks, and these into operational and implementational tasks. His systematic search for a design concept starts with a solution discovered more or less intuitively, and then this initial solution is modified in respect of structure, materials, and method of manufacturing to achieve the optimum solution by continuous evaluation and improvement.

R. Franke (1948), H. Franke (1984), and Rodenacker (1970) developed a solution methodology for mechanical power transmission systems using a logical–functional structure based on elements with different physical effects (electrical, mechanical, hydraulic effects for the same logical functions, guiding, coupling and separating).

Kesselring (1942, 1951, 1954) introduced a method of successive approximations and the evaluation of design variants according to a set of technical and economic criteria, together with the underlying engineering science principles and business constraints:

- Minimum production costs
- Minimum space requirement
- Minimum weight
- Minimum losses
- Optimum handling

Niemann (1950) assumed that a layout of the machine existed. He proceeded from a definition of the machine design task to a systematic variation of the elements of the initial solution for a formal selection of the optimum solution. He modernized the machine elements approach of Redtenbacher and Reuleaux and connected the design of machine elements with the design of the machine as a system.

Hansen (1956) and other members of the Ilmenau school introduced four working steps:

1. Analysis, critique, and specification of the task, leading to the basic solution concept, which encompasses the overall function that has been derived from the design task and the design requirements and constraints
2. A systematic search for solution elements and their combination into working concepts
3. Analysis of any shortcomings of the several working concepts with respect to the design goal and, if necessary, improvement
4. Evaluation of these improved working concepts to determine the optimum working concept for the task

Similarly, Mueller (1970), and Yoshikawa (1983) introduced a theoretical foundation of the design process. They offer essential foundations of design science.

Rodenacker (1970, 1991) further developed the method of systematic design of a working design concept through a step-by-step layout of logical, physical, and embodiment relationships. He applied a systematic identification of possible failures of the design, evaluating every early design step with quantity, quality, and cost criteria. He based his design methodology on logical function structures interrelated with binary logic, using a rigorous task decomposition.

Wachtler (1967) argued that creative design is the most complex form of the learning process. Learning represents a higher form of control, one that involves not only quantitative changes at constant quality (control), but changes in the quality itself. Moreover, design changes technical quantities as well as working principles. In structural terms, learning and control can, despite qualitative differences, be considered as comparable circular processes. Optimization means that the design process should not be treated statically, but dynamically as a control process in which the information feedback must be repeated until the information content has reached the level at which the optimum solution can be found. The learning process thus keeps increasing the level of information and hence facilitates the search for a solution (Pahl and Beitz 1991).

Roth (1986, 1994) introduced a set of integrated and logically structured design catalogues. A particular goal is the partitioning of the design process into small steps that can be described by algorithms. It should then be possible to process these steps completely and continuously using the computer. To that end, "product representing models" are defined that can be described unambiguously by "product defining data." Roth divided the design process into three phases and three corresponding models:

- *Problem formulation* phase, problem-representing models, requirements list to identify the main design tasks and the design requirements

Figure I.11. The *Concorde,* epitomizes all mechanical design achievements of the 20th Century. Yet it did not receive much attention. It was a solution looking for a problem.

- *Functional* phase, function-representing models that include logical function structures; relational function structures with material, energy, and information flow functions; physical function structures with physical, chemical, and other effects; and physical solution solutions
- *Embodiment* phase, embodiment-representing models with geometrical working structures based on predesigned modules and functional elements

A reasonably comprehensive model that still retains some clarity is that offered by Pahl and Beitz (1996). It is based on the following design stages:

- *Clarification of the task.* Collect information about the requirements to be embodied in the solution and also about the constraints.
- *Conceptual design.* Establish function structures, search for suitable solution principles, combine into concept variants.
- *Embodiment design.* Starting from the concept, determine the layout and forms and develop a technical product or system in accordance with technical and economic considerations.
- *Detail design.* Lay down arrangement, form, dimensions, and surface properties of all the individual parts; specify materials; recheck technical and economic feasibility; produce all drawings and other production documents.

Koller (1989) proposed an algorithmic description of the design process that will make it suitable for machine implementation and design automation (1989, 1994). He developed twelve functions and their inverses, which he called basic operations, to describe the flow of energy, matter, and information in a machine. In the *function synthesis,* logical and physical basic operations are linked together in a function structure. In the *qualitative synthesis,* physical effects, effect carriers, principle solutions, and system and embodiment variants are used. In the *quantitative synthesis,* dimensioned layouts and production documents are used.

Figure I.12. Harley Davidson 1200S. The motorcycle was never at the forefront of engineering. Yet it became very popular because it was a good solution to an important problem.

12. CONCLUSION

It appears that certain fundamental ideas return with some regularity to the methodology of machine design. It was pointed out above that group technology, symbolic manipulation, and open-ended design are not at all new ideas. The last concept, currently promoted vigorously by the Accreditation Board of Engineering and Technology, was practiced extensively by the engineers of ancient times, albeit in a different way. Machine design moves from machine elements to the design of a machine as a system. This transition was best described by Reuleaux (1876):

> Formerly, the fundamental idea of alteration or extension was improvement, a word which says much in itself of the nature of the process. Now, on the other hand, we have a direct production of new things, a sudden bringing into being of heretofore completed machines. We see the beginnings of a perception that will someday apparently be universal among those who have to do with all classes of machinery. Upon this growing sense I believe that our polytechnic machine-instruction should act with increasing certainty.

REFERENCES

Aristotle. 1936. *Mechanical Problems.* In *Minor Works,* trans, W. S. Hett. Loeb Classical Library. Cambridge, Mass.: Harvard University Press.

Benvenuto, E. 1991. *An Introduction to the History of Structural Mechanics.* New York: Springer-Verlag.

Cohen, M. R., and I. E. Drabkin. 1958. *A Source Book in Greek Science.* Cambridge, Mass. Harvard University Press.

Cotterell, B., and J. Kamminga. 1990. *Mechanics of Pre-industrial Technology.* Cambridge: Cambridge University Press.

De Camp, L. S. 1960. *The Ancient Engineers.* New York: Ballantine Books.

Dugas, R. 1955. *A History of Mechanics.* Neuchatel, Switzerland: Éditions du Griffon, and New York: Central Book Co.

Erkens, A. 1928. "Beiträge zur Konstruktionserziehung." *Z. VDI* 72:17–21.

Euler, L. 1736–1742. *Mechanica, sive, Motus scientia analytice exposita.*

Forbes. 1967. In *Technology in Western Civilization,* ed. M. Kranzberg and C. E. Pursell. New York: Oxford University Press.

Franke, H. M. 1984. *Der Lebenszyklus technischer Produkte.* VDI-Berichte, no. 512. Düsseldorf: VDI-Verlag.

Hansen, F. 1956. *Konstruktionssystematik.* Berlin: VEB-Verlag Technik.

Hartenberg, R. S., and J. Denavit. 1964. *Kinematic Synthesis of Linkages.* New York: Mc-Graw-Hill.

Heinrich, W. 1992. "Kreatives Problemlösen in der Konstruktion." *Konstruktion* 44:57–63.

Hill, D. 1984. *A History of Engineering in Classical and Medieval Times.* LaSalle, Ill.: Open Court.

Ibn Shakir, Ahmad bin Musá, and Hasan bin Musá ibn Shakir. 1979. *The Book of Ingenious Devices (Kitob al-hiyal).* D. R. Hill (tr.) Dordrecht: D. Reidel.

Jazari, Isma'il ibn al-Razzaz. 1974. *The Book of Knowledge of Ingenious Mechanical Devices,* trans. D. R. Hill. Dordrecht: D. Reidel.

Kesselring, F. 1942. "Die Starke Konstruktion." *Z VDI* 86:321–330, 749–752.

———. 1951. *Bewertung von Konstruktionen.* Düsseldorf: VDI-Verlag.

———. 1954. *Technische Kompositionslehre.* Berlin: Springer.

Kirby, R. S., S. Withington, A. B. Darling, and F. G. Kilgour. 1956. *Engineering in History.* New York: McGraw Hill.

Koller, R. 1989. *CAD—Automatisiertes Zeichnen, Darstellen und Konstruieren.* Berlin: Springer-Verlag.

———. 1994. *Konstruktionslehre für den Maschinenbau,* 3d ed. Berlin: Springer-Verlag.

Landels, J. G. 1978. *Engineering in the Ancient World.* Berkeley: University of California Press.

Laudien, K. 1931. *Die Maschinenelemente.* Leipzig: Dr. Max Junecke Verlagsbuchhandlung.

Liddell, H. G., and R. Scott. 1843. *A Greek-English Lexicon Based on the German Work of Francis Passow.* Oxford: Oxford University Press.

Mueller, I. 1970. *Grundlagen der Systematischen Heuristik.* Berlin: Dietz.

Niemann, G. 1950. *Maschinenelemente,* 1st ed., vol. 1. Berlin: Springer.

Pahl, G., and W. Beitz. 1996. *Konstruktionslehre.* Berlin: Springer-Verlag.

Reuleaux, F. 1876. *The Kinematics of Machinery,* trans. B. W. Kennedy. London: Macmillan.

Riedler, A. 1913. *Das Maschinen-Zeichnen,* 2nd rev. ed. Berlin: J. Springer.

Rodenacker, W. G. 1970. *Methodisches Konstruieren.* Konstruktionsbücher, no. 27. Berlin: Springer.

Roetscher, F. 1927. *Die Maschinenelemente.* Berlin: Springer.

Roth, K., 1986. "Modellbildung für das methodische Konstruieren ohne und mit Rechnerunterstützung." *VDI-Z,* 21–25.

———. 1994. Konstruieren mit Konstruktionskatalogen, 2nd ed. Berlin: Springer-Verlag.

Schopenhauer, A. 1911. *Die Welt als Wille und Vorstellung,* ed. P. Deussen. Munich.

Timoshenko, S. P. *History of the Strength of Materials.* New York: McGraw-Hill.

Wachtler, R. "Beitrag zur Theorie des Entwickelns (Konstruierens)." *Feinwerktechnik* 71 (1967): 353–58.

Wögerbauer, H. 1943. *Die Technik des Konstruierens,* 2nd ed. München: Oldenbourg.

Yoshikawa, H. 1983. "Automation in Thinking in Design." In International IFIP Conference on Computer Applications in Production and Engineering, Amsterdam, The Netherlands; 1982, *Computer Applications in Production and Engineering.* Amsterdam: North-Holland.

ADDITIONAL READING

Babbage, C. *A Method of Expressing by Signs the Action of Machinery.* London: W. Nicol, 1826.

Bach, C. *Die Maschinenelemente,* 1st ed. Stuttgart: A. Bergstrasser, 1880.

———. *Die Maschinen-Elemente,* 12th ed. Stuttgart: A. Bergstrasser, 1920.

Beitz, W. "Systemtechnik in der Konstruktion." *DIN-Mitteilungen* 49 (1970): 295–302.

———. *Systemtechnik im Ingenieurbereich.* VDI-Berichte, no. 174. Düsseldorf: VDI-Verlag, 1971.

Bischoff, W., and F. Hansen. *Rationelles Konstruieren.* Konstruktionsbücher, no. 5. Berlin: VEB-Verlag Technik, 1953.

Bock, A. "Konstruktionssystematik—die Methode der ordnenden Gesichtspunkte." *Feingerätetechnik* 4:4 (1955).

Chestnut, H. *Systems Engineering Tools.* New York: John Wiley & Sons, 1965.

Cross, N. *Engineering Design Methods.* Chichester: John Wiley & Sons, 1989.

Dilke, O. A. W. *The Roman Land Surveyors.* David and Charles, Newton Abbot: England, 1971.

Dimarogonas, A. D. *History of Technology.* Athens: Symmetry, 1978. [In Greek]

———. "The Origins of Vibration Theory," *Journal of Sound and Vibration* 140(2) (1990): 181–189.

———. "Mechanisms of the Ancient Greek Theater." Paper presented at ASME Mechanisms Conference, Phoenix, Ariz., 1992.

———. "Machines in Homer's Poetry." Paper presented at 4th World Congress for the Theory of Machines and Mechanisms, Houlu, Finland, 1999.

———. "Standards and Quality Control in Antiquity." Paper presented at *4th World Congress for the Theory of Machines and Mechanisms,* Houlu, Finland, 1999.

Dixon, J. R. *Design Engineering: Inventiveness, Analysis, and Decision Making.* New York: McGraw-Hill, 1966.

Euler, L. *Theoria motus corporum,* 1765, and 1775, Nov. Comm. Petrop., 1790.

Farrington, B. *Greek Science.* London: Harmondsworth, 1944.

Franke, R. *Vom Aufbau der Getriebe.* Düsseldorf: VDI-Verlag, 1948/51.

French, M. J. *Conceptual Design for Engineers.* London: Design Council, and Berlin: Springer-Verlag, 1985.

———. *Invention and Evolution: Design in Nature and Engineering.* Cambridge: Cambridge University Press, 1988.

Gladett, M. *Greek Science in Antiquity.* New York: Aberland-Schuman, 1955.

Glegg, G. L. *The Design of Design.* Cambridge: Cambridge University Press, 1969.

————. *The Development of Design.* Cambridge: Cambridge University Press, 1981.

Gregory, S.A. *Creativity in Engineering.* London: Butterworths, 1970.

Haselberger, L. "The Construction Plans for the Temple of Apollo at Didyma." *Scientific American* (December 1985): 126–132.

Heinrich, W. "Eine systematische Betrachtung der konstruktiven Entwicklung technischer Erzeugnisse." Habilitationsschrift Teknische Universität Dresden, 1976.

Hill, D. R. "Mechanical Engineering in the Medieval Near East," *Scientific American* (May 1991): 100–105.

Hubka, V. *Theorie Technischer Systeme.* Berlin: Springer-Verlag, 1984.

Hubka, V., and W. E. Eder. *Theory of Technical Systems : A Total Concept Theory for Engineering Design.* Berlin: Springer-Verlag, 1988.

Keller, W. *The Bible as History,* new ed. New York: Hodder & Stoughton, 1974.

————. *The Bible as History,* rev. ed. New York: Bantom Books, 1982.

Koller, R. *Konstruktionslehre für den Maschinenbau: Grundlagen, Arbeitsschritte, Prinziplösungen,* 1st ed. Berlin: Springer-Verlag, 1985.

————. "Entwicklung und Systematik der Bauweisen technischer Systeme—ein Beitrag zur Konstruktionsmethodik." *Konstruktion* 38:1–7, 1986.

Leupold, J. *Theatrum Machinarum.* 1724.

Mayr, O. *The Origins of Feedback Control.* Cambridge, Mass.: MIT Press, 1970.

Moll, C. L., and F. Reuleaux. *Konstruktionslehre für den Maschinenbau.* Braunschweig: J. Vieweg, 1854.

Mueller, J. *Arbeitsmethoden der Technikwissenschaften : Systematik, Heuristik, Kreativität.* Berlin: Springer-Verlag, 1990.

Nadler, G. *The Planning and Design Approach.* New York: John Wiley & Sons, 1981.

Needham, J., and L. Wang. *Science and Civilization in China.* Cambridge: Cambridge University Press, 1959.

Niemann, G. *Maschinenelemente.* Berlin: Springer, 1965.

Niemann, G., and M. Hirt. *Maschinenelemente,* 2nd rev. ed. Berlin: Springer, 1975.

Ostrofsky, B. *Design, Planning, and Development Methodology.* Englewood Cliffs, N. J.: Prentice-Hall, 1977.

Pahl, G. "Entwurfsingenieur und Konstruktionslehre unterstützen die moderne Konstruktionsarbeit." *Konstruktion* 19(1967): 337–344.

————. *Engineering Design : A Systematic Approach,* trans., ed. London: Design Council, and Berlin: Springer-Verlag, 1988.

Pahl, G. and W. Beitz. "Für die Konstruktionspraxis." *Aufsatzreihe in der Konstruktion* 24 (1972).

————. "Für die Konstruktionspraxis." *Aufsatzreihe in der Konstruktion* 25 (1973).

————. "Für die Konstruktionspraxis." *Aufsatzreihe in der Konstruktion* 26 (1974).

Proclus Diadochos. *Commentary on Euclids Elements I,* ed. G. Friedlein. Leipzig, 1873.

Pugh, S. *Total Design; Integrated Methods for Successful Product Engineering.* Reading, Mass.: Addison-Wesley, 1990.

Redtenbacher, F. *Prinzipien der Mechanik und des Maschinen-Baues.* Mannheim: Bassermann, 1852, 257–90.

Reuleaux, F. *Der Konstructeur.* Braunschweig: J. Vieweg, 1872.

Reuleaux, F., and C. Moll. *Konstruktionslehre für den Maschinenbau.* Braunschweig: J. Vieweg, 1854.

Rodenacker, W. G. *Methodisches Konstruieren,* 2nd ed. Berlin: Springer, 1976.

———. *Methodisches Konstruieren,* 3d ed. Berlin: Springer, 1984.

———. *Methodisches Konstruieren,* 4th ed. Berlin: Springer, 1991.

———. "Neue Gedanken zur Konstruktionsmethodik." *Konstruktion* 43 (1991): 330–334.

Rodenacker, W. G., and U. Claussen. *Regeln des Methodisches Konstruierens.* Mainz: Krausskopf, 1973/74.

Roth, K. *Konstruieren mit Konstruktionskatalogen,* 1st ed. Berlin: Springer-Verlag, 1982.

Schröedinger, E. *Nature and the Greeks.* Cambridge: Cambridge University Press, 1954.

Soedel, W., and V. Foley. "Ancient Catapults." *Scientific American* (1979): 120–28.

Strandh, S. *The History of the Machine.* New York: Dorset Press, 1989.

Suh, N. P. *The Principles of Design.* Oxford: Oxford University Press, 1988.

Taguchi, G. *Introduction to Quality Engineering.* New York: Unipub, 1986.

Theon of Smyrna. [2nd Century AD.] In M. R. Cohen and I. E. Drabkin. *A Source Book in Greek Science.* Cambridge: Cambridge University Press, 1958.

Ullman, D. G. "A Taxonomy for Mechanical Design." *Res. Eng. Des.* 3 (1992): 179–89.

———. *The Mechanical Design Process.* New York: McGraw-Hill, 1992.

Vitruvius, Marcus Pollio. *On Architecture.* Cambridge, Mass.: Harvard University Press, 1914.

CHAPTER 1

MACHINE DESIGN METHODOLOGY

Dodge Intrepid. One of the first machines to be designed and manufactured with the concurrent engineering procedure.

OVERVIEW

Designing a machine is a long road. Its beginning is a vague initial description of the machine design problem to be solved. Its end, is a concrete and unambiguous description of the machine, usually in the form of detailed design drawings of the whole machine and every single part of it, including a parts list describing the materials and the manufacturing processes by which every part will be made. This description enables a manufacturing facility to make the machine exactly as the designer intended it.

This chapter shows how we can develop a multitude of possible solutions and identify the most promising solution concept, described in generic form, usually in the form of a schematic that describes the machine's operation and a set of detailed design specifications.

1.1. THE ART AND SCIENCE OF MACHINE DESIGN

To an ever-increasing extent, our immediate visual environment is dominated by machinery and industrial products. In the home and workplace, in schools, factories, offices, shops, even recreation areas, in public streets and transport systems, machines constitute the visible cultural landscape of everyday life, comprising in their totality a complex pattern of function and meaning in which our perception of the world, our attitudes and sense of relationship to it, are closely interrelated.

Because machines are themselves products of other machines, with a precision the human hand cannot match, their space and design rarely yield to the layman any indication of the participation and personality of the designer and manufacturer. They are all, however, a manifestation of a process of a human design, of conception, judgment, and specification.

The precise nature of the design process is infinitely varied and therefore difficult to summarize in a simple design formula, or book of formulas, or a precise definition.

- It can be the work of a person.
- It can be the effort of a team.
- It may emanate from creative intuition, from executive decisions based upon market research, or from a calculated judgment.
- It may be constrained by resource availability or organizational, political, social and aesthetic considerations, aiming of course at acceptance by the end user, the customer.

Whatever the particular situation, machine design, as defined in the previous chapter, is a process of creation, invention, and definition, involving an eventual synthesis of contributory and often conflicting factors into a three-dimensional form capable of multiple reproduction, at marketable price, with acceptable quality of products and with specified reliability (Orlov 1976).

Machine design is an applied science relying heavily on engineering science because no machine can defy Newton's Law or the strength limits of the materials the machine is made of.

Machine design is also an art (Dimarogonas 1997). A strength of materials problem usually has one solution. A machine design problem usually has an infinite number of solutions. In rough mathematical terms, the number of unknown parameters is usually orders of magnitude greater than available data or performance specifications, that is, equations. This uncertainty can be eliminated to a small extent by optimization methods and to the full extent by good technical judgment by the designer. What is "good" and what is "bad" technical judgment? Though experience has accumulated rules, the final judge is the end user and the quantitative measure of acceptance is market success.

With the dissemination of information the world is experiencing in our times, competing designers have at their disposal about the same amount of information. Proprietary information is usually much less in reality than advertised. Why is one product (and the manufacturer) successful and the other not? In most cases the main reason is the decisions made at design stage, not of course those based on engineering science but those based on engineering (and business) judgment.

As stated above, this cannot be corrected with concrete rules. There are, however, guidelines that can help the designer with the decisions he has to make for a successful design. A product designed "by the book" is not guaranteed success. It is almost certain, however, not to be a total failure.

1.1.1. Classification

A *machine* is a device for doing useful work, such as manufacturing and transporting goods. Depending on the nature of its useful function, a machine may be simple or complex, fragile or strong. Before the industrial age, machines were simple and low-powered, most of them very crude devices designed to relieve people of backbreaking burdens such as lifting heavy loads, as in mining operations and milling grain. Humans and animals, sails, windmills, and waterwheels provided power for these early machines. The power produced by these sources were clearly adequate for the easy tasks of the time. Thus, the lack of sources of highly concentrated mechanical power provided little incentive for anyone to make machines that were faster or more powerful. We have seen in the Historical Introduction that crude steam and heat engines were invented by Hero of Alexandria about two millennia ago, but the needs and means of production at that time could not make use of these inventions. Thus, technical progress was slow, with few exceptions, in the fields of mechanics and metallurgy. The development of the steam engine and the harnessing of the chemical energy in coal, just prior to the Industrial Revolution, provided the incentive needed for humans to build more complex and powerful machines.

Invention of the modem steam engine by James Watt in 1769 gave people the mechanical power required for large-scale industry and mass transportation. Use of the steam engine for mine pumping and railroads led to other and much improved power sources, such as the current internal combustion engines, the turbine, the jet engine, and the rocket. In 200 years (1769–1969), humanity advanced technically from a primitive steam engine to a powerful rocket engine that would carry people to the moon. The invention of the steam engine was one of the most important events in history.

The explosive nature of the Industrial Revolution underscores the fundamental fact that industrial production is much more dependent on available sources of power than on supplies of raw materials and production tools.

The most dramatic development in the history of the machine came, no doubt, with the introduction and rapid development of automatic operations, machines directed by machines instead of by humans. Termed *automation,* this interrelationship among machines initially depended on linkages and cams, then on punch cards, and finally on magnetic tape and electronic equipment to direct and control mechanical processes without the constant attendance of operators. At the heart of automation is the computer, a cybernetic machine for processing, storing, and retrieving information.

Another way of looking at the machine is that it enables work to be done with greater ease or speed than would be obtainable without its use. The term is most commonly applied to mechanisms used in the industrial arts for shaping and joining materials. Machines are frequently named from their use (e.g., screw cutter) or from the product made (e.g., bolt maker). *Compound* machines are formed when two or more simple machines are combined. *Tools* are the simplest implements of the industrial art. *Machines,* on the other hand, are more complicated in structure.

When machines act with great power, they generally take the name of engines (e.g., internal combustion, steam, or aircraft).

A machine is therefore essentially a structure consisting of a frame with various fixed and moving parts. The parts are rigid or resistant and are relatively constrained so that they can transmit power, modify force or motion, and do useful work. In short, a machine is a device that transmits and changes the application of energy.

Machinery is a derived term used to represent (1) the internal working parts of a complex assembly, usually of large size, and (2) a grouping, such as the machinery of a plant or mill. Machines can be classified broadly as *basic* or *simple* and *complex* or *compound*. A *simple* machine is a device with at least one mechanically actuated member (lever or screw) (Figure 1.1). A *complex* machine is merely a combination of simple machines, as exemplified by a typewriter (Hinhede et al. 1983).

Complex machines may be classified as *stationary*, *portable*, or *mobile*. As the name implies, mobile machines, such as self-propelled combines or street sweepers, move themselves in doing useful work. This group includes all means of transportation, farm machinery, and construction equipment. Stationary machines include most factory production machinery. Portable machines are primarily items such as power tools, chain saws, and vacuum cleaners; that is, the user can carry them.

Complex machines are also classified as *prime* movers, *secondary* movers, and *power-driven*. Any machine that utilizes a natural source of energy to produce power for other machines is a prime mover. An automobile engine is a prime mover that converts the chemical energy of the fuel into mechanical energy. This in turn is used to propel and direct the entire vehicle along some selected path. An electric motor, by contrast, is a secondary mover because it receives energy directly or indirectly from a generator driven by a prime mover, usually a steam turbine. A power-driven machine or power absorption unit utilizes energy to do useful work. Typical power-driven machines are pumps, machine tools, and air compressors.

Machines are thus *dual* in their makeup: they all have a power source separate from the parts doing useful work. The power supplied to a machine is called *input* power, while the power delivered by a machine is called *output* power. Thus, the

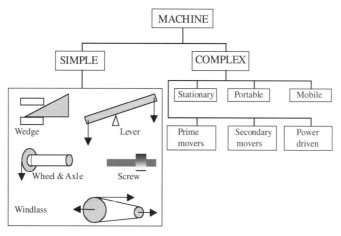

Figure 1.1. Classification of machines.

output power of one machine becomes the input power of the driven machine. It is common to have a single place for power input and another for output power. Some machines, however (e.g., trucks and tractors), have two output shafts, one to propel the vehicle and another to drive accessories.

Five simple machines have been known since the time of Hero of Alexandria[1]: *lever, windlass, pulley and axle, wedge, screw* (Figure 1.1). Simple machines provide a mechanical advantage (MA) by increasing force at the expense of speed; herein lies their primary importance. Sometimes, however, they are used in reverse to augment speed at the expense of force (MA < 1). The high speeds of vehicles, for example, are obtained by using the wheel and axle with MA < 1.

1.1.2. The Traditional Machine Design Procedure

Machine design is a systematic process. Even if a new machine was conceived by invention, systematic machine design is needed to transform the invented concept into a working system that users will appreciate.

George Sandor (1913–1997). Born and educated in Hungary, he came to the United States at a young age to work in industry before starting his graduate work at Columbia University, where he received his doctorate in 1959. He then became a professor of Mechanical Design at Yale University and, in 1967, at the Rensselaer Polytechnic Institute. He wrote hundreds of books and technical papers, received six patents, and was awarded membership in several academies. An avid flyer, sailor, musician and poet, Sandor was one of the first who introduced computer methods in mechanism analysis and design and of computer-aided design in general. His book *Mechanism Design,* which he coauthored with A. G. Erdman, is a classic text in the field.

The machine design process is subject to a large number of variations. In every text on the subject of engineering design, a different division of the design process in distinct stages is proposed. They all make sense, although they seem very different from one another. This reflects the complex nature of the design process and the fact that every design problem requires a special treatment. This process cannot be exactly specified by an equation or an algorithm. A systematic approach is useful only to the extent that the designer is presented with a strategy that he can use as a basis for planning the required design strategy for the problem at hand.

This strategy, proposed by Sandor (1964), is described in the flowchart of Figure 1.2. The flowchart is arranged in a Y-shaped structure.

1. The two upper branches of the Y represent, on one hand, the evolution of the design task, and on the other hand, the development of the available, applicable engineering background.

---

[1] Hero of Alexandria (1st Century AD) considered seven simple machines as the seven machines then known to lift a weight, the most important machine design problem of that time.

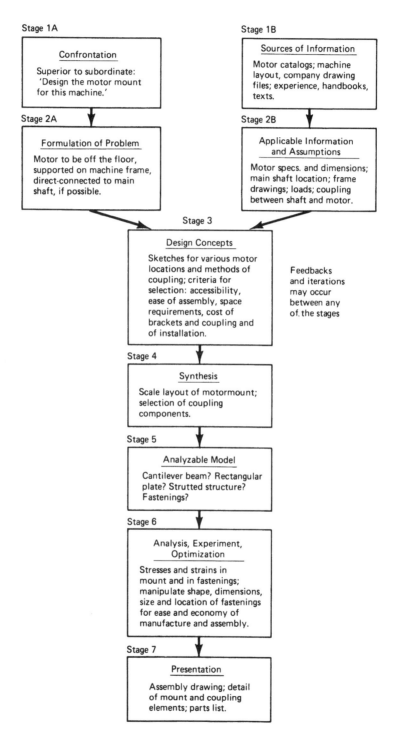

Figure 1.2. Sequential design procedure. (Reprinted from G. N. Sandor, "The Seven Stages of Engineering Design," *Mechanical Engineering,* 86(4) (1964):21, by permission of the American Society of Mechanical Engineers.)

2. The junction of the Y stands for the merging of these branches: generation of design concepts.

3. The leg of the Y is the guideline toward the completion of the design, based on the selected concept.

The flowchart implies, but is not encumbered by, the feedbacks and iterations that are essential and inevitable in the creative process.

Stage 1A: Confrontation. The "confrontation" is not a mere problem statement; it is the actual encounter of the engineer with a need to take action. It usually lacks sufficient information and often demands more background and experience than the engineer possesses at the time. Furthermore, the "real need" may not be obvious from this first encounter with an "undesirable situation."

Stage 1B: Sources of information. The sources of information available to the engineer encompass all human knowledge. Perhaps the best source is other people with expertise in the same or related fields.

Stage 2A: Formulation of problem. Because confrontation is often so indefinite, the designers must clarify the problem that is to be solved: they must recognize and ferret out the "real need" and then define it in concrete, quantitative terms suitable for engineering action.

Stage 2B: Preparation of information and assumptions. From the vast variety of sources of information the designer must select the applicable areas, including theoretical and empirical knowledge, and, where information is lacking, fill the gap with sound engineering assumptions.

Stage 3: Generation and selection of design concepts. Here the background developed by the foregoing preparation is brought to bear on the problem as it was just formulated, and all conceivable design concepts are prepared in schematic skeletal form, drawing on related fields as much as possible.

It should be remembered that creativity is largely a matter of diligence. If the designer lists all the ideas that can be generated or assimilated, workable design alternatives are bound to develop, and the most promising can be selected in the light of requirements and constraints.

Stage 4: Synthesis. The selected design concept is a skeleton. We must give its substance: fill in the blanks with concrete parameters with the use of systematic design methods guided by intuition. Compatibility with interfacing systems is essential. In many areas, advanced analytical, graphical, and combined, computer-aided methods have become available. However, intuition, guided by experience, is the traditional approach.

Stage 5: Analyzable model. Even the simplest physical system or component is usually too complex for direct analysis. It must be represented by a model amenable to analytic or empirical evaluation. In abstracting such a model, the engineer must strive to represent as many of the significant characteristics of the real system as possible and necessary, commensurate with the available time, methods, and means of analysis or experimental techniques. Typical models: simplified physical version, free-body diagrams, and kinematic skeletal diagrams.

Stage 6: Experiment, analysis, optimization. Here the objective is to determine and improve the expected performance of the proposed design.

1. Design-oriented experiment, on either a physical model or its analog, must take the place of analysis where the latter is not feasible.

2. Analysis or test of the representative model aims to establish the adequacy and responses of the physical system under the entire range of operating conditions.

3. In optimizing a system or a component (getting the most out of it—see Chapter 7), the engineer must decide three questions in advance: (a) With respect to what criterion or weighted combination of criteria should he or she optimize? (b) What system parameters can be manipulated? (c) What are the limits on these parameters; that is, what constraints is the system subject to?

Although systematic optimization techniques have been and are being worked out (see Chapter 7), this stage is largely dependent on the engineer's intuition and judgment. The amount of optimizing effort should be commensurate with the importance of the function or the system component and/or quantity involved.

"Experiment," "analysis," and "optimization" form one integral "close-loop" stage in the design process. Their results may give rise to feedbacks and iterations involving any or all of the previous stages, including a possible switch to another design concept.

Stage 7: Presentation. No design can be considered complete until it has been presented to (and accepted by) two groups of people:

1. Those who will utilize it
2. Those who will make it

   The engineer's presentation must therefore be understandable to the prospective user and contain all the details necessary to allow manufacture and construction by the builder.

1.1.3. The Concurrent Engineering Machine Design Procedure

In recent years it has been observed that the sequential approach to the design procedure, as outlined in the previous section, has some implementation advantages from an organization standpoint, such as:

• Simple organization, well-defined responsibilities of each of the sections of the organization.

• Easy and simple communication: at the end of each step, information is simply "thrown over the wall to the next section."

This method, however, has several shortcomings:

• Because the duration of the product cycle development is the sum of the durations of the different design steps, the product cycle from confrontation to production is very long.

- During the process there is a need for several iterations due to the late input from the sections down the serial structure and the development cost and time increase.
- The input of the manufacturer and the user arrives too late or is not considered at all so that the final product is of inferior quality.

The solution of the problem is the formation of *multifunctional design teams* that include personnel of all sections-activities of the *design/manufacturing/ production/distribution/finance/marketing* cycle so that the development of the design is *concurrent* in all fronts (Figure 1.3).

Concurrent engineering is the ideal of planning and implementing all machine development steps, from early product conceptualization to delivery and service, as early as possible. Team members are responsible for each step, working together throughout.

In the traditional machine development process, each step is conceived of as a unit with clear inputs and outputs. Steps further downstream, such as manufacturing process development, are not supposed to start until the results of previous steps, such as component design, are well defined. This production-line view of the development process assumes that time is wasted in downstream steps if upstream steps have not yet been completed, with plans solidified.

Although it is true that downstream work must take upstream decisions into account, the major problem not dealt with in the traditional model is that upstream steps may arrive at results that are unrealistic, impractical, or not optimal for downstream implementation. For example, the engineer may unwittingly choose a design that is unnecessarily difficult to manufacture or is expensive to repair in the field. Automobiles have been made that required removal of the engine for changing the spare plugs.

Manufacturing and field support interaction with the designers could influence those decisions and thereby lower the company's downstream costs.

The best approach to making upstream and downstream decisions is an interactive one, in which representatives from all functions collaborate, sharing their decision-making processes throughout. In such a dynamic, give-and-take scenario, realistic decisions can be made that achieve the best, most workable results for all functions.

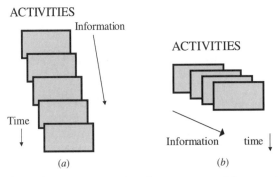

Figure 1.3. Sequential and concurrent engineering processes: (*a*) sequential; (*b*) concurrent.

The concurrent engineering model aims at starting all development process steps as early as possible, even simultaneously (Figure 1.3). Its success comes from *each step influencing the other* as the development process moves forward. With sufficient communication between people responsible for each step, practical and optimal results are more likely for all steps.

The method by which communication can occur frequently enough and at a detailed enough level to achieve these desirable results is to treat the people responsible for each development step as a single "multifunctional" team, located together, and all with the same objective—the success of their jointly developed product. More and more U.S. companies are adopting this style of product development, often referred to as "project," rather than "functional," organization. The objectives of such "project" organization are (Clausing 1994):

- Earlier start of each development step, leading to earlier completion of all steps
- Optimization of decisions at every step through dialogue and collaboration by representatives from all disciplines, leading to optimal product design and low production and delivery costs
- Overall lower development cost

1.1.4. Axiomatic Foundation and Fundamental Rules of Machine Design

Since very early times, designers have been confronted with the question of the feasibility of a design theory, that is, the existence of a set of concrete principles upon which design can be developed as a science using mathematics and reason.

In the Historical Introduction we saw that F. Redtenbacher introduced design synthesis (*Konstruktion*), a systematic methodology for finding design solutions to machine design problems, and a set of general mechanical design principles:

- Sufficient strength
- Sufficient resistance to wear
- Sufficiently low friction
- Optimum use of materials
- Easy manufacturing
- Easy maintenance
- Simplicity

Redtenbacher's principles are too contradictory and overlapping to some extent to become a formal system of design axioms. A more abstract set of design principles was introduced by his student Reuleaux (1854), professor at Zurich and Berlin, who separately addressed the function and the form considerations in the following two ground rules (*Grundsätze*) of machine design (Reuleaux 1876):

Franz Reuleaux was born to a family of master builders. His father was a designer and builder of steam engines. He studied engineering in Karlsruhe, natural sciences in Berlin, and arts in Bonn. He taught a few years at the ETH in Zurich and moved to Berlin Technical University. Known already from his 1854 book *Machine Elements* and 1861 handbook *The Designer,* Reuleaux published in 1875 *Theoretische Kinematik,* which was translated immediately into English (1874–1876) by Kennedy under the title *Kinematics of Machinery.* In this work, he presented a complete treatment of the kinematics of mechanism. It was intended to be part of a trilogy, but only one more book was published, in 1900. Reuleaux, who was equally known in cultural circles for his lectures and articles on "Culture and Technology," made extensive studies on the application of kinematics to human body mechanics, translated poetry, wrote essays on art and books on a variety of subjects, and had a great effect on the development of kinematics.

Franz Reuleaux, portrait in the Deutsches Museum, Munich (by permission).

1. (*Function rule*): The design must provide a uniform satisfaction of the design requirements.
2. (*Form rule*): The form of the design embodiment must have the highest possible symmetry.

The first principle can be regarded as an optimization principle. It means that all the (nonreplaceable) parts of the machine should fail at the maximum value of the machine loading. In other words, if a part fails, every part should fail. There is no point in letting half of the machine fail while the rest remains intact. The part that did not fail was overdesigned and thus more expensive than needed.

The second principle is essentially a minimum information principle and aims at minimizing the resources needed for the manufacturing of the machine.

In recent years, the need for a rigorous foundation of the design and manufacturing sciences prompted Suh (1988) to introduce a modified form of Reuleaux's design principles as design axioms, which stirred a very productive discussion on the subject:

Axiom I (*function rule*): The design should independently satisfy the design requirements.
Axiom II (*form rule*): The design should have minimum information content.

While axiom II is essentially the same as Reuleaux's ground rule 2, axiom II sounds similar to Reuleaux's ground rule 1 but is not quite the same. The mathematical form of axiom I is that the every functional requirement of the design is related with only one design parameter and one design equation.

The next problem is whether we should speak about rules or axioms. Aristotle (384–322), in his *Posterior Analytics,* defined an axiom as "truth par excellence."

However, the least one can say about the design axioms is that they are not truths par excellence in the Aristotelian sense. Indeed, there are a myriad of designs that clearly violate these design axioms yet are considered very successful. One might find utility in such propositions as design rules, however.

Yoshikawa (1981) used a topological approach to propose an axiomatic foundation of design. Taguchi (1988) gave a very general definition of the quality of the design, including both the function and the form, setting the design goal as the total societal satisfaction from the product to be designed. The celebrated example of the application of his principle is the design of an electronic circuit for a TV tube.

It seems that one can reconcile the Reuleaux design rules and the Taguchi principle in the form:

Design rule 1 (*function rule*): Among many feasible designs the preferable is the one that is closer to the design functional goal.
Design rule 2 (*form rule*): The embodiment of a design that satisfies rule 1 is best when it leads to the minimum path of manufacturing resources.

Definition 1.1: A *feasible design* is one that satisfies all functional requirements; that is, feasible designs are the link of demand and capacity, all the elements of the demand parameters set that are included in the elements of the respective capacity parameters set.

Definition 1.2: *Closeness* is measured with a rule, which is an integral part of the functional goal.

Corollary 1.1: The design that is closer to the design goal satisfies uniformly the design requirements.

Corollary 1.2: The design that is closer to the design goal satisfies the Taguchi principle with the proper definition of the design goal.

Definition 2.1: The definition of the value of the path of the manufacturing process is an integral part of the manufacturing process(es) to be used.

Corollary 2.1: The embodiment of a design that satisfies rule 2 has the maximum possible degree of symmetry.

Corollary 2.2: The embodiment of a design that satisfies rule 2 has the minimum information content.

Corollary 3.1: The embodiment of a design that satisfies rule 1 and rule 2 is unique.

Corollary 3.1 has a very important consequence: the design that satisfies the design rules is not open, but it is unique. It might not be possible to find, but its existence and uniqueness is guaranteed.

Existence of a design solution is not *a priori* certain, because of the limitations that nature imposes, such as material strength. It is obvious that the designers of the ziggurat in ancient Sumeria that would reach the sky, the Tower of Babel, were not aware of this observation.

Example 1.1 With the limited knowledge of what goes into the design of a Boeing 747 aircraft, discuss the application for that design of (a) the Reuleaux design rule, (b) the Suh design axioms, (c) the revised design rules.

Solution   The Reuleaux design rules: Rule 1 appears to be satisfied because there are no reports of failures of one particular system of the aircraft while the others are yet intact. This is partly due to the fact that the FAA demands certain parts to be changed at regular intervals. Rule 2 is nearly satisfied, with one notable exception: the fuselage is not exactly a solid of revolution. Symmetry is tradeoff for other operational requirements.

The Suh axioms: Axiom 1 is not satisfied because, for example, the number of passengers specified is not related with one only design parameter (i.e., length) but with many others (dimensions, engine thrust, etc.). Axiom 2 is nearly satisfied in the sense of satisfaction of Reuleaux rule 2, discussed above.

The modified design rules: Both rules appear to be satisfied; better, from what is commonly known, there is no indication of violation of these rules.

1.2. CONCEPTUAL MACHINE DESIGN

1.2.1. Clarifying the Design Task

A designer has to start from some basic information when confronted with a design task. Sources of the information and the assignment might be:

- An assignment by a planning organization or customer; such an assignment generally specifies the required parameters of the machine and the field and conditions of its use
- A technical suggestion initiated by a designing organization or a group of designers
- A research work or an experimental model based on the research
- An invention proposition or an invention prototype
- An existing machine prototype, which has to be reproduced with modifications and alterations

As stated in the previous section, design starts with the confrontation with a particular problem, a *task* associated with a number of (usually loosely) defined *specifications* or *requirements*. A phase of further data collection must then be initiated (Pahl and Beitz 1996). One starts with some initial questions:

- What is the problem really about?
- What implicit wishes and expectations are involved?
- Do the specified constraints actually exist?
- What paths are open for development?

Any idea for a solution should be subject to some initial *scrutinizing questions:*

- What objectives is the intended solution expected to satisfy?
- What properties must it have?
- What properties must it *not* have?

Design requirements are distinguished as *demands* and *wishes:*

- *Demands* are requirements that must be met under all circumstances, requirements without which the solution is not acceptable: e.g., for the design of a car, gasoline consumption better than 20 miles to a gallon, satisfaction of federal standards of safety.
- *Wishes* are requirements that should be satisfied if possible, sometimes further classified as of *major, medium,* or *minor importance:* e.g., for the design of a certain automobile, major importance wishes are gasoline consumption better than 25 miles to a gallon and a certain maximum noise level and minor importance wishes might be gasoline consumption better than 35 miles to a gallon or indication of low tire pressure.

The list of demands and wishes should be distinguished as *quantitative* and *qualitative:*

- *Quantitative:* All data involving numbers and magnitudes, such as, maximum weight, power output, throughput rate, volume flow rate
- *Qualitative:* All requirements that cannot be quantified, such as, water proof, corrosion proof, aesthetically pleasing

In the search for solutions, *abstraction* is used. This means ignoring what is particular or incidental and emphasizing what is general and essential for the design of a particular machine leading directly to the most important aspects of the design task.

TABLE 1.1. Requirement List for a Motor-Driven Water Pump

Requirement	Numerical Value	D(emand), W(ish)
Basic Design Parameters:	–	–
Motor mains	3PH, ~60 Hz, 120/220 V	D
Water flow	30 m^3/hr	D
Pump head	50 m	D
Fluid	city water	D
Operation:	–	–
Noise and vibration	(Low, residential building)	W
Fully automatic operation	–	D
Motor overheat alarm	–	W
No water in suction alarm	–	D
Maintenance:	–	
Lubrication permanent, no attendance	1 time/year	W
Filter change	in operation	W

Consider, for example, the design of a water pump system. The task is described by means of an abstract sketch (Figure 1.4) where only the critical parameters are shown and not the noncritical ones, such as color. In addition are a requirements list and the formulation of the goal to be achieved—for example, the water pressure and flow and the characteristics of the power source available. The designer should also ask about other requirements, such as:

- Improving quality, e.g., efficiency, durability, reliability
- Reducing weight or space
- Lowering costs
- Shortening delivery time
- Improving manufacturing methods

The next step is to analyze the requirements list in respect of the required function and essential constraints of the problem. The functional relationships contained in the requirements list should be formulated explicitly and arranged in order of their importance.

This analysis, coupled with a step-by-step abstraction, will reveal the general aspects and essential features of the task, as follows:

- Eliminate personal preferences.
- Omit requirements that have no direct bearing on the function and the essential constraints.
- Transform quantitative into qualitative data and reduce them to essential statements.
- Generalize the results of the previous step.
- Formulate the problem in solution-neutral terms.

Depending on either the nature of the task or the size of the requirements list, or both, certain steps may be omitted. For example, let us examine the task of improving the method of filling, storing and loading bags of animal feed (Krick 1969, Pahl and Beitz 1996). An analysis gave the situation shown in Figure 1.5.

Some of the possible formulations would be:

- Filling, weighing, closing, and stacking bags of feed
- Transferring feed from the mixing silo to stacked bags in the warehouse

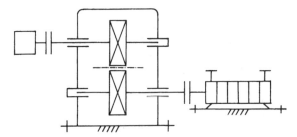

Figure 1.4. Abstract sketch of a water pump system.

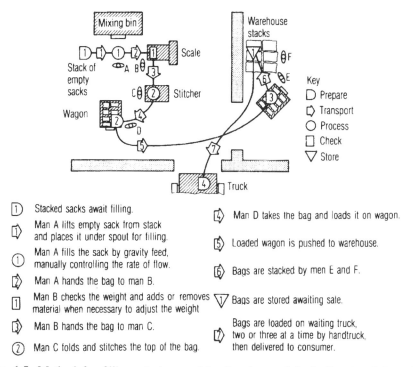

Stack of
empty
sacks

Wagon

Key
D Prepare
⇨ Transport
○ Process
□ Check
▽ Store

1̄	Stacked sacks await filling.	
⇨	Man A lifts empty sack from stack and places it under spout for filling.	
1	Man A fills the sack by gravity feed, manually controlling the rate of flow.	
⇨	Man A hands the bag to man B.	
1̄	Man B checks the weight and adds or removes material when necessary to adjust the weight	
⇨	Man B hands the bag to man C.	
2	Man C folds and stitches the top of the bag.	

⇨ Man D takes the bag and loads it on wagon.

⇨ Loaded wagon is pushed to warehouse.

⇨ Bags are stacked by men E and F.

▽ Bags are stored awaiting sale.

⇨ Bags are loaded on waiting truck, two or three at a time by handtruck, then delivered to consumer.

Figure 1.5. Method for filling, storing, and loading bags of feed. (Reprinted from E. V. Krick, *An Introduction to Engineering and Engineering Design,* 1969, by permission of John Wiley & Sons.)

- Transferring feed from the mixing silo to bags on the delivery truck
- Transferring feed from the mixing silo to the delivery truck
- Transferring feed from the mixing silo to a delivery system
- Transferring feed from the mixing silo to the consumer's storage bins
- Transferring feed from ingredient containers to the consumer's storage bins
- Transferring feed ingredients from their source to the consumer

Krick (1969) incorporated some of these formulations in a diagram shown in Figure 1.6. The problem formulation was systematically made as broad as possible in successive steps.

Some examples of abstraction and broad problem formulation are (Pahl and Beitz 1996):

- Do not design a garage door, but look for means of securing a garage in such a way that the car is protected from thieves and the weather.
- Do not design a keyed shaft, but look for the best way of connecting gear wheel and shaft.
- Do not design a packing machine, but look for the best way of dispatching a product safely or, if the constraints are genuine, of packing a product compactly and automatically.

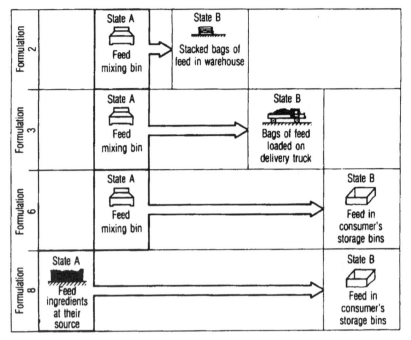

Figure 1.6. Alternative methods for filling, storing, and loading bags of feed. (Reprinted from E. V. Krick, *An Introduction to Engineering and Engineering Design,* 1969, by permission of John Wiley & Sons.)

- Do not design a clamping device, but look for a means of keeping the workpiece firmly fixed.

The final formulation can be derived in a way that does not prejudice the solution, that is, is *solution-neutral,* and at the same time turns it into a *function,* rather than a *device:*

- "Seal shaft without contact" instead of "Design a labyrinth seal."
- "Measure quantity of fluid continuously" instead of "Gauge height of liquid with a float."
- " Measure out feed" instead of "Weigh feed in sacks."

1.2.2. Design Specifications—Quality Function Deployment

The *requirements list* is the initial starting point for the design of a machine. However, the design group needs to reformulate these requirements in much more concrete terms that will be recognizable by all members of the team and will be more quantitative than qualitative so that the results of evaluating different designs will be more concrete.

A successful machine design effort many times indicates good planning, which starts with a clear definition of the design objectives and particular specifications.

We distinguish here the specific design objectives and design specifications from the clarification of the task of the previous section.

Definition of the design objectives and specifications will provide guidance during the course of design, but it will also eventually be used to judge the result of the design effort.

The design objectives have the form of clearly stated specifications for the machine to be designed. Specifications for the machine are imposed from the task and also for safety reasons, availability of certain raw material, cooperation with other machines in production, and so on. Specifications for the product of the machine are always imposed.

Usually the starting objective is cost of the machine itself and of production. Many other objectives may be stated for the machine or its products without an immediate cost interpretation, such as appearance, surface texture, noise, and environmental impact.

The machine should not be overspecified, that is, the design objectives should always be the minima of the acceptable limits. Overspecified machines have an unnecessarily high cost. A friend of mine, a layman, told me once that he would file a lawsuit against Omega because his wristwatch was accurate only to 1/10 of a second while at the time of purchase they told him that it was accurate to 1/100 of a second. Obviously in this case the watch was overspecified.

Overspecification also reduces creativity because it imposes restrictions and removes freedom from the designer's imagination and inventiveness.

A systematic way in developing the design objectives with the broadest possible input has been the main achievement of *quality function deployment* (QFD). QFD was first developed in Japan to ensure that the customer's requirements are met throughout the design process and also in the design of production systems. It was originated at Mitsubishi's Kobe shipyard site in 1972, based on works of Yoji Akao, a professor of industrial engineering in Tokyo. Toyota started using QFD in the mid-1970s and achieved impressive results. In the launch of their new van (January 1977 to October 1979), they achieved a 20% reduction in startup costs, and by 1982, startup costs had fallen 38% and by 1984, 61%. Development time fell by one third, and quality improved markedly. The first U.S. implementations began in 1986. Ford, GM, and Xerox started training in 1986, and by 1988 many companies were using QFD in design.

The QFD process involves constructing one or more *requirements development matrices* (RDM) (sometimes called *quality tables*). The first of these matrices is called by many authors the *house of quality* (HOQ). It displays the *requirements list* (the *voice of the customer* or the *whats*) along the left (*A*) and the development team's technical response to meeting those *wants, the design technical specifications,* along the top (*C*). The requirements list is related to the technical specifications by way of a correlation matrix (*D*) that might consist of several sections or submatrices joined together in various ways, each containing information related to the others (Clausing 1994) (Figure 1.7).

Each of the labeled sections, *A* through *F*, is a structured, systematic expression of a product or process development team's understanding of an aspect of the overall planning process for a new product, service, or process. The lettering sequence suggests one logical sequence for filling in the matrix.

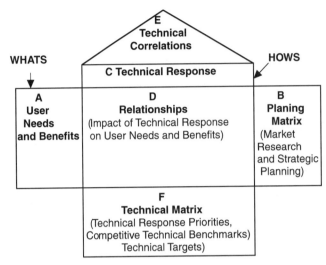

Figure 1.7. House of quality. (Reprinted from D. Clausing, *Total Quality Development,* 1994, by permission of ASME Press.)

QFD authorities differ somewhat in the terminology. In general:

- Section *A* contains a structured list of customer/user wants and needs. The structure is usually determined by qualitative market research. Usually this is the set of initial specifications that are the result of the clarification of the task.
- Section *B* contains three main types of information:
  1. Quantitative market data indicating the relative importance of the wants and needs to the customer/user and the customer/user's satisfaction levels with the organization's and its competition's current offerings
  2. Strategic goal-setting for the new machine
  3. Computations for rank ordering the customer/user wants and needs
- Section *C* contains, in the designer's technical language, a high-level description of the machine they plan to develop. Normally this technical description is generated (deployed) from the customer's wants and needs in Section *A*.
- Section *D* contains the development team's judgments of the strength of the relationship between each element of their technical response and each customer's wants and needs.
- Section *E*, technical correlations, is half of a square matrix, split along its diagonal and rotated 45°. Because it resembles the roof of a house, the term house of quality has been applied to the entire matrix and has become the standard designation for the matrix structure. We shall use here the term *system requirements development matrix* (System RDM). Section *E* contains the development team's assessments of the implementation interrelationships between elements of the technical response.
- Section *F* contains three types of information:

1. The computed rank ordering of the technical responses, based on the rank ordering of customer wants and needs from Section *B* and the relationships in Section *D*
2. Comparative information on the competition's technical performance
3. Technical performance targets

Beyond the RDM, QFD optionally involves constructing additional matrices for the machine subsystems and components, which further guide the detailed decisions that must be made throughout the machine development process (Table 1.1). Figure 1.8 illustrates one possible configuration of a collection of interrelated matrices. It also illustrates a standard QFD technique for carrying information from one matrix into another. In Figure 1.8 we start with the system RDM. We place the requirements list (whats) on the left of the matrix. *Whats* is a term often used to denote benefits or objectives we want to achieve. The design technical specifications (hows) are correlated with the whats at this stage, but they become whats in the next stage, usually the RDM for a subsystem at a lower hierarchical level in the task decomposition scheme. This process continues until designers reach the component level, as shown in Figure 1.8.

Other multiple-matrix QFD schemes are considerably more elaborate than the three-matrix scheme described in Figure 1.8. Some QFD matrix schemes involve as many as 30 matrices that use the *voice of the customer* (VOC) priorities to plan multiple levels of design detail, quality improvement plans, process planning, manufacturing equipment planning, and various value engineering plans.

QFD is a central tool in support of concurrent engineering. It brings the multifunctional team together in the first place to develop the top-level system requirements development matrix. At each step in the process, it helps keep the team focused on customer/user satisfaction, the primary ingredient of product success.

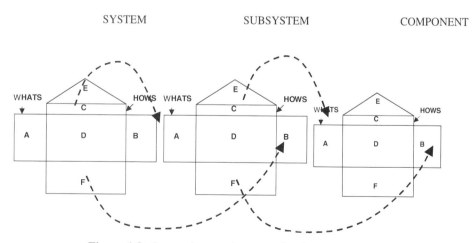

Figure 1.8. Successive requirements development matrices.

TABLE 1.2. Typical Model for QFD

Matrix	What	How
Requirements Development matrix	Requirements list	Technical Performance Measures, Technical Specifications
Subsystem design matrix	Technical performance measures	Piece-part characteristics
Piece part design matrix	Piece-part characteristics	Process parameters
Process design matrix	Process parameters	Production operations

---

Example 1.2 Build a system RDM for the design of a hand-held power tool for making holes for concrete, with the following requirements list:

a. Holes for the usual steel anchors for concrete
b. Fast drilling
c. One-man handling
d. Reliable
e. Safe
f. Low noise
g. Reasonable cost

Compare it with model *T* of the same company and model *X* of the main competitor.

Solution The requirements list given will be listed in room *A* of the HOQ. We assign importance (0–10) and units. No unit appears in this list.

The technical performance measures (technical specifications) for room *C* of the HOQ, technical hows, are:

- Capacity for 6–25 mm diameter holes
- Drilling speed 0.2 mm/sec minimum
- Maximum weight 200 N
- MTBF[2] 6,000 hours
- Should comply with applicable UL[3] and other standards.
- The noise level at 3 m distance should not be more than 70 DB above background noise.
- Manufacturing cost should be less than $150.
- Should not exceed the ISO[4] allowable vibration levels at the hands of the user.

We assign numerical values, units, direction of improvement (higher–lower) and importance (1–10).

---

[2] *Mean time between failures.* A very important measure of machine durability. Usually time is meant as operation time, but sometimes the number of operations is used as a durability index.
[3] Underwriters Laboratories. An industry-supported laboratory to certify the safety of equipment, mainly electrical equipment and electrical and electronic consumer products.
[4] International Standards Organization. Its U.S. member organization is NIST, the National Institute of Standards and Technology.

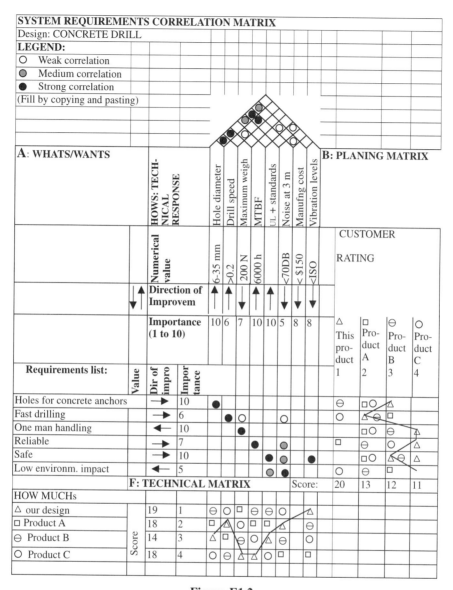

Figure E1.2

Now we fill room D with the relationships of the users and the engineering metrics. Then we specify the correlation of the engineering metrics in room E (roof).

Finally we do the benchmarking. In room B we compare the new design specifications with model T of the same company and model X of the best competitor. In room B the benchmarking is made on the basis of the user's perception of quality, in room F on the basis of an engineering assessment of quality. Now we multiply importance factors with ratings in both rooms B and F for the three designs T, X, N.

In room B we see that the quality sum for the new machine N is 155, slightly less than machine T of the same company and better than machine X of the competitor.

In room F we see that the quality sum for the new machine N is 259, much greater than machine T of the same company and better than machine X of the competitor.

The new design has specifications better than those of the older machine of the company and much better than those of the competing machine. Should this not be the case, one can see very easily what specification improvements will result in a better competitive standing of the new machine, observing also the organizational difficulty row (top of room *F*), because some improvements can be made more easily than others in a certain organization.

1.2.3. The Machine Functional Concept

The design requirements do not necessarily determine the function, that is, the relationship between the inputs and outputs of a machine. A *machine functional concept* needs to be developed first.

The designer should always generalize the task as much as possible because:

· There might be another way to accomplish the task.
· The specifications might be widened to include other possible uses of the machine, without leading to a substantial cost increase, thus widening the market for the product.

Designers can use several methods to generate functional concepts:

1.2.3.1. Conventional Methods

1. *Literature and design records survey.* In the open literature and in the designer's organization a substantial body of information may be available on how people in the past have solved the design problem. With computerized searches available today one can accumulate a substantial number of existing solutions. Links for searching journal articles, conference proceedings, libraries, and so on are available through most libraries and organizations.

2. *Patent survey.* Right after the victorious conclusion of the War of Independence, the U.S. Congress established the procedure for patenting and invention. In fact, one of the first inventors to receive a patent was George Washington. This was already the practice in some European countries at that time and today is a global practice for awarding patents for the protection of intellectual property. According to the Global Association for Tarifs and Trade (GATT) agreement, a patent protects the owner from infringement[5] for 20 years after the date of submission of the application. Unfortunately, each country issues its own patents, so a thorough patent search is a very complex operation. The U.S. depository of patents, however, includes nearly 6,000,000 patents (year 1998), and most major inventions have been patented in the United States due to its huge internal market. Depositories of patents exist in most major U.S. cities. Searching used to be quite difficult and was the job of professionals. Today, however, with the available computerized

---

[5] Unauthorized production and sale of a product that includes a patented design or method or procedure.

searches,[6] a patent search is not so difficult and any designer can perform it from the design office through the Internet.

3. *Asking experts and experienced people in the trade.* One has to be careful, however, using this resource because the more expertise one has in a certain field, the less creative one is because past experience causes a bias toward using the solutions that are well known.

1.2.3.2. Intuitive Methods. We will not discuss here why people are creative. It suffices to say that people can generate ideas with a conscious and subconscious mental operation that we commonly call intuition. These ideas can be completely new or can be initiated by reflection on other ideas of solutions for the problem at hand. Several such methods, such as *brainstorming, method 635,* the *gallery method,* the *Delphi method,* and *synectics,* are described in more detail in books on design methodology, such as Pahl and Beitz (1996). In general, the problem is given to a small number of participants, 5–8, possibly with the participation of a team leader or moderator, and they start generating ideas on how to solve the problem at hand and enhancing these ideas by sharing them with other members of the group. This is done in one or more successive sessions until a number of alternatives are generated. In general, as they generate the ideas they do not judge them immediately because one nonfeasible idea might help the generation down the line of better, feasible ideas. In some methods (brainstorming, method 635), the participants need to have very diverse backgrounds; in others (Delphi), the participants need to be experts in the field of the problem at hand.

One method that a designer can use without involving an organization to pursue intuitive methods is the bottom-up development of the functional concept. To this end we start from the end goal, for example, of a power tool: drill holes in concrete. By stating "power tool" in the problem statement we have already decided that the solution will involve a drill with some kind of a motor. Even if the problem statement has a solution bias (which many times seems logical—the Black & Decker Co. is highly unlikely to pursue a solution radically different from an electric motor coupled to a drill through a shaft or gear train), we need to generalize the problem so that we will be able to identify other solution principles that might surface later if and when the competition or the company starts doing research on them for a possible future product. Sometimes this helps a company to identify that a certain technology is becoming obsolete and search for alternatives.

1.2.3.3. The Systematic Method. We start by listing possible physical effects that can be used to achieve the task: for example, drilling holes in concrete, or removing material from a mass of concrete in a focused way, that is, forming a hole of uniform diameter, for only such holes can be used to locate anchors in the concrete. Searches in the literature, the patent depository, and the market and in-

---

[6]Free patent search sites can be found at the U.S. Patent and Trademark Office web site (http://www.uspto.gov/patft/index.html) and an IBM server (http://patent.womplex.ibm.com/ibm.html). Unfortunately, they cover only a little more than the past 20 years. For the years before 1976, one currently has to do the search the hard way, in a patent directory.

terviews with experts in the field have identified the following three possible physical effects that can be used to drill holes in concrete:

1. Cutting away the concrete
2. Burning the concrete
3. Chemically altering the concrete to a gas- or water-soluble substance

The physical principles that can accomplish these physical effects are now specific enough to be quantified by application of the laws of physics and mathematical
methods. For the case at hand, the physical principles could be as in the second
column in Table 1.3. Further, each physical principle can be implemented with a
physical device, as shown in the third column of Table 1.3. The result is that there
are at least eight physical devices that can, in principle, perform the function of
making holes in concrete. At this point, a decision should be made on which
physical device should be selected.

To this end, we construct a decision matrix (Table 1.4). In the first column, rows
2 on, we place the requirements list. In the next column we place the "Importance
Factor" from room C of the system RDM, the engineering requirements. We then
assign a "degree of satisfaction" for each requirement by each physical device in
the range 1–10 (could be any range). Then for every physical device we calculate
the total score by summing the products of the elements of the respective column
times the importance factor and placing the sum at the bottom cell. The bottom
row (weighted total) now gives the merit of every one of the solution functional
concepts, in this case the helicoidal impact drill. Now, however, we have explored
all possible solution principles and in fact can identify solution 7 as a possible,
though somewhat distant, alternative.

**TABLE 1.3. Systematic Development of a Machine Functional Concept for a
Concrete Drill**

Physical Effect	Physical Principle	Physical Device
a. Cutting away the concrete:	Shearing the concrete by rotational shear	Hollow cylindrical cutting tool Helicoidal impact drill
	Shearing the concrete in the direction of the hole	Cylindrical punch
	Impacting with a sharp solid to generate high, nonuniform compressive stresses, thus shear	Speeding bullet Jackhammer
b. Burning the concrete:	By high-temperature flame By high-speed friction and wear	Oxyflame burner "Diamond"-head rotating rod
c. Chemically altering the concrete to a gas or water soluble substance:	By flow of an acid	Sitrring acid flow through an orifice

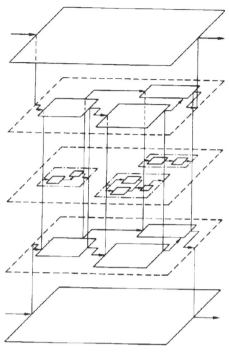

Figure 1.9. Task decomposition (VDI standard 2221, 1993, by permission of the Society of German Engineers (VDI).

In our bottom-up procedure, we now need to identify other components of the power tool that will complete the functional concept. Because we use as a basis an impact drill, a search in our sources of information (our experience, literature, patents, experts, etc.) will reveal that we need a mechanism to generate the rotary motion of the drill, the periodic impact, the static force that must be applied to the drill, and a power source that will power all of the above.

1.2.4. Design Alternatives: Task Decomposition

In the previous section, we outlined the method for finding one (possibly more) machine functional concept. The selected concept can be embodied in a variety of ways, called *design alternatives* or *design variants.* For example, in the power tool design, the functional concept is a high-speed motor connected through a speed reducer to a drill holder. Now, there are many different types of motors, speed reducers, and tool holders. Moreover, there might be solutions combining the above basic elements—for example, a variable speed motor that might not need a speed reducer. The search for design alternatives is more systematic and capable for machine implementation than the highly intuitive search for a basic functional concept.

Depending on the complexity of the problem, the resulting functional concept will in turn be more or less complex. By complexity we mean the relative lack of transparency of the relationships between inputs and outputs, the relative intricacy

TABLE 1.4. Decision Table, Concrete Drill Power Tool Functional Design Concept

Requirements List	Importance	Solution Number							
	–	1	2	3	4	5	6	7	8
Capacity for 6–25 mm diameter holes	10	10	8	1	2	5	2	10	1
Drilling speed 0.2 mm/s minimum	8	8	8	10	10	10	2	7	1
Maximum weight 200 N	10	10	10	3	4	10	7	10	8
MTBF 6000 hours	7	8	8	3	5	10	7	8	2
Should comply with applicable UL and other standards	8	6	6	6	6	5	7	7	2
The noise level at 3 m distance should not be more than 70 db above background noise	5	1	1	5	5	1	10	2	10
Manufacturing cost should be less than $150	6	8	4	5	5	5	10	2	10
Should not exceed the ISO[a] allowable vibration levels at the hands of the user	4	5	5	5	5	5	5	5	5
Weighed total	–	441	397	258	298	395	341	410	258

[a]International Standards Organizations. Its U.S. member organization is NIST, the National Institute of Standards and Technology.

of the necessary physical processes, and the relatively large number of assemblies and components involved.

The methodology for systematic development of the machine functional concept through the generation of design alternatives is a top-down technique. Just as a technical system can be divided into subsystems and elements, a complex implementation of the machine functional concept can be broken down into subfunctions of lower complexity. This process is termed *task decomposition* (Woegebauer 1943). The combination of individual subfunctions results in a *function structure* representing the overall function (Rodenacker 1970).

The object of breaking down complex functions is:

• The determination of subfunctions facilitating the subsequent search for solutions
• The combination of these subfunctions into a simple and unambiguous function structure

A function structure allows a clear definition of existing subsystems or of those to be newly developed, so that they can be dealt with separately. As a result, function structures may lead to a much better design.

We start by determining the main flow of matter, energy, and information in the machine. The auxiliary flows should only be considered later. When a basic function structure, including the most important links, has been found, we then consider the auxiliary flows with their subfunctions.

Thus we develop a function structure schematic as in Figure 1.9. In this example, a three-level task decomposition is shown in the top three layers from the overall function to subfunctions to elementary functions (*analysis*). As we assign solutions to the elementary problems corresponding to the elementary functions, we start combining them in the levels below to form subunits until we form the overall machine solution, as in the bottom level (*synthesis*).

The lower three (or more) levels in the function structure might eventually correspond to a structural decomposition of the machine, such as the machine in Figure 1.10.

To formalize the process, we consider the machine or any module or part of it, and also the function and each subfunction as some kind of processor of matter, energy, and information (Figure 1.11).

In setting up a function structure, one should (Pahl and Beitz 1996):

1. First derive a rough function structure with a few subfunctions from what functional relationships one can identify in the requirements list, and then break this rough structure down, step by step, by the resolution of complex subfunctions. This is much simpler than starting out with more complicated structures. In certain circumstances, it may be helpful to substitute a first solution idea for the rough structure and then, by analysis of that first idea, derive other important subfunctions. It is also possible to begin with subfunctions whose inputs and outputs cross the assumed system boundary. From these we can then determine the inputs and outputs for the neighboring functions—in other words, work from the system boundary inwards.

2. If no clear relationship between the subfunctions can be identified, the search for a first solution principle may, under certain circumstances, be based on the mere enumeration of important subfunctions without logical or physical relationships but, if possible, arranged according to the extent to which they have been realized.

3. Logical relationships may lead to function structures through which the logical elements of various working principles (mechanical, electrical, etc.) can be anticipated.

4. Function structures are not complete unless the existing or expected flow of energy, matter, and information can be specified. Nevertheless, it is useful

Level: 1   2 3   4

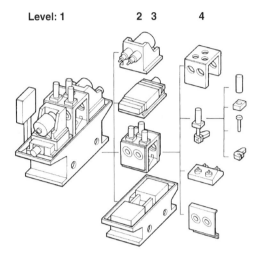

Figure 1.10. Machine structural decomposition.

to begin by focusing attention on the main flow because, as a rule, it determines the design and is more easily derived from the requirements. The auxiliary flows then help in further elaborating the design, coping with faults, and dealing with problems of power transmission, control, etc. The complete function structure, composed of all flows and their relationships, can be obtained by iteration, that is, by looking first for the structure of the main flow, completing that structure by taking the auxiliary flows into account, and then establishing the overall structure.

5. In setting up function structures, it is helpful to know that in the conversion of energy, material, and information, several subfunctions recur in most structures and should therefore be introduced first. Essentially, these are the generally valid functions of Figure 1.12, and they can prove extremely helpful in the search for task-specific functions.
  - *Conversion of energy:*
    - Changing energy, e.g., electrical into mechanical energy
    - Varying energy components, e.g., amplifying torque

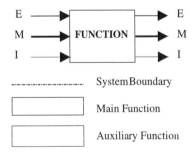

Figure 1.11. The machine, as a processor of energy, matter, and information.

- Connecting energy with a signal, e.g., switching on electrical energy
- Channeling energy, e.g., transferring power
- Storing energy, e.g., storing kinetic energy
- *Conversion of material:*
  - Changing matter, e.g., liquefying a gas
  - Varying material dimensions, e.g., rolling sheet metal
  - Connecting matter with energy, e.g., moving parts
  - Connecting matter with signal, e.g., cutting off steam
  - Connecting materials of different type, e.g., mixing or separating materials
  - Channeling material, e.g., mining coal
  - Storing material, e.g., keeping grain in a silo
- *Conversion of information:*
  - Changing information, e.g., changing a mechanical into an electrical signal or a continuous into an intermittent signal
  - Varying signal magnitudes, e.g., increasing a signal's amplitude
  - Connecting information with energy, e.g., amplifying measurements
  - Connecting information with matter, e.g., marking materials
  - Connecting information with information, e.g., comparing target values with actual values
  - Channeling information, e.g., transferring data
  - Storing information, e.g., in data banks

7. From a rough structure, or from a function structure obtained by the analysis of known systems, it is possible to derive further *variants,* and hence to optimize the solution, by:

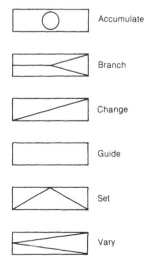

Figure 1.12. Function subsystems.

- Breaking down or combining individual subfunctions
- Changing the arrangement of individual subfunctions
- Changing the type of switching used (series switching, parallel switching, or bridge switching)
- Shifting the system boundary

   Because varying the function structure introduces distinct solutions, setting up function structures constitutes a first step in the search for solutions.

8. Function structures should be kept as simple as possible, so as to lead to simple and economical solutions. To this end, it is also advisable to aim at the combination of functions for the purpose of obtaining integrated function carriers. There are, however, some problems in which discrete functions must be assigned to discrete function carriers; for instance, when the requirements demand separation or when there is a need for extreme loading and quality.

9. In the search for a solution, none but promising function structures should be introduced, for which purpose a selection procedure should be employed, even at this early stage.

10. For the representation of function structures it is best to use the simple and informative symbols shown in Figures 1.12 and 1.13, supplemented with task-specific verbal clarifications.

11. An analysis of the function structure leads to the identification of those subfunctions for which new working principles have to be found and of those for which known solutions can be used. This encourages an efficient approach. The search for solutions then focuses on the subfunctions that are essential for the solution and on which the solutions of other subfunctions depend.

Function structures are intended to facilitate the discovery of solutions; they are not ends in themselves. To what degree they are detailed depends very much on the novelty of the task and the experience of designers.

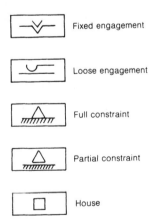

Figure 1.13. Elementary functions.

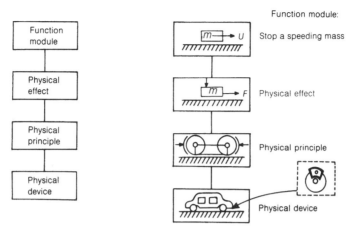

Figure 1.14. Evolution of a function module.

Moreover, it should be remembered that function structures are seldom completely free of physical or formal presuppositions, which means that the number of possible solutions is inevitably restricted to some extent. Hence it is perfectly legitimate to conceive a preliminary solution and then abstract this by developing and completing the function structure by a process of iteration.

1.2.5. Development of Alternative System Design Concepts

By the combination of different physical devices that perform the same operation for each elemental function, design alternatives can start being produced.

Every combination of possible physical devices under every elemental function constitutes an alternative design concept. A tool that can facilitate this selection by providing a visual basis is the *morphological table* (Zwick 1966–71, Pahl and Beitz 1996) (Figure 1.15). Every row corresponds to an elemental function, and, from column 2 and higher, different physical devices appear that correspond to different devices embodying the elemental function on the left at column 1. If we connect

	SOLUTIONS			
Function	1	2	3	4
Provide rotary motion	DC-Motor	AC Motor	Air motor-piston	Air motor-turbine
Speed change	Parallel gears	Bevel gears	Belt drive	Friction drive
Impact	Cam	Air piston		
Hold drill	Chuck	Magnetic	Tapered hole	Coupler
ON/OFF	Switch			
Speed control	Variable voltage	Mechanical	Variable no. of motor poles	

Figure 1.15. Morphological table for the design of a power drill.

with a broken line any physical device in the first row with any physical device in the second row, and so on until we reach the last row, this broken line corresponds to a design alternative.

The result will be a list, hopefully an extensive one, of design alternatives.

1.2.6. Preliminary Evaluation of System Design Alternatives

The several alternatives found in the previous section need to be evaluated in order to select the best. This process is highly subjective, since it is not usually feasible to design all alternatives, build prototypes and test them, because of the cost and time involved. Thus, one has to use a systematic process to select the preferable alternative with the best information available. We shall discuss here two methods of alternatives evaluation: the decision matrix and the Pugh's comparison matrix.

1.2.6.1. Decision Matrix. To this end we construct a matrix, as we did in Table 1.3. In the first column, rows 2 on, we place the requirements list from section C of the system requirements development matrix (Figure 1.7). In the next column we place the importance factor, also from section C of the system RDM (Figure 1.7). In the first row, from column 2 on, we list the design alternatives by number, as listed in the previous section, for example for the power tool of the previous sections from 1–8. We then assign a "degree of satisfaction" for each design alternative in the range 1–10 (could be any range). Then for every design alternative we calculate the total score by summing the products of the elements of the respective column by the importance factor and placing the sum at the bottom cell. The bottom row now gives the merit of every one of the design alternative, in this case the power tool as originally suggested in the problem statement. Now we have explored all possible solution principles, and in fact we can identify design alternative 1 as the most preferable and design alternative 2 as the one to watch as a possible alternative.

1.2.6.2. Pugh's Comparison Matrix. Pugh (1981) introduced a visual, comparative process for design alternatives evaluation and for the concurrent generation of new alternatives (Clausing 1994).

As in the decision matrix, the design evaluation criteria are entered in the first column of a matrix (see Table CS1.1.2., Case Study 1.1), and are taken from the engineering requirements list from section C of the system RDM (Figure 1.7). In the first row, from column 2 on, we list the design alternatives by number, as listed in the previous section, for example for the power tool of the previous sections (Table 1.5). The last one is a datum design, which can be the most important competing design or the previous design of the same company. For a completely new design, datum will be the first choice of the designer, and as the alternatives are improved and reevaluated, the datum design might change. Typically 15–20 criteria are used. The concepts are posted as the column headings. Often there are more than 15 concepts, indicative of vigorous concept generation.

The design team then judges each concept relative to the datum in its ability to satisfy the quality criteria that are included in the engineering requirements list. If a concept is clearly superior, it is given a plus mark. If clearly inferior, it is given a minus mark. If it is neither clearly superior nor clearly inferior, it is given an S for "the same." The team judges all concepts on the basis of the engineering requirements list. As this evaluation is being done, the designers develop much new insight because the weak points of the different alternatives are identified. Commonly there are differing initial perceptions about the evaluations, and the discussions among the design/development team add greatly to clear understanding of the concepts and the criteria. Frequently a team member will observe that the combination of, say, concepts 2 and 6 is a superior concept. This hybrid is added to the matrix as a new concept.

After several hours, the matrix evaluations are complete. The team then computes the total score for each concept. However, the scores are not nearly as important as the insight that has been gained by the team. The team now has a much clearer understanding of the types of concept features that will be responsive to the customer.

The team agrees on additional work and comes back together to "run" the matrix again in a few weeks. Typically the team has dropped roughly half of the original concepts, but new and better concepts have been generated and added to the matrix. The team iterates in this fashion until it converges on the dominant concept. The Pugh concept selection process greatly improves the capability of the team to move forward with a strong concept and thus avoid much rework later in the development process. Generally, the final machine version is chosen from several versions that have been carefully evaluated and compared from all possible points of view for perfection of kinematic and force characteristics, cost of manufacture, energy consumption, cost of labor, reliable operation, size, metal content, weight, suitability for industrial production, unitizing, servicing, assembly–disassembly, inspection, setting, and adjustment. The original design must assure future development and subsequent improvement.

Not always, even during the most careful search, is a successful solution found that fully answers the posed demands. Only in the exceptional designing case is a solution a success in all respects. The reason is sometimes not the lack of inventiveness, but rather conflicting requirements being imposed. Under such circumstances it is necessary to find a compromise solution, waiving some requirements that are not of prime importance in the given conditions of the machine application. Often a variant is chosen not because it incorporates more advantages, but because it has fewer disadvantages in comparison with others.

After selection of the working concept and the basic parameters of the machine, a sketch is made, and then a general arrangement drawing (layout), on the basis of which the initial, technical and working designs are composed. The development of design alternatives is not a matter of individual preference or inspiration of the designer, but a regular design method aimed at seeking the most rational solution.

1.2.7. Empirical Improvement: Continuation, Inversion

The designer has to utilize the previous experience gained in the given branch of mechanical design, as well as in the allied branches, and introduce into the particular design useful features of the existing machines.

Nearly every machine is the result of the work of several generations of designers. The original model of a machine is gradually improved, equipped with new units and mechanisms, and made better as a result of new solutions, the results of creative efforts of many people. Some design solutions die due to the development of more rational ones, new technological procedures, or higher operational demands, while some, proved exceptional, remain in force for very long periods, although at times being slightly modified. The Volkswagen Beetle is an example.

In the course of time, machine specifications, both technical and economic, become more and more demanding, their power and productive capacities are enhanced, automation is more extensively introduced, and reliability and durability are bettered. Alongside the improvements to existing machines, new machines of identical purpose are devised, which are principally based on other design schemes. More progressive and viable designs win the competition.

When reviewing the history of mechanical engineering or visiting any patent office, one sees that a huge number of miscellaneous design schemes were tried before. Many of them disappeared and were completely forgotten, then were revived after many years on a new technological basis and again returned to use. Such historical reviews help the designer to avoid errors and the repetitions of the stages traversed and allow him to choose more prospective methods of machine development.

In particular, for new inventions, the inventor or designer must first conduct a thorough patent search to avoid reinventing the wheel or spending effort to solve problems other people have solved before.

It is useful to plot graphs showing the tendency of basic machine parameters over the years. Tendencies of constructional forms are well expressed by graphs showing as percentages the occurrence of various designs. An analysis of such graphs and their extrapolation make it possible for a clear picture to be drawn of future machine parameters and design trends.

Of particular importance is the study of the starting data when working on a new design. Here the main task is to choose machine parameters correctly. Partial design errors can be eliminated during machine manufacture and tests, but errors in basic parameters are often irreparable and may lead to an unsuccessful design.

The selection of parameters must be preceded by a comprehensive study of all factors determining the machine's life. It is necessary to study carefully the experience gained in the operation of foreign and domestic machines, correctly analyze their advantages and disadvantages, choose a correct prototype, and make clear the tendencies of the development and requirements of the branch of engineering for which the machine is intended.

An important prerequisite for correct design is the availability of pertinent references on materials. Apart from archives of their own products, design organizations should have at their disposal design catalogues from manufacturers.

In his particular sphere, the designer must be aware of research in his field. In addition to the research carried out in its own branch, every designer or design organization should make wide use of the experiences of other, sometimes rather alien branches of mechanical engineering. This widens the outlook of the designer and enriches his store of constructional means and devices.

It is especially useful to study the experience of advanced branches of engineering, in which the design and technological concepts, forced by the demands for the product quality (aviation) and mass production (automotive industry) continually develop new designs, methods of increasing strength, reliability, and durability, and better production techniques.

The concept of design continuation does not impose limitations on initiative. The designing of any machine presents an unlimited field of creativity to the designer, who must not invent anything already invented and must not forget the rule formulated early in the 20th Century by Gueldner: *weniger erfinden, mehr konstruiren* (invent less, design more).

The process of continuous machine improvement, caused by the ever-growing demands of the industry, results in the development of the design thought. The striving for a perfect design penetrates the flesh and blood of the designer; it becomes his life. However, one should remember that the enemy of the "good enough" is the "perfect."

The designer must constantly improve and enrich his store of design solutions. An experienced designer will invariably note and mentally take a snapshot of interesting solutions, even in machines foreign to him.

The designer must always be aware of all the latest technological processes, including physical, electrophysical, and electrochemical methods of processing (spark-erosion, electron-beam, laser, ultrasonic, dimensional electrochemical etching, processing with blast, electrohydraulic impact, electromagnetic impulses, etc.). Otherwise he will be rather restricted in his choice of the most rational shapes for parts and unable to build into the design the conditions for productive manufacture.

Among the large number of methods devised to facilitate the design insight is the method of inversion. Its essence consists of inverting the function, shape, and position of parts. In assemblies it is occasionally preferable to invert the functions of parts—for example, to turn a driving element into a driven one, to turn a leader into a follower, to make a female component out of a male one, or to convert a stationary part into a movable element.

Sometimes it becomes advisable to invert shapes of parts—for example, to change a female taper to a male one or a convex spherical surface to a concave one. In other cases it may be more suitable to shift some elements from one part to another. Each time the design will acquire new qualities. The designer must carefully evaluate all the advantages and drawbacks of the original and inverted alternatives, taking into account durability, size, suitability for industrial production, and convenience of operation, and then choose the most suitable version.

Some typical examples of inversion in mechanical engineering are shown in Figure 1.16.

- *Rod drive* (Figure 1.16*a*). In sketch I the rod is actuated by a fork lever acting through the aid of the rod pin. In inverted design II the pin is transferred to the lever. This results in a reduced transverse force acting on the rod. However, the design shown in sketch I has smaller size.
- *Tappet drive* (Figure 1.16*b*). In design I the striker of the rocker arm is flat and the tappet plate spherical. In inverted scheme II the rocker arm is spherical and the tappet flat. Moreover, the striker can be made cylindrical, thus ensuring a linear contact in the joint, whereas in design I the contact is a point.
- *Connecting rod and gudgeon pin* (Figure 1.16*c*). In scheme I the gudgeon pin is secured to the connecting rod and runs in the fork bearings; in inverted design II the pin is fixed to the fork and the bearing is in the connecting rod. In this case the inversion reduces overall dimensions and lightens the work of the bearing acquiring greater rigidity.

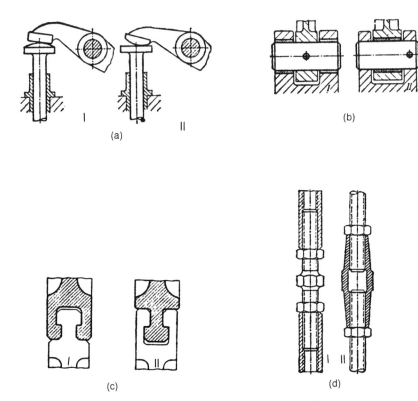

Figure 1.16. Examples of design inversion.

- *Nipple unions.* In sketch I (Figure 1.16*d*) the nipple is tightened by an internal captive nut, but in inverted version II an external nut is used. As a result, axial dimensions are decreased and radial dimensions increased. The alternative shown in sketch II is more convenient to tighten.

Example 1.3 Define the function structure and its symbolic form for the power train of an automobile

Solution A standard gearshift is assumed. The boundary of the system is *B*. The system is connected to the engine and to the ground and transmits and transforms energy between them. Also, it transmits and transforms information between the driver and the system (to clutch and gearbox). The resulting function and symbolic structure are shown in Figure E1.3.

Example 1.4 Develop design alternatives for a 90° angle speed reducer.

Solution

(a) Standard design
(b), (c) Bearings for one shaft are mounted directly on the casing instead of an intermediate sleeve

The casing is shorter in the respective direction.

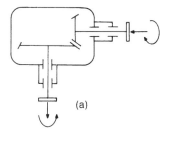

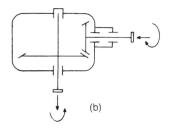

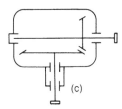

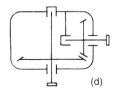

Figure E1.3

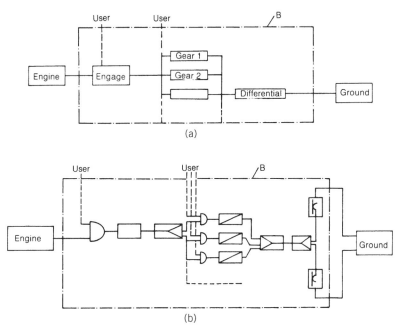

Figure E1.4. (*a*) Function structure of the automobile gear train; (*b*) Power train of Audi 80 Quatro (by permission of the FAG Co.).

Figure E1.4 (*Continued*)

(d) A more compact design. Second bearings for both shafts mounted on the casing without the use of sleeves. Smaller bearing reactions and smaller bearings

---

Design Procedure 1.1: Conceptual Design

Step 1: Define the problem. List user requirements. Perform a thorough search for codes, standards, and regulations applicable to the machine to be designed (see Section 1.4.1).

Step 2: Draw a house of quality sheet with the correlation of user wants and needs with the technical specifications and, further, the comparison with the previous or competing designs (see Sections 1.2.1, 1.2.2, 1.3.3).

Step 3: Do a thorough literature and patent survey to identify previous work in this area and intellectual property constraints (see Sections 1.2.1–1.2.3). Do a market survey to identify the user needs and market-imposed constraints.

Step 4: Develop the machine functional concept (see Section 1.2.3). Draw task decomposition diagram (see Section 1.2.4*).

Step 5: Define elementary tasks and develop solution principles for each.

Step 6. Draw morphological table. Develop design alternatives(see Section 1.2.5).

Step 7: Draw a decision matrix. Find preferred solution. Evaluate selected solution for feasibility (see Section 1.2.6).

Step 8: Apply empirical improvement: continuation, inversion. Using approximate analyses, approximately size the selected design concept (see Section 1.2.7).

Step 9: Compare with other designs of the same and other manufacturers. Use HOQ or Pugh's matrix. Finalize the selection of the design concept (see Section 1.2.6).

Step 10: Write a feasibility report.

CASE STUDY 1.1: Design a hand-held power tool to drill holes in concrete using design specifications as in Example E1.2.

The design engineer will immediately think of a motor, a gearbox for speed reduction, and a chuck to mount the drill—"business as usual." However, the designer should always generalize the task. For example, the power tool to be developed might work on principles of bullet-firing, a laser beam for drilling, chemical drilling, and so on. On the other hand, the tool might be applicable for drilling holes in lumber, metals, and other materials and could also be used as a sander, polisher, and so on.

We start by listing possible physical effects (Table 1.3) that can be used to remove material from a mass of concrete in a focused way, that is, forming a hole of uniform diameter, for only such holes can be used to locate anchors in the concrete. Searches in the literature, the patent depository, and the market and interviews with experts in the field have identified the following three possible physical effects that can be used to drill holes in concrete:

1. Cutting away the concrete
2. Burning the concrete
3. Chemically altering the concrete to a gas- or water-soluble substance

The physical principles that can accomplish these physical effects are now specific enough to be quantified by application of the laws of physics and mathematical methods. For the case at hand, the physical principles could be as in the second column in Table 1.3. Further, each physical principle can be implemented with a physical device, as shown in the third column of Table 1.3. The result is that there are at least eight physical devices that can, in principle, perform the function of making holes in concrete. At this point, a decision should be made for the physical device that should be selected.

To this end, we construct a decision matrix (Table 1.4). In the first column, rows 2 on, we place the requirements list. In the next column we place the "Importance Factor" from room C of the system RDM, the engineering requirements. We then assign a "degree of satisfaction" for each requirement by each physical device in the range 1–10 (could be any range). Then for every physical device we calculate the total score by summing the products of the elements of the respective column times the importance factor and placing the sum at the bottom cell. The bottom row (weighted total) now gives the merit of every one of the solution functional concepts, in this case the helicoidal impact drill. Now, however, we have explored all possible solution principles and in fact can identify solution 7 as a possible, though somewhat distant, alternative.

In our bottom-up procedure, we now need to identify other components of the power tool that will complete the functional concept. Because we use as a basis an impact drill, a search in our sources of information (our experience, literature, patents, experts, etc.) will reveal that we need a mechanism to generate the rotary motion of the drill, the periodic impact, the static force that must be applied to the drill, and a power source that will power all of the above.

From the machine functional concept we know that the power tool needs to run the drill with rotary motion and periodic impact. Then we need to turn the system on and off and provide speed control, if possible. Further, wc know that a drill will be a separate piece and not the exit shaft itself, because it has to bc made from a hard steel, which is not good for shafts and, because of rapid wear, has to be replaceable. Thus, we need to have some way of gripping the drill, in addition to using different diameter drills. Finally, it appears that there will be moving parts and we need to support and house them properly. A first attempt to organize this will be a function tree diagram as shown in Figure CS1.1.

We note that usually the different kinds of motors are high-speed, especially in portable equipment, because their mass is then small. Because drilling in concrete is known to require, in general, low rotational speed, it seems logical to require some kind of speed change between the motor and the drill. Moreover, it seems prudent to provide for some kind of speed control. Finally, it seems logical to incorporate support to the individual components within themselves, that is, bearings of the motor will be considered an integral part of the motor. Thus, we modify the function tree to include these features (Figure CS1.2).

Now we need to express the interconnections of the different subfunctions, that is, the exchange of energy, matter, and information. The *function structure* of the design is shown in Figure CS1.3 for the power tool design based on the function tree of Figure CS1.2. Each level of the function tree corresponds to one level of the *function decomposition* diagram. Finally, in the third level we have a block diagram describing the *function structure*. This is shown in plane view in Figure CS1.4.

Finally, from the diagram of Figure CS1.4 we can construct a symbolic function structure as in Figure CS1.5 showing the logical interrelations of the different elements and the system control.

The function tree diagram (Figure CS1.2), the function decomposition diagram (Figure CS1.3), the function structure, (Figure CS1.4), and the symbolic function structure, (Figure CS1.5), are diagrams useful in dealing with the problem of finding design alternatives. During the phase of finding design alternatives, these diagrams are under continuous scrutiny because improvements can be made along the way.

In the bottom row of Figure CS1.2, the bottom plane of Figure CS1.3, and in Figures CS1.4 and CS1.5, if one can find a number of physical devices performing the same operation for each elemental function, then design alternatives can start being produced.

In Figure CS1.2, we can start listing under every subfunction box in the bottom row, called *elemental functions,* a number of possible alternative solutions (Figure CS1.6).

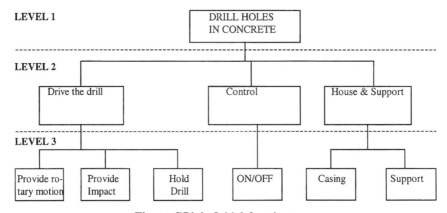

Figure CS1.1. Initial function tree.

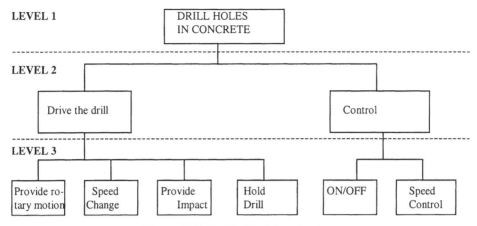

Figure CS1.2. Modified function tree.

Physical devices have to be found for the various bottom row subfunctions, and these devices must eventually be combined into a working structure. The concretization of the working structure will lead to the solution concept. For every element of the bottom row, we need to develop the sequence ⟨physical effect, physical principle, physical device⟩. A physical device must reflect the physical effect needed for the fulfillment of a given function.

Thus, one or more physical devices can be found for every elemental function. Every combination of possible physical devices under every elemental function constitutes an alternative design concept. For example, one such design alter-

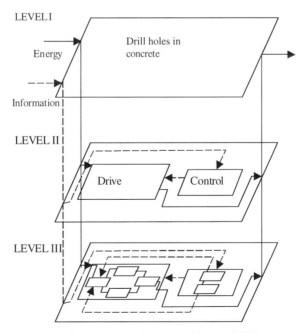

Figure CS1.3. Function decomposition, machine to drill holes in concrete.

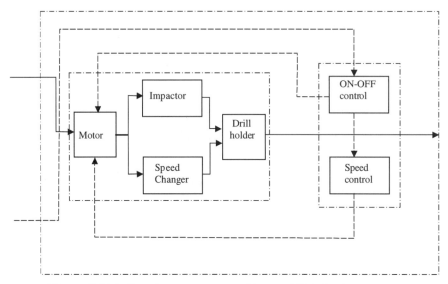

Figure CS1.4. Function structure, machine to drill holes in concrete.

native would be DC motor, belt drive, cam, tapered hole, switch, and mechanical speed changer, as shown in Figure CS1.6.

   We are not concerned at this point with the quality of each alternative; our aim is to develop as many alternatives as possible. As we form alternatives, however, we discard the obviously inconsistent. For example, if in one alternative we select an electric motor, then for an impactor we cannot select an air piston because one cannot ask for two sources of power, electricity and air. In forming the alternatives, one can combine functions, add new elements, or even change the task decomposition if a better overall concept comes out of this elaboration.

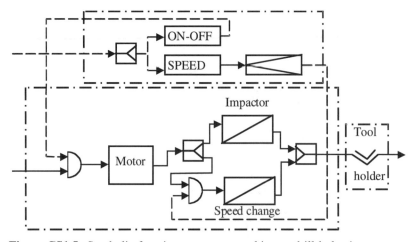

Figure CS1.5. Symbolic function structure, machine to drill holes in concrete.

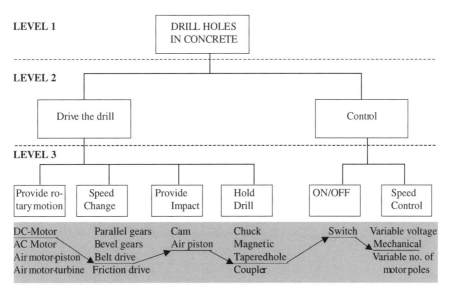

Figure CS1.6. Development of element physical devices.

In our bottom-up procedure, we need to identify other components of the power tool that will complete the functional concept. Because we use as a basis an impact drill, a search in our sources of information (our experience, literature, patents, experts, etc.) will reveal that we need a mechanism to generate the rotary motion of the drill, the periodic impact, the static force that must be applied to the drill, and a power source that will power all the above.

One can select alternatives directly from Figure CS1.6. This can be done in a systematic way in a morphological table (Figure 1.8). Every row corresponds to an elemental function, and, from column 2 and higher, different physical devices appear that correspond to different devices embodying the elemental function on the left at column 1. If we connect with a broken line any physical device in the first row with any physical device in the second row, and so on until we reach the last row, this broken line corresponds to a design alternative.

The result will be a list, hopefully an extensive one, of design alternatives. For the power tool example, a limited (in the interest of space) list of alternatives would be:

1. AC motor with parallel gears, cam, chuck, switch, and mechanical speed changer

2. AC motor with bevel gears, cam, chuck, switch, and resistor speed changer

3. DC motor with parallel gears, cam, chuck, switch, and resistor speed changer

4. Air-piston motor with parallel gears, piston-driven impactor, chuck, switch, and mechanical speed changer

5. Air-piston motor with parallel gears, piston-driven impactor, chuck, switch, and air pressure-controlled speed changer

TABLE CS1.1. Decision Matrix for Conceptual Design of the Power Tool

Requirements List	Importance	Solution Concept Number							
		1	2	3	4	5	6	7	8
Capacity for 6–25 mm diameter holes	10	10	10	10	10	8	8	8	8
Drilling speed 0.2 mm/s minimum	8	8	8	6	8	8	8	8	8
Maximum weight 200 N	10	8	8	7	7	4	4	4	4
MTBF 6000 hours	7	10	6	5	5	6	8	8	8
Should comply with applicable UL and other standards	8	8	6	6	6	8	8	8	8
The noise level at 3 m distance should not be more than 70 DB above-background noise	5	6	6	6	6	4	6	6	6
Manufacturing cost should be less than $150	6	10	6	4	4	4	2	4	2
Should not exceed the ISO[a] allowable vibration levels at hands of user	6	10	6	4	4	4	2	4	2
Weighed total		468	412	401	361	348	356	368	356

[a]International Standards Organization. Its U.S. member organization is NIST, the National Institute of Standards and Technology.

6. Air turbine motor with parallel gears, cam, chuck, switch, and mechanical speed changer
7. Air turbine motor with parallel gears, cam, chuck, switch, and air pressure-controlled speed changer

Preliminary Evaluation of System Design Alternatives. The several alternatives found in the previous section need to be evaluated in order to select the best. The decision matrix and the Pugh's matrix are shown in Tables CS1.2 and CS1.3, respectively.

It is apparent that concept 1 is the winning concept: AC motor with parallel gears, a cam, a chuck, switch, and mechanical speed changer. The conceptual

TABLE CS1.2. Pugh's Comparison Matrix for Conceptual Design of the Power Tool

Requirements List	Solution Concept Number							Datum = 1
	2	3	4	5	6	7	8	
Capacity for 6–25 mm diameter holes				−	−	−	−	
Drilling speed 0.2 mm/s minimum		−						
Maximum weight 200 N		−	−	−	−	−	−	
MTBF 6000 hours	−	−	−	−	−	−	−	
Should comply with applicable UL and other standards	−	−	−					
The noise level at 3 m distance should not be more than 70 DB above-background noise				−				
Manufacturing cost should be less than $150	−	−	−	−	−	−	−	
Should not exceed the ISO[a] allowable vibration levels at the hands of the user	−	−	−	−				
Weighed total	−4	−6	−5	−6	−4	−4	−4	

[a]International Standards Organization. Its U.S. member organization is NIST, the National Institute of Standards and Technology.

design is concluded with a concept sketch-layout (Figure CS1.7). It is to be noted that no dimensions appear on the concept sketch. Assigning materials, manufacturing methods, and dimensions to different elements, such as frame, shafts, bearings, and gears, will be the subject of embodiment design, which will be discussed over most of this book.   □

1.3. EMBODIMENT DESIGN

1.3.1. Design Synthesis

The design concept developed in the previous section is highly abstract. We have a sketch-layout that shows the machine operation and, perhaps, estimates of some of the machine design parameters. We need to complete the design of the machine by selecting all the critical design parameters.

We observed in Figure 1.9 the process of task decomposition, by which we take the overall task and systematically split into small, easy-to-implement subtasks, the function modules. After we select the physical devices that embody the function modules, we need to put them together into a working system, the machine. This process is called synthesis (in Greek, put together). During synthesis it is important to separate the critical design parameters from the secondary ones.

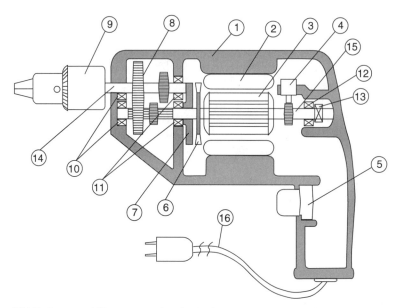

Figure CS1.7. Impact drill concept sketch. 1, frame; 2, armature; 3, motor rotor; 4, commutator assembly; 5, ON/OFF switch; 6, cooling fan; 7, impact cam; 8, two-speed gear assembly; 9, chuck; 10, front-end bearings; 11, intermediate bearings; 12, rear-end bearings; 13, thrust bearing; 14, main shaft; 15, motor shaft; 16, electric cord.

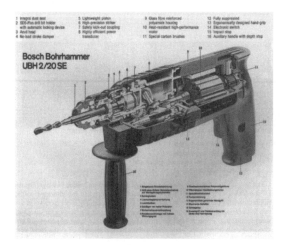

Figure CS1.8. A similar-function Bosch UBH 2/20 SE impact drill (by permission of the FAG Co.).

Attempting to conceive all elements of a design at one time is a typical error characteristic of young designers. Having finalized the design specifications, the designer often begins at once to draw the complete design in all its details. Such a method makes the design an irrational one because it forms a mechanical string of constructional elements and units, poorly arranged.

It is necessary to begin the synthesis process with the solution of the main design problems, that is, selection of kinematic and power trains and correct sizes and shapes of parts and determination of their most suitable relative positions. Any attempt to fully describe parts at this stage of design is not only useless, but harmful, because it draws the attention from the main problems of synthesis and confuses the logical course of developing the design.

Design synthesis must always be accompanied by calculations, if only tentative approximations. One should never rely merely on judgment when selecting dimensions and shape of parts. Of course, there are very skillful designers who almost without mistakes can establish sizes and cross sections ensuring stress levels acceptable for the given branch of engineering, but this is a doubtful virtue. On the other hand, to depend wholly upon calculations is wrong. The existing techniques of strength calculations do not consider many factors influencing the fitness of the design, and there are some parts (e.g., housings) that cannot be calculated except by tedious methods. Moreover, other factors besides strength affect the sizes of parts. For example, the design of cast parts in the first place depends on casting technology. For parts being mechanically machined it is necessary to consider their resistance to cutting forces and give them a certain rigidity. Thus, alongside calculations, the designer must refer to existing designs and introduce, if necessary, well-founded corrections.

Another prerequisite for correct designing is a continuous consideration of the manufacturing problems; at the very beginning the component should be given a technologically rational shape. A skilled designer makes a part easy to produce at the design stage. Young designers should constantly consult production engineers.

Design must be based on standard dimensions (fitting diameters, sizes of keyed and splined connections, diameters of threads, etc.) and maximum unification of standard elements as much as possible.

Design is not always an one-way road. Often problems overlooked in the first design must be corrected. Sometimes it is necessary to return to designs rejected earlier or to develop new ones. Design is an iterative process.

Some units are not always successfully designed from start. This should not discourage the designer who has to devise some tentative alternatives and raise the design to the required level in the process of further activities. In such cases it is useful, as Italians say, *dare al tempo il tempo* (to give time to time), that is, to take a breathing space, after which, as a result of subconscious thought, the problem is often solved. After a while the designer looks at the outline drawing in another light and sees the mistakes made at the first stage of the development of the main design idea.

Sometimes the designer unintentionally loses his objectivity and does not see the drawbacks of his favorite variant or the potentialities of other versions. In such cases impartial opinions of outsiders, the advice of senior designers and coworkers, and even their faultfinding criticism turn out to be most helpful. Moreover, the sharper the criticism, the greater the benefit derived by the designer. Of course, the rule "perfect design is the main enemy of the good design" must be always observed.

A general rule is: the wider the discussion of the design and the more attention the designer pays to useful advice, the better the design becomes and the more perfect the design.

The cost of designing is only a small portion of the machine manufacturing expenditures (excluding machines produced individually or on small production). In the final analysis, the greater the development work on the design, the more the savings in the machine cost, time of manufacture, and finishing, the better its quality, and the greater the economic gains over the machine's service life.

If the overall dimensions of a project allow, its design is best drawn to a 1:1 scale. This facilitates selection of required dimensions and sections and enables a realistic presentation of machine part proportions, their strength and rigidity, to be individually and as a whole obtained. In addition, such a scale obviates the necessity for a large number of dimensional specifications and simplifies subsequent stages of design, in particular detail drawings, because such dimensions can be taken directly from the drawing.

Tracing to a reduced scale, particularly at reductions exceeding 1:2, strongly obstructs the design process and distorts proportions and clarity of the drawing representation.

If the dimensions of design deny the use of a 1:1 scale, then at least individual units and groups should be composed to natural size. Design of simple components may be developed in one projection if the drawing explains it sufficiently well and clearly. The form of the construction in a cross-sectional plane will be filled by a three-dimensional imagination. However, when applied to more sophisticated designs, this method may cause serious errors; therefore, the design must be developed in all necessary projections and cross-sections. Computers can play a major role here, as it will be shown in Chapter 2.

The fulfillment of an arrangement drawing is a continuous process of research, trial, approximation, development and comparison of alternatives, and rejection of unsuitable versions. Sketches, such as the one in Figure 1.14, should be lightly traced with a pencil during the design. Corrections, follow one another, which means that an eraser is applied more often than a pencil.

Machine design synthesis is further exemplified in Chapter 7.

1.3.2. Design Analysis and Evaluation

Only in a very few cases it is possible with preliminary calculations to design a final version of a machine or an element. In most cases, the designer, using his experiences and simple design calculations, makes a first draft of his design in the form of free-hand sketches with some basic dimensions. This is the product of his experience, creativity, imagination, and calculations. It has to be thoroughly analyzed to verify that:

1. It meets the design objectives.
2. It has adequate strength, stability, durability, safety, etc. This process of design analysis is essential to any machine design process. Since most of this

book is devoted to design analysis of the machines, this subject will come up quite often in the chapters to follow.

To evaluate a specific design, one must have a complete design analysis. Among other reasons, he has to verify not only that the machine designed meets the specifications but also that it is close to them and is not overdesigned. A design analysis is also an invaluable tool in solving problems at later times during the development or operation of the machine.

1.4. MACHINE DETAIL DESIGN

Codes and Standards

Long experience in machine design, manufacturing, and operation has accumulated valuable data. Such data have been summarized and evaluated in many areas by government or industry standards organizations.

Engineering standards have many purposes, such as:

1. *Utilization of experience*. The designer should not reinvent the wheel every time he faces a problem with rotary motion.
2. *Interchangeability of components*. This need became apparent (where else) at time of the first mass production of firearms and ammunition. Every bullet ought to fit properly with every firearm. Now, if you buy a ball bearing of a certain type, it can replace a defective bearing of the same type regardless of its manufacturer.
3. *Setting safety and quality standards* for a range of products. Such standards are adopted many times by governments to form mandatory codes.

The need for the designer to conform to such standards is apparent. However, he should not simply design "to available standards" but use them with continuous criticism, wherever possible *and allowable*. The reason is that many times some standards are superconservative because in one simple standard they have to include a variety of different cases, and the safety margins are not uniform. Dimensioning standards are the most universally accepted. For this case, the International Standards Organization has issued several standards, which can be obtained from the national affiliate of ISO in every country. Some of the national and international standards organizations mentioned in this book are listed in Table 1.6.

1.5. PHILOSOPHICAL AND LEGAL ASPECTS OF MACHINE DESIGN

1.5.1. Aesthetics in Machine Design

An aesthetically pleasing appearance is by no means a luxury for a machine. Many times, for machines of established principle in particular, appearance is the only factor that counts. And for every machine it is a definite advantage.

A simple number cannot express the quality of a machine. Several factors, often contradictory, are included in the overall judgment of the quality, such as cost, production rate, reliability, and durability. Most times not all necessary information is available. The judgment of quality then becomes a relatively subjective affair, which can be greatly influenced by the personal impression of the machine on the end user. Moreover, it is now verified that productivity and rate of accidents are greatly affected by the work environment, where the appearance of the workplace is not of the least value.

One basic design rule for the form, though not universally accepted, is Sullivan's rule "form follows function." The first impression the machine or product has on the observer must be closely related to the machine function.

Sergio Pininfarina, the noted car designer, wrote on the function versus appearance question for car design (Pininfarina 1978):

> Recent years have seen traditional design parameters joined by others arising out of matters of safety and the energy crisis, constraints which have brought with them a hole series of new functions.
>
> As one of the mind's main categories, ancient Greek philosophy pointed to the Beautiful, meaning by this a concept inclusive of functional aesthetic (Useful) and even ethical (the Good) implications. The Beautiful was thus an extremely important achievement, laboriously or miraculously (with the help of the gods) arrived at by way of the convergence of varying and, at times, contrasting requirements.
>
> Modern philosophers, especially the idealists and their followers, particularly Croce, have attempted to split the Beautiful and the Useful, and render them autonomous.
>
> In art, these two interpretations and classifications are readily observed in so-called classical and modern art: the eternal dilemma or contrast between Form and Content (Function) has set its imprint on almost all human production, both figurative and literary.
>
> But what is the situation in other not strictly artistic fields? Can we speak of the Beautiful and the Useful, of form and content, of Appearance and Function?
>
> I believe that where these definitions concern creative activities the reply is positive, the question or questions are legitimate. The design of a car is certainly creative in so far as it is a speculative mental activity: creativity applied to an object produced by a modern industrial society, which is more complex precisely for that reason.

And he concluded:

> By contrast, artistic creativity is and will always remain the work of just one or two people and the resources he or they need to express their art, however much technological sophistication is deployed, will never be comparable with what we have been talking about, as the major contribution comes from the mind, eye, and hand. The very definition of Man, however, means that this invaluable contribution will not fall by the wayside; at the most it might become sublimated in a higher and more pregnant sphere than the one in which it has lived until now.
>
> To conclude, then, I would not talk about "Function versus Appearance" but "Function in Harmony with Appearance."

In many cases, colors indicate a particular material or process, such as pipe colors (red for steam, green for water, etc.)

A very important aspect of form is style, which can be roughly defined as the generally accepted appearance at a certain time. An egg-shaped car would have looked ugly 30 years ago but today looks beautiful. An old Remington typewriter would not look right in a modern office—it is out of style.

Marketing demands greatly influence appearance. Producers for a large internal market do not put great emphasis on appearance. Industries, which mostly export and thus have greater competition, are very careful on form. A typical example again is U.S. automobiles on the one hand and Japanese and European on the other. Moreover, Italian-made machinery has a very carefully designed appearance.

1.5.2. Machine Design Ethics

Ethics is the division of philosophy that studies the rules that should govern the free activities of man. We can find ethical discussions in the written and oral tradition since the beginning of civilization. Ancient Greek philosophers discussed ethics, notably Socrates and Plato, and ethics became a part of the philosophical discourse after Aristotle founded it as a science with his books *Ethica Nicomachea, Ethica Eudemia,* and *Magna Moralia.* Two of the main questions of philosophical ethics are:

1. What are the rules that should govern the free activities of man—in Aristotle's terminology, the "good life" in the sense of the "right way of life."
2. Do these rules exist *a priori,* or do they reflect the contemporary ideas of a given society for rules that should govern the free activities of man?

The first problem will be discussed in conjunction with the second problem, which is central to the philosophical discussion of ethics and has very important implications for the ethics of machine design.

Aristotle is not very clear about the epistemological status of ethical rules. He seems to assume that the principles of the good life are self-evident, but the purpose of ethics is to facilitate the life of people in society.

Descartes maintained that intellect and logic are capable of distinguishing between good and bad. Kant believed that ethical rules were given to people *a priori,* together with free will, from the beginning of man and thus ethical rules are invariant in time.

The utilitarian school believed that the intellect can indeed judge the good and bad and that what is good may vary in place and time. They recognized, however, that among different times and places the similarities are more notable than the differences.

Most religions maintain that God revealed the rules of ethics to people. As such, they are invariant through time, subject to changes that may be made by God.

Engineering ethics is a direct extension of the medical ethics or legal ethics to which the medical and legal professions expose their practitioners. It has been a practice for quite some time to legitimize the establishment of a profession in the eyes of the public by introducing and enforcing a code of ethics for the members of that profession. This practice has its roots in medieval and even ancient times, and there were good reasons to establish such practices.

The physician has an important function to perform. There is no way for the layman to judge whether a physician is indeed a physician, let alone even whether he is a good physician. Because he is trusting the physician with his life, and death is a nonreversible process, he needs some external assurance of the physician's competence. In the early days of the historical period, given the prevailing superstition and the lack of detailed regulation and adequate enforcement, the oath was the main means of establishing one's credibility. Thus the Hippocratic Oath became the main vehicle for giving confidence to the public, at least as to the good will and diligence of the physician. From medieval times on, regulation became necessary, and the professional societies started qualifying professionals by several means:

1. Testing the competence of professionals as they entered the profession
2. Having and enforcing a strict code of ethics
3. Establishing a legal basis for prohibiting practice of the profession for nonmembers of the professional society.

Further, the well being of their members and their material remuneration for their services would be proportional to the esteem in which they were held by the society, and the code of ethics was a proof that members of the profession were held to the highest of ethical standards. Thus, clients could trust them with their lives (physicians or lawyers) or their property and freedom (lawyers) and pay higher fees for their services.

Engineers adopted their practices in England and continental Europe later, establishing professional societies that had substantial regulatory powers. In most European countries, they were given the legal standing to administer tests or establish regulations for admission to the society and the profession and make membership a prerequisite for practicing the profession.

The American engineering societies followed later, but owing perhaps to a more liberal enterprise system did not have the legal standing to establish mandatory professional rules and thus only introduced codes of ethics, like their European counterparts.

Engineering design ethics change over time for many reasons:

1. Codes of ethics were established when professionals (mainly physicians and lawyers) worked alone and had total responsibility over the professional services they were offering to the public. Engineers formerly worked as individual designers or consultants, and some still do, but the vast majority of engineers are employed by large companies that have separation of responsibilities. Thus, each engineer knows only a more or less small part of the whole picture and is thus unable to judge the social implications of the final design, simply because he usually does not know exactly what the final design will be. Therefore, the scope of the code of ethics is practically that of a code of conduct and is limited to the conduct of the engineer in relation to the society, his colleagues, and his employer.
2. Most of the articles of the codes of ethics are now part of the common law, such as bidding practices and plagiarism.

3. Enforcement of the codes of ethics is now very difficult in the United States due to the litigation explosion. Professional societies are now hesitant about enforcement because of the danger of long and very expensive lawsuits that those accused of professional misconduct will most probably initiate.

4. Professional engineering societies in the United States established themselves as open scientific societies rather than closed professional societies. For example, there is no true restriction that a member of the ASME be a mechanical engineer at all. Thus, the code of ethics offers very little comfort to someone who wants to hire a mechanical engineer: he might be an ethical person, but he might have no relation to mechanical engineering whatsoever.

5. Until about mid-century, engineering activities were totally negligible in comparison with the global forces of nature. Even today, the total energy produced by all humans and all technological processes is about four orders of magnitude less than the solar energy alone that the earth receives. It is apparent that man cannot even make a dent in the global process of nature. However, on a smaller scale, the results of technology can be devastating. Examples are the floods of the Mississippi River due to the irrational flood control systems, the extensive pollution of air and water, and unsafe transportation systems. Though some of these problems have been exaggerated for political reasons, they can get out of hand if not addressed early. The present approach of the professional societies is merely euchological: *The engineer must hold paramount the public safety.* The solution of the problems is left to legislators, who are concerned mostly with the legal aspects being mostly lawyers.

Decisions in machine design are no longer made by a single person or even a single organization but, in most cases, by society as a whole. For example, better automobiles and highways can be built to reduce the number of automotive accident deaths in the United States from about 50,000 per year to, say, 10,000 per year. The society, through its elected representatives, every year makes the decision that "the cost involved does not justify saving 40,000 lives per year." Thus, one cannot assign full ethical responsibility to the design engineer who did not design better cars or better highways.

1.5.3. The Legal Constraints: Product Liability

It is essential that a machine design engineer be familiar with the issues of safety and liability, as they are an integral part of human–product interaction and greatly affect the perceived quality of the machine. Moreover, liability suits increased from 100,000 in 1966 to more than one million in 1998 (Pechner 1983). A substantial portion of those refer to machines, such as automobiles, household appliances, and industrial machinery. Fortunately, lawyers sue the manufacturers, from whom they hope to recover substantial money, and thus they do not sue individual machine designers, but this courtesy may not last for long. Machine designers should try to build safety in the machines they design, for the protection of their employer, of themselves, and of society at large.

Design for safety means ensuring that the product will not cause injury or loss. Two issues must be considered in designing a safe product:

1. Who or what is to be protected from injury or loss during the operation of the product?
2. How is the protection actually implemented in the product?

The main consideration in designing a machine for safety is the protection of humans from injury, the loss of other property affected by the machine and the impact on the environment in case of failure.

The designer needs to consider the effect the machine can have on other machines or systems, either during normal operation or upon failure. For example, the manufacturer of a fuse for a circuit breaker that fails to cut the current flow to a device may be liable because the fuse did not perform as designed and caused loss of or injury to a different machine or device. There must also be concern for the machine's effect on the environment.

There are three ways to design a machine for safety:

1. Design the machine against failure, that is, so that the device poses no inherent danger during normal operation or in case of failure. The main vehicle for this is redundancy. For example, a control circuit for regulating the maximum speed of a machine is not adequate protection for overspeeding. The designer needs to provide independent means of protection against overspeed, that is, a centrifugal brake, a high electric current switch, a sound or red light warning to the operator, and so on.
2. Addition of protective devices to the machine, such as shields around rotating parts, crash-protective structures (as in automobile body design), light array switches, that stop the machine upon human proximity, and "dead man's switches" that automatically turn a device off (or on) if there is no human contact.
3. Use of a warning to point out dangers inherent in the use of a product, such as labels, loud sounds, and flashing lights.

In layman's terms, "For safety, *design* for it. If you cannot design, *protect*. If you cannot protect, *warn*."

It is very important that the machine be designed for safety. It is very difficult to design protective shields that are foolproof, and warning labels do not absolve the designer from liability in case of an accident. The only truly safe machine is one with safety designed into it.

Product liability is the name of the special branch of law dealing with alleged personal injury or property or environmental damage resulting from a defect in a product. It is important that design engineers know the extent of their responsibility in the design of a product. If, for example, a worker is injured while using a machine, the designer(s) of the machine and/or the manufacturer may be sued to compensate the worker and the employer for the losses incurred.

A products liability suit is a common legal action. It involves the plaintiff (the party alleging injury and suing to recover damages) and the defendant (the party being sued). Lawyers are using technical experts, both plaintiff and defense, to testify about the operation of the machine that allegedly caused the injury.

The product liability issues might be:

1. *The product was defective by design.* Typical charges include the failure to use state-of-the-art design considerations, that improper calculations were made, that poor materials were specified, that insufficient performance testing was carried out before the machine was released for the market, and that commonly accepted standards were not followed. In order to protect themselves from these charges, designers must, in addition to designing a safe machine:

   - Keep good records to show all that was considered during the design process. These include records of calculations made, standards considered, results of tests, and all other information that demonstrates how the product evolved.
   - Use commonly accepted standards when available. "Standards" are either voluntary or mandatory requirements for the product or the workplace; they often provide significant guidance during the design process. They are issued by governmental agencies, engineering societies, industry associations or laboratories, or the international and national standards organizations (see Section 1.4.1).
   - Use state-of-the-art evaluation techniques for proving the quality of the design before it goes into production.
   - Follow a rational design process (such as that outlined in this book) so that the reasoning behind design decisions can be defended.

2. *The design did not include proper safety provisions.* It is essential that the design engineers foresee all reasonable safety-compromising aspects of the machine during the design process.

3. *The designer did not foresee possible alternative uses of the product.* In many cases, courts have demanded that machines must be "idiot-proof."

Other cases of product liability might be that:

- The machine was defectively manufactured by use of poor workmanship, materials of inferior quality, or improper manufacturing processes or the designer's drawings not being followed.

TABLE 1.5. Hazard Frequency of Occurrence (MIL-STD 8828)

Description	Level	Individual Item	Inventory
Frequent	A	Likely to occur frequently	Continuously experienced
Probable	B	Will occur several times in life of an item	Will occur frequently
Occasional	C	Likely to occur sometime in life of an item	Will occur several times
Remote	D	Unlikely, but possible to occur in life of an item	Unlikely, but can reasonably be expected to occur
Improbable	E	So unlikely, it can be assumed that occurrence may not be experienced	Unlikely to occur, but possible

· The product was improperly advertised.
· Instructions for safe use of the product were not provided for the user.

Because safety is such an important concern in military operations, the Armed Services have a standard—MIL-STD 882B, System Safety Program Requirements—focused specifically on ensuring safety in military equipment and facilities. This document gives a simple method for dealing with any hazard, which is defined as a situation that, if not corrected, might result in death, injury, or illness to personnel or damage to or loss of equipment.

MIL-STD 882B defines two measures of a hazard: the likelihood or frequency of its occurring and the consequence if it does occur. Five levels of frequency of occurrence are given in Table 9.5, ranging from "Improbable" to "Frequent." Table 9.6 lists four categories of consequence of occurrence. These categories are based on the results expected if the hazard does occur. Finally, in Table 1.7, frequency and consequence of recurrence are combined in a hazard-assessment matrix. By considering the level of frequency and category of consequence, we find a hazard-risk index. This index gives guidance on how to deal with the hazard. For example, say that during the design of a power lawnmower, the possibility of using the mower as a hedge trimmer was indeed considered.

What action should be taken? First, using Table 1.5, we decide that the frequency of occurrence is either remote (D) or improbable (E). Most likely it is improbable. Next, using Table 1.6, we rate the consequence of the occurrence as critical, category 11, because severe injury may occur. Then, using the hazard-assessment matrix (Table 1.7), we find an index of 10 or 15. This value implies that the risk of this hazard is acceptable, with review. Thus, the possibility of the hazard should not be dismissed without review by others with design responsibility. If the potential for seriousness of injury had been less, the hazard could have been dismissed without further concern. The very fact that the hazard was considered, an analysis performed according to accepted standards, and the concern documented might sway the results of a products liability suit.

1.5.4. Societal Constraints: Green Design, Recyclability

There is evidence of ethical concern for environmental issues in very early times. Hippocrates (5th Century BC) wrote a treatise, *Airs, Waters and Places,* in which he discussed the problem of water pollution (Landels 1978).

TABLE 1.6. Hazard-Consequence of Occurrence (MIL-STD 8828)

Description	Category	Mishap Definition
Catastrophic	I	Death or system loss
Critical damage	II	Severe injury, minor occupational illness, or major system damage
Marginal damage	III	Minor injury, minor occupational illness, or minor system
Negligible	IV	Less than minor injury, occupational illness, or system damage

TABLE 1.7. Hazard-Assessment Matrix (MIL-STD 8828)

Frequency of Occurrence	Hazard Consequence Categories			
	I	II	III	IV
	Catastrophic	Critical	Marginal	Negligible
A. Frequent	1	3	7	13
B. Probable	2	5	9	16
C. Occasional	4	6	11	18
D. Remote	8	10	14	19
E. Improbable	12	15	17	20

Hazard-risk index	Criteria
1–5	Unacceptable
6–9	Undesirable
10–17	Acceptable with review
18–20	Acceptable without review

In the design of the Roman aqueducts, elaborate measures were taken to guard against water pollution. Vitruvius's *De Architectura* advised against the use of lead pipes, to protect the public from lead poisoning that was apparent in the workers in lead-processing shops. He recognized that lead oxide (*cerussa*) was the toxic substance responsible for lead poisoning. It is also known that noise pollution was a problem in ancient Rome, and vehicles were not allowed to operate in the center of the city during certain hours.

In this chapter, we have seen the machine as a processor of energy, matter, and information. Some of the output of the machine might not be intended for the machine's mission but a byproduct of the processes used in the machine. Table 1.8 lists some of these unwanted byproducts.

Materials recycling plays a major role in design for a clean environment. It reduces the environmental impact not only of materials that otherwise would have had to be produced, but of the energy production that would have been needed.

Materials can be recycled in different ways (Figure 1.17):

TABLE 1.8. Types of Pollution by Different Machines

Hazardous byproduct	Examples of polluting machines
Noise	Motor vehicles, construction equipment, HVAC equipment
Radiation	Nuclear reactors, x-ray equipment for weld inspection, TV sets
Air pollution	Internal combustion engines, refrigeration equipment, boilers
Water pollution	Metal processing machinery, chemical process machinery

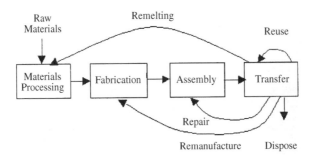

Figure 1.17. Recyclability.

1. By *reuse* of parts or systems that are needed only at some phase of the production and they can be reused in production.

2. By *repair* and reuse of parts that are replaced because of wear.

3. By *remanufacture,* the complete disassembly, repair, or replacement of worn parts and reassembly with the quality control procedures for the manufacturing of a new machine or part.

4. By *remelting,* the reduction of machines or parts to a raw material, usually called scrap, for producing new parts or machines.

Thus, design for recyclability has to be viewed in the above broad sense and provision should be made at the design stage to facilitate all four levels of recyclability.

Nonferrous materials such as aluminum, copper, and copper alloys are prime candidates for recycling because they can become almost completely recycled. The recovery rate for aluminum cans is about 65% for nondeposit states and 90% for deposit states (Lindbeck and Wygant 1994). *Iron and steel* scrap accounts for nearly 35% of all steel production in the United States. Some processes can reduce drastically the efficacy of remelting, such as zinc plating (used in automotive body production). *Glass,* although the cost of recycling is high compared with the cost of glass produced from raw materials, is a prime source of recyclable material. *Plastics* can be recycled, depending on the type of plastic. Thermoplastics can be remelted, but thermosetting plastics cannot, and the selection of plastic materials in design has to consider this property.

Figure 1.18. Depending on the design of each particular scrap metal item, this scrapyard can be a wasteland or a valuable resource for raw materials.

Design Procedure 1.2: Embodiment and Detail Machine Design

Step 1: Perform design procedure 1.1, conceptual design.

Step 2: Perform design synthesis (see Section 1.4, also Chapter 7).

Step 3: Perform machine sizing using the machine macromodel (Chapter 7).

Step 4: Assign tasks to each component (see Chapters 3 and 7).

Step 5: Select materials and manufacturing processes to each component (see Chapter 4).

Step 6. Perform a kinematic and a loading analysis (see Chapters 2, 3)

Figure 1.18. Depending on the design of each particular scrap metal item, this scrapyard can be a wasteland or a valuable resource of raw materials.

Step 7: Perform machine component design (see Chapters 5, 8–14).

Step 8: Prepare detail drawings, and a parts list (see Section 1.4).

Step 9: Perform a final design analysis to verify satisfaction of design requirements.

Step 10: Review the design for intellectual property protection, potential ethical problems and product liability, manufacturability, recyclability (see Sections 1.4, 1.5, and Chapter 4).

Design Procedure 1.3: Machine Development

Step 1: Design procedure 1.1, conceptual design.

Step 2: Design procedure 1.2, embodiment and detail design.

Step 3: Production planning, including selection of manufacturing processes and quality control procedures.

Step 4: Design of the production facility, if not already available.

Step 5: Cost analysis, financial analysis, estimation of production cost for the machine.

Step 6: Preparation of a manual that will include installation, operation, maintenance and repair procedures.

Step 7: Formal presentation to company management.

REFERENCES

Clausing, D. 1994. *Total Quality Development*. New York: ASME Press.

Deming, W. E. *Out of the Crisis,* Cambridge, Mass.: MIT Center for Advanced Engineering Study, 1982.

Dimarogonas, A. D. 1997. "Philosophical Issues in Engineering Design." *Journal of Integrated Design and Process Science* 1(1):94–113.

Hinhede, U., et al. 1983. *Machine Design Fundamentals.* New York: John Wiley & Sons.

Hubka, V. 1982. *Principles of Engineering Design,* trans. W. E. Eder. London: Butterworth Scientific.

Krick, E. V. 1969. *An Introduction to Engineering and Engineering Design.* New York: John Wiley & Sons.

Landels, J. G. 1978. *Engineering in the Ancient World.* Berkeley: University of California Press.

Lindbeck, J. R., and R. M. Wygant. 1994. *Product Design and Manufacture.* Englewood Cliffs, N.J.: Prentice Hall.

Orlov, P. 1976. *Fundamentals of Machine Design.* Moscow: Mir.

Pahl, G., and W. Beitz. 1996. *Konstruktionslehre.* Berlin: Springer-Verlag.

Pechner, D. "Products Liability and the Engineer." *Automotive Engineering* (November 1983):33–38.

Pininfarina, S. 1978. "Future Trends." In IME Proceedings, October, *Function versus Appearance in Vehicle Design.*

Reuleaux, F. 1876. *The Kinematics of Machinery,* trans. B. W. Kennedy. London: Macmillan.

Rodenacker, W. G. 1970. *Methodisches Konstruieren,* 2nd ed. Berlin: Springer Verlag.

Sandor, G. N. 1964. "The Seven Stages of Engineering Design," *Mechanical Engineering,* 86(4):21.

Suh, N. 1988. *The Theory of Design.* New York: Oxford University Press.

Taguchi, G. 1988. "The Development of Quality Engineering," *ASI Journal* 1(1):5–29.

Yoshikawa, H. 1981. "General Design Theory and a CAD System." In IFIP W65.2–5.3 Working Conference on Man–Machine Communication in CAD/CAM, *Man–Machine Communication in CAD/CAM,* ed. T. Sata and E. Warman (Editors), North-Holland Publishing Company, 1–19.

Zwicky, F. 1966–1971. *Entdecken, Erfinden, Forschen im Morphologischen Weltbild.* Munich: Dromer-Knaur.

ADDITIONAL READING

Aristotle. *Ethica Nicomachea.* In *The Works of Aristotle,* ed. W. D. Ross. Oxford: Oxford University Press.

———. *Mechanical Problems.* In *Minor Works,* trans. W. S. Hett Loeb Classical Library. Cambridge, Mass.: Harvard University Press, 1936.

Cortes-Comerer, N. "Defensive Designing: On Guard against the Bizzare." *Mechanical Engineering* (August 1988): 40–42.

Dimarogonas, A. D. "On the Axiomatic Foundation of Design," In International Conference on Design Theory and Methodology, *Design Theory and Methodology, DTM '93.* New York: ASME, 1993, 253–258.

Hartenberg, R. S., and J. Denavit. *Kinematic Synthesis of Linkages.* New York: McGraw-Hill, 1964.

Leupold. *Theatrum Machinarum.* 1724.

Mayr, O. *The Origins of Feedback Control.* Cambridge, Mass.: MIT Press, 1970.

Moll, C. L., and F. Reuleaux. *Konstruktionslehre für den Maschinenbau.* Braunschweig: J. Vieweg, 1854.

Moody, J. A., W. L. Chapman, F. D. van Voorhees, and A. T. Bahill. *Metrics and Case Studies for Evaluating Engineering Designs.* Upper Saddle River, N.J.: Prentice Hall, 1997.

Redtenbacher, F. *Prinzipien der Mechanik und des Maschinenbaus.* Heidelberg: F. Basserman, 1852.

Reuleaux, F. *Der Konstructeur.* Braunschweig: J. Vieweg, 1872.

Ulrich, K. T., and S. D. Eppinger. *Product Design and Development.* New York: McGraw-Hill, 1995.

Willis, R. *Principles of Mechanism,* London: J. E. Parker, and Cambridge, J. & J. J. Deighton, 1841.

PROBLEMS

1.1. For a typical food processor for the kitchen:
- Is it a machine and why?
- Observing a food processor of your choice, does it satisfy the design rules?

1.2. For a typical gas engine for a lawnmower:
- Is it a machine and why?
- Observing a lawnmower of your choice, does it satisfy the design rules?

1.3. For a typical electric fan:
- Is it a machine and why?
- Observing a electric fan of your choice, does it satisfy the design rules?

1.4. For a typical electric typewriter:
- Is it a machine and why?
- Observing a typewriter of your choice, does it satisfy the design rules?

1.5. For a typical computer notebook:
- Is it a machine and why?
- Observing a notebook of your choice, does it satisfy the design rules?

1.6. For an electric refrigerator:
- Propose a set of customer "wants" and technical "hows."
- Develop a "house of quality."
- Find a machine functional concept.

1.7. For a disk player:
- Propose a set of customer "wants" and technical "hows."
- Develop a "house of quality."
- Find a machine functional concept.

1.8. For an electric heater:
- Propose a set of customer "wants" and technical "hows."
- Develop a "house of quality."
- Find a machine functional concept.

1.9. For a garbage disposal machine:
- Propose a set of customer "wants" and technical "hows."
- Develop a "house of quality."
- Find a machine functional concept.

1.10. For an automotive automatic wash machine:
- Propose a set of customer "wants" and technical "hows."
- Develop a "house of quality."
- Find a machine functional concept.

1.11. For the machine of Problem 1.6, using the machine functional concept developed in that problem, perform a task decomposition and develop five design alternatives.

1.12. For the machine of Problem 1.7, using the machine functional concept developed in that problem, perform a task decomposition and develop five design alternatives.

1.13. For the machine of Problem 1.8, using the machine functional concept developed in that problem, perform a task decomposition and develop five design alternatives.

1.14. For the machine of Problem 1.9, using the machine functional concept developed in that problem, perform a task decomposition and develop five design alternatives.

1.15. For the machine of Problem 1.10, using the machine functional concept developed in that problem, perform a task decomposition and develop five design alternatives.

1.16. For an electric refrigerator:
- Search for applicable codes and standards.
- Determine the frequency of hazard occurrence, consequences, and hazard-risk.
- Comment on the machine recyclability.

1.17. For an electric dryer:
- Search for applicable codes and standards.
- Determine the frequency of hazard occurrence, consequences, and hazard-risk.
- Comment on the machine recyclability.

1.18. For a room air conditioning unit:
- Search for applicable codes and standards.
- Determine the frequency of hazard occurrence, consequences, and the hazard risk.
- Comment on the machine recyclability.

1.19. For an automobile:
- Search for applicable codes and standards.

- Determine the frequency of hazard occurrence, consequences, and hazard-risk.
- Comment on the machine recyclability.

1.20. For a laptop computer:
- Search for applicable codes and standards.
- Determine the frequency of hazard occurrence, consequences, and hazard-risk.
- Comment on the machine recyclability.

LIST OF SUGGESTED MACHINE DESIGN PROJECTS

The projects that follow can be used to practice the procedures presented throughout this book. It is suggested that a plan of action be devised with any project, to be executed in the sequential or concurrent engineering approach (see Design Procedures 1.1, 1.2, and 1.3), which might include some of the following steps:

- A thorough search for codes, standards, and regulations applicable to the machine to be designed (see Section 1.4.1)
- A thorough literature and patent survey to identify previous work in this area and intellectual property constraints (see Sections 1.2.1–1.2.3)
- A market survey to identify the user needs and market-imposed constraints
- A complete set of final specifications, the technical response to the user needs and wants
- A "house of quality" sheet with the correlation of user wants and needs with the technical specifications and, further, the comparison with the previous or competing designs (see Sections 1.2.1–1.3.3)
- Development of a machine functional concept (see Section 1.2.3)
- A task decomposition (see Section 1.2.4)
- Development of alternative system design concepts (see Section 1.2.5)
- Preliminary evaluation of system design alternatives (see Section 1.2.6)
- Empirical improvement: continuation, inversion (see Section 1.2.7)
- Design synthesis (see Section 1.3.1)
- Design analysis and evaluation (see Section 1.3.2)
- Detail design, production of a complete set of drawings (hand sketches or formal drawings—see Section 1.4)

The complete set of drawings should include:

- An assembly drawing with numbering of all parts and the dimensions needed only for assembly, packaging, and shipment
- A parts list on the assembly drawing
- Individual part dimensioned drawings for parts to be manufactured

- Production planning, including selection of manufacturing processes and quality control procedures
- Design of the production facility, if not already available
- Cost analysis, financial analysis, estimation of production cost for the machine
- A manual that will include installation, operation, maintainance, and repair procedures
- A formal presentation to company management

If these projects are used for instruction, additional discussion, details, constraints, and other information might be provided by the instructor.

1. Design of a mousetrap for use in the house.
   Initial Specifications:
   - Should effectively kill household mice.
   - Should not kill household pets.
   - Should have maintenance-free operation for a long time.
   - Should not cost the consumer more than $20.
   - Should be safe for small children.

2. Design of a small air compressor for automotive tire inflation.
   Initial Specifications:
   - Should use 12 V automotive battery power.
   - Should supply pressures up to 0.4 MN/m^2.
   - Should be capable of inflating a mid-size car tire in 10 min.
   - Should not cost the consumer more than $35.
   - Should have maintenance-free operation for a long time.

3. Design of a machine that can harvest strawberries
   Initial Specifications:
   - Should move in a 1.2 m wide dirt road in the strawberry field.
   - Should be able to harvest 2 m on each side.
   - Should be able to recognize and pick up mature strawberries.
   - Maximum density of 200 strawberries per m^2.
   - Strawberry size from 1.5–4 cm diameter (approximate).
   - Should deposit strawberries in 3 layers maximum in wooden boxes 40 cm wide $\times$ 50 cm long $\times$ 15 cm high.
   - Should be driven and operated by one operator.
   - Should retract for transportation to a maximum width of 1.8 m.
   - Should not apply to the strawberries more than 3g acceleration.
   - Should be powered by a diesel motor.

4. Design of a machine that cuts almonds into slices.
   Initial specifications:
   - Should process almonds 1–2 cm long, 4–8 mm thick, and 8–14 mm wide.
   - Should cut the almonds in slices, 0.5–3 mm thick, along the major cross-section.

- Should process 50 kg almonds per hr.
- Should be powered by an electric 120/220 V, 60 Hz motor.
- Should comply with applicable safety and sanitary standards.
- Should be relatively free of maintenance.

5. Design of an almond-placing machine in chocolate production.

   A machine that places 8 almonds minimum to 10 almonds maximum in each chocolate bar while it is in the molten stage in a 10-bar mold on a belt moving at a speed of one row of chocolate bars per second.

   - Should process chocolate bars 10 cm long, 4 cm wide, and 8 mm thick.
   - Chocolate bars are formed in the mold lengthwise, distance between bars 1 cm.
   - Should process at least 2000 bars per hr.
   - Almonds may be 1–1.5 cm long, 4–6 mm thick, and 6–10 mm wide.
   - Should be powered by an electric 120/220 V, 60 Hz motor.
   - Should comply with applicable safety and sanitary standards.
   - Should be relatively maintenance-free.

6. Design of a mechanism to provide orientation of a solar collector towards the sun during the day.

   Initial specifications:

   - Dimensions of the solar collector 2 m × 1 m × 0.15 m.
   - Weight of the solar collector = 400 N.
   - Constant angle of longitudinal axis, the axis of rotation, in respect to the horizontal plane adjustable between 30° and 50°.
   - Electric motor 120/230 V, 60 Hz.
   - Must comply with applicable safety and industry standards.
   - Must have flexible connections to the (stationary) water storage tank.

7. Design of a welding table.

   A welding table (1 × 1 meters) that can have adjusted height 0.5 to 1.2 m and tilt up to 30° in any orientation and secured in the desired place (Hubka 1980).

   Initial Specifications (in addition to the ones in the problem statement):

   - Maximum weight of the welded work = 5,000 N.
   - Table must be made of some grade of steel for electrical conductivity and resistance to high-temperature molten metal drops.
   - Means for securing the workpiece must be provided.
   - Table must be safe for the welder.
   - Readjustment to a different position must be quick.
   - The table must have longevity and be relatively maintenance-free.

8. Design of a mechanism to handle brick transport.

   A mechanism to pick up bricks from a moving transport belt and deposit them onto another one moving at 90° angle. Both belts are horizontal and they transport one pair of bricks per second.

   Initial specifications:

   - Belt speed = 0.5 m/sec.

- Belt width = 0.4 m.
- Brick dimensions = 20 × 10 × 5 cm.
- Brick mass = 2.5 kg.
- Maximum acceleration on the brick = 2g.
- Bricks are oriented with their longitudinal axis perpendicular with the direction of motion and they lie flat on the belt in pairs with distance between bricks 20 cm in the direction of motion and 10 cm along the longitudinal axis of the bricks.
- Vertical distance between the two belts = 1 m.

9. Design a safer log splitter with the following characteristics.

Initial Specifications:

- Able to be towed at highway speeds behind a full-size pickup truck.
- An 8 hp gasoline engine drives a two-stage hydraulic pump that in turn pressurizes a hydraulic cylinder to split the log.
- Accommodates a 2 ft long log.
- Generates 15 tons of force on the log against a stationary splitting wedge.
- Has a safety cage that covers the log/wedge/cylinder area during splitting to prevent injury to the operator. This cage slides (manually) out of the way to load/unload logs and is interlocked such that it must be in place before the hydraulic cylinder will move.

10. Design a battery-powered, motorized shopping cart.

Initial Specifications:

Capable of carrying a 200 lb person plus 50 lb of groceries around the aisles of a supermarket. It should hold at least half the volume of foodstuffs of a conventional, manual shopping cart.

- Be speed-limited, safe against tipover, and require constant pressure on its control to run (i.e., a "dead man's" switch).
- When the power is cut, an automatic brake should stop it within 1 ft.
- Intended users are elderly or infirm shoppers.
- It should run for 1 hr between recharges.

11. Design a device to transfer a 100 kg paraplegic patient safely from bed to wheelchair and vice versa.

Initial Specifications:

- The patient has good upper body strength but no control of the lower extremities.
- Your design should be operable by the patient with minimal assistance.
- Safety is a paramount concern.

12. Design an indoor bicycle exerciser that will use a common bicycle and a system to absorb energy at the wheels.

Initial Specifications:

- The driver will propel the bicycle in the usual bicycle-driving manner.
- The bicycle will be stationary and the motion from the rear wheel will be transmitted to a proper mechanism for conversion of the user's power into some other form of energy that can be dissipated, stored, or transmitted.

· The bicycle should be kept in an upright, stable location.

· There should be indicators of speed and torque generated by the user.

· There should be a manual control of the resisting torque at the pedal.

· The exerciser should be safe for use in the presence of small children.

13. Design of a motorized wheelchair to climb on the curb.

Though most municipalities and building complexes throughout the world are building ramps for wheelchairs to move from the street to sidewalks, in many places such ramps still do not exist, impairing the mobility of people that have to use a motorized wheelchair. A motorized wheelchair needs to be designed to allow climbing a single step, facilitating movement from street to sidewalk without use of a ramp.

Initial Specifications:

· Is capable of carrying a 200 lb person.

· When the power is cut, an automatic brake should stop it within 1 ft.

· Intended users are elderly or infirm persons. It should run for 1 hr between recharges.

· Retrofitting an existing system to climb the sidewalk is desirable.

14. Design a paint-mixing machine for use in factories and retail stores.

The machine will be self-contained in a steel cabinet (for safety) and will accept either one five-gallon can of paint or four one-gallon cans, still in their cardboard shipping cartons. The machine will rotate the can (or cans) simultaneously about two axes. One possible concept for this design is envisioned as a cabinet that constitutes the main structural support of the

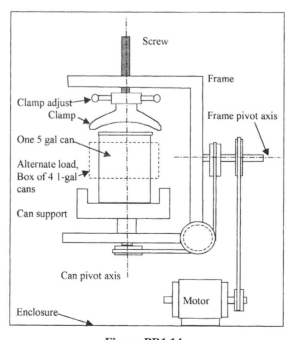

Figure PR1.14

machine with a rectangular (or open-sided c-shaped) frame inside that is pivoted from the cabinet at the center of two opposing sides (or at the closed side of a c-shaped frame). The other two sides of the frame would also contain pivots with their axis at 90° to the first pivot axis. One of these would have a platform on which the paint can or box is placed. The other would have an adjustable clamp that could be brought to bear on the top of the can or box, holding it in place during the mixing operation. The support and clamp will be arranged so that the center of gravity of the can(s) is at or near the crossing point of the two rotation axes.

A system of belts and sheaves would be provided so that the frame axes and a common motor could rotate both the can axes. One belt would drive the frame while a second belt from the same motor would drive the can clamp pivots. A system of belts similar to an old-fashioned dental drill would transfer this motion along the arms of the frame to the clamp pivots.

The design should minimize noise and vibration so that the machine does not have to be bolted down and hearing protection should not be required.

Initial Specifications:

- Motor 3/4 hp, 120 V, 60 Hz
- Machine size (maximum) 2 ft × 3ft × 4 ft
- Loading from front
- Safety features: door interlock switch
- Excessive vibration cutoff switch internal
- Controls: timer with start switch
- Emergency stop palm button
- Load capacity = 125 lb
- Cost to manufacture (approx.) $1,200

15. Design a solar-powered ventilator for a car.

A car parked in the sun can develop substantial temperatures during the day. This is uncomfortable for the driver when entering the car after a long parking time, but it also has detrimental effects on the car upholstery, wiring, controls, paint, etc. A small ventilator, powered with a solar cell array, could provide continuous ventilation and keep the temperature down.

The solar array should be located in the rear of the car, and the ventilator should be mounted in the trunk and pump air out through the area behind the rear seats. Air will enter the car through the front ventilation vents and the air ducts of the heating/AC system.

Initial specifications:

- Should not use battery power.
- Should provide ten air changes per hr for a typical mid-size car.
- Solar array should be difficult to dismount without trunk being opened.

16. Design a tapwater-powered can crusher for home use.

Recycling of aluminum cans becomes difficult sometimes because users do not like the substantial storage space that accumulating cans can take between collections. An aluminum can crusher needs to be designed, powered by the tapwater pressure.

Initial specifications:

- Activation with a turn of a valve or the push of a button.
- Deactivation should be affected automatically at the end of the stroke of the piston.
- After operation, the crushed can should drop in a bag and the piston return to the open position automatically.
- It should be absolutely safe for children.

17. Design a motor-driven mechanical hoist as shown.

    A schematic outline is shown in Figure PR1.17. A simple reduction gear reduces motor speed to that of the drum. Rotation of the drum, in turn, moves the load either up or down. Maximum lifting capacity is 49 kN. Drum diameter is 350 mm, total efficiency 0.78, and motor torque 200 Nm. The safety brake on the left has four bevel shoes for power input, each of mass 0.13 kg, guided radially, and balanced by four springs for a motor-driven mechanical hoist. When the four shoes simultaneously make contact with the outer rim (output), each spring force is 320 N, and the distance r to the center of gravity of each shoe is 120 mm. The brake must keep the speed below 15 m/s. Use 0.15 for the brake–shoe interface friction coefficient. Design the system.

18. Design of a platform elevator.

    One possible concept for a platform that can be elevated and lowered mechanically while carrying a heavy load F is using four screws—vertical poles near the four corners of the rectangular-horizontal platform. The weight of platform and load is 30 kN, concentrated at the center of table. The platform, assumed rigid, is to be lifted in the vertical direction at a speed $V = 0.5$ m/s. The spindles may be turned by bevel gears or other suitable means. Design the system.

Initial specifications:

- Platform weight and load = 30 kN.
- Motor: electric motor, 120/220 V, 60 Hz.
- Platform dimensions = 2 m × 2 m.

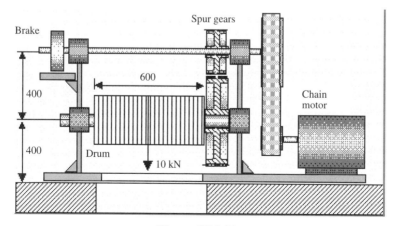

Figure PR1.17

- Motion should be coordinated at all four corners.
- Should comply with applicable safety standards.
- Should be durable and relatively maintenance-free.

19. Design of an improved snowblower.

A typical snowblower operates on the principle of the auger. In this concept, two power-driven augers of the same diameter and length, but of opposite hand, mounted on the same horizontal shaft pull the snow in from both sides and discharge it radially with great force. The engine, rated 5.5 kW at 2700 rpm, has its speed reduced in two steps. First, it is reduced by 1.5:1 in a belt drive; next, a chain cuts speed by a factor of 2.25:1. Design the snowblower.

Initial specifications:

- Motor: gas engine, 5.5 kW at 2700 rpm.
- Blower rotor speed = 800 rpm.
- Blower rotor external diameter D = 250 mm.
- Blower rotors length = 300 mm each.
- Should comply with UL and other applicable safety standards.
- Should be durable and relatively maintenance-free.
- Should stop when user stops holding the handle.

20. Design a single-lever faucet (Pahl and Beitz 1996).

Initial Specifications:

- Throughput (mixed flow) max 10 lit/min at 20 MPa pressure
- Maximum pressure 1 MPa (test pressure 1.5 MPa)
- Cold water temperature 15–25°C
- Hot water temperature 60–80°C
- Easy operation (for adults and children)
- No external energy
- Hard water supply (drinking water)
- No connection of cold/hot water lines when valve shut
- Handle not to heat above 35°C
- Obvious operation, simple and convenient handling
- Smooth, easily cleaned contours; no sharp edges
- Noiseless operation
- Service life 10 years at about 300,000 operations
- Easy maintenance and simple repairs; use standard spare parts
- Maximum manufacturing cost $15 (3000 units per month)
- Should comply with safety and sanitation standards

21. Design of an electric toothbrush.

Initial specifications:

- Clean teeth better than a hand brush, massage gums, reduce decay, hygienic family sharing, electrical and mechanical safety, etc.
- Sweeten breath and whiten teeth (symbolic needs for social acceptance); handle colors to match bathroom, etc.

- Autonomy in deciding when and how one's teeth are to be cared for, self-esteem from care of teeth, praise for effort, pleasure from giving or receiving a gift, etc.
- Diameter, length, brush size, amplitude, frequency, weight, running time, reliability, useful life, etc.
- $1 per person is the lowest cost alternative, but electric razors sell for 20 times the price of a manual razor, so the sale price should be $20 or lower.

22. Design of a hand-operated lifting device.

    Figure PR1.22 shows one design of a hand-operated lifting device to lift a load 8,000 N in the vertical direction while the device is mounted on a vertical wall. Design a device to perform this function.

    Initial specifications:

    - Maximum load = ±8,000 N.
    - Maximum force the user should deliver = ±300 N.
    - Maximum radius of the hand-crank = 300 mm.
    - There should be a locking device that will prevent the load from coming down by itself during any time.
    - Lowering the load should be done by applying a torque to the crank.
    - Crankshaft distance from ground = 1.50 m.
    - Maximum travel of load = 1.20 m.
    - Design should comply with applicable safety and quality standards.

23. Design a hand-operated winch.

    Figure PR1.23 shows one design of a hand-operated winch to lift a load 20,000 N in the vertical direction through a cable. Design a device to per-

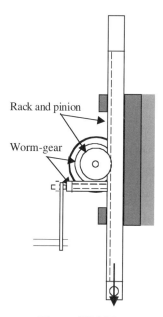

Rack and pinion

Worm-gear

Figure PR1.22

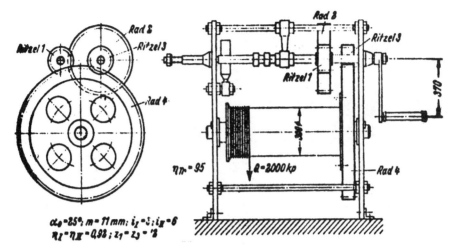

Figure PR1.23

form the same function.

Initial specifications:

· Maximum load = 20,000 N.

· Maximum force the user should deliver = ± 300 N.

· Maximum radius of the hand-crank = 370 mm.

· There should be an operator-controlled locking device that will prevent the load from coming down by itself at any time.

· Crankshaft distance from ground = 1.50 m.

· Maximum travel of load = 60 m.

· Design should comply with applicable safety and quality standards.

24. Design a machine to remove pieces of iron from an iron foundry broken slug.

Figure PR1.24 shows one design of a machine to remove pieces of iron from an iron foundry broken slug. The slug is carried on a transport belt, and above the transported slug is placed a magnet powerful enough to attract the iron from the slug. A belt operates around the magnet; thus the pieces of iron are carried away, and when they are outside the magnetic field they fall into a hopper. Design a device to perform the same function.

Initial specifications:

· Belt speed = 0.45 m/s.

· Magnet dimensions, length = 1 m, width = 0.6 m, height = 1.0 m.

· Weight of the magnet = 25,000 N.

· An electric motor 120/230 V, 60 Hz should be used.

· Design should comply with applicable safety and quality standards.

25. Design of a steam valve.

Figure PR1.25 shows one design of a steam valve. It is hand-operated and works on the principle of a wedge-shaped gate sliding up and down to

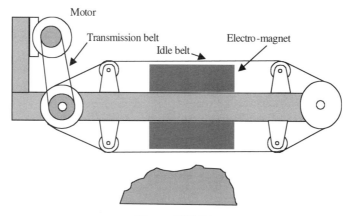

Figure PR1.24

control the flow. Design a device to perform the same function.

Initial specifications:

· Inner diameter of the pipe = 200 mm.
· Steam pressure = 160 MPa.
· Steam is saturated.
· Turning wheel radius = 170 mm and the maximum force an operator will put normally is 500 N.
· The valve should be safe against overloading by an operator tangential force = 1000 N.

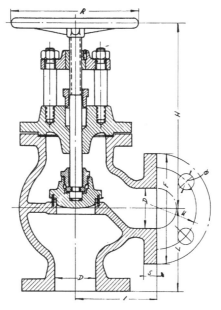

Figure PR1.25

- The valve stem should have a seal so that steam will not escape the valve.
- Design should comply with applicable safety and quality standards.

26. Design of a hand-operated lifting device.

Figure PR1.26 shows one design of a hand-operated lifting device to lift a car for tire change. Design a device to perform the same function.

Initial specifications:

- Maximum load = 10,000 N.
- Maximum force the user should deliver = ± 300 N.
- Maximum radius of the hand-crank = 300 mm.
- The device should not allow the load to come down by itself during any time.
- Lowering the load should be done by applying a torque to the crank.
- Minimum height = 250 mm.
- Maximum travel 200 mm.
- Design should comply with applicable safety and quality standards.

27. Design of a rotating crane.

Figure PR1.27 shows one design of a rotating crane. Design a device to perform the same function.

Initial specifications:

- Maximum load = 20,000 N.
- Standard, off-the-shelf electric winch with lifting motion with an electric motor 120/230 V and 60 Hz.
- Manual radial motion.
- Maximum radius of the load = 2.5 m.

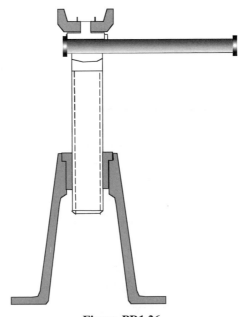

Figure PR1.26

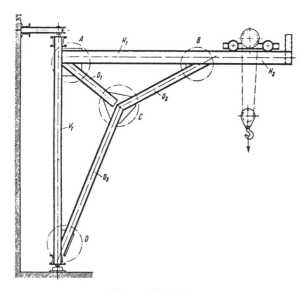

Figure PR1.27

- Maximum height of the load at the outer radius = 2.5 m.
- Ceiling height = 3.30 m.
- Design should comply with applicable safety and quality standards.

28. Design of a manually operated winch.

    Figure PR1.28 shows one design of a manually operated winch. Design a device to perform the same function.

    Initial specifications:

    - Maximum load = 10,000 N.
    - Maximum operator force = 400 N.
    - Maximum travel of hook = 3 m.
    - Design should comply with applicable safety and quality standards.

29. Design of a bookbinder's press.

    Figure PR1.29 shows one design of a bookbinder's press. Design a device to perform the same function.

    Initial specifications:

    - Maximum load = 2,000 N at operator's normal torque 300 Nm.
    - Crank total length = 500 mm.
    - Maximum torque the operator can apply = 250 Nm.
    - Maximum travel = 800 mm.
    - Dimensions of the press plate = 400 × 400 mm.
    - Design should comply with applicable safety and quality standards.

30. Design a Land Surveyor's Utility Cart (Jack Leffler, Department of Technological Studies, The College of New Jersey).

    Land surveyors working in the field have the need to carry quantities of heavy supplies throughout their job site. Sometimes this means carrying concrete markers or mix as well as wooden hubs and lath along with their

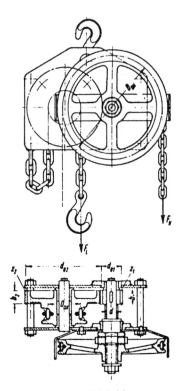

Figure PR1.28

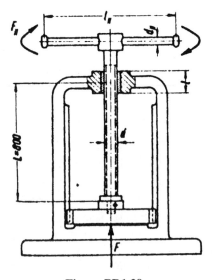

Figure PR1.29

surveying equipment over large distances.

The land surveyor covers all types of terrain, from pavement to mud construction areas, from dry ground to mountains, woods, and swamps. No matter the ground conditions, the land surveyor must carry a large amount of equipment in order to be able to work. The transportation of supplies requires a large output in time and energy and increases the danger of work-related strains and injuries. Each output of time and energy required in the transportation of tools and materials takes away from the productivity of the land surveyor.

Initial specifications. The system must:

- Transfer the weight of the material to the ground.
- Keep items in place (lashed down).
- Be able to transport 500 N over dry ground.
- Be able to transport instruments of length up to 1.60 m.
- Be buoyant while transporting 300 N.
- Be capable of sliding over snow.
- Contain water-resistant storage space for extra miscellaneous supplies.
- Able to withstand abusive treatment (bouncing over rocks).
- When empty, be easily lifted by two average people.
- Fit into, or on top of, a standard Chevrolet Suburban.

31. Design of an adjustable lawn sprinkler.

    Many lawns are irregularly shaped, and lawn sprinklers can have a circular or rectangular coverage. The result is underwatered corners of lawn or overwatered lawns and unwanted moistening of neighboring areas or structures. There is a market need for the design of a sprinkler for watering irregularly shaped lawns.

    Initial specifications:

    - Powered by the water mains.
    - Maximum pressure = 0.6 MPa.
    - Minimum pressure = 0.3 MPa.
    - Lawn shape: polygon, up to six straight edges.
    - Maximum dimension of the lawn in any direction = 10 m.
    - Minimum flow required = 0.5 m^3/hr.
    - Maximum cost = $30.

32. Design of a power riveting machine.

    Design a riveting machine with the following specifications:

    - The machine should have a U-structure (part *B*) to allow free space 1200 × 300 mm.
    - The riveting mechanism *A* should deliver 5 kN of maximum force.
    - The machine must be capable of forming one rivet every 2 seconds.
    - The riveting actuator *D* should be operated by a hydraulic piston, a slider crack mechanism, a cam, or any other suitable mechanism.
    - The operator will have the hands occupied, so that actuation should be done by a foot pedal.

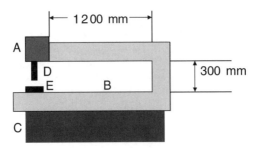

Figure PR1.32

- Sufficient safety should be built into the design.

  Milestones:
- Select the operating concept of the riveting head, select a concept for the cross section of the U-structure (solid, tubular, truss, double *T,* etc.) and prepare a working sketch of the machine.
- Design part *B* for strength.
- Design the riveting head *A*. If the concept you have selected has elements not covered in class yet, design them later.
- Design the operation and control system.
- Make drawings and write a final report.

  At the end of every milestone, submit a progress report that will also include all previous reports so that the progress report should be self-standing.

CHAPTER 2

KINEMATIC ANALYSIS OF MACHINES

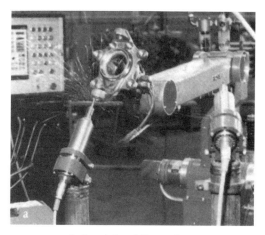

Industrial welding robot.

OVERVIEW

In Chapter 1 we saw how from a vague initial description of the machine design problem to be solved we can develop a multitude of possible solutions and identify the most promising solution concept. We have seen that the concept is described in generic form, usually in the form of a schematic that describes the machine operation and a set of detailed design specifications. From this point to having a complete set of drawings of the machine, there is a lot to be done. We need to describe the machine quantitatively, selecting materials and manufacturing processes for every part and selecting dimensions that will assure that the machine will operate in accordance with the specifications without failing.

Often many of the main dimensions of a machine can be found using purely kinematic considerations before it is necessary to discuss materials and their ability to carry loads. This facilitates considerably the design task and will be the subject of this chapter.

2.1. KINEMATIC ANALYSIS

As explained in Chapter 1, the existing detail design methods are component-based and require each component to be designed separately. However, the operation of each component is not independent, and the component is constrained to operate with the other parts of the machine as a system. Therefore, in order to design the particular component, we need to study the machine as a system and determine the *resource interface* (exchange of energy, matter, and information) of each component with the rest of the system. This is difficult, in general, because study of the system presumes knowledge of its components. The designer is always confronted with this vicious circle. Some ways to overcome this problem are:

1. If the design procedure for each component leads to a number of design equations relating the component design parameters and the component resource interfaces, if the design of the system can be expressed mathematically as a system of equations relating the system constraints, the resource interfaces among the components, and the design parameters, and if these equations can be solved, the design problem has a formal solution. This method will be discussed in Chapter 7.

2. Since the above ideal design situation is not always the case, the designer very often has to make some assumptions, based on previous experience, about some initial values of the component design parameters. The system is then designed to determine the component resource interfaces and design the components. If the resulting design parameters deviate considerably from the assumed ones, an iterative procedure must be used. This method requires considerable experience, and this is the main reason for specialization. It is very difficult for an aircraft designer to design a lawnmower this way.

3. Some of the resource interface parameters can be determined without the knowledge of all the component design parameters. An example is the determination of information interfaces, in particular system velocities.

The latter method can be employed in most machines. The possibility of isolating the kinematics of a machine (which depends only on input–output speeds and the main dimensions of the machine) from its statics and dynamics (which depends on the detail dimensions and the mass distribution) was recognized by the great kinematicians of the 18th and 19th Centuries (see Historical Introduction). In 1830, Ampère, in his *Essai sur la philosophie des sciences,* defined the machine "an instrument by the help of which the direction and velocity of a given motion can be altered." He excluded forces from kinematic analysis and gave to the study of the machine the name *kinematics.*

Ampère, André Marie (1775–1836), French scientist, known for his important contributions to the study of electrodynamics. Ampère, the son of a Lyons city official, was born in Polémieux-au-Mont-d'Or, near Lyons. His electrodynamic theory and his views on the relationship of electricity and magnetism were published in his *Recueil d'observations électrodynamiques* (*Collection of Observations on Electrodynamics,* 1822) and in his *Théorie des phénomènes électrodynamiques* (*Theory of Electrodynamic Phenomena,* 1826).

In the new science, the movements are to be studied as movements, independent of the power creating them, in the same way as we observe the static bodies around us, especially in such combinations as we call machines. . . . This science (kinematics) ought to include all that can be said with respect to motion in its different kinds, independently of the forces by which it is produced.

2.1.1. Kinematic Elements

Machines and mechanisms are distinguished from mere structures due to movements of their parts in relation to one another by magnitudes, which are comparable to the size of these parts. Moreover, deformations of the whole machine and/or its parts are much smaller than the motions of these parts and, for the purpose of kinematic analysis, can be neglected as a first approximation. This makes kinematic analysis possible as it can be performed without reference to the forces that produce the motion, stresses throughout the machine, and so on. Thus, we need to describe the machine not in every detail but in geometric terms sufficient to describe the motion of the machine. Such a description of a machine will be denoted here as *kinematics of mechanism.*

Since we do not consider element deformations, the exact shape of the machine parts is not important, only the geometric characteristics that affect the motion. Since we do not consider deformations, the first family of elements of the mechanism that we will consider are different forms of a *rigid body,* a concept we already know from mechanics. A rigid body is known to have, in general, six degrees of freedom, that is, to describe its motion we need to define six coordinates, usually three translations along the three axes of a Cartesian coordinate system and three rotations about these axes. A *mechanism element* will be defined as the geometrical form by which two parts of a mechanism are joined together so that the relative motion between these two parts is consistent. Such elements are shown in Table 2.1. Since a rigid body has six degrees of freedom for general spatial motion and three degrees of freedom for planar motion:

1. *Revolute joint* (R): allows relative rotation of the two parts about a point, one degree of freedom, and imposes five *constraints.* Typical revolute joints are *pins* and *hinges.*

TABLE 2.1. Mechanism Joints[a]

Name of Pair	Geometric Form	Schematic Representations	Degrees of Freedom
1. Revolute (R)			1
2. Cylinder (C)			2
3. Prism (P)			1
4. Sphere (S)			3
5. Helix (H)			1
6. Plane (P_L)			3

[a] From C. H. Suh and G. W. Radcliffe, Kinematics and Mechanical Design, 1978, by permission of John Wiley & Sons, Inc.

2. *Cylindrical joint (C)*: only rotation about its axis and translation along the axis are possible; allows two degrees of freedom, and imposes four constraints on the motion.

3. *Spherical joint (S)*: only three rotations about three axes are possible; allows three degrees of freedom and imposes three constraints on the motion.

4. *Prismatic joint (P)*: allows only rectilinear motion; only one degree of freedom and two constraints in planc motion, five constraints in space motion. A typical prismatic joint is a slider.

5. *Link (L)*: rigid body; has:

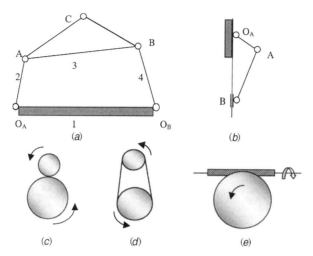

Figure 2.1. Miscellaneous mechanisms.

- Three degrees of freedom if it is confined to move on a plane (*planar link*), two translations and one rotation;
- Six degrees of freedom if it has general spatial motion (*planar link*), three translations and three rotations.

Typical links are the robot arm and the crank and connecting rod of a piston engine. For mechanisms consisting only of links and joints we use the term *linkages*.

A mechanism consists of several elements. For example, the planar four-bar linkage (Figure 2.1a) consists of four links, 1–4, and four revolute joints, A, B, O_A, O_B.

It is interesting to study the *mobility* of a mechanism. We note that one link is usually fixed to the ground, which removes three degrees of freedom. Since every link has six degrees of freedom and every revolute joint removes five degrees of freedom, the number of degrees of freedom of the mechanism will be, in general,

$$M = 6(n - 1) - 5m \tag{2.1a}$$

where n is the number of links and m the number of revolute joints. Equation (2.1a) is called *Gruebler's equation*. For a planar mechanism,

$$M_{\text{planar}} = 3(n - 1) - 2m \tag{2.1b}$$

Equations (2.1a) and (b) should be used with care because they can give erroneous results for some mechanisms.[1]

---

[1] For more on this, see Erdman and Sandor, 1997.

Example 2.1   Investigate the mobility of the four-bar linkage of Figure 2.1*a*.

Solution   Gruebler's equation (Equation (2.1*b*), for $n = 4$ and $m = 4$, yields:

$$M = 3(4 - 1) - 2 \times 4 = 1$$

We have assumed here that the coupler 3 is one rigid triangular link. Let us drop this assumption and consider the three coupler links separately. There are six links ($n = 6$), and there seem to be five revolute joints. However, the joints at A and B are double because they constrain two pairs of links each; thus, $m = 7$. Therefore:

$$M = 3(5 - 1) - 2 \times 7 = 1$$

The four-bar linkage has one degree of freedom.

2.1.2. Kinematic Layout

At this stage in the design of the machine, the design concept has already been defined in the form of a process layout, a sketch showing all the operation sequences and the paths of flow of the system resources, mass, energy, and information. From the process layout, a *kinematic layout* can be produced. On it, the known motions (input, output, and sometimes motions of specific components) and the dimensions that affect the kinematic behavior are shown. The kinematic layout need not be to scale. It is usually presented in plane or in axonometric view. Substantial abstraction is allowed in the form of a *schematic* or *skeleton diagram:* gear pairs are represented as circles in contact, a connecting rod as a line connecting the two pins, etc. (see Figure 2.1).

Example 2.2   For the design of an electric drill, it was decided during the conceptual design stage that the motor would be a 4000 rpm Universal Motor, the output speed at the drill would be 300 or 600 rpm, selectable, a first stage of speed reduction would be with a pair of helical gears to 1000 rpm, and the final speed would be obtained with another stage with a double pair of spur gears with a two-gear pinion sliding in a spline. Draw a kinematic layout.

Solution

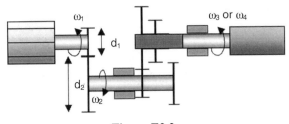

Figure E2.2

2.1.3. Fixed Kinematic Pairs

Transmission of speed involves two parts, the *driving* and the *driven* parts, that we shall term a *kinematic pair*. Some kinematic pairs have a relationship between input and output motions that is invariable with time, termed *fixed kinematic pairs*. Others have a relationship between input and output motions that is changing with time, termed *variable kinematic pairs*.

The most common fixed kinematic pairs are speed reducers. Motors and prime movers, in general, work most efficiently at speeds higher than the ones we need to run our machines. To match the speeds, speed reducers are used in between. The most common of them, with their kinematic relationships, are:

1. *Gears* (Figure 2.2a): Gears are cylindrical, mostly, elements with toothed cylindrical surfaces that engage to the mating gear preventing slip. Input is the angular velocity ω_1, output is the angular velocity ω_2. The input–output relationship, because the peripheral velocities of both gears $V = \omega_1 d_1/2$ and $V = \omega_2 d_2/2$ are equal (the no-slip condition), is:

$$\frac{\omega_2}{\omega_1} = \frac{d_1}{d_2} = R_{12} \tag{2.2}$$

where ω is the angular velocity (rad/s), d is the nominal diameter of the gear, R_{12} is the *gear ratio,* or *speed ratio,* and 1 and 2 refer to the driving (input) and the driven (output) gears, respectively.

The *speed ratio* (gear ratio) R_{12} is exact and remains constant at all times. The direction of rotation is reversed, in general.

Alternative expressions for the angular velocity are:

Frequency:

$$f = \frac{\omega}{2\pi} \quad \text{(cps or Hz)}$$

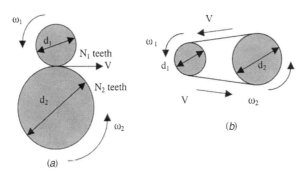

Figure 2.2. Fixed (speed ratio) kinematic pairs.

Rotating speed:

$$n = \frac{60\omega}{2\pi} \quad \text{(rpm)}$$

In all computations, unless otherwise specifically stated, should be used.[2] For multiple gear pairs connected in series, the total speed ratio is $R_{1n} = \omega_1/\omega_n = R_{12}R_{23}R_{34} \cdots R_{(n-1)n}$.

2. *Belts* (Figure 2.2*b*): A belt wrapped around the periphery of two cylindrical elements (called *pulleys*) allows for transmission of power from one pulley to the other at different speeds. Input is the angular velocity ω_1, output is the angular velocity ω_2. The input–output relationship, because the peripheral velocities of both pulleys $\omega_1 d_1/2$ and $\omega_2 d_2/2$ are equal (though approximately, due to some slipping; see Chapter 12), is:

$$\frac{\omega_2}{\omega_1} \approx \frac{0.985 d_1}{d_2} \tag{2.3}$$

where ω is the angular velocity (rad/s), d is the pulley diameter, 1 and 2 refer to the driving (input) and driven (output) pulleys, respectively, and 0.985 is an approximate *slippage factor* for preliminary design purposes. More on the slippage factor will be discussed in Chapter 12.

The velocity ratio is not exact, due to slipping, and does not remain constant at all times. The direction of rotation is preserved, in general.

Example 2.3   For the kinematic layout shown, find the angular velocities of the shafts A, B, C, D, if the motor speed (input) at point A is 3500 rpm and the dimensions are $d_1 = 100$, $d_2 = 300$, $d_3 = 120$, $d_4 = 340$, $d_5 = 150$, $d_6 = 300$ (Figure E2.3).

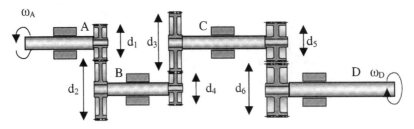

Figure E2.3

---

[2] The units should be rad/s, unless otherwise prompted in some empirical formulas.

Solution

$$\omega_A = 3500 \frac{2\pi}{60} = 366.5 \text{ rad/s}$$

$$\omega_B = \frac{\omega_A d_1}{d_2} = 366.5 \frac{100}{300} = 122.2 \text{ rad/s}$$

$$\omega_C = \frac{\omega_B d_3}{d_4} = 122.2 \frac{120}{340} = 43.1 \text{ rad/s}$$

$$\omega_D = \frac{0.985 \omega_C d_5}{d_6} = 0.985 \times 43.1 \frac{150}{300} = 21.2 \text{ rad/s}$$

2.2. VARIABLE SPEED KINEMATIC PAIRS—CLOSED LOOP LINKAGES

Linkages are often used in machines to achieve:

1. Change of reciprocating motion to rotary motion, and the inverse
2. Change of a constant speed rotary motion to one of non-constant speed
3. Change of a reciprocating motion to another such motion with different characteristics

Usually, one link of a linkage, the *input link,* is connected with a prime mover that provides the power, another link, the *output link,* is connected with a part that provides the useful function of the linkage, and at least one link, a *base link,* is fixed in space (usually on the machine frame). We quantify the relation of input to output by assigning a value to the input or output, such as the angle of rotation of each link in respect to a predefined datum, or the motion of one or more points on a link.

2.2.1. The Four-Bar Linkage

For a four-bar linkage (Figure 2.3*a*) input usually is the rotation of the crank (link 2) and output the rotation of the follower (link 4). Link *AB* is the *coupler* (link 3), while link $O_A O_B$ is the base link (1). In order that one link of the four-bar linkage can make a full revolution, geometric compatibility demands that the sum of the lengths of the shortest link *s* and the longest *l* should be less than the sum of the lengths *p* and *q* of the other two links of intermediate length (*Grashof's law*):

$$l + s \leq p + q \tag{2.4}$$

Depending on the selection of the input link, the four-bar linkage can have the following modes of operation (Shigley and Vicker 1980):

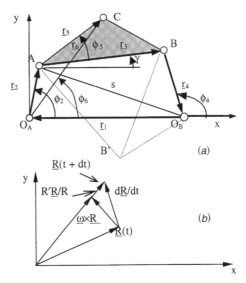

Figure 2.3. Geometry and kinematics of a four-bar linkage.

1. If the shortest link is attached to the base link as in Figure 2.3*a*, we obtain the *crank and rocker* linkage. The crank executes a full rotation while the rocker oscillates only between two limits.
2. If the shortest link is the base link, we obtain the *drag-link* or *double-crank* linkage. The crank and the follower both execute full rotations.
3. If the opposite to the shortest link is the base link, we obtain the *double rocker* linkage. Both the crank and the follower oscillate only between two limits.

Franz Grashof, born in Germany in 1826, succeeded Redtenbacher in the Chair of Mechanics at the Technical University of Karlsruhe. In 1866, he wrote *Theorie der Elasticität und Festigkeit* (*Theory of Elasticity and Strength*). He was among the founders of the VDI, the German Engineering Society, and was founding editor of its journal. He died in 1893.

Displacement analysis: The first problem in linkage analysis is to find the geometric configuration for different locations of the input links; in the case of a four-bar linkage (Figure 2.3*a*), the output angle ϕ_4 for different values of the input angle ϕ_2, if the lengths of the links are known.

The geometric closure equation for the vector polygon in Figure 2.3*a* is:

$$r_1 + r_2 + r_3 + r_4 = 0 \qquad (2.5)$$

This is a vector equation in a coordinate system (x, y) and is equivalent with two scalar equations:

$$r_1 + r_2 \cos \phi_2 + r_3 \cos \phi_3 + r_4 \cos \phi_4 = 0 \qquad (2.6a)$$

$$r_2 \sin \phi_2 + r_3 \sin \phi_3 - r_4 \sin \phi_4 = 0 \qquad (2.6b)$$

where:

$$r_1 = \text{length of } base\ link$$

$$r_2 = \text{length of } crank$$

$$r_3 = \text{length of } coupler$$

$$r_4 = \text{length of } follower$$

The system of two scalar nonlinear Equations (2.6a) and (b) has two scalar unknowns, ϕ_3 and ϕ_4. A numerical procedure for solving such equations using EXCEL will be discussed later in this chapter. Here we can obtain a closed form solution as follows:

We solve Equation (2.6a) and (b) for $\cos \phi_3$ and $\sin \phi_3$, respectively:

$$\cos \phi_3 = \frac{r_1 - r_2 \cos \phi_2 - r_4 \cos \phi_4}{r_3} \qquad (2.7a)$$

$$\sin \phi_3 = \frac{-r_2 \sin \phi_2 + r_4 \sin \phi_4}{r_3} \qquad (2.7b)$$

Using the identity $\sin^2 \phi_3 + \cos^2 \phi_3 = 1$, we obtain one equation on the angle ϕ_4:

$$(r_1 + r_2 \cos \phi_2 + r_4 \cos \phi_4)^2 + (-r_2 \sin \phi_2 + r_4 \sin \phi_4)^2 = r_3^2 \qquad (2.8)$$

Using the half angle trigonometric identities $\sin \phi_4 = 2 \tan \phi_4/2/(1 + \tan^2 \phi_4/2)$, $\cos \phi_4 = (1 - \tan^2 \phi_4/2)/2/(1 + \tan^2 \phi_4/2)$, we obtain one equation in the unknown $\tan \phi_4/2$ and then calculate $\phi_4/2$. Thus, the relations of the input–output angles (*link rotations* or *angular displacements* ϕ_2 and ϕ_4) are:

$$\phi_4 = \pi - \arccos \frac{r_1^2 + s^2 - r_2^2}{2r_1 s} \pm \arccos \frac{r_4^2 + s^2 - r_3^2}{2r_4 s} \qquad (2.9a)$$

$$\phi_3 = \phi_4 - \gamma \qquad (2.9b)$$

where:

$$\gamma = \pm \arccos \frac{r_3^2 - s^2 + r_4^2}{2r_3r_4}$$

$$s = (r_1^2 + r_2^2 - 2r_1r_2 \cos \phi_2)^{1/2}$$

The $\pm$ sign in Equation (2.9a) points to the existence of two solutions. Indeed, in Figure 2.3a we observe that for the same input angle ϕ_2 links 3 and 4 can be assembled in two positions, ABO_B and $AB'O_B$. Both solutions are valid; the one that the designer will use will depend on other considerations of machine function, constraints, and geometry.

To trace the motion of the linkage for changing values of the input angle ϕ_2, one can use Equation (2.9a) and (b) to compute the output. In more complex linkages, obtaining closed form solutions is not always possible and solving numerically the closure equations is computationally intensive and sometimes difficult to obtain. The differential method is used to find the change in mechanism configuration if the initial configuration is known, and the mechanism input is given a small increment, $\Delta \phi_2$ in the case of a four-bar linkage. To this end, in the equations of closure in scalar form 2.6 we give a small increment to the link angles to obtain

$$r_1 + r_2 \cos(\phi_2 + \Delta\phi_2) + r_3 \cos(\phi_3 + \Delta\phi_3) + r_4 \cos(\phi_4 + \Delta\phi_4) = 0 \quad (2.10a)$$

$$r_2 \sin(\phi_2 + \Delta\phi_2) + r_3 \sin(\phi_3 + \Delta\phi_3) - r_4 \sin(\phi_4 + \Delta\phi_4) = 0 \quad (2.10b)$$

Using the trigonometric identities $\sin(a + b) = \sin a \cos b + \cos a \sin b$, $\cos(a + b) = \cos a \cos b - \sin a \sin b$ and that for small angles Δa, $\sin \Delta a \approx \Delta a$, $\cos \Delta a \approx 1$, we obtain from Equation (2.10a) and (b):

$$-r_2 \sin \phi_2 \Delta\phi_2 - r_3 \sin \phi_3 \Delta\phi_3 + r_4 \sin \phi_4 \Delta\phi_4 = 0 \quad (2.10c)$$

$$r_2 \cos \phi_2 \Delta\phi_2 + r_3 \cos \phi_3 \Delta\phi_3 + r_4 \cos \phi_4 \Delta\phi_4 = 0 \quad (2.10d)$$

or

$$-r_3 \sin \phi_3 \Delta\phi_3 + r_4 \sin \phi_4 \Delta\phi_4 = r_2 \sin \phi_2 \Delta\phi_2 \quad (2.10e)$$

$$r_3 \cos \phi_3 \Delta\phi_3 + r_4 \cos \phi_4 \Delta\phi_4 = -r_2 \cos \phi_2 \Delta\phi_2 \quad (2.10f)$$

These are two linear algebraic equations that yield the output change $\Delta\phi_3$, $\Delta\phi_4$ for input change $\Delta\phi_2$:

$$\Delta\phi_3 = \frac{-(r_2 \sin \phi_2 r_4 \cos \phi_4 + r_2 \sin \phi_2 r_4 \sin \phi_4) \Delta\phi_2}{r_3 \sin \phi_3 r_4 \cos \phi_4 + r_3 \cos \phi_3 r_4 \sin \phi_4} \quad (2.10g)$$

$$\Delta\phi_4 = \frac{-(r_3 \sin \phi_3 r_2 \sin \phi_2 - r_3 \sin \phi_3 r_2 \cos \phi_2) \Delta\phi_2}{r_3 \sin \phi_3 r_4 \cos \phi_4 + r_3 \cos \phi_3 r_4 \sin \phi_4} \quad (2.10h)$$

The new configuration is used as a starting position for a new small increment and so on until the full range of change of the input angle is achieved. This numerical method has the added advantage that the designer selects the initial configuration of the mechanism and the incremental procedure maintains, in general, this configuration, while the algebraic method or numerical solution of the equation of closure might give the symmetric configuration of the linkage in an uncontrollable way. Moreover, the incremental method can be easily generalized for different and more complex mechanism configurations by writing the appropriate equations of closure of the different loops.

For many applications, the coupler is not just a bar but a rigid body with considerable transverse dimensions, and the motion of points that lie outside the centerline AB of the coupler is important. In Figure 2.3a, point C is defined by the length of a vector r_5 and the angle $\alpha = \phi_5$. Since Equations (2.9) fully define the locations of points A and B, simple trigonometric determination of the coordinates of point C gives the position vector of point C (*coupler point*):

$$r_6 = \sqrt{r_2^2 + r_5^2 + 2r_2r_5 \cos(\phi_3 + \alpha - \phi_2)}$$

(2.10i)

$$\theta_6 = \tan^{-1} \frac{r_2 \sin \phi_2 + r_5 \sin(\phi_3 + \alpha)}{r_2 \cos \phi_2 + r_5 \cos(\phi_3 + \alpha)}$$

Example 2.4 A four-bar linkage has link lengths $r_1 = 1.4$ m, $r_2 = 0.436$ m, $r_3 = 1.044$ m, $r_4 = 0.831$ m. At the starting position, the input angle $\phi_2 = 1.25$ rad. Find the output angle ϕ_4. For nomenclature, see Figure E2.4.

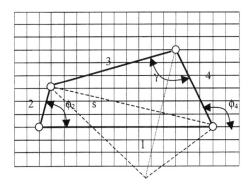

Figure E2.4

Solution
Data:

Description	EXCEL Symbol	Value	Units
Base link	r_1	1.4	m
Crank	r_2	0.436	m
Coupler	r_3	1.044	m
Follower	r_4	0.831	m
Input angle ϕ_2	f_2	1.25	rad
Input velocity ω_2	om_2	50	rad/s

From Equation (2.9a) and (*b*) we obtain:

Auxiliary distance $s = 1.32859$ m

Auxiliary angle $\gamma = 1.561952$ rad

Follower angle $\phi_4 = (+)3.728831$ rad and $(-)1.920968$ rad

Coupler angle $\phi_3 = 2.16688$ rad $- 1.561952$ rad

The configuration shown in Figure E2.4 corresponds to the second values of the follower and coupler angles, 1.92097 and 0.35902, respectively. The first angles, 3.72883 and 2.166883, correspond to follower and coupler angles of the assembly denoted by a dashed line.

2.2.1.1. Velocity Analysis. To obtain the angular velocities, if the input angle ϕ_2 changes with time, one can differentiate the equations for the angular displacements 2.9*a* and 2.9*b* in respect with the time. An alternative way to proceed is by vector analysis. Let a vector $R(t)$ (Figure 2.3*b*) be displaced at time $t + \Delta t$ to a new position $R(t + \Delta t)$. The first and second derivatives of the vector in respect to time are, if the length of the vector is constant:

$$\frac{d\mathbf{R}}{dt} = \lim_{\Delta t \to 0} \frac{\mathbf{R}(t + \Delta t) - \mathbf{R}(t)}{\Delta t} = \frac{\mathbf{k} \times \mathbf{R}d\phi}{dt} = \boldsymbol{\omega} \times \mathbf{R} \qquad (2.11)$$

$$\frac{d^2\mathbf{R}}{dt^2} = \boldsymbol{\alpha} \times \mathbf{R} + (\boldsymbol{\omega} \times \boldsymbol{\omega}) \times \mathbf{R}. \qquad (2.12)$$

where $\mathbf{k}$ is the unit vector along the z axis, perpendicular with the plane of the paper, $\omega = d\phi/dt$ the angular velocity and $\alpha = d^2\phi/dt^2$ the angular acceleration about the z axis. If the length of the vector changes,

$$\frac{d\mathbf{R}}{dt} = d\left(R\frac{\mathbf{R}}{R}dt\right) = \boldsymbol{\omega} \times \mathbf{R} + \dot{R}\frac{\mathbf{R}}{R} \qquad (2.13)$$

where $\mathbf{R}/R$ is the unit vector along the vector direction and a dot denotes the time derivative. The second derivative is:

$$\frac{d^2\mathbf{R}}{dt^2} = \boldsymbol{\alpha} \times \mathbf{R} + (\boldsymbol{\omega} \times \boldsymbol{\omega}) \times \mathbf{R} + \dot{R}\frac{\mathbf{R}}{R} + \dot{R}\boldsymbol{\omega} \times \frac{\mathbf{R}}{R} - \dot{R}\frac{\mathbf{R}}{R^2} \qquad (2.14)$$

The geometric closure equation (Equation (2.5)) for the vector polygon in Figure 2.3*a* is

$$\mathbf{r}_1 + \mathbf{r}_2 + \mathbf{r}_3 + \mathbf{r}_4 = 0 \qquad (2.15)$$

The time derivative of this equation, using Equation (2.13), gives:

$$\boldsymbol{\omega}_2 \times \mathbf{r}_2 + \boldsymbol{\omega}_3 \times \mathbf{r}_3 + \boldsymbol{\omega}_4 \times \mathbf{r}_4 = 0 \qquad (2.16)$$

Since $|a \times b| = ab \sin(\angle a, b)$, this vector equation is equivalent to the scalar equations,

$$-r_2 \sin \phi_2 \omega_2 - r_3 \sin \phi_3 \omega_3 - r_4 \sin \phi_4 \omega_4 = 0 \qquad (2.17a)$$

$$r_2 \cos \phi_2 \omega_2 + r_3 \cos \phi_3 \omega_3 + r_4 \cos \phi_4 \omega_4 = 0 \qquad (2.17b)$$

This is a linear system of equations in ω_3 and ω_4. The solution is a function of ϕ_2, ϕ_3, ϕ_4, ω_2:

$$\omega_4 = \frac{r_2 \sin(\phi_3 + \phi_2)}{r_4 \sin(\phi_4 - \phi_3)} \omega_2 \qquad (2.18a)$$

$$\omega_3 = \frac{r_2 \sin(\phi_4 + \phi_2)}{r_3 \sin(\phi_4 - \phi_3)} \omega_2 \qquad (2.18b)$$

Example 2.5 A four-bar linkage of Example 2.3a has link lengths $r_1 = 1.4$ m, $r_2 = 0.436$ m, $r_3 = 1.044$ m, $r_4 = 0.831$ m. At the starting position, the input angle $\phi_2 = 1.25$ rad and the angular velocity of the input link is $\omega_2 = 50$ rad/s. Find the angular velocities of the coupler ω_3 and of the follower ω_4 at this position of the linkage. For nomenclature, see Figure 2.3.

	A	B	C	D	E
1	KINEMATIC DIAGRAMS FOR A SLIDER-CRANK LINKAGE				
2	A. Definition of the system parameters				
3	Rod L=	0,16			
4	Crank Rcr =	0,08			
5	Crank ω_{cr}= Om =	377,00			
6	B. Spreadsheet				
7	Angle ϕ	θ	Displacement s	Velocity v/Om	Acceleration/Om^2
8	0	0	0,24	0	-0,12
9	0,314159265	0,155129957	0,234163155	-0,036619949	-0,109734137
10	0,628318531	0,298296775	0,21765551	-0,066922763	-0,080242527
11	0,942477796	0,416441332	0,193348299	-0,085520067	-0,03646333
12	1,256637061	0,495564432	0,165473431	-0,089447826	0,010795921
13	1,570796327	0,523598776	0,138564065	-0,08	0,046188022
14	1,884955592	0,495564432	0,116030712	-0,062721217	0,06023864
15	2,199114858	0,416441332	0,099302658	-0,043922652	0,057582311
16	2,513274123	0,298296775	0,088212791	-0,027122877	0,049200193
17	2,827433388	0,155129957	0,081994112	-0,01282277	0,042434906
18	3,141592654	6,12574E-17	0,08	-4,90059E-18	0,04
19	3,455751919	-0,155129957	0,081994112	0,01282277	0,042434906
20	3,769911184	-0,298296775	0,088212791	0,027122877	0,049200193
21	4,08407045	-0,416441332	0,099302658	0,043922652	0,057582311
22	4,398229715	-0,495564432	0,116030712	0,062721217	0,06023864
23	4,71238898	-0,523598776	0,138564065	0,08	0,046188022
24	5,026548246	-0,495564432	0,165473431	0,089447826	0,010795921
25	5,340707511	-0,416441332	0,193348299	0,085520067	-0,03646333
26	5,654866776	-0,298296775	0,21765551	0,066922763	-0,080242527
27	5,969026042	-0,155129957	0,234163155	0,036619949	-0,109734137
28	6,283185307	-1,22515E-16	0,24	2,94036E-17	-0,12
29	Formula:	ASIN((Rcr/L)*SIN(A8))	Rcr*COS(A8)+L*COS(B8)	Rcr*(-SIN(A8)-SIN(B8)*COS(A8)/COS(B8))	-Rcr*(COS(A8)-SIN(A8)*TAN(B8)+(Rcr/L)*(COS(A8))^2/(COS(B8))^3)

Figure E2.5

Solution

Using the values of the angles found in Example 2.3a and using Equation (2.12), we obtain from Equation (2.18a) and (b):

$$\omega_4 = \frac{r_2 \sin(\phi_3 + \phi_2)}{r_3 \sin(\phi_4 - \phi_3)} \omega_2 = 26.21532 \text{ rad/s}$$

$$\omega_3 = \frac{r_2 \sin(\phi_4 + \phi_2)}{r_3 \sin(\phi_4 - \phi_3)} \omega_2 = -0.770524 \text{ rad/s}$$

2.2.1.2. Acceleration Analysis. One further differentiation of Equation (2.13) yields the acceleration equation:

$$\alpha_2 \times r_2 + (\omega_2 \times \omega_2) \times r_2 + \alpha_3 \times r_3 + (\omega_3 \times \omega_3) \times r_3 + \alpha_4 \times r_4$$

$$+ (\omega_4 \times \omega_4) \times r_4 = 0 \tag{2.19a}$$

or

$$\alpha_3 \times r_3 + \alpha_4 \times r_4 = -\alpha_2 \times r_2 - (\omega_2 \times \omega_2) \times r_2 - (\omega_3 \times \omega_3)$$

$$\times r_3 - (\omega_4 \times \omega_4) \times r_4 \tag{2.19b}$$

The vector equation for the accelerations 2.19 is equivalent to the two scalar equations, in the Cartesian coordinate system (x, y, z):

$$-r_3 \sin \phi_3 \alpha_3 - r_4 \sin \phi_4 \alpha_4 = f_1 \tag{2.20a}$$

$$r_3 \cos \phi_3 \alpha_3 + r_4 \cos \phi_4 \alpha_4 = f_2 \tag{2.20b}$$

where:

$$f_1 = r_2 \sin \phi_2 \alpha_2 + r_2 \sin \phi_2 \omega_2^2 + r_3 \sin \phi_3 \omega_3^2 + r_4 \sin \phi_4 \omega_4^2$$

$$f_2 = -r_2 \cos \phi_2 \alpha_2 + r_2 \cos \phi_2 \omega_2^2 + r_3 \cos \phi_3 \omega_3^2 + r_4 \cos \phi_4 \omega_4^2$$

Solving the system of linear equations 2.20 for the unknown angular accelerations,

$$\alpha_3 = \frac{f_1 \cos \phi_4 + f_2 \sin \phi_4}{r_3 \sin(\phi_4 - \phi_3)} \tag{2.21a}$$

$$\alpha_4 = \frac{-f_2 \sin \phi_3 + f_1 \cos \phi_3}{r_4 \sin(\phi_4 - \phi_3)} \tag{2.21b}$$

The geometric interpretation of Equations (2.15), (2.16), and (2.19) follows:

1. Equation (2.15) shows that the location of every point can be found if to the location of another point we add a vector connecting the two points. For

example, in Figure 2.3a, since the coordinates of point O_A are (0, 0), the location of point D is the sum of vectors $\mathbf{r}_2$, $\mathbf{r}_3$, and $\mathbf{r}_4$.

2. The velocity of point A is $\mathbf{v}_A = \boldsymbol{\omega}_2 \times \mathbf{r}_2$, of magnitude $\omega_2 r_2$ and direction perpendicular with $O_A A$. The velocity of point B is $\mathbf{v}_B = \mathbf{v}_a + \boldsymbol{\omega}_3 \times \mathbf{r}_3$. The velocity of any other point C will be:

$$\mathbf{v}_C = \mathbf{v}_A + \boldsymbol{\omega}_3 \times \mathbf{r}_5 \tag{2.22}$$

The term $\boldsymbol{\omega}_3 \times \mathbf{r}_5$ is the relative velocity of point C in respect to A.

3. The acceleration of point A is $\boldsymbol{\alpha}_2 \times \mathbf{r}_2 + \boldsymbol{\omega}_2 \times (\boldsymbol{\omega}_2 \times \mathbf{r}_2)$. The first term is due to the angular acceleration perpendicular to $O_A A$, and the second term is the centripetal acceleration due to the angular velocity of magnitude $\omega_2^2 r_2$ and directed from A to O_A. At any point C:

$$\mathbf{a}_C = \mathbf{a}_A + \boldsymbol{\alpha}_3 \times \mathbf{r}_5 + \boldsymbol{\omega}_3 \times (\boldsymbol{\omega}_3 \times \mathbf{r}_5) \tag{2.23}$$

The term $\boldsymbol{\alpha}_3 \times \mathbf{r}_5 + (\boldsymbol{\omega}_3 \times \boldsymbol{\omega}_3) \times \mathbf{r}_5$ is the relative acceleration of point C in respect to A.

Example 2.6   A four-bar linkage of Example 2.5 has link lengths $r_1 = 1.4$ m, $r_2 = 0.436$ m, $r_3 = 1.044$ m, $r_4 = 0.831$ m. At the starting position, the input angle $\phi_2 = 1.25$ rad and constant angular velocity of the input link $\omega_2 = 50$ rad/s.

• Find the angular acceleration of the coupler α_3 and of the follower α_4 at this position of the linkage.
• Find the acceleration of the coupler point C shown in Figure E2.3c.

Solution

1. Using the values of the angles found in Example 2.4, the output angular velocities are (Example 2.5) $\omega_4 = 26.21532$ rad/s, $\omega_3 = -0.770524$ rad/s. The angular acceleration $\alpha_2 = 0$ because the angular velocity is constant. Therefore, from Equations (2.21):

$$f_1 = r_2 \sin \phi_2 \alpha_2 + r_2 \sin \phi_2 \omega_2^2 + r_3 \sin \phi_3 \omega_3^2$$
$$+ r_4 \sin \phi_4 \omega_4^2 = 1571.052 \text{ m} - \text{rad/s}^2$$

$$f_2 = -r_2 \cos \phi_2 \alpha_2 + r_2 \cos \phi_2 \omega_2^2 + r_3 \cos \phi_3 \omega_3^2$$
$$+ r_4 \cos \phi_4 \omega_4^2 = 148.3612 \text{ m} - \text{rad/s}^2$$

$$\alpha_3 = \frac{f_1 \cos \phi_4 + f_2 \sin \phi_4}{r_3 \sin(\phi_4 - \phi_3)} = -382.7786 \text{ rad/s}^2$$

$$\alpha_4 = -\frac{f_2 \sin \phi_3 + f_1 \cos \phi_3}{r_4 \sin(\phi_4 - \phi_3)} = 1832.82 \text{ rad/s}^2$$

Using unit vectors $\mathbf{i}$, $\mathbf{j}$ along the x and y axis, respectively:

$$\mathbf{r}_1 = 0.1\mathbf{i} + 0.3\mathbf{j}, \ \mathbf{r}_5 = 0.3\mathbf{i} + 0.5\mathbf{j}$$

$$\omega_2 = 50\mathbf{k} \text{ rad/s}, \ \omega_4 = 26.21532\mathbf{k} \text{ rad/s}, \ \omega_3 = -0.770524\mathbf{k} \text{ rad/s}$$

$$\alpha_3 = -382.7786\mathbf{k} \text{ rad/s}^2, \ \alpha_4 = 1832.82\mathbf{k} \text{ rad/s}^2$$

Using Equation (2.23):

$$\mathbf{a}_A = \mathbf{r}_1 \times \omega_2^2 = (0.1\mathbf{i} + 0.3\mathbf{j}) \times 50\mathbf{k} = -5\mathbf{j} + 15\mathbf{i}$$

$$\mathbf{a}_C = \mathbf{a}_A + \alpha_3 \times \mathbf{r}_5 + \omega_3 \times (\omega_3 \times \mathbf{r}_5)$$

$$= 15\mathbf{i} - 5\mathbf{j} + 382.7786\mathbf{k} \times (0.3\mathbf{i} + 0.5\mathbf{j}) + 0.770524\mathbf{k}$$

$$\times [0.770524\mathbf{k} \times (0.3\mathbf{i} + 0.5\mathbf{j})]$$

$$= 205.9\mathbf{i} + 109.7\mathbf{j}$$

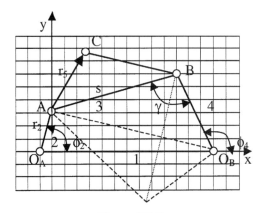

Figure E2.6

2.2.2. The Slider–Crank Linkage

For the slider–crank mechanism (Figure 2.4), we obtain similar expressions: input is the *angle of rotation* or *angular displacement* of the crank ϕ and output is the distance s traveled by the slider (the piston in a piston engine), point B.

The vector equation for the closure is:

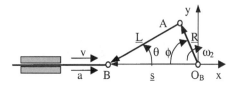

Figure 2.4. The slider–crank linkage.

$$\mathbf{R} + \mathbf{L} - \mathbf{s} = 0 \tag{2.24}$$

This is equivalent to the scalar equations

$$R \cos \phi + L \cos \theta - s = 0 \tag{2.25}$$

$$R \sin \phi - L \sin \theta = 0 \tag{2.26}$$

Determination of the unknowns θ and s can be done by a solution of the above Equations (2.25) and (2.26):

$$L \sin \theta = R \sin \phi \tag{2.27}$$

$$s = R \cos \phi + L \cos \theta = R \cos \phi + L \sqrt{1 - (R/L) \sin^2 \phi} \tag{2.28}$$

The time differentiation of the equation of closure 2.24 yields the velocity equation

$$\boldsymbol{\omega}_R \times \mathbf{R} + \boldsymbol{\omega}_L \times \mathbf{L} - v\mathbf{i} = 0 \tag{2.29}$$

where $\mathbf{i}$ is the unit vector along the x axis, $\omega_R = d\phi/dt$, the angular velocity of the crank, $\omega_L = d\theta/dt$, the angular velocity of the connecting rod, and v is the velocity of point B. The corresponding scalar equations yield:

$$\omega_L = \frac{\omega_R(R/L)\cos \phi}{\cos \theta} \tag{2.30}$$

$$v = -\omega_R R \sin \phi - \omega_L L \sin \theta = -R\omega_R(\sin \phi + \cos \phi \tan \theta) \tag{2.31}$$

The time derivative of the velocity equation gives:

$$\boldsymbol{\alpha}_R \times \mathbf{R} + (\boldsymbol{\omega}_R \times \boldsymbol{\omega}_R) \times \mathbf{R} + (\boldsymbol{\omega}_L \times \boldsymbol{\omega}_L) \times \mathbf{L} + \boldsymbol{\alpha}_L \times \mathbf{L} + a\mathbf{i} = 0 \tag{2.32}$$

The corresponding scalar equations yield:

$$\alpha_L = -\alpha_R \frac{R}{L} \frac{\cos\phi}{\cos\theta} + \omega_R^2 \frac{R}{L} \frac{-\sin \phi \cos \theta - \cos \phi \sin \theta}{\cos^2 \theta} \tag{2.33}$$

$$a = -\alpha_R \sin\phi - \omega_R^2 R \cos \phi - \alpha_L L \sin \theta + \omega_L^2 R \cos \theta$$

$$= -R\alpha_R \left(\sin \phi + \frac{\cos \phi}{\tan \theta} \right)$$

$$- R\omega_R^2 \left(\cos \phi - \sin \phi \tan \theta + R\omega_R^2 \frac{R \cos^2 \phi}{L \cos^3 \theta} \right) \tag{2.34}$$

where R is the crank length, L is the follower length, ϕ is the input angle, s, v, and a are the displacement, velocity, and acceleration of the slider, ω and α are the angular velocities and angular accelerations, and subscripts R and L designate crank and follower, respectively.

Example 2.7   Determine the velocity of the head slider S of the quick return mechanism of the shaper machine shown if the crank C rotates with constant angular velocity and the other dimensions are as shown in Figure E2.7.

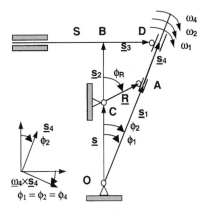

Figure E2.7

Solution   Let the vectors $\mathbf{s} = (\vec{OC})$, $\mathbf{s}_1 = (\vec{OA})$, $\mathbf{s}_2 = (\vec{OB})$, $\mathbf{s}_3 = (\vec{BD})$, $\mathbf{s}_4 = (\vec{OD})$. The equation of closure for the triangle OAC is:

$$\mathbf{s} + \mathbf{R} \quad \mathbf{s}_1 = 0$$

In scalar form:

$$\mathbf{s} + R \cos \phi_R - s_1 \cos \phi_1 = 0$$

$$R \sin \phi_R - s_1 \sin \phi_1 = 0$$

The solution is:

$$\phi_1 = \tan^{-1} \frac{R \sin \phi_R}{s + R \cos \phi_R}$$

$$s_1 = \frac{R \sin \phi_R}{\sin \phi_1}$$

The time derivative of the equation of closure is:

$$\boldsymbol{\omega} \times \mathbf{R} - \boldsymbol{\omega}_1 \times \mathbf{s}_1 - \frac{\dot{s}_1 \mathbf{s}_1}{s_1} = 0$$

In scalar form:

$$\omega R \sin \phi_R - \omega_1 s_1 \sin \phi_1 - \frac{\dot{s}_1}{s_1} s_1 \cos\phi_1 = 0$$

$$-\omega R \cos \phi_R + \omega_1 s_1 \cos \phi_1 - \frac{\dot{s}_1}{s_1} s_1 \sin\phi_1 = 0$$

The solution is:

$$\omega_1 = \frac{\omega(\sin \phi_R \sin \phi_1 + \cos \phi_R \cos \phi_1)R}{s_1} = \frac{\omega \cos(\phi_R - \phi_1)R}{s_1}$$

$$\dot{s}_1 = (-\omega_R R \sin \phi_R - \omega_1 s_1 \sin \phi_1)/\cos\phi_1$$

Now we write the equation of closure for the triangle *OBD*:

$$\mathbf{s}_2 - \mathbf{s}_4 + \mathbf{s}_3 = 0$$

Time differentiation gives:

$$-\boldsymbol{\omega}_4 \times \mathbf{s}_4 - \frac{\dot{s}_4 \mathbf{s}_4}{s_4} + \frac{\dot{s}_3 \mathbf{s}_3}{s_3} = 0$$

since $\omega_2 = 0$, $\omega_3 = 0$.

The velocity equation yields the scalar equations:

$$-\omega_4 s_4 \cos \phi_2 - \dot{s}_4 \sin \phi_2 + \dot{s}_3 = 0 \text{ (horizontal direction)}$$

$$\omega_4 s_4 \sin \phi_2 + \dot{s}_4 \cos \phi_2 = 0 \text{ (vertical direction)}$$

Therefore, since $\phi_2 = \phi_1$, $\omega_4 = \omega_2 = \omega_1$:

$$\dot{s}_4 = -\omega_4 s_4 \tan \phi_1$$

$$\dot{s}_3 = \omega_1 s_4 \cos \phi_1 + \dot{s}_4 \sin \phi_1 = s_2 \omega_1 (1 - \tan^2 \phi_1)$$

Example 2.8 Kinematic Diagram for a Slider–Crank Linkage

The slider–crank mechanism for an automotive engine (Figure 2.4) consists of a crank of length $R = 80$ mm and a connecting rod of length $L = 160$ mm. At constant rotating speed $n = 3000$ rpm, calculate and plot the velocity and acceleration of the piston over one cycle of rotation.

Solution The angular velocity of rotation of the crank is $\omega = 2\pi n/60 = 2\pi \times 3600/60 = 377$ rad/s. The piston displacement, velocity and acceleration (Figure 2.4) are s, v, and a, respectively. The input angular displacement and angular velocity are the crank angle ϕ and angular velocity ω_R, respectively. The angular acceleration of the crank α_R is zero because the rotating speed is constant. The respective equations are, from Equations (2.28), (2.31), (2.34):

$$s = R \cos \phi + L\sqrt{1 - (R/L)^2 \sin^2 \phi} \qquad (a)$$

$$\sin \theta = \frac{R}{L} \sin \phi, \quad \cos \theta = \sqrt{1 - \left(\frac{R}{L}\right)^2 \sin^2 \phi} \qquad (b)$$

$$v = -R\omega_R \left(\sin\phi + \frac{\cos \phi}{\tan \theta} \right) \qquad (c)$$

$$a = -R\alpha_R \left(\sin \phi + \frac{\cos \phi}{\tan \theta} \right) - R\omega_R^2 \left(\cos \phi - \sin \phi \tan \theta + R\omega_R^2 \frac{R \cos^2 \phi}{L \cos^2 \theta} \right) \qquad (d)$$

Equations (*a–d*) can be conveniently plotted using a spreadsheet, such as EXCEL (Figure

E2.5). Since the angular velocity usually is $\omega_R \gg 1$, the numerical values of the velocity v will be much greater than the values of the angular displacement ϕ, and the accelerations a will be even larger. To plot them in a single diagram, we plot ϕ, v/ω_R and a/ω_R^2, instead of ϕ, v, and a, for 24 values of the input angle $0 < \phi < 2\pi$. In cells $A1$, $A2$, $C2$, we define variables Crank, Rod, Om for R, L, and ω_R, respectively. We assign respective values 0.08, 0.16, and 377. To simplify the equations we included a column for the angle θ.

The parameters Crank, Rod, and Om (for R, L and ω_{cr}, respectively) were defined in cells $B3$–$B5$ where the numerical values have also been assigned. Then in cells $B8$–$E8$ the expressions we written for the results shown in cells $B7$–$E7$, respectively. The rest of the columns were filled by dragging each one down by the cursor at the rightmost lower corner of the cells $B8$–$E8$. For column A, cell $A7$ was set to zero and cell $A8$ was set $= A7 + PI(\ )/24$. Then the cell was dragged down to generate cells $A9$–$A28$. For the convenience of the reader, in line 29 we placed the equations used in cells $B8$–$E8$.

The plot is shown in Fig. E2.8.

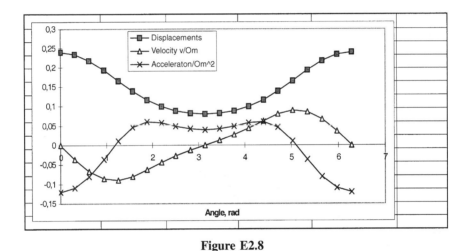

Figure E2.8

2.2.3. Instantaneous Center of Rotation

Instantaneous center of rotation for a point on a moving rigid body is defined in mechanics as the center of the circular arc that coincides with the path of the point over an infinitesimal part of the path about the point. It is obvious then that this center will lie on a straight line that is perpendicular to the velocity vector (which is always tangent to the path) and on the plane that is tangent to the path. That plane is the linkage plane in the case of planar linkage (Figure 2.5). At the position of the linkage shown, the coupler AB instantaneous center of rotation C must lie on the extension of the links O_AA and O_BB because they are perpendicular to the velocities of points A and B, respectively.

For an automotive suspension system, the double wishbone axle, the instantaneous center of rotation of the coupler, which is the wheel in this case, is called *roll center* and needs to be at a height from the ground as close as possible to the height of the mass center of the vehicle and as far from the vehicle as possible to improve the transverse stability of the car and reduce roll when cornering. The roll center is in the intersection of crank and follower, (Figures 2.6, 2.7).

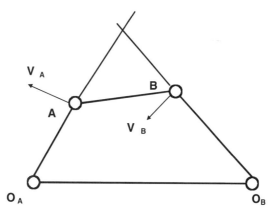

Figure 2.5. Instantaneous center of rotation of the coupler for a four-bar linkage.

Design Procedure 2.1: Kinematic Analysis of a Closed Loop Linkage

Step 1: Determine the number of degrees of freedom M (Equation (2.1)), typically $M = 1$.

Step 2: Sketch linkage in the starting position, sketch as many closed loops as necessary to sketch the linkage completely.

Step 3: On the loops, assign a vector to every bar.

Step 4: Write the vector equations of closure for each loop, as many as the number of closed loops (such as Equations (2.5)).

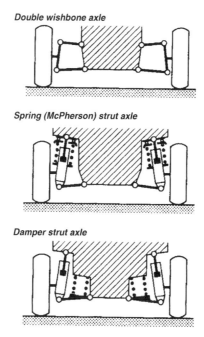

Figure 2.6. Automotive suspension systems. (Reprinted from Bosch, *Automotive Handbook*, 1987, by permission of VDI-Verlag.)

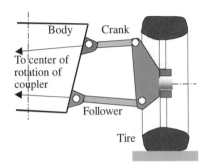

Figure 2.7. Double wishbone axle suspension.

Step 5: Split each vector equation into two scalar equations (such as Equations (2.6)).

Step 6: Identify input (given) quantities (their number = M), i.e., ϕ_2, and output (unknown) quantities, i.e., ϕ_3, ϕ_4, for the four-bar linkage.

Step 7: Solve scalar forms of closure equations for the unknown quantities, analytically, if possible (such as Equations (2.7) to (2.9)), or numerically, for example using Solver of EXCEL (see next section).

Step 8: Take the time derivatives of the vector equations of closure and repeat steps 5–7 for the velocities.

Step 9: Take the time derivatives of the vector equations for the velocities and repeat steps 5–7 for the accelerations.

Step 10: Plot displacements, velocities, accelerations of links vs. input (i.e., input angle).

2.3. VARIABLE SPEED KINEMATIC PAIRS—OPEN LOOP LINKAGES

2.3.1. Coordinate Transformations

Consider two coordinate systems (X, Y, Z) and (x, y, z) (Figure 2.8). The coordinates of a point P in the coordinate system (x, y, z) can be transformed into coordinates of the same point in the coordinate system (X, Y, Z) with the transformation:

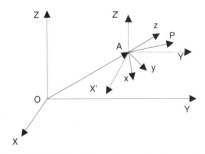

Figure 2.8. Definition of coordinate systems.

$$P(X, Y, Z) = GP(x, y, z) \tag{2.35}$$

$$\text{where} = \begin{bmatrix} l_x & m_x & n_x & t_x \\ l_y & m_y & n_y & t_y \\ l_z & m_z & n_z & t_z \\ 0 & 0 & 0 & 1 \end{bmatrix}$$

t_x, t_y, t_z are the components of translation of the origin vector OA in the coordinate system (X, Y, Z), and l_x, l_y, l_z, m_x, m_y, m_z, n_x, n_y, n_z are direction cosines, that is, the cosines of the angles of the axes Ax, Ay, Az with the axes AX' or AY' or AZ', respectively, as indicated by the subscript, of a coordinate system (X', Y', Z') which is parallel with the (X, Y, Z) but has origin at A.

Consider now a kinematic pair of two rigid links i and j connected by way of a revolute pin joint at point B (Figure 2.9). These links might be also connected to other links with revolute pins at points A and C. We shall study the relationships of the coordinates at different coordinate systems along the chain of links having various origins and orientations.

The coordinate system $(A, x_{iA}, y_{iA}, z_{iA})$ has origin at A and the x_{iA} axis oriented along the line AB. A similarly oriented coordinate system with origin at point B is denoted as (x_{iB}, y_{iB}, z_{iB}). The relationship among the coordinates of any point P in the two coordinate systems is as in Equation (2.35), with $l_x = a_i$, where a_i is the distance (AB) and all the direction cosines are 0 or 1 because the two coordinate systems are parallel. This relationship can be written in a symbolic form (from Equation (2.35), as:

$$P_{iA} = T_{AB}P_{iB} \tag{2.36}$$

where:[3]

$$P_{iA} = \{x_{iA}, y_{iA}, z_{iA}, 1\}, \quad T_{AB} = \begin{bmatrix} 1 & 0 & 0 & a_i \\ 0 & 1 & 0 & 0 \\ 0 & 0 & 1 & 0 \\ 0 & 0 & 0 & 1 \end{bmatrix}, \quad P_{iB} = \{x_{iB}, y_{iB}, z_{iB}, 1\}$$

Matrix T_{AB} is called a *translation matrix* (member of a class of *homogeneous*

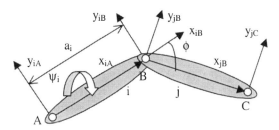

Figure 2.9. Coordinate systems on two rigid links connected through a revolute pin.

[3]Curly brackets denote column vectors, i.e., $\{a, b\} = [a \, b]^T$.

transformations) because it refers to a translation of a coordinate system from one origin to another retaining the same orientation. Thus, *the coordinate vector in one coordinate system is the product of the vector in another, parallel, coordinate system multiplied from the left by a translation matrix* T.

Consider now a coordinate system with origin at point B but rotated in respect to (x_{iB}, B, y_{iB}) by an angle φ so that its x axis coincides with line BC of link j, where φ is the angle between lines AB and BC, as shown. This system is denoted $(B, x_{jB}, y_{jB}, z_{jB})$.

The relationship between the coordinates of a point P in the two coordinate systems can be written in a symbolic form, from Equation (2.35), as:

$$P_{iB} = R_\phi P_{jB} \tag{2.37}$$

where:

$$P_{iB} = \{x_{iB}, y_{iB}, z_{iB}, 1\}, \quad R_\varphi = \begin{bmatrix} \cos\phi & \sin\phi & 0 & 0 \\ -\sin\phi & \cos\phi & 0 & 0 \\ 0 & 1 & 1 & 0 \\ 0 & 0 & 0 & 1 \end{bmatrix}, \quad P_{jB} = \{x_{jB}, y_{jB}, z_{jB}, 1\}$$

because $\cos(90 + \phi) = -\sin\phi$, $\cos(90 - \phi) = \sin\phi$.

Matrix R_φ is called a *rotation matrix* because it refers to a rotation of a coordinate system about the z axis through the origin by an angle φ. Thus, *the coordinate vector in one coordinate system is the product of the vector in another, rotated, coordinate system multiplied from the left by a rotation matrix* R.

If the arm i is rotated about its longitudinal axis (about the x axis) by an angle ψ_i, the x coordinates in the system $\{x_{iB}, y_{iB}, z_{iB}, 1\}$ will remain the same while the y and z coordinates will change according to Equations (2.35). The corresponding transformation matrix is:

$$R_\psi = \begin{bmatrix} 1 & 0 & 0 & 0 \\ 0 & \cos\psi & \sin\psi & 0 \\ 0 & -\sin\psi & \cos\psi & 0 \\ 0 & 0 & 0 & 1 \end{bmatrix} \tag{2.38}$$

It follows from Equations (2.36) and (2.37) that:

$$P_{iA} = T_{AB} R_\psi R_\varphi P_{jB} \tag{2.39}$$

Thus, *successive coordinate transformations can be represented by coordinate transformation matrix multiplications.*

2.3.2. Robot Arms and Manipulators

2.3.2.1. Direct (Forward) Kinematics. Unlike the closed loop linkages studied in Section 2.3, robot arms and manipulators have several degrees of freedom and, in the robot arm of Figure 2.10a for example, each link can rotate about the

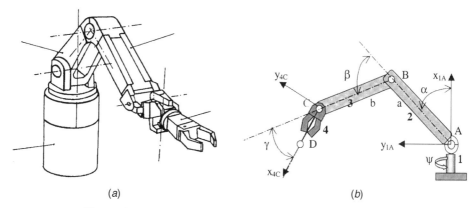

(a)   *(b)*

Figure 2.10. (*a*) Four-axes robot; (*b*) schematic of the robot.

respective joint independently by way of motors that can turn the links by angles α, β, γ in respect to ground link 1, as shown in Figure 2.10*b*.

The link rotations about their longitudinal axes are assumed to be zero, except for link 1, which rotates about the vertical axis by angle ψ. In an open loop mechanism, the joint rotation variables are independent of one another, so that all joint rotation variables must be specified in order to define the particular configuration of the system and location of the *end effector,* the grip or tool that the robot arm manipulates.

The system consists of four links connected serially by three revolute joints. Thus, with successive applications of Equations (2.36) and (2.37), we can obtain an expression for the coordinates in a coordinate system attached to *base link* (x_{1D}, y_{1D}, z_{1D}), or *base frame*, in terms of the three revolute joint angles α, β, γ, the base rotation angle ψ, and the point coordinates in the coordinate system attached to the *end effector frame* or the *end effector link* (x_{4D}, y_{4D}, z_{4D}). This has the matrix form

$$P_{1A} = A_{1-4}P_{4C} \tag{2.40}$$

where $A_{1-4} = R_\psi R_\alpha T_a R_\beta T_b R_\gamma$ is a Denavit–Hartenberg transformation matrix (Denavit and Hartenberg 1955, Hartenberg and Denavit 1964).

Generalization of Equation (2.40) for more links is apparent. One can perform the algebraic operations in Equation (2.40) or obtain explicitly with simple trigonometry, for the three-arm robot with base rotation ψ:

$$x_{1A} = x_{4c} \cos(\alpha + \beta + \gamma) - y_{4c} \sin(\alpha + \beta + \gamma) \tag{2.41a}$$

$$+ b \cos(\alpha + \beta) + a \cos \alpha$$

$$y_{1A} = y^* \cos \psi + z_{4c} \sin \psi \tag{2.41b}$$

$$z_{1A} = -y^* \sin \psi + z_{4c} \cos \psi \tag{2.41c}$$

where $y^* = x_{4c} \sin(\alpha + \beta + \gamma) + y_{4c} \cos(\alpha + \beta + \gamma) + b \sin(\alpha + \beta) + a \sin \alpha$.

Velocities and accelerations can be readily obtained by differentiation.

Design Procedure 2.2: Direct Kinematics of an Open Loop Linkage

Step 1: Determine the number of degrees of freedom M (Equation (2.1a)), typically $M = 2$–6.

Step 2: Sketch the robot arm in the starting position, identify end effector and base coordinate systems and the point of interest on the end effector.

Step 3: Identify position vectors of the point of interest on the end effector and base coordinate systems.

Step 4: Identify the input quantities (link lengths, input rotations, position vector of the point of interest on the end effector-based coordinate system).

Step 5: Multiply successively the position vector on the end effector-based coordinate system by the translation and rotation matrices of the links, from the end effector to the base, to find the position vector of the point of interest on the base coordinate system.

Step 6: Find values of the coordinates of the point of interest on the base coordinate system

Step 7: Take the time derivatives of the vector equations of closure and repeat steps 5 and 6 for the velocities.

Step 8: Take the time derivatives of the vector equations for the velocities found in step 7 and repeat steps 5 and 6 for the accelerations.

Step 9: Plot displacements, velocities, and accelerations of the point of interest vs. input (i.e., input angles or time).

Example 2.9 A three-arm robot such as the one shown in Figure 2.10 has $a = 1$ m, $b = 1$ m. The end effector is holding a cylindrical shaft on the arm centerline at distance 0.2 m from the pivot C. At a certain time, the arm configuration is at $\alpha = 30°$, $\beta = 30°$, $\gamma = 30°$, and $\psi = 0$ remaining constant, the angular velocities of the three arms are $\dot{\alpha} = 25$ rad/s, $\dot{\beta} = 80$ rad/s, $\dot{\gamma} = -100$ rad/s, while the angular accelerations of the three arms are $\ddot{\alpha} = 40$ rad/s^2, $\ddot{\beta} = -50$ rad/s^2, $\ddot{\gamma} = 30$ rad/s^2. Find the location, velocity, and acceleration of end point D.

Solution We first need to determine the initial location of point D in the coordinate system (x_{1A}, y_{1A}, z_{1A}) using Equations (2.41) with $x_{4C} = 0.2$ m, $y_{4C} = 0$, $z_{4C} = 0$:

$$x_{1A} = x_{4c} \cos(\alpha + \beta + \gamma) - y_{4c} \sin(\alpha + \beta + \gamma) + b \cos(\alpha + \beta) + a \cos \alpha \qquad (a)$$

$$y_{1A} = x_{4c} \sin(\alpha + \beta + \gamma) + y_{4c} \cos(\alpha + \beta + \gamma) + b \sin(\alpha + \beta) + a \sin \alpha \qquad (b)$$

The first derivative yields the velocities of point D:

$$\dot{x}_{1A} = [-x_{4c} \sin(\alpha + \beta + \gamma) - y_{4c} \cos(\alpha + \beta + \gamma)] (\dot{\alpha} + \dot{\beta} + \dot{\gamma}) \qquad (c)$$
$$- b \sin(\alpha + \beta) (\dot{\alpha} + \dot{b}) - a \sin \alpha \dot{\alpha}$$

$$\dot{y}_{1A} = [x_{4c} \cos(\alpha + \beta + \gamma) - y_{4c} \sin(\alpha + \beta + \gamma)] (\dot{\alpha} + \dot{\beta} + \dot{\gamma}) \qquad (d)$$
$$+ b \cos(\alpha + \beta) (\dot{\alpha} + \dot{\beta}) + a \cos \alpha \dot{\alpha}$$

Another derivative yields the accelerations of point D:

$$\ddot{x}_{1A} = [-x_{4c} \cos(\alpha + \beta + \gamma) + y_{4c} \sin(\alpha + \beta + \gamma)] (\dot{\alpha} + \dot{\beta} + \dot{\gamma})^2 \qquad (e)$$
$$+ [-x_{4c} \sin(\alpha + \beta + \gamma) - y_{4c} \cos(\alpha + \beta + \gamma)] (\ddot{\alpha} + \ddot{\beta} + \ddot{\gamma})$$
$$- b \cos(\alpha + \beta)(\dot{\alpha} + \dot{\beta})^2 - b \sin(\alpha + \beta)(\ddot{\alpha} + \ddot{\beta}) - a \cos \alpha \dot{\alpha}^2$$
$$- a \sin \alpha \ddot{\alpha}$$

$$\ddot{y}_{1A} = [-x_{4c} \sin(\alpha + \beta + \gamma) - y_{4c} \cos(\alpha + \beta + \gamma)](\dot{\alpha} + \dot{\beta} + \dot{\gamma})^2 \qquad (f)$$
$$+ [x_{4c} \cos(\alpha + \beta + \gamma) - y_{4c} \sin(\alpha + \beta + \gamma)] (\ddot{\alpha} + \ddot{\beta} + \ddot{\gamma})$$
$$- b \sin(\alpha + \beta))(\dot{\alpha} + \dot{\beta})^2 + b \cos(\alpha + \beta)(\ddot{\alpha} + \ddot{\beta})$$
$$- a \sin \alpha \dot{\alpha}^2 + a \cos \alpha \ddot{\alpha}$$

Substituting:

$$x_{1A} = 0.2 \cos 90° + 1.0 \cos 60° + 1.0 \cos 30°$$
$$= 0 + 0.5 + 0.866 = \underline{1.366 \text{ m}} \qquad (g)$$

$$y_{1A} = 0.2 \sin 90° + 1.0 \sin 60° + 1.0 \sin 30° = 0.2 + 0.866 + 0.5$$
$$= \underline{1.566 \text{ m}} \qquad (h)$$

$$\dot{x}_{1A} = -0.2 \sin(90)(5) - 1.0 \sin(60)(105) - 1.0 \sin 30 \times 25$$
$$= \underline{-104.4 \text{ m/s}} \qquad (i)$$

$$\dot{y}_{1A} = 1.0 \cos(60) \times 105 + 1.0 \cos 30 \times 25$$
$$= \underline{74.1 \text{ m/s}} \qquad (j)$$

$$\ddot{x}_{1A} = -0.2 \cos(90)(-5)^2 - 0.2 \sin(90)(20) - 1.0 \cos 60 \times 105^2$$
$$- 1.0 \sin 60(-10) - 1.0 \cos 30 \times 25^2 - 1.0 \sin 30 \times 40$$
$$= \underline{6069.1 \text{ m/s}^2} \qquad (k)$$

$$\ddot{y}_{1A} = -0.2 \sin(90)(-5)^2 + 0.2 \cos(90) \times 20 + 1.0 \sin(60)(105)^2$$
$$+ 1.0 \cos(60)(-10) - 1.0 \sin 30 \times 25^2 + 1.0 \cos 30 \times 40$$
$$= \underline{9358.8 \text{ m/s}^2} \qquad (l)$$

2.3.2.2. *Inverse Kinematics.* The problem of determining the position of the end effector for a given set of robot control parameters (Equation (2.40)) is termed the *direct kinematics problem*. Of interest for the operator of the robot is the inverse problem: for a given set of coordinates P_{1A} of the end effector in the base frame coordinate system, find the robot control parameters, for example α, β, γ, a, b and ψ in the above example. This is called the *inverse kinematics* problem.

Since vectors P_{1A} and P_{4A} are known, the geometric transformation that relates them can be obtained directly from the coordinates of the origin of the coordinate system, which is fixed on the end effector, in respect to the base coordinate system t_x, t_y, t_z, the direction cosines of the respective axes of the two coordinate systems

$l(l_x, l_y, l_z)$, $m(m_x, m_y, m_z)$, $n(n_x, n_y, n_z)$ that relate the coordinates of a point in the two coordinate systems as in Equation (2.35).

If we are interested only in the position of point P, the three scalar equations that result from Equation (2.40) suggest that only with a three-degree-of-freedom robot arm do we have a unique solution of this inverse kinematic problem.

If, however, we are interested not only in the location of point P but also in the orientation of the end effector, if the latter needs to have a given orientation, matrix G in Equation (2.35) is known. Indeed, the information it contains consists of the location of the origin of the end effector-fixed coordinate system in respect to the base-fixed coordinate system t_x, t_y, t_z, and the direction cosines that define the orientation of the axes of the end effector-fixed coordinate system in respect with the base-fixed one l_x, l_y, l_z, m_x, m_y, m_z, n_x, n_y, n_z.

On the other hand, Equation (2.40) relates exactly the same vectors as Equation (2.35) and thus the two matrices G_{1-4} (which contains only known geometric information) and A_{1-4} (which contains only robot control information) are equal and their elements are equal, one to one. Therefore, we possess twelve equations to determine the six unknowns in the example discussed. Great care should be exercised, however, since some of the twelve equations are obviously interdependent or identities (i.e., $0 = 0$, $1 = 1$), and to fully define the location and orientation of the end effector in respect to the coordinate system fixed at the machine frame we need only six coordinates. Thus, robots with less than six degrees of freedom cannot achieve full location and orientation of the end effector and robots with more than six degrees of freedom can have multiple solutions of the inverse kinematic problem, that is, there are many sets of robot control parameters that can achieve the same location and orientation of the end effector. An example is the human arm, with seven degrees of freedom, that can assemble the same part from different angles.

Though in some cases closed-form solutions can be found, in general this is by no means a trivial problem, and numerical methods should be used in general; see Example 2.7. Closed-form solutions for most cases of typical robots can be found in the literature (Shabinpoor 1987, Asada and Slotine 1986).

Design Procedure 2.3: Inverse Kinematics of an Open Loop Linkage

Step 1: Determine the number of degrees of freedom M (Equation (2.1a)), typically $M = 2$–6.

Step 2: Sketch robot arm in the starting position, identify end effector and base coordinate systems and the point of interest on the end effector.

Step 3: Identify position vectors of the point of interest on the end effector and base coordinate systems.

Step 4: Identify the input quantities (link lengths, input rotations, position vector of the point of interest on the end effector-based coordinate system).

Step 5: Multiply successively the position vector on the end effector-based coordinate system by the translation and rotation matrices of the links, from the end effector to the base, to find the position vector of the point of interest on the base coordinate system.

Step 6: If possible, obtain scalar equations from the vector equation developed at step 5.

Step 7: Solve for the unknown input quantities determining the unknown arm configuration (i.e., link rotations or extensions) analytically, if possible, or numerically.

Example 2.10 A three-arm robot such as the one shown in Figure 2.10 has $a = 1$ m, $b = 1$ m. The end effector is holding a cylindrical shaft on the arm centerline at distance 0.2 m from the pivot C. Initially, the arm configuration is at $\alpha = 30°$, $\beta = 30°$, $\gamma = 30°$, $\psi = 0$. We need to move the cylinder horizontally to another location at distance -0.5 m in the direction of the y_{1a} axis, keeping $\gamma = 30°$, $\psi = 0$. Find the values of the rotation angles that will accomplish this.

Solution We first need to determine the initial location of point D in the coordinate system (x_{1A}, y_{1A}, z_{1A}) using Equations (2.41) with $x_{4C} = 0.2$ m, $y_{4C} = 0$, $z_{4C} = 0$:

$$x_{1A} = 0.2 \cos 90° + 1.0 \cos 60° + 1.0 \cos 30° = 0 + 0.5 + 0.866 = 1.366 \text{ m} \qquad (a)$$

$$y_{1A} = 0.2 \sin 90° + 1.0 \sin 60° + 1.0 \sin 30° = 0.2 + 0.866 + 0.5 = 1.566 \text{ m} \qquad (b)$$

To find the desired configuration we note that in the new position $x_{1A} = 1.366$ again but $y_{1A} = 1.566 - 0.5 = 1.066$ m. We now apply again Equations (2.41):

$$1.366 = 0.2 \cos(\alpha + \beta + 30°) + 1.0 \cos(\alpha + \beta) + 1.0 \cos \alpha \qquad (c)$$

$$1.066 = 0.2 \sin(\alpha + \beta + 30°) + 1.0 \sin(\alpha + \beta) + 1.0 \sin \alpha \qquad (d)$$

We rewrite the equations in the form

$$0 = 0.2 \cos(\alpha + \beta + 30°) + 1.0 \cos(\alpha + \beta) + \cos \alpha - 1.366 \qquad (e)$$

$$0 = 0.2 \sin(\alpha + \beta + 30°) + 1.0 \sin(\alpha + \beta) + \sin \alpha - 1.066 \qquad (f)$$

The solution can be found using the Solver function of EXCEL. Equation (e) is used as a quantity to be set to value 0 and Equation (f) is used as a constraint. Angles α and β are the unknowns, set initially to 30° or $\pi/6$, and are found to have values $\alpha = -0.05$ rad or $-3.02°$, $\beta = 1.22$ rad or 70.0°. A symmetric solution $\alpha = 1.38$ rad or 79.0°, $\beta = -1.39$ rad or $-79.7°$ can be found using initial values $\alpha = 1$ rad and $\beta = -1$ rad. Both solutions are valid, though the former requires less relative motion and probably is preferable.

2.3.3. Infinitesimal Motion Analysis

Equations (2.40) and (2.41) are nonlinear in the control variables α, β, γ, ψ, and solving, for example, for the angular displacements necessary to transport the robot arm grip to a new position at a finite distance is computationally intensive. An alternative is to compose the finite motion of a number of very small motions that can be treated mathematically as infinitesimal motions, making the mathematical model linear. To this end, we assume that the new position in the base coordinate system will be $P_{1A} + \Delta P_{1A}$, where $\Delta P_{1A} = \{\Delta x_{iA}, \Delta y_{iA}, \Delta z_{iA}, 1\}$, a small step of motion of the robot grip. If $\Delta P_{1A} \to 0$:

$$\Delta P_{1A} = (R_{\psi,\psi}R_\alpha T_a R_\beta T_b R_\gamma \Delta\psi + R_\psi R_{\alpha,\alpha}T_a R_\beta T_b R_\gamma \Delta\alpha$$

$$+ R_\psi R_\alpha T_{a,a}R_\beta T_b R_\gamma \Delta a + R_\psi R_\alpha T_a R_{\beta,\beta}T_b R_\gamma \Delta\beta$$

$$+ R_\psi R_\alpha T_a R_\beta T_{b,b}R_\gamma \Delta b + R_\psi R_\alpha T_a R_\beta T_b R_{\gamma,\gamma}\Delta\gamma)P_{4C} \qquad (2.42)$$

where the partial derivatives are denoted as $X_{y,y} = \partial X_y / \partial y$.

Equations (2.42) are three linear equations relating the three components of the grip displacement Δx_{1A}, Δy_{1A}, Δz_{1A} with the six variables $\Delta\psi$, $\Delta\alpha$, Δa, $\Delta\beta$, Δb, $\Delta\gamma$, where Δa and Δb indicate the change of the respective lengths a and b, if the links have, in addition, controllable length. If three of these variables are prescribed, the other three can be computed with the system of linear equations (2.42). For example, if $\psi = 0$, $\gamma = $ constant, $z = 0$ (planar motion), and a and b are constant, thus $\Delta\gamma = \Delta a = \Delta b = 0$, Equations (2.42) give:

$$\Delta x_{1A} = [-x_{4c}\sin(\alpha + \beta + \gamma) - y_{4c}\cos(\alpha + \beta + \gamma) - b\sin(\alpha + \beta)$$

$$- a\sin\alpha]\Delta\alpha + [-x_{4c}\sin(\alpha + \beta + \gamma) - y_{4c}\cos(\alpha + \beta + \gamma)$$

$$- b\sin(\alpha + \beta)]\Delta\beta \qquad (2.43a)$$

$$\Delta y_{1A} = [x_{4c}\cos(\alpha + \beta + \gamma) - y_{4c}\sin(\alpha + \beta + \gamma) + b\cos(\alpha + \beta)$$

$$+ a\cos\alpha]\Delta\alpha + [x_{4c}\cos(\alpha + \beta + \gamma) - y_{4c}\sin(\alpha + \beta + \gamma)$$

$$+ b\cos(\alpha + \beta)]\Delta\beta \qquad (2.43b)$$

These are two linear equations in $\Delta\alpha$ and $\Delta\beta$, and they can be readily solved to yield the robot displacement parameters needed to proceed along the desired path by a small step (Δx_{1A}, Δy_{1A}). After every step, the values of the robot parameters are updated:

$$\alpha^{(1)} = \alpha^{(0)} + \Delta\alpha, \quad \beta^{(1)} = \beta^{(0)} + \Delta\beta \qquad (2.43c)$$

where the superscript in parentheses indicates the step number. This facility, however, is not without undesired effects. Since we have to use a small but not truly infinitesimal step, we commit a small position error at every step, which in a long operation might accumulate. Most robot controllers allow for a correction after one or more finite operations of the robot arm by having the grip hit one or more switches that determine the position of the robot and allow the controller to start again with an accurate initial position.

2.4. VARIABLE SPEED KINEMATIC PAIRS—CAMS AND FOLLOWERS

2.4.1. Cam and Follower Function and Classification

The class of functions that closed loop mechanisms can produce as input–output relationship functions is limited by their finite number of possible configurations and the functions they generate. Thus, for approximate function generation they are many times adequate. There are applications, however, requiring a very accurate output function, and *cam and follower* mechanisms are often used for this purpose. For example, for lifting the valves in an automotive engine, cams (lobes mounted on the camshaft) and followers (the valve pushing rods) are used (Figure 2.11). The geometry of the curve bounding the cam (termed *cam profile* or *contour*) determines the relationship of the engine angular velocity with the displacement, velocity, and acceleration of the pushing rods and, in turn, of the valves themselves (Figure 2.12).

A cam mechanism consists of two parts:

1. The diving link: the *cam,* which is attached to a *camshaft,* usually rotating with a constant angular velocity
2. The driven link: the *follower,* which takes motion from the cam and transfers it to the part we want to move at a desired motion (rotational or rectilinear), usually a nonlinear function of the angular rotation of the crankshaft

Typical cams are shown in Figure 2.13 and typical followers in Figure 2.14.

A *disk cam* (Figure 2.13*a*) is attached to a camshaft and a follower slides or rolls over its outer surface, producing the desired motion. A *translation* cam (Figure 2.13*b*) is attached to a part moving along a straight line and a follower slides or rolls inside a face slot that can push the follower in either direction. A *face cam* (Figure 2.13*c*) is attached to a camshaft and a follower slides or rolls inside a face slot that can push the follower in either direction. A *cylindrical cam* (Figure 2.13*d*) is attached to a camshaft and a follower slides or rolls inside a cylindrical face slot that can push the follower in either direction.

A *translation follower* (Figure 2.14*a,b,c*) has an oscillatory rectilinear motion inside a guide and contacts the cam through a sharp edge (*a*), or a roller (*b*) or a flat part (*c*). A *rotational follower* (Figure 2.14*d,e,f*) has an oscillatory rotary mo-

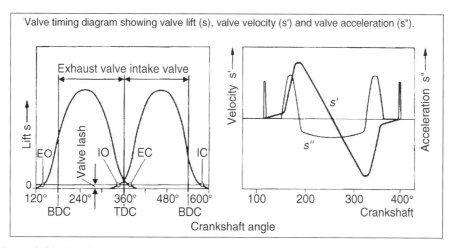

Figure 2.11. Displacement, velocity and acceleration diagrams for the valves of an automotive engine. (Reprinted from Bosch, *Automotive Handbook,* 1987, by permission of VDI-Verlag.)

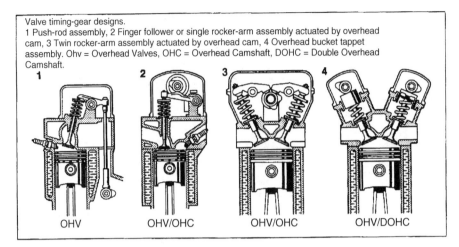

Figure 2.12. Alternative camshaft train designs for automotive engines. (Reprinted from Bosch, *Automotive Handbook,* 1987, by permission of VDI-Verlag.)

tion inside a guide and contacts the cam through a sharp edge (*d*), a roller (*e*), or a flat part (*f*). Every one of the cams shown in Figure 2.13 can drive nearly any one of the followers shown in Figure 2.14, a total of 30 possible different combinations of cam–follower mechanisms.

2.4.2. Follower Kinematics

The purpose of the cam and follower mechanism is to force the follower in a motion that is related with the rotation of the cam with some given function. The follower motion relation to cam rotation function (*cam motion profile*) can be described in a diagram such as in Figure 2.15. The follower is at position zero while the cam rotates from angle zero to angle ϕ_1. This phase is termed *dwell*. From angle ϕ_1 to angle ϕ_2 the follower moves to a new position by a distance s_1 for a linear follower or an angle for a rotational follower. This phase is termed *rise* or *lift,* and the

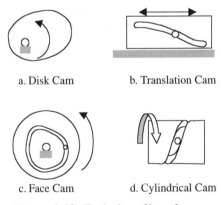

Figure 2.13. Typical profiles of cams.

Linear reciprocating followers

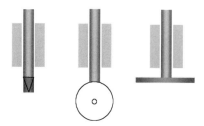

Rotational Oscillating Followers

Figure 2.14. Typical profiles of followers.

function that connects the points (ϕ_1, 0) and (ϕ_2, s_1) is of paramount importance for the operation of the mechanism. In Figure 2.15, this function is a straight line, a function that will be shown to be very undesirable. Indeed, the velocity of the follower, since $\phi = \omega t$, where ω is the angular velocity of the cam and will be assumed constant, is $v = s_1/(t_2 - t_1) = \omega s_1/(\phi_2 - \phi_1)$ and has a finite value that increases as the duration of the rise decreases. The velocity is shown in Figure 2.15 with dashed line.

We observe that the velocity jumps at $\phi = \phi_1$ and $\phi = \phi_2$, and thus acceleration at these two points jumps to infinite values while everywhere else is zero. This is a very dangerous situation because high acceleration is associated with high inertia forces (Newton's law) and subsequently with noise, vibration, and wear, unwanted features in machine operation. Thus, the rise function needs to be selected in a way that will lead to a smooth motion of the follower.

An additional requirement is that the third derivative of the follower displacement, termed *jerk,* should have finite values. The reason is that high values of the jerk mean rapidly changing acceleration. This usually means that acceleration might change sign suddenly. Because machines are designed with operating tolerances

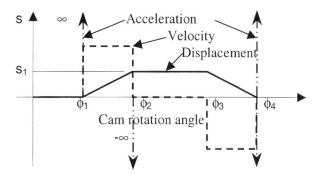

Figure 2.15. Follower motion.

(see Chapter 4), change in the direction of acceleration means change in the direction of inertia forces. This in turn might lead to impacts in the machine clearances and subsequently to noise, vibration, fatigue, and wear.

Further rotation of the cam between angles ϕ_2 and ϕ_3 is with the follower remaining at position s (second dwell). Further, from angle ϕ_3 to angle ϕ_4 the follower returns to the original position, a phase termed *fall*. During the fall of the follower, displacement, velocity and acceleration are shown in Figure 2.15.

Before the availability of computers, several special functions were used for the rise and the fall of the follower, such as sinusoidal (all derivatives have finite values), parabolic, and cycloid, and extensive tables have been compiled of such suggested functions (see, e.g., Erdman and Sander 1997). In contemporary designs, a high order polynomial is frequently used and its constant coefficients are adjusted to obtain all desired characteristics.

To this end, we shall assume without loss in generality that the cam and follower mechanism consists of a face cam with a linear follower, and we set up the relationship between input angle ϕ and the output displacement s (*cam motion profile*) as, normalizing with division by s_1:

$$\frac{s}{s_1} = \sigma = a_n\phi^n + a_{n-1}\phi^{n-1} + \cdots + a_2\phi^2 + a_1\phi + a_0 \tag{2.44}$$

where $a_0, a_1, \ldots, a_n$ are yet undetermined constants. The velocity, acceleration, and jerk are the first, second, and third derivatives of the displacement, respectively:

$$v = \frac{ds}{dt} = \frac{ds}{d\phi}\frac{d\phi}{dt} = \omega s_1[na_n\phi^{n-1} + (n-1)a_{n-1}\phi^{n-2} + \cdots + 2a_2\phi + a_1] \tag{2.45}$$

$$a = \frac{dv}{dt} = \frac{dv}{d\phi}\frac{d\phi}{dt} = \omega^2 s_1[n(n-1)a_n\phi^{n-2} + (n-2)(n-1)a_{n-1}\phi^{n-3}$$
$$+ \cdots + 6a_3\phi + 2a_2] \tag{2.46}$$

$$j = \frac{da}{dt} = \frac{da}{d\phi}\frac{d\phi}{dt} = \omega^3 s_1[(n-2)(n-1)na_n\phi^{n-3}$$
$$+ (n-3)(n-2)(n-1)a_{n-1}\phi^{n-4} + \cdots + 6a_3] \tag{2.47}$$

In a normalized form, the equations for the follower motion will be:

$$\frac{s}{s_1} = f_s(\phi) = a_n\phi^n + a_{n-1}\phi^{n-1} + \cdots + a_2\phi^2 + a_1\phi + a_0 \tag{2.48}$$

$$\frac{v}{\omega s_1} = f_v(\phi) = na_n\phi^{n-1} + (n-1)a_{n-1}\phi^{n-2} + \cdots + 2a_2\phi + a_1 \tag{2.49}$$

$$\frac{a}{\omega^2 s_1} = f_a(\phi)$$

$$= n(n-1)a_n\phi^{n-2} + (n-2)(n-1)a_{n-1}\phi^{n-3} + \cdots + 6a_3\phi + 2a_2$$

$$(2.50)$$

$$\frac{j}{\omega^3 s_1} = f_j(\phi)$$

$$= (n-2)(n-1)na_n\phi^{n-3} + (n-3)(n-2)(n-1)a_{n-1}\phi^{n-4} + \cdots + 6a_3$$

$$(2.51)$$

The *boundary conditions* for smooth operation of the cam–follower mechanism are:

$$f_s(\phi_1) = 0, \quad f_v(\phi_1) = 0, \quad f_a(\phi_1) = 0, \quad f_j(\phi_1) = 0 \qquad (2.52)$$

$$f_s(\phi_2) = 1, \quad f_v(\phi_2) = 0, \quad f_a(\phi_2) = 0, \quad f_j(\phi_2) = 0 \qquad (2.53)$$

If a 7th degree of the polynomial is selected, the unknown coefficients are eight $(a_7 \ldots a_0)$ and application of the eight boundary conditions (Equations (2.52) and 2.53)) will yield a system of eight linear equations in eight unknowns $a_7 \ldots a_0$. The problem is simplified considerably if the independent variable ϕ is normalized in the form $\xi = (\phi - \phi_1)/(\phi_2 - \phi_1)$ and the follower function is set in the form:

$$F_s(\xi) = \xi^4(b_3 \xi^3 + b_2 \xi^2 + b_1\xi + b_0) \qquad (2.54)$$

If the follower function has the form as in Equation (2.54), the boundary conditions (2.52) are always satisfied and it remains to find four coefficients $b_i, i = 0 \ldots 3$, using the boundary conditions in Equations (2.53) as normalized in Equations (2.55):

$$F_s(1) = 1, \quad F'_s(1) = 0, \quad F''_s(1) = 0, \quad F'''_s(1) = 0 \qquad (2.55)$$

Thus, Equations (2.54) and (2.55) yield:

$$F_s(1) = b_3 + b_2 + b_1 + b_0 = 1 \qquad (2.56)$$

$$F'_s(1) = 7b_3 + 6b_2 + 5b_1 + 4b_0 = 0 \qquad (2.57)$$

$$F''_s(1) = 42b_3 + 30b_2 + 20b_1 + 12b_0 = 0 \qquad (2.58)$$

$$F'''_s(1) = 210b_3 + 120b_2 + 60b_1 + 24b_0 = 0 \qquad (2.59)$$

For the solution, LINEQ was selected from the MELAB 2.0 menu. Input–output is shown in Figure 2.16. The result is $b_3 = -20$, $b_2 = 70$, $b_1 = -84$, $b_0 = 35$. The follower function is then:

```
EDIT DATA
Problem Identification: <3x3 system of linear equations>?
4x4 system of linear equations
Number of Equations    < 4 >? 4
Enter Coefficients Matrix:
Enter element 1 1 of matrix < 0 >? 1
Enter element 1 2 of matrix < 0 >? 1
Enter element 1 3 of matrix < 0 >? 1
Enter element 1 4 of matrix < 0 >? 1
Enter element 2 1 of matrix < 0 >? 7
Enter element 2 2 of matrix < 0 >? 6
Enter element 2 3 of matrix < 0 >? 5
Enter element 2 4 of matrix < 0 >? 4
Enter element 3 1 of matrix < 0 >? 42
Enter element 3 2 of matrix < 0 >? 30
Enter element 3 3 of matrix < 0 >? 20
Enter element 3 4 of matrix < 0 >? 12
Enter element 4 1 of matrix < 0 >? 210
Enter element 4 2 of matrix < 0 >? 120
Enter element 4 3 of matrix < 0 >? 60
Enter element 4 4 of matrix < 0 >? 24
Enter Constant vector:
Enter element 1 of vector < 0 >? 1
Enter element 2 of vector < 0 >? 0
Enter element 3 of vector < 0 >? 0
Enter element 4 of vector < 0 >? 0

Are the data correct (Y/N)? ▮
```

```
DATA:
Coefficient matrix
1     1     1     1
7     6     5     4
42    30    20    12
210   120   60    24
Constant vector
1     0     0     0
```

```
Results from LINEQ
-20.00004    70.00014    -84.00017    35.00006
```

Figure 2.16. LINEQ solution of the linear equations.

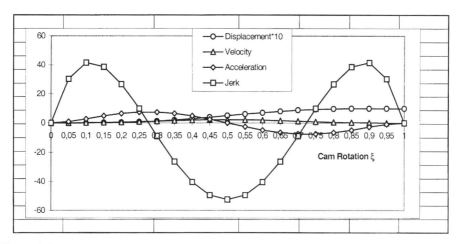

Figure 2.17. Normalized displacement ($\times 10$), velocity, acceleration and, jerk of a 7th degree polynomial cam follower.

$$F_s(\xi) = -20\xi^7 + 70\xi^6 - 84\xi^5 + 35\xi^4 \tag{2.60}$$

The follower function and its derivatives for velocity, acceleration, and jerk are plotted with EXCEL in Figure 2.17 for the rise.

In Example 2.8, a 6th order polynomial will be used, dropping the conditions on the jerk. We observe somewhat lower values of the jerk but the differences from the seventh degree polynomial follower function are in general small (see Figure E2.8a). Thus, the designer might want to relax the requirements for zero jerk at the two ends of the follower rise interval and impose other requirements, such as maximum value of the acceleration and the jerk and minimum rise time. In this case, the conditions (2.52) and (2.53) need to be reformulated according to the design task.

Design Procedure 2.4: Design of a Polynomial Cam Motion Profile

Step 1: Select a polynomial degree, typically 7.

Step 2: Write the normalized form of the polynomial, such as Equation (2.48).

Step 3: Take the time derivatives to find velocity, acceleration, and jerk, such as in Equations (2.49) to (2.51), respectively.

Step 4: Identify the design data, i.e., rise, fall magnitude, and timing.

Step 5: Write boundary conditions for displacement, velocities, acceleration, and jerk, such as Equations (2.52) and (2.53).

Step 6: Apply boundary conditions to equations for displacement, velocity, acceleration, and jerk to obtain linear algebraic equations for the unknown coefficients of the cam profile polynomial.

Step 7: Solve for the unknown coefficients of the polynomial.

Example 2.11 Find an appropriate follower function to lift the follower by L during a rise from 60° to 90°, and a fall between 150° and 180°, having zero velocity and acceleration at the beginning and end of the rise and fall.

Solution   The polynomial needs to be of order 5 because there are six conditions to satisfy. The polynomial

$$F_s(\xi) = b_2 \, \xi^5 + b_1 \, \xi^4 + b_0 \, \xi^4$$

satisfies automatically the conditions

$$F_s(0) = 0, \quad F'_s (0) = 0, \quad F''_s (0) = 0$$

Therefore, the remaining conditions are:

$$F_s(1) = 1, \quad F'_s (1) = 0, \quad F''_s (1) = 0$$

Therefore, application to polynomial equation and its derivatives yields

$$F_s(1) = b_2 + b_1 + b_0 = 1$$
$$F'_s(1) = 5b_2 + 4b_1 + 3b_0 = 0$$
$$F''_s(1) = 20b_2 + 12b_1 + 6b_0 = 0$$

The solution of the above system of three linear equations in three unknowns was obtained with LINEQ, $b_2 = 6$, $b_1 = -15$, $b_0 = 10$, and the polynomial is:

$$F_s(\xi) = 6\xi^5 - 15\xi^4 + 10\xi^3$$

This polynomial applies for both the rise (60–90°) and the fall (150–180°). The rest of the time the follower is in dwell (0–60, 90–150, 180–360°).

The definition of the normalized angle of the cam is $\xi = (\phi - 60)/(90 - 60)$, ϕ in degrees.

For the fall, the equation will be:

$$F_s(\xi) = 1 - (6\xi^5 - 15\xi^4 + 10\xi^3)$$

because the displacement starts from the value of 1 and goes down to 0 at 180° angle and the boundary conditions are satisfied. Here, the definition of the normalized angle of the cam is $\xi = (\phi - 150)/(180 - 150)$, ϕ in degrees.

The corresponding plot is shown in Figure E2.11.

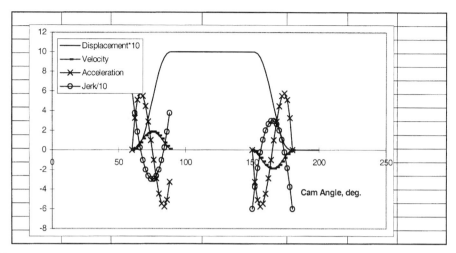

Figure E2.11 Normalized displacement ($\times 10$), velocity, acceleration and jerk/(10) of a 5th degree polynomial cam-follower.

2.4.3. Cam Profile Design

In the past, design of the cam profile was a tedious job requiring application of graphical or analytical methods. In most applications today, a cam profile is generated by a numerically controlled milling cutter that can be programmed to cut the cam profile so that the desired follower motion will be generated without explicitly determining the cam profile first. This will be demonstrated by way of an example, the cam and rectilinear (translating) flat follower (Figure 2.18).

A cam rotates about point O_A and a flat follower is always moving so that its geometric axis is always on the line $O_A B$. A reference line fixed on the cam has an angle ϕ to the follower axis $O_A B$ at time t. The follower contacts the cam in this position at A. A milling cutter has its center at O_c and radius R_c. The distance $L(\phi)$ is the displacement of the follower and is a function of the angle of rotation ϕ. In a milling cutter, the cam stock can rotate about O_A by a prescribed angle at a time and for each angle, and the center of the milling cutter can be programmed to have such a position that the displacement of the follower will be prescribed by the function $L(\phi)$. Thus, given are the radius of the milling cutter R_c and the location of the cam center of rotation O_A, and for each angle ϕ we need to know the location of the center of the milling cutter O_c. To this end, we assume a small incremental rotation of the cam $d\phi$ about O_A, but we observe it from the cam, so that we see the centerline of the follower to rotate counterclockwise by $d\phi$. The dashed line shows the new position of the follower and the change of location dL = $ad\phi$ Therefore, the angle of the vector R with the centerline θ is (Erdman and Sandor 1997):

$$\tan \theta = \frac{a}{L} = \frac{dL}{Ld\phi} \tag{2.61}$$

and the coordinates of the center O_c of the milling cutter are:

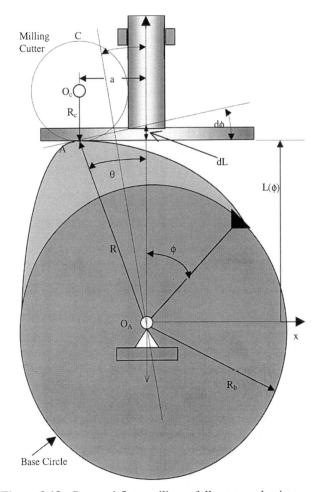

Figure 2.18. Cam and flat rectilinear follower mechanism.

$$x_c = -L(\phi)\tan \theta = \frac{dL}{Ld\phi}, \quad y_c = L(\phi)\cot \theta + R_c = \frac{L^2(\phi)}{dL/d\phi} + R_c \quad (2.62)$$

Though the cam profile itself is not needed for the manufacturing process, one can obtain the polar coordinates of point A in the form:

$$R = \sqrt{L^2(\phi) + a^2} = \sqrt{L^2(\phi) + \left(\frac{dl}{d\phi}\right)^2}, \quad \theta = \arctan \frac{dL}{Ld\phi} \quad (2.63)$$

The profile of the cam sometimes needs to be plotted at the design stage because there are some designs that are not acceptable, though they might lead to correct follower motion. For example, Figure 2.19a shows a cam profile with a cusp. The point contact will lead to excessive wear at that point due in part to high contact stresses and inadequate lubrication (see Chapters 10 and 11). In Figure 2.19b, there

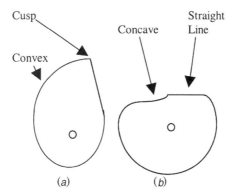

Figure 2.19. Nonfeasible cam designs.

is an area with outward curvature (concave) that a flat follower cannot reach. A roller follower would be acceptable if its radius is smaller than the smallest radius of curvature on the concave section.

The latter form is related to the necessity to start the displacement function not from zero, because then the profile will touch the center of rotation, which is impracticable, but from a certain value that is usually the radius of the cam at the arc corresponding to the lowermost position of the follower at the dwell. It is apparent that the profile of the cam that corresponds to the dwell should be a circular segment so that the distance of the follower from the center of rotation of the cam will remain constant during the dwell. We call this circle *base circle*, of radius R_b in Figure 2.18, and its size plays a role of paramount importance for the design of a cam. The designer must select a radius R_b big enough to avoid concave sections for flat followers or sharp concave sections for roller followers, and on the other hand small enough to yield the smallest possible cam size. This point will be illustrated in Example 2.9.

Some of the most common other configurations of cam–follower systems with the respective milling cutter coordinates and the cam profile equations are included in Table 2.2.

Design Procedure 2.5: Design of a Cam

Step 1: Find the cam motion profile (Design Procedure 2.4).

Step 2: Select the type of cam from Table 2.2 and the necessary geometric data, i.e., diameter of the base circle, location of follower, etc.

Step 3: From Table 2.2, for the cam type selected, determine the required geometric quantities, i.e., center of milling cutter, profile equations.

Step 4: For a sequence of angles of rotation, typically from 0–360°, determine the cam profile coordinates and the milling cutter center of rotation coordinates.

Step 5: Plot the cam profile.

Step 6: Inspect the cam profile for cusps and other abnormalitics.

Step 7: Repeat steps 2–5, if at step 6 it is found that the cam profile is not acceptable.

TABLE 2.2. Design Equations for Cam–Follower Mechanisms

Mechanism Type	Drawing	Equations[a]
Translating, flat face follower	Figure 2.18	2.63 to 2.65
Oscillating, Flat-Face Follower		$\theta = \arctan\left[\dfrac{(d\zeta/d\theta)}{1 - (d\zeta/d\phi)}\dfrac{m\cos\zeta}{f + m\sin\zeta}\right]$ *Profile:* $R = \dfrac{f + m\zeta}{\cos\theta}$, angle $(\phi + \psi + \theta)$ $\psi = \pi/2 - \zeta$ *Center of milling cutter:* $y_c = R\cos(\pi/2 - \theta - \psi) + R_c\cos(\pi/2 - \psi)$ $x_c = R\sin(\pi/2 - \theta - \psi) + R_c\sin(\pi/2 - \psi)$

156

Offset Translating Roller Follower

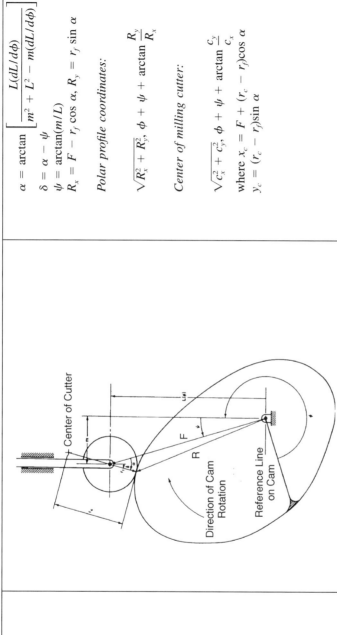

$$\alpha = \arctan\left[\frac{L(dL/d\phi)}{m^2 + L^2 - m(dL/d\phi)}\right]$$

$$\delta = \alpha - \psi$$

$$\psi = \arctan(m/L)$$

$$R_x = F - r_f \cos \alpha, \ R_y = r_f \sin \alpha$$

Polar profile coordinates:

$$\sqrt{R_x^2 + R_y^2}, \ \phi + \psi + \arctan \frac{R_y}{R_x}$$

Center of milling cutter:

$$\sqrt{c_x^2 + c_y^2}, \ \phi + \psi + \arctan \frac{c_y}{c_x}$$

where $x_c = F + (r_c - r_f)\cos \alpha$
$y_c = (r_c - r_f)\sin \alpha$

TABLE 2.2. (*Continued*)

Mechanism Type	Drawing	Equations[a]
Oscillating Roller Follower	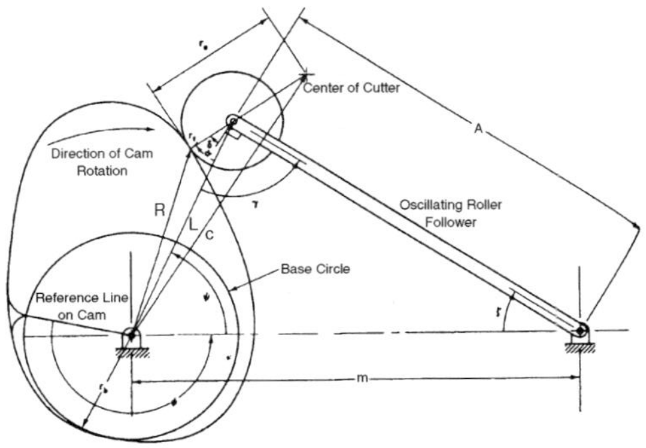	$\alpha = \arctan\left[\dfrac{A\sin\gamma(d\zeta/d\phi)}{L - A\cos\gamma(d\zeta/d\phi)}\right]$ $\delta = \gamma + \alpha - \pi/2$ $R_x = L - r_f\cos\alpha,\ R_y = r_f\sin\alpha$ *Profile polar coordinates:* $\sqrt{R_x^2 + R_y^2},\ \phi + \psi + \arctan\dfrac{R_y}{R_x}$ *Center of milling cutter:* $\sqrt{c_x^2 + c_y^2},\ \phi + \psi + \arctan\dfrac{c_y}{c_x}$ where $x_c = L + (r_c - r_f)\cos\alpha$ $y_c = (r_c - r_f)\sin\alpha$

[a]From Erdman and Sandor 1997.

158

Example 2.12 Find the cam profile for the cam in Example 2.11 for a flat translating follower, assuming that the maximum rise of the follower is $s = 1$ cm.

Solution   The follower rise function was found to be

$$L_r(\xi) = sF_s(\xi) = s(6\xi^5 - 15\xi^4 + 10\xi^3)$$

with $\xi = (\phi - \phi_0)/\Delta\phi$, where $\phi_0 = 60°$, $\Delta\phi = 30°$. If we add a base circle of radius R_b,

$$L_r(\xi) = R_b + s(6\xi^5 - 15\xi^4 + 10\xi^3)$$

The follower fall function will be:

$$L_r(\xi) = R_b + s[1 - (6\xi^5 - 15\xi^4 + 10\xi^3)]$$

with $\xi = (\phi - \phi_0)/\Delta\phi$, where $\phi_0 = 150°$, $\Delta\phi = 30°$.

Equations (2.61) and (2.63) give the polar description for the cam profile:

$$\tan\theta = \frac{dL/d\phi}{L}$$

$$R = \sqrt{L^2(\phi) + \left(\frac{dL}{d\phi}\right)^2}, \quad \theta = \arctan\frac{dL}{Ld\phi}$$

Plotting will be done in EXCEL. We construct a spreadsheet with columns ϕ(deg), ξ, $L(\phi)$, $dL/d\phi$, θ, $\theta + \phi$, R, and ask for the polar plot of the last two columns. First, we start with a radius of base circle $R_b = 1$ cm. The result is shown in Figure E2.12a. We observe areas of odd, concave shape that cannot work with a flat follower.

Now we try $R_b = 10$ cm. The resulting profile is shown in Figure E2.12b. This profile is now much better.

2.5. PROFILE CONTACT MECHANISMS: GEARS

2.5.1. Classification of Gearing

A gear is a mechanism that, by means of meshing teeth, transmits or converts motion, changing the angular velocity and torque between two moving systems. It is a member of the family of mechanical drives (Figure 2.20).

Toothed gearings convert and transmit rotary motion between shafts with parallel, intersecting, and nonparallel, nonintersecting (crossed) axes, and also convert rotary motion into translational motion and vice versa.

Toothed gearing between parallel shafts is accomplished by spur, helical, and herringbone (double-helical) gears (Figure 2.21a,b,c,d). Gearing between intersecting axes is usually accomplished by straight and spiral bevel gears (Figure 2.21f and h) and less often by skew bevel gears (Figure 2.21g). Gearing to convert rotary motion into translational motion or vice versa is accomplished by a rack and pinion (Figure 2.21e).

Crossed helical gears, worm gears, and hypoid gears are used to transmit rotation between shafts with nonparallel, crossed axes.

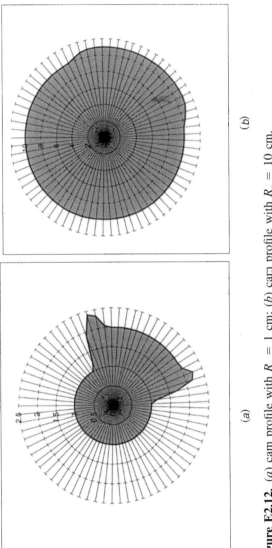

Figure E2.12. (*a*) cam profile with $R_c = 1$ cm; (*b*) cam profile with $R_c = 10$ cm.

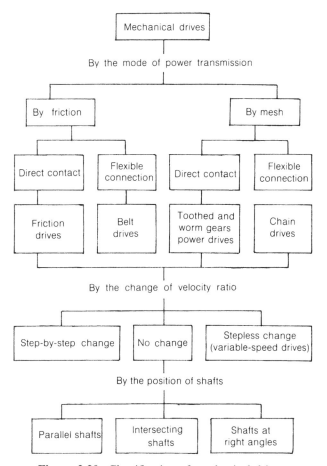

Figure 2.20. Classification of mechanical drives.

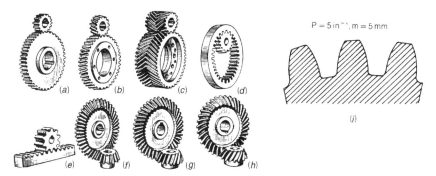

Figure 2.21. Gear drives.

Toothed gearing is the most widely used and most important form of mechanical drive. Gears are used in many fields and under a wide range of conditions, from watches and instruments to the heaviest and most powerful machinery.

Peripheral forces from decimals of a gram to thousands of tons, torques up to a thousand ton-meters, and power from negligibly small values to tens of thousands of kilowatts are transmitted using gears of diameter from a fraction of a millimeter to ten or more meters.

In comparison to other mechanical drives, toothed gearing has essential advantages, namely: (a) small overall size, (b) high efficiency, (c) long service life and high reliability, (d) constant speed ratio owing to the absence of peripheral slipping, and (e) the possibility of being applied to a wide range of torques, speeds, and ratios.

Shortcomings of toothed gearing are the noise generated at high speeds and the relatively high cost.

Gears have been known since antiquity. A mechanical computer of the First Century BC found on the Mediterranean island of Antikythyra employed a large number of steel gears to reproduce the motion of the planets of our solar system (Figure 2.22), according to one interpretation.

2.5.2. Kinematics of Gears

The basic kinematic condition that must be satisfied by the gear tooth profiles is the constancy of the instantaneous velocity ratio of the gearing. Various classes of curves can meet this requirement. To ensure efficiency, strength, and a long service life, the profile should also provide for low sliding velocity and a sufficiently large radius of curvature at the points of contact. The profile should be easy to manufacture. In particular, it should be feasible to cut gears with various numbers of teeth with the same simple tool. The involute teeth meet all of these requirements.

Consider two cams revolving about points O_1 and O_2 with contact at point A (Figure 2.23). In order to maintain contact, point A on either cam must have the same velocity along the line perpendicular to the surfaces at the contact point. Therefore:

$$u_1 = \omega_1 r_1 = u_2 = \omega_2 r_2, \qquad \frac{\omega_2}{\omega_1} = \frac{r_1}{r_2} = \frac{O_1 P}{O_2 P} = R_{12} \qquad (2.64)$$

Equation (2.64) implies that in order to maintain a constant ratio R of angular velocities, the point P where the line perpendicular to the surfaces at contact meets the centerline is constant and its location depends only upon the ratio of the two angular velocities. Moreover, this line must be stationary on the plane, at some angle $\pi/2 - \varphi$ with the centerline. Angle φ is known as the *pressure angle*. To an observer on either cam, the tooth profile appears to be traced by point A along tangent N–N on the circle; therefore, it is an *involute*. The circles with radii r_1 and r_2 have a ratio given by Equation (2.64). They are called *rolling circles* and are fully determined by the velocity ratio and the center distance.

At some rotation angle, the two cams will lose contact. This will not happen if the involute profiles are repeated around the cam at equal peripheral distances. This forms an *involute gear*.

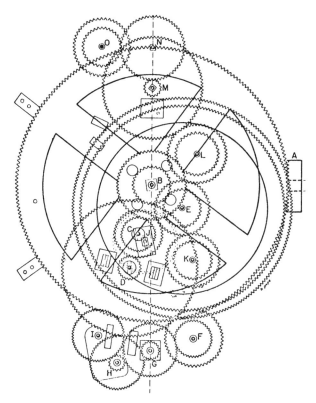

Figure 2.22. The Antikythyra Computer (courtesy of late Derek de Solla Price, Yale University).

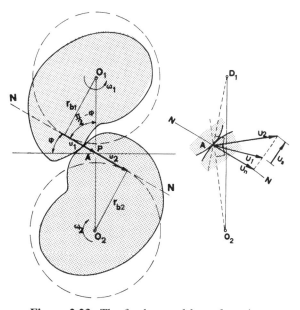

Figure 2.23. The fundamental law of gearing.

Conventionally, the smaller gear is called the *pinion* and the larger simply the *gear*. Indices 1 and 2 will correspond in the following to pinion and gear, respectively.

2.5.3. Geometry of Spur Gears

The meshing of gears is equivalent to the rolling without slipping of rolling circles of diameters $d_1 = 2(O_1P)$ and $d_2 = 2(O_2P)$ (Figure 2.23), called the *rolling* or *pitch* diameters. The circles are known as the rolling or pitch circles. As straight line NN rolls without slipping around the base circles of diameters $d_{b1} = d_1 \cos \varphi$ and $d_{b2} = d_2 \cos \varphi$ (where φ is the pressure angle), points of this straight line describe involutes on each of the gears. The gears themselves have considerable slipping on the mating surfaces, as one can see from Figure 2.24, observing carefully the mating points as the gears rotate and their relative position.

The *circular pitch p* is the distance between like profiles of adjacent teeth measured along an arc of the pitch circle of the gear: $p = \pi d/N$, where N is the number of teeth and d is the diameter of the rolling circle. Because the pitch is found on the rolling circle, the latter is also called *pitch circle*. The circular pitch p is a multiple of the number π and is therefore inconvenient to standardize as a basic parameter of gearing. Therefore, in the metric system the *module m = p/π* is defined and in the English system the *diametral pitch*

$$P = \frac{\pi}{p} \tag{2.65}$$

is used. Further, the module

$$m = \frac{d}{N}, \quad m \text{ (mm)} = \frac{25.4}{P} \text{ (in.)} \tag{2.66}$$

The module and the diametral pitch have been standardized. Table 2.3 has the most common modules (those of the first series are preferable to those of the second).

Additional modules that can be used for reducing gears are 1.6, 3.15, 6.3, and 12.5 mm.

As the number of teeth is increased to infinity, the gear becomes a gear rack and the involute tooth profile becomes a straight line, which is convenient in manufacture and inspection. The basic rack (Figure 2.25a) completely determines the

TABLE 2.3. Standard Gear Modules *m* mm

1st series	1	1.25	1.5	2	2.5	3	4	5	
2nd series	1.125	1.375	1.75	2.25	2.75	3.5	4.5	5.5	
1st series	6	8	10	12	16	20	25	32	40
2nd series	7	9	11	14	18	22	28	36	45

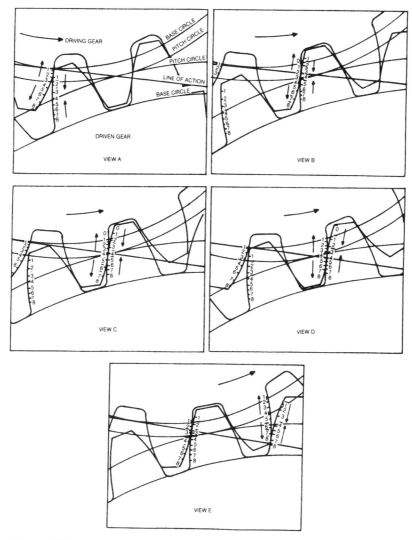

Figure 2.24. Meshing of spur gears, contact line (courtesy Mobil Oil Corp.).

tooth profiles of all normal gearing, enabling all gears to mesh with one another. The parameters of the initial contour have been standardized, and the profile angle is usually $\varphi = 20°$. The outer circle defining the top of the teeth (*addendum circle*) is at radial distance m from the rolling circle, for standard gears, while the circle defining the bottom of the teeth is at radial distance $1.25m$ from the rolling circle; thus the working depth of the teeth is $h = 2.25m$; the radial clearance between the bottom of tooth space and the top of the tooth of the mating gear is $c = 0.25m$ (or up to $0.35m$ in cutting gears with sharper cutters); the fillet radius at the bottom of the tooth of the spur and helical gears is, according to most international standards, $r = 0.38m$.

To reduce the force of impact in high-speed gearing when the teeth come into and go out of mesh and to reduce noise, the faces of the teeth are modified by

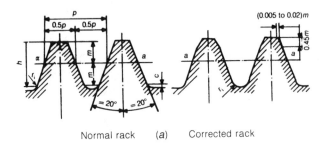

Normal rack   (a)   Corrected rack

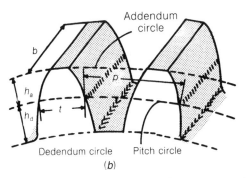

(b)

Figure 2.25. Forms of spur gear teeth.

flanking, which is a deliberate deviation from the involute profile at the top of the tooth (over a portion of the face) into the body of the tooth. This procedure is known as *correction*.

The standard principal geometric relationships for uncorrected gearing are listed in Table 2.4. The names and designations of the elements of toothed gearing are illustrated in Figures 2.25 and 2.26.

Gear ratios $\omega_1/\omega_2 = N_2/N_1$ for spur gears might only have values that are ratios of two integers.

The basic principle of the involute curve implies that contact between teeth will always take place along the pressure line NN. On the other hand, no contact can take place outside the addendum circle of the gear. Therefore, contact will take place on the portion of the pressure line that is inside both addendum circles, segment BC on the pressure line (Figure 2.26). The length of this segment is of particular importance for the operation of the gears. The longer this segment is, the more teeth are in contact at the same time. The sum of the arcs determined by the angles AO_2B and AO_2C is the arc of contact, which divided by the circular pitch will yield the number of teeth in contact. A measure of this is the contact ratio, the ratio of the arc length (BC) divided by the circular pitch $p = \pi m$. From Figure 2.26,

$$m_c = \frac{(BC)}{\pi m} = \frac{(BE - AE) + (CD - AD)}{\pi m} \tag{2.67}$$

For standard teeth, the radius of the addendum circle is $r + m$. (BE) and (CD) are computed from the triangles O_2BE and O_1CD. Using also the relations

TABLE 2.4. Geometric Relationships for Spur Gears[a]

Parameter	Formula
Circular pitch	$p = \pi d/N = \pi d_1/N_1 = \pi d_2/N_2$
Diametral pitch	$P = \pi/p = N/d = N_1/d_1 = N_2/d_2$
Module	$m = p/\pi = 1/P = d/N = d_1/N_1 = d_2/N_2$
Center-to-center distance a	$a = \dfrac{(N_1 + N_2)m}{2} = 0.5(N_1 + N_2)m$
Whole depth of teeth h	$h = 2.5\ m$
Addendum (head) h_a	$h_a = m$
Dedendum (foot) h_f	$h_f = 1.25m$
Radial clearance c	$c = 0.25m$
Rolling circle diameter d	$d_1 = mN_1,\ d_2 = mN_2$
Outside (addendum circle) diameter d_a	$d_{a1} = d_1 + 2m$
	$d_{a2} = d_2 + 2m$
Foot (dedendum circle) diameter d_f	$d_{f1} = d_1 - 2.5m - 2c$
	$d_{f2} = d_2 - 2.5m - 2c$

[a]d = rolling circle diameter, N = number of teeth, c = radial clearance.

$$r_1 = mN_1/2, \qquad r_2 = mN_2/2, \qquad r_b = R\cos\phi \qquad (2.68)$$

the contact ratio is obtained:

$$m_c = \frac{[(N_2 + 2)^2 - N_2^2 \cos^2\phi]^{1/2} - N_2\sin\phi + [(N_1 + 2)^2 - N_1^2 \cos^2\phi]^{1/2} - N_1\sin\phi}{2\pi}$$

$$(2.69)$$

Thus, pressure angle affects the contact ratio as follows:

Pressure angle ϕ	Contact ratio m_c
14.5°	1.7–2.5
20°	1.45–1.85
25°	1.2–1.5

The higher values correspond to the higher number of teeth of the gear, in the limit to the basic rack. The lower values correspond to equal diameters, gear ratio 1. Smaller pressure angle results in higher contact ratio, thus more teeth in contact at the same time

The basic circle has diameter $d_b = d\cos\phi$. Therefore, the distance between the basic and the pitch circles is $r - r_b = r(1 - \cos\phi)$. If the size of the teeth is such that the tooth and the contact extend below the basic circle, that is, when the depth of the tooth below the pitch circle $1.25m < r(1 - \cos\phi)$, since involute does not exist below the basic circle, a noninvolute shape will continue up to the dedendum circle.

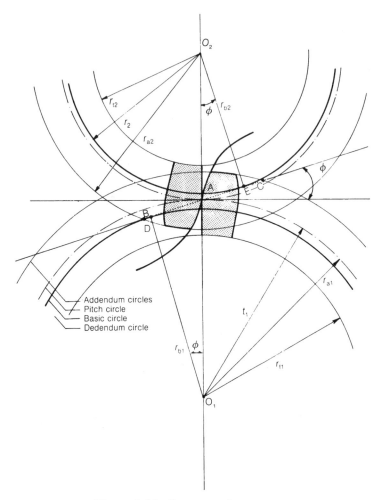

Figure 2.26. Geometry of spur gears.

A radial direction is usually selected because the involute is perpendicular to the basic circle and so is the radius. At this portion of the tooth, the basic law of the gearing does not hold and therefore the speed ratio is not constant. This means change in velocity, acceleration, and additional dynamic loads, which quickly wear this portion of the tooth. The tooth becomes thinner at the high-stress area and breakage eventually will occur. This phenomenon is known as *interference* and the weakening of the tooth as *undercutting,* situations that obviously must be avoided. It is apparent that the larger the pressure angle, the smaller is the diameter of the basic circle and the less likely it is to have interference.

Selection of the pressure angle is based on the trade-off between low pressure angle/high contact ratio and high pressure angle/less interference. The compromise $\phi = 20°$ is used in most gears. Gear displacement (correction) and additional machining of the top of the tooth (Figure 2.25a) have been used to reduce interference. However, the main measure to avoid interference is to keep the number of teeth above a minimum value. Indeed, the distance between pitch and basic circle depends only on the pressure angle and not the number of teeth. The depth of the tooth depends on the number of teeth for a given diameter. The larger the number of teeth, the smaller the tooth and the less likely it is to have interference.

To quantify this, Figure 2.27 shows the contact of the top of the tooth of a rack with the bottom of the tooth of the mating pinion, point D. It is evident that if point D is on the basic circle and above, there will be no interference. In the limit, from the triangle AKD, because $(OK) = (OA) - (AK) = r - m$

$$m = r_1 - r_{b1} \cos \phi = r_1 - r_1 \cos^2 \phi = r_1 \sin^2 \phi \qquad (2.70a)$$

$$N_{\min} = \frac{2r_1}{m} = \frac{2}{\sin^2 \phi} \qquad (2.70b)$$

In general, if the mating gear is not a rack but a gear, as in Figure 2.27, if O_2 is the center of the gear and the contact at D, from the triangle O_1KD, in the most favorable case where $r_1 = r_2 = r$:

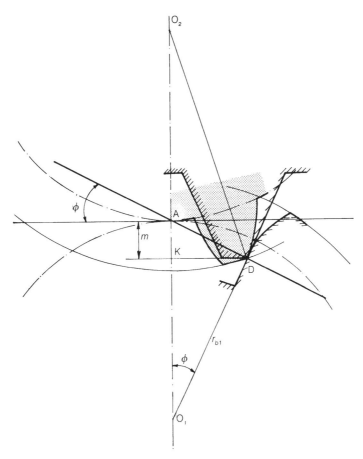

Figure 2.27. Generation of the involute.

$$(O_2K)^2 + (KD)^2 = (O_2D)^2$$

$$(O_2K) = (O_2A) + (O_1A) - (O_1K) = r(1 + \sin^2 \varphi) \tag{2.71}$$

$$(KD) = r_{b1} \sin \varphi = r \sin \varphi \cos \varphi$$

$$(O_2D) = r + m$$

Therefore, because $N = 2r/m$:

$$N_{\min} = \frac{2}{[1 + 3 \sin^2 \varphi]^{1/2} - 1} \tag{2.72}$$

For the usual pressure angles, the minimum number of teeth to avoid interference is related with the pressure angle to Equations (2.70b) and (2.72) as in Table 2.5.

For certain applications, such as gear-pumping of liquids, a small number of teeth is required. For such purposes, unconventional tooth profiles have been used, such as cycloidal shapes of different forms.

Example 2.13 Geometry of Gearing

A spur gear reducer was measured to have a pinion with 20 teeth and external diameter 110 mm and a gear with 40 teeth and external diameter 210 mm. The distance of the centers was measured to be 150 mm. It is supposed that the gears have standard involute teeth. Determine the module, pitch diameters, the contact ratio, and whether interference will be observed.

Solution   For standard teeth, the addendum (external) circle has diameter $d + 2m$. For the pinion:

$$d_1 + 2m = 110 \text{ mm}$$

Because $d_1 = mN_1$, $mN_1 + 2m = 110$. Therefore, $m = 110/(N_1 + 2) = 5$ mm. The pitch diameters $d_1 = mN_1 = 100$ mm, $d_2 = mN_2 = 200$ mm. The contact ratio for $\phi = 20°$, from Equation (2.69):

$$m_c = \frac{[(40 + 2)^2 - 40^2 \cos^2 20°]^{1/2} - 40 \sin 20° + [(20 + 2)^2 - 20^2 \cos^2 20°]^{1/2}}{2\pi} = 1.537$$

From Equation (2.70b), the minimum number of teeth to avoid interference, the most

---

TABLE 2.5. Minimum Number of Teeth for Gears to Avoid Interference

Pressure Angle ϕ	Minimum Number of Teeth $N_{\min}$ gear-rack, $N_2 = \infty$ Equation (2.70b)	Minimum Number of Teeth $N_{\min}$ equal gears, $N_2 = N_1$ Equation (2.70b)
14.5°	32	23
20°	17	12
25°	12	9

unfavorable case is 17, for $\varphi = 20°$ pressure angle. Therefore, no interference will be encountered.

Example 2.14 Geometry of Gearing

A spur gear reducer was designed to have module $m = 5$ mm, gear ratio 2 and the center distance of the two gears $a = 150$ mm. It is supposed that the gears have standard involute teeth. Determine the pitch and external diameters, the contact ratio, and whether interference will be observed.

Solution The center distance $a = (d_1 + d_2)/2 = 150$ mm and the gear ratio $R_{12} = d_2/d_1 = 2$. These are two equations in two unknowns d_1 and d_2, and they yield $d_1 = 100$ mm, $d_2 = 200$ mm, $N_1 = d_1/m = 100/5 = 20$, $N_2 = d_2/m = 200/5 = 40$.

For standard teeth, the addendum (external) circle has diameter $d + 2m$. For the pinion:

$$d_{1a} = d_1 + 2m = 110 \text{ mm}$$

For the gear, $d_{2a} = d_2 + 2m = 210$ mm.

The contact ratio for $\varphi = 20°$, from Equation (2.69):

$$m_c = \frac{[(40 + 2)^2 - 40^2 \cos^2 20°]^{1/2} - 40 \sin 20° + [(20 + 2)^2 - 20^2 \cos^2 20°]^{1/2}}{2\pi} = 1.537$$

From Equation (2.70*b*), the minimum number of teeth to avoid interference, the most unfavorable case is 17, for $\varphi = 20°$ pressure angle. Therefore, no interference will be encountered.

CASE STUDY 2.1: Kinematic Design of a Metal Planer–Shaper

The Cincinnati Tool Co., Inc. wants to design a planer–shaper machine with the following specifications:

· Machine functional concept: The quick-return mechanism
· Maximum stroke: 500 mm
· Maximum linear speed: 10 m/s
· Forward to return time ratio: 2

Find the major dimensions of the machine by kinematic design.
Solution

A similar machine previously made by the company is shown in Figure CS2.1.

The crank has maximum length R and rotates with constant angular velocity ω. Let $(OC) = h$ and $(OB) = L$. Since w is constant, the time for point A to move on a circle with center C and radius R is proportional to the rotation of the crank:

$$\frac{AA_3A_1}{AA_2A_1} = 2$$

But $(AA_3A_1) + (AA_2A_1) = 2\pi$. Therefore:

Figure CS2.1. An older model of the Cincinnati planer–shaper.

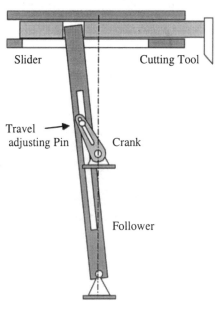

Figure CS2.2. First kinematic layout.

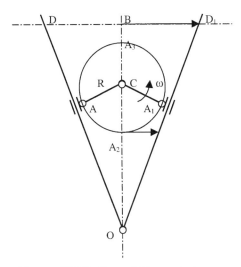

Figure CS2.3. Second kinematic layout.

$$(AA_3A_1) = \frac{4\pi}{3}, \; (AA_2A_1) = \frac{2\pi}{3}$$

Thus, angle $(OAC) = 60°$ and $(OC) = h = R/\sin 60° = 2R$.

When the oscillating follower is in the vertical position and the crank is at CA_2, the velocity of point A_2 is $v_2 = \omega R$ while the velocity of the follower at point B is $V_B = v_2 L/(h - R)$. The problem statement limits this velocity to 10 m/s. Therefore (because we found $h = 2R$):

$$V_B = \frac{\omega RL}{h - R} = 10 \text{ m/s}$$

The maximum stroke is $(DD_1) = 0.5$ m. But $(DD1) = 2 \times L \tan 30° = 0.5$ m. Therefore, $L = 0.5/(2 \tan 30°) = 0.433$ m.

The maximum crank length R is selectable by the designer so that $h < L$. Let $h = 0.3$ m, thus $R = h/2 = 0.15$ m. Then the crank angular velocity is $\omega = 10(h - R)/(RL) = 10 \times (0.3 - 0.15)/(0.15 \times 0.433) = 23.1$ rad/sec. The kinematic sizing of the shaper is now complete with the following results:

- Crank maximum angular velocity $\omega = 23.1$ rad/s = 221 rpm
- Maximum crank length $R = 150$ mm
- Pivot distance $(OC) = h = 300$ mm
- Mechanism height $(OB) = L = 433$ mm

To conclude the kinematic design of the shaper, we need to tabulate the angular velocity of the follower and the linear velocity of the slider. To this end we refer to Example 2.7, from which we obtain:

Crank angle: ϕ_R, from 0 to 2π

Follower angle: $\phi_1 = \tan^{-1} \dfrac{(h + R \sin \phi_R)}{R \cos \phi_R}$ (here $s = h$)

$$s_1 = \frac{R \cos \phi_R}{\cos \phi_1}$$

Follower angular velocity $\omega_1 = \dfrac{\omega \cos(\phi_R - \phi_1)R}{s_1}$

Therefore, since $\phi_2 = \phi_1$, $\omega_4 = \omega_2 = \omega_1$:

Slider displacement: $s_3 = L \tan \phi_1$

Slider velocity: $v_3 = \dot{s}_3 = L\dot{\phi}_1 \dfrac{1 - \sin^2 \phi_1}{\cos^2 \phi_1}$

Accelerations can be found by further differentiation in respect to time. We obtain:

Angular acceleration of the follower

$$\alpha_1 = \frac{d\omega_1}{dt}$$

$$= \frac{\omega[\omega_1(\sin(\phi_R - \phi_1)\sin \phi_1 + \cos(\phi_R - \phi_1)\cos \phi_1] + \omega \cos(\phi_R - \phi_1)\sin \phi_1 \cos \phi_R}{\sin^2 \phi_R}$$

Linear acceleration of the head:

$$a_3 = L \frac{a_1(1 - \tan^2 \phi_1) - 2\omega_1^2 \sin \phi_1}{\cos^3 \phi_1}$$

These results can be tabulated for later use in the detail design effort, using the spreadsheet of Figure CS2.4. The last two columns were normalized by ω for plotting convenience.

CASE STUDY 2.2: Computer-simulated Kinematics of Mechanisms

Using Working Model, simulate the operation of the aircraft nose landing gear (see Figure P2.16) for constant angular velocity of the motor.

Solution   There are several other kinematic and dynamic analysis packages, such as Automatic Dynamic Analysis of Mechanical Systems (ADAMS) and Dynamic Analysis of Dynamical Systems (DADS). The definition of the problem via the user's interface is similar in all those programs. Here we shall use Working Model. We do not provide detailed instructions, since these programs frequently change user interface and it is better for the reader to consult the appropriate user's manual.

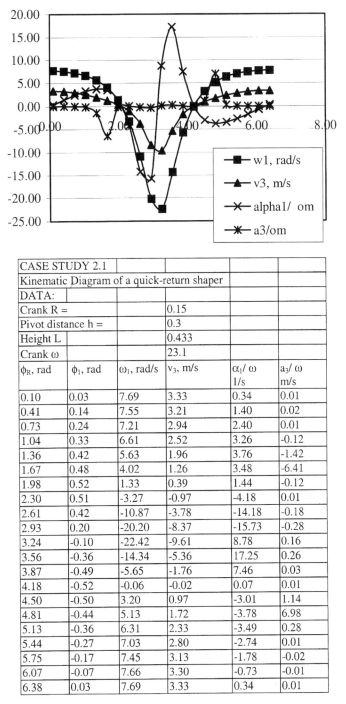

CASE STUDY 2.1					
Kinematic Diagram of a quick-return shaper					
DATA:					
Crank R =		0.15			
Pivot distance h =		0.3			
Height L		0.433			
Crank ω		23.1			
ϕ_R, rad	ϕ_1, rad	ω_1, rad/s	v_3, m/s	α_1/ω 1/s	a_3/ω m/s
0.10	0.03	7.69	3.33	0.34	0.01
0.41	0.14	7.55	3.21	1.40	0.02
0.73	0.24	7.21	2.94	2.40	0.01
1.04	0.33	6.61	2.52	3.26	-0.12
1.36	0.42	5.63	1.96	3.76	-1.42
1.67	0.48	4.02	1.26	3.48	-6.41
1.98	0.52	1.33	0.39	1.44	-0.12
2.30	0.51	-3.27	-0.97	-4.18	0.01
2.61	0.42	-10.87	-3.78	-14.18	-0.18
2.93	0.20	-20.20	-8.37	-15.73	-0.28
3.24	-0.10	-22.42	-9.61	8.78	0.16
3.56	-0.36	-14.34	-5.36	17.25	0.26
3.87	-0.49	-5.65	-1.76	7.46	0.03
4.18	-0.52	-0.06	-0.02	0.07	0.01
4.50	-0.50	3.20	0.97	-3.01	1.14
4.81	-0.44	5.13	1.72	-3.78	6.98
5.13	-0.36	6.31	2.33	-3.49	0.28
5.44	-0.27	7.03	2.80	-2.74	0.01
5.75	-0.17	7.45	3.13	-1.78	-0.02
6.07	-0.07	7.66	3.30	-0.73	-0.01
6.38	0.03	7.69	3.33	0.34	0.01

Figure CS2.4. Kinematic analysis of the planer–shaper of Case Study 2.1.

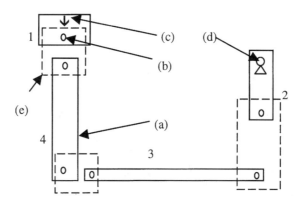

Figure CS2.5. Input preparation for the landing gear simulation in Working Model.

1. Using the rectangle tool, make four rectangles, the three links 2, 3, 4 and the base link 1, in the vicinity we want (the exact location will be determined later).
2. Click twice in the joint element (small circle button °) and then click on the six locations of the revolute joints (excluding the pivot of the crank).
3. Click on the anchor button (↓) and then click on the base link to ground it.
4. Click on the motor button (lower left corner) and then click on the location of the pivot point of the crank.
5. Click on the pointer tool (↖) and click and drag the mouse to create the four rectangles with the dashed line, one by one. Each rectangle must contain the joint points that we want to form a joint. After each rectangle is formed, click on the **Joint** button.

Now you are ready for the first trial. Just click on the **Run** button and the crank will start rotating. To use the exact dimensions, click on every one of the elements

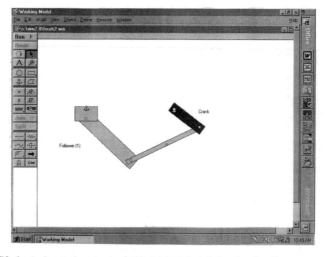

Figure CS2.6. Animated output of Working Model for the landing gear simulation.

above and then click on **Window/Geometry** or **Window/Properties** to enter the exact numerical values.

The output window is shown in Figure CS2.6.

To observe the variation of any parameter, for example the follower rotation, click on the follower and then click on **Measure/Position/Rotation** graph in the menu. A window will open with a time graph of the rotation of the follower vs. time.

REFERENCES

Asada, H., and J.-J. E. Slotine. 1986, *Robot Analysis and Control.* New York: John Wiley & Sons.

Denavit, J., and R. S. Hartenberg. 1955. "A Kinematic Notation for Lower Pair Mechanisms Based on Matrices." *J. Appl. Mech.* 22.

Erdman, A. G., and G. N. Sandor. 1984. *Mechanism Design*, 1st ed. Englewood Cliffs, N.J.: Prentice Hall.

———. *Mechanism Design*, 2nd ed. 1991. Englewood Cliffs, N.J.: Prentice Hall.

———. *Mechanism Design*, 3rd ed. 1997. Upper Saddle River, N.J.: Prentice Hall.

Hartenberg, R. S., and J. Denavit. 1964. *Kinematic Synthesis of Linkages.* New York: McGraw-Hill.

Pfeiffer, F., and E. Reithmeier. 1987. *Roboterdynamik.* Stuttgart: Teubner.

Shahinpoor, M. 1987. *A Robot Engineering Textbook.* New York: Harper & Row.

Shigley, J. E., and J. J. Uicker Jr. 1980. *Theory of Machines and Mechanisms.* New York: McGraw-Hill.

Suh, C. H., and C. W. Radcliffe. *Kinematics and Mechanical Design.* New York: John Wiley & Sons, 1978.

ADDITIONAL READING

Adler, U., ed. *Automotive Handbook*, 2nd ed. Detroit: Society of Automotive Engineers, 1984.

Dudley, D. W., ed. *Gear Handbook.* New York: McGraw-Hill, 1962.

Reuleaux, F., *The Kinematics of Machinery*, trans. B. W. Kennedy. London: MacMillan, 1876.

Rooney, J., and P. Steadman. *Computer-Aided Design.* London: Pitman, 1987.

PROBLEMS[4]

2.1. [T] For a pair of pliers, if one leg is kept fixed:

　　1. Identify all the kinematic elements.

　　2. Draw a skeleton diagram.

　　3. Find the number of degrees of freedom.

---

[4] [C] = certification problem, [D] = design problem problem, [N] = numerical problem, [T] = theoretical problem.

2.2. [T] For the crank–follower–piston mechanism of a single-cylinder internal combustion engine, if the crank revolves about a fixed point on the engine block and the piston slides inside a cylinder which is also fixed on the engine block:

1. Identify all the kinematic elements.
2. Draw a skeleton diagram.
3. Find the number of degrees of freedom.

2.3. [T] For the hood closing linkage shown (Figure P2.3):

1. Identify all the kinematic elements.
2. Draw a skeleton diagram.
3. Find the number of degrees of freedom.

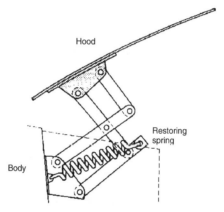

Figure P2.3 (Adapted from A. G. Erdman and G. N. Sandor, *Mechanism Design,* 1984, by permission of Prentice Hall, Inc.)

2.4. [T] For the robot arm of Figure 2.10*a*:

1. Identify all the kinematic elements.
2. Draw a skeleton diagram.
3. Find the number of degrees of freedom.

2.5. [T] For the car suspension system shown in Figure P2.5:

1. Identify all the kinematic elements.
2. Draw a skeleton diagram.
3. Find the number of degrees of freedom.

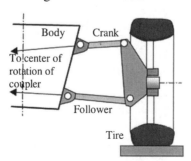

Figure P2.5

2.6. [C] An electric motor has speed 3500 rpm and power $P = 5$ kW. It transmits the motion to a shaft of a lathe that has speed of rotation 1200 rpm.

1. Draw a kinematic layout.
2. Calculate the torques on the motor shaft and the shaft of the lathe.
3. Draw a kinematic diagram showing the speeds and torques in the system.

2.7. [C] An electric motor has speed 1750 rpm. It is directly connected to a shaft that transmits motion through two different belt drives to two oil pumps, the first, which is located closer to the motor, at 800 rpm and 3.5 hp, the second at 600 rpm and 8 hp.

1. Draw a kinematic layout.
2. Calculate the torques on the motor shaft and the shafts of the pumps.
3. Draw a kinematic diagram showing the speeds and torques in the system.

2.8. [C] An automotive engine has maximum torque 250 Nm at 3500 rpm speed. A four-forward speeds gear box is used with the shaft connected with the engine transmitting motion to an intermediate shaft through two selectable gear pairs of gear ratio 3.8 and 5.5. The intermediate shaft transmits motion to the exit shaft, which is connected with the differential through two selectable gear pairs of gear ratio 3.8 and 4.4.

1. Draw a kinematic layout.
2. Calculate the torques on the motor shaft, the intermediate shaft, and the exit shaft at all possible speeds.
3. Draw a kinematic diagram showing the speeds and torques in the system.

2.9. [C] An electric motor has speed 3500 rpm and power $P = 6$ kW. It transmits the motion to an intermediate shaft that has speed of rotation 1750 rpm through a V-belt drive. This shaft in turn powers four textile spindles of approximately 1.5 kW, each through gear boxes of gear ratio 2.5.

1. Draw a kinematic layout.
2. Calculate the torques on the motor shaft and the shaft of the lathe.
3. Draw a kinematic diagram showing the speeds and torques in the system.

2.10. [C] A 6000 kW ship propulsion steam turbine has speed of rotation 6000 rpm and the propeller shaft has speed 200 rpm. The gear speed reducer consists of a number of intermediate shafts and pairs of gears to successively reduce the speed from 6000 to 200 rpm. The gear ratio, for efficiency and reliability reasons, should not exceed 4.

1. Select number of speed reduction steps assuming that all gear ratios are the same.
2. Draw a kinematic layout.
3. Calculate the torques on all shafts.
4. Draw a kinematic diagram showing the speeds and torques in the system.

2.11. [C] A four-bar linkage as in Figure 2.3a has link lengths $r_1 = 450$ mm, $r_2 = 150$ mm, $r_3 = 250$ mm, $r_4 = 300$ mm. Initially, the angle ϕ_2 is 60° and the angular velocity of link 2 is 40 rad/sec, clockwise. Determine the output angle ϕ_4 and the follower angular velocity ω_4 at this position of the linkage.

2.12. [C] A four-bar linkage as in Figure 2.3a has link lengths $r_1 = 500$ mm, $r_2 = 100$ mm, $r_3 = 200$ mm, $r_4 = 400$ mm. For values of the input angle ϕ_2 60°, 80°, 100°, 1200°, determine the output angle ϕ_4 of the follower and its angular velocity and angular acceleration.

2.13. [C] A four-bar linkage as in Figure 2.3a has link lengths $r_1 = 450$ mm, $r_2 = 150$ mm, $r_3 = 250$ mm, $r_4 = 300$ mm. Determine the input (crank) angle ϕ_2 and the input angular velocity ω_2 so that the output (follower) angle $\phi_4 = 110^0$ and the output angular velocity $\omega_4 = 35$ rad/sec at this position of the linkage.

2.14. [C] A four-bar linkage as in Figure 2.3a has link lengths $r_1 = 450$ mm, $r_2 = 150$ mm, $r_3 = 250$ mm, $r_4 = 300$ mm. Initially, the angle ϕ_2 is 60° and the angular velocity of link 2 is 40 rad/sec, clockwise. The coupler AB is extended beyond B at point C such that $(AB) = (BC)$. Find at this position of the linkage the location and velocity of the coupler point C.

2.15. [C] A four-bar linkage as in Figure 2.3a has link lengths $r_1 = 450$ mm, $r_2 = 150$ mm, $r_3 = 250$ mm, $r_4 = 300$ mm. Initially, the angle ϕ_2 is 60° and the angular velocity of link 2 is 40 rad/sec, clockwise. The coupler AB is replaced by a triangle ABC, having three equal sides $(AB) = (BC) = (CA)$, with point C above line AB. Find at this position of the linkage the location and velocity of the coupler point C.

2.16. (Adapted from Erdman and Sandor 1991) [C] The aircraft nose landing gear shown in Figure P2.16 is a crank and follower mechanism. The motor turns the crank and the follower supports the wheel as shown.
   1. Draw a skeleton diagram of the linkage and assume that the drawing is to scale. Prove that it is a crank and follower mechanism.
   2. Find the maximum and minimum oscillation angle of the follower for the two extreme locations of the control arm, assuming that the mechanism was drawn to scale in Figure P2.16.

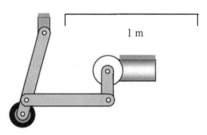

Figure P2.16

2.17. [C] The mechanism shown in Figure P2.17 is a crane that is supposed to deliver a nearly straight horizontal trajectory of coupler point P. Assuming that the linkage is drawn to 1:200 scale:
   1. Find the type of the linkage, applying the Grashof criteria. Draw a skeleton diagram of the linkage.
   2. Determine the location of P for angles of deviation 0 (the position shown), ± 5°, ± 10° from the location shown of the input rocker, link 2.

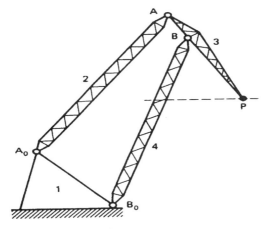

Figure P2.17

2.18. [C] The mechanism in Figure P2.18 is the Geneva mechanism, an intermitted motion mechanism in which at a full rotation of the disk-crank the four-slot follower wheel rotates by one quarter of a revolution.

1. Draw a skeleton diagram of the mechanism that will describe the operation for one quarter revolution of the crank and draw a skeleton diagram.
2. If the crank rotates with angular velocity 120 rpm, compute the maximum angular velocity and angular acceleration of the four-slot wheel.

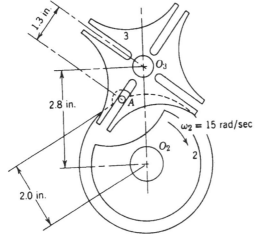

Figure P2.18

2.19. [C] The dump mechanism of the dump truck mechanism shown in Figure P2.19 is equivalent to a four-bar linkage. Assuming that the drawing is to 1: 100 scale:

1. Find the type of the mechanism using the Grashof criteria and draw a skeleton diagram.
2. Find the desired total free travel of the hydraulic piston so that the track bed should have a maximum slope of $45°$, if distance $AD = 0.80$ m and the sketch is to scale.

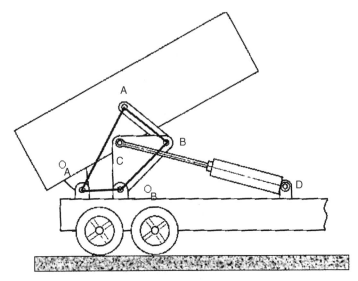

Figure P2.19

2.20. [C] The mechanism of drum beating shown in Figure P2.20 is equivalent to a four-bar linkage. Assuming that the drawing is to 1:25 scale:

1. Find the type of the mechanism using the Grashof criteria and draw a skeleton diagram.

2. Find the total angle of rotation of foot pedal that will move the drum-hitting follower to the vertical position.

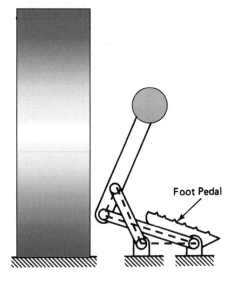

Figure P2.20. Dotted line denotes the four-bar linkage. (Adapted from A. G. Erdman and G. N. Sandor, *Mechanism Design,* 1984, by permission of Prentice Hall, Inc.)

2.21. [T] For the robot shown (Figure P2.21), with three cylindrical joints plus base and grip rotation, determine the forward kinematic relations.

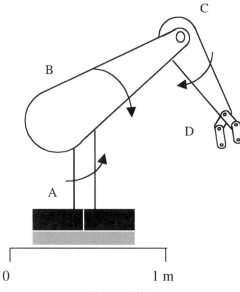

Figure P2.21

2.22. [T] For the cylindrical robot shown (Shahinpoor 1987), base and grip rotation, one cylindrical joint, two sliding arms, determine the forward kinematic relations.

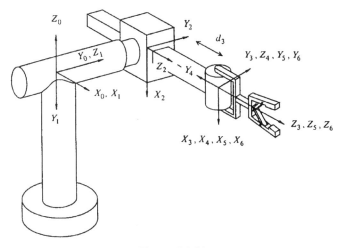

Figure P2.22

2.23. [T] For the Microbot robot shown (Shahinpoor 1987), with three cylindrical joints and grip rotation, determine the forward kinematic relations.

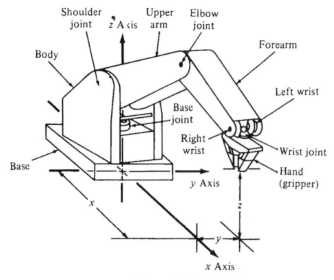

Figure P2.23

2.24. [T] For the cylindrical robot shown (Shahinpoor 1987), with base and arm rotation, two sliding arms and a sliding grip, determine the forward kinematic relations.

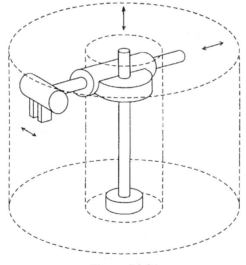

Figure P2.24

2.25. [T] For the cylindrical robot shown (Pfeiffer and Reithmeir 1987), with base and grip rotation and horizontal arm sliding, determine the forward kinematic relations.

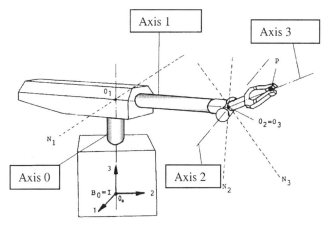

Figure P2.25

2.26. [C] A follower dwells for 120°, then rises by 20 mm in 45°, then dwells again for 120°, then falls by 20 mm in 45°, then dwells again to 360°. Assume a 7th degree polynomial and continuity conditions at the beginning and end of the rise and fall in displacement, velocity, acceleration and jerk. Design and draw the follower time function. If a computer and a spreadsheet are available, draw the diagrams for the normalized displacement, velocity, acceleration, and jerk and determine the maximum normalized acceleration.

2.27. [C] A follower dwells for 120°, then rises by 20 mm in 45°, then dwells again for 120°, then falls by 20 mm in 45°, then dwells again to 360°. Assume a 5th degree polynomial and continuity conditions at the beginning and end of the rise and fall in displacement, velocity, and acceleration. Design and draw the follower time function. If a computer and a spreadsheet are available, draw the diagrams for the normalized displacement, velocity, acceleration, and jerk and determine the maximum normalized acceleration.

2.28. [C] A follower dwells for 120°, then rises by 20 mm in 45°, then dwells again for 120°, then falls by 20 mm in 45°, then dwells again to 360°. Assume a sinusoidal function $a_1 + a_2 \cos a_3\phi$, where a_1, a_2, a_3 constants to be determined from the continuity conditions at the beginning and end of the rise and fall in displacement and velocity. Design and draw the follower time function. If a computer and a spreadsheet are available, draw the diagrams for the normalized displacement, velocity, acceleration, and jerk and determine the maximum normalized acceleration.

2.29. [C] A follower dwells for 120°, then rises by 20 mm in 30° with constant acceleration, then rises further by 20 mm in 30° with constant deceleration, equal to the previous constant acceleration, then dwells again for 120°, then falls by 20 mm in two steps of constant deceleration and accelerations as for the rise, then dwells again to 360°. Design and draw the follower time function. If a computer and a spreadsheet are available, draw the diagrams for the normalized displacement, velocity, acceleration, and jerk.

2.30. [C] A follower dwells for 120°, then rises by 20 mm in 45°, then dwells again for 120°, then falls by 20 mm in 45°, then dwells again to 360°. Assume a sinusoidal function $a_1 + a_2 \cos a_3\phi$, where a_1, a_2, a_3 constants to be determined from the continuity conditions at the beginning and end of the rise and fall in displacement and velocity. Design and draw the follower time function. If a computer and a spreadsheet are available, draw the diagrams for the normalized displacement, velocity, acceleration, and jerk and determine the maximum normalized acceleration.

2.31. [D] A follower dwells for 120°, then rises by 20 mm in 45°, then dwells again for 120°, then falls by 20 mm in 45°, then dwells again to 360°. Assume a 7th degree polynomial and continuity conditions at the beginning and end of the rise and fall in displacement, velocity, acceleration, and jerk. Design and draw the cam profile for radius of the base circle $R_b = 80$ mm for flat, translating follower.

2.32. [D] A follower dwells for 120°, then rises by 20 mm in 45°, then dwells again for 120°, then falls by 20 mm in 45°, then dwells again to 360°. Assume a 5th degree polynomial and continuity conditions at the beginning and end of the rise and fall in displacement, velocity, and acceleration. Design and draw the cam profile for radius of the base circle $R_b = 80$ mm for flat, translating follower.

2.33. [D] A follower dwells for 120°, then rises by 20 mm in 45°, then dwells again for 120°, then falls by 20 mm in 45°, then dwells again to 360°. Assume a sinusoidal function $a_1 + a_2 \cos a_3\phi$, where a_1, a_2, a_3 constants to be determined from the continuity conditions at the beginning and end of the rise and fall in displacement and velocity. Design and draw the cam profile for radius of the base circle $R_b = 80$ mm for flat, translating follower.

2.34. [D] A follower dwells for 120°, then rises by 20 mm in 30° with constant acceleration, then rises farther by 20 mm in 30° with constant deceleration, equal to the previous constant acceleration, then dwells again for 120°, then falls by 20 mm in two steps of constant deceleration and accelerations as for the rise, then dwells again to 360°. Design and draw the cam profile for radius of the base circle $R_b = 80$ mm for flat, translating follower.

2.35. [D] A follower dwells for 120°, then rises by 20 mm in 45°, then dwells again for 120°, then falls by 20 mm in 45°, then dwells again to 360°. Assume a sinusoidal function $a_1\phi + a_2 \cos a_3\phi$, where a_1, a_2, a_3 constants to be determined from the continuity conditions at the beginning and end of the rise and fall in displacement and velocity. Design and draw the cam profile for radius of the base circle $R_b = 80$ mm for flat, translating follower.

2.36. [D] A follower dwells for 120°, then rises by 20 mm in 45°, then dwells again for 120°, then falls by 20 mm in 45°, then dwells again to 360°. Assume a 7th degree polynomial and continuity conditions at the beginning and end of the rise and fall in displacement, velocity, acceleration, and jerk. Design and draw the cam profile for radius of the base circle $R_b = 80$ mm for an oscillating, flat-face follower, if $f = 3$ cm and $m = 20$ cm.

2.37. [D] A follower dwells for 120°, then rises by 20 mm in 45°, then dwells again for 120°, then falls by 20 mm in 45°, then dwells again to 360°. Assume a 5th degree polynomial and continuity conditions at the beginning and end of the rise and fall in displacement, velocity, and acceleration. Design and draw the cam profile for radius of the base circle R_b = 80 mm for an oscillating, flat-face follower, if f = 3 cm and m = 20 cm.

2.38. [D] A follower dwells for 120°, then rises by 20 mm in 45°, then dwells again for 120°, then falls by 20 mm in 45°, then dwells again to 360°. Assume a 7th degree polynomial and continuity conditions at the beginning and end of the rise and fall in displacement, velocity, acceleration, and jerk. Design and draw the cam profile for radius of the base circle R_b = 80 mm for an offset translating roller follower, if r_f = 3 cm and m = 2 cm.

2.39. [D] A follower dwells for 120°, then rises by 20 mm in 45°, then dwells again for 120°, then falls by 20 mm in 45°, then dwells again to 360°. Assume a 5th degree polynomial and continuity conditions at the beginning and end of the rise and fall in displacement, velocity, and acceleration. Design and draw the cam profile for radius of the base circle R_b = 80 mm for an offset translating roller follower, if r_f = 3 cm and m = 2 cm.

2.40. [D] A follower dwells for 120°, then rises by 20 mm in 45°, then dwells again for 120°, then falls by 20 mm in 45°, then dwells again to 360°. Assume a 7th degree polynomial and continuity conditions at the beginning and end of the rise and fall in displacement, velocity, acceleration, and jerk. Design and draw the cam profile for radius of the base circle R_b = 80 mm for an oscillating roller follower, if r_f = 3 cm and m = 15 cm.

2.41. [C] A standard 20° involute pair of gears have number of teeth 17 and 40 with module m = 6 mm. Determine the geometric characteristics of the gearing, diameters, center distance, and contact ratio. Sketch the main lines in the vicinity of two mating teeth and verify graphically the computed contact ratio.

2.42. [C] A standard 20° involute pair of gears has center distance a = 300 mm, gear ratio R = 4 with module m = 6 mm. Determine the geometric characteristics of the gearing, diameters, numbers of teeth, and contact ratio. Sketch the main lines in the vicinity of two mating teeth and verify graphically the computed contact ratio.

2.43. [C] A standard 20° involute pair of gears has addendum diameters 135 and 385 mm and numbers of teeth 25 and 75, respectively. Determine the geometric characteristics of the gearing, diameters, center distance, and contact ratio. Sketch the main lines defining the profile in the vicinity of two mating teeth and verify graphically the computed contact ratio.

2.44. [C] A standard 20° involute pair of gears has pitch diameter of the pinion 96 mm, gear ratio 3 with module m = 4 mm. Determine the geometric characteristics of the gearing, diameters, center distance, and contact ratio. Sketch the main lines in the vicinity of two mating tccth and verify graphically the computed contact ratio.

2.45. [C] Two standard 20° involute pairs of gears have parallel geometric axes of rotation and center distance $a = 125$, approximately, while the speed ratio $N_1/N_2 = 6.25$ and the module $m = 2$ mm. Determine the geometric characteristics of the gearing, diameters, center distance, and contact ratio. Sketch the main lines in the vicinity of two mating teeth and verify graphically the computed contact ratio.

CHAPTER 3

ANALYSIS OF MACHINE LOADS

Cranes are designed to lift very high loads. Thus, knowledge of the load distribution is of paramount importance for design.

OVERVIEW

In Chapter 1 we saw how from a vague initial description of the machine design problem to be solved we can develop a multitude of possible solutions and identify the most promising solution concept, described in generic form, usually in the form of a schematic that describes the machine operation and a set of detailed design specifications.

In Chapter 2 we saw that often many of the main dimensions of a machine can be found using purely kinematic considerations before it is necessary to discuss materials and their ability to carry loads. For example, in Case Study 2.1 we found the main dimensions of a shaper–planer by kinematic methods. We can observe, however, that for the machine members for which we determined dimensions, at most we found a length but nothing for their lateral dimensions. These dimensions need to be sufficient that each member will not fail (e.g., by fracture) during normal operation.

One of the main reasons for failure of machine members is the limited capacity of materials to carry loads. It is apparent, therefore, that before sizing the machine components for strength and strength-related considerations, we need to find the loads that each machine member sustains during operation. This will be the subject of the present chapter.

3.1. POWER TRANSMISSION IN A MACHINE

3.1.1. Work and Energy

To do useful work, machines have to process energy. They receive the energy through a prime mover that converts the primary source of energy (electricity, chemical energy of fuels, kinetic energy of wind, etc.) into mechanical work, which is converted in the machine into forces or torques that are applied to moving parts. The first law of thermodynamics states that a system conserves its energy, that is:

$$E_{in} - E_{out} = \Delta E \tag{3.1}$$

where E_{out} is the energy leaving the system, E_{in} the energy entering the system, ΔE the change in the amount of energy stored in the system. The last can have many different forms, as follows.

3.1.1.1. Kinetic Energy. If the mass of a solid part is m and the part is accelerated from speed v_1 to speed v_2, the kinetic energy stored in the part is (Beer and Johnston 1981):

$$KE = 0.5m(v_2^2 - v_1^2) \tag{3.2}$$

For rotary motion of a wheel about its geometric axis with initial and final angular velocities ω_1 and ω_2:

$$KE = 0.5J(\omega_2^2 - \omega_1^2) \tag{3.3}$$

where J is a measure of the distance of the mass of the part from the axis of rotation, termed *mass moment of inertia* and defined as:

$$J = \int_m r^2 \, dm \tag{3.4}$$

and r is the distance of the elementary mass dm from the axis of reference (here the axis of rotation) and the integral is taken over the mass m of the solid part. Mass moments of inertia for common solids are tabulated in Appendix F.

The mass moment of inertia of a composite solid about an axis is the sum of the mass moments of inertia of the component solids about the same axis. Moreover, the *theorem of de Moivre* states that the mass moment of inertia J_x of a solid of mass m about an axis x is related with the mass moment of inertia $J_{x'}$ of the solid about another axis x', parallel with the axis x and at distance a, as

$$J_x = J_{x'} + ma^2 \tag{3.5}$$

3.1.1.2. Potential Energy. The potential energy of raising a weight W from a height h_1 to height h_2 is:

$$PE = W(h_2 - h_1) \tag{3.6}$$

We shall encounter other forms of potential energy, such as energy of elastic deformation, later on in this book.

3.1.1.3. Mechanical Work. If a force F or torque T is applied on a solid and the point of application of the force or the solid moves with a velocity V or angular velocity ω, respectively, the mechanical work produced is:[1]

$$W = \int_{t_2}^{t_2} F \cdot V \, dt \tag{3.7}$$

$$W = \int_{t_1}^{t_2} T \cdot \omega \, dt \tag{3.8}$$

3.1.1.4. Frictional Heat. In sliding friction, if the friction force F is collinear with the direction of the velocity V of the point of application, the amount of the friction heat generated while the force was applied from time t_1 to time t_2 is:

$$Q = \int_{t_1}^{t_2} F \cdot V \, dt = \int_{t_1}^{t_2} FV \, dt \tag{3.9}$$

If the force and velocity remain constant in time,

$$Q = FV(t_2 - t_1) = F \, \Delta s \tag{3.10}$$

where $\Delta s = V(t_2 - t_1)$ is the distance traveled with constant velocity V. *Amonton's law* demands in turn that the friction force is:

$$F = fF_n \tag{3.11}$$

where F_n is the normal force compressing the two rubbing solids towards one another and f is a constant, called the *friction coefficient,* that (as a first approximation) depends only on the materials and the surface conditions of the rubbing solids and not on the applied normal load or the velocity.

Similarly, if constant friction torque T is applied to a solid rotating at an angular velocity ω, during a time interval $t_2 - t_1$ or angle of rotation $\Delta\theta = \omega(t_2 - t_1)$, the heat generated is:

---

[1] Large dot indicates an inner product of two vectors, i.e., $x \cdot y = xy \sin (\angle x, y)$

$$Q = T\omega(t_2 - t_1) = T \, \Delta\theta \tag{3.12}$$

3.1.1.5. Stored Heat. A solid of mass m, to change its temperature from T_1 to T_2, needs to absorb heat:

$$\Delta Q = mc(T_2 - T_1) \tag{3.13}$$

where c is a material constant called *heat capacity*.

Example 3.1   An automobile weighs 10,000 N and is designed to stop in 60 m when the disk brakes are applied while the automobile cruises at speed 40 MPH. Determine:

1. The braking force applied by the tires to the automobile, assumed constant.
2. The minimum required coefficient of friction between the tires and the road.
3. The temperature rise of the disk brakes if the disks have total mass 25 kg and the heat capacity of the disk material is 500 J/kg°C.

Solution

1. The automobile speed is 40 MPH $= 40 \times 1609.3/3600 = 17.88$ m/s. The mass is $10,000/g = 10,000/9.81$ kg and the kinetic energy is (Equation (3.2)):

$$KE = 0.5m(v_2^2 - v_1^2) = 0.5 \times \frac{10,000}{9.81} \times 17.88^2 = 162,943 \text{ Nm (joules)}$$

   Equations (3.7) and (3.10) give the braking force F from $KE = W = F \, \Delta s$, $F = KE/\Delta s = 162,943/60 = \underline{2,715 \text{ N}}$.

2. The minimum coefficient of friction is $f = F/N = 2,715/10000 = 0.27$.
3. The temperature rise of the disks will be found from Equation (3.13), observing that the heat stored in the disks equals the mechanical work of braking, equal in turn to the kinetic energy of the automobile, $\Delta Q = mc(T_2 - T_1) = KE$:

$$T_2 - T_1 = \frac{KE}{mc} = \frac{162,943}{25} \times 500 = \underline{13.04°\text{C}}$$

3.1.2. Power Flow Analysis

Along the line of a kinematic chain the speed changes, as we saw in the previous chapter. Power is transmitted, too, along the kinematic chain. At every stage of speed change, the power is reduced somewhat due to the mechanical losses. This is expressed by way of the *mechanical efficiency:*

$$\eta = \frac{\text{power out}}{\text{power in}} \tag{3.14}$$

Later in this book we will discuss efficiency of different machine components in detail. For preliminary analysis, one can assume the following efficiencies for different speed change mechanisms:

1.	Spur, helical and bevel gears	97%
2.	Worm gears (varies)	75%
3.	Belts	95%
5.	Linkages:	
	four-bar	97%
	slider-crank	95%

The available power H_j at stage j will be:

$$H_j = \eta_1 \eta_2 \ldots \eta_j H_0 \tag{3.15}$$

where H_0 is the input power (the power of a prime mover, for example).

The transmitted force F or torque T at any step of speed change is a function of the power H and the velocity, linear V or angular ω, respectively, at this step:

$$F = \frac{H}{v} \tag{3.16}$$

$$T = \frac{H}{\omega} \tag{3.17}$$

where v is the linear speed and ω is the angular velocity.

As the speed increases, since power remains nearly constant, the force or torque will decrease, and the inverse. At step j, the force or torque will be:

$$F_j = \eta_1 \eta_2 \eta_3 \ldots \eta_j \frac{H_0}{v_j} \tag{3.18}$$

$$T_j = \eta_1 \eta_2 \eta_3 \ldots \eta_j \frac{H_0}{\omega_j} \tag{3.19}$$

For further design calculations, it is useful to draw diagrams of speed, force/torque, and power along the transmission system, called *kinematic diagrams,* as shown in Figure E3.2, Example 3.2.

Design Procedure 3.1: Kinematic and Power Flow Analysis

Step 1: Identify input power, force or torque, speed. Two of them should be specified and the third computed. Change all data to one system of units.

Step 2: Identify solid components of the system. Draw free-body diagrams for each one.

Step 3: Find kinematic relations for the relative motion between solid components.

Step 4: Compute the power transmitted through each solid component, assuming an efficiency of transmission from the previous component.

Step 5: Compute the linear and/or angular velocity of each solid component using the kinematic relationships with the previous component. If the kinematic relationship is a function of time (in linkages, for example) the velocity of the solid component will be, in general, a function of time.

Step 6: Compute the force or torque transmitted by each solid component using the kinematic relationships with the previous component and Equations (3.16) and (3.17). If the kinematic relationship is a function of time (in linkages for example), the force or torque through the solid component will generally be functions of time.

Example 3.2 A 3 kW, 3500 rpm electric motor moves an intermediate shaft at 1750 rpm through a belt drive. This shaft moves a pump at 1000 rpm through a one-stage gear drive. Determine the torques at the different parts of the system and draw the kinematic diagram.

Solution We first change all data to the SI system:

$$H_0 = 3 \times 1000 = 3000 \text{ W}$$

$$\omega_0 = 3500 \times \frac{2\pi}{60} = 366.5 \text{ rad/s}$$

$$\omega_1 = 1750 \times \frac{2\pi}{60} - 183.2 \text{ rad/s}$$

$$\omega_2 = 1000 \times \frac{2\pi}{60} = 104.7 \text{ rad/s}$$

Assume 95% efficiency for the belt drive and 97.5% for the gear drive. The power of each stage is:

$$H_0 = 3000 \text{ W}$$

$$H_1 = 0.95 \times 3000 = 2850 \text{ W}$$

$$H_2 = 0.975 \times 2850 = 2788.7 \text{ W}$$

The torques are:

$$T_0 = \frac{H_0}{\omega_0} = \frac{3000}{366.5} = \underline{8.19 \text{ Nm}}$$

$$T_1 = \frac{H_1}{\omega_1} = \frac{2850}{183.2} = \underline{15.5 \text{ Nm}}$$

$$T_2 = \frac{H_2}{\omega_2} = \frac{2788.7}{104.7} = \underline{26.6 \text{ Nm}}$$

The power and torque flow diagram is shown in Figure E3.2.

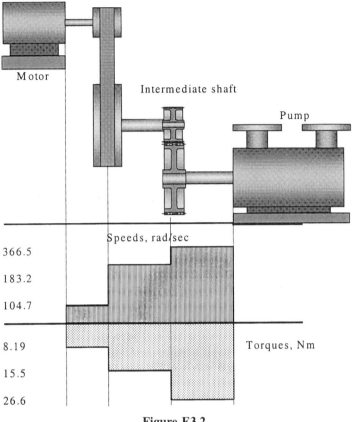

Figure E3.2

3.2. FORCE TRANSMISSION

3.2.1. Sources of Forces in Machines

Machines transform energy through the transmission of forces that, as we saw in the previous section, produce (or consume) mechanical work. Therefore, static loads are inherent in a machine and some of the mechanisms of generation of static loads are:

3.2.1.1. Loads Due to the Transmission of Power. These are torques and peripheral forces in gears, belts, etc., that generate the torques. For example, if a certain torque T is transmitted at a gear stage, a peripheral force is needed:

$$F_t = \frac{T}{d/2} = \frac{2H}{\omega d} \tag{3.20}$$

where d is the gear pitch diameter and H is the power transmitted. This force is statically equivalent to a static torque T and an equal force F_t acting at the center of the gear.

As the gears rotate, the point of contact between mating teeth moves along the profile, as one can observe in Figure 2.24. If the friction force is neglected, it can be assumed that the interaction force is perpendicular to the tooth profile all the times. It will be assumed at this point that only one tooth at a time carries the load, owing to cutting errors and high tooth stiffness. Multiple contact gear teeth will be discussed later (Chapter 13).

We saw in Chapter 2 that the normal to the gear teeth surfaces at the point of contact is always at angle ϕ (the pressure angle) with the common tangent of the pitch circles. The normal force can be analyzed in tangential and radial components (Figure 3.1):

$$F_t = F_n \cos \phi \tag{3.21}$$

$$F_r = F_n \sin \phi = F_t \tan \phi \tag{3.22}$$

Different types of gears, such as helical, bevel, and worm gears, have a more complex system of forces. They will be discussed in Chapter 13.

With a belt drive, the two belt sections transmit different forces, F_1 and F_2. Their difference must equal the net peripheral force, as in Equation (3.20).

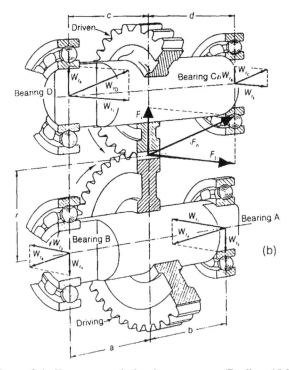

Figure 3.1. Force transmission in spur gears (Dudley, 1962).

$$F_1 - F_2 = F_t = \frac{T}{d/2} = \frac{2H}{\omega d} \qquad (3.23)$$

The ratio $(F_1 - F_2)/(F_1 + F_2)$ is called *pull ratio* and is usually 0.5 for flat belts and 0.8 for V-belts (see Chapter 12 for ways to calculate the pull ratio). Therefore, forces F_1 and F_2 can be computed. They are equivalent to the vector sum of the two forces plus the torque, applied to the center of the pulley.

3.2.1.2. Pressure Loads. Due to pressures in fluid-flow machinery static forces are applied on the rotor and the stator. For example, axial forces are usually found in axial fans, pumps, turbines, and so on.

3.2.1.3. Loads Due to Dead Weight. These are important in large machines only. This is because pressure loads increase with the square of the linear dimensions while dead loads increase with the cube of the linear dimensions.

3.2.1.4. Inertia Loads. If parts of a machine accelerate, inertia forces might be significant. We shall discuss these loads later in this chapter.

A *solid element* will be defined as a one-piece element that deforms only elastically under loads and there is no other type of relative motion among its points. Moreover, the deformation of the element due to the service loads is very small compared with the motion of the element.

A machine consists of solid elements that are linked together with connections. The machine base can be considered as one element attached to the machine foundation to form a solid element with it.

As discussed in Chapter 2, every solid element has six degrees of freedom: three translations and three rotations. A *connection* or *restraint* restricts the relative motion between two elements in j_i directions, $i = 1$ to m, if there are n elements, one fixed on the machine base, and m connections. Then:

If $6(n - 1) - \sum_{i=1}^{m} j_i > 0$, the system has mobility and is a mechanism.
If $6(n - 1) - \sum_{i=1}^{m} j_i = 0$, the system is statically determinate.
If $6(n - 1) - \sum_{i=1}^{m} j_i < 0$, the system is statically indeterminate, or redundant.

One has to be careful in using the above criterion because some elements might be overconstrained so that the result might show less mobility than the actual one.

A *free body* is an assembly of solid elements or a solid element or part of it that is acted upon by a number of forces (see Example 3.3). These forces can be external forces, forces from connections to other elements, or body forces (i.e., weight). A *free body diagram,* a sketch showing the geometry and the forces that act on the rigid body, is very useful for force analysis of machines.

3.2.2. Force Analysis, Static Equilibrium

For every solid element, we can apply six equations of equilibrium of forces and moments, a total of $6(n - 1)$ equations if n is the number of solid elements and one of them is the ground element. There are also unknown forces of reaction along the connection constraints. For statically determinate systems, the number of equations of equilibrium is equal to the number of unknowns, and therefore the unknown forces can be found by a system of linear equations.

For mechanisms, additional input forces are needed and then the unknown forces can be computed as outlined above.

For redundant systems, additional *compatibility equations* for the displacements are needed. In machines, however, we very seldom design redundant systems, for two reasons:

1. Redundant systems generate very large internal stresses due to thermal expansion, a very common condition in machines.
2. Redundant systems also generate very large internal stresses due to manufacturing or assembly errors, also a common condition in machines.

Therefore we can assume that, in general, a machine is a statically determinate system or a mechanism possessing mobility.

To develop the force equilibrium equations, we first need to understand the concept of a *rigid body*. All materials in nature are deformable under force. In a statically determinate system, in most cases the deformations of machine members are very small as compared with the overall dimensions of the members themselves and do not affect the operation of the machine. One can see this in the opposite way: a machine is designed so that the deformations are not only significantly smaller than the dimensions of the machine members but smaller than operating clearances as well, because otherwise the operation of the machine would be affected. Therefore, we shall define a *rigid body* as a solid element whose change of shape under the forces it carries is very small and does not affect significantly the system geometry and force distribution. The latter is no true in the case of redundant systems where deformation can alter decisively the distribution of forces.

The first of Newton's laws of motion can be interpreted as that in order for a rigid body to be in steady motion, the external forces and moments have to be in equilibrium (Figure 3.2). It can be expressed as:

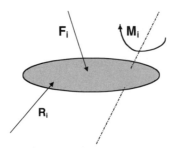

Figure 3.2. A free-body diagram of a solid element.

$$\sum_{i=1}^{m} F_i = 0, \quad \sum_{i=1}^{m} M_i + R_i \times F_i = 0 \qquad (3.24)$$

where F_i are the m external forces acting on the rigid body and $M_i + R_i \times F_i$ are the moments applied to it, where R_i is a vector from any point of reference to the line of application of the force F_i (Figure 3.2). The scalar form of Equations (3.24) are:

$$\sum_{i=1}^{m} F_{xi} = 0, \quad \sum_{i=1}^{m} F_{yi} = 0, \quad \sum_{i=1}^{m} F_{zi} = 0$$

$$\sum_{i=1}^{m} [M_{xi} + a_{xi}F_{xi}] = 0, \quad \sum_{i=1}^{m} [M_{yi} + a_{yi}F_{yi}] = 0, \quad \sum_{i=1}^{m} [M_{zi} + a_{zi}F_{zi}] = 0 \quad (3.25)$$

where a_{xi}, a_{yi}, a_{zi} are the distances of the projection of the point of reference to the projection of the forces F_i on the planes $x = 0$, $y = 0$, $z = 0$, respectively.

Force analysis is a systematic process and can be performed with the following algorithm:

Design Procedure 3.2: Force Analysis in General Determinate Systems

Step 1: Select a suitable coordinate system.

Step 2: Identify and number all individual $(n - 1)$ solid elements.

Step 3: Identify all m connections and the unknown forces of constraint associated with these connections. Assign arbitrary directions.

Step 4: Check if $6(n - 1) - ? = 0$. If > 0 (mechanism), identify the specified external loads that make the system statically determinate.

Alternative 1:

Step 5: Write down the $6(n - 1)$ equations of equilibrium of forces and moments, 6 for every solid element (in 4 and 5 use 3 instead of 6 for plane problems).

Step 6: Solve linear equations and determine unknown forces.

Alternative 2:

Step 5: In many machine design problems, it is possible to identify for a solid element two planes perpendicular to one another with their intersection coinciding with a line of symmetry (a straight beam or a shaft, for example) (Figure 3.3). In this case, we split the three-dimensional problem into two plane problems. To this end, we develop separate force diagrams in two separate planes xy and xz, taking in each plane the projections of all the lateral forces on that plane. Forces along the axis that coincide with the plane intersection need to be considered on one of the two planes only.

Step 6: We determine the components of the unknown reactions on the two planes separately by solving two plane equilibrium problems. If the vector force at the supports is needed, it will be the resultant of the reactions along the two planes.

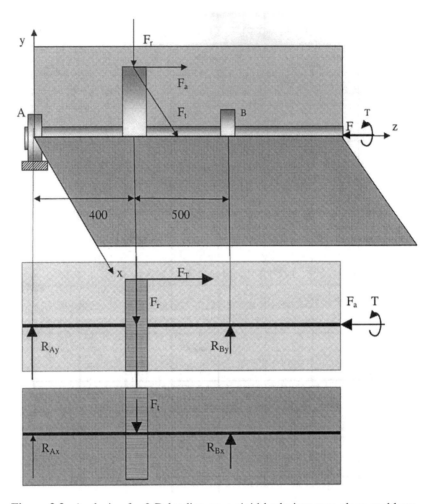

Figure 3.3. Analysis of a 3-D loading on a rigid body into two plane problems.

Example 3.3 A riveting machine is operated by a hydraulic cylinder and an angular lever, as shown in Figure E3.3. The riveting force $F_1 = 20$ kN and the diameter of the piston is $d = 100$ mm. If $L = 300$ mm and $H = 500$ mm, find the pressure p needed in the cylinder.

Solution We start first with a free-body representation of the system. To this end, we dis-assemble the system into solid elements and on each one we designate the applied forces. At the points of connections among solid elements, we assign (usually unknown) equal and opposite interaction forces. Newton's third law requires that the forces applied to two different elements at the common point of contact be equal and opposite. Thus, we represent the system with a number of compatible free-body diagrams. For each free-body, application of the equations of equilibrium will yield a sufficient number of equations to determine the unknown forces

Thus, for the riveting machine at hand, we have two solid elements, the piston and the angular bracket. We represent them separately by two free-body diagrams, Figure E3.3.

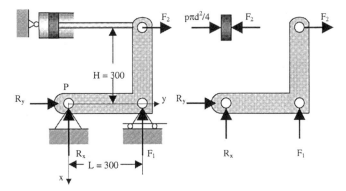

Figure E3.3

A simple way to solve the problem is to apply the equilibrium of the moments about the pivot P:

$$-F_1L + F_2H = 0, \quad F_2 = \frac{F_1L}{H} = 20{,}000 \times \frac{300}{500} = 12{,}000 \text{ N}$$

The force on the piston is:

$$P_A = p \times \frac{\pi d^2}{4} = F_2, \quad p = \frac{4F_2}{\pi d^2} = 4 \times \frac{12{,}000}{\pi \times 0.100^2} = 1.53 \times 10^6 \text{ Pa}$$

It should be observed that the unknown reactions R_x and R_y did not enter the calculations due to the lucky selection of the point P at the intersection of the unknown reactions to apply the moment equilibrium. This trick is used in many cases, but it is not always applicable.

To apply Equations (3.24) of force equilibrium:

1. We select the coordinate system (x, y) as shown.
2. We identify two solid elements, the body and the lever.
3. We identify one connection and two unknown forces of constraint associated with this connection. We assign arbitrary directions, say along the axes.
4. $6(n - 1) - \Sigma_{i=1}^{m} j_i = 6 \times (2 - 1) - 1 \times 5 = 1$ (mechanism). We identify the specified external load F_1 that makes the system statically determinate.
5. We write down the $3 \times (2 - 1) = 3$ equations of equilibrium of forces and moments for the solid element:

$$\sum F_x = R_x + F_1 = 0$$

$$\sum F_y = R_y + F_2 = 0$$

$$\sum M_P = -F_1L + F_2H = 0$$

6. These are the equations in three unknowns R_x, R_y, F_2. From the third, we find $F_2 =$

$F_1L/H = 20,000 \times 300/500 = 12,000$ N. From the second, $R_y = -F_2 = -12,000$ N. From the first, $R_x = -F_1 = -12,000$ N.

Example 3.4   Find the reactions at the bearings A and B in the system of Figure 3.3, if the radius of the gear is 300 mm and the forces shown are $F_t = 3$ kN, $F_r = 500$ N, and $F_a = 1000$ N.

Solution

1. The sum of the moments about bearing B are zero:

$$\sum M_B = R_{Ay} \times 900 + 1000 \times 300 - 500 \times 500 = 0$$

$$R_{Ay} = \frac{(-1000 \times 300 + 500 \times 500)}{900} = -55.556 \text{ N}$$

2. Equilibrium in the vertical direction gives:

$$R_{Ay} - F_r + R_{By} = 0, \quad -55.6 - 500 - R_{By} = 0, \quad R_{By} = 500 + 55.6 = 555.6 \text{ N}$$

Equilibrium at plane x–z (Figure 3.3c) demands that:

1. The sum of the moments about bearing B is zero:

$$\sum M_B = R_{ay} \times 900 - 3000 \times 500 = 0$$

$$R_{Ax} = \frac{(3000 \cdot 500)}{900} = 1667 \text{ N}$$

2. Equilibrium in the x direction gives:

$$-R_{Ax} - F_t + R_{Bx} = 0, \quad 1667 - 3000 + R_{Bx} = 0, \quad R_{Bx} = -1667 + 3000 = 1.333 \cdot 10^3$$

Therefore, the reactions are:

$$R_A = \sqrt{1667^2 + 55.6^2} = 1.668 \times 10^3 \text{ N}$$

$$R_B = \sqrt{556.6^2 + 1333^2} = 1.445 \times 10^3 \text{ N}$$

Example 3.5   On coupler 3 of a four-bar linkage shown in Figure E3.5, there is a vertical load at point C, $W = 1000$ N. On crank 2 a torque M is applied so that the linkage is in equilibrium. The link lengths are $r_1 = 1$ m, $r_2 = 0.2$ m, $r_3 = 0.5$ m, $r_4 = 0.7$ m, the distances $a = 0.18$ m, $b = 0.47$ m, and the input–output angles are $\phi = 80°$ and $\psi = 140°$. Find the unknown torque M and the forces that load the crank and the follower.

Solution   On a free-body diagram of the coupler, we sketch the forces acting on it. At A there is a general force (due to the torque), that we resolve into a radial (along $O_A A$) and tangential components T_A and R_A, respectively. At B there is a force along the direction of the follower R_B.

Equilibrium in the vertical and horizontal directions and moments about A give:

$$\sum F_y = W + T_A \cos \phi + R_A \sin \phi + R_B \sin \psi = 0 \qquad (a)$$

$$\sum F_x = -T_A \sin \phi + R_A \cos \phi + R_B \cos \psi = 0 \qquad (b)$$

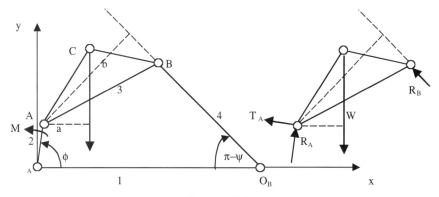

Figure E3.5

$$\sum M_A = Wa - R_B b = 0, \quad \text{or } R_B = \frac{Wa}{\mathbf{b}} \tag{c}$$

Substituting (c) into (a) and (b) and solving for T_A and R_A:

$$T_A = W\left(1 - \frac{a}{b}\sin\psi\right)\cos\phi + W\frac{a}{b}\cos\psi\sin\phi \tag{d}$$

$$R_A = W\left(1 - \frac{a}{b}\sin\psi\right)\sin\phi - W\frac{a}{b}\cos\psi\cos\phi \tag{e}$$

Substituting $T_A = 1000[1 - (0.18/0.5)\sin 140°]\cos 80° + 1000(0.18/0.5)\cos 140° \sin 80° = -138$ N:

$$R_A = 1000\left(1 - \frac{0.18}{0.5}\sin 140°\right)\sin 80° - 1000\frac{0.18}{0.5}\cos 140° \cos 80° = 805 \text{ N}$$

$$R_B = 1000\frac{0.18}{0.5} = 360 \text{ N}$$

Moment equilibrium about $O_A A$ for the crank gives $M = T_A r_2 = -138 \times 0.2 = -27.6$ Nm.

3.2.3. Force Analysis: Principle of Virtual Work

If a kinematic analysis of a machine has already been performed, the computation of the work-producing forces can be done easily on the basis of work power considerations. Indeed, conservation of energy demands that, neglecting the losses, for a mechanical system:

$$\sum_{i=1}^{m} F_i \cdot \Delta x_i + \sum_{i=1}^{m} M_i \cdot \Delta\phi_i = 0 \tag{3.26}$$

$$\sum_{i=1}^{m} F_i \cdot v_i + \sum_{i=1}^{m} M_i \cdot \omega_i = 0 \tag{3.27}$$

where F_i and M_i are forces and moments *external* to the system, Δx_i and $\Delta\phi_i$ are small displacements, termed *virtual,* that is, possible displacements that do not violate the kinematic constraints of the machine. Furthermore, v_i and ω_i are the velocities of the point of application of the forces and the angular velocities of the rigid bodies on which the moments are applied (and in the same direction). Equation (3.26) expresses the conservation of the mechanical work, Equation (3.27) the conservation of the rate of change of energy (power).

Example 3.6 Solve Example 3.3 using the kinematic method.

Solution The only external forces on the angle lever are F_1 and F_2. If the angular velocity of the lever is ω, then $v_1 = \omega R_1 = \omega L$, $v_2 = \omega R_2 = \omega H$, therefore, from Equation (3.27), observing that $F_1 v_1 = F_1 v_1 \cos \pi = -F_1 v_1$:

$$-F_1 v_1 + F_2 v_2 = -F_1 \omega L + F_2 \omega H = 0, \quad F_2 = \frac{F_1 L}{H} = 20{,}000 \times \frac{300}{500} = 12{,}000 \text{ N}$$

The force on the piston is:

$$P_A = \frac{p \pi d^2}{4} = F_2, \quad p = \frac{4 F_2}{\pi d^2} = 4 \times \frac{12{,}000}{\pi \times 0.100^2} = 1.53 \times 10^6 \text{ Pa}$$

3.2.4. Force Analysis: Dynamic Loads

Substantial inertia loads might be the result of variable speed elements in high-speed machinery. If this is the case, a thorough kinematic analysis is needed first to compute the displacement, velocity, and acceleration of every element of the machine. The inertia forces may be then computed by application of Newton's second law, which can be expressed as:

$$m \ddot{r} = \sum_{i=1}^{m} F_i \tag{3.28a}$$

$$J \ddot{\varphi} = \sum_{i=1}^{m} [M_i + R_i \times F_i] \tag{3.28b}$$

where F_i are the m external forces acting on the rigid body and $M_i + R_i \times F_i$ are the moments applied to it, R_i is a vector from any point of reference to the line of application of the force F_i (Figure 3.2), m is the mass of element, J is the mass moment of inertia tensor (see Beer and Johnston 1981), r is the position vector of the mass center, and ϕ is the rotation vector. In principle, if the accelerations are known from kinematic analysis, application of Equations (3.28) is straightforward. Since most designers have a better grasp of static analysis, the kinetostatic method is used, employing the D'Alembert principle in the form:

$$\sum_{i=1}^{m} F_i - m\ddot{r} = 0 \qquad (3.29a)$$

$$\sum_{i=1}^{m} [M_i + R_i \times F_i] - J\ddot{\varphi} = 0 \qquad (3.29b)$$

Equations (3.29) are equivalent to the static equilibrium Equations (3.24), if we apply at the center of mass of the rigid element D'Alembert forces and moments $-m_i\ddot{r}_i$, $-J_i\ddot{\varphi}_i$. This transforms a dynamic problem to a static one. For a plane problem, in particular:

$$\sum_{i=1}^{m} F_{xi} - m_i\ddot{x}_i = 0, \quad \sum_{i=1}^{m} F_{yi} - m_i\ddot{y}_i = 0, \quad \sum_{i=1}^{m} [M_i + a_iF_i] - J_{pi}\ddot{\varphi} = 0 \quad (3.30)$$

where x_i, y_i are the coordinates of the mass center of the element i, a_i is the distance of the force F_i from the mass center, and ϕ_i and M_i are the rotation and moment about the mass center of the element i.

Jean Le Rond d'Alembert (1717–1783). Abandoned as an infant at the doorsteps of a church, Le Rond managed to get a very broad education and collaborated with Diderot in the 28 volumes of the *Encyclopédie*. He played an important role in the ideological preparation of the French Revolution. At the age of 24 he was elected to the Académie des Sciences and in 1772 he became its permanent secretary. His principle appeared in his treatise *Traité de dynamique* in 1743. He made important contributions to many branches of mathematics and mechanics.

Example 3.7   To increase traction in the driving system of a locomotive, driving wheel 1 is in turn driving a second wheel 2 through a coupler *AB* (Figure E3.7a). The center of mass of coupler G is in the middle of it, the crank and follower lengths are 300 mm, the coupler length is 1.0 m, and the coupler mass is 100 kg and is uniformly distributed. Find the forces at bearings O_A and O_B for a full rotation at 1000 rpm rotating speed if for every wheel the transmitted traction force is constant $F_T = 150$ kN and the static load is 500 kN.

Solution   The angular velocity of the wheels is $\phi = 1000 \times 2\pi/60 = 104.7$ rad/s.

Point G has rotary motion about point O_G, in the middle of the distance $(O_A O_B)$. There is no tangential acceleration because the speed of rotation is constant and the radial acceleration is $a_r = \omega^2 R = 104.7^2 \times 0.3 = 3{,}290$ m/s². Thus, the inertia force is $ma_r = 100 \times 3{,}290 = 329{,}000$ N, directed along direction $O_G G$. Because $\angle GO_G O_B = \angle O_A O_B = \phi_2 - \omega t$, equilibrium of the rigid body demands that:

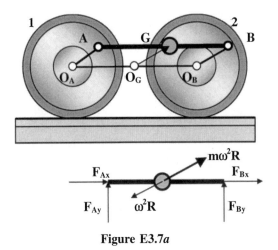

Figure E3.7a

$$\sum M_A = 1.0 \times B_y - m\omega^2 R \frac{AB}{2} \sin \phi_2 = 0$$

$$B_y = m\omega^2 R \frac{AB}{2} \sin \phi_2 = 164{,}500 \times \sin 104.7t = 164{,}500 \times \sin 104.7t$$

$$\sum F_y = F_{Ay} + F_{By} - m\omega^2 R \sin \phi_2$$

$$F_{Ay} = -F_{By} + m\omega^2 R \sin \phi_2 = -98{,}700 \times \sin 104.7t + 329{,}000 \times \sin 104.7t$$

$$= 164{,}500 \sin 104.7t \text{ N}$$

The horizontal component of the inertia force will be transferred to the bearing that is in the direction of that force. Thus:

$$F_{Ax} = m\omega^2 R \cos \phi_2 = 164{,}500 \times \cos 104.7t$$

$$F_{Bx} = m\omega^2 R \cos \phi_2 = 164{,}500 \times \cos 104.7t$$

The bearing force F is the vector sum of the transmitted traction force and the horizontal and vertical components due to the inertia force:

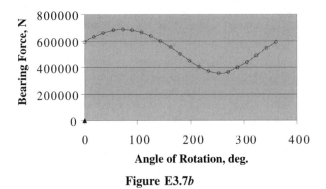

Figure E3.7b

$$F_A = [(F_{Ax} + F_T)^2 + (F_{Ay} + W)^2]^{1/2}$$

$$= [(164{,}500 \times \cos 104.7t + 150{,}000)^2 + (164{,}500 \times \sin^2 104.7t + 500{,}000)^2]^{1/2} \text{ N}$$

$$F_B = [(F_{Bx} + F_T)^2 + (F_{By} + W)^2]^{1/2}$$

$$= [(164{,}500 \times \cos 104.7t + 150{,}000)^2 + (164{,}500 \times \sin^2 104.7t + 500{,}000)^2]^{1/2} \text{ N}$$

This force is plotted in Figure E3.7*b*. One can see that the contribution of the inertia forces to the bearing load can be substantial.

3.3. INTERNAL LOADING OF MACHINE ELEMENTS

3.3.1. Equilibrium of a Sectioned Member

The external loads on a machine member are generally not directly related to member failure due to material failure. The latter is usually due to internal loads caused by the external loads the member carries. The determination of the internal loads and the stresses and strains they cause is the aim of *stress analysis.*

Every member of the machine that is fairly inflexible, that is, whose deformation is very small compared with the motions of the machine, can be considered a rigid body for the purpose of stress analysis. As we saw in the previous section, a rigid body diagram is drawn with all known external forces applied to the rigid body. Unknown applied forces, such as bearing reactions and forces transmitted from one member to another, are also drawn but with an unknown magnitude. Then the unknown forces are found by application of the equilibrium equations for each rigid body.

Figure 3.4 shows a rigid element sectioned into two parts, each represented by way of a free-body diagram. The sign convention is also shown in this figure. Forces in the direction of the *y* axis are positive, otherwise negative.

3.3.2. Internal Loading Diagrams

Once all external forces upon the rigid body, that is, the particular machine member, have been determined, the forces throughout the member are determined by the method of sections. If we assume a hypothetical cut of the member with a plane

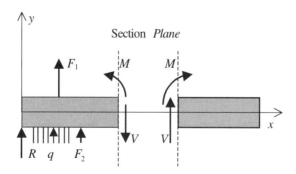

Figure 3.4. Sectioned member and sign convention for forces.

(dashed line) (Figure 3.4), to maintain equilibrium, we need to apply an appropriate force V and a moment M on the section. These forces are the *internal forces* on the member. Since equilibrium should be maintained, with the sign convention in Figure 3.4 (positive forces and distributed loads point upwards):

$$\sum_{i=1}^{m} F_i + V = 0, \qquad \sum_{i=1}^{m} [M_i + R_i \times F_i] + M = 0 \qquad (3.31)$$

Therefore:

$$V = -\sum_{i=1}^{m} F_i, \qquad M = -\sum_{i=1}^{m} [M_i + R_i \times F_i] \qquad (3.32)$$

Since, by virtue of Newton's third law, action equals reaction, the section forces on both sides of the section are equal in magnitude and opposite in direction. Therefore, we can apply Equations (3.32) for any one of the two parts in which the element is divided by the section, whichever is computationally more convenient.

Thus, at each point along the structural member of the machine, one can find the section shear force and bending moment and plot them. It is known from mechanics of deformable bodies (Beer and Johnston 1981) that equilibrium of a part between two infinitely close sections demands that the change in bending moment equal the local shear:

$$V = \frac{dM}{dx} \qquad (3.33)$$

Similarly, the change in shear equals the local distributed load per unit length q:

$$q = \frac{dV}{dx} = \frac{d^2M}{dx^2} \qquad (3.34)$$

Equations (3.33) and (3.34) show that:

1. When the external loads are distinct forces, the shear diagram between adjacent forces is a horizontal line and the moment diagram is a straight line (Equation (3.33)).
2. When the external load is distributed over the extent of the distributed load, the shear diagram is a straight line and the moment diagram is a parabola (Equation (3.34)).

For most problems, the designer can quickly draw internal force diagrams without resorting to extensive computations. For example, for the beam-like member of Figure 3.5:

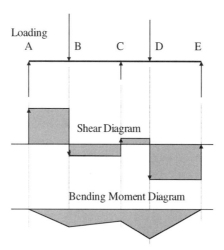

Figure 3.5. Internal loading diagrams.

1. The shear diagram is constructed by moving from left to right, and each time we encounter a force we move in the direction of this force at a distance equal to a certain scale with the force magnitude.

2. The bending moment diagram is constructed by first determining at every point of interest (here, at points where forces are applied) the bending moment of all forces on the left of that point, plotting these values and connecting the vertices with straight lines.

3. In a similar manner, we can construct the diagrams for the distribution of torques and axial forces, usually horizontal lines between points where torques or axial forces are applied.

4. If distributed loads are present, one can approximately lump them as a small number of distinct forces or use analytical methods, integrating Equations (3.33) and (3.34). This is encountered very seldom in machine design because external forces are usually much greater than body forces. The internal loading diagrams are very important in design because the designer can easily identify the maximum values of the internal loads and design the section properly. Moreover, one can identify input errors by inspecting the loading diagrams and avoiding design errors. For example, if bending is the dominant mechanism of failure, simple inspection shows that point D is the one most dangerous one, while one might make the wrong assumption that point C is the most dangerous because it is closer to the mid-span of the beam, this usually being the case.

In most problems all the external loads are not known, such as the support reactions in the case of loaded beams. For 3-D loading, it is usually simpler to work separately on two perpendicular planes such that their line of intersection coincides with the longitudinal axis of the member (Figure 3.3). Then on each plane there are generally three unknown reactions and three equilibrium equations. In this case, Internal Force Analysis Design Procedure 3.3 applies.

Design Procedure 3.3: Internal Loading Analysis of a General Element in Two Dimensions

Step 1: Select a suitable Cartesian coordinate system with the x axis along the length of the element and the planes (x, y), (x, z) on the planes of symmetry of the member, if possible.

Step 2: Identify all external forces and moments applied to the element. Find the components of these forces and moments along the two planes 1 and 2.

Step 3: Select plane 1. Identify all connections or supports and assign arbitrary directions to the unknown forces of constraint associated with these connections.

Step 4: Check if $6(n - 1) - ? = 0$. If >0 (mechanism), identify the specified external loads that make the system statically determinate.

Step 5: Write down the three equations of equilibrium of forces along x and y and moments about any point.

Step 6: Solve three linear equations and determine unknown forces.

Repeat the same for plane 2, steps 3–6.

Step 7: Compute internal loads at points of interest and draw force diagrams.

Example 3.8 A 3 kW, 3500 rpm electric motor (Figure E3.8) moves an intermediate shaft at 1750 rpm through a V-belt drive. This shaft moves a pump at 1000 rpm through a one-stage gear drive. The belt transmission is vertical, with the motor above the intermediate shaft, the angle of the two branches of the belt is 60°, the branch on the left has tension twice the tension of the branch on the right, and the gear transmission is horizontal with a

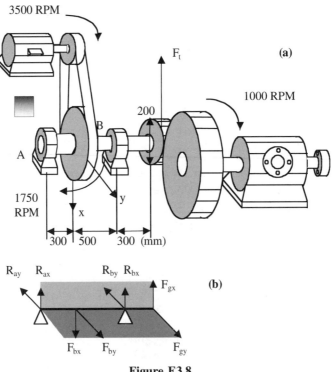

Figure E3.8

peripheral force F_t and radial force $F_t \tan 20°$. If the diameter of the pulley is 300 mm and the gear pitch diameter is 200 mm:

1. Find the torques in the different parts of the system.
2. Determine the bearing reactions on the intermediate shaft.
3. Draw torque, shear force, and bending moment diagrams.

Solution

Step 1: Select x—vertical, y—horizontal, z—along the shaft. Select planes xz and yz as shown.

Step 2: We now change all data to the SI system:

$$H_0 = 3 \times 1000 = 3000 \text{ W}$$

$$\omega_0 = 3500 \times \frac{2\pi}{60} = 366.5 \text{ rad/s}$$

$$\omega_1 = 1750 \times \frac{2\pi}{60} = 183.2 \text{ rad/s}$$

$$\omega_2 = 1000 \times \frac{2\pi}{60} = 104.7 \text{ rad/s}$$

Assume 95% efficiency for the belt drive and 97.5% for the gear drive. The power at the intermediate shaft is:

$$H_1 = 0.95 \times 3000 = 2850 \text{ W}$$

The torque at the shaft is:

$$T_1 = \frac{H_1}{\omega_1} = \frac{2850}{183.2} = \underline{15.5 \text{ Nm}}$$

The power at the pump shaft is:

$$H_2 = 0.975H_1 = 0.975 \times 2850 = 2779 \text{ W}$$

The torque is:

$$T_2 = \frac{H_2}{\omega_2} = \frac{2779}{104.7} = \underline{26.5 \text{ Nm}}$$

If F_1 and F_2 are the tensions of the two branches of the belt, it was given that $F_1 = 2F_2$ and the peripheral force at the belt is found from the torque equilibrium about the axis of rotation:

$$F_t = F_1 - F_2 = F_2 = \frac{T_1}{d/2} = \frac{2H_1}{\omega_1 d} = \frac{2 \times 2850}{183.2 \times 0.300} = 103.7 \text{ N}$$

Then, $F_1 = 2F_2 = 207.4$ N. These two forces have vertical and horizontal components

$$F_v = (F_1 + F_2)\cos 30° = (207.4 + 103.7)\cos 30° = 269.4 \text{ N}$$

$$F_h = (F_1 - F_2)\sin 30° = 75.6 \sin 30° = 51.9 \text{ N}$$

At the gear, the peripheral force is:

$$F_t = \frac{2H_1}{\omega_1 d} = \frac{2 \times 2850}{183.2 \times 0.200} = 155.5 \text{ N}$$

The radial force is:

$$F_r = F_t \tan 20° = 155.5 \tan 20° = 56.6 \text{ N}$$

Step 3: Referring to the free-body diagram (Figure E3.8*b*), the vertical external force components on the shaft are:

$$\text{Belt:} \quad F_{bx} = -269.4 \text{ N}$$

$$\text{Gear:} \quad F_{gx} = F_t = 155.5 \text{ N}$$

Step 4: There are two unknown reactions, R_{Ax} and R_{Bx}, and no axial force.

Steps 5 and 6: Moments about the y axis through A:

$$269.4 \times 0.300 - R_{bx} \times 0.800 + 155.5 \times 1.100 = 0$$

$$R_{bx} = \underline{314.8 \text{ N}}$$

$$\Sigma F_x = R_{ax} + R_{ax} + 269.4 + 155.5 = 0$$

$$R_{ax} = \underline{-110.1 \text{ N}}$$

Steps 3–6 are repeated for the horizontal forces:

Step 3: The horizontal external force components on the shaft are:

$$\text{Belt:} \ F_{by} = 51.9 \text{ N}$$

$$\text{Gear:} \ F_{gy} = F_r = 11.34 \tan 20° = 56.6 \text{ N}$$

Step 4: There are two unknown reactions, R_{Ay} and R_{By}, and no axial force.

Steps 5 and 6: Moments about A:

$$51.9 \times 0.300 - R_{by} \times 0.800 + 56.6 \times 1.1 = 0$$

$$R_{By} = \underline{-58.4 \text{ N}}$$

$$\Sigma F_y = 0: R_{Ay} + R_{By} - 51.9 + 56.6 = 0$$

$$R_{yA} = \underline{53.7 \text{ N}}$$

3.3.3. Computer Methods

Internal loading diagrams can be produced in a spreadsheet. To this end, Equations (3.33) and (3.34) are integrated along the length of the beam. To do this, we need to express the loads, generally not continuous functions, as functions of x, the coordinate along the length of the beam. We introduce the singularity function $\langle x - a \rangle^n$ (Macauley 1919) which takes the values:

$$x \text{ for } x \leq a \tag{3.35}$$

$$0 \text{ for } x > a$$

These functions can be integrated with the known integration rules. Thus, the following common loading conditions have the corresponding singularity functions and integrals:

Concentrated moment M at point $x = a$:

$$M \langle x - a \rangle^{-2}, \quad \int_{-\infty}^{x} M \langle x - a \rangle^{-2} \, dx = M \langle x - a \rangle^{-1}$$

Concentrated force F at point $x = a$:

$$F \langle x - a \rangle^{-1}, \quad \int_{-\infty}^{x} F \langle x - a \rangle^{-1} \, dx = F \langle x - a \rangle^{0}$$

Distributed force q starting at point $x = a$:

$$q \langle x - a \rangle^{-1}, \quad \int_{-\infty}^{x} q \langle x - a \rangle^{0} \, dx = q \langle x - a \rangle^{1}$$

From Equations (3.33) and (3.34):

$$V = \int_{0}^{L} q \, dx \tag{3.36}$$

$$M = \int_{0}^{L} V \, dx \tag{3.37}$$

Starting from the leftmost ($x = 0$) point in the beam, we express the external loads with singularity functions, and we integrate using Equations (3.36) and (3.37) and the properties above of the singularity functions. For example, for a simply supported beam of length L, with simple supports at $x = 0$ and $x = L$, with a transverse load F at mid-span, the differential equation for the moment is (Equation 3.34):

$$\frac{d^2 M}{dx^2} = q = -F \left\langle x - \frac{L}{2} \right\rangle^{-1} \tag{3.38}$$

Integrating twice:

$$M = -F \left\langle x - \frac{L}{2} \right\rangle^1 + c_1 x + c_2 \tag{3.39}$$

where c_1 and c_2 are constants of integration to be determined from the boundary conditions $M(0) = 0$ and $M(L) = 0$ (free ends, no bending moment there). Application to Equation (3.39) yields $c_2 = 0$ and $c_1 = F/2$. Therefore:

$$M = -F \left\langle x - \frac{L}{2} \right\rangle^1 + \frac{Fx}{2} \tag{3.40}$$

Equation (3.40) describes a triangular shape for the moment diagram with zero values at $x = 0$ and $x = L$, and maximum value $FL/4$ at the point of application of the force, as expected.

Example 3.9 Internal Loading Analysis of a Beam on a Spreadsheet
Use a spreadsheet to determine internal loading diagrams for the vertical loading of the shaft of Example 3.8.

Solution We shall use a direct numerical integration method instead of the singularity functions. We first divide the length of the beam in n small segments of length Δx_i (no need to be equal). We rewrite Equations (3.33) and (3.34) in the approximate form (replacing the differentials with small finite changes):

$$\Delta V_{i+1} = V_{i+1} - V_i = q_i \Delta x_i + F_i$$
$$\Delta M_{i+1} = M_{i+1} - M_i = V_i \Delta x_i + m_i$$

where F_i and m_i are concentrated forces and moments at station i, q_i the distributed load between stations i and $i + 1$. These equations can be rewritten as:

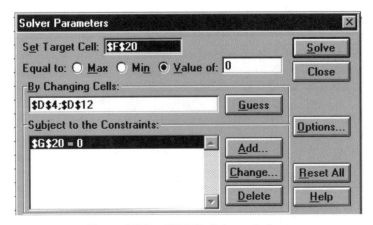

Figure E3.9a. EXCEL Solver window.

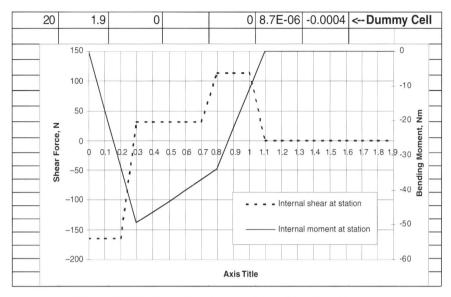

| 20 | 1.9 | 0 | | 0 | 8.7E-06 | -0.0004 | **<--Dummy Cell** |

Figure E3.9*b*. Spreadsheet-drawn internal loading diagram.

$$V_{i+1} = V_i + q_i \, \Delta x_i + F_i$$

$$M_{i+1} = M_i + V_i \, \Delta x_i + m_i$$

Successive application of the last two equations from left to right allows for the determination of shear forces and bending moments along the beam.

The shaded area of Figure 3.9*c* (columns *B–E*) is the input area where the user specifies distance of each station from the left end (column *B*), distributed load just right of the station (column *C*), external force at the station (column *D*), external moment at the station (column *E*).

At the support stations, we initially assign arbitrary reactions. Then the EXCEL Tools/ Solver has to be used to solve for the two unknown reactions by making zero the shear and bending moment at a dummy station on the right end of the beam, because in this case the boundary conditions demand a right end of the beam free of forces.

The Solver window appears and we specify to set the target cell F20 to zero (zero shear at the very right end of the beam) and by clicking Add we specify the second condition (as a constraint) G20 = 0 (zero bending moment at the very right end of the beam). Then we specify solution by changing cells D4:D12 (the unknown reactions) and hit Solve. Because the problem is linear, Solver always returns with a unique solution and updates the spreadsheet and the diagram (Figure E3.9*b* and *c*).

The reader can make changes in the geometry and loading and obtain the diagram for a particular problem by using Solver. All input cells need to be filled, with zero if there is no input for the particular cell. Because this spreadsheet uses a number of columns greater than the number of stations for plotting purposes, the user is advised to keep it that way and simply enter zeroes in the input cells for the artificial station.

Sometimes the scale might be such that shears and moments cannot be read on the same diagram. In this case, the user can copy and paste a second graph and then eliminate from the first graph the moment diagram and from the second the shear. EXCEL then adjusts the

	A	B	C	D	E	F	G	H	
2	Example 3.9								
3	BEAM INTERNAL LOADING DIAGRAMS								
4	INPUT					OUTPUT			
5	Station Number	Distance from left end	External distributed load at right	External force at station	External moment at station	Internal shear at station	Internal moment at station		
6	1	0	0	-165.3	0	-165.3	0		
7	2	0.1	0	0	0	-165.3	-16.53		
8	3	0.2	0	0	0	-165.3	-33.06		
9	4	0.3	0	196.4	0	31.124	-49.58		
10	5	0.4	0	0	0	31.124	-46.47		
11	6	0.5	0	0	0	31.124	-43.36		
12	7	0.6	0	0	0	31.124	-40.25		
13	8	0.7	0	0	0	31.124	-37.13		
14	9	0.8	0	82.276	0	113.4	-34.02		
15	10	0.9	0	0	0	113.4	-22.68		
16	11	1	0	0	0	113.4	-11.34		
17	12	1.1	0	-113.4	0	9E-06	-4E-04		
18	13	1.2	0		0	9E-06	-4E-04		
19	14	1.3	0		0	9E-06	-4E-04		
20	15	1.4	0		0	9E-06	-4E-04		
21	16	1.5	0		0	9E-06	-4E-04		
22	17	1.6	0		0	9E-06	-4E-04		
23	18	1.7	0		0	9E-06	-4E-04		
24	19	1.8	0		0	9E-06	-4E-04		
25	20	1.9	0		0	9E-06	-4E-04	← Dummy Cell	

Figure E3.9c

scale in each diagram automatically. Elimination of one of the two graphs can be made by clicking on the curve to be eliminated so that it will be highlighted and then hitting the Delete key.

The shear diagram has the transition lines across single forces not vertical but at a slightly inclined slope. This is due to the numerical method used and can be corrected if we take two stations very close to the point forces. However, since the results of engineering significance are correct, there is no need for this if the designer understands the reason for the slope. To check the results, we compute the maximum bending moment at point B (stations 4):

$$M_B = 165.27 \times 0.300 = 49.58$$

which agrees to four significant digits with the one calculated in the spreadsheet (cell G7).

One can use this example as template and simply change the data to obtain a solution to another problem. As one changes the data, EXCEL updates the results and the graph. Moreover, one can eliminate stations simply by deleting rows or add stations by adding rows. One should not delete the first and last rows, nor add a new row before the first or after the last row. If rows are added, the result cells above the added rows have to be dragged down to activate the result cells of the new rows. Finally, Solver needs to be always used to find the correct reaction forces.

3.4. DEFORMATION OF MACHINE ELEMENTS

3.4.1. Deformation of Simple Shapes

Deformation of machine members under loading is important in design for many reasons, such as:

1. Deformed members might not fit properly with mating parts, for example screws with nuts.
2. Deformed moving members might interfere with stationary components as they move—for example, rotating shafts in seals.
3. Easily deformable members might be an indication of instability, a situation that will be discussed in a later chapter.

Deformation of simple members, such as rods and simple beams, is discussed in detail in textbooks on the mechanics of deformable bodies. Some pertinent results are included in Table 3.1, where E is the modulus of elasticity and G the shear modulus of the materials.

Deformation for multiple loading of simple members can be found with the *principle of superposition:* the deformation of any point of a linear system under the action of many loads is the sum of all deformations due to each load acting alone.

3.4.2. Deformation of Complex Shapes

From the mechanics of deformable bodies (Beer and Johnston 1981) it is known that for small deformations, the lateral deflection y of the beam satisfies the equation:

$$\frac{d^2y}{dx^2} = \frac{M}{EI}, \qquad \frac{d\theta}{dx} = \frac{M}{EI} \tag{3.41}$$

where θ is the slope of the deflected beam dy/dx, M is the bending moment, I is the cross-sectional moment of inertia, E is the Young's modulus, and all can be functions of x. Therefore, using Equation (3.34):

$$
\begin{aligned}
y &= \int\left(\int \frac{M}{EI}\, dx\right) dx + c_1 x + c_2 \\
&= \int\left\{\int\left[\int\left(\int q\, dx\right) dx\right] dx\right\} dx + \int\left(\int c_1 x\, dx\right) dx \\
&\quad + \int\left(\int c_2\, dx\right) dx + c_3 x + c_4
\end{aligned}
\tag{3.42}
$$

where c_1, c_2, c_3, c_4 are integration constants that can be determined from the support

TABLE 3.1. Deformation of Simple Structural Members

Case	Schematic	Deformation
Uniform rod in tension of cross-sectional area A, length L		Elongation of the rod due to end force F: $$\Delta L = FL/AE$$
Uniform cylindrical rod in torsion, of cross-sectional area polar moment of inertia J_p, length L		Angle of twist of one end relative to the other due to end torsion T: $$\Delta\phi = TL/J_pG$$
Simply supported beam, constant sectional area moment of inertia I		Lateral deflection y at x due to: a) Concentrated lateral force F at B $y(x < a) = Fbx(L^2-b^2-x^2)/6LEI$, $y(x > a) = Fbc[L^2 - b^2 - (L - x)^2]/6EIL$ b) Distributed load q from A to C: $y = q(x^4 - 2Lx^3 + L^3x)/24EIL$. c) Bending Moment M at C: $y = M(x^3 - L^2x)/6EIL$
Cantilever beam, constant sectional area moment of inertia I		Lateral deflection y at x due to: a) Concentrated force F at $x = L$: $y = F(3Lx^2 - x^3)$ b) Distributed load q from A to B: $y = q(x^4 - 4Lx^3 + 6L^2x^2)/24EI$ c) Bending moment at B: $y = Mx^2/2EI$

(boundary) conditions. In the case of redundant beams, the reactions are treated as undetermined constants and the boundary conditions together with the equations of equilibrium are sufficient to determine the undetermined constants of integration and the unknown support reactions.

Deformation diagrams can be produced in a spreadsheet (see Example 3.9) or by direct integration. For the latter, Equation (3.42) can be used with appropriate definition of the forcing function q using singularity functions (Equation (3.35)). To do this, we need to express the loads, generally not continuous functions, as functions of x, the coordinate along the length of the beam.

Starting from the leftmost ($x = 0$) point in the beam, we express the external loads with singularity functions and integrate using Equations (3.42) and the properties of the singularity functions. For example, for a simply supported beam of length L, with simple supports at $x = 0$ and $x = L$, with a transverse load F at mid-span, the differential equation for the moment is (Equations (3.40) and (3.41)):

$$EI \frac{d^2y}{dx^2} = M = \frac{Fx}{2} - F \left\langle x - \frac{L}{2} \right\rangle^1 \tag{3.43}$$

Integrating twice,

$$EIy = \frac{Fx^3}{12} - F \left\langle \frac{x}{6} - \frac{L}{12} \right\rangle^3 + c_1 x + c_2 \tag{3.44}$$

where c_1 and c_2 are constants of integration to be determined from the boundary conditions $y(0) = 0$ and $y(L) = 0$. Application to Equation (3.44) yields $c_2 = 0$ and $c_1 = 3FL^2/48$. Therefore:

$$y = \frac{F}{6EI} \left[\frac{x^2}{2} - \frac{3L^2}{8} - \left\langle x - \frac{L}{2} \right\rangle^3 \right] \tag{3.45}$$

Example 3.10 Deflection Analysis of a Beam on a Spreadsheet

Use a spreadsheet to find the deflection for the vertical loading of the shaft of Example 3.8.

Solution We shall use a direct numerical integration method instead of the singularity functions. We first divide the length of the beam in n small segments of length Δx_i (no need for them to be equal). We rewrite Equations (3.41) in the approximate form (replacing the differentials with small finite changes):

$$\Delta \theta_{i+1} = \theta_{i+1} - \theta_i = \frac{M_i \, \Delta x_i}{EI_i}$$

$$\Delta y_{i+1} = y_{i+1} - y_i = \theta_i \, \Delta x_i$$

where M_i are the bending moments at station i. These equations can be rewritten as:

$$\theta_{i+1} = \theta_i + \frac{M_i \, \Delta x_i}{EI_i}$$

$$y_{i+1} = y_i + \theta_i \, \Delta x_i$$

Successive application of the last two equations from left to right allows for the determination of slopes and deflections along the beam.

This procedure is implemented in EXCEL, and the spreadsheet is shown in Figure E3.10a.

The input area (columns B–E) is where the user specifies distance of each station from the left end (column B), distributed load just right of the station (column C), external force at the station (column D), and external moment at the station (column E). At the support stations, we initially assign arbitrary reactions.

The Solver procedure must be used again for the deflection. To this end, we note that we have to start the spreadsheet with arbitrary slope and deflection at station 1. Deflection (cell J4) is zero because of the support. We assign an arbitrary slope (cell I4), say 1, at station 1. Then we write the recursive calculation formulas at cells I5 and J5 and then drag them down to the bottom of the spreadsheet. The deflection at bearing B ought to be 0, but in general it is not (cell J12) due to the arbitrary selection of the slope at station 1. Therefore, we use Solver and specify to set target cell J12 to zero (deflection at b) and target cells F16 and G16 also to zero (shear and moment zero past the right end; see Example 3.8). Then we specify solution by changing cells I4 (the unknown slope), D4, and D12 (the unknown

BEAM DEFLECTION

	A	B	C	D	E	F	G	H	I	J
			EXTERNAL LOADS			INTERNAL LOADS				
3	Station No	Distance from left end	Distributed load at right	Force at station	Moment at station	Shear at station	Moment at station	Flexural rigidity EI at station	Slope	Deflection
4	1	0	0	-165,3	0	-165,3	0	133596,228	0,00012	0
5	2	0,1	0	0	0	-165,3	-16,53	133596,228	0,00010763	1E-05
6	3	0,2	0	0	0	-165,3	-33,06	133596,228	8,2888E-05	2E-05
7	4	0,3	0	196,4	0	31,124	-49,58	133596,228	4,5775E-05	2E-05
8	5	0,4	0	0	0	31,124	-46,47	133596,228	1,099E-05	2E-05
9	6	0,5	0	0	0	31,124	-43,36	133596,228	-2,1464E-05	2E-05
10	7	0,6	0	0	0	31,124	-40,25	133596,228	-5,1588E-05	2E-05
11	8	0,7	0	0	0	31,124	-37,13	133596,228	-7,9383E-05	9E-06
12	9	0,8	0	82,276	0	113,4	-34,02	133596,228	-0,00010485	-1E-06
13	10	0,9	0	0	0	113,4	-22,68	133596,228	-0,00012183	-1E-05
14	11	1	0	0	0	113,4	-11,34	133596,228	-0,00013031	-3E-05
15	12	1,1	0	-113,4	0	9E-06	-4E-04	133596,228	-0,00013031	-4E-05
16	13	1,2	0		0	9E-06	-4E-04	133596,228	-0,00013031	-5E-05
17	14	1,3	0		0	9E-06	-4E-04	133596,228	-0,00013031	-7E-05
18	15	1,4	0		0	9E-06	-4E-04	133596,228	-0,00013032	-8E-05
19	16	1,5	0		0	9E-06	-4E-04	133596,228	-0,00013032	-9E-05
20	17	1,6	0		0	9E-06	-4E-04	133596,228	-0,00013032	-1E-04
21	18	1,7	0		0	9E-06	-4E-04	133596,228	-0,00013032	-1E-04
22	19	1,8	0		0	9E-06	-4E-04	133596,228	-0,00013032	-1E-04
23	20	1,9	0		0	9E-06	-4E-04	133596,228	-0,00013032	-1E-04

Figure E3.10a. Spreadsheet for constructing beam deflection diagram.

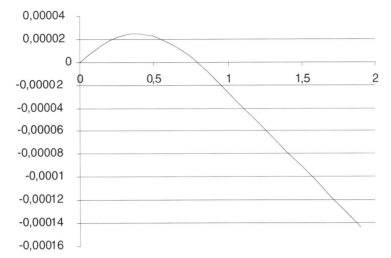

Figure E3.10*b*. Spreadsheet-drawn beam deflection diagram.

bearing reaction) and hit Solve. Because the problem is linear, Solver always returns with a unique solution and updates the spreadsheet and the diagram.

If the beam is redundant, there will be additional bearing reactions and an equal number of added constraints due to the additional bearing boundary conditions.

The reader can make changes in the geometry and loading and obtain the diagram for a particular problem by using Solver. All input cells need to be filled, with zero if there is no input for the particular cell. Because this spreadsheet uses a number of columns greater than the number of stations for plotting purposes, the user is advised to keep it that way and simply enter zeroes in the input cells for the artificial station.

The shear diagram is plotted with slanted lines at the force station due to the discretization. The larger the number of stations, the more the diagram approaches the correct one.

Sometimes the scale might be such that shears and moments cannot be read on the same diagram. In this case, the user can copy and paste a second graph and then eliminate from the first graph the moment diagram and from the second the shear. EXCEL then adjusts the scale in each diagram automatically. Elimination of one of the two graphs can be made by clicking on the curve to be eliminated so that it will be highlighted and then hitting the Delete key.

One can use this example as a template and simply change the data to obtain a solution to another problem. As one changes the data, EXCEL updates the results and the graph. Moreover, one can eliminate stations simply by deleting rows or add stations by adding rows. One should not delete the first and last rows, nor add a new row before the first or after the last row. If rows are added, the result cells above the added rows have to be dragged down to activate the result cells of the new rows.

CASE STUDY 3.1: George Furlow vs. Toledo Conveyors, Inc. and Belleville Machines, Inc.

The case:

In 1968, George Furlow, a farmer in Belleville, Illinois, purchased a grain conveyor for his farm to load grain in his truck from Toledo Conveyors, Inc. The grain was fed into a hopper and was conveyed by way of a rotating auger.

In 1981, Furlow acquired a new truck that had a taller carriage. He asked Belleville Machines, Inc., in Belleville, Illinois, if they could augment the conveyor to deliver grain at a higher elevation. The answer was affirmative, and Belleville Machines modified the conveyor, adding an extension of the auger and supporting structure.

Furlow testified that to secure the machine, he always pushed the conveyor forward against the truck so that it would rest on the carriage of the truck. After a few times the modified machine was used, and while Furlow's assistant was driving the truck away, the conveyor overturned to the right and seriously injured Furlow.

Investigation

By inspection, measuring and weighing, it was found that the weight of the original machine, with the auger filled with grain and the gas tank filled, was W_1 = 5500 N and the weight of the addition was W_2 = 1700 N. The mass centers of the two G_1 and G_2 are as shown in Figure CS3.1.

Let R_A and R_B the ground reactions at the two wheels A and B, respectively. Equilibrium of moments about point B yields:

$$2.3R_A - 0.8W_1 + 2.5W_2 = 2.3R_A - 0.8 \times 5500 + 2.5 \times 1800 = 0$$

Thus:

$$R_A = \frac{0.8 \times 5500 - 2.5 \times 1700}{2.3} = 150 \text{ N}$$

There is upwards-directed reaction on wheel A, meaning that the machine is

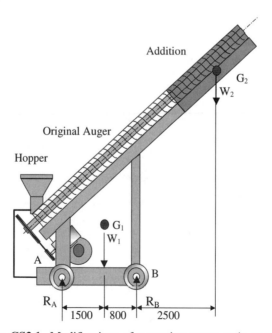

Figure CS3.1. Modifications of a rotating auger grain conveyor.

stable. The reaction of 150 N, however, indicates only marginal stability that can be affected by several uncertainties. For example, if only 150 N of grain (33 lb) remain in the extension of the auger, the reaction will be:

$$R_A = \frac{0.8 \times 5500 - 2.5 \times 1700 - 2.5 \times 150}{2.3} = -13 \text{ N}$$

and the conveyor now becomes unstable. Amount of gas in the gas tank, leveling, and other factors can adversely affect stability because it is so marginal.

The extension of the conveyor was based on a defective and unsafe design. Since the mass center G_1 of the original machine was well within the wheel span, the original design was reasonably safe.

REFERENCES

Beer, F. P., Johnston, E. R. 1981. *Mechanics of Materials.* New York: McGraw-Hill.

Macauley, W. H. "Note on the Deflection of Beams." *Messenger of Mathematics* 48 (1919): 129–30.

ADDITIONAL READINGS

American Society for Metals (ASM). "Failure Analysis and Prevention." In *Metals Handbook,* 8th ed. Materials Park, Oh.: ASM, 1975.

Baumeister, T., ed. *Marks' Standard Handbook for Mechanical Engineers,* 7th ed. New York: McGraw-Hill, 1967.

Boresi, A. P., et al. *Advanced Mechanics of Materials.* New York: John Wiley & Sons, 1978.

Dudley, D. W., ed. *Gear Handbook.* New York: McGraw-Hill, 1962.

Popov, E. P. *Mechanics of Materials,* 2nd ed. Englewood Cliffs, N.J.: Prentice-Hall, 1976.

Roark, R. J., and Young, W. C. *Formulas for Stress and Strain,* 5th ed. New York: McGraw-Hill, 1975.

PROBLEMS[2]

3.1. [C] An electric motor has speed 3500 rpm and power $H = 5$ kW. It transmits the motion to a shaft of a lathe that has speed of rotation 1200 rpm.

1. Calculate the torques on the motor shaft and the shaft of the lathe.

2. Draw a kinematic diagram showing the speeds and torques in the system.

3.2. [C] An electric motor has speed 1750 rpm. It is directly connected to a shaft that transmits motion through two different belt drives to two oil pumps, the first, which is located closer to the motor, at 800 rpm and 3.5 hp, the second at 600 rpm and 8 hp.

---

[2][C] = certification, [D] = design, [N] = numerical, [T] = theoretical problem.

1. Calculate the torques on the motor shaft and the shafts of the lathe.

2. Draw a kinematic diagram showing the speeds and torques in the system.

3.3. [C] An automotive engine has maximum output torque 250 Nm at 3500 rpm speed. A four-forward-speeds gearbox is used with the shaft connected with the engine transmitting motion to an intermediate shaft through two selectable gear pairs of gear ratio 3.8 and 5.5. The intermediate shaft transmits motion to the exit shaft which is connected with the differential through two selectable gear pairs of gear ratio 3.8 and 4.4.

1. Calculate the torques on the motor shaft, the intermediate shaft, and the exit shaft at all possible speeds.

2. Draw a kinematic diagram showing the speeds and torques in the system.

3.4. [C] An electric motor has speed 3500 rpm and power $H = 6$ kW. It transmits the motion to an intermediate shaft that has speed of rotation 1750 rpm through a V-belt drive. This shaft in turn powers four textile spindles of 1.5 kW approximately each through gear boxes of gear ratio 3.2.

1. Calculate the torques on the motor shaft and the shaft of the lathe.

2. Draw a kinematic diagram showing the speeds and torques in the system.

3.5. [D] A ship propulsion steam turbine has speed of rotation 6000 rpm and the propeller shaft has speed 200 rpm. The gear speed reducer consists of a number of intermediate shafts and pairs of gears to successively reduce the speed from 6000 to 200 rpm. The gear ratio, for efficiency reasons, should not exceed 4.

1. Select number of speed reduction steps assuming that all gear ratios are the same.

2. Calculate the torques on all shafts

3. Draw a kinematic diagram showing the speeds and torques in the system.

3.6. [C] The mechanism shown in Figure P2.17 is a crane that is designed to deliver a nearly straight horizontal trajectory of the coupler point P. Assuming that the linkage is drawn to scale with $A_0B_0 = 2$ m, that a vertical load 5 kN is applied at point P, and that a moment is applied on link A_0A about point A_0 to assure equilibrium of the crane:

1. Find, at the position shown, the applied moment and bearing reactions at pivots A_0 and B_0.

2. Draw rigid-body diagrams for all links and find the forces transmitted through the revolute joints.

3.7. [C] The dump mechanism of the dump truck shown in Figure P2.19 is equivalent with a four-bar linkage. Assuming that the drawing is to scale, that the distance between the centerline of the wheels is 1.5 m, and that the dumping platform is loaded with a 100 kN vertical load at point A, find the force of the hydraulic cylinder at the position.

3.8. [C] The PUMA robot shown in Figure P2.21 is loaded at point C with a 5 kN weight. Assuming that the joint motors apply sufficient torques on the respective arms for equilibrium:

1. Draw free-body diagrams for all rigid links of the robot.

2. Calculate all joint loads at the positions shown.

3.9. [C] A four-bar linkage as in Figure 2.3a has link lengths r_1 = 450 mm, r_2 = 150 mm, r_3 = 250 mm, r_4 = 300 mm. Initially the angle ϕ_2 is 60°. Torque T_2 = 100 Nm is applied on link 2, and another torque T_4 is applied on link 4 in counterclockwise direction about pivots O_A and O_B, respectively, so that the linkage will be in equilibrium. Find:

1. Torque T_4
2. The reactions at pivots O_A and O_B
3. The loads on coupler link 3

3.10. [C] A four-bar linkage as in Figure 2.3a has link lengths r_1 = 450 mm, r_2 = 150 mm, r_3 = 250 mm, r_4 = 300 mm. Initially angle ϕ_2 is 60°. The coupler AB is replaced by a triangle ABC having three equal sides (AB) = (BC) = (CA), with point C above the line AB, and horizontal force F = 2 kN is applied on point C with direction from left to right. Torque T_2 is applied on link 2 about pivot O_A in a counterclockwise direction to counterbalance force F so that the mechanism will be in equilibrium. Determine:

1. Torque T_2
2. The reactions at pivots $A_A A$ and $O_B B$
3. The loads on coupler link 3

3.11. [C] The mechanism shown in Figure P2.17 is a crane that is designed to deliver a nearly straight horizontal trajectory of coupler point C. Assuming that the linkage is drawn to scale with $A_0 B_0$ = 2 m, that a 50 kN weight is applied at point P, and that a moment is applied on link $A_0 A$ about point A_0 so that link $A_0 A$ has constant angular velocity 1.5 rad/s at the position shown:

1. Find the dynamic load on the weight at P.
2. Find the bearing reactions at pivots A_0 and B_0.
3. Draw rigid-body diagrams for all links and find the forces transmitted through the revolute joints.

3.12. [C] The dump mechanism of the dump truck shown in Figure P2.19 is equivalent to a four-bar linkage. Assuming that the drawing is to scale, that the distance between the centerline of the wheels is 1.5 m, and that the dumping platform is loaded with a 100 kN weight at point A, find the force of the hydraulic cylinder in the position shown, if the angular acceleration of the damping platform is 1 rad/s^2 and the angular velocity is 2 rad/s.

3.13. [C] The PUMA robot shown in Figure P2.21 is loaded at point D with a 5 kN weight. Assuming that the joint motors apply sufficient torques on the respective arms for equilibrium and that the angular velocity of rotation about the vertical axis of rotation is constant 6 rad/s:

1. Calculate the dynamic load at the weight.
2. Draw free-body diagrams for all rigid links of the robot, assuming that inertia loads from links and pivot motors are negligible.
3. Calculate all joint loads at the positions shown.

3.14. [C] A four-bar linkage as in Figure 2.3a has link lengths r_1 = 450 mm, r_2 = 150 mm, r_3 = 250 mm, r_4 = 300 mm. Initially angle ϕ_2 is 60°. Torque T_2 = 100 Nm is applied on link 2 in a counterclockwise direction about pivot O_A. If the coupler is a bar of uniform mass m = 300 kg, and the crank and follower have negligible mass, find at the position shown:

1. The angular accelerations α_2, α_3, α_4.
2. The reactions at pivots A_A and O_B.
3. The loads on coupler link 3.

3.15. [C] A four-bar linkage as in Figure 2.3a has link lengths r_1 = 450 mm, r_2 = 150 mm, r_3 = 250 mm, r_4 = 300 mm. Initially angle ϕ_2 is 60°. Coupler AB is replaced by a triangle ABC having three equal sides $(AB) = (BC) = (CA)$, with point C above line AB, a mass m = 200 kg is fixed on point C, and the masses of the links are negligible. Torque T_2 is applied on link 2 about pivot O_A in a counterclockwise direction to counterbalance the weight mg and make the mechanism move. If, at the position shown, the angular velocity of the crank ω_2 = 40 rad/s and the angular acceleration α_2 = 20 rad/s², determine:

1. Torque T_2
2. The reactions at pivots A_AA and O_BB
3. The loads on follower link 3.

3.16. [C] The crank of a bicycle, shown in Figure P3.16, is loaded at the horizontal position with a vertical force F = 1000 N. The length of the rectangular cross-section bar is 150 mm and the total length of the circular cross-section bar is 80 mm. Draw internal loading diagrams throughout the crank.

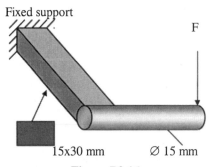

Fixed support

F

15x30 mm   Ø 15 mm

Figure P3.16

3.17. [C] The structure of the portal crane shown is loaded by the payload plus the crane hardware as shown, with F = 20 kN, L = 2 m, H = 2.5 m. Draw internal loading diagrams for the beam and the columns. Assume that friction at the floor is negligible, and thus the reactions at the floor are vertical only.

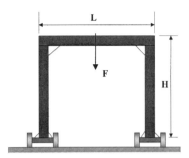

Figure P3.17

3.18. [C] Figure P3.18 shows a riveting machine that is loaded at the riveting head with a maximum force $F_1 = 20$ kN that acts between the riveting head and the base plate. Draw the bending moment and shear diagrams for the U-shaped main structure of the machine.

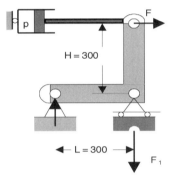

Figure P3.18

3.19. [C] For the railroad axle shown, $l = 1.2$ m, $d_1 = 80$ mm, $d_2 = 100$ mm, $d_3 = 110$ mm, $d_4 = 130$ mm, and the rail reaction at the rim can be assumed at the middle of the hub length (100 mm). Draw the internal loading diagrams.

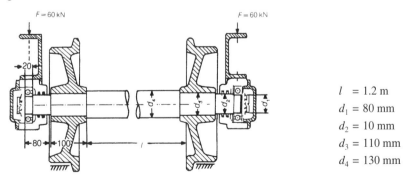

$$l = 1.2 \text{ m}$$
$$d_1 = 80 \text{ mm}$$
$$d_2 = 10 \text{ mm}$$
$$d_3 = 110 \text{ mm}$$
$$d_4 = 130 \text{ mm}$$

Figure P3.19

3.20. [C] The actuation arm of a pad/drum brake is actuated by force $F = 500$ N at the right end, as shown. Draw the internal loading diagrams for the arm, assuming that the connection of the arm to the pad is through a revolute joint that can be assumed to transmit only vertical forces.

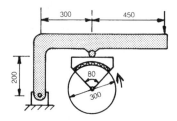

Figure P3.20

3.21. [C] A shaft of diameter $d = 60$ mm has a gear of radius $r = 200$ mm. At the gear there is a radial force $R = 3$ kN and an axial force $F = 2$ kN. The shaft transmits 10 kW at 3000 rpm.

1. Find the reactions at the two bearings A and B.
2. Draw shear force and bending moment diagrams.

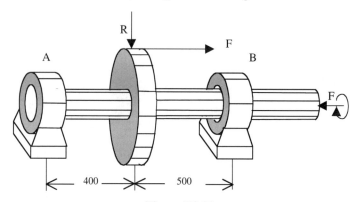

Figure P3.21

3.22 to 3.30. [C] The shaft of Figure P3.22 rotates with angular velocity ω(rad/ s). It is driven from left with power W (kW) that is transmitted partly through two pulleys 1 and 2, shown only schematically, powers P_1 and P_2 respectively, and the remaining power is transmitted to a driven machine at the right end with a very flexible coupling. Assuming no shear forces and bending moments transmitted to left and right ends,

TABLE P3.1

Problem	ω, rad/s	P_1 (kW)	P_2 (kW)	F_{r1} (N)	F_{r2} (N)	F_{t1} (N)	F_{t2} (N)	F_{a1} (N)	F_{a2} (N)
P3.22	200	4	4	1000	2000	0	0	0	0
P3.23	250	4	2	1000	2000	2000	2000	0	0
P3.24	300	2	0	2000	2000	2000	2000	0	0
P3.25	350	2	4	0	0	0	0	0	0
P3.26	200	2	4	2000	2000	0	0	0	0
P3.27	250	2	4	0	0	2000	2000	1500	1000
P3.28	300	6	0	3000	1000	0	0	0	0
P3.29	350	4	4	1000	2000	0	0	0	0
P3.30	400	4	2	1000	2000	2000	2000	1500	1000

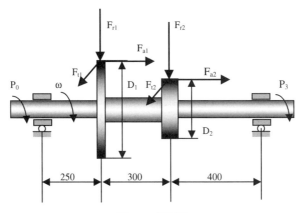

Figure P3.22

1. Determine the bearing reactions,
2. Draw internal force diagrams, torque, shear force, axial force, and bending moment.

The data are given in Table P3.1. F_{t1} and F_{t2} are lateral horizontal forces in the direction perpendicular to the plane of paper at pulleys 1 and 2, F_{a1} and F_{a2} are axial forces, shown in Figure P3.47. For all problems, P_0 = 8 kW, ω = 100 rad/s, D_1 = 250 mm, D_2 = 180 mm.

CHAPTER 4

MACHINE DESIGN MATERIALS
AND MANUFACTURE

An I-beam emerges red-hot from a rolling mill.

OVERVIEW

In Chapter 2 we saw that often many of the main dimensions of a machine can be found using purely kinematic considerations before it is necessary to discuss materials and their ability to carry loads.

In Chapter 3 we found the forces that are loading the machine components. Since one of the main reasons for failure of machine members is the limited capacity of materials to carry loads, before we assess this capacity we need to specify the materials to be used. Moreover, because the properties of the materials are sometimes dependent on the manufacturing process used for the part, we need to specify that process for every part. These will be the subjects of the present chapter.

4.1. PROPERTIES OF MATERIALS FOR MACHINE DESIGN

4.1.1. The Tensile Strength Test

Capacity of a machine component is related to the most severe condition it can sustain without a change that will prevent the component from continuing its intended function. In most cases, sustaining loads is the main manifestation of capacity. To assess the load-carrying capacity of a machine component, the maximum unit load (stress) has to be compared with the appropriate material property. *Stress* is the *force density,* defined as $\lim(\Delta F/\Delta A)$ over a small part of the area cross-section ΔA as $\Delta A \rightarrow 0$, while strain is the elongation ratio $\Delta L/L$. Depending on the orientation of the transmitted force in respect to the cross section under examination, stress can be *normal* σ (force perpendicular with the cross section) or *shear* τ (force parallel to the cross section), while normal stress can be *tensile* σ_i or *compressive* σ_c. Finding stresses in a machine component under service loading will be discussed in a later chapter. Here we shall discuss the capacity of engineering materials to sustain service loads.

Thomas Young (1773–1829). Born in Somerset to a Quaker family. Fluent in Latin, Greek Arabic, Persian, and Hebrew at age 14, he studied medicine and did research in the theory of light and sound. In 1802, he was elected to the Royal Society and installed as professor of natural philosophy (physics) at the Royal Institution, where he continued his research on mechanics of materials. He resigned the following year and his short tenure was considered a failure. Lord Rayleigh remarked, "Young . . . from various causes did not succeed in gaining due attention from his contemporaries. Positions which he had already occupied were in more than one instance re-conquered by his successors at a great expense of intellectual energy."

A most informative material test is the simple tensile test. A specimen in the form of a cylindrical bar, machined to a certain specification as shown in Figure 4.1, is *slowly* loaded in a tensile testing machine and the axial tensile load F and displacement (extension) ΔL over a length L are recorded. The resulting load–displacement curve is shown in Figure 4.2 for a typical low-carbon (also termed *mild*) steel. Assuming constant cross section of the rod, the same curve relates stress $\sigma = F/A$ and strain $\varepsilon = \Delta L/L$.

Figure 4.1. Tensile strength test specimen.

The initial part of the curve is a straight line, meaning that stress and strain are related by way of *Hooke's*[1] *Law* $\sigma = \varepsilon E$, where the constant of proportionality E is termed *modulus of elasticity* or *Young's modulus*.

The curve of this figure does not correspond exactly to the real stress–strain relationship because of the way it is made; that is, we measure force and displacement and interpret them as stress and strain based on the initial length and the initial cross section of the test specimen. These properties continuously change during the experiment, and therefore the results of such a test will have only a conventional value.

From this diagram, which is typical for most materials used in machine design, we observe the linear stress–strain relationship $\sigma = \varepsilon E$, which extends up to some stress S_p, and shortly thereafter we can observe an increasing deformation without a proportional increase in the load and the stress. This roughly corresponds to the point at which we start having an appreciable yielding of the material, and we call this the *yield point*. The corresponding stress (S_y) is called the *yield strength* of the material.

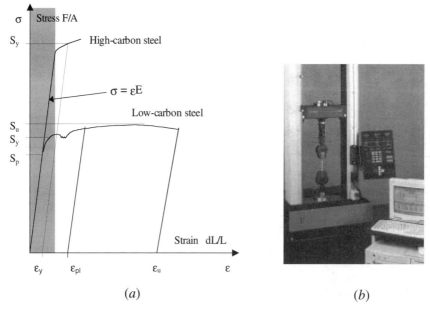

(a)   (b)

Figure 4.2. (*a*) Stress–strain diagram for steel; (*b*) Instron materials-testing machine.

---

[1] Robert Hooke (1635–1703) was Curator of Experiments at the Royal Society and a professor of geometry at Gresham College.

We have been intentionally vague in precisely defining the yield point because in most materials it is not possible to identify a single point where the transition occurs from elastic to plastic behavior. In fact, in most engineering materials, this transition is not abrupt and it is a matter of definition or convention to specify the yield point. Usually we define the yield point as the point where a certain percentage of plastic deformation remains after loading, usually 0.2% or 0.5%, and designate it as $S_{0.2}$ or $S_{0.5}$, respectively.

Further loading of the specimen results in *rupture,* physical separation into two parts. *Ultimate tensile strength* (S_u) is the maximum nominal stress that can be observed into the stress–strain diagram before rupture, which corresponds to the maximum nominal stress that the material can sustain, the ratio of maximum load to the original cross-sectional area.[2]

Materials with substantial plastic deformation before rupture, such as low-carbon steel (Figure 4.2) are termed *ductile,* while materials with minimal plastic deformation before fracture, such as high-carbon steel (Figure 4.2) are termed *brittle.* In ductile materials there is a substantial difference between yield and tensile strength. In brittle materials the difference between these two values is usually small (Figure 4.2).

An important measure of ductility is the energy of plastic deformation for rupture. Since the product *force* × *displacement* is energy, the product $\sigma\varepsilon$ is energy per unit volume. Therefore, the area under the stress–strain curve is energy per unit volume, (mostly) energy of elastic deformation before the yield point, and energy of plastic deformation beyond the yield point (Figure 4.3):

$$u_e \approx \frac{1}{2} S_y \varepsilon_y = \frac{S_y^2}{2E} \tag{4.1}$$

$$u_p \approx S_u \varepsilon_u \tag{4.2}$$

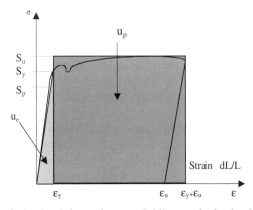

Figure 4.3. Energy of elastic deformation at yielding and plastic deformation energy at rupture for mild steel.

---

[2] We shall denote stress properties of the materials with capital Latin letters, i.e., S_{xx}, and actual stresses due to loads in machine components with lowercase Greek letters, i.e., σ_{xx}, τ_{xx}.

Thus, for a specimen of cross-sectional area A and length L, the energy of elastic deformation at yielding and the energy of plastic deformation at rupture are:

$$U_c = \frac{1}{2}(S_y A)(\varepsilon_y L) = \frac{S_y^2}{2E} \tag{4.3}$$

$$U_p = (S_y A)\Delta l = \frac{S_y V \delta}{100} \tag{4.4}$$

where δ is the percent plastic deformation (elongation) at rupture, $\approx 100\varepsilon_u$.

In most materials the strength is the same in tension and in compression. Some materials however, such as cast iron, have very different values of strength in tension and compression. As we shall see later (Chapter 5), micro-cracks exist in the structure of this material and give rise to high-stress concentration during tensile loading, while in compressive loading, for geometrical reasons, these micro-cracks are ineffective and therefore the material can sustain much higher loads. In such materials the strength in tension and in compression have to be recorded independently.

In material property tables, especially for design purposes, we will also observe different strengths in tension and bending. We might observe similar differences in shear and torsion. Although it appears that no matter what loads cause the stresses, the strength to a particular type of stress must be the same, this is not always the case. Take, for example, the strength in pure tension and in bending. In both cases the direction of stress is the same, namely tensile stress. In flexure, however, only the outer fibers of the material have high stress, while the stresses diminish as we move towards the neutral line. Since, as shall be seen later (Chapter 5), the micro-cracks already mentioned are uniformly distributed in the material, the probability of having a micro-crack in the area of high stress is smaller in flexure than in pure tension, and therefore the strength in flexure is generally greater than in pure tension. A similar situation exists for strengths in torsion and direct shear.

As already mentioned, the yield point in ductile materials is usually well defined. In cases where there is no pronounced yield point in the diagram, the yield strength is defined as the stress at which the permanent set $\varepsilon_y = 0.002$ or 0.2% (Figure 4.2). In some cases the yield strength is established for $\varepsilon_y = 0.5\%$. To distinguish between the yield point in tension and in compression, an additional subscript t or c is introduced in the notation when it is necessary in some materials. Thus, we obtain the symbols S_{yt} and S_{yc} for the yield point.

Another deviation of the stress–strain curve from reality must be pointed out: the horizontal scale is usually arbitrarily nonlinear because elastic deformation at small strains is very small compared with plastic deformation. For this reason, the strain scale is enlarged for small strains (shaded region in Figure 4.2). If yielding occurs, unloading of the specimen will result in returning not to the origin but to a point at the abscissa at distance ε_{pl} from the origin (Figure 4.3). This is the plastic deformation. The slope of the line of return will be the same with the one stress–strain line during loading. Subsequent loading will follow the new line (Figure 4.4). In this case we note that the yield point will be higher. This is termed *work hardening* or *strain hardening* because plastic deformation has increased the yield

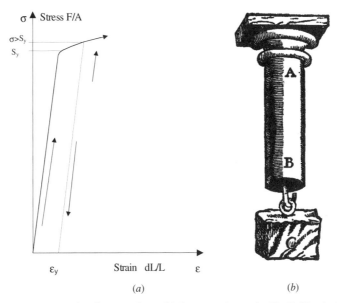

Figure 4.4. (*a*) Stress–strain diagram for a high-strength steel; (*b*) Galileo's illustration of the tensile test.

strength of the material. This property is extensively used in improving the strength of ductile materials.

The yield point is one of the main mechanical characteristics of a material in machine design, where yielding is mostly unwanted, even if the structural integrity of the component is maintained, because of geometric or other implications of yielding that might affect the machine operation.

As noted above, the ratio of the maximum force that the specimen is capable of sustaining to its original cross-sectional area is termed *the ultimate tensile strength* and is denoted S_u. It is important to note that S_u is not the stress at which the specimen fractures. If the tensile force is referred to the minimum section at a given point in time rather than to the original cross-sectional area of the specimen, it may be observed that the average stress on the narrowest section of the specimen before rupture is appreciably greater than S_u. Thus, the ultimate tensile strength is also a conventional quantity. Owing to the convenience and simplicity of its determination, the ultimate tensile strength is widely used in design practice as a basic comparative characteristic of the strength properties of materials. However, for machine component design sizing calculations, the yield strength is used because yielding of the machine parts usually has adverse effects on machine operation.

Another important characteristic of a material design is determined from tension tests, *the percentage elongation at rupture,* which is the average permanent deformation produced in a specified standard length of the specimen at the moment of rupture. The ability of a material to acquire large permanent deformations without fracture is known as *ductility.* The property of ductility is of prime importance in such manufacturing processes as extrusion, drawing, and bending. The measure of ductility δ is the percentage elongation at rupture $\delta = 100\varepsilon_u$ (Figure 4.3). The greater δ is, the more ductile is the material. Highly ductile materials include annealed copper, aluminum, brass, and low-carbon steel. Duralumin and bronze are less ductile. Low-ductility materials include many alloy steels, cast iron, and ceramics.

A property opposite to ductility is brittleness, that is, the tendency of a material to fracture without any appreciable permanent deformation. Materials possessing this property are called *brittle*. For such materials, the amount of elongation at rupture does not exceed 2–5%, and in some cases it is expressed by a fraction of 1%. Brittle materials include cast iron, high-carbon tool steel, glass, brick, and stone. The tension test diagram for brittle materials has no yield point or work hardening zone (Figure 4.4).

Ductile and brittle materials behave differently in compression tests as well. The compression test is conducted on short cylindrical specimens placed between parallel plates. The compression test diagram for low-carbon steel has a shape such as represented in Figure 4.5. Here, as for tension, the yield point can be observed with subsequent transition to the zone of strain hardening. Thereafter, however, the load does not fall, as in tension, but increases abruptly. This is due to the fact that the cross-sectional area of the compressed specimen increases; the specimen takes a barrel-like shape owing to friction at the ends (Figure 4.5). It is practically impossible to bring the specimen of a ductile material to fracture. The test cylinder is compressed into a thin disk (see Figure 4.5), and further testing is limited by the capacity of the machine. Hence, the ultimate compressive strength cannot be found for materials of this kind.

Brittle materials also behave in a different way in compression tests. The compression test diagram for these materials retains the qualitative features of the tension test diagram. The ultimate compressive strength of a brittle material is determined in the same way as in tension. The fracture of the specimen occurs with cracks forming on inclined or longitudinal planes (Figure 4.6).

A comparison of the ultimate tensile strength S_{ut} and the ultimate compressive strength S_{uc} of brittle materials shows that these materials possess, as a rule, higher strength in compression than in tension. The magnitude of the ratio $\kappa = S_{ut}/S_{uc}$ for

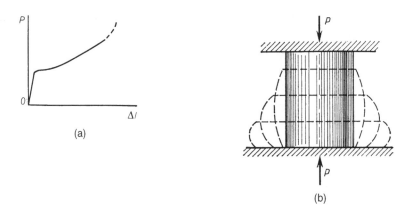

Figure 4.5. (*a*) Stress–strain diagram for a ductile material in compression; (*b*) Compression of ductile materials (after Feodosyev 1973).

Figure 4.6. Compression of brittle materials (after Feodosyev 1973).

cast iron ranges from 0.2 to 0.4. For ceramic materials, $\kappa = 0.1$ to 0.2. For ductile materials, usually $S_{ut} \approx S_{uc}$ and $\kappa \approx 1$.

Fiber-reinforced plastics may sustain larger loads in tension than in compression. Some metals such as magnesium also possess this property.

The division of materials into ductile and brittle is purely conventional because there is no sharp dividing line between them as regards the index κ. Many brittle materials may behave as ductile materials and vice versa, depending on the conditions of testing. For example, a cast iron specimen fails by yielding in a tension test under high pressure of the surrounding medium ($p > 400$ MPa).

The duration of loading and the temperature affect ductility and brittleness. Under rapid loading, brittleness is displayed more sharply, while under prolonged loading, ductility is more pronounced. For example, brittle glass is capable of developing permanent deformations under sustained loading at normal temperature. Ductile materials, such as low-carbon steel, exhibit brittle properties under sudden impact loading.

Tension and compression tests give an objective assessment of material properties. In industry, however, this method of testing is often very inconvenient where the quality of manufactured parts is desired. For example, it is difficult to measure the results of heat treatment on finished parts by tension and compression tests. It is therefore common practice to resort to a comparative assessment of material properties by hardness test. *Hardness* is the capacity of material to resist mechanical penetration of sharp objects. The most commonly used are the Brinell and Rockwell hardness tests. In the first case a steel ball 10 mm in diameter and in the second case a pointed diamond indenter is pressed against the surface of the part. The hardness of the material is determined by measuring the resulting indentation. In the Shore hardness test, hardness is measured as the material resistance to scratching.

Experiments have provided empirical conversion tables for estimating the ultimate tensile strength of a material from hardness indices, as in Figure 4.7, for example, for steel. Thus, nondestructive hardness tests provide means of measuring the strength indices of the material.

4.1.2. Environmental and Time Factors

The foregoing discussion of material properties applies to tests under normal conditions, that is, at room temperature and at relatively small rates of change of loads and elongations.

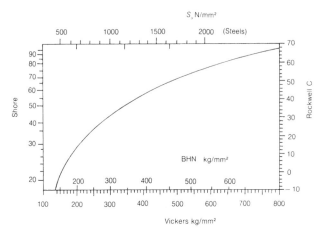

Figure 4.7. Hardness scales.

The range of temperatures over which structural materials actually operate extends far beyond the limits of the above normal conditions. In some structures the material is subject to extremely high temperatures, as for example in the walls of combustion chambers of aircraft engines. In other structures, in contrast, working temperatures are extremely low, such as in elements of refrigerating plants and tanks containing liquefied natural gases.

The speeds of loading and the duration of external forces also vary over a wide range. There are loads varying very slowly and other loads varying very rapidly. Some loads act for years and others for a fraction of a second. It is clear that the mechanical properties of materials will differ depending on the environment and the loading conditions.

A general analysis of material properties taking into account the temperature and time effects is very complicated and cannot be confined to simple experimental curves similar to tension test diagrams. Therefore, temperature and time effects are treated at present with reference to particular types of problems.

Many machine loads are slowly varying, or static, loads. The rate of change of such loads in time is so small that the kinetic energy that is acquired by moving particles of a deformed body is a negligible fraction of the work done by external forces. In other words, the work done by external forces is transformed only into elastic energy and also into irreversible thermal energy as a result of plastic deformation of the material. The testing of materials in so-called normal conditions is performed under static loads.

Figure 4.8 shows the relation between temperature and the modulus of elasticity E, the yield point S_y, the ultimate tensile strength S_u, and the percentage elongation at rupture δ for a low-carbon steel over a range from 0–500°C. As can be seen from these curves, the modulus of elasticity practically does not vary with temperature up to 300°C. The quantities S_{yt} and δ undergo more substantial changes; so-called embrittlement of the steel takes place—that is, the percentage elongation at rupture is reduced. With further increase of temperature, the ductile properties of the steel are recovered while the strength indices decrease rapidly.

Embrittlement at elevated temperatures is encountered mostly in low-carbon steel. Alloy steels and nonferrous alloys mainly exhibit a monotonic increase of δ

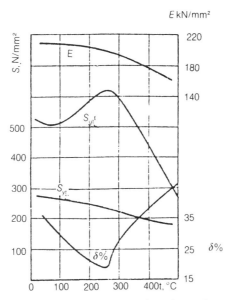

Figure 4.8. Properties of a mild steel at elevated temperatures.

and a similar monotonic reduction of S_{yt} and S_{ut} with increase in temperature. Figure 4.9 represents the corresponding curves for a chrome–manganese steel.

The variation in time of strains and stresses induced in a loaded part is called *creep*. A particular case of creep is a growth of irreversible strains under a constant stress, as on the disks and blades of a gas turbine subjected to large centrifugal forces and high temperatures. This increase in dimensions is irreversible and usually occurs after long operation. A result of creep is relaxation: a redistribution of stresses in time under a constant strain. An example of relaxation is the loosening of bolt connections operating under high-temperature conditions.

Creep lends itself most readily to experimental studies. If a specimen is loaded with a constantly acting force (Figure 4.10) and the variation of its length at a fixed temperature is observed, it is possible to obtain creep diagrams (Figure 4.11) giving the strain versus time relationship at various values of stress. As can be seen from these curves, the strains grow very rapidly at the start. Then the process settles

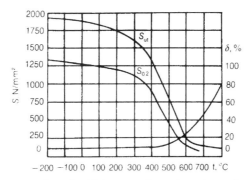

Figure 4.9. Properties of a high-strength steel at elevated temperatures.

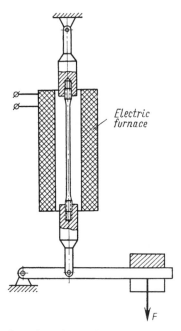

Figure 4.10. Testing of steel at elevated temperatures (after Feodosyev 1973).

down and strains increase at a constant rate. As in ordinary testing, the specimen begins to neck down through time. Shortly before rupture a rapid increase in local strains occurs as a consequence of the reduction in cross-sectional area. At higher temperatures, the variation of strains with time takes place more rapidly.

For a given material, it is possible to convert creep diagrams to relaxation diagrams by using the methods of the theory of creep. Relaxation diagrams can, however, be obtained experimentally. This demands more complex apparatus because it is necessary to measure changes in magnitude of the tensile force while maintaining the elongation of the specimen. Relaxation diagrams giving the stress–time relation are shown in Figure 4.12.

The basic mechanical characteristics of a material under creep deformation are the *creep–rupture strength* and the creep limit. The creep–rupture strength is defined as the ratio of the load at which a tension specimen fails in a given length

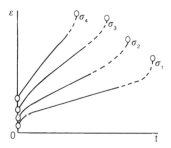

Figure 4.11. Creep diagrams at different stress levels.

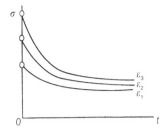

Figure 4.12. Relaxation diagrams at different strain levels.

of time to the original cross-sectional area. Thus, the creep–rupture strength depends on a given time to rupture. The latter is chosen equal to the service life of a part and ranges from minutes in rockets to 30–40 years in large steam turbines.

The *creep limit* is defined as the stress at which plastic strain reaches a given value in a given time. To determine the creep limit it is necessary to assign a time interval (which depends on the service life of a part) and a range of permissible strains (which depends on the service conditions of a part).

The creep–rupture strength and the creep limit are greatly affected by temperature. As temperature increases, they obviously decrease.

Periodically varying loads are of particular importance in machine design and are associated with the concepts of *endurance* or *fatigue* of materials. These problems will be discussed in detail in Chapter 5.

Some loads vary quite rapidly, producing appreciable velocities of the particles of a deformed body. These velocities are so high that the overall kinetic energy of moving masses is now a considerable fraction of the total work done by external forces. On the other hand, the rate of change of a load may be related to the rate of development of plastic deformations. A load may be considered as rapidly varying if plastic deformations cannot fully develop during the process of loading. This materially affects the character of observed stress–strain relations.

It is quite apparent that because the development of plastic deformations cannot be fully accomplished under rapid loading the material becomes more brittle (Figure 4.13). Strain rate effects become significantly more important at high temperatures.

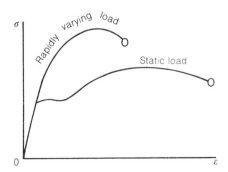

Figure 4.13. Effect of rapid loading.

The last of the three kinds of load under consideration is loads varying very rapidly in time. Their rate of change is so great that the work done by external forces is almost completely transformed into kinetic energy of moving particles of a body while the elastic and plastic strain energy is relatively small.

Very rapidly varying loads are produced by impact of bodies moving with velocities of several hundred meters per second and higher. These loads are dealt with in the study of problems of armor piercing, the assessment of the destructive action of a blast, and the investigation of the penetrating power of interplanetary dust encountered by spacecraft.

Because the strain energy is relatively small under conditions of very high rates of loading, the elastic properties of a solid are of minor importance in the present case. However, problems involving very high rates of loading lie outside the framework of machine design.

Some representative mechanical properties of engineering materials are tabulated in Appendix E.

Example 4.1   A carbon steel AISI 1020 rod is subject to a compressive axial impact. Calculate the elastic energy up to yielding and the plastic energy up to rupture, absorbed per unit volume of the material. Comment on the ductile or brittle nature of this material. Neglect dynamic effects.

Solution   From the material tables (Appendix E), we obtain $S_y = 340$ N/mm^2, $\delta = 39\%$, $E = 2.1 \times 10^5$ N/mm^2. In the elastic region, the elastic energy is the area under the straight portion of the stress–strain curve. The plastic energy is the area under the yielding curve at (approximate) constant stress S_y up to the point of rupture $\varepsilon_p = \delta$ (Equations (4.3) and (4.4)):

$$U_c = \frac{1}{2}(S_y A)(\varepsilon_y L) = \frac{S_y^2 V}{2E}$$

$$U_p = (S_y A)\Delta l = \frac{S_y V \delta}{100}$$

where A is the cross section, L the length, and V is the volume of the bar. Therefore, per unit volume:

$$u_e = \frac{S_y^2}{2E} = 0.275 \text{ N-mm/mm}^3$$

$$u_p = \frac{S_y}{100} = 123.6 \text{ N-mm/mm}^3$$

We observe that the energy of elastic deformation at yielding is more than three orders of magnitude smaller than the energy of plastic deformation at rupture, and thus the material is ductile.

4.2. MATERIAL PROCESSING

4.2.1. Classification

Materials are commercially available in standard forms and almost invariably, will be processed to the designed form. Not all materials, however, are processed in the same way. That will depend on the material, the type and size of the element to be manufactured, the particular service conditions of the element, and the available manufacturing facilities; last but not least is the production cost. Therefore, the designer has to be aware of available methods of material processing and their main features. Many times, the manufacturing requirements for a component have decisive influence on the design. For example, wall thickness of a cast part cannot be less than castability requires for the particular material and size, even if strength is assured with much smaller thickness. With the advent of manufacturing automation, many of our design philosophies have to change to facilitate the application of new technology.

The selection of the proper processing method in the process of designing a given machine element requires a knowledge of all possible production methods. Factors that must be considered are volume of production, quality of the finished product, and the advantages and limitations of the various types of equipment capable of doing the work. Most parts can be produced by several methods, but one method will be the most economical.

Metal working may be classified according to various types of process, many of which, with some modifications, can also be applied to most nonmetallic materials (De Garmo 1974):

1. Processes used to change the shape of material:
    (a) Casting
    (b) Hot and cold working
    (c) Power metallurgy forming
    (d) Plastics molding
2. Processes used for machining parts to a fixed dimension:
    (a) Traditional machining, chip removal
    (b) Nontraditional machining
3. Processes for obtaining a surface finish:
    (a) Metal removal
    (b) Polishing
    (c) Coating
4. Processes used for joining parts or materials (welding)
5. Processes used to change the physical properties (heat treatment)

4.2.2. Change of Shape

Most metal products originate as an ingot casting or continuous casting. Molten metal is poured into metal or graphite molds to form ingots of convenient size and shape for further processing.

Processes used primarily to change the shape of metals include the following:

1. Casting	8. Piercing	15. Torch cutting
2. Forging	9. Swaging	16. Explosive forming
3. Extruding	10. Bending	17. Electrohydraulic forming
4. Rolling	11. Shearing	18. Magnetic forming
5. Drawing	12. Spinning	19. Electroforming
6. Squeezing	13. Stretch forming	20. Powder metal forming
7. Crushing	14. Roll forming	21. Plastics molding

In this group of processes, material is changed into its primary form for some selected part. Sometimes the parts are suitably finished for commercial use, as in metal spinning, cold rolling of shafts, die casting, stretch forming of sheet metal, and drawing wire. Other times, neither the dimensions nor the surface finish is satisfactory for the final product and further work on the part is necessary. It should be noted that the last three processes, electroforming, the forming of powder metal parts, and plastic molding, do not start as a casting. Electroformed parts are produced by electrolytic deposition of metal onto a conductive performed pattern. Metal is supplied from the electrolyte and a bar of pure metal that acts as an anode. This process can make parts of controlled thickness, having high precision.

The method used in the production of powder metal parts is essentially a pressing operation. Metal powders are placed in a metal mold and compacted under great pressure. Most powder metal products also require a heating operation to assist in bonding the particles together.

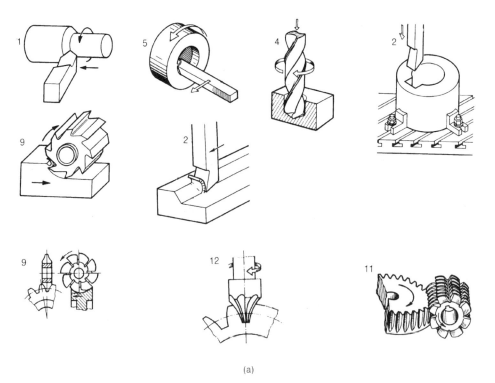

(a)

Figure 4.14(a). Traditional chip-removal process.

Honing

Lapping

Grinding

Roll burnishing

Electrolytic grinding

Burnishing

Jet electrochemical machining

Electric-spark machining

Ultrasonic machining

Electric-pulse machining

Electrochemical
rotary cathode machining

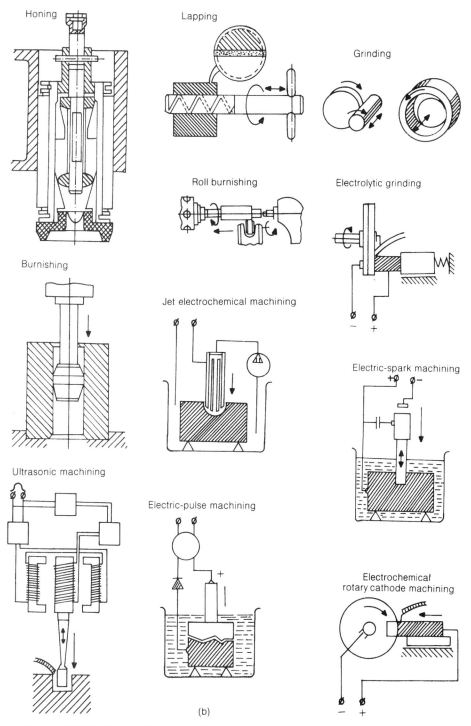

(b)

Figure 4.14(*b*). Nontraditional machining processes: fine machining and polishing.

Plastics are molded under heat and/or pressure to conform to the configuration of a mold.

Explosive, electrohydraulic, and magnetic forming are high-energy rate processes in which parts are formed very rapidly by extremely high pressures.

4.2.3. Machining

While some forming processes can deliver components in final form within acceptable tolerances, in most cases some form of machining is required to bring the material stock or semifinished component to its final dimensions by removing the excess material.

Traditional chip-removal processes include (Figure 4.14*a*):

1. Turning	5. Boring	9. Milling
2. Planing	6. Reaming	10. Grinding
3. Shaping	7. Sawing	11. Hobbing
4. Drilling	8. Broaching	12. Routing

Nontraditional machining processes include (Figure 4.14*b*):

1. Ultrasonic	5. Electrochemical
2. Electrical discharge	6. Chemical milling
3. Electro-arc	7. Abrasive jet cutting
4. Optical lasers	8. Electron beam machining
	9. Plasma-arc machining

In these secondary operations, which are necessary for many components and require close dimensional accuracy, metal is removed from the parts in small chips. Such operations are performed on machine tools, which include the various power-driven machines used for cutting metal. All of these operate on either a reciprocating or a rotary-type principle: either the tool or the work reciprocates or rotates, as indicated in Figure 4.13. The planer is an example of a reciprocating machine, because the work reciprocates past the tool, which is held in a stationary position. In other machines, such as the shaper, the work is stationary and the tool reciprocates. The rotary machine is the lathe, which has the work rotating and the tool stationary. In the drill press it is the tool that rotates.

In ultrasonic machining, metal is removed by abrasive grains, which are carried in a liquid and attack the work surface at high velocity by means of an ultrasonic generator. For electrical discharge and electro-arc machining, special arcs are generated that can be used to machine any conducting material. The optical laser is a strong beam of photons that can be used to generate extremely high temperatures and thus cut or weld metal. Chemical machining is done either by attacking the metal chemically or by using a reverse plating process.

4.2.4. Surface Finish

Surface finishing operations are used to ensure a smooth surface, great accuracy, an esthetically pleasing appearance, or a protective coating. Some processes used are:

1. Polishing	5. Honing	9. Inorganic coating
2. Abrasive belt grinding	6. Lapping	10. Parkerizing
3. Barrel tumbling	7. Superfinishing	11. Anodizing
4. Electroplating	8. Metal spraying	12. Sheradizing

In this group are some processes that cause little change in dimension and result primarily in finishing the surface. Other processes, such as grinding, remove some metal to the designed dimension in addition to giving it a good finish. Processes such as honing, lapping, and polishing consist of removing small scratches with little change in dimension. Superfinishing is also a surface-improving process that removes undesirable fragmented metal, leaving a base of solid crystalline metal. Plating and similar processes, used to obtain friction-, wear-, and corrosion-resisting surfaces or just to give a better appearance, do not change dimensions materially.

4.2.5. Improvement of Material Strength

There are a number of processes in which the physical properties of the material are changed by the application of an elevated temperature or by rapid or repeated stressing of the material. Processes in which properties are changed include:

1. Heat treatment
2. Hot working
3. Cold working
4. Shot peening

Heat treating includes a number of processes that result in changing the properties and structure of metals. Although both hot and cold working are primarily processes for changing the shape of metals, these processes have considerable influence on both the structure and the properties of the metal. Shot peening renders many small parts, such as springs, resistant to fatigue failure.

Because strength of metals and alloys can be drastically improved by heat treatment, it has become one of the most important and commonly used processes, in particular for steel. Steel and cast iron consist basically of iron and carbon in varying composition and structure. The crystal and chemical structure of these two elements in equilibrium, even for fixed composition, largely depends on the temperature history. Therefore, heat treatment is mostly the application of a temperature scenario suitable for the particular purpose.

The several heat-treatment methods can be identified on the phase equilibrium diagram of steel, which shows the phases, and structure of the iron–carbon equilibrium for various compositions and temperatures (Figure 4.15).

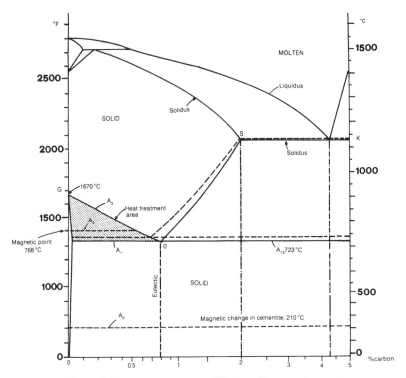

Figure 4.15. Iron–carbon equilibrium diagram for steel.

Since ancient times it has been known that hot iron immersed in water becomes hard and brittle, and this process was used to make stronger weapons. This process, known as *quenching,* improves strength and hardness and requires heating to about 20–50°C above the line GOSK in Figure 4.15 and rapid cooling rate. It is shown schematically in Figure 4.16 together with some of the other important heat treatments. All processes are described in terms of the two temperatures corresponding to the intersection of the carbon content vertical and the curves A_1 and A_3 of Figure 4.15, which bound the transformation.

Tempering is applied to quenched steel to reduce internal stresses and improve ductility and toughness. *Stress relieving* in turn relieves internal stresses. *Normalizing* is used to produce a uniform structure. *Annealing* is used to bring the material in its softest state to facilitate processing such as forming or welding.

Surface treatments involve the diffusion on a surface layer of carbon, nitrogen, and other substances to give higher strength to this layer and improve resistance to fatigue and contact stresses while the bulk of the material retains its ductility. For components of small size, the surface treatment can fully penetrate the component.

4.2.6. Applicable Manufacturing Processes

Once the material is selected for an application, the designer has a choice of manufacturing process for the material stock in order to produce the final product. The equilibrium diagram (Figure 4.15) is very helpful for steels.

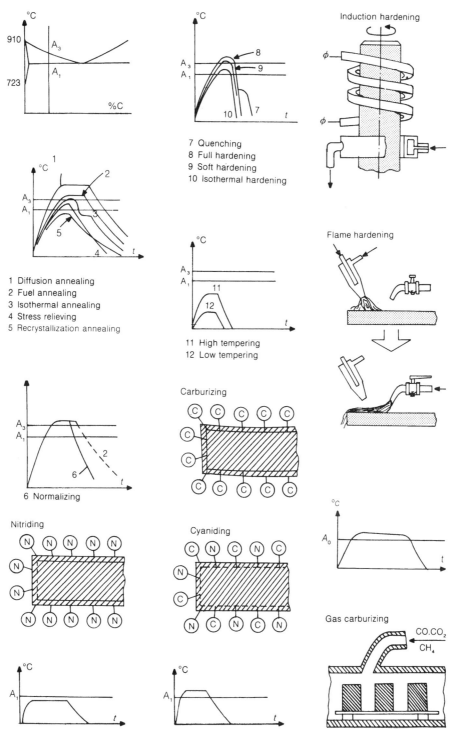

Figure 4.16. Heat treatment processes for steel.

Areas *A* and *C* correspond to low-carbon, ductile steels that can be formed by making them yield by the application of a force to the desired form. For small thicknesses this can be done in room temperature (area *C*) with cold working. This process will yield a material with improved strength but also sensitive to high temperatures because then it will undergo annealing and it will loose the strength benefits from cold working. All steel sheet metal and wire stock with thickness less than 6 mm are normally produced that way.

For higher thickness, the force needed for cold working is prohibitive, which renders cold rolling very expensive or impossible. In this case, the material is heated to a point where the yield strength is drastically reduced and then hot forming (rolling, punching, forging) is applied (area *C*).

For high-carbon steel, plastic deformation processes are not applicable, not only because of the prohibitively high force required but also because of crack formation during plastic deformation. The metal is molten and used in casting at temperatures obviously above the melting point (area *B*). This method is easy to apply and economical for a small number of parts but yields generally rough surfaces, and due to temperature-related contraction the final dimensions are within a relatively wide range. If dimensional accuracy and good surface quality are needed, machining methods are used (area *D*). Such methods are usually expensive, and for mass production a compromise is the powder metallurgy. Powder of the base metal is pressurized at high temperature in a mold and forms the part, many times in a final form.

Example 4.2   For the indexing mechanism shown in Figure E4.2, suggest manufacturing processes to be used for the parts shown.

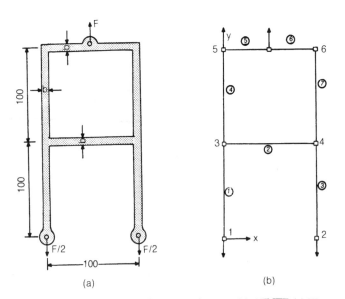

Figure E4.2. Analysis of a plane frame with FINFRAME.

DATA (From file FRAME42)

Node	x	y	z
		NODE COORDINATES	
1	50	0	0
2	50	99.99999	0
3	150	99.99999	0
4	150	0	0
5	50.00001	200	0
6	100	200	0
7	150	200	0

Hit RETURN to continue?

Node	RX	RY	RZ	FX	FY	FZ
			NODE RELEASES			
1	0	0	1	0	0	0
4	0	0	1	0	0	0

Hit RETURN to continue?

Member	Node 1	Node 2
	ELEMENT INCIDENCES	
1	1	2
2	2	3
3	3	4
4	2	5
5	5	6
6	6	7
7	7	3

Hit RETURN to continue?

No	Material	Area	Iy	Iz	Iyz	Roll angle
			ELEMENT PROPERTIES			
1	1	60	80	1125	1205	0
2	1	60	80	1125	1205	0
3	1	60	80	1125	1205	0
4	1	60	80	1125	1205	0
5	1	60	80	1125	1205	0
6	1	60	80	1125	1205	0
7	1	60	80	1125	1205	0

Hit RETURN to continue?

Node	Fx	Fy	Fz	Mx	My	Mz
		LOADING				
6	0	5000	0	0	0	0

Hit RETURN to continue?

NODE DISPLACEMENTS

Element	ux	uy	uz	rotx	roty	rotz
1	0.00E+00	0.00E+00	0.00E+00	0.00E+00	0.00E+00	0.00E+00
2	2.60E-03	1.98E-02	0.00E+00	0.00E+00	0.00E+00	-9.41E-04
3	-2.60E-03	1.98E-02	0.00E+00	0.00E+00	0.00E+00	9.41E-04
4	0.00E+00	0.00E+00	0.00E+00	0.00E+00	0.00E+00	0.00E+00
5	-2.10E-03	3.97E-02	0.00E+00	0.00E+00	0.00E+00	4.77E-03
6	-1.61E-07	2.69E-01	0.00E+00	0.00E+00	0.00E+00	4.66E-10
7	2.10E-03	3.97E-02	0.00E+00	0.00E+00	0.00E+00	-4.77E-03

Figure E4.2. (*Continued*)

```
ELEMENT FORCES
------------------

Element   Fx         Fy          Fz          Mx          My          Mz
   1    1.26E+02   -2.50E+03   0.00E+00    0.00E+00    0.00E+00   -4.08E+03
   2    6.55E+02    1.53E-04   0.00E+00    0.00E+00    0.00E+00   -4.45E+03
   3    1.26E+02    2.50E+03   0.00E+00    0.00E+00    0.00E+00    8.53E+03
   4   -5.29E+02   -2.50E+03   0.00E+00    0.00E+00    0.00E+00    1.30E+04
   5   -5.29E+02   -2.50E+03   0.00E+00    0.00E+00    0.00E+00   -4.00E+04
   6   -5.29E+02    2.50E+03   0.00E+00    0.00E+00    0.00E+00    8.50E+04
   7   -5.29E+02    2.50E+03   0.00E+00    0.00E+00    0.00E+00   -4.00E+04

Normal termination. Press any key.
```

(c)

Figure E4.2. (*Continued*)

Solution

- *Shaft:* Long piece, should be made of steel stock, hot rolled to be machined to the final dimensions.
- *Slider:* Complex shape, to be cast and machined to final dimensions.
- *Spring:* Long shape, must be hard. To be cold drawn and, perhaps, shot peened due to repeated loading.
- *Cam, gear:* Hot-rolled steel stock, machined to shape, surface hardened, heat treated to relieve residual stresses, finally ground to precise shape.

4.3. DESIGN FOR MANUFACTURING

4.3.1. Dimensioning: Preferred Numbers (ASME 1995A)

The old system of dimensioning in inch fractions was in effect a form of preferred numbers system. Standard sizes, however, were used long ago. In ancient Rome, standard-sized pipes were used for water supply. Much later, during the 1870s, Charles Renard, a French army captain, was successful in reducing the number of dimensions of rope for military balloons from 425 to 17 using known geometric series of numbers. The sequence of such numbers is obtained from 1 by multiplication by a constant number. In fact, for the series designated as R5, R10, R20, R40 in accordance with ISO 3-1973, the constants are, respectively, $\sqrt[5]{10}$, $\sqrt[10]{10}$, $\sqrt[20]{10}$, $\sqrt[40]{10}$, giving:

```
R5  1    1.60   2.5
R10 1    1.25   1.60   2.00   2.5    ...
R20 1    1.12   1.25   1.40   1.60   1.80   2.00   2.24   2.5  ...
    ...
```

Derived series are fractions of the standard series, such as R5/2, R10/3, R5/4, ... Using preferred numbers is very convenient in design because:

1. It reduces inventory for many machine components, such as bolts, pipes, bearings, materials (e.g., steels of various compositions and strength), and manufacturing tools, such as drills and cutters,

2. It results in product line simplification when planning model sizes in respect to capacity, speed, power rating, etc.

4.3.2. Dimensioning: Fits and Tolerances (ASME 1995B)

In machine production, exact numbers are only the integer quantities, such as number of teeth of a gear or number of bolts of a coupling. Dimensions are never exact, for a variety of reasons:

- Measuring errors
- Manufacturing errors due to wear of tools, changes in temperature, etc.
- Intended deviations from the nominal dimension for functional purposes.

For example, mass-produced shafts of diameter $d = 40$ mm have actual diameters very close to 40 mm but not exactly that. In accepting or not accepting a shaft piece, one has to set some limits to the deviation from 40 mm, the *nominal* dimension, because this shaft will have to work with other elements, such as bearings and gears, or will have to rotate within a confined annulus. For this reason, quality control sets up limits to the deviations from the nominal diameter for acceptance. These limits have to be decided by the designer because many times they influence the function or strength of a machine component.

Because interchangeability of parts imposes restrictions on the acceptable limits, one of the first standards was on those deviations, called *tolerances*. In particular, if two components must work together, their tolerances must be interrelated. In this case, the components form a *fit*.

Eli Whitney (1765–1825) was born in Westboro, Mass. Whitney's father was a respected farmer who served as justice of the peace. In May 1789, Whitney entered Yale College. After graduation in the fall of 1792, Whitney was disappointed twice in promised teaching posts. The second offer was in Georgia, where, stranded, without employment, short of cash, and far from home, he was befriended by one Catherine Greene. Then he invented the cotton gin and went into business manufacturing and servicing the new gins. He designed machine tools to make muskets. Any part would fit any musket of that design. He had grasped the concept of interchangeable parts, fits, and tolerances. This was the inauguration of the American system of mass production.

Engraving of Eli Whitney, by Samuel F. B. Morse, ca. 1822. New Haven Colony Historical Society.

It is obvious that the tolerance range for a certain dimension depends on the manufacturing procedure used. ISO specifies 18 basic tolerance grades, numbered as IT01, IT0, IT1, IT2, . . . , IT18, from finest to larger tolerances. Machine components used for fits are manufactured usually to grades IT5 to IT11. The tolerance unit is a function of the *nominal dimension D*, for a shaft–hub fit an approximate diameter both for the shaft and the hole, usually an integer number of millimeters:

$$i = 0.45 \sqrt[3]{D} + 0.001D \qquad (4.5)$$

Allowed tolerances for the different nominal size ranges are given in Table 4.1.

Machining quality defines the tolerance range but cannot give information on the location of this range in respect to the nominal dimension.

ISO has specified 27 fundamental deviations of the tolerance range from the nominal dimension. Each deviation is designated with a letter, lower case a, b, c, . . . for shaft dimensions, upper case A, B, C, . . . for hole dimensions. Figure 4.17. illustrates the relative position of tolerance ranges for the standard fundamental deviations (ISO 286) in respect to the nominal diameter[3] (zero line) for shafts and holes. It can be seen that category H or h starts the tolerance range on nominal diameter. It is apparent that the specific tolerance range and deviation for the shaft and hole will determine the operation of the fit. In fact, we usually distinguish three types of fits:

TABLE 4.1. Allowable Tolerance Ranges it or IT and Fundamental Deviations δ_F for the Hole-Basis System

Quality	Tolerance (μm)	Basic Hole System (D in mm)		
		Category	Minimum Clearance a_o (μm)	Minimum Interference a_u (μm)
IT 5	$7i$	a	$(265 + 1.3D)$, $D \le 120$	
IT 6	$10i$		$3.5D$, $D > 120$	
IT 7	$16i$	b	$(140 + 0.85D)$, $D \le 160$	
IT 8	$25i$		$1.8D$, $D > 160$	
IT 9	$40i$	c	$52D^{0.2}$, $D \le 40$	
IT 10	$64i$		$(95 + 0.8D)$, $D > 40$	
IT 11	$100i$	d	$16D^{0.44}$	
IT 12	$160i$	e	$11D^{0.41}$	
IT 13	$250i$	f	$5.5D^{0.41}$	
IT 14	$400i$	g	$2.5D^{0.34}$	
IT 15	$640i$	k		$-0.6\sqrt[3]{D}$, $3 < IT < 8$ else, 0
IT 16	$1000i$	m		$-IT7 + IT6$

Limit dimensions (basic hole)

Hole

$D_{0.0}^{IT}$

Shaft

$d_{-a_o-it}^{-a_o} = d_{-a_u}^{-a_u+it}$

Category	Minimum Clearance a_o (μm)	Minimum Interference a_u (μm)
n		$-5D^{0.34}$
p		$-IT7 - 0...5$
r		Average of p and s
s		$-IT8 - 1...4$, $D \le 50$
		$-IT7 - 0.4D$, $D > 50$
t		$-IT7 - 0.63D$
u		$-IT8 - D$
v		$-IT7 - 1.25D$
x		$-IT7 - 1.6D$
y		$-IT7 - 2.0D$
z		$-IT7 - 2.5D$

[3] We shall limit our discussion to shafts and holes, but the discussion is valid for any type of fit.

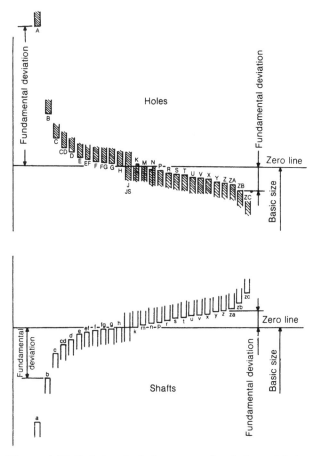

Figure 4.17. Relative deviation ranges for shafts and holes.

1. *Clearance,* or *running,* fits, classes *RC*1 to *RC*9. All the tolerance range of the shaft is below the tolerance range of the hole (Figure 4.18*a*).

2. *Interference,* or *shrink,* fits, classes *FN*1 to *FN*5. All tolerance range of the shaft is above the tolerance range of the hole (Figure 4.18*c*). All hole diameters will be smaller than any of the shaft diameters, and the fit should be "forced." Such fits are used to achieve solid connections transmitting force and torque. They will be discussed further in a later chapter.

3. *Transition* fits, classes *LC*1 to *LC*11. If the two tolerance ranges overlap, some pairs will have clearance fit and some others interference fit (Figure 4.18*b*).

The same effect can be achieved for any location of the two tolerance ranges in respect to the nominal size, provided that their relative distance remains the same. For standardization purposes, ISO specifies two alternatives:

1. Shaft-basis. The shaft has deviation *h* and the deviation of the hole depends on the type of fit desired.

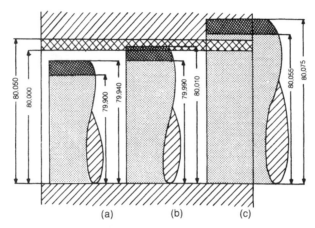

Figure 4.18. Clearance (*a*), transition (*b*), and interference (*c*) fits for a 80-mm diameter shaft.

2. Hole-basis. The hole has deviation *H* and the shaft deviation depends on the desired fit.

For shaft and hole, therefore, a deviation (letter) is specified followed by a digit designating tolerance, such as *b*8, *H*6, *O*7. A fit needs the specification of both shaft and hole, such as *H*8/*f*7, *P*7/*h*6. Preferred fits per ANSI B4.2 (American National Standards Institute) are shown in Table 4.2 with suggested application.

Figures 4.19. and 4.20. show preferred hole-basis and shaft-basis fits according to ANSI B4.2.

ISO and ANSI have published detailed tables with dimensions of different fits. These tables, for every combination of deviation and tolerance (i.e., *h*6, *D*7), give maximum and minimum deviations from the nominal dimension (see Figure 4.9):

Δ_u = upper deviation of the hole
Δ_l = lower deviation of the hole
δ_u = upper deviation of the shaft
δ_l = lower deviation of the shaft

The smallest deviation is called *fundamental deviation,* Δ_F for the hole, δ_F for the shaft, and is tabulated in Table 4.1*b*. For running fits, categories *a–h*, the fundamental deviation is termed *minimum clearance.* For interference fits, categories *k–z*, the fundamental deviation is termed *minimum interference.*

All deviations and the limit dimensions can be computed from the quality (number) and category (letter) of the fit and the nominal dimension. With the terminology of Figure 4.21:

• For running fits:

$$\delta_u = \delta_F, \qquad \delta_l = \delta_F + it, \qquad \Delta_l = \Delta_F, \qquad \Delta_u = \Delta_F + IT \qquad (4.6)$$

TABLE 4.2. Preferred Fits (ANSI B4.2)

	ISO Symbol Hole Basis	ISO Symbol Shaft Basis	Description	
↑ Clearance fits	H11/c11	C11/h11	*Loose running* fit for wide commercial tolerances or allowances on external numbers.	↑ More clearance
	H9/d9	D9/h9	*Free running* fit not for use where accuracy is essential, but good for large temperature variations, high running speeds, or heavy journal pressures.	
↓	H8/f7	F8/h7	*Close running* fit for running on accurate machines and for accurate location at moderate speeds and journal pressures.	
↑ Transition fits	H7/g6	G7/h6	*Sliding* fit not intended to run freely, but to move and turn freely and locate accurately.	
	H7/h6	H7/h6	*Locational clearance* fit provides snug fit for locating stationary parts, but can be freely assembled and disassembled.	
↓	H7/k6	K7/h6	*Locational transition* fit for accurate location, a compromise between clearance and interference.	
	H7/n6	N7/h6	*Locational transition* fit for more accurate location where greater interference is permissible.	
↑ Interference fits	H7/p6	P7/h6	*Locational interference* fit for parts requiring rigidity and alignment with prime accuracy of location but without special bore pressure requirements.	More interference
	H7/s6	S7/h6	*Medium drive* fit for ordinary steel parts or shrink fits on light sections, the tightest fit usable with cast iron.	
↓	H7/u6	U7/h6	*Force* fit suitable for parts which can be highly stressed or for shrink fits where the heavy pressing forces required are impractical.	↓

- For interference fits:

$$\delta_l = \delta_F, \qquad \delta_u = \delta_F + it, \qquad \Delta_u = \Delta_F, \qquad \Delta_1 = \Delta_F + IT \tag{4.7}$$

The limit dimensions are then:

- For running fits (*a–h*) (Figure 4.21*a*):

Fundamental deviations: $\delta_F = \delta_u, \quad \Delta_F = \Delta_l$

$$\text{Shaft:} \quad d_{\max} = D - \delta_F, \quad d_{\min} = D - \delta_F - it \tag{4.8}$$

$$\text{Hole:} \quad D_{\max} = D + \Delta_F + IT, \quad D_{\min} = D + \Delta_F \tag{4.9}$$

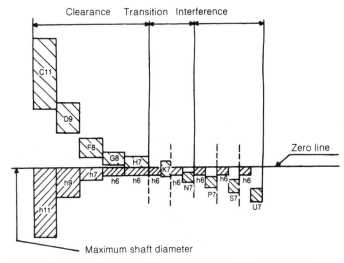

Figure 4.19. Standard fits, shaft base (ANSI B4.2), to scale for 25 mm diameter.

- For interference fits $(k–z)$ (Figure 4.21b):

Fundamental deviations: $\delta_F = \delta_1, \quad \Delta_F = \Delta_u$

$$\text{Shaft:} \quad d_{\max} = D + \delta_F + it, \quad d_{\min} = D + \delta_F \quad (4.10)$$

$$\text{Hole:} \quad D_{\max} = D - \Delta_F, \quad D_{\min} = D - \Delta_F - IT \quad (4.11)$$

It is apparent that the maximum and minimum clearances between shaft and hole are:

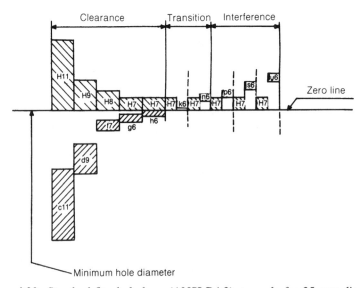

Figure 4.20. Standard fits, hole base (ANSI B4.2), to scale for 25 mm diameter.

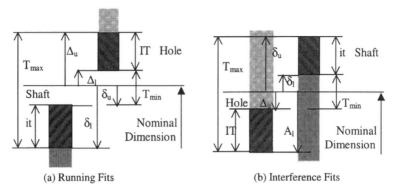

(a) Running Fits   (b) Interference Fits

Figure 4.21. Definition of basic fit dimensions (gray designates the solid).

- Running fits:

$$T_{\max} = \delta_F + \Delta_u + it \qquad T_{\min} = \delta_F + \Delta_l \tag{4.12}$$

- Interference fits:

$$T_{\max} = \delta_u + \Delta_F + IT \qquad T_{\min} = \Delta u + \delta_l \tag{4.13}$$

Negative clearances mean interference fits.

The computation procedure starts with selection of the machining method for shaft and hole. This yields the tolerance ranges IT and it. In usual applications $it = IT + 1$ because machining of holes is more difficult than machining of shafts. Then hole-basis or shaft-basis is selected. This is mostly arbitrary, and machine manufacturers select one or the other on the basis of cost and availability of measuring tools. The hole-basis system is the most common.

Furthermore, from the preferred fits (Table 4.2), depending on function, one is selected. The limit dimensions and the minimum deviations δ_f and Δ_F are found from Table 4.1. For hole-basis, $\Delta_F = 0$. For shaft-basis, $\delta_F = 0$. This is found from Table 4.1a and Equation (4.5).

Nominal dimensions are in millimeters and tolerance limits in micrometers (μm). Most important for the designer are the interference fits (or shrink fits) because they induce stresses on the fitted components that must be taken into account at component design. This will be discussed in Chapter 5.

Design Procedure 4.1: Fit Design

Step 1: Select shaft-basis or hole-basis system (designer's choice).
Step 2: Identify nominal diameter D. From Table 4.2, select preferred fit.
Step 3: Find the tolerance unit from Equation (4.5).
Step 4: Find tolerances it and IT for shaft and hole, respectively, from Table 4.1a.
Step 5: Find fundamental deviation from Table 4.1b.
Step 6: Find limit dimensions using Equations (4.14) and (4.15).

Step 7: For shrink fits, strength of the hub and transmission of force/torque need to be determined (see Chapter 5.

Example 4.3:   Limit Dimensions of a Fit

A journal-bearing fit has nominal diameter 16 mm and fit $H7/f7$. Determine the limit dimensions.

Solution   Since $f < H$, it is a running fit.

The nominal diameter is $D = 16$ mm. The tolerance unit is (Equation (4.5)):

$$i = 0.45D^{1/3} + 0.001D = 1.15 \ \mu m$$

For quality 7, both for the shaft and the hole:

$$it7 = 16i = 16 \times 1.15 = 18.4 \ \mu m, \quad IT7 = it7 = 18.4 \ \mu m$$

The fundamental deviation for the hole is $\Delta_F = 0$ (category H). The minimum clearance for the shaft (category f) is (from Table 4.1):

$$\delta_F = 5.5D^{0.41} = 5.5 \times 16^{0.41} = 17.142 \ \mu m$$

The maximum clearance for the shaft is:

$$\delta u = \delta_F + it7 = 17.142 + 18.4 = 35.5 \ \mu m$$

The maximum clearance for the hole is:

$$\Delta u = \Delta_F + IT7 = 0 + 18.4 = 18.4 \ mm$$

Thus, the limit dimensions for the shaft are 16^{-35}_{-17} ($15.965 < d < 15.983$) and the limit dimensions for the hole 16^{18}_0 ($0 < D < 16.184$).

The problem can also be solved with MELAB 2.0. From the main menu we select Fasteners, then Fits. The input–output session is in Figure E4.3.

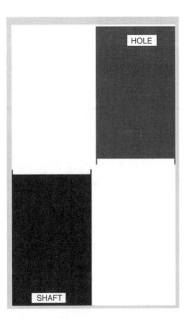

```
COMPUTER PROGRAMS FOR MECHANICAL ENGINEERS
Problem . . . . . . . . . . . . Default example?     16    Hi
Enter Rlsil: Dimension (m)  . . . . . . . . . . . . . . .  XG   Y    1
Enter Quality of Shaft it  . . . . . . . . . . . . . . . .   7    ?
Enter Quality of Hole IT  . . . . . . . . . . . . . . . .   8    ?    7
Enter Category of Shaft (Hole H) . . . . . . . . .   9    Y    H

RESULTS:

    10  11  7  /r  7    fit

XHAFT:    14.96122   <  d  <   14.98286
UNIT:       10   <  n  <   10.01814
enax:       1.714162E 02
enin:       5.342735E 02

ENTER  <C> to change fit
           <P> for another problem
           <V> to quit

Y ▮
```

Figure E4.3

4.3.3. Design for Manufacturability

Designing a machine that meets all the performance requirements is not enough in a competitive world. The machine should be designed for manufacturing with a minimum amount of manufacturing resources, according to the second fundamental rule of design. To this end, designers use *design for X* (DFX) methodologies, where X may correspond to criteria such as reliability, robustness, serviceability, environmental impact, or manufacturability. The most common of these methodologies is *design for manufacturing* (DFM), which is of primary importance because it directly addresses manufacturing costs (Ulrich and Eppinger 1995).

DFM begins during the concept development phase. When product specifications are refined, the designers make trade-offs between desired performance characteristics—for example, weight reduction may increase or decrease manufacturing costs. The DFM methodology consists of five parts:

1. Estimate the manufacturing costs.
2. Reduce the costs of components.
3. Reduce the costs of assembly.
4. Reduce the costs of supporting production.
5. Consider the impact of DFM decisions on other factors.

This process is iterative. Several times the manufacturing cost is recomputed in an iterative procedure until the design is acceptable. As long as the product design is improving, these DFM iterations may continue even until pilot production begins. At some point the design is "frozen" (or "released"), and any further modifications are considered formal "engineering changes" or become part of the next generation of the product.

Manufacturing cost of a machine consists of costs in three categories (Ulrich and Eppinger 1995):

1. *Component costs:* the *components* of a machine might be *standard parts,* purchased from suppliers, or *custom parts,* made according to the manufacturer's design from raw materials.
2. *Assembly costs:* machines are generally assembled from parts. The process of assembling almost always incurs labor costs and may also incur considerable costs for equipment and tooling.
3. *Overhead costs:* overhead is the category used to encompass all of the other costs. We find it useful to distinguish between two types of overhead: *support costs* and other *indirect allocations.* Support costs are the costs associated with materials handling, quality assurance, purchasing, shipping, receiving, facilities, and equipment/tooling maintenance (among others). Indirect allocations are the costs of manufacturing that cannot be directly linked to a particular product but that must be paid for so a company can be in business. For example, the salary of the security guard and the cost of maintenance to the building and grounds are indirect costs because these activities are shared among several different products and are difficult to allocate directly to a specific product.

Another way to distinguish manufacturing costs is to divide them into *fixed costs* and *variable costs*. Fixed costs are those incurred in a predetermined amount, regardless of how many units of the product are manufactured, such as purchasing of a machine for the manufacturing of the specific part. This cost is independent of the number of units produced. However, because machines have a maximum capacity, exceeding that capacity requires another machine, making this cost variable.

Variable costs are those incurred in direct proportion to the number of units produced, such as the cost of raw materials, which is directly proportional to how many parts are produced and therefore to how many machines are made.

Henry Ford (1863–1947) was born on a farm near Greenfield, Michigan. He became a machinist's apprentice in Detroit. In 1891, he became an engineer at the Edison Illumination Company in Detroit. Henry built his first car, the Quadricycle Runabout, in the summer of 1896. It had a 4 hp engine and could reach 20 MPH. He sold the car for $200 to finance his second car, which was completed around early 1898. He incorporated the Ford Motor Company on June 16, 1903. It was capitalized for $100,000 with 12 stock holders. Ford produced 1708 cars the first year. In 1908, he brought out the Model T, which sold 15,000,000 units. Henry Ford is the undisputed father of the automobile and efficient manufacturing. He was the first to introduce the idea of design for manufacturability.

Henry Ford, photograph, courtesy of the Henry Ford museum.

- *Standard components:* The costs of standard components are estimated by either (1) comparing each part to a substantially similar part the firm is already producing or purchasing in comparable volumes or (2) soliciting price quotes from vendors or suppliers. Vendors for most standard components can be found in the *Thomas Register of American Manufacturers*[4] in the United States and *Botin* in Europe, or by looking for company names on components used in related products. To obtain a price quote, first request a catalog or product literature. Then either choose a part number or, if a custom component will be used, write a one-page description of the requirements of the component. Next, telephone the vendor, ask to speak to someone in sales, and request price information. Make sure to inform vendors that the information is for estimation purposes only; otherwise they may claim they do not have enough information to determine exact prices.
- *Custom components:* Custom components, which are parts designed especially for the product, are made by the manufacturer or by a supplier. We estimate the cost of a custom component by adding up the costs of raw materials, processing, and tooling.

---

[4] Available on-line at http://www.thomasregister.com/

- *Assembly:* Products made of more than one part require assembly. For products made in quantities of less than several hundred thousand units per year, this assembly is almost always performed manually. One exception to this generalization is the assembly of electronic circuit boards, which is now almost always done automatically, even at relatively low volumes. There will likely be more exceptions in the coming years as flexible, precision automation becomes more common. Manual assembly costs can be estimated by summing the estimated time of each assembly operation and multiplying by a labor rate.
- *Overhead:* Accurately estimating overhead costs for a new product is difficult, and industry practices are not uniform. Overhead rates are used as a convenient way to account for overhead costs, but this scheme can yield inaccurate estimates of the true costs experienced by the manufacturer to support production.

Several techniques can be used to reduce production cost:

- Redesign components to eliminate processing steps.
- Choose the appropriate economic scale for the part process.
- Standardize components and processes.
- Adhere to "black box" component procurement.
- Reduce assembly costs.
- Integrate parts.
- Reduce the support costs.
- Minimize systemic complexity.

4.4. MATERIALS SELECTION IN MACHINE DESIGN

Materials selection plays a very important role in machine design. The cost of materials in any machine is a good portion of the cost of the machine. More than the cost is the fact that materials are always a very decisive factor in a good design. The materials in design have to be very carefully selected according to the specific requirements posted on the several machine components because these components will operate in a certain environment. This will frequently be far from the usual environment and many times at high temperatures or in the presence of oxidizing conditions.

The choice of the particular material for the machine member depends on the particular purpose and the mode of operation of the machine component, as well as on the expected mode of failure of this component, as will be discussed in Chapter 5.

4.4.1. High-Strength Materials in Machine Design

If the expected mode of failure is fracture, the material must have high-strength features and usually must be made of structurally improved or hardened steels and high-strength cast irons. Furthermore, one must differentiate between the several modes of fracture. Because in machine members general yielding is not allowed, machine operation usually being very seriously affected in such case, the basis of selection is the yield strength of the particular material. From this point on, the particular mode of fracture expected will dictate more features of the selected material. If the part has space or weight limitations, such as the parts used in aircraft and the aerospace industry, very high-strength steels have to be used, or similar high-strength materials. In this case, besides the yield strength, other considerations have to be taken into account, such as resistance to crack propagation, brittle fracture, and creep.

If the main consideration is rigidity, the material has to be selected on the basis of modulus of elasticity. In this case steels and cast irons can be selected. In particular, parts where we expect large elastic deformations, such as springs, have to be made out of high-hardness materials such as hardened steel or out of non-metallic materials such as rubber or of plastics that have a very high ratio of ultimate strength to modulus of elasticity.

If contact stresses are the main load considered in a machine part, as in antifriction bearings and gears, materials must be selected that must be able to take surface hardening and similar treatments. Structurally improved steel or hardened steel must be utilized.

Many machine parts move relative to others, and their wear characteristics are very important. In this case one of the two materials has to be made of very hard steel or cast iron and the other material must be a softer one with good antifriction properties. The harder material is for the larger part, which is not easily replaceable, while the *antifriction material* is for the part that can be replaced easier. Antifriction materials such as bronzes, babbitt metals, and other nonferrous alloys, antifriction plastics, etc. have low coefficient of friction, high wear resistance, good resistance to sizing, good running-in properties, and low wear on the mating part. Sliding couples have to be made from very dissimilar materials because similar materials have a tendency to form local bonds, which increase wear and friction. In such couples where lubrication is also present, the effect this might have on oil circulation and the disposal of the wear debris should be considered.

In other cases, where we want a high coefficient of friction, such as in brakes and clutches, *friction materials* are used. These materials have high coefficient of friction and wear rate and also present high resistance to heat. They must also have a coefficient of friction that does not change very much with temperature and environmental conditions, and they must also have low wear on the mating parts under dry or lubricated conditions. In modern machinery operating in high temperatures, such as aircraft engines and gas turbines, materials must have very high wear resistance, and in this case heat-resistant superalloys are used.

Many times, when we design for high-contact stresses or for friction and wear resistance, it is possible to use surface hardening or surface-coating techniques. In this case we do not have to use a very strong and usually very expensive material for the whole part, but instead can use a material that gives adequate strength for the body of the part and can be surface-strengthened or coated with the proper harder or softer (depending on the case) layer.

As noted above, cost of the material is a primary consideration. Therefore, in selecting a material, the mere strength value cannot be the absolute criterion. For this purpose we use specific material indices, such as strength divided by the unit price, strength divided by density if the weight of the material is of importance, and other indices of this type, depending on particular situations. Many times, of course, material selection is limited due to material availability or manufacturing limitations in the particular locality.

Because stocks are expensive, every manufacturing facility tries as hard as possible to reduce the number of available materials on these stocks. This is a usual limitation on the available materials for design, and many times the designer must look through such a list of available materials. For machine parts that will be produced in very large quantities, this is not a limitation because the materials will be purchased for the production of the particular component.

When the material does not contribute greatly to the cost of the machine, or for a new design of the machine, materials of high quality must always be selected. This is because we usually design machines and their members based on the major modes of expected failure without considering secondary effects (which in many cases are much greater than we think). Having a good material will help in many aspects in this direction. Later, when the machine has to be redesigned, economizing with the use of lower-cost materials can be considered, based on the experience gained during the period of its initial use.

Materials frequently impose technological limitations on developing larger and more powerful machines and structures. For example, we can design a simply supported beam with some load up to a certain span. Beyond that span, if we try to increase the strength of the beam by strengthening the section of the beam, the additional weight we put on the beam is higher than the additional strength of the beam. To make a beam with longer span, we must use a material with a higher strength-to-density ratio or a different design of the section of the beam to increase the strength without simultaneously increasing the weight per unit length. Similar limitations are imposed by the materials on the development of more powerful aircraft engines. To have more power in an engine, we have to have longer blades in the compressor and turbine section. Longer blades mean higher loads and therefore higher stresses as a result. To make an engine with higher power, we have to use a material with a higher strength-to-density density ratio. However, for the available materials, there is a limit to this ratio beyond which, at the present state of technology, we cannot go. A breakthrough in material development is usually followed by other technological breakthroughs in areas where this material can be utilized.

Physical properties of the materials are not influenced substantially by small variations in composition and can be represented as in Table 4.3. Strength, on the other hand, is greatly influenced by composition. Therefore, for every type of material, a separate table needs to be compiled. This was done in Appendix E.

It must be emphasized that the material properties given are average values or ranges. Some wide ranges in strength properties indicate the influence of heat treatment. Fatigue strength given in the tables is discussed in Chapter 5.

4.4.2. Designing with Cast Irons

Irons with carbon content above 1.2% are usually referred to as *cast irons*. The carbon appears mostly as graphite inclusions in spherical or other forms, which impose on cast iron certain unique properties, such as a great difference between

TABLE 4.3. Physical Properties of Some Engineering Materials

Material No.	Material Name	Specific Weight	Young's Modulus (GPa)	Shear Modulus (GPa)	Therm Exp. Coeff. $(\times 10^{-})/°C$	Sp. Heat Al (kJ/kg°C)	Thermal Cond. (kJ/mh°C)	Electrical Resistance $\mu(\Omega m)$	Poisson Ratio
1	Aluminum	2.70	62.1	23.3	22.2	0.921	775	0.027	0.34
2	Wrought Al alloys	2.72	74	28	22	0.921	500	0.045	0.34
3	Cast Al alloys	2.7	68	28	23	0.921	560	0.053	0.34
4	Structural steels	7.85	210	85	11.45	0.477	190	0.17	0.27
5	Alloy steels	7.85	210	84	11.4	0.510	120	0.7	0.27
6	Stainless steels	7.7	200	86	18	0.5	45	0.7	0.29
7	Heat res. steels	7.83	210	82	11.45	0.4	45	0.8	0.35
8	Copper	8.97	117	50	16	0.385	1400	0.017	0.295
9	Bronze	8.5	112	41	17	0.385	600	0.045	0.295
10	Brass	8.5	109	40	17	0.377	245	0.08	0.295
11	Cast iron	7.5	66–170	9.6–28	10	0.586	180	0–9	0.2
12	Cast steel	7.83	207	77	12.5	0.48	134	1	0.31
13	Mg alloys	1.8	45	16.6	26	1.05	300	0.14	0.3
14	Titanium	4.51	107	41	8~.5	0.469	50	0.12	0.34

tensile and compressive strengths, high internal friction, low coefficient of friction, and high rigidity.

4.4.2.1. Gray Cast Iron.
Gray cast iron is the principal material for casting larger shapes. It has very good castability, average strength, small elongation at fracture (which means limited impact strength), good wear resistance, and high internal friction with low sensitivity to heating. Gray cast iron has low tensile strength due to the graphite inclusions, which result in local stress concentrations, which in turn lower the strength of the material. In compression this factor does not appear, and therefore the strength of cast iron in compression is comparable to the strength of steel.

Gray cast iron is used mainly for parts of relatively complicated shape, where casting is easier than fabricating. They are used mostly for stationary parts of machines like machine housing and have good resistance to wear; for this purpose they are used for parts of machines over which other parts are sliding.

According to the ISO, gray cast irons are identified by two letters followed by two numbers representing the tensile strength and the bending strength in kilograms per square millimeter.

The modulus of elasticity of cast iron increases with the tensile strength.

Many times the size of the machine component made out of gray cast iron is determined not on the basis of strength but on the minimum thickness that a castable part can have because of casting limitations. The higher the percentage of carbon, the lower the strength and the better the castability, and therefore the smaller the permissible dimensions of the component.

1. Low-strength castings are employed for parts that are subject to low loads and no severe sliding wear.
2. Medium-strength castings are used for parts that have medium loads and work at low sliding speeds and low pressures. These castings are the ones most extensively used, as for example in most machine housing and supporting parts.
3. High-strength castings are used for parts that are subject to high loads and stresses or work at high speeds and pressures such as crankshafts, drive components, and heavily loaded guideways.

The properties of cast iron can be improved by alloying or adding particular forms of carbon before casting. For extremely thin-wall castings of cast iron, we use a high carbon content, up to 4.6%, and silicon up to 2.8%, also with increased phosphorus content.

For good antifriction properties we alloy the cast iron with nickel (0.3–0.4%) and chromium (0.2–0.35%). Nickel and chromium together with a low percentage of silicon and phosphorus are used for heat- and wear-resistance castings. In high-strength cast irons, magnesium and other materials are added, causing the graphite to precipitate as the cast iron solidifies in the form of spherical nodules and having the effect of reducing the internal stress concentrations. The modulus of elasticity of such cast irons ranges from 160 to 190 GPa, while the endurance limit for cross sections of medium size is approximately equal to that of medium-carbon steel.

Such cast irons are frequently used instead of steel because although they might have somewhat lower strength, very complicated parts can be made and the strength improved by making the parts lighter.

4.4.2.2. White Cast Irons. White cast irons are very hard and can be machined only with special tools. For this reason they have very high wear and heat resistance and also high resistance to corrosion. They are used as:

1. Parts with high wear, such as brake shoes, grinding and crushing parts, and pumps for abrasive particle-carrying liquids
2. Parts that are subject to flame and heat environments
3. Parts operating in chemical environments

White cast irons can be also alloyed with the addition of elements such as nickel and boron for wear resistance, chromium for wear and heat resistance, and silicon for acid resistance.

4.4.2.3. Malleable Cast Irons. Malleable cast irons are used for parts of somewhat complicated shape but with certain impact loads. The word "malleable" is used only by convention; these cast irons cannot be subjected to plastic working as the name suggests.

The properties of the several types of cast irons and steels are shown in Appendix E.

4.4.3. Designing with Structural Steels and Steel Alloys

Steel is by far the most widely used material in machine construction. In general, it has very high strength per unit price. Beyond that it has many other advantages over other structural materials.

Structural steels are classified on the bases of their carbon content and alloying elements. With respect to their carbon content and type of heat treatment, steels are classified as:

1. Low-carbon carburizing steels with a carbon content up to 0.25%
2. Medium-carbon structurally improvable and hardening steels with carbon content between 0.25 and 0.6%
3. High-carbon hardening steels with a carbon content over 0.6%

Castings are made out of steel with relatively high carbon content. Such steels are somewhat inferior to rolled or forged steels and have lower machineability. Structural section steels are supplied in various types and size ranges as rounds, squares, and several other forms. Steels of ordinary quality are used for parts that are not to be heat treated.

Some uses of carbon steels are:

1. Low-carbon structural steels are used for parts that during manufacturing will have to undergo plastic deformation or machining that is also based on plastic deformation.
2. Medium-carbon steels are used for parts that have low loads and will not be heat treated after machining. Such steels are generally not used for sliding parts.
3. High-carbon structural steels are used for parts that have medium stresses and are subject to heat treatment.

Carbon steels with manganese content are used for larger parts to improve strength and wear resistance. They are known for their hardenability with special heat treatments. Carbon steels with an increased sulfur content are called *free-cutting steels* and are used for parts where smooth surfaces and high machineability are required.

As mentioned above, medium-carbon steels can be heat treated and hardened with special methods over their surface to assume good friction properties, while their body can take high loads, especially impact loads. When high strength and special surface properties are required, alloyed steels are used, which can also in general be heat treated. In particular, chromium steels have high strength and high resistance to wear and corrosion. However, since they can be hardened near the surface they are generally used for parts with small cross-section.

Under high-temperature conditions, chromium–nickel steels can be used. They have good machineability, a property that makes them especially useful for gears. Critical parts with large cross-sections can be made out of chromium–nickel steels, which have high hardenability, high strength and wear resistance, and high toughness. For even higher mechanical and processing properties, steels with molybdenum or tungsten additions are used in very critical parts of machines such as gas turbines and aircraft engines.

In small quantities, titanium has advantageous effects on the grain size of the steel, raising its hardenability. It is also used for critical parts of machines.

Construction steel is a low-alloyed steel with a carbon content up to 0.18% and having an addition of manganese, silicon, chromium, nickel, and copper.

Corrosion-resisting, stainless, and acid-resisting steels are alloyed with chromium, nickel, and manganese.

Heat-resisting steels are used for temperatures above 700°C and consist of low- and medium-alloy chromium steels.

Strength properties of steels are shown in Tables E1 to E13, Appendix E. They are summarized in Table 4.4.

4.4.4. Designing with Nonferrous Alloys

Copper alloys are very widely used in machine construction. They have two main advantages: good friction properties and high resistance to corrosion.

Brasses are copper alloys with zinc as the main alloying element (up to 50%). All other copper alloys are called *bronzes*. With respect to the main component besides copper, bronzes are classified as tin, lead, aluminum, beryllium, silicon,

TABLE 4.4. Summary of Mechanical Properties of Materials for Machines[a]

Name	ISO designation	S_u (MPa)	S_y (MPa)	S'_n (MPa–Rotating bending)	$\delta\%$	Use
Iron: Cast	185 GR 10 to 40	100–1400	100–450	70–160	0.37–0.33	Machine frames
Cast iron/steel	1083 GR 38 to 70	370–800	230–450	170–270	12–27	Cast parts with impact loads
Austenitic cast	S-alloying elements	120–390	170–240	70–150	7–25	Cast parts with impact loads (higher $\delta\%$)
Malleable	94x GR A to D	290–690	190–540	200–310	2–12	Small, lightly loaded parts
Steel: Cast	3755 GR 90 to 30	400–1030	200–860	240–410	9–25	Large, heavily loaded machine frames and parts
Structural	630–Fe 37 to 70	360–690	230–360	170–320	26–11	General use for machine components
Carbon	C10 to C60	460–980	355–600	190–400	25–11	General use for machine components
Stainless (Fer-Mart)	R683 PART xx GR yy	880–1420	690–1039	350–630	12–9	General use for machine components in corrosive environment
Free-cutting	R683/9 GR 1–10	490–840	390–450	290–350	7–14	Machine parts, heat treated
Nitriding, case hard	R683/10,11 GR 1–15	490–1620	290–1080	290–410	7–14	Components resistant to friction and wear
Stainless, austenitic	R683/13 GR 10–23	440–690	180–220	270–290	35–40	Impact loaded parts
Spring	R683/14 GR 1–14	1100–1370	880–1180	550–650	5–6	Springs
Flame, induction hardening	R683/12 GR 1–11	620–1270	400–800	240–350	17–10	Components resistant to wear, high contact stresses, gears, rolling wheels

Material					Application
Titanium	500	340	280	18	Turbine blades, bio-implants
Ti alloys	1000	950	500	12	Turbine, aircraft, bio-mechanical parts
Aluminum	17	55		15	Vehicle parts, high-speed parts
Al–Mg alloys	220	110	70	20	Automotive, chemical machine parts
Cast Mg alloys	200	160	70	5	Aircraft, automotive parts
Wrought Mg alloys	270	170	60	15	High-strength parts
Al–Zn–Mg alloys	240	130	120	12	High-strength parts
Cast Al–Ti–Cu alloys	400	300	80	10	High-strength parts
Cast Al–Si–Mg alloys	180	160	110	3	High-strength parts
Copper	200	40		35	Electric conductors
Cu–Sn bronze	320	160	150	40	Sliding friction parts
Cu–Ni alloys	310	120	100	35	Corrosion-resistant parts
Cu–Al alloys	450	180	150	20	Corrosion-resistant parts
Wrought Cu–Zn brass	370	180	155	25	Corrosion-resistant parts
Cast Cu–Zn brass	450	170	150	12	Corrosion-resistant parts
Acetal	60				Structural and mechanical parts
Glass reinforced	133				"
Nylon 6/12	66				"
Glass reinforced	150				"
Polyester	58				"
Glass reinforced	133				"

[a]For detailed tables, see Appendix E.

271

and others. In general, bronzes have high antifriction properties and high corrosion resistance and are conducive to several processing methods, such as casting or machining. Because of the above properties, bronzes are used for bearings, guides, gears, and nuts of power screws and for fitting and parts in corroding environments.

Bronzes with tin contents between 4 and 12% are widely used with smaller contents of lead, zinc, and phosphorus, resulting in high corrosion resistance. Because tin is expensive, these bronzes are also expensive.

Lead bronzes contain 27–32% lead and have good antifriction properties, being used mostly for bearings. Because of their low hardness they require that the mating material be surface-hardened and have a good surface finish.

Aluminum bronzes are used also as antifriction material at high pressures, but only at low and medium sliding speeds. The requirements on the mating parts are as for lead bronzes.

Brasses have good resistance to corrosion, high electrical conductivity, sufficient strength, and especially good processing properties. Brasses can be casted in the foundry, but they can also be subject to cold working and rolled into thin sheets and wires. With the exception of higher grades, brasses can be machined at high speeds, and they give a high-class surface finish. Because of these properties, brasses are widely used for tubing, sheets, wires, fittings, instruments, and electrical machinery.

Babbitt metals, or white metals, are alloys with soft metals such as tin, lead, and calcium. They are very good antifriction materials and also have good running-in properties in bearings operating at high speeds and high pressures.

Lightweight alloys are based on aluminum or magnesium and have specific weights below 4.5. They can be made both in castings or rolled stock. These alloys have high strength-to-weight ratios and are used for parts that have high speed and intermittent motion, such as reciprocating machines, and in general where dynamic loads are high. They are also used for rapidly rotating parts to reduce the centrifugal forces and for housings of engines and machines used in the aircraft industry. In mass production they are particularly useful because they can be machine-casted to precise dimensions, thereby eliminating the need for expensive machining.

4.4.5. Designing with Engineering Plastics

Plastics are materials consisting of high-molecular-weight organic compounds, usually synthetic materials. To improve mechanical properties, fillers are used, as well as small amounts of additives. The fillers must be in the form of cloth fabric, paper, glass or graphite fibers, or small particles.

Due to their good properties, plastics have been developed to a very high degree during the last few decades. Their use is very economical because usually their price is low and they can be manufactured to a great variety of forms with a very small amount of material loss during manufacturing.

Generally, plastics have low density, high heat and electrical insulation, chemical stability, very high damping capacity, good appearance, and last but not least, high strength. In fact, some plastics have strength comparable to that of very high-quality steels. Limitations of plastics are their low heat resistance and some tendency to change dimensions and shape and sometimes degrade with time or exposure to heat or water or other factors.

Depending on the resins used, plastics can be *thermosetting* or *thermoplastic.* Thermosetting materials undergo a chemical change owing to high temperature during manufacturing. They cure to an infusible shape not permitting any reforming. Thermoplastic materials soften with heat and harden on cooling. They can be resoftened again by heating to be formed to a different shape.

Plastic parts are produced by hot- or cold-pressure molding, injection molding, transfer molding, and machining. Thermosetting plastics can be laminated with a fabric base with a filler of cotton cloth in sheets, plates, and fibers. Moreover, hardened paper, laminated asbestos fabric, wood laminate, and fiberglass can be used. The last produces a plastic that has very high strength, elasticity, low notch sensitivity, high heat resistance, and good electrical insulating properties. It is among the materials with the highest strength per unit mass.

Thermoplastic materials are also very extensively used; they have somewhat lower mechanical properties than thermosetting plastics but have good manufacturing properties. Such thermoplastic materials are plexiglas, polyethylene, and polyvinyl chloride.

Mechanical properties of some nonferrous materials are shown in Table E13, Appendix E.

4.4.6. Material Designations and Selection

In Table 4.4 and the material tables of Appendix E, the ISO material designations are used together with the most common designations in the United States and Germany. Widely adapted are the designations of the American Society of Testing and Materials (ASTM 1991), which include many of the AISI (American Institute of Steel Construction) and the SAE (Society of Automotive Engineers) designations in a common standard Unified Numbering System (UNS). This system designates metals and alloys in the form

Xabccd

where X stands for the particular metal and its alloys as follows:

A = aluminum
C = copper
D = steels
F = cast iron and cast steel
G = carbon and alloy steel
J = cast steels
K = miscellaneous steels and ferrous alloys
L = low-melting-point metals and alloys
M = miscellaneous nonferrous
N = nickel
S = stainless steels
T = tool steels
Z = zinc

abccd stands for the alloy composition:

  a for the alloy group, the major alloying element (1 = carbon steel, 2 = nickel
  steel, 3 = nickel–chromium steel, 4 = molybdenum steel, 5 = chromium
  steel, 6 = chromium-vanadium steel, 8 = chromium–nickel-molybdenum,
  9 = silicone–manganese steel)

  b for the alloying element content (roughly) % ($\times$ 100)

  cc for the approximate carbon content % ($\times$ 100)

  d for other designations

Example 4.4 The indexing mechanism shown (Figure E4.4) has a rotating gear with a cam
profile on its top horizontal surface that pushes the slider, also having a cam profile, upwards
once per revolution. The slider moves along a stationary shaft and it is pressed downwards
by a helical spring. Select the appropriate materials.

Solution The shaft should be made of carbon steel. The slider has sliding friction against
the shaft and the gear. A good selection is a grade of surface-hardened steel or cast iron.
Surface-hardened steel needs to be selected for the gear. The bushing between gear and shaft
has to be made from a very dissimilar material, such as cast iron or phosphor–bronze, with
good antifriction properties. One of the spring steels grade is selected for the springs.

   Design calculations are needed to establish the specific material grade for each component.

4.5. FACTOR OF SAFETY: RELIABILITY

4.5.1. The Empirical Method

As a result of tension and compression tests, we obtain basic data on the mechanical
properties of a material. Let us now see how the results so obtained can be used
in machine design.

   The fundamental and most commonly used method of machine component sizing is the method based on stresses. The components must perform under loading
without failure. According to this method, the design is based on the maximum
stress developed at some point of a loaded structure. This stress is called the *maximum working stress.* It must not exceed a certain value, the *allowable stress,*
characteristic of a given material and the service conditions of the structure.

   Design based on stresses uses a relation of the form:

$$\sigma_{\max} = \frac{S_L}{N} \tag{4.14}$$

where S_L is a certain limiting stress for a given material and N is a number greater
than unity, called the *factor of safety.* The dimensions of a structure are usually
already known and assigned, for example, from service or technological considerations. In this case the value of $\sigma_{\max}$ is calculated to determine the existing factor
of safety:

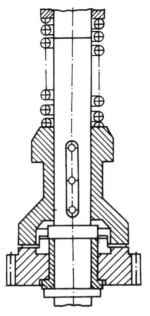

Figure E4.4

$$N = \frac{S_L}{\sigma_{max}} \tag{4.15}$$

If this factor of safety satisfies the designer, it is concluded that the design is acceptable.

When a structure is in the design stage and some characteristic dimensions are to be assigned directly from strength requirements, the magnitude of N is prescribed beforehand. The required dimensions are obtained from a relation of the form:

$$\sigma_{max} \leq \frac{S_L}{N} \tag{4.16}$$

The quantity on the right is called *the allowable stress*. It remains to be decided what stress is to be taken as a limiting stress S_L and how to assign N.

In order to avoid appreciable permanent deformation of a functioning structure it is customary in machine design to take the yield strength as S_L for ductile materials. Then the maximum working stress is S_y/N (Figure 4.22). Here the factor of safety is denoted by N and is called the factor of safety with respect to yielding. For brittle materials and in some cases for moderately ductile materials, S_L is taken to be the ultimate tensile strength S_u. We then obtain:

$$N_u = \frac{S_u}{S_{max}} \tag{4.17}$$

where N_u is the factor of safety with respect to fracture.

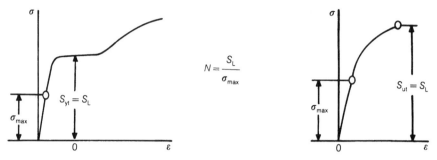

Figure 4.22. Definition of the safety factor.

For design based on any limiting load, it is possible to introduce similarly the factor of safety based on the limiting load:

$$N = \frac{P_L}{P_w} \qquad (4.18)$$

where P_L and P_w are the limiting and working loads. In design based on stiffness:

$$N = \frac{\delta_L}{\delta_w} \qquad (4.19)$$

where δ_L and δ_w are the limiting and working displacements, respectively.

The choice of N is made in two ways: analytically, based on reliability analysis as shown in Section 4.5.2, and empirically, as shown below.

The factor of safety assigned is based on experience with the specific service conditions for the structure being designed. The factor N is virtually determined by past practice and by the state of the art at the particular moment. Each field of engineering has its own traditions, requirements, methods, and specificity of designs, and the factor of safety is assigned accordingly. Thus, for example, in the designing of stationary engineering structures intended for prolonged service, the factors of safety are rather large ($N = 2$–5). In aircraft engineering, where severe weight restrictions are imposed on the structures, they are in the range 1.5–2. In view of the high reliability requirements, it has become the practice in this field to conduct obligatory static tests on individual components and complete aircraft for direct determination of the limiting loads.

The choice of the factor of safety depends on the methods of stress analysis, the degree of accuracy of these methods, and the gravity of consequences that the failure of a part may entail. The factor of safety depends also on the properties of the material. In the case of a ductile material, the factor of safety with respect to yielding may be lower than for a part made of brittle material. This is quite evident since a brittle material is more sensitive to accidental damage and unexpected manufacturing defects. Moreover, any accidental increase of stresses may cause only small permanent deformations in a ductile material, whereas for a brittle material this may result in failure. The safety factor is not, as is commonly thought, a fudge factor to give us a margin of safety, that is, secured higher strength than the maximum expected load. The proper choice of the factor of safety depends to a considerable extent on the judgment, experience, and ingenuity of the analyst and the designer, but also on the degree of uncertainty about material properties, methods of analysis, and service conditions.

If the above factors were known *exactly,* then the safety factor would be 1.

The differential method for the estimation of the factor of safety assumes this factor as a product of several sub-factors;

$$N = N_1 N_2 N_3 N_4 \ldots \tag{4.20}$$

where:

- N_1 reflects the reliability with which design loads can be determined. It takes values between 1 and 1.5. $N_1 = 1$ when rated loads are determined with unquestionable accuracy, or where safety devices protect the machine from overloading, such as pressure vessels with relief valves; $N_1 = 1.5$ when the load is determined from questionable data, such as wind generators.

- N_2 reflects the reliability of the material properties. It is taken to be 1.2–1.5 for rolled or forged steel and 1.5–2.5 for cast iron and brittle material, partly owing to easier inspection of the rolled steel and smaller effect of imperfection on ductile materials.

- N_3 depends on the consequences of a failure. It is taken to be 1 when failure will not affect anything else, or ≥ 1.5 when failure results in total loss of the machine, environmental damage, or danger to operators.

- N_4 refers to starting and accidental overloads, if the calculations are based on the rated load of the machine, to frequent starts, to operation with or without shocks, and so on. This subfactor N_4, also called *service factor,* depends on the particular application and can sometimes reach high values. N_4 can be taken up to a value of 2 for smooth operation and low starting torque, and to 5 for rough operation, high starting torque, and frequent start-ups. Of course, it is taken to be 1 when the exact operating loads have been used in the calculations. Design is based on the yield strength.

As a guide only, and if design is based on both the yield and fatigue strengths, N_4 is taken (see Decker 1973) to be:

1.3–1.5 for static loads, e.g., turbines and electric machines

1.4–1.6 for regularly varying loads, e.g., for reciprocating machines and machine tools

1.5–1.8 for irregularly varying loads, e.g., for punching and pressing machines

1.6–2.0 for impact load, e.g., stone-crushing machines and steel mills

---

Example 4.5 Select an empirical safety factor for the shaft of an electric motor driving an air compressor for paint spray, if the design is based on the rated load, with the differential method.

Solution Rated loads are very accurately determined because there are electrical load limit switches and pressure relief valves on the air side. Therefore, assuming that safety devices operate 10% above the rated load, $N_1 = 1.1$.

Rolled steel used for shafts is reliable and therefore $N_2 = 1.2$. For the application envisaged, $N_3 = 1$.

Smooth compressor operation and frequent start-ups are expected, requiring a rather high subfactor $N_4 = 1.6$.

Finally, $N = 1.1 \times 1.2 \times 1.0 \times 1.6 = 2.11$.

4.5.2. Statistical Character of Strength: Design for Reliability

The foregoing discussion might have left the reader with the impression that the purpose of the safety factor is to ensure that the strength of the structure is greater than the expected loading by a fixed percentage. This is of course not true. If the expected loads were known exactly (loading) and the material properties (strength) were also known exactly, then, as we have already explained, the safety factor would be exactly 1. There would be no reason at all for it to be otherwise. However, loads and material properties have a certain degree of uncertainty. Take, for example, the design of the shaft for an elevator motor. The maximum load is usually specified by the maximum allowed number of passengers. For the designer, this is translated into load, multiplying the number of passengers by the average weight of each person. However, people's weights are different. It we have the weights of a great number of people, we can plot the fraction $p(L)$ of the number of people with weight within a specified step range L, say between five successive kilogram integers (60–65, 65–70, 70–75, etc.) (Figure 4.23).

Such a histogram represents the distribution of weights of people of a certain type (locality, sex, age, etc.). If the sample is large enough and the spacing of steps small, this distribution appears as a continuous curve (Figure 4.24), the probability of the load curve. This curve has the property that the area under the curve has a value 1 because the sum of the loads that belong to all load ranges equals that total sample.

There is a value $\overline{L}$ that is the average of the sample population. This point is called the *mean* and can be computed as:

$$\overline{L} = \sum_{i=1}^{n} \frac{L_i}{n} \tag{4.21}$$

where L_i is the value of L of the sample i and n is the total number of sample loads. The closer the distribution curve is to the mean, the less is the deviation of the sample from the mean.

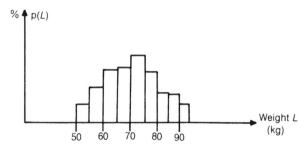

Figure 4.23. Histogram of weight distribution of people.

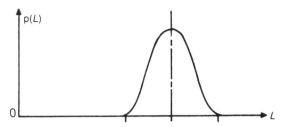

Figure 4.24. Load distribution.

To quantify that, we sum all the squares of the differences between each step range and the mean:

$$S_L^2 = \sum_{i=1}^{n} \frac{(L_i - \bar{L})^2}{n - 1} \tag{4.22}$$

We call this quantity *variance* and its square root (S) the *standard deviation* and denote it by σ if n is very large. Many physical quantities closely follow a specific type of curve, called the *normal distribution* (Figure 4.25), which is defined by:

$$p(L) = \frac{1}{S_L(2\pi)^{1/2}} \, e^{(-L-\bar{L})^2/2S_L^2} \tag{4.23}$$

Introducing the new variable:

$$t = \frac{L - \bar{L}}{S_L} \tag{4.24}$$

Equation (4.21) gives:

$$0 = \sum_{i=1}^{n} \frac{L_i - \bar{L}}{n} = S_L \sum_{i=1}^{n} \frac{L_i - \bar{L}}{nS_L} = S_L \bar{t} \tag{4.25}$$

which means that $\bar{t} = 0$. Equation (4.22) now gives:

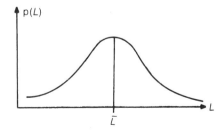

Figure 4.25. Normal distribution.

$$S_L^2 = \sum_{i=1}^{n} \frac{(L_i - \overline{L})^2}{n - 1} = \sum_{i=1}^{n} \frac{t_i^2 S_L^2}{n - 1} \qquad (4.26)$$

and

$$\sum_{i=1}^{n} \frac{t_i^2}{n - 1} = 1, \quad S_t = 1 \qquad (4.27)$$

which means that the standard deviation of t is unity. The normal distribution becomes

$$S_L p(L) = \overline{p}(t) = \frac{1}{(2\pi)^{1/2}} e^{-t^2} \qquad (4.28)$$

The integral

$$A_{t_1}^{t_2} = \int_{t_1}^{t_2} p(t)\, dt$$

gives the probability that the load is between t_1 and t_2. If, for example, between loads $L = 60$ and 70 kg the integral is 0.3, that means that 30% of the people of the sample weigh between 60 and 70 kg. For the normal distribution, tables exist that give the above integral. The following relation is an obvious identity:

$$A_{-\infty}^{t_1} + A_{t_1}^{t_2} + A_{t_2}^{\infty} = 1 \qquad (4.29)$$

The same behavior is exhibited by the material properties. The tabulated properties are actually *median* values, and it is expected that during repeated tensile tests half the specimens will fail below the mean strength and half above it. The strength has also a standard deviation S, which with most engineering standards should not exceed 8% of the value of the mean.

Suppose now that we plot in the same diagram the load, in the form of stress L on a particular section and the material capacity, in the form of the strength c (Figure 4.26). It is obvious that if overlap exists, some of the designs will fail,

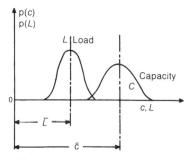

Figure 4.26. Distributions of load and capacity.

though the mean of the capacity (material strength) is greater than the mean of the load (stress).

The ratio

$$N = \frac{\bar{c}}{\bar{L}} \qquad (4.30)$$

is, by what was indicated in Section 4.5.1, the safety factor, expressing the degree of uncertainty of the expected loads and material strengths. The greater the safety factor, the less probability of failure (load exceeds capacity if $N < 1$).

To quantify this, let us form the difference for each design $D = c - L$. It is known from probability theory that:

$$\bar{D} = \bar{c} - \bar{L} \qquad (4.31)$$

$$S_D^2 = S_c^2 + S_L^2 \qquad (4.32)$$

Therefore, if the statistical properties, mean and standard deviation, of the load and the capacity are known, so are the statistical properties of the difference D. The distribution of D will also be normal (Figure 4.27). We will have failure when $D < 0$. Therefore, the probability of failure will be:

$$A_{-\infty}^0 = \int_{-\infty}^0 \frac{\exp[-t^2/2]dt}{\sqrt{2\pi}} \qquad (4.33)$$

where

$$t = \frac{0 - (\bar{c} - \bar{L})}{S_D} \quad \text{or} \quad t = \frac{-\bar{c} + \bar{L}}{[S_c^2 + S_L^2]^{1/2}}$$

Since $N = \bar{c}/\bar{L}$, dividing by $\bar{c}$:

$$t = \frac{1 - N}{[(S_c/\bar{c})^2 + (S_L/\bar{L})^2/N^2]^{1/2}} \qquad (4.34)$$

Therefore, if the relative standard deviations $S_c/\bar{c}$ and $S_L/\bar{L}$ are known, from a certain value of the safety factor N one can compute t and from a table of the

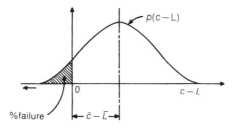

Figure 4.27. Distribution of the over capacity (capacity − load).

standard distribution compute the probability of failure. Therefore, a function $A = f(N)$ can be plotted (Figure 4.28). Since such function has been established for a given probability of failure, the appropriate safety factor N can be found.

The probability of failure is not purely a technical matter. For the Apollo mission, for example, it must have been close to zero. For home appliances, experience shows that it is higher than zero. In general, customer demands and economic factors determine the allowable probability of failure. The penalty for lower probability of failure is a higher safety factor, which leads to heavier sections and more expensive product. N is computed in the program SAFFAC (below). A probability of failure A is assigned and the corresponding safety factor N is to be computed. Given also are the ratios and $S_L/\overline{L}$ and $S_c/\overline{c}$.

The program starts with an initial guess N_0 and computes the probability of failure A_0, not equal in general to the given A. Then it iterates with the Newton-Raphson method on N until it comes close enough to $A - A_0 = 0$. A guess of N, (N_0), is assumed and t is then computed with Equation (4.23). The integral

$$A = A^t_{-\infty} = \int_{-\infty}^{t} \frac{\exp[-t^2/2]dt}{(2\pi)^{1/2}} \tag{4.35}$$

is then evaluated numerically from a large negative value of t to t_1. This can be done conveniently with the trapezoidal rule.

Let A_1 be the value of this integral for $N = N_1$. Given a small variation in N, say ΔN, this results in a change, ΔA. A numerical value of the derivative dA/dN is $\Delta A/\Delta N$. From geometry of Figure 4.28:

$$N_2 = N_1 - \frac{A_1 - A_0}{\Delta A/\Delta N} \tag{4.36}$$

In the same way the sequence of points N_3, N_4, . . . , is found, which converges, in general, at the desired point N.

A useful nomogram is shown in Figure 4.29. It shows the probability of failure on the left as a function of the ratio

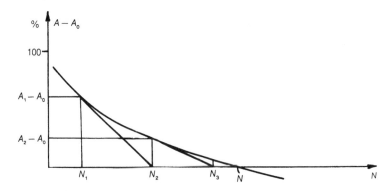

Figure 4.28. Probability of failure versus safety factor.

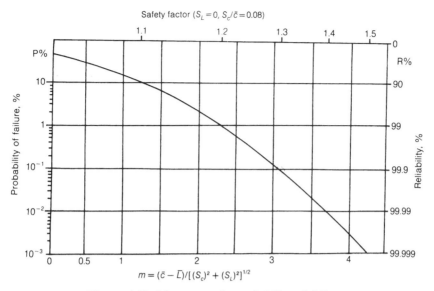

Figure 4.29. Nomogram for probability of failure.

$$m = \frac{\bar{c} - \bar{L}}{[S_c^2 + S_L^2]^{1/2}} = \frac{N - 1}{[(S_c/\bar{c})^2 N^2 + (S_L/\bar{L})^2]^{1/2}} \qquad (4.37)$$

When the allowed probability of failure and the distribution of the load and capacity are known, equation (4.37) can be solved for the safety factor N. This has been done in the module SAFFAC of MELAB 2.0 or using the function Solver of EXCEL.

Considering material uncertainty only $(S_L = 0)$ $m = (N - 1)/0.08N$ (for standard deviation of the material 8%), the safety factor N is shown in the upper horizontal.

Example 4.6 The cable for a high-rise building elevator is made of high-strength steel with standard deviation 8% of the yield strength. The standard deviation for the weight of the town's population per capita is 15 kg while the average weight is 75 kg. Determine the safety factor for a reliability of 99.99%:

1. With the statistical method and Figure 4.29.

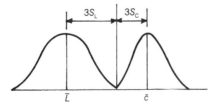

Figure E4.6

2. Assuming that all the population (both for materials and passengers) is 100% within 3 standard deviations.
3. With the SAFFAC program.

Solution

1. It is $S_c/\bar{c} = 0.08$, $S_L/\bar{L} = 15/75 = 0.2$. Using the nomogram of Figure 4.29, $P = 0.01\%$, $m = 3.7$. Therefore:

$$3.7 = \frac{N-1}{[(S_c/\bar{c})^2 N^2 + (S_L/\bar{L})^2]^{1/2}} = \frac{N-1}{[0.08^2\, N^2 + 0.2^2]^{1/2}}$$

Solution of the resulting quadratic equation yields $N = 1.936$.
2. The situation is shown in Figure E4.6. Obviously,

$$\bar{c} - \bar{L} = 3(S_L + S_c)$$

Dividing by L:

$$N - 1 = 3\left(\frac{S_L}{\bar{L}} + \frac{S_c N}{\bar{c}}\right)$$

$$N = \frac{1 + 3(S_L/\bar{L})}{1 - 3(S_c/\bar{c})} = \frac{1 + 3 \times 0.2}{1 - 3 \times 0.08} = 2.1$$

3. Run of SAFFAC gives $N = 1.936$.

CASE STUDY 4.1: Leroy Matisse vs. Ladders Unlimited, Inc.[5]

Case Summary

Leroy Matisse is a painting contractor in Mt. Vernon, Illinois. In 1972, he purchased a ladder standoff manufactured by Ladders Unlimited, Inc. of Buffalo, N.Y. On September 12, 1983, he was painting the exterior of a house in Mt. Vernon, Illinois, and was on the ladder when suddenly the ladder standoff aluminum pipe structure broke at the upper left bolt of the top step and Mr. Matisse fell from a 14 ft height. He was seriously injured.

Investigation

The ladder standoff in question was manufactured by Ladders Unlimited, Inc. and consists of a tubular aluminum structure as shown in Figure CS4.1. The structure consists of two 1 in. diameter, $1/16$ in. thick aluminum tubes bent and connected as shown by way of two braces of $3/4 \times 1/8$ in. aluminum strip and two $1/16$ in. thick tubular aluminum steps, as shown. The ladder standoff is placed on the top of an aluminum ladder by inserting the vertical tubes over the ladder tubes. It keeps the ladder at some distance from the wall to facilitate the work.

[5] Courtesy of Professor Wallace Diboll, Consulting Engineer and Professor of Mechanical Design, Washington University, St. Louis, Missouri.

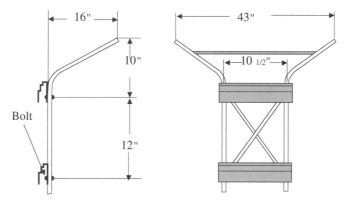

Figure CS4.1. A ladder standoff.

Each of the two tubular aluminum steps is secured on the aluminum structure by way of one ¼ in. bolt on each side that penetrates the aluminum tube through an 8 mm diameter hole.

The 1 in. aluminum tube fracture was right through that hole on the top left hook (see Figure CS4.1). The broken standoff is shown in Figure CS4.2. A close-up of the broken structure is in Figure CS4.3. By inspection of Figure CS4.3, one can see that it was a brittle fracture without appreciable plastic deformation.

The material of the tubes was 2011 tempered aluminum with plastic strain at fracture $\varepsilon_u = 0.10$, while a low-carbon steel (AISI 1018) has $\varepsilon_u = 1.05$, the ultimate strengths are 324 and 341 MPa, respectively. Thus the plastic energy at fracture is, respectively, 32.4 and 358 Nm/m^3 (see Equation (4.2)).

Figure CS4.2

Figure CS4.3

It was concluded that the fracture was due to a poor choice of material (brittle material in an application that anticipates impact) in combination with the improper placement of the 8 mm hole that weakened the aluminum tube.

A consulting engineer, a world-renowned pioneer in fracture mechanics, testified as an expert witness and concurred with the above conclusions.

REFERENCES

American Society of Mechanical Engineers (ASME). 1995A. *Dimensioning and Tolerancing: ANSI Standard V14.5M.* New York: ASME Press.

———. 1995B. *Preferred Limits and Fits: ANSI Standards B4.1-1967 and B4.2-1978.* New York: ASME Press.

American Society for Testing and Materials (ASTM, 1991). *Standard Practice for Numbering Metals and Alloys.* Standard No. E527-83. West Conshocken, Pa.: ASTM.

Decker, K.-H. 1973. *Maschinenelemente.* Munich: C. Hanser Verlag.

De Garmo, E. P. 1974. *Materials and Processes in Manufacturing.* New York: Macmillan.

Feodosyev, V., 1973. *Strength of Materials.* Mason: Mir Publishers.

Ulrich, K. T., and S. D. Eppinger. 1995. *Product Design and Development.* New York: McGraw-Hill.

ADDITIONAL READING

Clausing, D. *Total Quality Development.* New York: ASME Press, 1994.

Kverneland, K. O. *World Metric Standards for Engineering.* New York: Industrial Press, 1978.

Machine Design 53(6) (1981). Materials Issue.

American Society for Metals (ASM). *Metals Handbook.* Materials Park, Oh.: ASM, 1974.

Popov, E. P. *Mechanics of Materials,* 2nd ed. Englewood Cliffs, N.J.: Prentice-Hall, 1976.

Shigley, J. E. *Mechanical Engineering Design.* New York: McGraw-Hill, 1977.

PROBLEMS[6]

4.1. [T] Determine and compare the elastic energy up to yielding and the plastic energy up to rupture for two bars of identical geometry loaded by axial impact, per unit volume of the bar. The cross-section is circular and the materials of the two bars have deformation at rupture $\delta\% = 35$ and 22, yield strength $S_y = 220$ MPa and 400 MPa, and ultimate strength $S_u = 550$ MPa and 700 MPa. Comment on which bar can better take an axial impact.

4.2. [T] Solve Problem 4.1 for a tapered square cross-section of side a at mid-span and $0.8a$ at the supports. Assume that the impact is sufficient to cause general yielding through the length of the bar.

4.3. [T] Solve Problem 4.1 if the bar has a double T-section, the impact makes the flanges of the beam to yield between $\frac{1}{3}$ and $\frac{2}{3}$ of the length, the energy of the web is negligible, the beam is simply supported, and the impact is at mid-span by a lateral load in the direction of the plane of the web.

4.4. [T] Determine and compare the elastic energy up to yielding and the plastic energy up to rupture for two hollow cylinders of inner diameter $d = 100$ mm and thickness 10 mm of identical geometry loaded by torsional impact, per unit volume of the bar. The materials of the two bars have angular deformation at rupture $\gamma\% = 35$ and 22, shear properties, yield strength $S_{sy} = 100$ MPa and 200 MPa, and ultimate strength $S_{su} = 300$ MPa and 600 MPa. Comment on which bar can better take an axial impact. Consider uniform yielding along the radius since the thickness is small as compared with the radius.

4.5. [T] Determine and compare the elastic energy up to yielding and the plastic energy up to rupture for two full cylinders of diameter $d = 100$ mm of identical geometry loaded by torsional impact, per unit volume of the bar. The materials of the two bars have angular deformation at rupture $\gamma\% = 35$ and 22, shear properties, yield strength $S_{sy} = 100$ MPa and 200 MPa, and ultimate strength $S_{su} = 300$ MPa and 600 MPa. Comment on which bar can better take an axial impact. Consider the case that the impact is just enough for rupture at the outer fiber of the cylinder.

4.6. [D] For the fluid (steam) relief valve shown in Figure P4.6, 1 is an adjusting nut, 2 a spherical seat, 3 a guide shaft, 4 a helical spring, 5 the body. Prepare a parts list.

---

[6][C] = certification, [D] = design, [N] = numerical, [T] = theoretical problem.

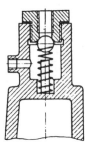

Figure P4.6

4.7. [D] For the fluid (steam) relief valve shown in Figure P4.7, 1 is an adjusting nut, 2 a spring seat, 3 a helical spring, 4 a guide shaft, 5 the body. Prepare a parts list.

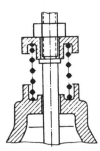

Figure P4.7

4.8. [D] For the fluid (steam) relief valve shown in Figure P4.8, 1 is the body, 2 a valve seat, 3 a helical spring, 3 a guide shaft, 4 a closing nut, 5 an adjusting nut. Prepare a parts list.

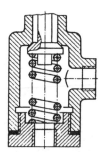

Figure P4.8

4.9. [D] For the indexing mechanism shown in Figure P4.9, 1 is a spring holder, 2 a cam, 3 a support plate, 4 the body, 5 a support bolt, 7 a helical spring. Prepare a parts list.

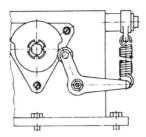

Figure P4.9

4.10. [D] For the steam safety valve shown in Figure P4.10:

  1. Explain the operation of the valve.

  2. Identify the numbered components.

  3. Prepare a parts list.

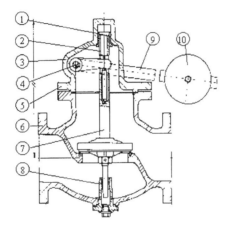

Figure P4.10

4.11. [D] For the bearing shown in exploded view in Figure P4.11:

  1. Explain the operation of the bearing.

  2. Identify and number the components.

  3. Prepare a parts list.

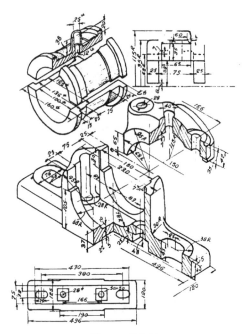

Figure P4.11

4.12. [D] For the steam valve shown in Figure P4.12:
1. Explain the operation of the valve.
2. Identify and number the components.
3. Prepare a parts list.

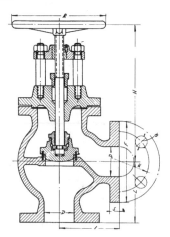

Figure P4.12

4.13. [D] For the check valve shown in Figure P4.13, which allows flow of the fluid only from the left to the right:
1. Explain the operation of the check valve.
2. Identify and number the components.
3. Prepare a parts list.

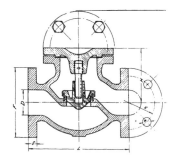

Figure P4.13

4.14. [D] For the shaft coupling shown in Figure P4.14:
1. Explain the operation of the check valve.
2. Identify and number the components.
3. Prepare a parts list.

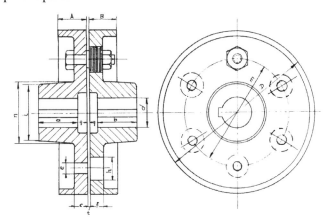

Figure P4.14

4.15. [D] For the fluid (steam) relief valve shown in Figure P4.6, Problem 4.6, the adjusting nut is made of low-carbon steel and the body of brass. What processes will be used for manufacturing of the two components?

4.16. [D] For the fluid (steam) relief valve shown in Figure P4.7, Problem 4.7, the spring seat is made of cast iron and the guide shaft of high-carbon steel. What processes will be used for manufacturing of the two components?

4.17. [D] For the fluid (steam) relief valve shown in Figure P4.8, Problem 4.8, the body is made of cast iron and the closing nut of bronze. What processes will be used for manufacturing of the two components?

4.18. [D] For the indexing mechanism shown in Figure P4.9, Problem 4.9, the support plate is made of low-carbon steel and the support bolt is low-carbon steel. What processes will be used for manufacturing of the two components?

4.19. [D] For the steam safety valve shown in Figure P4.10, Problem 4.10, the body is made of stainless steel and the valve assembly of high-carbon steel. What processes will be used for manufacturing of the two components?

4.20. [D] For the bearing shown in exploded view in Figure P4.11, Problem 4.11, the body of the bearing is made of cast iron and the bearing sleeve is phosphor bronze. What processes will be used for manufacturing of the two components?

4.21. [D] For the steam valve shown in Figure P4.12, Problem 4.12, the valve seat is made of low-carbon steel and the screw of high-carbon steel. What processes will be used for manufacturing of the two components?

4.22. [D] For the check valve shown in Figure P4.13, Problem 4.12, the top flange is made of cast iron and the valve seat of copper–nickel alloy. What processes will be used for manufacturing of the two components?

4.23. [D] For the shaft coupling shown in Figure P4.14, Problem 4.14, the two halves of the coupling are made of cast steel and the bolts of low-carbon steel. What processes will be used for manufacturing of the two components?

4.24. [D] The crankshaft of an automobile engine revolves on main bearings of nominal diameter 42 mm. Find the diameter range of bearing and shaft if the fit is hole-basis $H8F7$. Then determine the same range if the equivalent shaft-basis is selected.

4.25. [D] Select proper fits for the following applications:
   1. Valve stem in the valve guide of an automobile engine
   2. Head journal-bearing for a precision lathe
   3. Interference fit of coupling and shaft, if the torque is transmitted through a key
   4. A fit to locate a ball bearing on a shaft

4.26. [D] Identify on a lawnmower:
   1. A clearance fit
   2. A transition fit
   3. An interference fit

4.27. [D] Draw a sketch of an electric motor and a sectional view along the axis of rotation and identify all the fits that need to be specified.

4.28. [D] Determine the standard fit that describes a production lot of shafts of nominal diameters 30 and 72 mm with limit dimensions:

$$-0.007 \qquad 0.000$$
$$30 \qquad\qquad 72$$
$$-0.020 \qquad -0.019$$

4.29.–4.33. [D] For the fits indicated in Figures P4.29 to P4.33, respectively, determine the limit dimensions for shaft and hole, selecting appropriate standard fits wherever applicable.

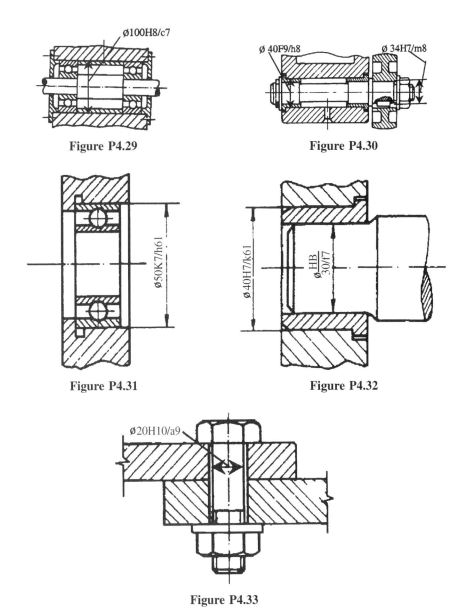

Figure P4.29

Figure P4.30

Figure P4.31

Figure P4.32

Figure P4.33

4.34. [D] For the fluid (steam) relief valve shown in Figure P4.6, Problem 4.6, continue the solution of that problem by entering the parts list, the material, and the manufacturing process you suggest for the part. Justify your selection.

4.35. [D] For the fluid (steam) relief valve shown in Figure P4.7, Problem 4.7, continue the solution of that problem by entering the parts list, the material, and the manufacturing process you suggest for the part. Justify your selection.

4.36. [D] For the fluid (steam) relief valve shown in Figure P4.8, Problem 4.8, continue the solution of that problem by entering the parts list, the material, and the manufacturing process you suggest for the part. Justify your selection.

4.37. [D] For the indexing mechanism shown in Figure P4.9, Problem 4.9, continue the solution of that problem by entering the parts list, the material, and the manufacturing process you suggest for the part. Justify your selection.

4.38. [D] For the steam safety valve shown in Figure P4.10, Problem 4.10, continue the solution of that problem by entering the parts list, the material, and the manufacturing process you suggest for the part. Justify your selection.

4.39. [D] For the bearing shown in exploded view in Figure P4.11, Problem 4.11, continue the solution of that problem by entering the parts list, the material, and the manufacturing process you suggest for the part. Justify your selection.

4.40. [D] For the steam valve shown in Figure P4.12, Problem 4.12, continue the solution of that problem by entering the parts list, the material, and the manufacturing process you suggest for the part. Justify your selection.

4.41. [D] For the check valve shown in Figure P4.13, Problem 4.13, continue the solution of that problem by entering the parts list, the material, and the manufacturing process you suggest for the part. Justify your selection.

4.42. [D] For the shaft coupling shown in Figure P4.14, Problem 4.14, continue the solution of that problem by entering the parts list, the material, and the manufacturing process you suggest for the part. Justify your selection.

Select the appropriate safety factors by the empirical method for the following elements:

4.43. [D] Main shaft of a lawnmower.

4.44. [D] Valve stem for an automotive engine.

4.45. [D] Head bolt for a diesel engine.

4.46. [D] Anchoring cables for a TV tower.

4.47. [D] Main shaft of a steel mill.

4.48. [D] Main propeller shaft of a helicopter.

4.49. [D] On a small gasoline engine, measurements were conducted at full throttle and at rated speed. The 8% standard deviation materials, for 99% reliability, are:

1	2	3	4	5	6	7	8	9	10
3.5	3.65	3.32	3.35	3.74	3.61	3.6	3.52	3.56	3.41

If the computations are based on the rated horsepower 4.5 HP, find the proper safety factor for 8% standard deviation materials and 99% reliability.

4.50. [D] An air compressor of 10 kW was designed for standard deviations of the load 15% and the material 10% for a 95% reliability. Subsequently, the marketing department suggested that 99% reliability was essential. It was decided to lower the rated power in order to achieve the required reliability. Calculate the reduced rated power.

4.51. [D] Measurements of stresses owing to pressure were conducted at rated load on a boiling-water reactor pressure vessel. It was found that standard deviation was 10%. Thermal stresses could not be measured with accuracy. Temperature measurements were made, showing a standard deviation of 18%. It was then assumed that thermal stresses would have a similar behavior. Assuming that mean values are additive while standard deviations have a square root of sum of the squares law, determine the safety factor for standard material 8% if the design stresses are the sum (pressure + thermal) stresses.

4.52. [D] During the initial production of a new gas turbine blade design, it was found that the rate of failures at a given period of time was 3%, more than the competitor's rate, which was 2%. The design was based on 10% standard deviation of the load and 8% of the material.

1. Is it possible, using nondestructive test methods, to screen the materials in order to isolate a new lot of materials with smaller standard deviation to achieve the 2% failure rate?

2. In this case, what percentage of the purchased materials will be rejected, assuming that 100% of the population is within 3 standard deviations of the mean?

4.53. [D] Develop a table giving the safety factor for different probabilities of failure for 8% standard deviation material for four different values of the relative standard deviation of the load in steps of 2.5%, starting at 2.5%.

CHAPTER 5

SIZING MACHINE COMPONENTS FOR STRENGTH

A Budd Manufacturing Co.-designed locomotive. Failures of locomotive axles prompted the early investigations into machine element static and fatigue strength.

OVERVIEW

In the foregoing chapters, we saw how we can analyze the machine as a system, find the motion of each component, find the forces that this component carries, and finally find the internal forces and moments that the different sections of the component carry. Moreover, in Chapter 4 we saw that each section, due to material limitations, can carry but a limited load to preserve the structural integrity of the respective component.

In this chapter, we shall see how the sizing of each component can be performed by comparing the forces that each section must carry to the capacity of the material it is made of to carry forces without failing.

5.1. SIMPLE STATES OF STRESS

5.1.1. Design for Strength Procedure

Every section of a structural member is capable of carrying a maximum load, whose value depends on the material strength and the size of the section. On the other hand, this limiting load should not exceed the maximum working load expected during the operation of the machine. This leads to the most common procedure for sizing a machine part so that its structural integrity will be maintained during operation:

1. *Identification of the mode of failure.* From past experience or from theoretical considerations we identify one or more possible mechanisms of failure. For example:
   - A crankshaft can fail due to the transmitted torque causing excessive shear stress.
   - A steel cable of a crane may fail due to excessive load that the cable carries.
   - A blade of a lawn mower can break due to excessive stresses if it strikes a stone.
2. *Quantification of failure.* As we have seen in the previous chapter, the basic method of quantification of failure is to observe that failure occurs when *demand exceeds capacity*. In terms of stress, failure occurs when the working stress due to the operating conditions of the machine σ reaches the allowable stress S/N, the ratio of some applicable material strength S divided by a quantity N termed safety factor (see Chapter 4). In machine design, permanent deformation of a machine part is not acceptable, in general, and the applicable material strength is the yield strength S_y. The safety factor reflects the fact that the material property S_y was obtained in the laboratory under close control of the material and the applied load, while in a real application there are many uncertainties. Therefore, the *design equation* is:

$$\sigma = S_{\text{all}} = \frac{S_y}{N} \tag{5.1a}$$

3. *Sizing the part.* Equation (5.1a) is a design equation that relates stress and strength. If, for example, the stress is due to tension of a cylindrical bar of diameter d, this stress (force/area) can be compared with the allowable stress that we find if we divide the yield strength of the material S_y by the safety factor N:

$$\sigma = \frac{F}{\pi d^2/4} = \frac{S_y}{N} \tag{5.1b}$$

Equation (5.1b) can be solved for the unknown design parameter (dimension) d if the operating conditions (force F) and the material properties (S_y) are known.

In Chapter 4, we saw that stress *capacity* is the allowable stress, the material strength divided by the safety factor, and stress *demand* is the stress σ due to the machine load. Failure happens when *demand exceeds capacity*. Of course, this relationship has meaning if the stress σ is the same type of stress applied to a material specimen to obtain the yield strength, if the latter is the material strength we use.

Certification. In some cases, we need to find the safety factor of an existing machine, typical case is the acceptance test for a machine or a machine part. In Equation (5.1*a*), the operating conditions, the geometry, and the materials are known and unknown is the safety factor. Therefore, the *certification equation* is:

$$N = \frac{S_y}{\sigma} \tag{5.2}$$

Almost any realistic loading of a machine element results in a general distribution of stresses and strains within the element. Detailed analysis is in principle possible, especially in view of the computer analysis methods now available. However, engineers are always trying to isolate in a problem the most important features for the specific requirements and disregard factors that, in their best judgment, do not contribute to the design goal in proportion to the effort needed for taking these factors into account. It seems strange that although sufficient computation facilities exist to analyze most conceivable loading situations, we will concentrate on the important parts of the problem disregarding minor factors. But the following have to be considered:

1. Many times, analytical or computational effort to include a certain factor in design might cost much more than the additional cost that will result from accepting a somewhat greater safety factor instead.
2. Calculations of high complexity sometimes lead to a loss of the general understanding of the problem.
3. Even in the case of a justified complete computation effort, simplified analyses are needed for preliminary design, to establish design solutions that subsequently might be further analyzed and optimized.
4. Long design practice has accumulated invaluable experience. Such experience is difficult to be transferred and incorporated in a short time into the new methods.

Simplified and modern methods therefore have to coexist for a long time to come. When can a certain factor be considered as unimportant in a machine design effort? There is no simple answer to such a question. The designer is not simply a computer; experience and judgment cannot be easily replaced.

5.1.2. Tensile–Compressive Loading

This is the simplest, yet the most useful, design loading situation in machines because direct use of the results of the standard material tests can be made. In fact, most material properties are tabulated for such loading. Therefore, the design equation for members that are loaded significantly only by tension–compression is Equation (5.1).

This equation is applicable to long prismatic elements, but approximations for other shapes can be effectively made. Usually, the force is given, a material is selected, and Equation (5.2) yields the required section dimension.

Equation (5.1) is simple but requires care in its application. It presumes a uniform distribution of stresses over the cross-section, which might not be true in certain cases, such as the following:

1. Near the point of application of forces and the points of supports: For this reason, axial members such as bolts and rivets are designed with an increased cross-section near such points.
2. Eccentric application of load or eccentric reaction of the support: In such cases, bending moments are introduced that result in bending stresses which might be much higher than pure tensile stresses.
3. Misalignment of axially loaded members: This again can introduce shear or bending loads and can be avoided with proper design and manufacturing to ensure alignment.
4. Holes, notches, stress raisers in general: Such field disturbances are most dangerous in pure tensile loading because parts of the section of the element are under maximum stress.

In addition, only limiting the stress below the allowable limits does not necessarily mean a good design. Other requirements must be met, such as rigidity and stability, resistance to fatigue and fracture, surface strength, and resistance to wear.

Example 5.1   The cylindrical rod of a hydraulic piston is subject to a compressive force F due to the piston $p = 10$ MPa in a cylinder of inner diameter $D = 100$ mm. Design the rod. It is to be made of low-carbon steel with $S_y = 150$ MPa, and the design should have a safety factor $N = 2$.

Solution

Figure E5.1

TABLE E5.1

DATA:			
Description	Symbol	Value	Units
Cylinder diameter	D	0.1	m
Pressure	p	1.00E+07	Pa
Yield strength	S_y	1.50E+08	Pa
Safety factor	N	2	
Cylindrical rod, diameter d_r			
The total axial piston force is	$F = p(\pi D^2/4) =$	78539.81634 N	(Pressure force)
Average tensile stress at rod	$\sigma = F/(\pi d_r^2/4) = S_y/N$		
Solving for d_r	$d_r^2 = 4FN/\pi S_y =$	0.001333333	
Required rod diameter	$d_r =$	0.036514837 m, or	36.148 mm
Final Design Decision		38 mm rod diameter is selected	

Design Critique:

The solution is acceptable. Buckling analysis might be needed, see Chapter 9. We must pay attention to the detail design at the two ends of the rod not to create conditions that will invalidate the assumption of the average stress in the section.

5.1.3. Shear Loading

Pure shear is a state of stress in which the faces of an isolated element are acted upon only by shearing stresses. Such a state of stress is characterized by parallel displacements of two parallel faces of the element relative to each other.

If the shearing stresses are considered to be uniformly distributed over the area of their action, then the shear stress is:

$$\tau = \frac{V}{A} \tag{5.3}$$

and the design equation for pure shear is:

$$\tau = \frac{V}{A} \leq \frac{S_{sy}}{N} \tag{5.4}$$

where S_{sy} is the yield strength in shear.

The same reservations for application of the design equation (5.4) exist as for pure axial loading. In particular, nonaligned application of the punch (Figure 5.1b) results in nonuniform shear stress over section A. This is more pronounced in brittle materials, while in ductile materials it might be partly relieved by local yielding.

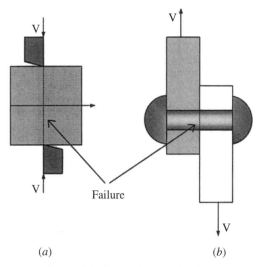

Figure 5.1. Pure shearing loading.

Example 5.2 The pin of the connecting rod shown in Figure E5.2 is loaded with a maximum force $V = 25$ kN. The pin material is low carbon steel with yield strength in shear $S_{sy} = 220$ MPa. For design based on the average shear stress, the safety factor is $N = 2$. Find the diameter of the pin.

Solution The stress area is $2\pi d^2/4$ because there are two sections of the pin to shear in the case of shear failure. The design equation is, using the failure criterion of the pure shear,

$$\tau_{av} = \frac{V}{A} = \frac{V}{2\pi d^2/4} = \frac{S_{sy}}{N}$$

$$4 \times \frac{25,000}{2\pi d^2} = 220 \times \frac{10^6}{2}$$

Therefore:

$$d^2 = \frac{2 \times 4 \times 25,000}{2\pi \times 220 \times 10^6} = 0.000145, \ d = 0.012 \text{ m or } 12 \text{ mm}$$

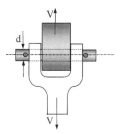

Figure E5.2

Design Critique

The solution looks reasonable. Because it was obtained on the assumption that two sections take the shear load, uniform distribution of the load on the pin is essential. Bending load was assumed to yield negligible contribution to failure. If the axial clearance is substantial, this assumption might not be valid.

5.1.4. Torsion

Prismatic elements are many times mainly loaded by torque, such as in shafts and torsional springs. The relation between torque T due to power transmitted in Newton-meters (Nm), the angular speed of rotation ω in rad/s and power P in watts is:

$$T = \frac{P}{\omega} \tag{5.5}$$

For the rotational speed n (rpm) and the power P, appropriate conversion of units leads to:

$$M = 7121 \frac{P(\text{hp})}{n} \text{ (Nm)} \tag{5.6}$$

$$M = 9549 \frac{P(\text{kW})}{n} \text{ (Nm)} \tag{5.7}$$

The shearing stress τ, at an arbitrary point at distance r from the center in a cross-section, is (Beer and Johnston 1981):

$$\tau = \frac{Tr}{J_p} \tag{5.8}$$

in which J_p is the polar moment of inertia of a circular cross-section of diameter $d = 2R$, $J_p = \pi d^4/32 \approx 0.1d^4$.

The maximum shearing stress, at the point most remote from the center, is:

$$\tau_{\max} = \frac{TR}{J_p} \leq \frac{S_{sy}}{N} \tag{5.9}$$

In Table 5.1, the maximum shear stress due to torsion has been compiled for various cross-sections. Polar moments of inertia of various cross-sections have been compiled in Appendix F. For the polar moments of inertia of compound cross-sections, Steiner's rule applies:

$$J_p = \sum_{i=1}^{n} [J_{pi} + r_i^2 A_i] \tag{5.10}$$

where J_{pi} is the polar moment of inertia about the centrode of the component section

TABLE 5.1. Torsional Loading of Sections

Section	Maximum stress	Twist angle per unit length
	$\tau_A = \dfrac{2T}{\pi hb^2}$	$\phi = \dfrac{h^2 + b^2}{\pi h^3 b^3} \cdot \dfrac{T}{G}$
	$\tau_A = \dfrac{20T}{b^3}$	$\phi = \dfrac{80}{b^4\sqrt{3}} \cdot \dfrac{T}{G} = \dfrac{46.2}{b^4} \dfrac{T}{G}$
	$\tau_A = \dfrac{T}{\alpha bh^2}$ $\alpha = 8/(3a + 1.8b)$ $\beta = 16/3 - 3.36\dfrac{b}{a}\left[1 - \dfrac{b^4}{12a^4}\right]$	$\phi = \dfrac{1}{\beta bh^3} \cdot \dfrac{T}{G}$
	$\tau = \dfrac{3T}{2\pi rt^2} = \dfrac{3r}{t} \cdot \dfrac{T}{2\pi r^2 t}$	$\phi = \dfrac{3}{2\pi rt^3} \cdot \dfrac{T}{G}$ $= \dfrac{3r^2}{t^2} \cdot \dfrac{1}{2\pi r^3 t} \cdot \dfrac{T}{G}$
	$\tau = \dfrac{Tr}{J_p} = \dfrac{T}{2\pi r^2 t}$	$\phi = \dfrac{T}{J_p G} = \dfrac{1}{2\pi r^3 t} \cdot \dfrac{T}{G}$
	$\tau_A = \dfrac{T}{2\pi bht}$	$\phi = \dfrac{\sqrt{2(b^2 + h^2)}}{4\pi b^2 h^2 t} \cdot \dfrac{T}{G}$
	$\tau_A = \dfrac{T}{2bht_1}$ $\tau_B = \dfrac{T}{2bht}$	$\phi = \dfrac{bt + ht_1}{2tt_1 b^2 h^2} \cdot \dfrac{T}{G}$

i, r_i is the distance of that centrode from the centrode of the compound section, and A_i is the area of the component section i.

Example 5.3 An electric motor has rated power 3 kW at 1650 rpm and it is expected that if failure occurs, this will be by torsional failure of the shaft. The shaft material is low-carbon steel with yield strength in shear S_{sy} = 220 MPa. For design based on the maximum shear stress due to torsion and the rated load, the safety factor is N = 2.5. Find the diameter of the shaft.

Solution The rated torque is (Equation (5.5)) $T = P/\omega = 3000/(2\pi \times 1650/60) = 9736 \times (3/1650) = 17.36$ Nm. The design equation is (Equation (5.9)):

$$\tau_{max} = \frac{TR}{J_p} \leq \frac{S_{sy}}{N}$$

$$\frac{17.36R}{\pi(2R)^4/32} = 220 \times \frac{10^6}{2.5}$$

Therefore:

$$R^3 = 17.36 \times 32 \times 2.5/220/\pi/16/10^6 = 0.126 \times 10^{-6}$$

$$d = 2R = 2 \times 0.0050 = 0.010 \text{ m or 10 mm}$$

Design Critique

The result looks reasonable, on the basis of experience from similar motors. The design was based on pure torsional stress contribution to failure. Notches, keyways, bending and other factors that might also affect stress failure must be avoided.

5.1.5. Pure Bending of Beams

We saw in Chapter 3 that long elements such as beams and shafts can have substantial internal loads in the form of section bending moment and shear. If only bending moment is present in a straight beam, for an isotropic, linear material with a symmetric cross-section, experimental observations have shown that during deformation due to application of a bending moment, plane cross-sections of the beam remain plane during deformation, that is, section ab will change into another plane section $a'b'$ under bending moment M (Figure 5.2a). Since the material is linear, Hooke's law states that stress is proportional to strain, and thus the distribution of the stress is linear, (Figure 5.2b).

The stress at any point y will be (Girard 1798, de St. Venant 1837):

$$\sigma = a + by \tag{5.11}$$

where a and b are constants. Let us observe the equilibrium of the portion of the beam on the left of the section. Equilibrium in the x direction demands that:

$$\int_A \sigma w(y)\, dy = \int_A (a + by)w(y)\, dy = a \int_A w(y)\, dy + b \int_A yw(y)\, dy = 0 \tag{5.12}$$

Therefore, since $\int_A w(y)\, dy = A$, the area of the section, the only way to satisfy Equation (5.12) is to be $a = 0$ and $\int_A yw(y)\, dy = 0$. The integral $\int_A yw(y)\, dy$ is the first moment of the area about the origin and is zero at some point of the section. At this point, $y = 0$ and the stress $\sigma = 0$ according with Equation (5.11). We call this point the *centroid* and the geometric axis of the beam through the centroid the *neutral axis*.

Equilibrium also demands that the moment of the normal forces over the area about the origin should equal the bending moment:

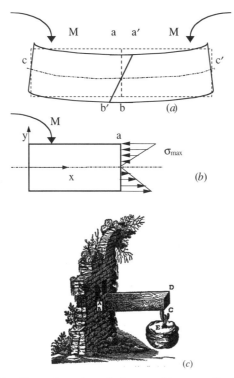

Figure 5.2. (*a*) Pure bending of a prismatic beam; (*b*) stress distribution; (*c*) Galileo's illustration of a bending strength test.

$$M = \int_A \sigma y w(y) \, dy = \int_A b y^2 w(y) \, dy = b \int_A y^2 w(y) \, dy$$

$$b = \frac{M}{\displaystyle\int_A y^2 w(y) \, dy} = \frac{M}{I} \tag{5.13}$$

where $I = \int_A y^2 w(y) \, dy$ is the second moment of the area, or *area moment of inertia,* of the section about the z-axis. For a circular cross-section, the centrode is at the center of the circle and $I = \pi d^4/64$. For different geometries of cross-sections, first and second section moments of areas are tabulated in Appendix F.

For compound cross-sections, *Steiner's rule* applies: The location of the centrode can be found by using first an arbitrary x-axis along the beam and locating the centrode at:

$$Y_c = \frac{\int_A y_i w(y_i) \, dy}{A} = \frac{\sum_{i=1}^n y_i A_i}{\sum_{i=1}^n A_i} \tag{5.14}$$

$$I = \sum_{i=1}^n (I_i + y_i^2 A_i) \tag{5.15}$$

where I_i is the moment of inertia about an axis through the centrode of the component section i, y_i is the distance of that centrode from the x-axis and A_i the area of the component section i. Thus, Equation (5.11) becomes:

$$\sigma = \frac{My}{I} \tag{5.16}$$

and the maximum stress is:

$$\sigma_{max} = \frac{Mc}{I} \le \frac{S_y}{N} \tag{5.17}$$

where c is the maximum vertical distance of any point on the perimeter of the section from the neutral axis. For example, for a circular cross-section, $c = R$. Equation (5.17) is the basic design equation for bending of beams.

Example 5.4 A steel beam of length $L = 2$ m, of square cross-section, simply supported at the two ends, supports at mid-span a hoist with maximum load capacity $F = 30$ kN. The beam material is structural steel with yield strength $S_y = 250$ MPa. For design based on the maximum bending stress and the rated load, the safety factor is $N = 3$. Find the size of the beam.

Solution   The support reactions are:

$$R_1 = R_2 = \frac{F}{2} = \frac{30,000}{2} = 15,000$$

The rated maximum bending moment is:

$$M_{max} = R_1 \times \frac{L}{2} = 15,000 \times \frac{2}{2} = 15,000 \text{ Nm}$$

The area moment of inertia for rectangular cross-section of side a is $a^4/12$ and $c = a/2$. The design equation (Equation (5.17)) is:

$$\sigma_{max} = \frac{M_{max}c}{I} \le \frac{S_y}{N}$$

$$15,000 \times \frac{a/2}{a^4/12} = 250 \times \frac{10^6}{3}$$

Therefore, solving for a:

$$a^3 = \frac{15,000 \times 12 \times 3}{250 \times 10^6 \times 2} = 0.000108, \ a = 0.1026 \text{ m}$$

$a = 104$ mm is selected.

Design Critique

The result looks reasonable, on the basis of experience from similar designs. The design was based on pure bending stress contribution to failure. Notches, holes, torsion due to unequal distribution of the load, and other factors that might also affect stress failure must be avoided.

5.1.6. Bending and Shear in Beams

If a beam is subject to both shear and bending moment, as usually happens, there is a shear stress developing over the cross-section. If we take a piece of the beam (Figure 5.3) from location $y = y_1$ and above, of length dx, at x the loads are M and V and at $x + dx$ the internal loads change to $V + dV$ and $M + dM$, respectively. At edge $(x + dx, y_1)$ stresses $\tau_{xy} = \tau_{yx}$ and equilibrium of the piece of beam A under the influence of the normal stresses, the shear stresses, and given that $V = dM/dx$ (see Chapter 2), yields (Beer and Johnston 1981):[1]

$$\tau = \frac{VQ}{Ib}, \quad Q = \int_{y_1}^{c} y \, dA \qquad (5.18)$$

where $dA = w(y)dy$, $w(y)$ the width at y and Q a section parameter. The shear stress given by Equation (5.18) is obviously zero at the top and bottom because the free surfaces are free of shear stress. In between there is a maximum, generally at the centrode for most practical sections, having the value, for different sections:

$$\text{Rectangular: } \tau_{\max} = \frac{3V}{2A}$$

$$\text{Circular: } \tau_{\max} = \frac{4V}{3A}$$

$$\text{Hollow circular: } \tau_{\max} = \frac{2V}{A} \qquad (5.19)$$

$$\text{I-beam: } \tau_{\max} = \frac{V}{A_{\text{web}}}$$

Example 5.5 The pin of the connecting rod shown in Figure E5.2 is loaded with a maximum force $V = 25$ kN. The pin material is low-carbon steel with yield strength in shear $S_{sy} = 220$ MPa. For design based on the maximum shear stress, the safety factor is $N = 2$. Find the diameter of the pin.

Solution The stress area is $2\pi d^2/4$ because there are two sections of the pin to shear in the case of shear failure. The design equation is, using the failure criterion of the maximum distortion strain energy and Equations (5.19) for the maximum stress:

---

[1] Developed originally by D. J. Jourawski (1821–1891) for the design of wooden bridges.

$$\tau_{av} = \frac{4V}{3A} = \frac{4V}{2} \times \frac{3\pi d^2}{4} = \frac{S_{sy}}{N}$$

$$\frac{4 \times 4 \times 25{,}000}{2 \times 3 \times \pi d^2} = 220 \times \frac{10^6}{2}$$

Therefore:

$$d^2 = \frac{2 \times 4 \times 4 \times 25{,}000}{2 \times 3\pi \times 220 \times 10^6} = 0.000193, \; d = 0.0139 \text{ m}$$

$d = 14$ mm is selected. It is 16% greater than the one computed on the basis of average stress in Example 5.2.

Design Critique

The result looks reasonable, on the basis of experience from similar motors. The design was based on pure shear stress contribution to failure. Notches, keyways, bending, and other factors that might also affect stress failure must be avoided in this design.

5.1.7. Bending of Curved Elements

Many machine parts have the form of curved beams, such as circular rings, springs, etc. The equations of the previous section are not valid because they were derived under the assumption of a straight beam.

The curvature is measured on the neutral axis x, which for curved beams does not pass through the center of area (centroid) of the cross-section.

The bending moment M determines the normal stress σ developed at a cross-section of a curved bar. Normal stresses σ are distributed over the cross-sectional area A, according to a hyperbolic law (Bresse 1854):

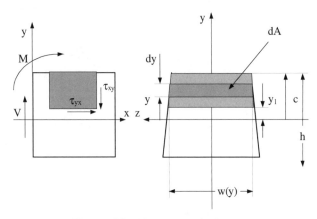

Figure 5.3. Shear stress in beams.

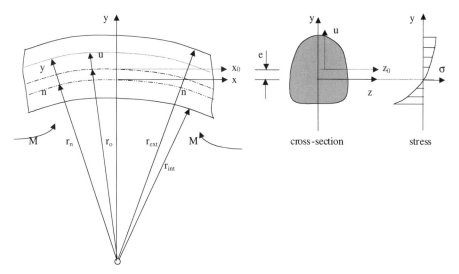

Figure 5.4. Bending stresses in a curved beam.

$$\sigma_m = \frac{M}{S}\frac{y}{(r_n + y)} \tag{5.20}$$

in which $S = Ae$ is the static moment of area A about the neutral axis z, r_n is the radius of curvature of the neutral line nn, and y is the coordinate of the point being considered from the z-axis (Figure 5.3). The neutral line nn is displaced with respect to the geometric axis of the beam towards the center of its curvature by the distance $e = r_0 - r_n$, in which r_0 is the radius of curvature of the geometric axis of the beam.

The radius of curvature of the neutral line of the beam for any shape of its cross-section is found from the equation (Beer and Johnston 1981):

$$r_n = \frac{F}{\int_F dF/(r_0 + u)} \tag{5.21}$$

in which u is the coordinate of the point being considered in the cross-section from the centroidal axis x_0 (Figure 5.2) and F is the cross-sectional area.

For a rectangular cross-section of width b:

$$r_n = \frac{b}{\ln(r_{ext}/r_{int})} \tag{5.22}$$

For a circular cross-section of diameter d:

$$r_n = \frac{r_0 + [r_0^2 - (d/2)^2]^{1/2}}{z} \tag{5.23}$$

For a trapezoidal cross-section:

$$r_n = \frac{A}{[b_{ext} + r_{ext}(b_{int} - b_{ext})/h]\ln(r_{ext}/r_{int}) - (b_{int} - b_{ext})} \quad (5.24)$$

where A is the area of the cross-section, r_{ext}, r_{int}, b_{ext}, and b_{int} are the radii of curvature and the widths of the external and internal fibers of the cross-section, respectively.

Finally, from Equation (5.20):

$$\sigma_m = \frac{M}{A(r_0 - r_n)}\frac{y}{(r_n + y)} \quad (5.25)$$

For certain other shapes of cross-sections, the values of r_n are given in textbooks on the strength of materials.

Since $r_n + y = r_0 + u$, where u is the distance of the point being considered in the cross-section from the centroidal axis x_0, Equation (5.25) can be written as:

$$\sigma = \frac{M}{a^2(r_0 - r_n)}\frac{r_0 + u - r_n}{(r_0 + u)} \quad (5.26)$$

The maximum stress occurs at the inner fiber, $u = -u_{int}$ and is:

$$\sigma_{max} = \frac{M}{a^2(r_0 - r_n)}\frac{r_0 - u_{int} - r_n}{(r_0 - u_{int})} \quad (5.27)$$

and the minimum stress at the outer fiber $u = u_{ext}$ and is:

$$\sigma_{min} = \frac{M}{a^2(r_0 - r_n)}\frac{r_0 + u_{ext} - r_n}{(r_0 + u_{ext})} \quad (5.28)$$

Equation (5.25) shows the nonlinear law of distribution of σ_y over the cross-section and is dependent on the shape of the latter and on the initial curvature of the beam. If $h/r_0 \leq 0.1$, where h is the height of the cross-section, then σ_y can be calculated from the equation for a straight beam, $\sigma = Mc/I$.

Example 5.6   A proving ring[2] shown in Figure 5.6a is rated at maximum force $F = 25$ kN and has square cross-section of side $2a$ (thickness) and mean radius $\rho = 200$ mm. The ring material is low-carbon steel with yield strength $S_y = 220$ MPa. For design based on the maximum bending stress, the safety factor is $N = 2$. Find the required thickness of the ring.

---

[2] Used in material testing to measure the applied force.

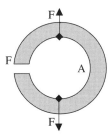

Figure E5.6

Solution   The radius of the centroid is $r_0 = 0.200$ m. The bending moment at the centroid at section A is $M_{max} = 25{,}000 \times 0.200 = 5{,}000$ Nm. The radius at the neutral line for rectangular cross-section (Equation (5.22)) is:

$$r_n = \frac{b}{\ln(r_{ext}/r_{int})} = \frac{a}{\ln[(r_0 + a)/(r_0 - a)]} = \frac{a}{\ln[(0.2 + a)/(0.2 - a)]} \qquad (a)$$

The maximum stress (Equation (5.27)) is:

$$\sigma_{max} = \frac{M_{max}}{a^2\,(r_0 - r_n)}\,\frac{r_0 - u_{int} - r_n}{(r_0 - u_{int})} = \frac{S_y}{N} \qquad (b)$$

$$\frac{5000}{a^2\,(0.2 - r_n)}\,\frac{0.2 - a - r_n}{(0.2 - a)} = -\frac{220 \times 10^6}{2} \qquad (c)$$

Equations (a) and (c) need to be solved simultaneously. This can be done by trial and error or by using Solver of EXCEL. The latter yielded the result $a = 14.5$ mm. The spreadsheet is shown in Table E5.6. The thickness found is $2a = 2 \times 0.014 = 0.028$ m or 28 mm.

TABLE E5.6

	A	B	C	D	E	F	G	H
1	**Solution of the nonlinear equation in Example 5.6**							
2	Moment $M =$		5000		$S_y =$	2.2E + 08		
3	Unknown a				0.0144775			
4								
5	Radius at the neutral line			$r_n =$	0.0998251			
6	Stress equation		$(E\_y/2) - (M/(a^2*(0, 2 - r\_n)))*$ $(0, 2 - a - r\_n)/(0, 2 - a) =$					0
7								
8	Cell $E5$:		$=a/LN((0.2 + a)/$ $(0.2 - a))$					

Design Critique

The result looks reasonable, on the basis of experience from similar proof rings. The design was based on pure shear stress contribution to failure. Notches, torsion due to unequal distribution of the load, and other factors that might also affect stress failure must be avoided in this design.

5.1.8. Design Methodology for Sizing for Strength

On the basis of element strength, the design procedure is to apply the stress–strength equations for comparison of the maximum stresses occurring on the element to the applicable material properties. Many times, more than one test must be made at different locations because of different loading at different areas of the element.

The stresses are computed with simple strength of materials equations if possible. For complex geometry, numerical methods are usually employed.

For sizing, the unknown is the geometry that enters as section properties in the computation of stresses. In this case, the design equation(s) is (are) solved for the unknown parameter(s).

Sometimes this solution can be obtained in a closed form. If this is not possible, one has to use a numerical method, such as Solver of EXCEL. Alternatively, one can assume values for the unknown geometric parameters and compute the stresses. If the design stresses are substantially higher or lower than the material strength, the geometry has to be adjusted and the computation repeated.

Design Procedure 5.1: Design for Strength: Sizing

Step 1: Select materials. Basis: engineering experience and judgment.

Step 2: Identify unknown parameters defining geometry.

Step 3: Express design requirements (material limits, operating features, etc.) in the form of design equations.

Step 4: Identify parameters of design equations that can be derived from data.

Step 5: Based on engineering experience and judgment, select superfluous parameters to leave undetermined as many design parameters as design equations.

Step 6: Solve design equations to determine unknowns.

Step 7: Check results against common sense, experience.

Step 8: Optimize by an educated better selection of superfluous parameters at step 5.

Step 9: Make detailed sketches of final design.

Step 10: Perform final design analysis to verify design.

5.2. COMBINED STRESSES: FAILURE THEORIES

5.2.1. Principal Stresses

In the previous section, we developed design equations by comparing demand (stress) to capacity (material strength). We have already seen stresses that are not compatible with the published material strength, which is usually limited to tensile

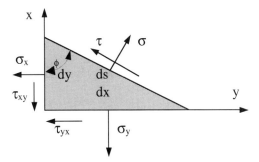

Figure 5.5. Plane stress.

strength, for example shear stress due to torsion. Moreover, there are components where sections are stressed by several loads, such as bending moment, shear, torsion, or axial forces. In this case, direct comparison of stresses and strength is not possible. Failure theories have been proposed to compare a multitude of stresses acting on a section of a machine component to the material tensile strength.

A solid in a general state of plane stress might have at a section normal and shear stresses acting simultaneously along the x and y directions, as shown. We will try to explore the possibility that along a certain direction there is only normal stress and no shear stress. We assume a plane perpendicular with the $x - y$ plane that cuts a triangular prism of base (dx, dy, ds). On the inclined side of length $ds = \sqrt{dx^2 + dy^2}$ there are normal and shear stresses σ and τ (Mohr 1882). Equilibrium in the x and the y direction of the forces $\sigma_x w \, dx$, $\sigma_y w \, dy$, . . . , etc., where w is the thickness of the element triangular shape shown, gives:

$$\sigma = \frac{\sigma_x + \sigma_y}{2} + \frac{\sigma_x - \sigma_y}{2} \cos 2\phi + \tau_{xy} \sin 2\phi \qquad (5.29)$$

$$\tau = \frac{\sigma_x - \sigma_y}{2} \sin 2\phi + \tau_{xy} \cos 2\phi \qquad (5.30)$$

Equation (5.30) shows that it is possible to have $\tau = 0$ for

$$\tan 2\phi = \frac{2\tau_{xy}}{\sigma_x - \sigma_y} \qquad (5.31)$$

At plane sections defined by angle ϕ given by Equation (5.31) and angle $\phi + \pi/2$, the shear stress is zero. Substituting the value of the angle ϕ into Equation (5.29) and solving for σ, we obtain:

$$\sigma_{1,2} = \frac{\sigma_x + \sigma_y}{2} \pm \sqrt{\left(\frac{\sigma_x - \sigma_y}{2}\right)^2 + \tau_{xy}^2} \qquad (5.32)$$

There are two stresses σ_1 and σ_2, termed *principal stresses,* at the planes where the shear stress is zero.

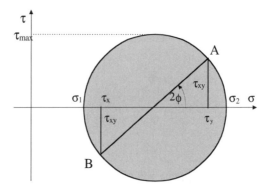

Figure 5.6. Construction of the Mohr circle for plane stress.

There is a simple geometric interpretation of Equations (5.29) and (5.30), shown in Figure 5.6. In a $(\tau - \sigma)$ coordinate system, we find points $A(\sigma_y, \tau_{xy})$ and $B(\sigma_x, -\tau_{xy})$ and we draw the circle with diameter AB, called the *Mohr circle*. Every point $(\tau - \sigma)$ on this circle has coordinates as in Equations (5.29) and (5.30). We observe that the two principal stresses σ_1 and σ_2 are found at the intersection of the Mohr circle with the σ axis. There the shear stress is zero, and σ_2 and σ_1 are the maximum and the minimum normal stresses. The maximum shear stress is at the top of the circle:

$$\tau_{\max} = \frac{(\sigma_2 - \sigma_1)}{2} \tag{5.33}$$

In machine design, there are many parts with complex, even three-dimensional, states of stresses. Mohr circles can be drawn for the general, three-dimensional state of stress in order to determine the principal stresses and the maximum shear stress, but this is rather tedious. It is preferable to do it algebraically. To this end, suppose that the state of stress in respect to three mutually perpendicular planes

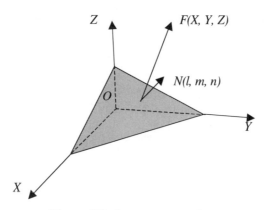

Figure 5.7. Stresses at a wedge.

forming the Cartesian system (*Oxyz*) is known, given by the three normal and three shear stresses, components of the stress tensor at point O (Cauchy 1822). At a plane inclined in respect to *Oxyz* by way of three direction cosines (see Chapter 2) *l, m, n,* the stress vector has components along *x, y*:

$$\begin{Bmatrix} X \\ Y \\ Z \end{Bmatrix} = \begin{bmatrix} \sigma_x & \tau_{xy} & \tau_{xz} \\ \tau_{yx} & \sigma_y & \tau_{yz} \\ \tau_{zx} & \tau_{zy} & \sigma_z \end{bmatrix} \begin{Bmatrix} l \\ m \\ n \end{Bmatrix} \tag{5.34}$$

This plane is *principal* if the stress vector is perpendicular to the plane, that is, vectors *F* and *n* are collinear:

$$X = \sigma l \quad Y = \sigma m, \quad Z = \sigma n \tag{5.35}$$

where σ is the (unknown) principal stress. Therefore, Equations (5.34) and (5.35) form a homogeneous system of linear algebraic equations in the unknown direction cosines, *l, m, n*. Solution exists if:

$$\begin{bmatrix} \sigma_x - \sigma & \tau_{xy} & \tau_{xz} \\ \tau_{yx} & \sigma_y - \sigma & \tau_{yz} \\ \tau_{zx} & \tau_{zy} & \sigma_z - \sigma \end{bmatrix} = 0 \tag{5.36}$$

Equation (5.36) is, in general, a cubic algebraic equation in the unknown principal stress σ admitting, in general, three solutions, the three principal stresses.

A general method for the solution of the above problem, the eigenvalue problem, is the Jacobi method, which is used in the program COMLOAD of MELAB 2.0. The same can be done in a spreadsheet (see Example 5.8).

The general Mohr diagram is as shown in Figure 5.8. The three principal stresses $\sigma_1, \sigma_2, \sigma_3$ define the three Mohr circles for the general state of stress. The extremal shear stresses are equal to the radii of the Mohr circles:

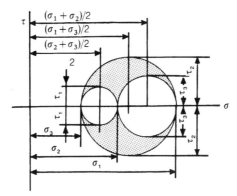

Figure 5.8. Mohr circles for the general state of stress.

Otto Mohr (1835–1918). Born in Wesselburen, Germany, studied engineering in the Technical University of Hanover. Had designed some of the first steel trusses in Germany as a railroad engineer when he became Professor of Engineering Mechanics at Stuttgart Polytechnic and later at Dresden Polytechnic. One of the pioneers in structural mechanics and a famous teacher. Students from all the other German technical universities attended his lectures. His contributions were in strength of materials and structural analysis.

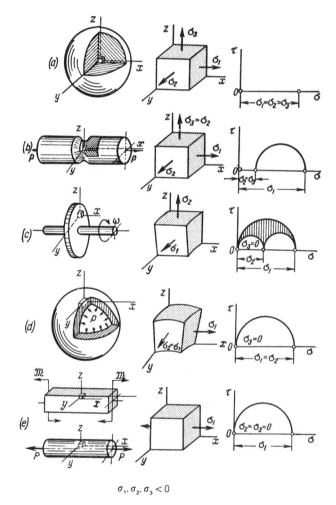

$$\sigma_1, \sigma_2, \sigma_3 < 0$$

Figure 5.9. Complex states of stress—principal tensile stresses only: (*a*) Sphere under hydrostatic pressure in a center cavity; (*b*) notched cylinder under axial load; (*c*) rotating disk; (*d*) hollow sphere under internal pressure; (*e*) pure bending or axial loading of a prismatic beam or bar.

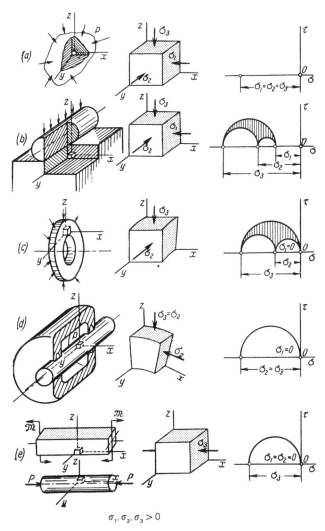

Figure 5.10. Complex states of stress—principal compressive stresses only: (*a*) Solid under external hydrostatic; (*b*) cylinder pressed against flat solid; (*c*) ring under external radial pressure; (*d*) cylinder under external radial pressure; (*e*) compressive axial loading of a prismatic bar.

$$\tau_1 = \pm\frac{\sigma_2 - \sigma_3}{2}$$

$$\tau_2 = \pm\frac{\sigma_1 - \sigma_3}{2} \tag{5.37}$$

$$\tau_3 = \pm\frac{\sigma_1 - \sigma_2}{2}$$

of which τ_2 is the maximum in this particular case.

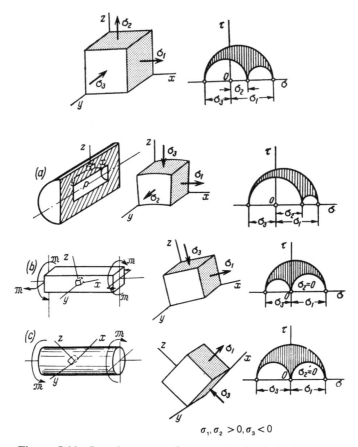

Figure 5.11. Complex states of stress, mixed principal stresses.

These stresses occur in planes inclined at 45° to the direction of the principal stresses: τ_1 in two mutually perpendicular planes parallel to axis x; τ_2 in two mutually perpendicular planes parallel to axis y; and τ_3 in two mutually perpendicular planes parallel to axis z.

All the equations relating to the general state of stress can also be applied to the planar state of stress if one principal stress is equated to zero, as well as to the linear state of stress if two principal stresses are equated to zero.

Several examples of the general state of stress in design applications are shown in Figures 5.9–5.11.

Example 5.7   At a certain point of a machine component, strain gage measurements of stress resulted to $\sigma_x = 100$ MPa, $\sigma_y = -50$ MPa, $\tau_{xy} = 50$ MPa. Find the principal stresses and the maximum shear stress and construct the Mohr circle:

1. Analytically
2. Using a spreadsheet

Solution

1. Equation 5.32 gives:

$$\sigma_{1,2} = \frac{\sigma_x + \sigma_y}{2} \pm \sqrt{\left(\frac{\sigma_x - \sigma_y}{2}\right)^2 + \tau_{xy}^2}$$

$$= \frac{100 - 50}{2} \pm \sqrt{\left(\frac{100 + 50}{2}\right)^2 + 50^2}$$

$$\sigma_1 = 115.1 \text{ MPa}, \ \sigma_2 = -65.1 \text{ MPa}, \ \tau_{max} = \frac{\sigma_1 - \sigma_2}{2} = \frac{115.1 + 65.1}{2} = 90.1 \text{ MPa}$$

To construct the Mohr circle, in a (τ, σ) Cartesian coordinate system we draw vectors $\sigma_x = 100$ and $\sigma_y = -50$ from the origin in the σ direction and then from their tips vectors $\tau_{xy} = -50$ and $\tau_{xy} = 50$ in the τ-direction as shown. The two tips of the shear vectors define the diameter of the Mohr circle (dotted line). The intersection of the circle with the σ axis defines the principal stresses, about 115 and −65 MPa. The top point of the circle defines the maximum shear stress, about 90 MPa.

2. The spreadsheet is straightforward, as shown below. Cells E4 and E5 are computed with Equation (5.32). The Mohr circle is plotted in Table E5.7 in EXCEL.

To plot the Mohr circle, 20 points were used to calculate the normal and shear stress using equations (5.29) and (5.30).

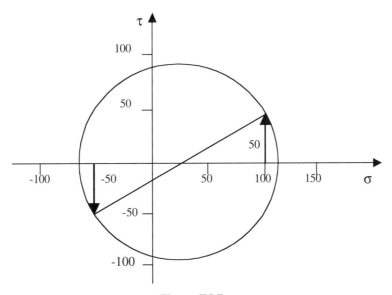

Figure E5.7*a*

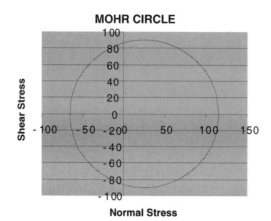

Figure E5.7b

TABLE E5.7

	A	B	C	D	E	F	G	H
1	GENERAL STATE OF STRESS—PLANE STRESS							
2								
3	INPUT			OUTPUT—Principal Stress				
4				Equation 5.32:				
5	$\sigma_x =$	100		$\sigma_1 =$	115.1388		$\sigma_{max} =$	115.1388
6	$\sigma_y =$	-50		$\sigma_2 =$	-65.1388		$\sigma_{min} =$	-65.1388
7	$\tau_{xy} =$	50					$\tau_{max} =$	90.13878

Example 5.8   In a part of a machine, a finite element analysis gave the following stresses (MPa):

$$\sigma_x = 122, \ \sigma_y = -87, \ \sigma_z = 56, \ \tau_{xy} = 89, \ \tau_{xz} = 76, \ \tau_{yz} = 103.$$

Find the principal stresses and the maximum shear stress.

Solution   A closed form solution of the cubic Equation (5.36) exists for this problem (Spiegel 1968).

1. We first calculate the three *invariants of the stress tensor:*

$$a_1 = -\sigma_x - \sigma_y - \sigma_z$$

$$a_2 = \sigma_x \sigma_y + \sigma_y \sigma_z + \sigma_y \sigma_z - \tau_{xy}^2 - \tau_{xz}^2 - \tau_{yz}^2$$

$$a_3 = \sigma_x \sigma_y \sigma_z - \tau_{xy}^2 \sigma_z - \tau_{xz}^2 \sigma_y - \tau_{yz}^2 \sigma_x + 2\tau_{xy}\tau_{xz}\tau_{yz}$$

2. We compute the auxiliary parameters:

$$Q = \frac{(1/3)a_1 a_2 - a_3 - 2(a_1^3)}{27}$$

$$R = \frac{a_1^2}{3} - a_2$$

$$T = \left(\frac{1}{27} R_3\right)^{0.5}$$

$$S = \left(\frac{R}{3}\right)^{0.5}$$

$$\alpha = a \cos \frac{Q}{2T}$$

Then the roots or the cubic equation—the principal stresses—are:

$$\sigma_1 = \left(\frac{1}{27} R^3\right)^{0.5}$$

$$\sigma_2 = 2S \left[\cos\left(\frac{\alpha}{3} + \frac{2\pi}{3}\right)\right] + \frac{a_1}{3}$$

$$\sigma_3 = 2S \left[\cos\left(\frac{\alpha}{3} + \frac{\pi}{3}\right)\right] + \frac{a_1}{3}$$

The spreadsheet that performs the above calculations follows in Table E5.8.

5.2.2. Failure Theories

The various failure theories propose criteria that determine the strength of an element of a material subject to complex stress conditions. According to these criteria, equivalent, or reduced, stresses are established, that is, stresses due to uniaxial tension of an element of a material that makes its stressed state equivalent for failure to a given complex state of stress. In other words, since only uniaxial tests are available, the failure theories interpret the failure under general state of stress to failure under uniaxial stress, which is better understood.

5.2.2.1. Coulomb[3] or Maximum Normal Stress Theory. According to the maximum normal stress theory, failure occurs when the maximum principal stress equals the uniaxial stress of failure. Therefore:

---

[3] Charles Augustin de Coulomb, French physicist (1736–1806).

TABLE E5.8

	A	B	C	D	E	F	G
1	**EXAMPLE E5.8**						
2	**GENERAL STATE OF STRESS-PRINCIPLE STRESSES**						
3							
4		**INPUT**		**Stress Matrix**			
5	Stress	Value		122	89	76	
6	$\sigma_x =$	122		89	-87	103	
7	$\sigma_y =$	-87		76	103	56	
8	$\sigma_z =$	56					
9	$\tau_{xy} =$	89		**Cubic Equation Coefficients (stress tensor invariants)**			
10	$\tau_{xz} =$	76		$a\_1 =$	-91		
11	$\tau_{yz} =$	103		$a\_2 =$	-32960		
12				$a\_3 =$	436362		
13	**Auxiliary Coefficients**						
14		$R = a_1^2/3 - a_2 =$	35720.333	$Q = (1/3)a_1a_2 - a_3 - 2*(a_1^3)/27 =$			619244.7
15		$S = (R/3)^{0.5} =$	109.11818	$T = ((1/27)R_3)^{0.5} =$			1299245.
16		$\alpha = a\cos(-Q/(2T)) =$	1.8114209				
17							
18	**PRINCIPAL STRESSES:**						
19	$\sigma_1 =$	-149.3144905		**Maximum Shear Stress:**			
20	$\sigma_2 =$	227.466741		188.3906			
21	$\sigma_3 =$	12.84774954					

$$\sigma_{\text{eqco}} = \max |\sigma_1| \leq \frac{S_y}{N}, \qquad i = 1, 2, 3 \tag{5.38}$$

In machine design, the most usual measure of failure is the yield strength because, while after a machine component yields it might not rupture, in most cases the machine will not be operational after a component yields.

Maximum normal stress theory is applicable to brittle materials, such as cast iron, high carbon steels, and ceramics, with tension being the predominant load. In Figure 5.11, the limits that Equation (5.38) imposes are shown (light shaded square).

5.2.2.2. Tresca[4] Maximum Shear Stress Theory. According to the maximum shear stress theory, failure occurs when the maximum shear stress max $|\sigma_i - \sigma_j|/2$ reaches the value of the shear stress at failure for the uniaxial tension test. Since this is half the tension stress at failure for uniaxial tension ($\tau_{max} = \sigma/2 = S_y/2$):

$$\sigma_{eqII} = \max |\sigma_i - \sigma_j| \le \frac{S_y}{N}, \qquad i, j = 1, 2, 3 \qquad (5.39)$$

In Figure 5.11, the limits that Equation (5.39) imposes are shown (dark shaded polygon).

For plane stress, the maximum shear stress can be found from the Mohr circle:

$$\tau_{max} = \sqrt{\left(\frac{\sigma_x - \sigma_Y}{2}\right)^2 + \tau_{xy}^2} \qquad (5.40)$$

For the tension test, $\tau_{max} = \sigma_x/2$, therefore:

$$\sigma_{eqTR} = \sqrt{(\sigma_x - \sigma_y)^2 + 4\tau_{xy}^2} \le \frac{S_y}{N} \qquad (5.41)$$

Equation (5.41) suggests that for simple tension only, failure will occur when $\tau_{max} = S_y/2N$. Thus, the yield strength in shear is:

$$S_{sy} = 0.5S_y \qquad (5.42)$$

The maximum shear stress theory is applicable to ductile materials, especially when torsion is the predominant source of stress.

5.2.2.3. von Mises Distortion Energy Theory.   According to the distortion energy theory, failure occurs when the strain energy due to distortion equals the same energy for the uniaxial test. It uses the observation that solids do not fail under high hydrostatic pressure that exceeds the strength of the material. For example, if one drops a spherical stone into the sea at 10,000 m depth, the hydrostatic stress will be 100,000 N/m^2, yet the stone will not break no matter what its strength is. Otherwise the earth's crust would be a sphere of powder beyond a small depth. Thus, if from the elastic energy at the general state of stress (which changes both the volume and the form) we subtract the energy at uniform hydrostatic stress (which maintains the form and changes the volume), we obtain the energy of changing the shape without changing the volume. That is why it is called *distortion energy*.

In terms of the principal stresses (Beer and Johnston 1981), equating the distortion energy of the uniaxial stress state to the one for the general stress state yields:

---

[4] Henri Edouard Tresca, French engineer (1814–1855).

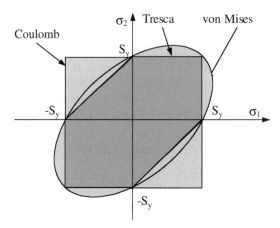

Figure 5.12. Comparison of the failure theories.

$$\sigma_{\text{eq}VM} = \left[\frac{1}{2}\left[(\sigma_1 - \sigma_2)^2 + (\sigma_2 - \sigma_3)^2 + (\sigma_3 - \sigma_1)^2\right]\right)^{1/2} \leq \frac{S_y}{N} \quad (5.43)$$

In terms of the normal and shear stresses:

$$\sigma_{\text{eq}VM} = \sqrt{\frac{1}{2}\left[(\sigma_x - \sigma_y)^2 + (\sigma_x - \sigma_z)^2 + (\sigma_y - \sigma_z)^2 + 6(\tau_{xy}^2 + \tau_{xz}^2 + \tau_{yz}^2)\right]} \leq \frac{S_y}{N} \quad (5.44)$$

For plane stress:

$$\sigma_{\text{eq}VM} = \sqrt{\left[(\sigma_x^2 + \sigma_y^2 - \sigma_x\sigma_y) + 3\tau_{xy}^2\right]} \leq \frac{S_y}{N} \quad (5.45)$$

If only shear is present:

$$\sigma_{\text{eq}VM-\text{shear}} = \sqrt{3}\,\tau_{xy} \quad (5.46)$$

Therefore, because for the simple tension test $\sigma_{\text{eq}VM-\text{tension}} = \sigma_x$, the shear strength is:

$$S_{sy} = \frac{S_y}{\sqrt{3}} \approx 0.577 S_y \quad (5.47)$$

For plane stress, in terms of the principal stresses, from Equation (5.43) for $\sigma_3 = 0$:

$$\sigma_{\text{eq}NM} = \sqrt{\sigma_1^2 + \sigma_2^2 - \sigma_1\sigma_2} \le \frac{S_y}{N} \tag{5.48}$$

Equation (5.48) is the equation of an ellipse shown in medium-dark shading in Figure 5.7. Comparing the Tresca and von Mises theories, we see that they do not differ very much, but since the von Mises theory is supported by experimental evidence, it is generally preferred by designers. However, the Tresca theory is more thus conservative and thus conservative designers favor it. With the increasing need for competitiveness in industry, the von Mises theory tends to prevail in most design applications.

5.2.2.4. Mohr Theory of Limiting States of Stress (Mohr 1882).[5] The theory of limiting states of stress is a modification of the shear stress theory for materials with different tension and compression strengths. Figure 5.13 shows a graphical illustration of the theory. If stress is positive, failure is at the yield strength in tension S_{yt} and the Mohr circle has center at $C(S_{yt}/2, 0)$ and diameter S_{yt}. If stress is negative, failure is at the yield strength in compression S_{yc} 15) and the Mohr circle has center at $A(S_{yc}/2, 0)$ and diameter S_{yc}. A combined loading with positive and negative principal stresses, say $\sigma_1 > 0$ and $\sigma_2 < 0$, is represented by the dashed Mohr circle with center at point $B(\sigma_1 + \sigma_2)/2, 0)$ and has diameter $\sigma_1 - \sigma_2$. Mohr's hypothesis is that failure will occur when any one of the Mohr circles of the general state of stress intersects the failure line $(a - a)$, tangent to the tension and compression failure circles or the (dashed) line $M_1M_2M_3$ that is defined by the strength in pure shear S_{sy}, point M_2. If the common tangent $(a–a)$ is accepted as failure line (*Coulomb–Mohr* or *internal friction* theory), similarity of the triangles ACD and BCE yields:

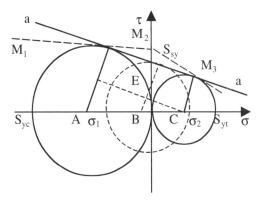

Figure 5.13. Illustration of the Mohr theory.

[5] In most texts, the strength used in this theory and the Coulomb theory is ultimate strength, while here we use yield strength. For brittle material to which these two theories are applicable, there is no appreciable difference between the two strengths. We preferred the use of yield strength in tune with the general direction of using yield strengths in machine design, for reasons explained earlier.

$$\sigma_{eqMO} = \sigma_1 - \frac{S_{yt}}{S_{yc}} \sigma_2 = \frac{S_y}{N} \qquad (5.49)$$

If design is based on the yield strength in tension S_{yt}, the equivalent stress will be then the maximum of all combinations of σ_1, σ_2, σ_3:

$$\sigma_{eqMO} = \max |\sigma_i - \kappa\sigma_j| \le \frac{S_y}{N}, \ i, j = 1, 2, 3 \qquad (5.50)$$

where $\kappa = S_{yt}/S_{yc}$.

In summary, in machine design theories III and IV are usually applied, namely the von Mises distortion strain energy theory for ductile materials and the Mohr theory of limiting states of stress for brittle materials of different yield strength in tension and compression. It is advisable for machine design applications to use the von Mises theory of distortion strain energy, especially for ductile materials or for brittle materials with the same yield strength in tension and compression. Alternatively, for a conservative design, the maximum shear stress theory could be used. For brittle materials exhibiting different strengths in tension and compression, the Mohr theory of limiting states of stress should be used. For computer-aided machine design, the von Mises theory has the added advantage that one can use the general stresses, (Equation (5.44)), and there is no need to compute the principal stresses.

Design Procedure 5.2: Design of a Part with Complex State of Stress for Static Strength—Sizing

Input: Loading, material properties, safety factor. *Output:* Part section dimension(s).

Step 1: From the internal loading diagram of the part, identify the point(s) where failure is more likely to occur.

Step 2: For the type of material, select a criterion of failure: Very brittle, Coulomb (I). Ductile and moderately brittle materials, von Mises (III). Brittle materials with different strength in tension and compression, Mohr (IV). Conservative design with moderately brittle or ductile materials, Tresca (II).

Step 3: Find all stress components at the point under consideration. Use algebraic symbols for the unknown geometric quantities.

Step 4: Apply Equations (5.38), (5.39), (5.44), or (5.50) if your chosen criterion of failure is Coulomb, Tresca, von Mises, or Mohr, respectively.

Step 5: Solve the equation for the unknown section dimension.

Design Procedure 5.3: Design of a Part with Complex State of Stress for Static Strength—Certification

Input: Loading, material properties, part section dimensions. *Output:* Safety factor.

Step 1: From the internal load diagram of the part, identify the point(s) where failure is more likely.

Step 2: For the type of material, select a criterion of failure: Very brittle, Coulomb (I). Ductile and moderately brittle materials, von Mises (III). Brittle materials with different strength in tension and compression, Mohr (IV). Conservative design with moderately brittle or ductile materials, Tresca (II).

Step 3: Find all stress components at the point under consideration. Use the algebraic symbol N for the unknown safety factor.

Step 4: Apply Equations (5.38), (5.39), (5.44), or (5.50) if your chosen criterion of failure is Coulomb, Tresca, von Mises or Mohr, respectively.

Step 5: Solve the equation for the unknown safety factor.

See also Examples 5.9, 5.10.

Example 5.9   For the stress state of Example 5.7, the material has yield strength $S_y = 220$ MPa. Find the safety factor with theories of failure I, II, III.

Solution   From Example 5.7, $\sigma_x = 100$ MPa, $\sigma_y = -50$ MPa, $\tau_{xy} = 50$ MPa, $\sigma_1 = 115.1$ MPa, $\sigma_2 = -65.1$ MPa.

Maximum normal stress theory:

$$N = \frac{S_y}{\sigma_{max}} = \frac{220}{115.1} = 1.91$$

Maximum shear stress theory:

$$N = \frac{S_{sy}}{\tau_{max}} = \frac{0.5 S_y}{(\sigma_1 - \sigma_2)/2} = \frac{0.5 \times 220 \times 2}{115.1 + 65.1} = 1.22$$

Distortion strain energy theory (Equation (5.48):

$$N^2 = \frac{S_y^2}{\sigma_1^2 + \sigma_2^2 - \sigma_1\sigma_2} = \frac{220^2}{115.1^2 + 65.1^2 + 115.1 \times 65.1} = 1.93, N = 1.391$$

It is to be noted that to apply the distortion strain energy theory, finding the principal stresses is unnecessary. Indeed, from equation (5.45):

$$N = \frac{S_y}{\sqrt{\sigma_x^2 + \sigma_y^2 - \sigma_x\sigma_y + 3\tau_{xy}^2}} = \frac{220}{\sqrt{100^2 + 50^2 + 100 \times 50 + 3 \times 50^2}} = 1.391$$

Example 5.10   For the stress state of Example 5.8, the material has yield strength in tension $S_{yt} = 400$ and in compression $S_{yc} = 600$ MPa. Find the safety factor with theories of failure I, II, III, IV.

Solution   From Example 5.8:

$$\sigma_x = 122, \sigma_y = -87, \sigma_z = 56, \tau_{xy} = 89, \tau_{xz} = 76, \tau_{yz} = 103$$

$$\sigma_1 = 227.4667, \sigma_2 = -149.314, \sigma_3 = 12.84775$$

Maximum normal stress theory:

$$N = \frac{S_y}{\sigma_{max}} = \frac{400}{227.5} = \underline{1.76}$$

Maximum shear stress theory:

$$N = \frac{S_{sy}}{\tau_{max}} = \frac{0.5 S_y}{(227.5 + 149.3)/2} = \frac{0.5 \times 400 \times 2}{(227.5 + 149.3)/2} = \underline{1.061}$$

Distortion strain energy theory (Equation (5.43)):

$$N = S_y \bigg/ \left[\frac{1}{2} \left[(\sigma_1 - \sigma_2)^2 + (\sigma_2 - \sigma_3)^2 + (\sigma_3 - \sigma_1)^2 \right] \right]^{1/2}$$

$$= \frac{400}{\{[(227.4 + 149.3)^2 + (-149.3 - 12.8)^2 + (12.8 - 227.4)^2]/2\}^{1/2}} = \underline{1.22}$$

Equation (5.44) could also be used.
Mohr theory of limiting states of stress:

$$\kappa = \frac{S_{yt}}{S_{yc}} = \frac{400}{600} = 0.66$$

It is apparent that the maximum difference in Equation (5.50) is max $|\sigma_i - k\sigma_j| = |\sigma_1 - k\sigma_2| = 227.4 + 0.66 \times 149.3 = 325.9$ MPa. Equation (5.50) gives:

$$N = \frac{S_{yt}}{\max |\sigma_i - k\sigma_j|} = \frac{400}{325.9} = \underline{1.22}$$

5.3. SIZING OF OTHER MACHINE COMPONENTS BY STRESS ANALYSIS

5.3.1. Thin Pressure Vessels, Membrane Stresses

Shells are curved members with a thickness much smaller than their other dimensions. They are capable of sustaining very high loads, mostly in the form of internal pressures in liquid or gas containing vessels, called *pressure vessels*. Therefore, many times they are very thin and their resistance to bending is very small. The loads are sustained by tensile stresses in the shell material. Such shells are called *membranes* or *thin pressure vessels* and the corresponding stresses are called *membrane stresses*.

Considering a differential element $ds_1 ds_2$ on the surface and assuming an internal pressure p, expressing the equilibrium of all forces along the direction perpendicular to the surface, the *Laplace equation* is obtained:

$$\frac{\sigma_t}{\rho_t} + \frac{\sigma_m}{\rho_m} = \frac{p}{h} \tag{5.51}$$

where σ_t and σ_m are *circumferential* (or *hoop*) and *meridional* (or *axial*, sometimes)

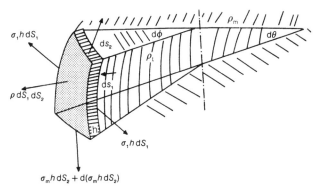

Figure 5.14. Equilibrium of a membrane element.

stresses respectively, ρ_t and ρ_m are the corresponding radii of curvature, p is the pressure, and h is the thickness. The Laplace equation is not enough to find the unknown stresses because there are two. In axisymmetric shells, however, we can obtain a second equation in most cases.

From Figure 5.15, summing up forces in the z-direction yields:

$$\sigma_m \, 2\pi Rh \, \cos \, \theta = \pi R^2 p \tag{5.52}$$

Thus, σ_m can be computed with Equation (5.52) and σ_t can then be computed from Equation (5.51). For a cylindrical pressure vessel of radius R, for example, $\theta = 0$, $\cos \, \theta = 1$, and Equation (5.51) yields:

$$\sigma_m = \frac{pR}{2h} \tag{5.52a}$$

and then Equation (5.51) gives for the hoop stress:

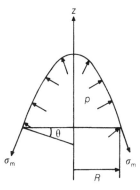

Figure 5.15. Axisymmetric membrane.

$$\sigma_t = \frac{pR}{h} \qquad (5.53)$$

The design equation then uses the distortion energy criterion of failure (Equation 5.48) with $\sigma_1 = \sigma_t$, $\sigma_2 = \sigma_m$ from Equations (5.52a) and (5.53):

$$\sqrt{\left(\frac{pR}{h}\right)^2 + \left(\frac{pR}{4h}\right)^2 - \left(\frac{pR}{2h}\right)^2} = \sqrt{3}\left(\frac{pR}{2h}\right) = \frac{S_y}{N} \qquad (5.54)$$

Usually the radius is known from the operating requirements (volume) and the design parameter is the thickness h, for which Equation (5.54) can be readily solved.

Design Procedure 5.4: Design of a Pressure Vessel—Sizing/Certification

Input: Loading, material properties, safety factor or thickness, geometry.
Output: Vessel thickness or safety factor.
Step 1: From equilibrium along axis of symmetry, find axial stress.
Step 2: From Laplace equation, determine tangential stress.
Step 3: Apply Procedure 5.2 for sizing, Procedure 5.3 for certification.
Step 4: Solve stress design equation for the thickness, if a sizing problem.
Step 5: Solve stress design equation for the safety factor, if a certification problem.

Example 5.11   A thin tube of inner diameter $d = 30$ mm, thickness $t = 5$ mm, is subject to internal pressure $p = 30$ N/mm^2 and a torque $T = 200$ Nm. Calculate the stresses at a point far from the ends, the principal stresses, and the equivalent stresses with (a) the maximum shear stress theory, (b) the equivalent distortion strain energy theory.

Solution   Circumferential stress:

$$\sigma_t = \frac{pR}{t} = 30 \times \frac{15}{5} = 90 \text{ N/mm}^2$$

Longitudinal stress:

$$\sigma_z = \frac{pR}{2t} = 45 \text{ N/mm}^2$$

Shear stress:

$$\tau_{tz} = \frac{TR}{J_p} = \frac{200 \times 10^3}{2\pi \times 15^2 \times 5} = 28 \text{ N/mm}^2$$

The program COMLOAD is used to determine the principal stresses and the equivalent stresses. The RUN of the program is shown in Figure E5.11c.

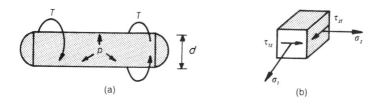

```
(c)                    Input normal stresses  Sx, Sy, Sz  ? 90,0,45
                       Input shear stresses   Txy,Txz,Tyz ? 0,28,0

     RESULTS:
     ----------------

     ----------Principal stresses-----------
        0            31.57996       103.4201
     ------------eigenvectors----------------
        0            1              0
      .4322089       0             -.9017736
      .9017736       0              .4322089

     Do you want the limiting states of stress theory? (Y/N) ? n
     Maximum shear stress theory,           Seq= 103.42
     Distortion strain energy theory,       Seq= 91.79
     Ok
```

Figure E5.11

Example 5.12 A communications satellite has the dimensions shown, axisymmetric geometry, thickness $h = 10$ mm, and internal pressure 0.2 MPa. Outside pressure is 0. Draw a stress map in the section shown in Figure E5.12a.

Solution
Point C: From Equation (5.51), due to symmetry:

$$\sigma_t = \sigma_m = \frac{pR}{2h} = \frac{0.2 \times 10^6 \times 0.5}{2 \times 0.01} = 0.5 \times 10^7 \text{ Pa} = 5 \text{ MPa}$$

Point F: Similarly:

$$\sigma_t = \sigma_m = \frac{pR}{2h} = \frac{0.2 \times 10^6 \times 0.3}{2 \times 0.01} = 0.3 \times 10^7 \text{ Pa} = 3 \text{ MPa}$$

Point D:

$$\tan \theta = \frac{250 - 150}{800}, \ \theta = 7.12°$$

$$\sigma_m = \frac{pR}{2h \cos \theta} = \frac{0.2 \times 10^6 \times 0.15}{2 \times 0.01 \times \cos 7.12} = 1.51 \text{ MPa}$$

$$\rho_t = 0.150 \times \cos 7.12 = 0.149 \text{ m}$$

$$\sigma_t = \left(\frac{p}{h} - \frac{\sigma_m}{\rho_m} \right) \quad \rho_t = \frac{p\rho_t}{h} = 2.98 \text{ MPa}, \ \rho_m = \infty$$

Point A:

$$\sigma_m = \frac{0.2 \times 10^6 \times 0.250}{2 \times 0.01 \times \cos 7.12} = 2.52 \text{ MPa}$$

$$\rho_t = 0.250 \cos 7.12 = 0.248 \text{ m}$$

$$\sigma_t = \frac{0.2 \times 0.248}{0.01} = 4.96 \text{ MPa}$$

The stress map is shown in Figure E5.12b.

5.3.2. Thick Pressure Vessels

Not all shells and pressure vessels can be considered thin, as in the previous section. Pressure vessels and pipes for high pressure and interference fit hubs are some examples. In such situations, the stress is not uniform across the section. Let us consider a thick cylindrical pressure vessel with inner radius r_1 and outer radius r_2 having internal pressure p_1 and external pressure p_2 (Figure 5.16). As a first approximation we will consider the material incompressible, and thus the displacement along a radius due to the stresses will be uniform. The circumference of a circle with radius r is:

$$c = 2\pi r \tag{5.55}$$

A change of radius δr results in a change in the circumference:

$$\delta c = 2\pi \delta r \tag{5.56}$$

Because of the incompressibility assumed, the change in radius δr is constant, and thus according to Equation (5.56) the change in the circumference δc is constant. The circumferential strain is:

$$\varepsilon_t = \frac{\delta c}{c} = \frac{2\pi \delta r}{2\pi r} = \frac{constant}{r} \tag{5.57}$$

Hooke's law for the material and Equation (5.57) yield for the stress in the circumferential direction (*tangential stress* σ_t):

$$\sigma_t = \varepsilon_t E = \frac{a}{r} \tag{5.58}$$

where a is a yet undetermined constant.

A rigid-body diagram of one half of the pressure vessel is shown in Figure 5.16. The forces applied are the resultants of the stresses F and the pressures over a length of the vessel L. Looking at an elementary segment of the hub (Figure 5.16), summing up the forces on the half ring gives:

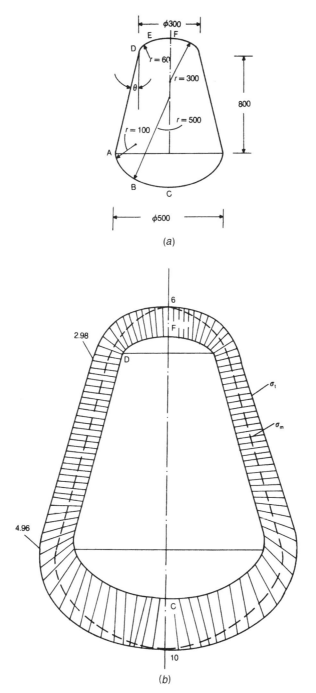

(a)

(b)

Figure E5.12

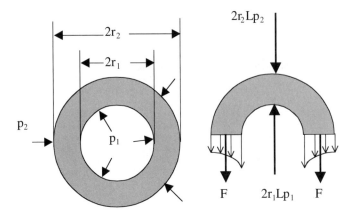

Figure 5.16. Thick pressure vessel.

$$2r_2Lp_2 - 2r_1Lp_1 + 2F = 2r_2Lp_2 - 2r_1Lp_1 + 2\int_{r_1}^{r_2} \sigma_t L \ dr = 0 \qquad (5.59)$$

Eliminating a between Equations (5.58) and (5.59) and solving for σ_t:

$$\sigma_t = \frac{p_1r_1^2 - p_2r_2^2 - r_1^2r_2^2 \ (p_1 - p_2)/r^2}{r_2^2 - r_1^2} \qquad (5.60)$$

Assuming plane strain and no axial stress on the cylinder (Beer and Johnston 1981):

$$\varepsilon_z = \frac{-v}{E} \ (\sigma_t + \sigma_r) \qquad (5.61)$$

Assuming constant strain in the z-direction:

$$\sigma_t = \text{constant} - \sigma_r = \alpha + \frac{\beta}{r} \qquad (5.62)$$

where α and β are yet-undetermined constants. But for $r = r_1$, $\sigma_r = -p_1$, for $r = r_2$, $\sigma_r = -p_2$, and constants a and b can be determined. Finally:

$$\sigma_r = \frac{p_1r_1^2 - p_2r_2^2 - r_1^2r_2^2 \ (p_1 - p_2)/r^2}{r_2^2 - r_1^2} \qquad (5.63)$$

The maximum value of σ_t is at the inner radius $r = r_1$:

$$(\sigma_t)_{max} = \frac{p_1(r_1^2 + r_2^2) - 2p_2r_2^2}{r_2^2 - r_1^2} \qquad (5.64)$$

$$(\sigma_r)_{max} = -p_1, \ if \ \ p_2 > p_1$$

With only internal pressure p:

$$(\sigma_t)_{max} = p\,\frac{r_2^2 + r_1^2}{r_2^2 - r_1^2} \tag{5.65}$$

The tangential strain is (Beer and Johnston 1981):

$$\varepsilon_t = \frac{1}{E}\,(\sigma_t - \sigma_r) \tag{5.66}$$

At a circle of radius ρ, the radius will be extended by u, such that:

$$2\pi u = 2\pi r \varepsilon_t \tag{5.67}$$

From Equations (5.66) and (5.67), the radial displacement is:

$$u = \frac{\rho}{E}\,(\sigma_t - v\sigma_r) \tag{5.68}$$

Equations (5.64) and (5.68) yield:

$$u(\rho) = \frac{(1 + \rho)(p_1 r_1^2) - (p_2 r_2^2)r + (1 - v)(p_1 - p_2)r_1^2 r_2^2/r}{E(r_2^2 - r_1^2)} \tag{5.69}$$

At the internal hub diameter, $r = r_1$:

$$u(r_1) = u_1 = \alpha_{11}p_1 + \alpha_{12}p_2 \tag{5.70}$$

where:

$$\alpha_{11} = [(1 + v)r_1^3 + (1 - v)r_1 r_2^2][E(r_2^2 - r_1^2)]$$

$$\alpha_{12} = \frac{-(1 + v)r_1 r_2^2 - (1 - v)r_1 r_2^2}{E(r_2^2 - r_1^2)}$$

At the external hub diameter, $\rho = r_2$, similarly:

$$u(r_2) = u_2 = \alpha_{21}p_1 + \alpha_{22}p_2 \tag{5.71}$$

where:

$$\alpha_{21} = \frac{(1 + v)r_1^2 r_2 + (1 - v)r_1^2 r_2}{E(r_2^2 - r_1^2)}$$

$$\alpha_{22} = \frac{-(1 + v)r_2^3 - (1 - v)r_1^2 r_2}{E(r_2^2 - r_1^2)}$$

For internal pressure only, $p_1 = p$, $p_2 = 0$, the increase of the inner radius will be:

$$u_1 \frac{r_1 p}{E} \left(\frac{r_2^2 + r_1^2}{r_2^2 - r_1^2} + v \right) \tag{5.72}$$

Equations (5.64), (5.68), and (5.69) are the stress and displacement equations for thick cylindrical pressure vessels. For such vessels the membrane stress equations (5.51) and (5.52) might be inaccurate if, as a rule of thumb, the inner diameter is less than 10 times the wall thickness, that is, $r_1 < 10(r_2 - r_1)$ or $r_2 > 1.1r_1$.

5.3.3. Interference Fits

Stress analysis in interference fit joints is usually performed with the assumption that a hub of orthogonal cross-section is press-fitted around a rigid cylindrical shaft. For other geometries, computer methods are used. Torque T and axial force F are transmitted.

Between the hub and the shaft there is uniform pressure p and the friction coefficient is f. The resulting friction moment is

$$M = fp \cdot Ld \frac{d}{2}$$

where L is the hub length. The axial friction force is $F = fp\pi Ld$. The minimum pressure p to secure the fit should provide sufficient friction for both the axial force and the torque transmitted and should be:

$$p = \left[\left(\frac{2M}{f\pi d^2 L} \right)^2 + \left(\frac{F}{f\pi Ld} \right)^2 \right]^{1/2} \tag{5.73}$$

The design requirements are:

1. For the fit with the minimum interference (see Chapter 4), the pressure must be at least p, as given in Equation (5.73).
2. At maximum interference, the hub must not fail under excessive stress, as in Equation (5.65).

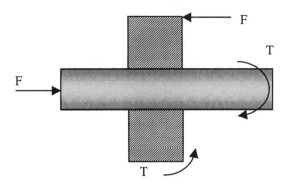

Figure 5.17. Loading of an interference fit.

Thus, an interference fit must be designed for:

1. *Minimum interference* e_{min}. The minimum pressure required for the fit is given by Equation (5.58a). Under this pressure p, the shaft diameter is reduced by (Equation (5.71)):

$$u_s = -\frac{(1 - v_s)r_2 p}{E_s} \qquad (5.74)$$

assuming zero internal diameter and only external pressure. The minimum interference e_{min} must consist of the reduction of the shaft diameter plus the extension of the internal hub diameter (Equation (5.72)), plus the roughness of the shaft R_s and hub R_h. Therefore:

$$e_{min} = \frac{pr_1}{E}\left(\frac{r_2^2 + r_1^2}{r_2^2 - r_1^2} + v\right) + \frac{(1 - v_s)r_1 p}{E_s} + R_h + R_s \qquad (5.75)$$

2. *Maximum interference* e_{max}. This is computed from the allowable stress of the hub material S_y/N. The pressure needed to result in tangential stress S_y/N is given by Equation (5.65), solved for p:

$$p_{max} = \left(\frac{r_2^2 - r_1^2}{r_2^2 + r_1^2}\right)\frac{S_y}{N} \qquad (5.76)$$

Then the corresponding extension of the internal hub diameter will be (Equation 5.72)):

$$e_{max} = \frac{r_1 p_{max}}{E}\left(\frac{r_2^2 + r_1^2}{r_2^2 - r_1^2} + v\right) \qquad (5.77)$$

Equations (5.75) and (5.77) are the design equations for the interference fit.

The above values assume 100% reliability. Since only a small percentage of fits have maximum interferences near e_{max}, if $e_m = (e_{max} + e_{min})/2$, the ranges $t_d = d_{max} - d_{min}$, $t_D = D_{max} - D_{min}$, analysis based on normal distribution yields the maximum interference, for $d = 2r_1$, $D = 2r_2$:

$$e_{max} = e_m + 0.5c[t_d^2 + t_D^2]^{1/2} \qquad (5.78)$$

where the factor c is related to the desired reliability P as:

P	0.99	0.95	0.9	0.5
c	0.78	0.55	0.43	0.0

For $P = 1$, $c = 1$. For lower reliability, the maximum interference will be substantially lower and so will be the maximum design pressure.

After the computation of e_{max} and e_{min}, the proper fit should be selected with the methods discussed in Chapter 4 (see also Example 5.13).

Design Procedure 5.5: Design of Interference Fits

Step 1: Identify hub width, shaft diameter, friction coefficient (see Appendix E), transmitted torque, and axial force.

Step 2: Use Equation (5.73) to find the minimum required interference pressure and Equation (5.60) to find the minimum required interference.

Step 3: Use Equation (5.76) to find the maximum allowed interference pressure and Equation (5.62) to find the maximum allowed interference.

Step 4: Use Equation (5.78) to find the maximum allowed interference accounting for surface roughness and desired reliability.

Step 5: Use Equation (4.8*b*) to find the fundamental deviation. Either $\Delta F = 0$ (hole-basis) or $\delta f = 0$ (shaft-basis).

Step 6: Select quality of shaft and hole to satisfy Equation (4.8*a*). Use Equation 4.5 and Table 4.1*a*.

Step 7: Use Equation (4.8*d*) to define the limit dimensions of the fit.

Example 5.13   A gear hub has inner diameter $d = 100$ mm, outer diameter $d = 200$ mm, and width $b = 40$ mm. It transmits to the shaft 15 hp at 140 rpm by an interference fit. The coefficient of friction between hub and shaft is $f = 0.4$, the hub material is steel ASTM A284 Grade D and has yield strength $220 N/mm^2$, and the required safety factor for strength is 1. Find the limit dimensions of the interference fit. The desired reliability is 95% and the surfaces can be machined to a roughness of 0.6 μm.

Solution   The minimum interference pressure required for torque transmission by friction (Equation (5.73)) is:

$$p_{min} = \frac{2M}{f\pi bd_1^2} = \frac{2 \times 7121 \times 15 \times 10^3}{140 \times 0.4\pi \times 4 \times 100} = 3.04 N/mm^2$$

Minimum interference (Equation (5.75)):

$$e_{min} = \frac{pd_1}{2E}\frac{d_2^2 + d_1^2}{d_2^2 - d_1^2} - v = 1 \ \mu m$$

Maximum pressure allowed by the hub strength (Equation (5.76)):

$$p_{max} = \frac{d_2^2 - d_1^2}{d_2^2 + d_1^2}S_y$$

Maximum interference (Equation (5.77)):

$$e_{max} = \frac{d_1 p_{max}}{2E}\left(\frac{d_2^2 + d_1^2}{d_2^2 + d_1^2} + v\right) = 60 \ \mu m$$

Therefore, the fit should be, taking into account the roughness of shaft and hub, using Equation (5.78) and the associated table:

$$e_{max} = 60 + 0.5 \times 0.55(2 \times 6^2)^{1/2} = 62.3 \; \mu m$$

We select the hole-basis system and thus $\Delta_F = 0$. From Equation (4.8b), $\delta_f = e_{min} = 1 \; \mu m$. From Equation (4.8$a$):

$$it + IT = e_{max} - \delta_f = 62.3 - 1 = 61.3 \; \mu m$$

The tolerance unit is $i = 0.45(100)^{1/3} + 0.001 \times 100 = 2.19 \; \mu m$. Tolerance 7 for the hole and 6 for the shaft, Table 4.1a give it $= 10 \times 2.19 = 21.9$, $IT = 16 \times 2.19 = 35$, which is acceptable. Therefore, from Equation (4.8d):

$$Hole: D = 100 \, {}^{0}_{-35}$$

$$Shaft: d = 100 \, {}^{23}_{1}$$

Example 5.14   A bearing consists of a bronze sleeve with $E_b = 1.1 \times 10^{11} N/mm^2$ and $S_{yb} = 70 N/mm^2$. It is interference fitted inside a steel bore of $S_y = 220 N/mm^2$. Find the maximum allowable interference with a safety factor 1. Then find the temperature of the steel ring necessary for the initial insertion of the bronze sleeve, if the coefficient of linear thermal expansion of steel is $\alpha = 12 \times 10^{-6}/°C$ and the Poisson ratio for both materials is 0.3. The surface is machined to a roughness of 0.8 μm.

Solution   With the dimensions shown in Figure E5.14, the maximum allowed interference pressure is $p = p_2$ for the bronze sleeve (bronze generally has lower strength than steel).

$$\sigma_t = \frac{-(2p \times 70^2)}{70^2 - 50^2} = S_{yb} = 70 N/mm^2; \quad p = 17.1 N/mm^2$$

The contraction of the bronze sleeve will be for $p_1 = 0$, $p_2 = 17.12 N/mm^2$ (Equation (5.71)). From Equation (5.69):

$$u_b = (1 + 0.3)(-17.1 \times 70) \times 70 \, \frac{+(1 - 0.3)(-17.1)50^2 \times 70^2/70}{1.1E11 \times (70^2 - 50^2)} = 0.05 \; mm$$

Similarly, the expansion of the steel sleeve will be for $p_1 = 17.1 N/mm^2$, $p_2 = 0$ (Equation (5.72)):

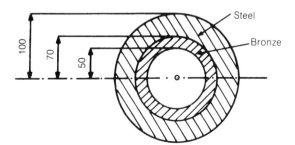

Figure E5.14

$$u_s = \frac{r_1 p}{E} \frac{r_2^2 + r_1^2}{r_2^2 - r_1^2} + v = 0.015 \text{ mm}$$

The maximum interference will be the sum:

$$e_{max} = u_b + u_s + 2(ra_b + Ra_s) = 50 + 15 + 2 \times 2 \times 0.8 = 68.2 \ \mu\text{m}$$

The radial strain should equal the thermal expansion $\varepsilon_r = e_{max}/r_2 = \alpha T$, therefore:

$$T = \frac{e_{max}}{\alpha r_2} = \frac{0.068}{(12 \times 10^{-6}) \times 70} = 80.9°\text{C}$$

over the room temperature.

5.3.4. Rotating Disks

Rotating disks are thick cylinders loaded also with inertia loads due to the rotation, commonly called *centrifugal loads*. These loads can produce substantial stresses and displacements in disks of large diameter at high speeds. To account for these loads, a centrifugal d' Alembert force is added to the volume differential dV, (Figure 5.6), of magnitude $\omega^2 r \rho dV$, where r is the radius at the differential volume, ω is the angular velocity of rotation, and ρ is the material density. The vertical resultant of the inertia forces need to added to the equation of equilibrium of forces (5.59) in the vertical direction:

$$2r_2 L p_2 - 2r_1 L p_1 + 2 \int_{r_1}^{r_2} \sigma_1 L dr - \int_{r_1}^{r_2} \int_0^p \omega^2 r \rho dr d\phi = 0 \qquad (5.79)$$

Following the method of Section 5.3.2, an additional term is added to the expression (5.69), namely:

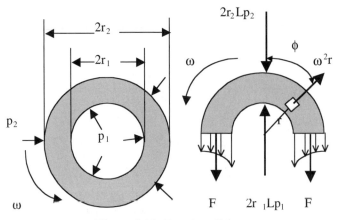

Figure 5.18. Rotating disk.

$$u_t = \frac{\gamma}{Eg} \omega^2 \rho \left(\frac{3+v}{8} \right) \left[(r_1^2 + r_2^2)(1-v) + \frac{r_1^2}{\rho^2}(1+v) - \frac{1-v^2}{3+v} \rho^2 \right] \quad (5.80)$$

and to the expressions for the stresses (Winkler 1860, Maxwell 1890) (Equations (5.75) and (5.78)):

$$\sigma_r = \rho\omega^2 \left(\frac{3+v}{8} \right) \left(r_1^2 + r_2^2 - \frac{r_1^2 r_2^2}{\rho^2} - \rho^2 \right)$$

$$\sigma_t = \rho\omega^2 \left(\frac{3+v}{8} \right) \left[r_1^2 + r_2^2 + \frac{r_1^2 r_2^2}{\rho^2} - \frac{1+3v}{3+v} \rho^2 \right] \quad (5.81)$$

Then we can write:

$$u(r_1) = u_1 = a_{11}p_1 + a_{12}p_2 + u_{1t}$$

$$u(r_2) = u_2 = a_{21}p_1 + a_{22}p_2 + u_{2t} \quad (5.82)$$

where from Equations (5.71) and (5.80):

$$u_{2t} = \frac{(\rho_0\omega^2/E)(3+v)r_2[(r_1^2 + r_2^2)(1-v) + r_1^2(1+v) - (1-v^2)r_2^2/(3+v)]}{8}$$

$$u_{1t} = \frac{(\rho_0\omega^2/E)(3+v)r_1[(r_1^2 + r_2^2)(1-v) + r_2^2(1+v) - (1-v^2)r_1^2/(3+v)]}{8}$$

The tangential (hoop) stress due to rotation has maximum value at the inner radius $r = r_1$:

$$\sigma_{t-\max} = \frac{3+v}{8} \frac{\rho\omega^2}{g} \left[r_1^2 + 2r_2^2 - \frac{1+3v}{3+v} r_1^2 \right] \quad (5.83)$$

Example 5.15   A turbine wheel has outer radius $r_2 = 450$ mm and inner radius $r_1 = 200$ mm, is made of CrMoV steel with density $\rho = 7800$ kg/m^3, Poisson's ratio $v = 0.3$, and yield strength $S_y = 450$ MPa. The wheel has uniform thickness and is fitted on the shaft with a free fit. Find the maximum safe speed for a factor of safety for strength $N = 3$.

Solution   There are no stresses due to fit pressure at the inner radius. We set up $\sigma_{t\max} = S_y/N$ in Equation (5.83) and solve for ω:

$$\omega_{\max} = \sqrt{\frac{8gS_y}{N(3+v)\rho\{r_1^2 + 2r_2^2 - [(1+3v)/(3+v)]r_1^2\}}}$$

$$= \sqrt{\frac{8 \times 9.81 \times 450 \times 10^6}{3(3+0.3)7800\{0.2^2 + 2 \times 0.45^2 - [(1+3\times 0.3)/(3+0.3)]0.2^2\}}}$$

$$\underline{\omega_{\max} = 1041 \text{ rad/s}} \text{ or } 1041 \times \frac{60}{2\pi} = 9,941 \text{ rpm}$$

5.4. STRESS CONCENTRATION

In strength of materials analysis, one usually assumes uniform distribution of stresses in tensile members or linear distribution of stresses and strains in beams. However, this is usually an approximation and is a reasonable approximation only under certain conditions. For example, stresses have to be considered far from point forces, supports, etc. Moreover, it is known that elastic materials have nonuniformity of stress distribution near geometric disturbances such as notches, holes, and keyways. Such occurrences are quite usual in machine members, which almost invariably are not simple cylinders, beams, or plates. In order to perform their function, they have to have such field disturbances, and the designer must consider the stress concentration near them.

Such a situation exists, for example, in a strip under uniform tension. As we know from strength of materials, far from the point of application of the forces the distribution of the stress everywhere in the section is uniform and equal to the applied stress. If, however, there is a hole in the strip, near the hole there is increased stress, as shown in Figure 5.19. For this case an analytical solution exists if the size of the hole is much smaller than the width of the strip and the hole has the shape of an ellipse with semi-axes a and b oriented as shown, and the maximum stress is (Beer and Johnston 1981):

$$\sigma_{\max} = \sigma_0 \left(1 + \frac{2a}{b} \right) \tag{5.84}$$

The ratio

$$K_t = \frac{\sigma_{\max}}{\sigma_0} \tag{5.85}$$

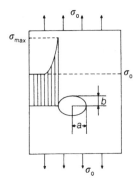

Figure 5.19. Stress concentration near an elliptical hole.

is called the *theoretical stress concentration factor* and expresses the intensity of the stress concentration due to the presence of the hole, if the material is purely elastic. Geometric disturbances that cause stress concentration are also referred to as *stress raisers*.

There are several methods for assessing the stress concentration. Many cases can be solved analytically, and there is a vast amount of literature on this subject. In more complicated cases one can apply numerical and experimental methods to estimate the stress fields and consequently the stress concentration factors. The finite-element method presented in Chapter 6 provides a very convenient tool for assessing the stress field in geometries encountered in engineering problems. For most practical situations, stress concentration factors are tabulated in the literature. In Appendix A.1, we have listed some results pertinent to machine design application.

Tabulated stress concentration factors (SCFs) are usually obtained for ideal elastic and homogeneous materials, and for this reason they are called *theoretical* or *geometric* SCFs.

It would seem that design on the basis of material strength has to be performed considering the maximum stresses at stress concentrations. This is not always the case. The reason is that in certain materials with high degree of ductility, the maximum stress may exceed the yield strength of the material. The local yielding in the area of stress concentration results in a redistribution of stresses such that a more uniform distribution will finally be achieved. This phenomenon is called *stress redistribution* or *stress relaxation* and is very common in ductile materials. In this case, failure will occur when the stress over the section is uniformly equal to the yield strength of the material. Thus, one should not consider stress concentration for such (highly ductile) materials. Local yielding near stress raisers in ductile materials results in residual stresses after unloading, a phenomenon common to most problems where partial yielding takes place.

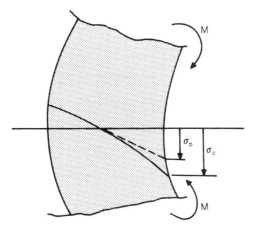

Figure 5.20. Stress concentration on a curbed beam.

(a)

(b)

Figure 5.21. (a) Photoelastic demonstration for stress concentration; (b) finite-element.

In brittle materials, however, stress concentration can be very dangerous because the yield point is almost indistinguishable from the point of fracture and yielding can be accompanied by the development of a crack, which might propagate rapidly and lead to final fracture. Therefore, for static loading, in general, stress concentrations have to be considered only in brittle materials. For dynamic loading the situation is quite different. This will be discussed later in this chapter. For some cases of static loading of brittle materials, stress concentration is not considered. This is the case when internal discontinuities such as microcracks exist that provide very high stress concentrations—higher than the sharpest geometrical stress raisers. Such a material is cast iron in tension, where this phenomenon is taken into account by assigning less strength in tension than in compression.

In designing machine members, one has to be careful where to put field disturbances such as holes and notches. If such disturbances have to be placed on a machine member (through the function of the machine), the designers must seek locations of low stress levels—if they have any choice at all. For example, in a member that is loaded by bending, holes and other disturbances must be located, if possible, on the neutral line and not near the outer fibers of the beam. A handbook of stress concentration factors is invariably a very good companion for every design office.

Photoelasticity is a useful method for demonstrating and measuring the stress concentration. Details can be found in any textbook on experimental stress analysis. Figure 5.21 shows a case of a stress raiser. High fringe concentration indicates stress concentration.

Theoretical stress concentration factors are tabulated, for some cases that are frequently encountered in machine design, in Appendix A, in two forms:

1. Approximate formulas found by approximating the test or numerical results. These could be used for synthesis or initial sizing only. They offer the convenience that they facilitate numerical computations because the stress concentration factors are functions of the geometry. Thus, to determine the geometry from design equations, one has to either use analytical expressions or resort to many iterations and longhand calculations.

2. Graphs with the experimental or numerical results are more accurate and are to be used for the final evaluation of a design or the certification of a machine.

Example 5.16   The nonrotating stub-shaft in the idle pulley assembly shown is loaded in bending. The transition from the 12 to 20 mm diameter is through a fillet of radius 3 mm. Find the theoretical stress concentration factor.

Solution   The ratio $D/d = 20/12 = 1.66$ and the ratio $r/d = 3/12 = 0.25$. From Appendix A, case 3, the theoretical stress concentration factor is $K_t = 1.4$.

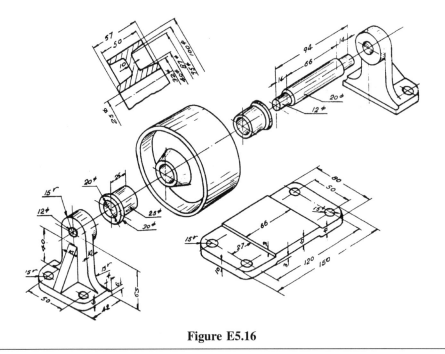

Figure E5.16

5.5. DYNAMIC LOADS: FATIGUE

5.5.1. Alternating Loads

Because most machine members move, many loads are not constant through time. Further, some constant loads can produce stresses that vary through time in a moving machine member. Take, for example, a rotating shaft that is loaded at midspan with a static load *W*, as shown in Figure 5.22*a*. If we measure the stress at fixed point *A* on the surface of the shaft rotating with it, we see that although the direction of the load on the shaft is constant, the produced stress at point *A* changes from tensile to compressive stress as point *A* moves with the shaft about the center of rotation. Therefore, in time the maximum tensile stress on the surface of the shaft follows a harmonic curve. Such stress is called *alternating stress*. The maximum value of such stress, which is equal to the semi-width of the stress variation, is called the *stress range* (Figure 5.22).

When an eccentric mass is rotating with the shaft, as shown in Figure 5.22*b*, the resulting inertia force (centrifugal force) produces a constant stress at the point *A* of the shaft, as shown, although the direction of the force varies continuously.

Next, consider the case of a hydraulic machine, an aircraft engine compressor for example, rotating at very high speed. If the static load and the unbalance give negligible effects, the turbulent flow of air around the shaft produces forces that are very irregular in time and appear as random. Such loading is called stochastic (Figure 5.22*c*).

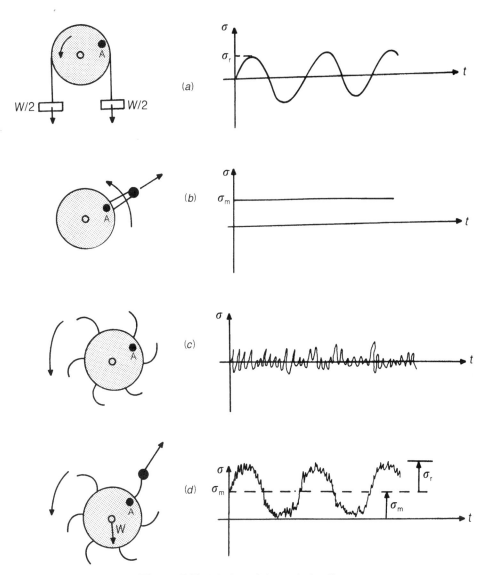

Figure 5.22. Modes of dynamic loading.

In most machines, all the types of loads appear simultaneously, as shown in Figure 5.22d, where the stress history in time is shown for a point on the periphery of the shaft. Disregarding the stochastic load, the effect of which will be discussed later, we have a periodic stress. Its axis of symmetry does not coincide with the time axis. Thus, we separate the mean *static stress* σ_m and the range of stresses σ_r. This is the most usual situation in machines, particularly in rotating shafts.

A. Woehler. (1819–1914). Born into the family of a school-master in Hanover, where he attended the Polytechnic Institute. As a distinguished student, he won a scholarship to go for training at Borsig, a locomotive manufacturer in Berlin involved in the building of Berlin railways. In 1843, he was sent to Belgium to study locomotive production. For the next 25 years he was in charge of railway machine shops, where he discovered the fatigue properties of materials while experimenting with railroad axles. They were unexplainable axle failures that Woehler should remedy. He designed and built various material testing machines, some of which are preserved in the Deutsches Museum. He had a paramount influence on materials testing in his time.

5.5.2. Fatigue

When one discusses static strength of materials, the assumption is that in the strength test the force is applied slowly and only once. In the 19th Century, the German engineer Woehler, experimenting with railroad axles, discovered that when he loaded them with forces producing stresses below the ultimate strength of the material, failure was observed after a number of full reversals of the stresses (revolutions in this case). Moreover, when the load was decreased, the number of cycles to failure increased. There was a stress level at which the material did not fail after any number of revolutions/stress reversals. This was called *continuous strength* or *fatigue strength*. The situation is shown in Figure 5.23 in a stress–time (number of cycles) diagram. The fatigue curve (called also a *Woehler curve*) levels for steel are usually between 5 and 10 million cycles. The asymptote to this curve would indicate the fatigue strength and is also called the *endurance limit*.

For steels and other materials, the fatigue curve can be approximated with the empirical formula $S^m N$ = constant, which is a straight line on semi-log paper. The exponent m is about 9 for steels. Assuming that continuous fatigue strength S'_n is achieved in a number of cycles N_0:

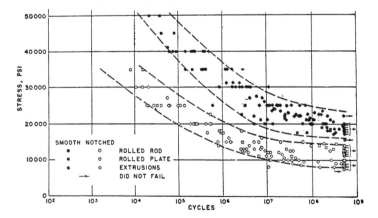

Figure 5.23. Fatigue curves (Woeler). (Courtesy ASME)

$$S_n^m \, N = S_n'^m \, N_0 \tag{5.86}$$

gives the fatigue strength S_n for any number of cycles $N < N_0$.

It is generally recognized that fatigue failure usually originates from a minute flaw discontinuity of the material after the loading results in local plastic deformation. Repeated deformation causes the flaw to gradually increase and form a gradually propagating crack, until the remaining section of the material can no longer sustain the load and the crack progresses rapidly, causing final failure. This phenomenon is called *high cycle fatigue,* to distinguish it from the fatigue (called *low cycle fatigue*) caused by more general yielding, which leads to failure after a much smaller number of cycles. This will be discussed in Section 5.5.4. In Figure 5.24, the effect of stress state on origin, appearance, and location of fatigue fracture is shown for tension and bending (*a*) and torsion (*b*).

High cycle fatigue strength, or simply fatigue strength, is usually tabulated with the other strength values of the materials (Appendix E), as discussed in Chapter 4. It must be emphasized, however, that the fatigue strength given in such tables usually has a rather high uncertainty because it is influenced by a great number of

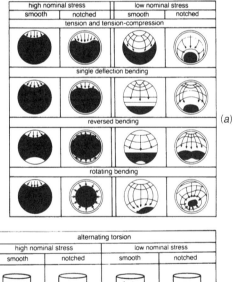

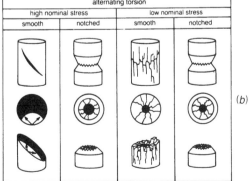

Figure 5.24. Effect of stress state on fatigue fracture origin, location, and appearance. (Adapted from L. Engel and H. Klingele, *An Atlas of Metal Damage,* Munich: Hanser Verlag, Munich, 1981.)

factors. From what was said for the procedure leading to fatigue failure, it is clear that fatigue strength depends on the size, orientation, and density of the material flaws in relation to the element size and on the mode of application of the load. For this reason, the tabulated values of the fatigue strength refer to very carefully performed experiments with a standard specimen. It has been found in practice that in actual machine elements the fatigue strength differs from the tabulated values, depending on the operating conditions and the geometry of the element. To obtain the design fatigue strength for the particular application, we usually multiply the theoretical fatigue strength by a number of *derating* factors. Each factor reflects a different mechanism of deviation from the ideal laboratory test. Here we shall use the expression (Marin 1962):

$$S_e = \frac{C_F C_S C_R C_H C_L S'_n}{K_f} \tag{5.87}$$

5.5.2.1. Surface Factor C_f. C_F is used to express the fact that different surface conditions result in different fatigue strength. This is due to the fact that rough surfaces have micro-cracks that with repeated loading might propagate and result in fracture. For structural steels this factor is given in Figure 5.25 (Karpov curves) for different levels of surface finishing.

5.5.2.2. Size Factor C_s. Fatigue tests are performed with standard diameters, usually 10 mm, and the variation of the diameter has an effect on fatigue strength.

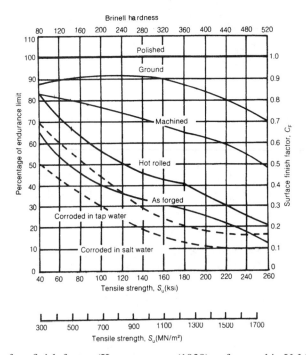

Figure 5.25. Surface finish factor. (Karpov curves (1939), referenced in V. M. Faires, *Design of Machine Elements,* 4th ed., New York: Macmillan, 1965.)

This is due to the fact (similar to the surface finish) that surface anomalies have a smaller effect on the bulk strength of larger parts than in smaller parts. On the other hand, if the dimensions become smaller, the probability of existence of internal cracks and other irregularities in the material is minimized. This is accounted for with factor C_s (Figure 5.26).

5.5.2.3. Reliability Factor C_R.
Published values of the fatigue strength are mean values of the fatigue strength, as was explained in Chapter 4. Therefore, the reliability of the component will be 50%. Higher reliability can be obtained by a proper safety factor (Chapter 4). If, however, this is not accounted for in the safety factor, the fatigue strength must be derated by a reliability factor, C_R, which is the inverse, $1/N$, of the safety factor obtained in the upper horizontal scale of Figure 4.29. If reliability is accounted for in the safety factor used in the design equations $\sigma_{\max} = S_e/N$, then $C_R = 1$.

5.5.2.4. Load Factor C_L.
The fatigue strength is influenced by the nature and density of the material flaws. It is then apparent that the fatigue strength of elements under pure tension–compression loading must be lower than that in reversed bending because for the latter, high stress occurs only at and near the outer fibers and therefore the probability of existence of a critical flaw in the highly stressed region is smaller.

It is therefore expected, and experimentally verified, that the fatigue strength in the tension–compression test is less than the values of fatigue strength that are obtained for cylindrical specimen in reversed (rotating) bending. For this case, the load derating factor C_L for tension–compression loading is taken to be 0.9, while for reversed bending and torsion (which has similar stress distribution to bending), it is taken to be 1.

5.5.2.5. Fatigue Stress Concentration Factor K_f.
It was pointed out earlier that the stress concentration factors do not have the same influence on ductile and brittle materials. In fatigue, the situation is a little more complicated. This is because even ductile materials under repeated loading and unloading, and therefore repeated yielding, might also fail. Experiments show that the proportion of the stress concentration that must be applied for design analysis is more or less a property of the material, and its effect is taken into account with the applied fatigue stress

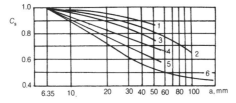

1–carbon steel, smoothly polished; 2–carbon steel, smoothly ground; 3–alloy steel, smoothly polished; 4–alloy steel, smoothly ground; carbon steel with stress concentration; 5–alloy steel with moderate stress concentration; 6–structural steel ($S_u < 65\,\text{kgf/mm}^2$ or $S_u < 650\,\text{MN/m}^2$), shaft with press-fitted part made of un-pressworked steel; at $d < 60\,\text{mm}$–alloy steel with intense stress concentration.

Figure 5.26. Size factor.

concentration factor K_f. This factor can be obtained from the theoretical stress concentration factor K_t using the Peterson equation:

$$K_f = 1 + q(K_t - 1) \tag{5.88}$$

Here the factor q is used to express the sensitivity of the material to stress concentrations. This factor is given for several materials and radii of notch in Figure 5.27. Such figures can be found in the literature for a variety of materials and forms of notch and loading.

The notch sensitivity factor q varies between 0 and 1. If it is 0, then the fatigue stress concentration factor is 1, indicating that the material is totally insensitive to fatigue due to stress concentration. Factor $q = 1$ make the fatigue stress concentration factor (K_f) equal to the theoretical stress concentration factor (K_t), and the material is then very sensitive to the stress concentration in fatigue.

5.5.2.6. Surface Hardening Factor C_H.

Neuber (1965) expressed the notch sensitivity factor q as a function of a material property a and the notch radius r, in the form $q = f(a/r)$. The material constant a was correlated with the material strength as shown in Figure 5.25.

Surface hardening has a strengthening effect, especially in small size elements. Not only does it strengthen a portion of the element section but it also strengthens the stress concentration areas near the surface, where a fatigue crack is most likely to originate. Therefore, an overrating factor C_H must be used, greater than 1. For the most common surface-hardening processes, C_H is given in Table 5.2.

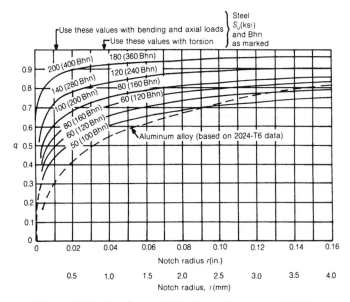

Figure 5.27. Notch sensitivity factor (courtesy ASME).

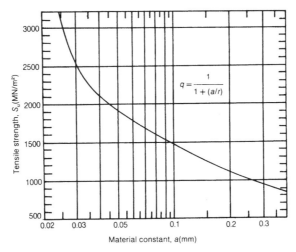

Figure 5.28. Notch sensitivity factor according to Neuber (courtesy of ASME).

Design Procedure 5.6: Design of a Part for Fatigue Strength— Reversed Stresses

Input: Loading, material properties, safety factor. **Output:** Part section dimension(s).

Step 1: From the internal load diagram of the part, identify the point(s) where failure is more likely.

Step 2: Find the reversing stress at the point under consideration. Use algebraic symbols for the unknown quantities.

Step 3: Find theoretical stress concentration factor (Appendix A.1) and the fatigue stress concentration factor (Equation (5.88)).

TABLE 5.2 Surface Strengthening Factor

		Strengthening Factor C_H		
Surface Hardening Procedure	S_u of Core (N/mm²)	Plain Shafts	Shafts with Low Stress Concentration $K_f = 1.5$	Shafts with High Stress Concentration $K_f = 1.8$ to 2
Induction surface	600–800	1.5–1.7	1.6–1.7	2.4–2.8
hardening	800–1000	1.3–1.5	—	—
Nitriding	900–1200	1.1–1.25	1.5–1.7	1.7–2.1
Carburizing and quenching	400–600	1.8–2.0	3	—
	700–800	1.4–1.5	—	—
	1000–1200	1.2–1.3	2	—
Work hardening by shot-peening	600–1500	1.1–1.25	1.5–1.6	1.7–2.1
Roll burnishing	—	1.2–1.3	1.5–1.6	1.8–2.0

Step 4: Find derating factors in Equation (5.87). Assume reasonable dimensions, if unknown, to find the derating factors.

Step 5: Use Equation (5.87) to find the applicable fatigue strength.

Step 6: Use stress equation $\sigma_r = Se/N$ to find either the safety factor (certification) or the unknown dimension (sizing).

Step 7: Go to step 3 and iterate again if the resulting dimensions (in sizing) are very different than the ones assumed in step 3.

See also Examples 5.17 and 5.18.

Example 5.17 (Fatigue, Certification) A circular shaft of length $L = 500$ mm, diameter $d = 40$ mm is made of AISI 1020 carbon steel, normalized, with $S_u = 440$, $S_y = 340$, $S'_n = 226$ N/mm^2. It has a normally machined surface and is integral with a wheel of diameter $D = 200$ mm weighing 100 kg at mid-span that results in bending stress. The shaft is connected to the wheel with a fillet of radius 5 mm. The shaft surface is not heat-treated. Find the effective safety factor for 99% reliability.

Solution The effective fatigue strength is:

$$S_e = \frac{C_F C_S C_R C_H C_L S'_n}{K_f}$$

where $C_F = 0.82$, from Figure 5.25; $C_L = 1$, bending is dominant; $C_s = 0.72$, from Figure 5.26, curve 4; $C_R = 1/1.2 = 0.83$ (Figure 4.29) for $R = 99\%$; $C_H = 1$, no surface hardening. For the stress concentration factor, $r/d = 5/40 = 0.125$, $D/d = 200/50 = 4$, therefore $K_t = 1.87$ (Appendix A). From Figure 5.27, $q = 0.7$ for $S_u = 440$ $N/\mathrm{mm}^2 = 62$ ksi, $r = 5$ mm, and $K_f = 1 + q (K_t - 1) = 1 + 0.7(1.87 - 1) = 1.609$. Therefore:

$$S_e = \frac{0.82 \times 0.72 \times 0.83 \times 1 \times 226}{1.609} = 68.8 \ N/\mathrm{mm}^2$$

The lateral force $W = 100 \times 9.81 = 981$ N and produces alternating stresses. The reactions are $981/2 = 460.5$ N and the bending moment at mid-span is $460.5 \times 0.25 = 115.2$ Nm. The maximum bending stress is:

$$\sigma_r = \frac{M}{W} = \frac{115.2}{\pi \times 0.040^3/32} = 18.3 \times 10^6 \ N/\mathrm{m}^2 = 18.3 \ N/\mathrm{mm}^2$$

$$N = \frac{S_e}{\sigma_r} = \frac{68.8}{18.3} = 3.76$$

Example 5.18 (Fatigue, Design) For the shaft of Example 5.17 it was decided that the safety factor was excessive and the shaft should be designed with a safety factor $N = 2.2$. Find the diameter.

Solution
First iteration: We assume first a diameter $d = 40$, as in Example 5.17, because we need the diameter to compute some of the factors involved in the shaft design, and then, as in Example 5.17, $S_e = 68.8$ N/mm^2. The stress equation for bending is:

$$\sigma_r = \frac{M}{W} = \frac{M}{\pi d^3/32} = \frac{S_e}{N}$$

Therefore:

$$d^3 = \frac{32MN}{\pi S_e} = \frac{32 \times 115.2 \times 2.2}{3.14159 \times 68.8 \times 10^6} = 3.752 \times 10^{-5} \text{ m}^3$$

$$\underline{d = 33.5 \text{ mm}}$$

Second iteration: The diameter is different than the one assumed, and thus the design calculations need to be iterated.

$C_F = 0.82$, from Figure 5.25; $C_L = 1$, bending is dominant; $C_s = 0.72$, from Figure 5.26, curve 4; $C_R = 1/1.2 = 0.83$ (Figure 4.29) for $R = 99\%$; $C_H = 1$, no surface hardening. For the stress concentration factor, $r/d = 5/33.5 = 0.15$, $D/d = 200/33.5 = 5.97$, therefore $K_t = 1.66$ (Appendix A). From Figure 5.27, $q = 0.7$ for $S_u = 440$ $N/mm^2 = 62$ ksi, $r = 5$ mm, and $K_f = 1 + q(K_t - 1) = 1 + 0.7(1.66 - 1) = 1.46$. Therefore:

$$S_e = \frac{0.82 \times 0.72 \times 0.83 \times 1 \times 226}{1.46} = 75.8 \text{ N/mm}^2$$

The diameter now is:

$$d^3 = \frac{32MN}{\pi S_e} = \frac{32 \times 115.2 \times 2.2}{3.14159 \times 75.8 \times 10^6} = 3.405 \times 10^{-5} \text{ m}^3, \underline{d = 32.4 \text{ mm}}$$

Third iteration: The diameter is again somewhat different than the one assumed, and thus we need another iteration.

For the stress concentration factor, $r/d = 5/32.4 = 0.154$, $D/d = 200/32.4 = 6.17$, therefore $K_t = 1.68$ (Appendix A). From Figure 5.27, $q = 0.7$ for $S_u = 440$ $N/mm^2 = 62$ ksi, $r = 5$ mm, and $K_f = 1 + q(K_t - 1) = 1 + 0.7(1.68 - 1) = 1.476$. Therefore:

$$S_e - \frac{0.82 \times 0.72 \times 0.83 \times 1 \times 226}{1.476} = 75.0 \text{ N/mm}^2$$

The diameter now is:

$$d^3 = \frac{32MN}{\pi S_e} = \frac{32 \times 115.2 \times 2.2}{3.14159 \times 75.0 \times 10^6} = 3.44 \times 10^{-5} \text{ m}^3, \underline{d = 32.5 \text{ mm}}$$

There is no need for further iteration and we will select the next available standard size, $d = 36$ mm.

Example 5.19 (Fatigue, Design, EXCEL) Solve Example 5.18 using EXCEL.

Solution The results-dependent parameter is K_t. From Appendix A, case 3:

TABLE E5.19

	AA	BB	CC	DD	EE	FF	GG	HH
1	EXAMPLE 5.19:							
2	**Design of a shaft with reversed bending stresses**							
3	*Data:*							
4	d_small =	0.032054	FilletR =	0.005	D_large =	0.1	S_N =	2.26E+08
5	Bending Moment =		115.2					
6	*Calculations:*							
7	r/d =	0.155986	c_F =	0.82	C_L =	1	C_S =	0.72
8	D/d =	3.119713	C_R =	0.83	C_H =	1		
9	q_fac =	0.7						
10	Coefficients for the analytical approximation of $K_t = 1 + (ca*r\_d^{\wedge}(-cb - cc*DC\_d))/(cd + ce/(DC\_d - 1)^{\wedge}cf)$							
11	ca	cb	cc	cd	ce	cf		
12	1.1169	0.693068	0.01069	3.976619	3.975696	0.237912		
13	Now we compute K_t and the effective fatigue stress concentration factor $k_f = 1 + q\_fac*(K\_t - 1)$							
14	K_t =	1.589868	K_ff =	1.412908				
15	The effective fatigue strength is		$S\_e = C\_F*C\_S*C\_R*C\_H*C$ $\_L*S\_N/K\_ff =$					78382506
16	The maximum bending stress is $s_{max} = Mc/I = B\_mom/$ $(PI()*d\_small^{\wedge}3/32) =$							35628425
17	The Safety Factor $N =$ $S_e/s_{max} =$			2.199999				
	Instructions: Use Tools/Solver to make N, cell DD17, equal to the desired value by changing d_small, cell BB4							

$$K_t = \frac{1 + 1.169(r/d)^{-0.693-0.107(D/d)}}{3.977 + 3.977/(D/d - 1)^{0.238}}$$

The spreadsheet is shown in Table E5.19. The diameter found is $d = 0.0032054$, the small deviation from the value found in Example 5.18 by longhand calculations is due to (a) the lesser error due to more iterations and (b) the greater error in the approximation for K_t.

5.5.3. Combined Loads

In Section 5.3, we saw already that for static loading the results of the tensile tests and the associated material properties can be extended to the more general loading of a material. An equivalent stress is defined as a function of the stresses acting on the material, and this equivalent stress is compared with the material strength. The related failure theories refer to the multiaxial static load on the material. With dynamic loading, the designers have to assess the strength of a component that is loaded by different mechanisms that produce stresses, both static and dynamic, and they then have to come up with some equivalent load that will subsequently be compared with the material strength.

5.5.3.1. Soderberg Line.
Assume uniaxial loading of a component with a combination of a static stress and an alternating stress as shown in Figure 5.29. The strength has to be evaluated based on the static and dynamic properties of the given material. To this end we note that for two combinations of static/dynamic load, the solution is already known. For zero dynamic load, the stress should not exceed the yield strength while for zero static load, the stress cannot exceed the fatigue strength of the material. If we now make a failure diagram on the (σ_m, σ_r) plane (Figure 5.30), there are two known points, the point S_e on the σ_r-axis and the point S_y on the σ_m-axis. The Soderberg hypothesis is that if these two points are connected with a straight line, this line separates the plane into safe and unsafe regions (Smith 1942). Below the *Soderberg line* the combination of static and dynamic load is assumed to be safe. Above the line the material will fail. Furthermore, if a factor of safety N is taken into account, the Soderberg line moves towards the origin by a factor of N, giving the line AB (Figure 5.30). In this diagram, every loading situation is represented by a point on the plane. In the limiting state, point C is on line AB. For this case, from the similar triangles AOB and CDB we obtain:

$$\frac{(S_y/N) - \sigma_m}{\sigma_r} = \frac{S_y}{S_e} \tag{5.89}$$

$$\frac{S_y}{N} = \sigma_m + \sigma_r \frac{S_y}{S_e} \tag{5.90}$$

It is apparent from Equation (5.90) that we can replace the combination of static and dynamic loads by an equivalent load given by Equation (5.90). The result can then be compared with the yield strength of the material:

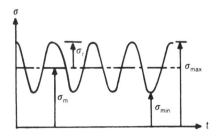

Figure 5.29. Static and alternating stress.

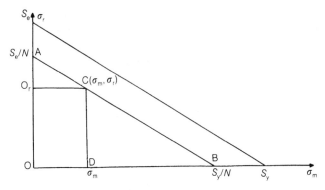

Figure 5.30. Soderberg diagram.

$$\sigma_{eq} = \frac{S_y}{N} = \sigma_m + \sigma_r \frac{S_y}{S_e} \tag{5.91}$$

The effective safety factor is then:

$$N = \frac{S_y}{\sigma_{eq}} = \frac{1}{(\sigma_m/S_y) + (\sigma_r/S_e)} \qquad \text{(Soderberg)} \tag{5.92}$$

In the above, only uniaxial static and dynamic loading is considered. In the case of more general loading, one may apply the static theories of failure (see Section 5.3) because the effect of dynamic loads is not taken into account in the equivalent static loads. We can write, for example, if only normal σ and shear stresses τ are present:

$$\sigma_{eq} = \left[\left(\sigma_m + \frac{S_y}{S_e} \sigma_r \right)^2 + a \left(\tau_m + \frac{S_{sy}}{S_{se}} \tau_r \right)^2 \right]^{1/2} \tag{5.93}$$

Equation (5.93) expresses the combined effect of static and dynamic loads acting as normal and shear stresses, with the Soderberg criterion. Again the factor a is 4 or 3, depending on application of the equivalent shear stress (Tresca) or equivalent distortion strain energy (von Mises) criteria.

5.5.3.2. Goodman Line. The Soderberg diagram is very convenient for design because of the facility of expressing the dynamic loads as equivalent static loads. It is a rather conservative criterion, however, because experimental observations show that the failure points are sometimes considerably above the Soderberg line. Experimental results show that the failure line should be taken as the one that connects the point on the σ_m-axis representing the ultimate strength with the point on the σ_r axis representing the fatigue strength. This is called the *Goodman line*. The tests show that the actual failure line is close to a parabola slightly above the Goodman line. Similar reasoning as for the Soderberg line yields for the Goodman line AD (Figure 5.31):

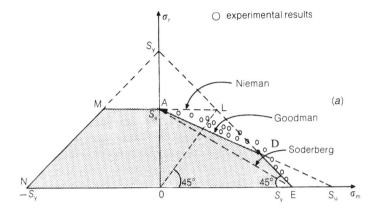

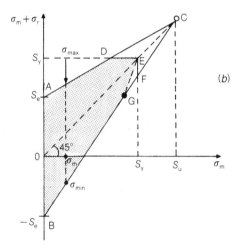

Figure 5.31. Goodman diagrams.

$$\sigma_{eq} = \frac{S_u}{N} = \sigma_m + \sigma_r \frac{S_u}{S_e} \qquad \text{(Goodman)} \qquad (5.94)$$

The effective safety factor for design based on the yield strength is:

$$N = \frac{S_y}{\sigma_{eq}} = \frac{S_y/S_u}{(\sigma_m/S_u) + (\sigma_r/S_e)} \qquad \text{(Goodman)} \qquad (5.95)$$

It must be observed that for machine-design purposes, static loads beyond the yield point are very rarely acceptable. Therefore, the sum of the static and dynamic load, that is, the maximum load that occurs at any time, cannot be above the yield line (Karpov 1936) (line *DE*):

$$\sigma_m + \sigma_r = S_y \qquad \text{(Yield line)} \qquad (5.96)$$

$$N = \frac{S_y}{\sigma_{eq}} = \frac{1}{(\sigma_m/S_y) + (\sigma_r/S_y)} \qquad \text{(Yield line)} \qquad (5.97)$$

This equation represents a $-45°$ straight line starting at point E (Figure 5.31). The intersection with the Goodman line completes the Goodman diagram: ADE is the failure line (Figure 5.31a). This situation may be also expressed in a diagram of maximum total stress versus static stress. Then the failure line is again ADE (Figure 5.31b).

John Goodman (1862–1935) was born in Royston, Hertfordshire. After a five-year engineering apprenticeship, he worked for the Brighton and South Coast Railway. He won the Whitworth scholarship and studied engineering at the University College, London. After his graduation in 1887, he worked for the Broadway Testing Works. In 1890, he was appointed to the Chair of Civil and Mechanical Engineering at Yorkshire College, Leeds. There he developed into a respected lecturer and researcher, and his book *Mechanics Applied to Engineering* had a lasting impact in engineering education. Most of his life he worked on tribological problems. His work on rolling bearings led him to the empirical evaluation of failure under combined static and reversing stresses. He retired in 1922 and shortly thereafter, due to friction with his successor, he was deprived of his experimental facilities at the university. He transferred his bearing testing machine to the Bradley's engineering garage where he continued his research until his death in 1935.

On the left of the diagram for negative static stresses (compression), the situation is a little different. Experiments show that most materials in static compression are rather insensitive to additional dynamic loads. Therefore, the Goodman line continues from point A to point M and then follows to point N, the safe static yielding line. This concludes the diagram.

There have been many attempts to define more realistic lines ADE, that better express the experimental data. Since points A and E are known, the in-between corner point has been the subject of different definitions. One such definition, very usual in continental Europe, is the pulsating strength of the material, which consists of subjecting the material to equal static stress and stress range. Obviously this point on the Goodman diagram must be on the 45° dichotomous of the AOE angle and, according to Niemann, on the yield line.

The foregoing discussion was based on the fatigue strength for a number of cycles greater than the number at which the material reaches the endurance limit. For steels it is about 10 million. Similar diagrams can be made for strengths at a lower number of cycles, and the corresponding fatigue strength at the particular number of cycles must be obtained from the Woehler curve, if available.

Design Procedure 5.7: Design for Fatigue Strength

Step 1: Compute all stresses at the section of interest. Identify static and alternating (dynamic) stresses

Step 2: Compute the effective fatigue strength S_e. If some of the parameters involved include unknown geometric parameters, make an educated guess, to be improved later.

Step 3: For the static stresses, use one of the failure theories to combine them into one single equivalent static stress one of (Equations (5.41)–(5.48)).

Step 4: For the dynamic stresses, use one of the failure theories to combine them into one single equivalent dynamic stress one of (Equations (5.41)–(5.48)).

Step 5: Use one of the theories of combined static and dynamic loads (Equation 5.92 or Equations (5.96) and (5.97)).

Step 6: For certification, solve for the safety factor N. If you use the Goodman theory, select the larger of the geometric quantities that result from Equations (5.96) and (5.97).

Step 7: For design, solve for the unknown geometric quantity (e.g., the diameter). If you use the Goodman theory, select the larger of the geometric quantities that result from Equations (5.96) and (5.97).

Step 8: If assumptions were made at step 2 for unknown geometric quantities, if the final design parameter found is substantially different, go back to step 2, use the new value found, and repeat the rest of the steps until convergence is achieved.

Example 5.20 (Fatigue, Certification, Combined Stresses)   A circular shaft of diameter $d = 40$ mm is made of AISI 1020 carbon steel, normalized, with $S_u = 440$, $S_y = 340$, $S'_n = 226$ N/mm^2. It has a normally machined surface and is integral with a 200 mm diameter wheel weighing 100 kg. There is also a constant thrust force of 2 tons. The shaft is connected to the wheel with a fillet of radius 5 mm. The shaft surface is not heat-treated. Find the effective safety factor for 99% reliability with the Soderberg and Goodman diagrams.

Solution   The effective fatigue strength (Equation (5.87)):

$$S_e = \frac{C_F C_S C_R C_H S'_n}{K}$$

where $C_F = 0.82$, from Figure 5.25; $C_L = 1$, bending is dominant; $C_s = 0.72$, from Figure 5.26, curve 4; $C_R = 1/1.2 = 0.83$ (Figure 4.29) for $R = 99\%$; $C_H = 1$, no surface hardening; $K_t = 1.7$ because $r/d = 5/40$, $D/d \gg 1$; $q = 0.7$ for $S_u = 440$ N/mm$^2 = 62$ ksi, $r = 5$ mm; and $K_t = 1 + q (K_t - 1) = 1.49$. Therefore:

$$S_e = \frac{0.82 \times 0.72 \times 0.83 \times 1 \times 226}{1.49} = 74.3 \text{ N/mm}^2$$

The axial force produces constant stress

$$\sigma_m = \frac{20,000}{\pi \times 40^2/4} = 15.9 \ \text{N/mm}^2$$

The lateral force produces alternating stresses:

$$\sigma_r = \frac{MR}{I} = \frac{1000 \times 500/4}{\pi \times 40^3/32} = 19.9 \ \text{N/mm}^2$$

The Soderberg formula (Equation (5.92)) gives:

$$N = 1[(\sigma_m/S_y) + (\sigma_r/S_e)] = 1/[(15.9/340) + (19.9/74.3)] = 3.17$$

The Goodman line (Equation (5.95)) gives:

$$N = \frac{S_u}{\sigma_m + (\sigma_r S_u/S_e)} = \frac{440}{15.9 + (19.9 \times 440/74.3)} = 3.29$$

The yield line (Equation (5.97)) gives:

$$N = \frac{S_y}{\sigma_m + \sigma_r S} = \frac{340}{15.9 + 19.9} = 9.49$$

The safety factor with the Soderberg theory is the smaller of the last two values, $N = 3.17$.

As expected, the Soderberg formula yields a smaller factor of safety, thus more conservative design. One more iteration is needed, to take into account factors that depend on the results (K_f, for example).

Example 5.21 Compute the shaft diameter in Example 5.2 if all data remain the same and the safety factor must be 2.5. Use the modified Goodman criterion.

Solution EXCEL is used. The spreadsheet is shown in Table E5.21. The safety factor was computed with both the Goodman criterion (5.95) and the yield line criterion (5.97). The smaller is the applicable safety factor. Solver is used to set this factor to 2.5 by changing the shaft diameter. The resulting diameter is 36.3 mm. The next standard diameter must be selected.

The problem could be solved by iteration (see Example 5.18).

5.5.4. Cumulative Fatigue Damage

In Section 5.3, it was assumed that the loads are constant during the life of the machine member. This might not be the case in an actual machine operation.

The machine by its nature might not operate all the time at the same load. On other occasions it is possible that the machine might operate accidentally for a period at loads higher than the design load. In cases where loads exceed, for part of the anticipated lifetime, the design load, we can deal with the values of either the remaining lifetime or the load level, in order for the machine to continue operation.

Palmgren, in the 1920s, made the assumption, which was restated later by Miner (1965), that operation at a certain level of dynamic loads for a certain number of

TABLE E5.21

	A	B	C	D	E	F	G	H
1	**EXAMPLE 5.21**							
2	**Design of a shaft with combined reversing + steady stresses**							
3	*Data:*							
4	d_small =	0.036335	Fillet$R$ =	0.005	D _large =	0.2	S _N =	2.26E+08
5	*Loads:*							
6	<u>Reversing</u>				<u>Steady</u>			
7	Bending Moment =		115.2		Axial Force		19620	
8	Material:							
9	S_u	4.40E+08	S_y =	3.40E+08	S'_n	2.26E+08		
10	*Calculations:*							
11	r/d =	0.137607	c_F =	0.82	C_L =	1		
12	D/d =	55.0428	C_R =	0.83	C_H =	1	C_Su =	0.7
13	q_fac =	0.7						
14	*Coefficients for the analytical approximation of K_r:*							
15	ca	cb	cc	cd	ce	cf		
16	1.1169	0.693068	0.01069	3.976619	3.975696	0.23791219		
17								
18	K _t =	1.734523	K_ff =	1.514166		S_e =	71109053	

TABLE E5.21 (Continued)

	A	B	C	D	E	F	G	H
19	Stress $\sigma_\tau =$	18921266		Stress $\sigma_m =$		18921266		
20	Safety Factor							
21	Goodman:		2.5		Yield Line:		8.9846	
22	Applicable Safety Factor:			2.5				
23								
24	Instructions: Use Tools/Solver to make cell $d22$ to the desired value by changing cell $B4$							
25	Cell B18: $=1 + (ca*B7^{\wedge}(-cb - cc*B8))/(cd + ce/(B8 - 1)^{\wedge}cf)$							
26	Cell D18: $=1 + q\_fac*(K\_t - 1)$							
27	Cell G18: $=C\_F*C\_Su*C\_R*C\_H*C\_L*S\_N/K\_ff$							

cycles has a cumulative effect on damage that, when summed up to a certain level, leads to the eventual failure of the material.

If we assume that the material is subjected to different loads, $\sigma_1, \sigma_2, \ldots$, at $n_{r1}, n_{r2}, \ldots$, cycles respectively, for each loading the ratio of the number of cycles over the number of cycles for failure $n_1, n_2, \ldots$, at that load (Figure 5.32) is a quantitative measure of fatigue damage incurred by the material. When the sum of all these damages reaches a certain limit, the material will fail. This is expressed as:

$$\frac{n_{r1}}{n_1} + \frac{n_{r2}}{n_2} + \cdots + \frac{n_{rm}}{n_n} \geq 1 \qquad (5.98)$$

In Equation (5.98), this limit is one.

Subsequently, experiments showed that this value differs from one for many materials and for many loading situations. There is a vast amount of experimental data for this sum. For softer steels, it is above one and can reach the value of three. For high-strength steels, it might fall slightly below one. It is a good practice to use the value of one, which results in more conservative design for lower-strength steels.

It must be pointed out that the above results are based on very strict assumptions, such as sinusoidal loading, constant load over a period of time, and so on. Machine members usually sustain loads of the type shown in Figure 5.22d. This suggests that the above results are only approximate and further experiments and assessment of the field-service experience are necessary for every relevant machine-design effort.

Design Procedure 5.8: Design for Cumulative Fatigue Damage

Step 1: Identify all different alternating loads $s_1, s_2, \ldots, s_{m-1}$, and their duration, $n_{r1}, n_{r2}, \ldots, n_{r(m-1)}$, in cycles.

Step 2: Use a semi-log paper (logarithmic abscissa) to develop the fatigue curve.

Step 3: For each of the alternating loads s_j, find the cycles to failure n_j from the fatigue diagram.

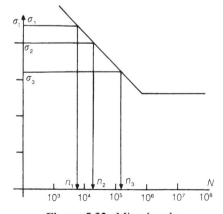

Figure 5.32. Miner's rule.

Step 4: Apply Equation (5.98). The last fraction on the left-hand side of the equation will have the remaining life n_{rm}, in number of cycles and the cycles to failure nm, at stress s_m of the remaining life.

Step 5: If the stress during the remaining life s_m is known, find the remaining life number of cycles n_{rm}.

Step 6: If the stress during the remaining life s_m is unknown and the remaining life number of cycles n_{rm} is known, find nm and enter the fatigue curve with the value found from Equation (5.98) to find the remaining life n_{rm}.

Example 5.22   Measurements were conducted on a mechanism component to determine the service conditions. A purely alternating stress was measured at a frequency of 40 cpm. The distribution of stresses and the relative frequency for 60 operations recorded were:

$\sigma_i(N/\text{mm}^2)$	140–150	130–140	120–130	110–120	100–110	90–100
n_{rj}	3	7	9	12	15	14

The material had an effective fatigue strength of 90 N/mm^2 reached at 10^7 cycles, while the strength at 1000 cycles was 440N/mm^2. Assuming a straight line of the Woehler curve between the two points on a (log–time) scale, determine the fatigue life of the component for stress 100 MPa.

Solution   From Figure E5.22, where the effective fatigue life curve is plotted with the component stress distribution, the fatigue life at each stress level is obtained:

σ_j (MPa)	140–150	130–140	120–130	110–120	100–110	90–100
n_j	10^3	7×10^3	3×10^4	1.2×10^5	8×10^5	4×10^6

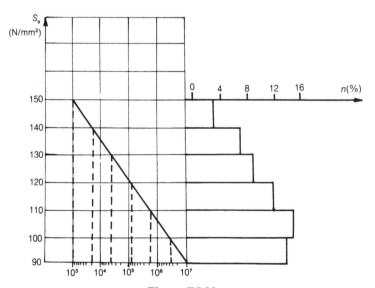

Figure E5.22

For 100 MPa, $n = 4 \times 10^6$. If n_{rm} is the remaining fatigue life, the Palmgren–Miner rule (Equation 5.98) gives:

$$\frac{n_{r1}}{n_1} + \frac{n_{r2}}{n_2} + \cdots + \frac{n_{rm}}{4 \times 10^6} = 1$$

$n_{rm} = 22,612$ cycles, which corresponds to 9.42 hours of operation at frequency 40 cpm.

5.5.5. Low Cycle Fatigue

In many service conditions, the material will only sustain a few thousand cycles of severe loading for which plastic strains are developed. Such situations exist many times in thermal machines where, during start-up, the thermal stresses developed lead to plastic flow. In other cases, such yielding comes from impact loads during start-up and in general where yielding is not present during normal operation of the machine, so that for the life of the machine the total number of occasions where yielding occurs is relatively low.

Several investigators have shown that for a finite life of less than about 100,000 cycles it is apparently the cyclic strain that is important for failure rather than the cyclic stress. For such situations it is usually impossible or irrelevant to measure or compute the stress under practical conditions of operation due to the plastic behavior of the material. For a number of metals, such as aluminum alloys, magnesium, low-carbon steels, and alloy steels, a fairly narrow scatterband was obtained when the test data were plotted in terms of maximum fiber strain versus endurance life in cycles, as shown in Figure 5.33.

As a further refinement, it has been observed that if the elastic strain is subtracted from the total strain measured in tests, a good straight-line relationship exists between the logarithm of plastic strain and the logarithm of cycles to failure. It has been shown that the experimental data are expressed with a relationship

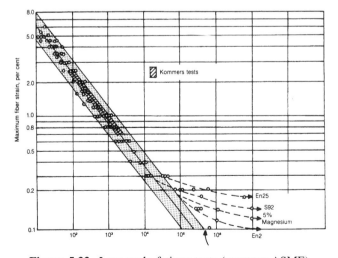

Figure 5.33. Low cycle fatigue tests (courtesy ASME).

$$N^{1/2}\Delta\varepsilon_p = C \tag{5.99}$$

where $\Delta\varepsilon_p$ is taken as the cyclic plastic strain, N is the fatigue life in cycles, and C is a constant of the material. The value of ε_p is determined by subtracting the elastic strain from the total observed strain. On log paper, strain lines are obtained such as in Figure 5.33.

It was found that the factor $C \cong 1/2$, the natural strain at fracture, which is defined as the logarithm of the ratio of the initial and final cross-sectional areas (area reduction at failure; for different materials see Appendix E) at fracture:

$$C = \ln \frac{A_0}{A} \tag{5.100}$$

For most steels, $C = 0.3$–0.8.

Example 5.23 For a typical test of mild steel for ductility, a strip of thickness $\alpha = 10$ mm is bent successively about two cylinders of diameter $2R = 200$ mm. It is assumed that plastic deformation prevails and elastic deformation can be neglected. The material is AISI 1015 steel with area reduction at failure 60%. Determine the number of deformation cycles to failure.

Solution The total strain at the outer fibers of the strip will be, at some arc length S:

$$\varepsilon = \frac{S_2 - S_1}{S_1} = \frac{S_2}{S_1} - 1 = \frac{R + \alpha}{R} - 1 = \frac{100 + 10}{100} - 1 = 0.1$$

It is assumed that elastic deformation is negligible, therefore $\Delta\varepsilon_p = 0.1$.

The low cycles fatigue constant C is:

$$C = \ln \frac{A_0}{A} = \ln \frac{1}{0.6} = 0.51$$

Equation (5.99) then yields:

$$N = \left(\frac{C}{\Delta\varepsilon p}\right)^2 = \left(\frac{0.51}{0.1}\right)^2 = 26 \text{ cycles}$$

5.6. BRITTLE FRACTURE

For many years, machine parts and engineering structures subjected to steady loads at moderate and low temperatures have been designed primarily on the basis of the yield strength. This method of design does not always produce satisfactory results, as large structures frequently fail at stresses below the yield strength, with disastrous consequences. For example, the tanker SS *Schenectady* failed while in dock, when the nominal stress in the deck was only 60 kN/mm^2. A liquid natural gas storage tank also failed in Cleveland, Ohio, on October 20, 1944. Brittle fracture of one large tank allowed the escape of gas into a heavily populated industrial area and over 130 people lost their lives. More recently, a number of large steam turbine and generator rotors have burst.

The aircraft and missiles industry frequently faces similar problems. During the initial operation of the B-29 airplane, the landing-gear assemblies failed with monotonous regularity, sometimes when the plane was sitting on the runway. These assemblies were made of high-strength (1400–1550 MPa) tensile strength steel forgings welded and sometimes plated.

Further failures occurred on welded pressure vessels used in the missiles program. These vessels must generally operate at very high stresses and sometimes at very low temperatures. Elevated-temperature "brittle fracture" may be basically the same phenomenon as that observed at lower temperatures, for it has many features in common with low-temperature brittle fracture.

Brittle fracture is the term commonly applied to the sudden failures described above. The use of the word "brittle" arises largely from the absence of any noticeable deformation preceding or accompanying the failure. Although the term *brittle fracture* is most commonly defined as a failure that was not preceded or accompanied by any noticeable plastic flow, this brittleness is not actually the primary cause for concern.

The terms *brittle* and *brittle crack propagation* both apply to failure under static *load,* as opposed to *fatigue* loading. For a very large number of completely reversed fatigue loading cycles (zero mean stress), the failure is always brittle; that is, there is no visible plastic deformation preceding crack initiation.

The following is a list of some of the more common criteria of brittleness (see Horger 1965):

1. Failure without any plastic flow preceding the initiation of a crack in an unnotched element.

2. Failure without any noticeable plastic flow accompanying the propagation of the crack. Such failure could also be called "failure with low crack propagation resistance."

3. Failures characterized by a "granular" or "crystalline" appearance, as contrasted with a "fibrous" or "silky" texture.

4. Failure characterized by flat fracture surfaces normal to the direction of stress, as contrasted with fracture surfaces that are inclined at 45° to the direction of stress.

5. Sudden failures, as contrasted with gradual tearing.

6. Failures in which the total energy absorbed is small compared to the energy that would be absorbed at a higher temperature.

7. Failures that occur before the load–deflection curve has deviated appreciably from a straight line.

8. Failures characterized by cleavage of the grains, as ascertained by a microsection through the fracture surface.

9. Failure at a load significantly less than that required for plastic flow to spread across the entire cross-section. For example, it is common practice in the case of annular grooves in cylindrical tension test bars to describe the behavior as "notch brittle" if the notch strength is less than the unnotched strength and "notch ductile" if the notch strength is greater than the unnotched strength.

All of the currently used or proposed approaches to engineering design are based on the principle that if one of the necessary conditions for fracturing is eliminated, or sufficiently reduced, failure can be avoided.

According to Griffith (1965), rapid crack propagation will commence whenever a small extension of the crack releases as much elastically stored energy as the energy used to form the additional crack surface.

The energy-release rate for the specimen is a function of nominal stress. It must be determined by either calculation or measurement. For example, for bending of a notched beam, the energy-release rate G is:

$$G = \frac{2h(1 - v^2)}{E} f\left(\frac{a}{d}\right) < G_c \tag{5.101}$$

where h is the net beam depth, (a/d) is the ratio of notch depth, and $f(a/d)$ is the function of the geometry only.

The critical energy release rate is a material property and is shown for some materials exhibiting brittle behavior in Table 5.3. For the particular geometry and loading, the energy-release rate can be computed for most design problems.

We note that Equation (5.84), for the stress concentration near an elliptical hole, gives infinite stress when the minor semi-axis tends to zero, as we can model a crack. In general, Irwin (1965) has shown that near the tip of a crack the stress field has the form

$$\sigma = Kf(\theta)[2\pi r]^{-1/2} \tag{5.102}$$

where r is the distance of the point from the crack tip, $f(\theta)$ is a function of the

TABLE 5.3 Critical Energy Release Rate for Some Materials[a]

Material	G_c(lb/in)
Dural	1.60×10^3
Key steel	5.71×10^2
Brass	3.43×10^2
Teak wood	68.5
Cast iron	45.7
Cellulose	22.8
Polystyrene	11.4
Polymethyl methacrylate	5.71
Epoxide resin	3.77
Polyester resin	2.51
Graphite	0.571–1.14
Alumina	0.457
Magnesia	0.114
Glass	4.57×10^{-2}

[a]From G. C. Sih et al., *Applications of Fracture Mechanics to Engineering Problems*, Bethlehem, Pa.: Lehigh University, 1972.

angle with the crack plane. K is a factor that depends on loading and geometry. For example, for a strip in uniform tension, perpendicular to the crack plane,

$$K = \sigma_0[\pi a]^{1/2} F(a) \tag{5.103}$$

where $2a$ is the crack length and $F(a)$ is a function that depends on the geometry. For an infinite strip, that is, if the strip boundary is very far from the crack, $F(a) = 1$.

The utility of the factor K, called the *stress intensity factor* (SIF), follows the observation that for every material, when this factor reaches a critical value, brittle fracture will follow. The critical value of the SIF is then a material property called *fracture toughness* (Table 5.4), which is known to be related to the critical energy release rate as

TABLE 5.4 Fracture Toughness for Some Materials[a]

Material	Ultimate Strength, S_u(ksi)	Critical Stress-Intensity Factor, $K_c$ksi(in)$^{1/2}$
A517F Steel (AM)	120	170
AISI 4130 Steel (AM)	170	100
AISI 4340 Steel (VAR)	300	40
AISI 4340 Steel (VAR)	280	40
AISI 4340 Steel (VAR)	260	45
AISI 4340 Steel (VAR)	240	60
AISI 4340 Steel (VAR)	220	75
300M Steel (VAR)	300	40
300M Steel (VAR)	280	40
300M Steel (VAR)	260	45
300M Steel (VAR)	240	60
300M Steel (VAR)	220	75
D6AC Steel (VAR)	240	40–90
H-ll Steel (VAR)	320	30
H-ll Steel (VAR)	300	40
H-ll Steel (VAR)	280	45
12Ni–5Cr–3Mo Steel (VAR)	190	220
18Ni (300) Maraging Steel (VAR)	290	50
18Ni (250) Maraging Steel (VAR)	260	85
18Ni (200) Maraging Steel (VAR)	210	120
18Ni (180) Maraging Steel (VAR)	195	160
9Ni–4Co–0.3C Steel (VAR)	260	60
Al 2014-T651	70	23
Al 2024-T851	65	23
Al 2219-T851	66	33
Al 2618-T651	64	32
Al 7001-T75	90	25
Al 7075-T651	83	26

[a]From G. C. Sih et al., *Applications of Fracture Mechanics to Engineering Problems,* Bethlehem, Pa.: Lehigh University, 1972.

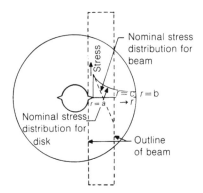

Figure 5.34. Congruency principle.

$$S_c = \frac{(1 + v)(1 - 2v)K_c^2}{2\pi E} \tag{5.104}$$

Therefore, for a given material with given either critical strain energy release rate or fracture toughness, an expression is needed giving the stress intensity factor of the energy release rate for the particular geometry and loading. The SIF has been tabulated in recent years in many publications for a variety of problems. Some useful results are given in Appendix B.

Failure criteria based purely on nominal stress are highly empirical in nature. They stem from the idea that fracturing is a local phenomenon and therefore is not influenced by conditions remote from the location of the fracture origin. For example, let us compare the behavior of an unnotched beam specimen with that of a rotating disk with a hole in the center (Figure 5.34). Suppose the beam depth is such that when the stress at the tension surface of the beam also equals the stress gradient at the bore of the disk. Now two elements of volume, one at the bore of the disk and the other at the tension surface of the beam, encounter the same stress situation and so should fail at the same stress and strain.

This concept, that failure will occur in two different objects when the local conditions are the same, can also be applied to notches (*congruency principle*).

Example 5.24 A wide strip of AISI 4340 steel with $S_u = 300$ ksi is under uniform tension of 50 N/mm². It is secured on the machine frame by a 10 mm diameter bolt and a 20 mm diameter nut. If a crack develops at the hole, it cannot be detected until it propagates outside the nut external diameter, that is, has a length of 5 mm. Assuming the crack to behave as an edge crack of a strip as shown with the same tension, accounting for stress concentration at the hole, find out if the strip is secured against brittle fracture when the crack has the length of 5 mm. Find also the critical crack length.

Solution With the approximation for the stress field shown, the stress concentration factor for the 10 mm bolt will be, from Appendix A, Table A, $K_t = 3$. Therefore, $\sigma_{max} = 3 \times 50 = 150$ N/mm². From Table 5.3, it is found that AISI steel, $S_u = 300$ ksi, or 2112 N/mm², and the fracture toughness $K_c = 40$ ksi (in.)$^{1/2}$ or 1419 N/mm² (mm)$^{1/2}$. The stress intensity factor $K = \sigma (\pi\alpha)^{1/2}$ for $f(\alpha/b) = 1$, $\alpha << b$ (Appendix B, case 2). Therefore:

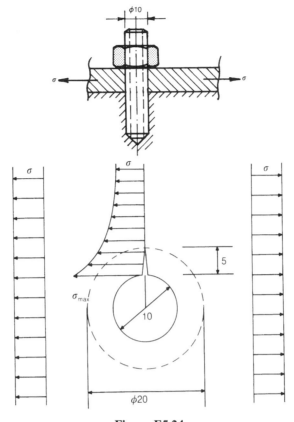

Figure E5.24

$$K = 150(\pi \times 5)^{1/2} = 594 \text{ N/mm}^2 \text{ (mm)}^{1/2} < K_c$$

Therefore, the strip is safe against brittle fracture. A more accurate value for $f(\alpha/b)$ without the edge crack of a strip approximation used here can be found in appropriate handbooks. The critical crack length will be, solving for α:

$$\alpha = \frac{(K_c/\sigma_{max})^2}{\pi} = 28.5 \text{ mm}$$

Example 5.25 A turbine wheel of uniform thickness has inner diameter 160 mm and outer 400 mm. It is made of 12Ni–5Cr–3Mo steel with $S_u = 1340$ N/mm^2 and $K_c = 7180$ N/mm^2 (mm)$^{1/2}$. It is press-fitted on the shaft. Find the maximum allowable pressure of the fit if the minimum observable crack has length 15 mm.

Solution The tangential stress is (from Equation (5.65)):

$$\sigma_t = \frac{(p_1 r_1^2 - p_2 r_2^2) + (r_2^2/r^2)(p_1 - p_2)}{r_2^2 - r_1^2}$$

where r_1 and r_2, p_1 and p_2 are inner and outer radii and pressures, respectively. Since $p_1 = p$, $p_2 = 0$.

$$\sigma_t = \frac{pr_1^2 + (r_1^2 r_2^2/r^2)p}{r_2^2 - r_1^2}$$

$$\sigma_t = pr_1^2 \frac{1 + r_2^2/r^2}{r_2^2 - r_1^2}$$

$$\frac{\partial \sigma_t}{\partial r} = \frac{-2pr_1^2 r_2^2/r^3}{r_2^2 - r_1^2}$$

The equivalent beam with height $2b$ and bending moment M, per unit width, would produce

$$\sigma_{max} = \frac{3M}{2b^2}$$

$$\frac{\partial \sigma}{\partial x} = \frac{\sigma_{max}}{b} = \frac{-3M}{2b^3}$$

Equating maximum stresses and slopes, to apply the congruency principle:

$$\frac{3M}{2b^2} = pr_1^2 \frac{1 + r_2^2/r_1^2}{r_2^2 - r_1^2}$$

$$= p80^2 \frac{1 + 200^2/80^2}{200^2 - 80^2} = 1.38p$$

and

$$\frac{-3M}{2b^3} = \frac{-2pr_2^2}{(r_2^2 - r_1^2)r_1} = -2.97 \times 10^{-2}p$$

Dividing the last equations,

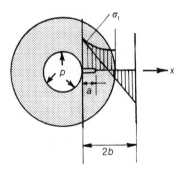

Figure E5.25

$$b = \frac{1.38}{2.97} \times 10^{-2} = 46.5 \text{ mm}$$

Then, $a/b = 46.5 = 0.32$. Then, from Appendix B, for an edge crack on a strip, $f(a/b) = 0.67$ and the stress intensity factor:

$$K = \frac{f(\alpha/b)M}{(b - \alpha)^{3/2}} = K_c$$

Replacing the moment $M = 1.38 \, p2b \, ^2\!/_3$

$$K_c = \frac{f(\alpha/b)5.52pb^2}{(b - \alpha)^{3/2}}$$

Solving for p:

$$p = \frac{K_c(b - \alpha)^{3/2}}{5.52f(\alpha/b)b^2} = 272 \ N/mm^2$$

REFERENCES

Beer, F. P., and E. R. Johnston. 1981. *Mechanics of Materials.* New York: McGraw-Hill.

Bresse, J. A. C. 1854. *Recherches analytiques sur la flexion et la réflexion des pièces courbes.* Paris.

Cauchy, A. 1822. *Acad. Sci. Paris.*

Engel, L., and H. Klingele. 1981. *An Atlas of Metal Damage.* Munich: K. Hansen Verlag.

Faires, V. M. 1965. *Design of Machine Elements,* 4th ed. New York: Macmillan.

Girard, P.-S. 1798. *Traité analytique de la résistance des solides et des solides d'égale résistance.* Paris: F. Didot.

Griffith. 1965. In O. J. Horger, ed., *ASME, Handbook.* New York: McGraw-Hill.

Horger, O. J., ed. 1965. *ASME Handbook.* New York: McGraw-Hill.

Irwin. 1965. In O. J. Horger, cd., *ASME Handbook.* New York: McGraw-Hill.

Karpov, A. V. 1936. "Modern Strength Theories," *Trans. ASCE* (October): 1127–53.

Marin, J. 1962. *Mechanical Behavior of Engineering Materials.* Upper Saddle River, N.J.: Prentice-Hall.

Maxwell, J. C. 1890. *The Scientific Papers of James Clerk Maxwell.* Cambridge: University Press.

Miner. 1965. In O. J. Horger, ed., *ASME Handbook.* New York: McGraw-Hill.

Mohr, O. 1882. *Civilingenieur.*

Neuber. 1965. In O. J. Horger, ed., *ASME Handbook.* New York: McGraw-Hill.

Sih, G. C., et al. 1972. *Applications of Fracture Mechanics to Engineering Problems.* Bethlehem, Pa.: Lehigh University.

Smith, J. O. 1942. "The Effect of the Range of Stress on the Fatigue Strength of Metals." *Univ. of Ill. Exp. Sta.,* Bulletin 334.

Spiegel, M. R. 1968. *Mathematical Handbook of Formulas and Tables.* Schaum's Outline Series in Mathematics. New York: McGraw-Hill.

St. Venant, M. de. 1837. *Leçons de Méchanique Appliquée.*

Winkler, E. 1860. *Civilingenieur* 6:325–62, 427–62.

ADDITIONAL READING

American Society for Metals (ASM). *Metals Handbook.* Materials Park, Oh.: ASM, 1974.

Duggan, T. V., and J. Byrne. 1977. *Fatigue as a Design Criterion.* London: Macmillan, 1977.

Hänchen, R., and K.-H. Decker. 1967. *Neue Festigkeitsberechnung fuer den Maschinenbau,* 3rd ed. Munich: K. Hanser Verlag.

Miner, M. A., "Cumulative Damage in Fatigue." *Trans. ASME, J. Appl. Mech.* 67 (1945): A-159.

Rooke, D. P., and D. J. Cartwright. *Stress Intensity Factors.* London: Ministry of Defence, 1976.

Tada, H., P, Paris, G. Irwin. *The Stress Analysis of Cracks Handbook,* Hellertown, Pa.: Del Research Corp., 1973.

PROBLEMS[6]

5.1. [C] The cylindrical stub-shaft shown is loaded with torque $T = 1000$ Nm and axial force $F = 1000$ N and has diameter $d = 80$ mm.

1. Determine the location of the most dangerous stress state, the principal stresses, the maximum shear stresses, and the equivalent stress with the failure theories of Tresca and von Mises.

2. Draw the appropriate Mohr circle.

Figure P5.1

5.2. [C] The cylindrical stub-shaft shown is loaded with torque $T = 1000$ Nm, bending moment $M = 800$ Nm and axial force $F = 1000$ N, and has diameter $d = 80$ mm.

1. Determine the location of the most dangerous stress state, the principal stresses, the maximum shear stresses, and the equivalent stress with the failure theories of Tresca and von Mises.

2. Draw the appropriate Mohr circle.

---

[6] [C] = certification, [D] = design, [N] = numerical, [T] = theoretical problem.

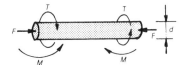

Figure P5.2

5.3. [C] The hollow–cylindrical hydraulic actuator shown is loaded with torque $T = 100$ Nm and inner pressure $p = 1.2$ MPa and has outer diameter $d = 80$ mm and thickness $t = 5$ mm.

1. Determine the location of the most dangerous stress state, the principal stresses, the maximum shear stresses, and the equivalent stress with the failure theories of Tresca and von Mises.
2. Draw the appropriate Mohr circle.

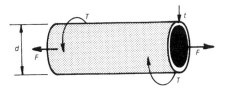

Figure P5.3

5.4. [C] The hollow–cylindrical hydraulic actuator shown is loaded wit h torque $T = 100$ Nm, bending moment $M = 80$ Nm, and inner pressure $p = 1.2$ MPa and has outer diameter $d = 80$ mm and thickness $t = 5$ mm.

1. Determine the location of the most dangerous stress state, the principal stresses, the maximum shear stresses, and the equivalent stress with the failure theories of Tresca and von Mises.
2. Draw the appropriate Mohr circle.

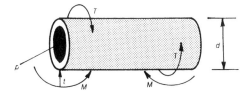

Figure P5.4

5.5. [C] A beam with square cross-section has dimensions and is loaded as shown in Figure P5.5. Determine the location of the most dangerous stress state, the principal stresses, the maximum shear stresses, and the equivalent stress with the failure theories of Tresca and von Mises.

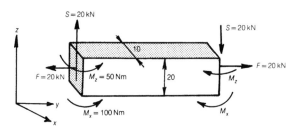

Figure P5.5

5.6. [C] The road sign shown has diameter $d = 1$ m and the wind loading p_{wind} $= 0.1$ MPa indicated in Figure P5.6. Determine the most dangerous stress state, the principal stresses, and the equivalent stresses with all the failure theories if the ratio tensile strength/compressive strength $= 0.5$.

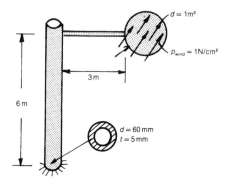

Figure P5.6

5.7. [C] A turbine wheel is loaded as shown in Figure P5.7. Find the state of stress near the wheel, at the most dangerous point, the principal stresses, and the equivalent stress with the failure theories of Coulomb, Tresca, and von Mises.

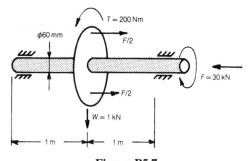

Figure P5.7

5.8. [C] A rotating shaft of diameter 50 mm for a water pump rotates at 200 rpm and transmits power of 10 kW. No other loads are acting on the shaft. After a number of failures an investigation was conducted, which found that bending moments of 100 Nm maximum were loading the shaft due to unbalance. It was decided to operate the pumps at lower loads to achieve the design safety factor. Find the new load rating.

5.9. [T] Hot rolling of round steel bars results in high local thermal stresses at the surface of the roll. These stresses are compressive and approximately equal the yield strength in the circumferential and axial direction $\sigma_t = \sigma_z = -S_y$. It was observed that internal pressure in the roll would improve roll strength because the resulting stress is tensile and reduces the stress σ_t, which is compressive. Using the distortion energy theory of failure, determine at the outer surface of the hollow cylinder (a) the equivalent stress with only the thermal stress and (b) the equivalent stress if there is an inner pressure of 50 MPa in the cylindrical cavity. The roll inner diameter is 100 mm, the outer diameter is 200 mm, and the material is low-carbon steel AISI 1020.

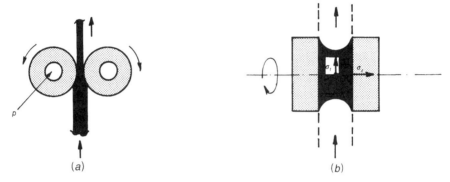

Figure P5.9

5.10. [D] Design the shaft of Problem 5.8 for the rated power, safety factor $N = 2.5$, material carbon steel AISI1025.

5.11. [D] The cylindrical pressure vessel of a hydraulic accumulator has inner diameter 250 mm and internal pressure 10 MPa. For material low-carbon steel AISI 1020 and safety factor $N = 3$, design the thickness of the vessel using the von Mises theory of failure.

5.12. [C] The shell of a boiler with hemispherical heads has diameter 2 m and internal pressure 1.5 MPa.
For material low-carbon steel AISI 1020 and safety factor $N = 3$:
1. Design the thickness of the boiler shell using the von Mises theory of failure.
2. Find the safety factor at the hemispherical end along the axis of the cylindrical shell.
3. Draw a stress map for a section with a plane through the axis of symmetry of the boiler.

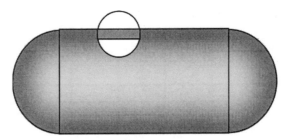

Figure P5.12

5.13. [D] The spherical shell of a liquefied natural gas tank has inner diameter 2 m and internal pressure 1.0 MPa.

1. For material low-carbon steel AISI 1020 and safety factor $N = 3$, design the thickness of the liquefied natural gas tank shell. Select an appropriate theory of failure considering the extremely low temperature of the liquefied gas and that the yield strength remains the same as at room temperature.

2. Draw a stress map for a section with a plane through the axis of symmetry of the spherical shell.

5.14. [D] The cylindrical shell of a heat exchanger has diameter 1 m and spherical heads of diameter > 1 m. The internal pressure is 1.5 MPa.

1. For material low-carbon steel AISI 1020 and safety factor $N = 3$, design the thickness of the shell using the von Mises theory of failure.

2. Find the diameter of the spherical end so that thickness of the spherical shell is the same as in the cylindrical shell, with the same safety factor.

3. Draw a stress map for a section with a plane through the axis of symmetry of the boiler.

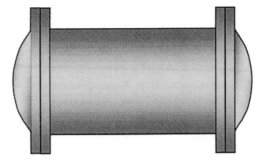

Figure P5.14

5.15. [D]The fuselage of an aircraft maintains an inside normal atmospheric pressure while at high altitudes the external pressure is very low. Considering the fuselage of an aircraft as a cylindrical shell, if the external pressure is 0.05 MPa, the internal pressure is 0.1 MPa, the diameter is 4 m, the material is an aluminum alloy with $S_y = 110$ MPa, density $\rho = 2700$ kg/m³, and the desired safety factor $N = 1.3$:

1. Determine the thickness.

2. Repeat the design for material carbon steel AISI 1030.

3. Compare the weight of the shell of the two designs per meter of aircraft length.

5.16. [D] The cylinder of a hydraulic accumulator has inner diameter 250 mm and internal pressure 100 MPa. For material low-carbon steel AISI 1020 and safety factor $N = 3$, design the thickness of the vessel using the von Mises theory of failure, considering it a thin pressure vessel.

5.17. [D] The shell of a high-pressure boiler accumulator with hemispherical heads has inner diameter 1 m and internal pressure 40 MPa.

1. For material low-carbon steel AISI 1020 and safety factor $N = 3$, design the thickness of the accumulator shell using the von Mises theory of failure and considering it a thick pressure vessel.

2. Repeat the design considering the accumulator a thin pressure vessel.

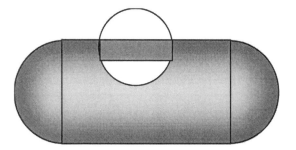

Figure P5.17

5.18. [C] The cylindrical shell of a liquefied natural gas tank has diameter 2 m and internal pressure 1.0 MPa. For material low-carbon steel AISI 1020 and safety factor $N = 3$, design the thickness of the liquefied natural gas tank shell. Select an appropriate theory of failure considering the extremely low temperature of the liquefied gas and that the yield strength remains the same as in room temperature.

5.19. [C] A high-pressure steel pipe has inner diameter 100 mm, outer diameter 150 mm, material steel with $S_y = 220$ MPa. Find the bursting pressure using the von Mises theory of failure, for safety factor $N = 1$:

1. Using the thin pressure vessel theory

2. Using the thick cylinder theory

5.20. [T] A composite thick pipe consists of an inner shell of inner diameter 200 mm and outer diameter 250 mm and an outer shell of outer diameter 300 mm.

1. Find the bursting pressure if the material is steel with $S_y = 300$ MPa.

2. If we replace the composite pipe with a single pipe of inner diameter 200 mm and outer diameter 300 mm, find the bursting pressure and compare with the one for the composite pipe.

(*Hint:* You have two thick cylinders. Treat the outer pressure of the inner cylinder and inner pressure of the outer cylinder as equals and unknown in the equation that results when you make equal the increase of the outer diameter of the inner cylinder and the inner diameter of the outer cylinder.)

5.21. [D] An ASTM A159-G4000 cast iron gear hub of width $w = 60$ mm and outer diameter 200 mm will be mounted on a shaft of diameter 60 mm with an interference fit. Find the limit dimensions and the closest fit if the transmitted power is 5 kW at 1800 rpm and the friction coefficient is $f = 0.3$.

5.22. [D] An ASTM A159-G4000 cast iron gear hub of width $w = 60$ mm and outer diameter 200 mm will be mounted on a shaft of diameter 60 mm with an interference fit. Find the limit dimensions and the closest fit if the torque transmitted is 50 Nm, the axial force of the gear is 2 kN, the radial force is 1.5 kN, and the friction coefficient is $f = 0.3$. (Hint: The friction force is the vector sum of the interference force and the radial force and has to overcome the vector sum of the tangential force and the axial force).

5.23. [C] Determine the maximum transmitted torque of an 80 mm diameter $H7/p8$ interference fit and the safety factor for the hub strength if the hub has outer diameter 220 mm, width 70 mm, material carbon steel C25, friction coefficient $f = 0.25$.

5.24. [C] In the system of Figure P4.32, determine the transmitted torque and the safety factor for the hub strength at the two extreme values of interference for the fit $H7/m6$ (on the right) of diameter 34 mm if the hub has outer diameter 100 mm, width 70 mm, material carbon steel C25, friction coefficient $f = 0.25$.

5.25. [D] In the fit $\varnothing 40H7/k6$ of the journal-bearing system of Figure P4.34, it was observed that the fit of the bearing on the steel plate became loose and it was decided to use an interference fit to locate it safely. Find the required fit if the hub web can be assumed to have rectangular section with outer diameter 100 mm, width 40 mm, material CuSn bronze. (Hint: no torque needs to be transmitted.)

5.26. [C] A flywheel for kinetic energy storage in a city bus has outer diameter 1.5 m, inner diameter 400 mm, uniform thickness 100 mm. The material is glass-reinforced acetal with $S_u = 150$ MPa and density 1500 kg/m^3. The safety factor should be $N = 3$.

1. Find the maximum speed and the energy stored at maximum speed.
2. Repeat the calculation for carbon steel C35 and compare the results.

5.27. [C] A gas turbine wheel has inner diameter 200 mm, outer diameter 600 mm, rotates at 3600 rpm, and on the periphery there are turbine blades of mass 0.1 kg/mm of peripheral length. The wheel is integral with the shaft and has thickness 80 mm, material AISI 4130 direct hardening CrMoV steel. Determine the safety factor.

5.28. [D] A gas turbine wheel has inner diameter 200 mm, outer diameter 600 mm, and on the periphery there are turbine blades of mass 0.1 kg/mm. The wheel is integral with the shaft. Determine the thickness for material AISI 4130 direct hardening CrMoV steel and safety factor $N = 2$.

5.29. [C] In Problem 5.23, determine the runaway speed, the one at which no torque can be transmitted.

5.30. [C] In Problem 5.24, determine the runaway speed, the one at which no torque can be transmitted.

5.31. [C] For the shaft of Figure P5.31, determine the theoretical stress concentration factor for axial loading (shown), for bending and torsion.

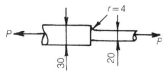

Material: AISI 1030 steel, machined

Figure P5.31

5.32. [C] For the shaft of Figure P5.32, if $D = 300$ mm, $d = 10$ mm, determine the theoretical stress concentration factor for axial loading (shown), bending and torsion.

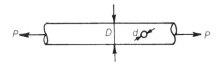

Figure P5.32

5.33. [C] For the steel bracket of Figure P5.33, determine the theoretical stress concentration factor for bending load (shown), axial loading, and torsion.

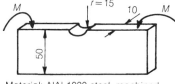

Material: AIAI 1030 steel, machined

Figure P5.33

5.34. [C] For the shaft of Figure P5.34, determine the theoretical stress concentration factor for torsional loading (shown), bending, and axial loading.

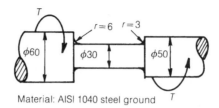

Material: AISI 1040 steel ground

Figure P5.34

5.35. [C] For the shaft of Figure P5.35, determine the theoretical stress concentration factor for torsion (shown), bending, and axial loading.

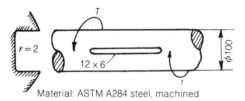

Material: ASTM A284 steel, machined

Figure P5.35

5.36. [C] For the steel bracket of Figure P5.36, determine the theoretical stress concentration factor for bending (shown), torsion, and axial loading.

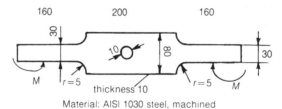

Material: AISI 1030 steel, machined

Figure P5.36

5.37. [C] For the retaining ring of Figure P5.37, determine the theoretical stress concentration factor.

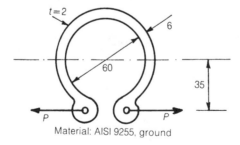

Material: AISI 9255, ground

Figure P5.37

5.38. [C] For the shaft of Figure P5.38, determine the theoretical stress concentration factor for torsion (shown), bending, and axial loading.

5.39. [C] For the shaft of Figure P5.39, determine the theoretical stress concentration factor for the lateral force (shown), bending, and torsion.

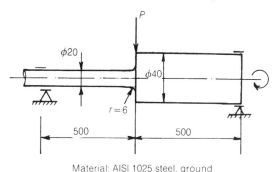

Material: AISI 1025 steel, ground

Figure P5.39

5.40. [C] For the body of the riveting machine of Figure P5.40, determine the theoretical stress concentration factor for the curved beam section.

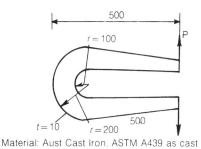

Material: Aust Cast Iron, ASTM A439 as cast

Figure P5.40

5.41–5.50. [C] Determine the maximum allowable alternating loads for the elements shown in Figures P5.31–P5.40 if there is a static load equal in magnitude to the dynamic load, for safety factor 1.8 including reliability. Assume all derating factors $C_i = 1$.

5.51–5.54. [C] Assume for these problems that the fatigue life curve is a straight line on semi-log paper with logarithmic time (cycles to failure) scale, connecting the ultimate strength $S_u = 350$ MPa at 1 cycle to the continuous fatigue strength $S'_n = 180$ MPa at 10^7 cycles.

1. Find the power law for fatigue strength vs. cycles to failure as in Equation (5.86).

2. Determine the applied alternating stress for continuous fatigue strength.

3. Determine the fatigue life if the applied stress does not have a constant amplitude but has the form indicated in Figures P5.51–P5.54, respectively, as percentages of the design alternating load for the above life of 10^6 cycles. Assume the Miner's rule sum to be $= 1$.

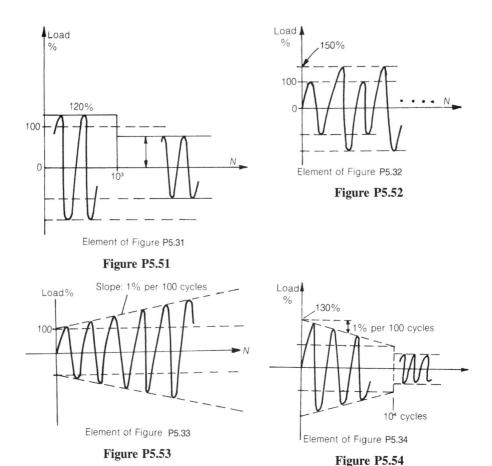

Element of Figure P5.31

Figure P5.51

Element of Figure P5.32

Figure P5.52

Element of Figure P5.33

Figure P5.53

Element of Figure P5.34

Figure P5.54

5.55. [C] On the element shown in Figure P5.38, with a load designed for 10^6 cycles for $N = 1$, $R = 99.9\%$, it was found by measurements that the actual load had the design mean value but had a standard distribution with deviation 5% of the mean, within two standard deviations. Under the same assumptions as in Problems 5.51–5.54, determine the number of cycles to failure.

5.56. [C] An AISI 1025 steel pipe is carrying a hot liquid at 480°C. Every time the hot liquid passes through the pipe, the pipe undergoes plastic deformation owing to thermal expansion because the pipe is fixed at the ends to a structure of constant temperature and very high rigidity. Plastic deformation occurs also at the cooling of the pipe. Determine the number of cycles up to the low cycle fatigue failure of the pipe. Room temperature is 30°C.

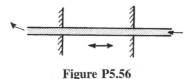

Figure P5.56

5.57. [C] A compound pipe consists of a thin internal layer of stainless steel AISI

321 and a thick external layer of AISI 4130 steel. The pipe is under high internal pressure with a chemical medium keeping the two layers in contact. The medium is gradually heated to 600°C and cooled to room temperature. Owing to differences in thermal expansion, the internal layer is deformed plastically during heating and cooling. Determine the number of cycles to low cycle fatigue rupture.

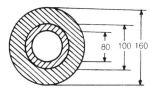

Figure P5.57

5.58. [C] Measurements on a turbine rotor revealed that during start-up the surface of the rotor is heated faster than the core, and a plastic flow of 0.008 was measured. The rotor is made of an AISI 5135 alloy steel and has normally one start-up per week. Determine its expected life considering low cycle fatigue.

5.59. [C] A safety pin to secure nuts has thickness 2 mm and is bent plastically to a curve of radius 5 mm. If it is made of ASTM A284, grade D steel, determine the number of times the pin can be used before it breaks.

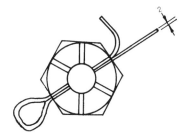

Figure P5.59

5.60. [C] A large electric generator shaft undergoes plastic deformation approximately 0.007 every time it is synchronized out of phase with the electric grid. If the shaft is made of carbon steel AISI 1025, determine its expected life if one incident of that type is expected every month.

5.61–5.63. [C] For cracked elements shown in Figures P6.61 to P5.63, respectively, determine the critical crack length.

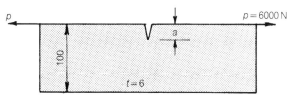

Material: Brass, machined

Figure P5.61

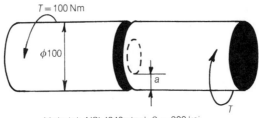

Material: AISI 4340 steel, $S_u = 300$ ksi

Figure P5.62

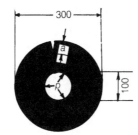

Material: Cast iron, $p = 100$ Mpa

Figure P5.63

5.64. A cylindrical pressure vessel is made of A517F steel. It has a diameter of 2 m, thickness of 20 mm. In the longitudinal welds, the inspection method can detect cracks down to 2 mm length. Determine the maximum allowable pressure to avoid brittle fracture.

5.65. In Problem 5.61, the plate is heated so that the upper side is kept at constant temperature 500°C while the lower is cooled to 20°C. The fracture toughness of the material is $K_c = 100 + 1.2T$, where the temperature T varies linearly along the width of the plate. Determine the expected length where the crack will stop propagating once it started (crack arrest) when it will reach a value where $K = K_c$, because K_c varies as the crack propagates.

CHAPTER 6

COMPUTER METHODS FOR MACHINE MODELING

A Boeing 777 aircraft. Developed with CAD and a system approach.

OVERVIEW

In the foregoing chapters, we saw how we can design the machine as a system (Chapter 1), find the motion of each component (Chapter 2), find the forces that this component carries and the internal forces and moments that the different sections of the component carry (Chapter 3), select appropriate material and manufacturing process (Chapter 4), and finally size the component by making capacity (material strength) equal to demand (working stress) (Chapter 5). In this process, we made many simplifying assumptions that might have considerable effect on the operation of the machine as a system. Moreover, the fact that the components maintain their structural integrity does not necessarily mean that the machine performs as intended. To study the overall performance of the machine, the designer should devise a model that will include all salient features of the design and the operation of the machine. Then the designer can assess the operation of the machine by applying mathematical and computer modeling. Modeling of machine components for design will be the subject of this chapter. System modeling of machines will be the subject of Chapter 7.

6.1. GEOMETRIC MODELING

6.1.1. Computer Aided Machine Design

In Chapters 2–5, analytical methods were mostly used for machine analysis and design. It became apparent that only for some simple shapes and low machine complexities can these methods be employed. For more complex shapes and higher machine complexities, the designer can resort to two alternative methodologies:

1. *Approximations.* One can approximate with simpler shapes or machines that, for the purpose of analysis or design, behave in a manner that closely represents the machine behavior. This is done to a great extent in machine design and pertinent examples will be shown throughout this book, but application of this methodology is still limited to parts and machines of low complexity.

2. *Computer methods.* Analytical approximations are also limited by the computational effort involved, and for fairly complex machine design problems, use of computers to perform the mathematical processing of the machine design problem has been a routine operation for many years. In machine design, computer methods are used to model complex parts and machines for analysis and design.

In the discussions in Chapters 2–5, modeling of machines and their parts involved:

- Defining the geometry
- Relating the machine capacity to the demand due to operating requirements for the particular geometric shape of the machine.

Thus, it becomes obvious that any attempt to utilize computer methods for machine design or analysis should start with the description of the geometry of the machine or a machine part in a complete and unambiguous manner by way of a set of computer-stored information that can be further utilized for analysis or design. This is termed *geometric modeling.*

Every design office is now equipped for computer aided design, geometric modeling in particular. A *design office* can be assumed to be:

- The room of a student who studies design
- The office of a consulting engineer involved in design
- The office of a company, small or large, where a design team works

The standard equipment of a design office must provide the essentials of drafting, although the drafting room is usually separate.

A standard design library should include manuals relevant to the design subject. Moreover, manufacturers' catalogs and electronic media (Internet access, CD-ROMS, etc.) on the same subject are the source for much essential information on any design.

Today the design office is equipped with computer aided design (CAD) facilities. This changes in many ways design and drafting practice and design philosophy.

The computer system employed depends very much on the scope and the amount of output of the design office. Three types of systems are available: microcomputers, minicomputers, and mainframes.

Microcomputers are essential, regardless of the size of the design office. They are simple to use and can be operated at any time, day or night, by personnel unskilled in using computers. Contemporary micros are equipped with main memory between 16 and 128 megabytes (increasing by the day), floppy disk drives, and hard disc storage devices with capacity between 2 and 8 gigabytes, even more (also increasing by the day). Most systems have graphics screens with high-resolution graphics, typically from 160×290 to 1000×1000 pixels (points on the screen).

Printers are available for text that are 80 or 132 characters wide and graphics using the dot matrix system or laser or other types of printers providing high-quality text and graphics printing. As a plotter device, one can use the printer, either by directly printing the graphics screen on paper or directly programming the printer for plotting. The plot then consists of many dots, with the resolution the spacing of the dots. There are, however, pen plotters for professional-quality plotting. Plotters can be of the flat-bed type (Figure 6.1a), which can plot on different sizes of paper, or the drum type (Figure 6.1b), which can plot on paper of the width of a drum and of any length.

Streaming tapes or floppy disk drives or high-capacity diskette drives are used to store information for back-up or transport purposes.

Interactive computer graphics are a two-way communication. The output is presented to the user on a CRT, a printer, or a plotter, or perhaps also stored in a data file. The input information often consists of a large amount of data that the user must transmit to the computer. There are several ways of doing this. Tracing an

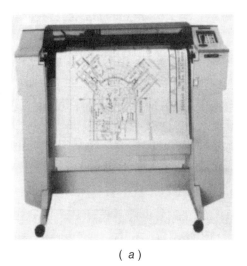

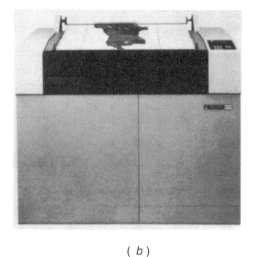

(a)   (b)

Figure 6.1. (a) Drum plotter (courtesy California Computer Co.); (b) flatbed x-y plotter (courtesy Hewlett-Packard Corp.)

Figure 6.2. A CAD workstation (courtesy of Autodesk, Inc.)

existing drawing or a sketch can be done by a digitizer or tablet (Figure 6.3). The drawing is placed on the tablet and a stylus or a cursor is traced over nodal points. Their $x - z$ coordinates are transmitted to the computer. The z coordinate can be transmitted through the keyboard or by tracing another view on the $x - z$ or $y - z$ plane. Device-handling procedures are hardware-dependent and instructions are included in the appropriate user manuals.

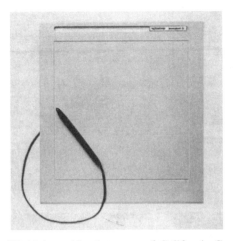

Figure 6.3. Digitizing tablet (courtesy of California Computer Co.)

6.1.2. Computer Aided versus Manual Drafting

It is assumed by newcomers to this field that computer drafting will eventually make manual drafting obsolete. For the immediate future at least, this is not so. A recent study in a particular company revealed that for simple drawings manual drafting was faster while for more complex drawings computer drafting was faster. Speed is not, therefore, the great advantage of computer drafting. The main advantage is that a computer-generated drawing can be stored away and recalled at any time. Furthermore, design is in many aspects an evolutionary process. Very few items are designed altogether new. Most designed components and machines are modifications and improvements of existing ones. In the manual mode, the drawing must be made from the beginning. In a computer drafting system, most of the time only the changes must be given and the new drawing is made by the system.

Charles Babbage (1791–1871) was born in Teignmouth, Devonshire, son of a banker. He entered Trinity College in 1810 and in 1817 received an MA from Cambridge. In 1816, he was elected a fellow of the Royal Society of London. He was instrumental in founding the Analytical (1812), Royal Astronomical (1820), and Statistical (1834) Societies. The idea of mechanically calculating mathematical tables first came to Babbage in 1812 or 1813. Its construction required the development of mechanical engineering techniques, to which Babbage of necessity devoted himself. In the meantime (1828–39) he served as Lucasian professor of mathematics at the University of Cambridge. During the mid-1830s Babbage developed plans for the so-called analytical engine, on the basis of instructions from punched cards, a memory unit in which to store numbers, sequential control, and most of the other basic elements of the present-day computer. The analytical engine, however, was never completed. Babbage's design was forgotten until his unpublished notebooks were discovered in 1937. He was also an accomplished mechanical designer and invented a type of speedometer and the locomotive cowcatcher.

Charles Babbage, detail of an oil painting by Samuel Lawrence, 1845, in the National Portrait Gallery, London

In many CAD systems, basic drawings are available, called *frames,* that the designer can modify to produce the desired drawings. Such frames can also be combined to yield more complex parts and assemblies. For most simple parts, computer programs are available that can make the drawing, using as input the basic dimensions. Furthermore, if a design system is available in the computer, the input is merely the service conditions of the part.

Long experience and the need for simpler, faster, and more economical drafting resulted in standard drafting practices that in many cases deviated considerably from representing a real object as a faithful photographic representation. Many conventional simplifications were merely for the designer's convenience, and these could be waived in CAD. In representing gears, for example, the draftsperson never draws the gear teeth in detail. For the computer, this is an easily programmable task. However, computer drafting has to comply in many aspects with conventional practices and therefore needs special treatment.

The purpose of this section is not to provide a detailed account of Computer graphics but to:

1. Serve as an introduction to computer graphics for the reader who is not familiar with the subject
2. Stress the points related to computer drafting
3. Help the reader who is using "canned" CAD programs to realize their structure, potential and limitations

6.2. SOLID MODELING

6.2.1. Wireframe Modeling

Creating a geometric model of a machine or a part facilitates the design work because the designer can work and experiment on it in the design office and perform analyses and changes at a minimal expense. Because machines consist of interrelated solids, it is very important for the machine designer to create a geometric model of the machine or the machine part, a *solid model,* in the form of an assembly of symbolic or arithmetic quantities capable of been manipulated in a digital computer. A *wireframe model* consists of a finite set of points (*vertices*) connected in pairs by straight lines (*edges*) or specific curves so that the three-dimensional form of a solid object can be visualized. This is a simple technique and can be accomplished with a minimum amount of computational resources. The data defining the vertices are termed *geometric data,* while the data defining the connectivity of the different vertices by way of edges are termed *topological data.*

The wireframe model has the advantage of simplicity but also has limitations:

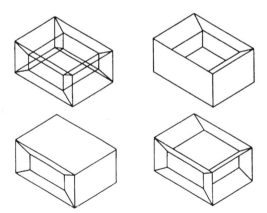

Figure 6.4. Axonometric wire drawing. (From J. Rooney and P. Stendman, *Computer-Aided Design,* 1987, by permission of Pitman Publishing.)

1. It has a certain degree of ambiguity, as one can see in the different possible interpretation of the wireframe model in the upper-left part of Figure 6.5.
2. It is transparent, meaning that the nonvisible lines in the real solid are visible in the model, which adds to the ambiguity and obstructs the fidelity of the model.

However, because a wireframe model can be easily generated, viewed, printed, and manipulated, it is a popular means for solid modeling (Figure 6.5).

A wireframe representation of a solid object consists of four lists:

1. The *vertex list* contains geometric data: the coordinates of the n vertices or other points of interest of the solid, e.g., V_1 (x_1, y_1, z_1), V_2 (x_2, y_2, z_2), ... V_n (x_n, y_n, z_n).
2. The *edge list* contains topological data: the numbers of the vertices defining each of the m edges, e.g., $E_1(V_a, V_b)$, $E_2(V_c, V_d)$, ..., $E_m(V_e, V_f)$, where a, ..., f can take values 1, 2, ..., n.

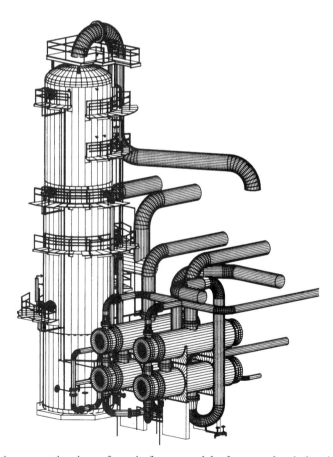

Figure 6.5. Axonometric view of a wireframe model of a complex industrial installation (courtesy of ASME)

3. The *edge type list* is an ordered list of types of edges, e.g., linear for straight lines, arc for circular arcs. Many solid modelers use straight line segments to approximate curves, and this list can be omitted in this case.

4. The *face list* is an ordered set of definitions of ℓ geometry polygons defining a small surface patch bounded by a number of edges, e.g., $F_1(V_a, V_b, \ldots)$, $F_2(V_a, V_b, \ldots), \ldots, F_\ell(V_a, V_b, \ldots)$, where $a, \ldots, f$ can take values 1, 2, \ldots, n.

It is apparent that upon projection onto the $x - y$ plane the density of the edges is proportional to the inclination of the patches in respect to the $x - y$ plane. This can be used to create a shading impression. The same thing can be achieved by painting the patches with a color of density proportional to the patch inclination. If the density of the mesh is high enough, a realistic 3-D picture can be created. Therefore, procedures to create solid primitives can be produced.

Furthermore, the solid data can be appended with the desired intersecting solids. The result, with the hidden line feature, is that any complex solid can be modeled with the available solid primitives. In fact, this is essentially the way most commercial graphics packages operate. Figure 6.5 shows an axonometric view of a wireframe model of a complex industrial installation with the hidden lines removed. Figures 6.6 and 6.7 show wireframe models of machine parts with the surface patch coloring method for generating a 3-D picture of the part.

Example 6.1   Define vertex, edge, and face lists for a cube of edge length $a = 1$ m with one vertex at the origin of a Cartesian coordinate system (x, y, z) and the edges parallel with the axes.

Figure 6.6. Wireframe model of turbine blade with the surface patch coloring method (Courtesy of Swanson Associates, by permission).

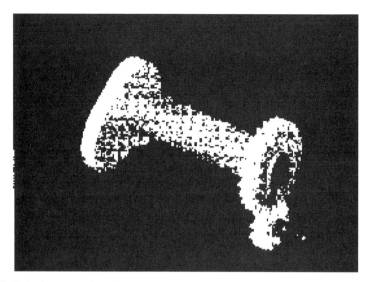

Figure 6.7. Wireframe model of machine part with the surface patch coloring method (courtesy of ASME).

Solution

Geometric data:

Vertex list:

$$V_1(0, 0, 0),\ V_2(1, 0, 0),\ V_3(1, 1, 0),\ V_4(0, 1, 0),\ V_5(0, 0, 1),$$

$$V_6(1, 0, 1),\ V_7(1, 1, 1),\ V_8(0, 1, 1)$$

Topologic data:

Edge list:

$$E_1(1, 2),\ E_2(2, 3),\ E_3(3, 4),\ E_4(4, 1),\ E_5(5, 6),\ E_6(6, 7),\ E_7(7, 8),\ E_8(8, 5),$$

$$E_9(1, 5),\ E_{10}(2, 6),\ E_{11}(3, 7),\ E_{12}(4, 8).$$

Face list:

$$F_1(1, 2, 3, 4),\ F_2(5, 6, 7, 8),\ F_3(1, 2, 6, 5),\ F_4(2, 3, 7, 6),\ F_5(3, 4, 8, 7),\ F_6(1, 4, 8, 5)$$

6.2.2. Surface Modeling

While wireframe modeling is very convenient and economical to apply, complex curved surfaces cannot be effectively represented by way of an assembly of plane surface polygons. For a long time, machine designers used surface modeling to represent such surfaces. In the simplest form, a curved surface was represented by way of a set of isocontour lines at different heights or different sections, in general, for example, to represent ship halls (Figure 6.8), propellers, aerofoils, automotive bodies, or aircraft fuselages.

This technique was derived from geographical mapping procedures. In geography, however, the surfaces are more or less random, while in machines, curved surfaces are generally rational ones that most of the time can be expressed in the form of, or approximated by, algebraic functions. With the availability of electronic computing power, surface modeling developed substantially for many applications.

We first observe that a straight line in space can be parametrically represented as:

$$\mathbf{p}(u) = (1 - u)\mathbf{p}_1 + u\mathbf{p}_2 \qquad (6.1)$$

where u is a parameter, $\mathbf{p}_1$ and $\mathbf{p}_2$ are column vectors with the coordinates of two points $\mathbf{p}_1(x_1, y_1, z_1)$ and $\mathbf{p}_2(x_2, y_2, z_2)$ that define a straight line in a Cartesian coordinate system (x, y, z), and $\mathbf{p}(x, y, z)$ is a column vector with the coordinates of a point p along the line. Indeed, for $u = 0$, $\mathbf{p} = \mathbf{p}_1$ and for $u = 1$, $\mathbf{p} = \mathbf{p}_2$. It is a straight line because it is a linear function of $\mathbf{p}_1$ and $\mathbf{p}_2$. Further, a plane surface can be represented as a linear combination of four coplanar points $\mathbf{p}_1$, $\mathbf{p}_2$, $\mathbf{p}_3$, and $\mathbf{p}_4$:

$$\mathbf{p}(u, v) = (1 - v)(1 - u)\mathbf{p}_1 + (1 - v)u\mathbf{p}_2 + v(1 - u)\mathbf{p}_3 + vu\mathbf{p}_4 \qquad (6.2)$$

The reader can easily verify that at $u = 0$ or 1 and $v = 0$ or 1, $\mathbf{p}$ coincides with one of $\mathbf{p}_1$, $\mathbf{p}_2$, $\mathbf{p}_3$, or $\mathbf{p}_4$.

To represent a curve, higher-order functions of several points need to be devised, called *blending functions*. For example (Figure 6.10), a *Bezier parametric curve* in respect of four points $\mathbf{p}_1$, $\mathbf{p}_2$, $\mathbf{q}_1$ and $\mathbf{q}_2$ is expressed as a cubic, passing through p_1 and p_2 and tangent at p_1 and p_2 with $p_1 q_1$ and $p_2 q_2$, respectively:

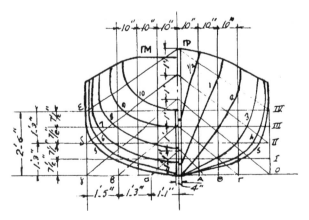

Figure 6.8. Isocontour lines of a ship hull.

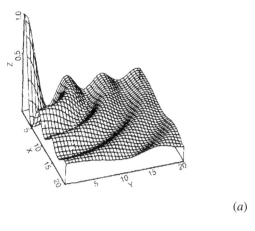

(a)

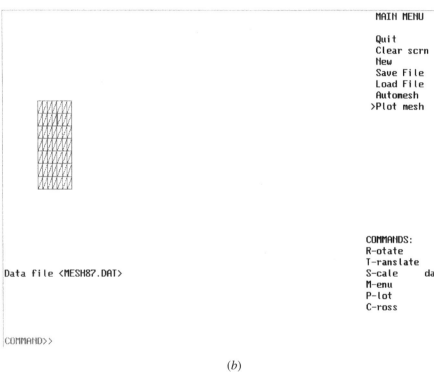

(b)

Figure 6.9. Surface representation of a function.

$$\mathbf{p}(u) = (1 - u)^3\mathbf{p}_1 + 3u(1 - u)^2\mathbf{p}_2 + 3u^2(1 - u)\mathbf{q}_1 + u^3\mathbf{q}_2 \qquad (6.3)$$

where p_1, p_2 are *collocation points* and q_1 and q_2 are *control points*. Thus, a curve can be specified through two points, the curvature being affected by the two control points.

As for the plane surfaces, curved surface regions (*surface patches*) can be described by higher-order functions, such as the Bezier surface curves. Considering

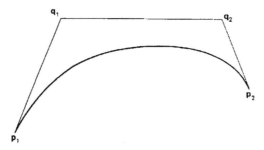

Figure 6.10. Bezier curve. (From J. Rooney and P. Stendman, *Computer-Aided Design,* 1987, by permission of Pitman Publishing.)

two parameters u and v and cubic function in these two parameters, the surface function can be set in the bicubic form:

$$p(u, v) = \sum_{i=0}^{3} \sum_{j=0}^{3} a_{ij} u^i v^j p_{ij} \qquad (6.4)$$

where a_{ij} are 16 coefficients to be determined so that the resulting function collocates with 16 points on the surface or is close to them with a predefined rule (Figure 6.11). Different schemes have been proposed (Faux and Pratt 1979) for the determination of the 16 coefficients a_{ij}.

Example 6.2: Parametric Surface Representation.

Determine the surface representation of a half cylinder of diameter $d = 200$ mm and length $L = 300$ mm in the form of a bicubic parametric patch.

Solution   The solution will be found using a spreadsheet in EXCEL. To this end, we assume that the half cylinder rests on the $x - y$ plane with one side on the y-axis, the one end on the x-axis while the cylinder is above the $x - y$ plane. The equation of the cylindrical surface

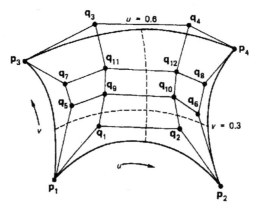

Figure 6.11. Bezier bicubic parametric patch. (From J. Rooney and P. Stendman, *Computer-Aided Design,* 1987, by permission of Pitman Publishing.)

is $z = \sqrt{(x - R)^2 + y^2}$ where $R = 200/2 = 100$ mm. We will specify 16 co-location points at the intersections of the lines $x = 0, 40, 160, 200$ mm and $y = 0, 100, 200, 300$ mm. They correspond to parameter values $u = 1, 2, 3, 4$ and $v = 1, 2, 3, 4$, respectively.

We now set the Solver of EXCEL to minimize the function

$$\varepsilon(u, v) = \sum_{u,v} \left[\left(\sum_{i=0}^{3} \sum_{j=0}^{3} a_{ij} u^i v^j z_{ij} \right) - z_{uv} \right]^2$$

where z_{ij} and z_{uv} are the z-coordinates of the grid points (the co-location points). The first term in the parenthesis is the term of the cubic function used for the surface representation, while the second term is the actual value at the grid points. If the sum of the squares of all these differences approaches zero, this means that the surface represented by the cubic function approaches the desired one at the co-location points.

The procedure is shown in the spreadsheet (Table E6.2), and the surface of the half cylinder is shown in Figure E6.2a, in wireframe representation. This shows the weakness of the latter, since it represents it as the nearest plane-patch bounded surface. A wireframe representation of higher fidelity requires the definition of progressively more geometric and topological data. For example, in Figure E6.2b, the same half-cylinder surface is shown for 12 divisions along the x axis while for the surface representation only a 4×4 matrix of the coefficients a_{ij} is needed for this example solid surface.

TABLE E6.2

Definition of surface co-location points of the surface patch,
   z-coordinate in the shaded area

x	0	100	y 200	300
0	0	0	0	0
40	80	80	80	80
160	80	80	80	80
200	0	0	0	0

Definition of the coefficient a_{ij} (Set initially to arbitrary value, say $= 1$, then use Solver to compute them for minimum sum)

	1	a_{ij} 2	3	4
1	1.0000	1.0000	1.0000	1.0000
2	0.5000	0.2500	0.0556	0.0078
3	0.1111	0.0556	0.0123	0.0017
4	0.0000	0.0000	0.0000	1.0000

Definition of the function to be minimized

$$\text{Sum} = \sum_{i=1}^{4} \sum_{j=1}^{4} (a_{ij} u^{i-1} v^{i-1} z_{ij})^2$$

5.46661E-09

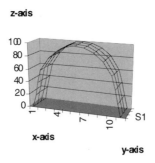

Figure E6.2*a*

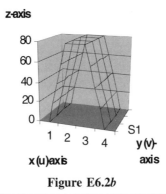

Figure E6.2*b*

6.2.3. Constructive Solid Geometry

A number of generic solids of simple shape can be defined by:

1. *Pure primitive instancing.* Solid shapes can be parametrically defined, such as a bolt (Figure 6.12).
2. *Sweeping of plane figures.* A plane figure upon rotation about a line, usually on the plane of the plane figure, or by displacement along a path can generate a solid (Figure 6.13).
3. *Constructive solid geometry.* A complex solid is composed of primitives, solids of simple geometry that can be catalogued beforehand and form the basis for a variety of different solid models. This is accomplished by applying

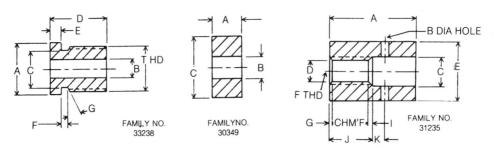

Figure 6.12. Parametric definition of a solid.

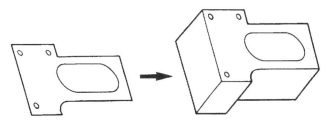

Figure 6.13. Solid model generation by sweep. (From J. Rooney and P. Stendman, *Computer-Aided Design,* 1987, by permission of Pitman Publishing.)

some Boolean operations on the solid primitives and the assembly, such as *union, intersection,* and *difference.*

Figure 6.14 shows the formation of a complex solid by union of cylinders and planes. Figure 6.15 shows the Boolean operations of union and difference of a rectangular block and a cylinder.

A solid modeler usually starts with the definition of some primitives, for example:

- A half space bounded by the plane $z = 0$, occupying the space $z < 0$ (HS)
- A cube of unit diameter and unit edge, based on the $x - y$ plane with the axis of symmetry along the positive z-axis (CUBE)
- A cylinder of unit diameter and unit length, based on the $x - y$ plane with the axis of symmetry along the positive z-axis (CYLINDER)
- A cone of unit base diameter and unit height, based on the $x - y$ plane with the axis of symmetry along the positive z-axis (CONE)
- A sphere of unit radius and centered at the origin (SPHERE)

Then it forms the solid by successive applications of Boolean operations and geometric transformations, which will be discussed in the next section, to bring them from the unit size of the primitives to the desired size of the solid.

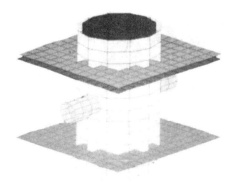

Figure 6.14. Solid modeling using intersections of cylinders and planes (courtesy of Swanson Associates).

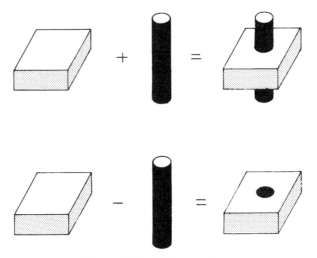

Figure 6.15. Solid operations.

6.3. GEOMETRIC TRANSFORMATIONS

Sometimes the same solid is defined in a different position, a different size, etc., or it is available in the form of a unit size primitive. One does not need to define the new solid from scratch; it is sufficient to transform the data so that they will reflect the desired change. For example, if a 10 mm diameter bolt of a certain type has been defined already in a machine and another bolt of the same geometry but of different size needs to be defined in another place in the machine, this can be done with a series of mathematical operations on the data, called *geometric transformations.*

In all the methods of solid modeling, a model was defined by way of geometric and topological data. In geometric transformations, one needs to change in most cases only the geometric data. Topological data (which usually are the bulk of the data, see Example 6.1) remain invariant during geometric transformations.

6.3.1. Plane Geometric Transformations

As in manual drafting, a drawing consists of lines, letters, numbers, and sometimes shades and colors. Therefore, a computer graphics system must have the capability of drawing on paper these essentials of visual communication. Although solid modeling is capable of presenting graphical images of models in three-dimensional views, even in three-dimensional perception with virtual reality methods, eventually this presentation is usually made on flat surfaces, such as paper or projection or computer screens. Therefore, plane graphics is of paramount importance for the graphical presentation of machine images. Since plane images are only abstractions of true three-dimensional images, they will be discussed separately.

6.3.1.1. Clipping. Many times, either accidentally or purposely, some node coordinates are outside the screen window. In other circumstances we want to plot part of a picture within a given window of the plotting area. Therefore we must remove the lines or part of lines outside this window.

If a (straight) line is plotted between a point x_1, y_1, within the window, to an outside point x_2, y_2, the coordinates x_2, y_2 must be replaced with x, y, that is, the intersection of the line with the screen window $x_b < x < x_a$, $y_b < y < y_a$. Simple geometry yields (Figure 6.16):

$$y = y_a \tag{6.5a}$$

$$x = x_1 + \frac{(x_2 - x_1)(y_a - y_1)}{(y_2 - y_1)} \tag{6.5b}$$

Similar results are obtained if there is intersection with the $x = x_a$, $x = x_b$, $y = y_b$ lines.

6.3.1.2. Scaling. A solid model consists of a set of vertex definitions $x(i)$, $y(i)$ and set of polygon definitions $C(N, N_1, N_2, \dots)$. Scaling onto the $x - y$ plane simply involves multiplication of the vertex coordinates by s_x and s_y, scale factors in the x and y directions, respectively (see figure in Table 6.1). Representing the transformed coordinates with capital letters:

$$X(i) = x(i)s_x \tag{6.6a}$$

$$Y(i) = y(i)s_y \tag{6.6b}$$

If $s_x = s_y$, this transformation is called *uniform*.

In matrix form, Equations (6.6) can be written as:

$$\begin{Bmatrix} X(i) \\ Y(i) \\ 1 \end{Bmatrix} = \begin{bmatrix} s_x & 0 & 0 \\ 0 & s_y & 0 \\ 0 & 0 & 1 \end{bmatrix} \begin{Bmatrix} x(i) \\ y(i) \\ 1 \end{Bmatrix} \tag{6.7}$$

The last equation is merely an identity, $1 = 1$, and is used for computational convenience. The matrix relating the transformed coordinates of a point (X, Y) to the original ones (x, y) is termed *transformation matrix*. We encountered such matrices in Chapter 2.

In Table 6.1 (pages 406–407), some plane geometric transformations are tabulated, together with the defining equations and an illustrative sketch.

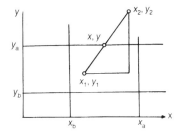

Figure 6.16. Straight line intersecting the window boundary.

TABLE 6.1. Plane Transformations

Transformation	Sketch	Equations	Transformation Matrix
Scaling by factors, s_x, s_y		$X(i) = x(i)s_x$ $Y(i) = y(i)s_y$	$\begin{bmatrix} s_x & 0 & 0 \\ 0 & s_y & 0 \\ 0 & 0 & 1 \end{bmatrix}$
Translation by t_x, t_y		$X(i) = x(i) + t_x$ $Y(i) = y(i) + t_y$	$\begin{bmatrix} 1 & 0 & t_x \\ 0 & 1 & t_y \\ 0 & 0 & 0 \end{bmatrix}$
Rotation about the origin by angle θ		$X = x \cos\theta + y \sin\theta$ $Y = -x \sin\theta + y \cos\theta$	$\begin{bmatrix} \cos\theta & \sin\theta & 0 \\ -\sin\theta & \cos\theta & 0 \\ 0 & 0 & 1 \end{bmatrix}$

Reflection in respect to the y-axis

$X(i) = -x(i)$
$Y(i) = y(i)$

$$\begin{bmatrix} -1 & 0 & 0 \\ 0 & 1 & 0 \\ 0 & 0 & 1 \end{bmatrix}$$

Zooming

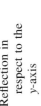

Translation . . . $t_x = -x_0$, $t_y = -y_0$
(center of detail to origin)

$$\begin{bmatrix} 1 & 0 & -x_0 \\ 0 & 1 & -y_0 \\ 0 & 0 & 1 \end{bmatrix}$$

Scaling . . . . $s_x = s_y = L_x/L$
Translation . . . $t_x =$
$L_x/2$, $t_y = L_y/2$ (center of detail
to center of frame)

$$\begin{bmatrix} s_v & 0 & 0 \\ 0 & s_v & 0 \\ 0 & 0 & 1 \end{bmatrix}$$

Clipping (to frame dimensions)

$$\begin{bmatrix} 1 & 0 & Lx/2 \\ 0 & 1 & L_v/2 \\ 0 & 0 & 1 \end{bmatrix}$$

6.3.2. Three-Dimensional Transformations

Since plotting devices are flat, only 2-D pictures can be plotted. One can however create 3-D objects and take several plane views, which is the standard practice in machine design. The description of the object again requires the coordinates x, y, z of each node and the definition of lines. Each line is defined by its beginning and end node.

As in plane graphics, transformations can be applied also on the node coordinates of the object, while the line definition remains unchanged, except for reflection, in which case more lines need to be defined.

In larger systems, the transformations are given in matrix form, a very convenient notation.

Scaling can be accomplished by the following transformation on the node coordinates:

$$X = s_x x, \quad Y = s_y y, \quad Z = s_z z \tag{6.8}$$

In matrix form:

$$\begin{Bmatrix} X \\ Y \\ Z \\ 1 \end{Bmatrix} = \begin{bmatrix} S_x & 0 & 0 & 0 \\ 0 & S_y & 0 & 0 \\ 0 & 0 & S_z & 0 \\ 0 & 0 & 0 & 1 \end{bmatrix} \begin{Bmatrix} x \\ y \\ z \\ 1 \end{Bmatrix} \tag{6.9}$$

where for a uniform transformation $s_x = s_y = s_z = s$.

Rotation is somewhat more complicated. In plane transformations, rotation is about a point. In space, rotation is about an axis (Figure 6.17). The rotation about an axis passing through the origin is defined by way of the three *direction cosines* of the axis:

$$n_1 = \cos(\varphi_x)$$

$$n_2 = \cos(\varphi_y) \tag{6.10}$$

$$n_3 = \cos(\varphi_z)$$

where φ_x, φ_y, φ_z are the angles of the axis of rotation with the three coordinate axes. Two of these need to be specified because it is known (Spiegel 1968) that:

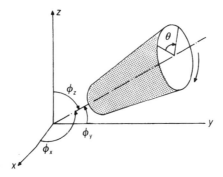

Figure 6.17. Rotation about the x axis by angle $\theta = \pi/2$.

$$n_1^2 + n_2^2 + n_3^2 = 1 \qquad (6.11)$$

The general rotation matrix in 3-D can be found in standard texts on analytic geometry for an angle of rotation θ about an axis with direction cosines n_1, n_2, n_3 passing through the origin. This and some other 3-D transformations are shown in Table 6.2.

For rotation about a general axis not passing through the origin, a translation is performed to bring the axis of rotation through the origin, then rotation, then translation again to the original position of the reference point.

Successive transformations can be performed by repeated application of the respective algorithms or by respective multiplication of the appropriate transformation matrices. For example, $P(i) = [R][T][S]\, p(i)$ means first scaling, then translation, then rotation. Transformations of this type are not commutative, that is, if applied in different order they will yield, in general, different results.

6.3.3. Mechanical Drawings

Mechanical drawings of an object are plane projections on the three planes $x - y$, $y - z$, $z - x$ and axonometric projections.

A solid can be described by a number of plane polygons in a wireframe representation. Curves can also be described by way of polygons with a sufficient number of straight line segments.

If the description of an object by way of the node coordinates and line definition is given, the projection on the $z = 0$ plane is trivial. A plane plot is performed with the given values of the node coordinates x and y, disregarding the z coordinate.

The projection on the $y = 0$ plane, view A, requires a rotation about the x-axis ($n_1 = 1$, $n_2 = 0$, $n_3 = 0$) by an angle $\theta = \pi/2$ and then plotting the new x, y coordinates (Figure 6.18). View B can be obtained with $\theta = -\pi/2$.

Axonometric views are obtained with proper selection of the axis and angle of rotation. An *axonometric isometric view* is obtained with a rotation such that the three orthogonal axes of symmetry of the solid, if such a situation exists, upon rotation are projected with equal angles 120° from one another. Such rotation is obtained with:

$$n_1 = -0.707, \; n_2 = 0.707, \; \theta = \frac{\pi}{4}$$

(See Figure 6.18).

6.3.4. Hidden Line Removal

In mechanical drawing, lines hidden behind visible planes are not shown on the drawing or are made with dashed lines. When the projection of an object on the $x - y$ plane is plotted by simply plotting all nodes and the connecting straight lines, the plotting resembles not a solid object but a wire mesh connecting the nodes. For machine drawings this is not acceptable, and the hidden lines must be identified and removed or plotted as dashed lines. This procedure takes a lot of computer time, and for microcomputers it can be considered a difficult task. Therefore, only

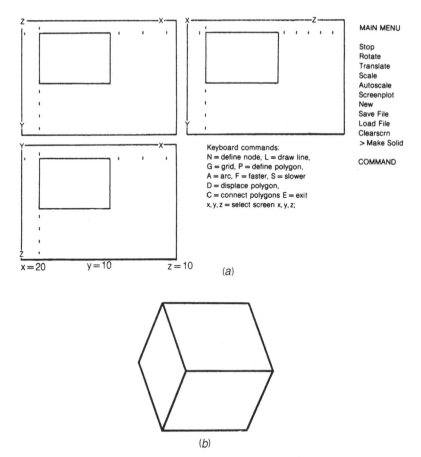

Figure 6.18. Views of a mechanical part.

in the final drawing can it be used because for large and complex objects it might take a long time to complete. Most commercial software is doing hidden line removal. Many methods have been proposed in the literature and used in the commercial software. A simple one will be described here.

Several changes have to be made to the plotting strategy to account for hidden lines. Lines cannot be plotted with one command; they have to be plotted point by point or in small segments. Each point has to be checked against all plane polygon surfaces of the object. Moreover, in addition to the definition of node coordinates and boundary lines, the bounding planes have to be defined by way of the node numbers that successively define each plane polygon, $N_1, N_2, \ldots, N_M, N_1$ and of course the number of nodes of each polygon. Before each point is plotted, it has to be tested against all bounding polygons. If it is found that the point is behind any one of them, then the point is not plotted. Therefore, the problem is reduced to:

Given a point X, Y and a plane polygon defined by a number of nodes $N_1(X_1, Y_1, Z_1)$, $N_2(X_2, Y_2, Z_2)$, $\ldots$, $N_m(X_m, Y_m, Z_m)$, find whether the point, for an observer at $Z = \infty$, is hidden behind the polygon (positive check) or not (negative check).

First, the equation for the plane of the polygon must be determined in the form:

$$z = ax + by + c \tag{6.12}$$

where a, b, c are yet undetermined constants. Application of this equation at the first three consecutive nodes (attention: they should not be on a straight line) yields:

$$ax_1 + by_1 + c = z_1$$
$$ax_2 + by_2 + c = z_2 \tag{6.13}$$
$$ax_3 + by_3 + c = z_3$$

Test 1

If the 3×3 coefficient matrix of the above system is singular, the plane is perpendicular to the $x - y$ plane and the check is negative. The next plane should be tested. Moreover, almost half of the polygons can be eliminated from the test with a simple trick. Suppose that the node numbering was given counterclockwise when the polygon was facing the surface from the outside. Upon projection on the $x - y$ plane, the polygon facing up maintains the counterclockwise direction of node numbering. Otherwise, for the polygon facing down, the direction is reversed. The polygons facing down should not be plotted and should not be checked during the hidden line check. To this end, the cross-product is formed $\mathbf{a} \times \mathbf{b}$, where $\mathbf{a}$ and $\mathbf{b}$ are vectors connecting the first node to the second and third, respectively (Figure 6.19). If the z-component $a_x b_y - a_y b_x$ of the cross-product $\mathbf{a} \times \mathbf{b}$ is positive, then the polygon is facing up and should be plotted and checked for hidden lines.

Test 2

If check 1 is positive, solution of the system (6.19) yields the constants a, b, c and the equation of the plane is known. For $x = X$, $y = Y$, then:

$$z = aX + bY + c \tag{6.14}$$

A straight line from point X, Y of the $x - y$ plane perpendicular to it meets the tested plane at z. If $Z > z$, the test point is above the test plane and the check is negative. Then we repeat from the beginning the procedure for the next plane. If $Z < z$, then the next test must be performed (Figure 6.20).

Test 3

Point X, Y, Z is below the test plane. It must be checked whether it is under the polygon or above it. From the geometry of Fig. 6.21, it is obvious that if a straight line emanating from the test point on the $x - y$ plane meets the boundary of the polygon at an odd number of points, the point is inside the polygon. Otherwise, it is outside.

TABLE 6.2. Some 3-D Transformation Matrices

Transformation	Sketch	Equations	Transformation Matrix
Scaling by factors s_x, s_y, s_z		$X(i) = x(i)s_x$ $Y(i) = y(i)s_y$ $Z(i) = z(i)s_z$	$\begin{bmatrix} S_x & 0 & 0 & 0 \\ 0 & S_y & 0 & 0 \\ 0 & 0 & S_z & 0 \\ 0 & 0 & 0 & 1 \end{bmatrix}$
Translation by t_x, t_y, t_z		$X(i) = x(i) + t_x$ $Y(i) = y(i) + t_y$ $Z(i) = z(i) + t_z$	$\begin{bmatrix} 1 & 0 & 0 & t_x \\ 0 & 1 & 0 & t_y \\ 0 & 0 & 1 & t_z \\ 0 & 0 & 0 & 1 \end{bmatrix}$

Rotation about an axis by angle θ

n_1, n_2, n_3 are direction cosines of the axis in respect to the x-, y-, z-axes

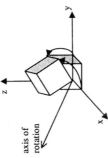

axis of rotation

See Chapter 2

See below

$$\begin{bmatrix} n_1^2 + (1 - n_1^2)\cos\theta & n_1 n_2(1 - \cos\theta) + \sin\theta & n_1 n_3(1 - \cos\theta) - n_2\sin\theta & 0 \\ n_1 n_2(1 - \cos\theta) - n_3\sin\theta & n_2^2 + (1 - n_2^2)\cos\theta & n_2 n_3(1 - \cos\theta) + n_1\sin\theta & 0 \\ n_1 n_3(1 - \cos\theta) + n_2\sin\theta & n_2 n_3(1 - \cos\theta) - n_1\sin\theta & n_3^2 + (1 - n_3^2)\cos\theta & 0 \\ 0 & 0 & 0 & 1 \end{bmatrix}$$

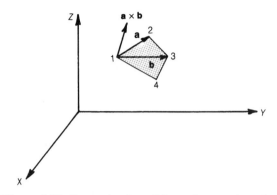

Figure 6.19. Determination of the polygon orientation.

For this test to be performed, such a straight line must be selected. There is no loss in generality if one assumes a parallel to the x-axis. The straight line between points x_1, y_1 and x_2, y_2 is defined on the $x - y$ plane by the equation:

$$x = x_1 + \frac{(y - y_1)(x_2 - x_1)}{(y_2 - y_1)} \tag{6.15}$$

The intersection of this line with the line $y = Y$ is at:

$$X = x_1 + \frac{(Y - y_1)(x_2 - x_1)}{(y_2 - y_1)} \tag{6.16}$$

Point X, Y is between points x_1, y_1, and x_2, y_2 if X is between x_1, x_2 and Y is between y_1, y_2, since plotting is performed in the first quadrant.

Polygons inside other polygons can be treated by combining them as indicated in Figure 6.22. Instead of polygons 1-2-3-4-5-6-1 and 7-8-9-10-1, a new polygon

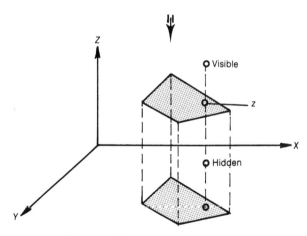

Figure 6.20. Hidden line detection.

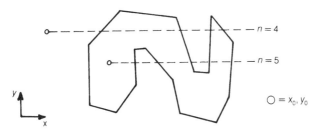

Figure 6.21. Polygon inclusion test.

is defined 1-2-3-4-5-6-7-8-9-10-7-6-1. Obviously, if a test for any plane polygon is positive (point hidden), there is no need for testing against the other planes. One must then proceed to the next point of the line.

All the lines of the solid have to be checked and plotted point by point. For a speedier but less precise plot, one can decide to plot a number of points at a time, say three, and consider them all hidden or visible on the basis of the test of one of them.

6.3.5. Sectioned Views

As mentioned above, machine design graphics deviates considerably from geometric computer graphics because of conventions and standards adopted during many years of design practice. To the extent that some of these conventions were adopted to simplify the manual work, with the applications of computer methods they are gradually fading out. Many conventions, however, are useful for many other purposes, such as manufacturing, assembly, and service, and they have to receive special consideration. One such feature is sectioned views.

A sectioned view of a solid is essentially a view of the part of a solid that is beyond some plane $z = V$ (Figure 6.23). This solid is projected as usual on the $x - y$ plane to give a sectional view of the solid in respect to the section plane. Practically, this means that:

1. All lines that are above this plane are not plotted.

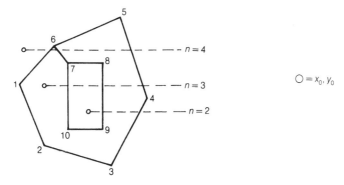

Figure 6.22. A multiple-connected polygon.

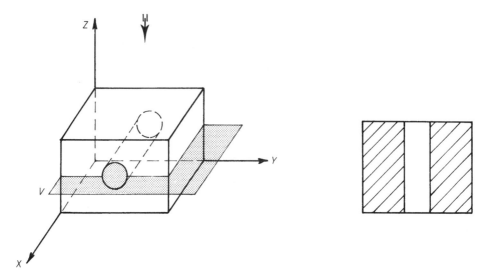

Figure 6.23. Sections of solids.

2. Polygons that are entirely above this plane do not hide lines that are below the section plane.
3. Lines that intersect with the section plane are plotted up to this plane.
4. For polygons that are intersecting with the section plane, their parts above the section plane have to be replaced with the line of intersection.

A typical sectioned axonometric view is shown in Figure 6.24 and a typical machine drawing is shown in Figure 6.25.

Example 6.3 Perform a rotation of the solid of Example 6.1 about an axis defined by the direction cosines $n_x = 0.5$, $n_y = 0.5$ and an angle of rotation $\phi = 0.7$ rad.

Solution The rotation matrix is (Table 6.2):

$$
\begin{bmatrix}
n_1^2 + (1 - n_1^2)\cos\theta & n_1 n_2(1 - \cos\theta) + \sin\theta \\
n_1 n_2(1 - \cos\theta) - n_3\sin\theta & n_2^2 + (1 - n_2^2)\cos\theta \\
n_1 n_3(1 - \cos\theta) + n_2\sin\theta & n_2 n_3(1 - \cos\theta) - n_1\sin\theta \\
0 & 0
\end{bmatrix}
$$

$$
\begin{matrix}
n_1 n_3(1 - \cos\theta) - n_2\sin\theta & 0 \\
n_2 n_3(1 - \cos\theta) + n_1\sin\theta & 0 \\
n_3^2 + (1 - n_3^2)\cos\theta & 0 \\
0 & 1
\end{matrix}
$$

The computations were performed in an EXCEL worksheet (Table E6.3, pages 418–419). In the upper right-hand corner is the resulting rotation matrix. On the left the vertex coordinates of the cube are shown while on the right are the transformed coordinates. The two solids were plotted in EXCEL (Figure E6.3, page 420). For this effect, each edge needs to be defined by the start and end vertex and be separated by a blank row from the other edges.

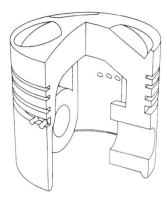

Figure 6.24. Sectioned wireframe drawing of a piston with hidden lines removed.

6.4. AUTOMATIC MESH GENERATION

6.4.1. Plane Mesh Generation

For many applications such as finite element modeling and solid modeling, a plane or curved surface must be divided into a number of triangles or rectangles or other shapes (Figure 6.9). This follows simple rules of geometry if the bounding curve has a simple shape.

The rectangular shape is of little use for most practical problems. In general, every plane shape can be divided into a number of curvilinear rectangles such as the one in Figure 6.26. It is assumed that the bounding curves can be expressed in

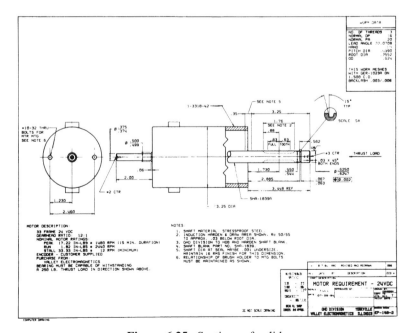

Figure 6.25. Sections of solids.

TABLE E6.3

	A	B	C	D	E	F	G	H
1	Example 6.3							
2	**SOLID TRANSFORMATIONS**				**Auxiliary computations:**			
3					Rotation Matrix			
4					0.823632	0.51432	−0.23897	0
5					−0.39674	0.823632	0.40525	0
6					0.40525	−0.23897	0.882421	0
7					0	0	0	1
8	Enter Data below: Direction cosines n_x, n_y, angle of rotation Theta							
9	$n_x =$	0.5	$n_y =$	0.5	$n_z =$	0.707107	Theta =	0.7
10	Cos(Theta)	0.76484	Sin(Theta)	0.644218				

ENTER BELOW INITIAL COORDINATES

RESULTS
Transformed Coordinates

VERTEX	x	y	z		x	y	z
start, 1	0	0	0		0	0	0
end	1	0	0		0.823632	−0.39674	0.40525
start, 2	0	0	0		0	0	0
end	0	1	0		0.51432	0.823632	−0.23897
start, 3	0	1	0		0.51432	0.823632	−0.23897
end	1	1	0		1.337952	0.42689	0.166282
start, 4	1	1	0		1.337952	0.42689	0.166282
end	1	0	0		0.823632	−0.39674	0.40525
start, 5	0	0	1		−0.23897	0.40525	0.882421
end	1	0	1		0.584664	0.008508	1.287671
start, 6	0	0	1		−0.23897	0.40525	0.882421
end	0	1	1		0.275352	1.228881	0.643453
start, 7	0	1	1		0.275352	1.228881	0.643453
end	1	1	1		1.098984	0.83214	1.048703
start, 8	1	1	1		1.098984	0.83214	1.048703
end	1	0	1		0.584664	0.008508	1.287671

start, 9	0	0	0	0	0
end	0	1	−0.23897	0.40525	0.882421
start, 10	1	0	0.823632	−0.39674	0.40525
end	1	1	0.584664	0.008508	1.287671
start, 11	1	0	1.337952	0.42689	0.166282
end	1	1	1.098984	0.83214	1.048703
start, 12	0	0	0.51432	0.823632	−0.23897
end	0	1	0.275352	1.228881	0.643453

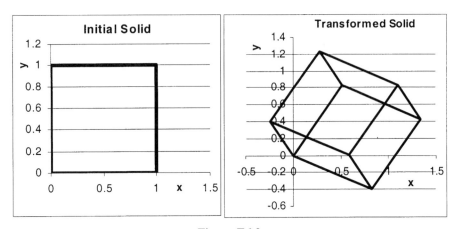

Figure E6.3

parametric form, so that the opposite curves can be assigned the parameter values of 1 to n of the parameters η and ξ. Thus, in the parameter plane η, ζ the curvilinear rectangle will be transformed into a square, which can be divided into triangles or rectangles. Therefore, the parametric equations relating x, y to η, ξ must now be developed. The parametric technique developed for surface representation (see Section 6.3) will be used.

For relatively smooth curves, a polynomial form is very practical. To this end, the parametric equations are set up, for quadratic polynomials, in the bi-quadratic form:

$$x = (a_1\xi^2 + a_2\xi + a_3)\eta^2 + (a_4\xi^2 + a_5\xi + a_6)\eta + (a_7\xi^2 + a_8\xi + a_9) \quad (6.17)$$

$$y = (b_1\xi^2 + b_2\xi + b_3)\eta^2 + (b_4\xi^2 + b_5\xi + b_6)\eta + (b_7\xi^2 + b_8\xi + b_9) \quad (6.18)$$

There are nine transformation constants a_i and nine b_i to be determined. For this purpose, nine points are specified on the curvilinear rectangle corresponding to the intersections of the two families of three curves for η, $\xi = 1$, $(1 + n)/2$, n. For each of these points there is a pair of values of η, ξ that correspond to known values of x, y. Therefore, they yield nine equations on the various a_i and nine equations on the various b_i. Solution of these equations will give the values of these a and b. Then the transformation is found.

Further, the rectangle in the η, ξ plane is divided with in an $n_x \times n_y$ mesh. Each pair of known η and ξ is transformed into a point x, y. Therefore, the mesh generation is completed, since the polygon definition does not change upon transformation.

6.4.2. Solid Mesh Generation

The plane mesh generated in the previous section can be supplemented by assigning z-coordinate values to yield a solid mesh. An example of this is shown in Figure 6.9, where the surface $z = \sin (R)/R$ is plotted in space. The rectangular facets of the resulting solid create an impression of solid surface. They are called *surface patches* and they play an important role in solid surface presentation.

Example 6.4   For the curved region shown, perform an automatic mesh generation with triangular mesh, three horizontal and six vertical layers. Dimensions are in mm.

Solution   The module Automesh of MELAB 2.0 is used. From the MELAB 2.0 menu we select Utilities, Automesh. We hit ENTER at the first page of Automesh and define triangular elements and the default screen width 640. We then select Automesh again in the menu. With the cursor we define first the three upper horizontal nodes ($x = 100, 250, 400$ at $y = 50$, in that sequence), and then we define the lower layer $x = 100$, $y = 100$, then $x = 250$, $y = 150$, then $x = 400$, $y = 100$. Then we hit A for Automesh and the program asks for the number of horizontal layers we defined, we enter 2, and the number of vertical layers, we enter 3. Then the program asks for the number of desired horizontal layers, we enter 3, and for the number of vertical layers, we enter 6. The program plots the mesh (Figure E6.4b). We hit M to return to the main menu of Automesh, and we can save the mesh in a file for further use.

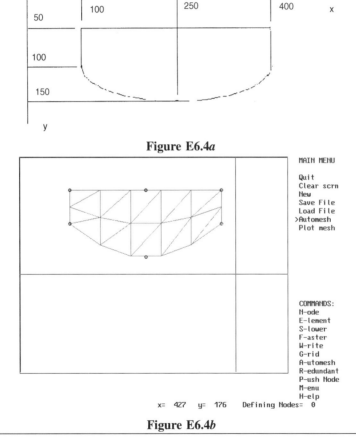

Figure E6.4a

Figure E6.4b

6.5. MODELING FOR STRESS ANALYSIS

6.5.1. Approximate Modeling

Because in machine design most of the failure modes originate from high loads, a stress analysis is almost invariably needed for every machine design effort. Stress analysis can be performed in three ways:

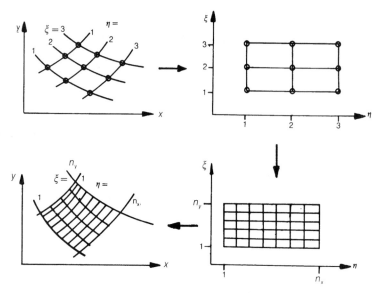

Figure 6.26. Curvilinear mesh transformations.

1. Analytically, using known principles of mechanics. Because analytical solutions exist only for very simple geometries, simplified assumptions must be made in almost all cases in machine design. As the geometry of machines and components becomes more and more complicated, analytical methods seem progressively crude. However, no matter what computer programs are available, such analyses must be always performed for order of magnitude estimations.

2. Experimentally, using several methods of experimental stress analysis.

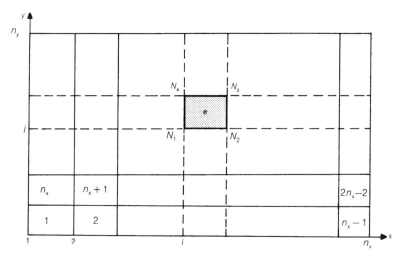

Figure 6.27. Rectangular mesh.

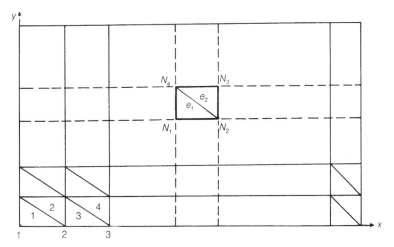

Figure 6.28. Triangular.

3. By computer. Several computer codes exist, and some computer methods are presented in this chapter.

Such methods can be used in very complicated geometries. Extreme caution, however, in using these methods will be emphasized continually in this book.

The reader will recall that strength of material methods are used in very specific and simple geometries, such as straight bars, cylindrical shafts with constant cross-section, and perfect trusses and frames. Such elements do not really exist in machines. Machine elements are usually much more complicated. Stress analysis in that geometry might be impossible analytically or difficult by way of experiment or computer analysis. The latter case might seem strange because, in principle, contemporary methods can analyze the most complicated geometries. True. But one should always bear in mind that cost, time, and reliability are essential ingredients in any machine design effort. The more complicated the computer analysis, the higher the cost. A longer time is needed and accumulation of errors in the computer arithmetic makes the results progressively more unreliable.

Once, in the early days of finite element analysis, we would have analyzed a thick pipe intersection by way of 3-D finite elements. It took an experienced engineer six weeks to prepare and verify the input only. It took a technician a day to do an experiment. The need always exists, therefore, to substitute for the real element a model that:

1. For the mode of failure under examination, will be expected to yield results to an engineering accuracy close enough to the results expected at the real element.
2. Will be manageable, i.e., will permit stress analysis to be performed with available resources: analytical methods, experimental capabilities, computer hardware and software.
3. Will have a cost within budget.

The next question is how. There are no equations for that. Experience, sound engineering judgment, and the trial-and-error process will be the guides.

In general, elements that will be manufactured in small quantities are analyzed with simpler models accompanied by higher safety factors. Elements manufactured in large quantities allow for detailed models with higher cost of stress analysis. As an example, consider the connecting rod shown in Figure 6.29, which is loaded by opposing loads F. This particular form can be analyzed with the finite element methods. However, results for preliminary design can be obtained with two different models, as indicated, depending on the relative dimensions.

- Model I assumes that the side members are very flexible and do not transmit moments to the members that bear the forces.
- Model II assumes the opposite. Therefore, the side members are beams loaded with eccentric loads.

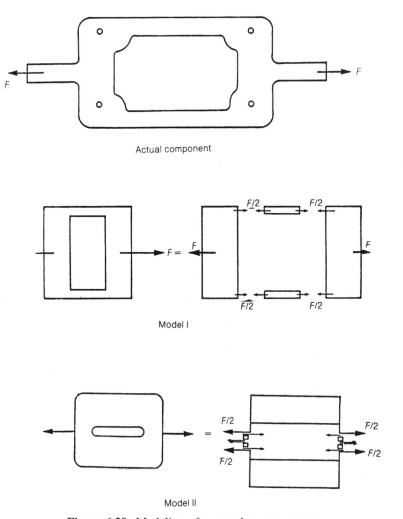

Actual component

Model I

Model II

Figure 6.29. Modeling of a complex component.

6.5.2. Computer Analysis of Line Members

A generally shaped machine member requires, in general, complicated procedures to produce the state of stress at each point due to the static forces and thermal and dynamic loads that the member sustains in service. This is a very complicated problem and the designer always seeks ways to achieve practical solutions for engineering accuracy. Fortunately, most machine components have a geometry that allows such solutions to be found by way of modeling this component approximately with simpler ones. One such simplification is possible when the dimensions of the component along one direction are much greater than along the ones perpendicular to it. We speak then for linear members that have the form of a rod when the length is several times greater than its maximum thickness. There is no specific rule to decide just when a long component can be considered a rod. It also depends on the form of loading and the accuracy required. In general, as a guide, a factor length/thickness of more than 10 can be considered adequate. Such components are usually shafts, bolts, power screws, hooks, springs, torsion bars, and so on. Stresses in some of them can be computed with strength of materials methods, such as prismatic bars in tension/compression, bending, shear, torsion. Stress analysis then is relatively simple.

In many applications, however, machine components have more complicated geometry and the designer needs adequate solutions that cannot be obtained with strength of materials methods. Fortunately, computer methods then come to the rescue and engineering solutions can be achieved for fairly complicated geometries. Examples are stepped shafts, machine frames, supports, and steel structures.

Some simple design analysis methods will now be discussed.

6.5.3. The Transfer Matrix Method

Stepped beams and shafts can be stress analyzed with computer methods that make use of the one-dimensional nature of the component, under some simplifying assumptions.

In strength of materials, the problems have solutions in a more or less exact form. Of course, strictly speaking, perfectly elastic solids, uniform beams, uniform loads, etc., do not really exist in nature. Furthermore, in mechanics we study systems with distributed mass and elasticity. There again, idealizations of beams with constant properties, linearity, boundary conditions of the rigid support type do not really exist. If one can conceive such systems, "exact" solutions exist in some form and, in certain cases, can be found.

This may have left us with a feeling of false security, however, because our solutions are, in any event, approximate. The first approximation is introduced in the modeling of the system itself, as we mentioned above. Second, only for very simple machine members with special and simple geometry can an exact solution be found. For more complicated systems, the very first difficulty is in their modeling. Moreover, the associated equations must be solved. In general, this has to be done numerically. Such situations exist for most moderately complicated machine members. Therefore, for these problems we do eventually resort to numerical methods. In this case, someone might argue, why bother with analysis at all; why not just work out the solution in the computer from the beginning? The answer to this is that we should always try to pursue the analytical solution as far as possible and then resort to numerical methods. One should not expect the computer to do all the work, if analysis to any extent is not cumbersome. Reliable computer results are much more difficult to obtain and ascertain than the enthusiastic inexperienced user might think.

A rather simple method that is almost exclusively used for rotating shafts and also lends itself for application to microcomputers will be discussed in the following.

The transfer matrix method was first introduced in an elementary form by Holzer (1921) for torsional vibration of rotating shafts. It makes use of the fact that in a large class of design problems, some structural member is designed along a line and the behavior at every point of the system is influenced by the behavior at neighboring points only. Typical examples are beams, shafts, and piping systems.

We shall start with the static lateral deflection of beams. We assume a beam, say simply supported, that consists of $n - 1$ beam sections of different moments of inertia (Figure 6.30a). Thus the beam has $n - 1$ elements with constant moments of inertia and n nodes, points (or planes) that define the beginning or end of a uniform beam element.

To fully describe the situation at each node, we need to know four quantities: the deflection y, the slope θ, the moment M, and the shear force V. The section j of a beam between nodes j and $j + 1$ is shown in Figure 6.30b with the usual, in statics, sign conventions. These four quantities can be arranged in a vector form $s = \{y \quad \theta \quad M \quad V\}$,[1] which, because it describes the state of affairs at node j is called *state vector*, and to designate node j we shall use a subscript j.

Let us suppose that at node 1 the state vector is:

$$\{z_1\} = \{y_1 \quad \theta_1 \quad M_1 \quad V_1\} \tag{6.19}$$

yet unknown. If no force is acting between nodes 1 and 2, the deflection, slope, moment, and shear at node 2, from simple beam theory, will be:

$$y_2 = y_1 + \ell_1\theta_1 + \frac{\ell_1^2}{2EI_1} M_1 + \frac{\ell_{31}}{6EI_1} V_1$$

$$\theta_2 = \theta_1 + \frac{\ell_1}{2EI_1} M_1 + \frac{\ell_{31}^2}{2EI_1} V_1$$

$$M_2 = M_1 + \ell_1 V_1 \tag{6.20}$$

$$V_2 = V_1$$

$$1 = 1$$

The third and fourth equations express the equilibrium of the forces and moments on the beam and the first two give the deflections due to these moments and forces. The last equation is an identity $1 = 1$.

---

[1] Curly brackets designate column vectors, i.e., $\{a \ b \ . . .\} = [a \ b \ . . .]^T$

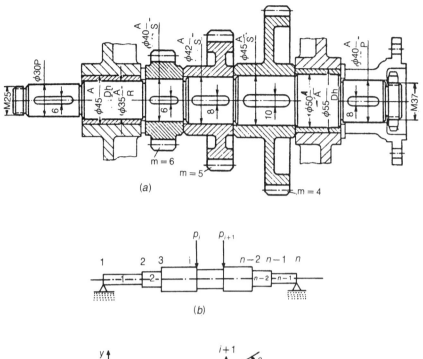

Figure 6.30. Modeling of a stepped shaft.

Equation (6.19) can be written in the matrix form:

$$\{z_2\} = [L_1]\{z_1\} \tag{6.21}$$

where a fifth component 1 is added to vectors $\{z\}$ and

$$[L_1] = \begin{bmatrix} 1 & \ell_1 & \dfrac{\ell_1^2}{2EI_1} & \dfrac{\ell_1^3}{6EI_1} & 0 \\[2mm] 0 & 1 & \dfrac{\ell_1}{EI_1} & \dfrac{\ell_1^2}{2EI_1} & 0 \\[2mm] 0 & 0 & 1 & \ell_1 & 0 \\[2mm] 0 & 0 & 0 & 1 & 0 \\[2mm] 0 & 0 & 0 & 0 & 1 \end{bmatrix} \tag{6.22}$$

The subscript 1 in the matrix $[L]$ indicates that the quantities ℓ, E, I are properties of the element number 1.

Equation (6.21) tells us that the state vector at node 2 is the state vector at node 1 multiplied by a square 5×5 matrix L, which depends on the element properties only and is well known if the part is well defined. This matrix transferred the state from node 1 to node 2 and therefore shall be called the *transfer matrix*. For every element of the beam there exists one known transfer matrix $[L]$. We can repeat the procedure for elements 2, 3, . . . , to obtain, using also the previous relations:

$$\{z_2\} = [L_1]\{z_1\}$$

$$\{z_3\} = [L_2]\{z_2\} = [L_2][L_1]\{z_1\} \tag{6.23}$$

$$\{z_4\} = [L_3]\{z_3\} = [L_3][L_2][L_1]\{z_1\}$$

At the nodes, the state vector as we approach the node from left and right is the same. However, if at the node we have a static force F, this is not true. In Figure 6.31 we show the situation. For a small length about the node, the deflection, slope, and moment remain unchanged, but in order to maintain equilibrium, we must have $V_j^R = V_j^L + F_j$, where superscript L refers to the situation at the left of the node and R to the situation at the right of the node.

We can write:

$$y_j^R = y_j^L$$

$$\theta_j^R = \theta_j^L$$

$$M_j^R = M_j^L \tag{6.24}$$

$$V_j^R = V_j^L + F_j$$

$$1 = 1$$

We can write this in matrix form as:

$$\{z_j^R\} = [P_j]\ \{z_j^L\} \tag{6.25}$$

where:

$$[P_j] = \begin{bmatrix} 1 & 0 & 0 & 0 & 0 \\ 0 & 1 & 0 & 0 & 0 \\ 0 & 0 & 1 & 0 & 0 \\ 0 & 0 & 0 & 1 & F_j \\ 0 & 0 & 0 & 0 & 0 \end{bmatrix} \tag{6.26}$$

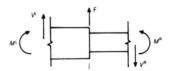

Figure 6.31. Node equilibrium.

The matrices $[L]$ refer to a beam element and are known as *field matrices*. Matrices $[P]$ refer to a nodal point and are called *nodal* or *point matrices*.

Finally, we can complete the sequence of Equations (6.23) as follows:

$$\{z_2^R\} = [P_2]\{z_2^L\} = [P_2][L_1][P_1]\{z_1^L\} \tag{6.27}$$

$$\{z_3^R\} = [P_3]\{z_3^L\} = [P_3][L_2]\{z_2^R\} = [P_3][L_2][P_2][L_1]\{z_1^R\} \tag{6.28}$$

$$\vdots$$

$$\{z_n^R\} = [A]\{z_1^L\} \tag{6.29}$$

where the 5×5 matrix $[A]$ is the product of all 5×5 matrices of the element and point matrices from left to right.

The first four of Equations (6.29) can be written as:

$$y_n = a_{11}y_1 + a_{12}\theta_1 + a_{13}M_1 + a_{14}v_1 + a_{15}$$

$$\theta_n = a_{21}y_1 + a_{22}\theta_1 + a_{23}M_1 + a_{24}v_1 + a_{25}$$

$$M_n = a_{31}y_1 + a_{32}\theta_1 + a_{33}M_1 + a_{34}v_1 + a_{35} \tag{6.30}$$

$$V_n = a_{41}y_1 + a_{42}\theta_1 + a_{43}M_1 + a_{44}v_1 + a_{45}$$

$$1 = 1$$

We have four equations with eight unknowns:

$$y_1 \ \theta_1 \ M_1 \ V_1; \ y_n \ \theta_n \ M_n \ V_n$$

However, because of the boundary conditions, we know four of these quantities. For example, for a simply supported beam, we shall have $y_1 = y_n = 0$ and $M_1 = M_n = 0$. Therefore, Equations (6.30) have only four unknowns: θ_1, V_1, θ_n, V_n.

After the computation of these unknowns, we can obtain the state vectors at the nodes, and thus the deflection of the beam, from Equations (6.27–30), applied successively from left to right.

What have we accomplished at this point? We have been able to compute the static deflection of the beam with only chain multiplications of 5×5 matrices and solution of a system of four algebraic equations. With the usual matrix inversion methods, we would have to invert a large matrix. In addition to the much greater computation effort, one would have to use a digital computer for this purpose. The line solution method can be used without a computer, but it is very suitable for machine computations. This is especially applicable with microcomputers.

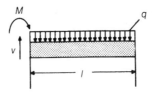

Figure 6.32. Field matrix for element with distributed load.

In the product of Equation (6.29), one can take into account any number of loads at the nodes by multiplying with all the respective point matrices. It is as easy to account for linear springs at the nodes, such as bearings and supports. Let such spring react with a force proportional to deflection (linear spring)$-ky$ and a moment proportional to rotation (torsional spring)$-k_T\theta$, in addition to external forces F_j and moments M_j at the node will be at this node:

$$M_j^R = M_j^L - k_{Tj} + M_j \qquad (6.31)$$

$$V_j^R = V_j^L - ky_j + F_j \qquad (6.32)$$

Of course, deflection and slope are the same before and after the spring. Therefore, at node j:

$$\begin{bmatrix} x \\ \theta \\ M \\ V \\ 1 \end{bmatrix}_j^R = \begin{bmatrix} 1 & 0 & 0 & 0 & 0 \\ 1 & 1 & 0 & 0 & 0 \\ 1 & -k_T & 1 & 0 & M \\ -k & 0 & 0 & 1 & F \\ 0 & 0 & 0 & 0 & 1 \end{bmatrix}_j \begin{bmatrix} x \\ \theta \\ M \\ V \\ 1 \end{bmatrix}_j^L \qquad (6.33)$$

In a similar way, one can obtain the transfer matrix for a beam element with a distributed static load q per unit length (Figure 6.32).

In general, nodes are designated at all points where some change takes place, such as changing sections, point forces, supports, and springs. For multi-supported shafts and beams, it is easy to account for intermediate supports, treating them as

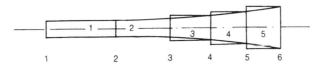

Figure 6.33. Modeling of a shaft with variable cross-section.

linear springs with large spring constant k, two orders of magnitude, or more, higher than some representative spring constant of the shaft $\ell/48EI$.

Members with continuously variable cross-section can be treated as in Figure 6.33 by taking enough nodes along the member and treating the in-between element as of constant section. A certain caution must be exercised, however. Increasing the number of nodes beyond a point will cause the solution to be in great error because the computer cannot handle more than a number of digits for every number. In many cases, a quantity is computed as a difference of two great numbers. If the difference appears beyond the significant digits, it will be zero. Fortunately, this is usually apparent in the solution because it will be shown as violation of the boundary conditions, erroneously large numbers, division by zero error message, etc.

The transfer matrix method is very convenient for design applications, especially on elements such as rotating shafts, and historically it has been one of the first applications in computer aided design of turbomachinery. In fact, despite the wide use of finite element methods, it is a method of choice for most shaft and rotor dynamic analyses because it can be made very fast and thus can be used for on-line simulation for machine control.

Example 6.5 A shaft is loaded as shown, dimensions in mm. The Young modulus $E = 2.1 \times 10\ N/mm^2$. Compute maximum deflection, reactions, shears, and bending moments.

Solution The input of program TRANSFER MATRIX (in SI units N, m) follows, together with the output. Maximum deflection is 0.00185 m (1.85 mm). It could be somewhat greater between stations. One can either draw the deflection curve or use more nodes along the beam to achieve more precision in the parameter tabulation. That, of course, depends on the particular problem and the required accuracy.

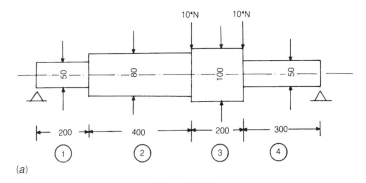

(a)

Figure E6.5a

```
            STATICS.TM              STATICS.TM
           STATICS.TM              STATICS.TM
          STATICS.TM              STATICS.TM
         STATICS.TM              STATICS.TM
        STATICS.TM              STATICS.TM
       STATICS.TM              STATICS.TM
      STATICS.TM              STATICS.TM
  A Transfer Matrix Program for Continuous Beam  analysis
           STATICS.TMSTATICS.TM
          STATICS.STATICS.TM
         STATICSTATICS.TM
        STATSTATICS.TM
       STSTATICS.TM
        STATICS.TM

      Type of section,<1> Circular, <2> General ? 1
      Number of Elements:                        ? 4
      Modulus of Elasticity                      ? 2.1e11
```

```
   Element data                          node data
   El/No
        length      Diameter      distr load   force        spring
    1  ? .2         ? .05         ?            ?            ?
    2  ? .4         ? .08         ?            ?            ?
    3  ? .2         ? .1          ?            ? 1e4        ?
    4  ? .3         ? .05         ?            ? 1e4        ?
       right end....                           ?            ?

   Enter no of solid supports? 2
   Enter numbers of support nodes:

   Support  1   at node ? 1
   Support  2   at node ? 5

   Are data correct (Y/N)  ? y
```

```
SOLUTION ...
========================

Node Deflection    Slope         moment        Shear        Reaction

1    1.470575E-06  5.561773E-03   0             0            7272.728

2    9.633144E-04  3.304113E-03  -1454.546     -7272.728

3    1.825637E-03  5.481806E-04  -4363.637     -7272.728

4    1.854139E-03 -2.45528E-04   -3818.182      12727.273

5    2.573361E-06 -9.135064E-03  -2.441406E-04  .7167969     12726.56
```

```
print return to continue? _
```

q-max= 0

(b)

Figure E6.5b

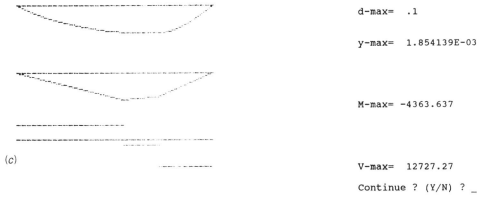

```
d-max=   .1

y-max=   1.854139E-03

M-max=  -4363.637

V-max=   12727.27

Continue ? (Y/N) ? _
```

Figure E6.5c

6.5.4. The Finite Element Method-Beam Element

The reader has been introduced already, with the transfer matrix method, to the idea of an elastic member whose state can be approximated if the state at some points on its boundary is known. For a piece of a prismatic beam, for example (Figure 6.34), if the conditions at the ends are known (deflection, slopes), the shape of the beam can be determined throughout this piece for static conditions. This can be done with methods of strength of materials, as in the previous section. However, one can observe that the equations of the deflection of the beam will be, by the simple beam theory:

$$\frac{d^2y}{dx^2} = \frac{M(x)}{EI}; \; V(x) = \frac{dM}{dx}; \; q(x) = \frac{d^2M}{dx^2} \tag{6.34}$$

If no load is assumed on the beam, $q(x) = 0$, differentiating twice the first equation and using the second, we obtain $d^4y/dx^4 = 0$, assuming that EI is constant. Its general solution is:

$$y(x) = c_3x^3 + c_2x^2 + c_1x + c_0 \tag{6.35}$$

where c_0, c_1, c_2, c_3 are constants to be determined from the boundary conditions

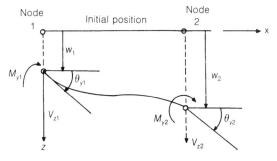

Figure 6.34. Deflected beam element.

$y_1, y_2, \theta_1, \theta_2$ and θ is the slope $= dy/dx$. Therefore, if the end conditions are known, the deflections in between are functions only of these conditions and independent of the beam properties E and I. In fact, it can be shown very easily that:

$$y(x) = f_1(x)y_1 + f_2(x)\theta_1 + f_3(x)y_2 + f_4(x)\theta_2 \tag{6.36}$$

where, for $s = x/\ell$ and ℓ the length of the element:

$$f_1(x) = 1 - 3s^2 + 2s^3; \; f_2(x) = \ell(s - 2s^2 + 2s^3); \; f_3(x)$$
$$= 3s^2 - 2s^3; \; f_4(x) = 1\,(-s^2 + s^3).$$

The slope $\theta = dy/dx$ is:

$$\theta(x) = f_1'(x)y_1 + f_2'(x)\theta_1 + f_3'(x)y_2 + f_4'(x)\theta_2 \tag{6.37}$$

where, for $s = x/\ell$ and ℓ the length of the element:

$$f_1'(x) = -6s + 6s^2; \; f_2'(x) = \ell\,(1 - 4s + 6s^2);$$
$$f_3'(x) = 6s - 6s^2; \; f_4'(x) = \ell\,(-2s + 3s^2)$$

Using Equations (6.34), one can compute the moment and shear by further differentiation of Equation (6.36). Therefore, the state at a distance x can be calculated once the end conditions are known, and instead of dealing with this beam segment, we can, in principle, deal only with the conditions at the two ends, already called *nodes*. Every piece of a continuum that can be described by way of the conditions at a finite number of points along its boundary will be called a *finite element*.

Another important property is that in finite elements the coordinates of the nodes (deflections and slopes) are related to the generalized forces at these nodes (forces and moments). In the sequel we shall use the terms *displacements* and *forces* respectively also for slopes and moments. For a prismatic bar in flexure, we saw already that the end conditions are related by way of a transfer matrix $[L]$. As we can see from Equations (6.20), we can solve them always for $y_1, \theta_1, y_2, \theta_2$ in terms of M_1, V_1, M_2, V_2 and the reverse. The result is:

$$\{F\} = [K]\{Y\}, \; \{Y\} = [K^{-1}]\{F\} \tag{6.38}$$

where:

$$[k] = \begin{bmatrix} 12/\ell^3 & 6/\ell^2 & -12/\ell^3 & 6/\ell^2 \\ 6/\ell^2 & 4/\ell & -6/\ell^2 & 2/\ell \\ -12/\ell^3 & -6/\ell^2 & 12/\ell^3 & -6/\ell^2 \\ 6/\ell^2 & 2/\ell & -6/\ell^2 & 4/\ell \end{bmatrix}$$

$$\{F\} = \{M_1, V_1, M_2, V_2\}, \; \{Y\} = \{y_1, \theta_1, y_2, \theta_2\}$$

The matrix $[K]$ is the stiffness matrix and its inverse $[A] = [K^{-1}]$ is the flexibility matrix. Both are symmetric.

The state of displacements at each node can have up to six components: three deflections and three slopes. Consequently, the forces at each node can have six components: three forces and three moments. Therefore, the stiffness and flexibility matrices can be of dimension 12×12, at the most, for two nodes. In general, for m nodes per element, the maximum dimension of these matrices will be $6m \times 6m$. In practical situations, we do not always use all the possible coordinates. For example, if only lateral motion of a beam on a plane is considered with the state vector $\{Y\}$ and force vector $\{F\}$ above, the force equilibrium is:

$$\{F\} = [K]\{Y\} \tag{6.39}$$

In general, this procedure can yield the desired solution for a continuous beam. Using the fact that nodes connect elements, and therefore the displacements of the same node have to be the same for two adjacent elements, we can write the equations for every element and use them as a system.

First, however, mention must be made of the nodal force F. This force is due partly to the external load and partly to the reaction on the node from the adjacent element. If we add the respective equations, only the sum of these nodal forces will appear, which is the known external force, because the internal forces at one node that belong to the two elements cancel each other upon addition. Therefore, in the equations of equilibrium, we have the external forces V and M. In a general system, this is called assembly of the elements, and it is shown in Equation (6.35), where we can observe that element stiffness matrices overlap each other in places where each element has common nodes with the others. This means addition of the respective equations.

6.5.5. The General Prismatic Element (Weaver 1967)

A beam element might have a more complex loading than the bending and shearing on one plane, as discussed in the previous section. Due to the assumed linearity,

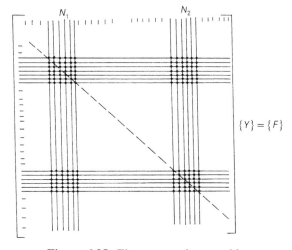

Figure 6.35. Element matrix assembly.

stress analysis can be performed independently on two planes, vertical $x - y$ and horizontal $x - z$ (Figure 6.34), analyzing the external loads to their components on these two planes. Moreover, axial load and torsion can also be considered separately.

Structures and machine components consist many times of interconnected prismatic elements under general space loading with all components of stress and deformation present and related to one another. In this case, for the two-node prismatic element already discussed, 6 degrees of freedom per node are needed, a total of 12 degrees of freedom per element, and the displacement vector will be for the beam element:

$$\{\delta^e\} = \{u_{x1}\ u_{y1}\ u_{z1}\ \theta_{x1}\ \theta_{y1}\ \theta_{z1}\ u_{x2}\ u_{y2}\ u_{z2}\ \theta_{x2}\ \theta_{y2}\ \theta_{z2}\}$$

with the force vector:

$$\{F\} = \{F_{x1}\ F_{y1}\ F_{z1}\ M_{x1}\ M_{y1}\ M_{z1}\ F_{x2}\ F_{y2}\ F_{z2}\ M_{x2}\ M_{y2}\ M_{z2}\}$$

The forces and moments in the vertical plane x–z, (Figure 6.34) are related to the forces and moments on the same plane by way of Equation (6.26). The same equation relates loads and displacements on the horizontal plane x–y.

Axial load F is related to the axial nodal displacements with:

$$F_{x2} = \left(\frac{AE}{L}\right) u_{x2} - \left(\frac{AE}{L}\right) u_{x1} \tag{6.39a}$$

$$F_{x1} = \left(\frac{AE}{L}\right) u_{x2} - \left(\frac{AE}{L}\right) u_{x1} \tag{6.39b}$$

Torsion M is related to the twist at the nodes as:

$$M_{x2} = \left(\frac{J_p G}{L}\right) u_{x2} - \left(\frac{J_p G}{L}\right) u_{x1} \tag{6.39c}$$

$$M_{x1} = \left(\frac{J_p G}{L}\right) u_{x1} - \left(\frac{J_p G}{L}\right) u_{x2} \tag{6.39d}$$

where A is the beam cross-section, G is the shear modulus, and J_p is the polar area moment of inertia.

The element force vector $\{F^e\}$ is related to the element displacement vector $\{\delta^e\}$ by way of Equations (6.38) and (6.39), properly applied. This relation can be written in matrix form as:

$$\{F^e\} = [K]\{\delta^e\} \tag{6.40}$$

For such an element the coordinate and node numbering system is indicated in Figure 6.37 with the y-axis normal to the plane of the paper. The stiffness matrix in Equation (6.40), obtained from Equation (6.38) and (6.39), by inspection, is:

$$[K] = \begin{bmatrix}
\frac{EA_x}{L} & 0 & 0 & 0 & 0 & 0 & -\frac{EA_x}{L} & 0 & 0 & 0 & 0 & 0 \\
0 & \frac{12EI_z}{L^3} & 0 & 0 & 0 & \frac{6EI_z}{L^2} & 0 & \frac{12EI_z}{L^3} & 0 & 0 & 0 & \frac{6EI_z}{L^2} \\
0 & 0 & \frac{12EI_y}{L^3} & 0 & -\frac{6EI_y}{L^2} & 0 & 0 & 0 & \frac{12EI_y}{L^3} & 0 & -\frac{6EI_y}{L^2} & 0 \\
0 & 0 & 0 & \frac{GI_x}{L} & 0 & 0 & 0 & 0 & 0 & -\frac{GI_x}{L} & 0 & 0 \\
0 & 0 & -\frac{6EI_y}{L^2} & 0 & \frac{4EI_y}{L} & 0 & 0 & 0 & \frac{6EI_y}{L^2} & 0 & \frac{2EI_y}{L} & 0 \\
0 & \frac{6EI_z}{L^2} & 0 & 0 & 0 & \frac{4EI_z}{L} & 0 & -\frac{6EI_z}{L^2} & 0 & 0 & 0 & \frac{2EI_z}{L} \\
-\frac{EA_x}{L} & 0 & 0 & 0 & 0 & 0 & \frac{EA_x}{L} & 0 & 0 & 0 & 0 & 0 \\
0 & -\frac{12EI_z}{L^3} & 0 & 0 & 0 & \frac{6EI_z}{L^2} & 0 & \frac{12EI_z}{L^3} & 0 & 0 & 0 & -\frac{6EI_z}{L^2} \\
0 & 0 & -\frac{12EI_y}{L^3} & 0 & \frac{6EI_y}{L^2} & 0 & 0 & 0 & \frac{12EI_y}{L^3} & 0 & \frac{6EI_y}{L^2} & 0 \\
0 & 0 & 0 & -\frac{GI_x}{L} & 0 & 0 & 0 & 0 & 0 & \frac{GI_x}{L} & 0 & 0 \\
0 & 0 & -\frac{6EI_y}{L^2} & 0 & \frac{2EI_y}{L} & 0 & 0 & 0 & \frac{6EI_y}{L^2} & 0 & \frac{4EI_y}{L} & 0 \\
0 & \frac{6EI_z}{L^2} & 0 & 0 & 0 & \frac{2EI_z}{L} & 0 & -\frac{6EI_z}{L^2} & 0 & 0 & 0 & \frac{4EI_z}{L}
\end{bmatrix}$$

$$(6.41)$$

This is not enough, however. The element was assumed on the x axis, which is not always the case. In general, it is convenient to compute the stiffness matrix in a local coordinate system coinciding with the natural axes of symmetry of the beam element and then modify it for a general orientation of the element.

The vectors $\{F^e\}$ and $\{\delta^e\}$ follow the rules of rotation discussed in Section 6.5. They have to be multiplied by a rotation matrix that accounts for:

1. The three direction cosines n_x, n_y, n_z of the axis of the beam in respect to the global coordinate system xyz.
2. The roll angle ϕ, i.e., the angle of rotation of the beam section about the x axis of the local coordinate system.

For each of the vectors $\{u_{x1}\ u_{y1}\ u_{z1}\}$, $\{\theta_{x1}\ \theta_{y1}\ \theta_{z1}\}$, $\{u_{x2}\ u_{y2}\ u_{z2}\}$, $\{\theta_{x2}\ \theta_{y2}\ \theta_{z2}\}$, $\{F_{x1}\ F_{y1}\ F_{z1}\}$, $\{M_{x1}\ M_{y1}\ M_{z1}\}$, $\{F_{x2}\ F_{y2}\ F_{z2}\}$, $\{M_{x2}\ M_{y2}\ M_{z2}\}$, the transformation matrix is:

$$[R] = \begin{bmatrix}
n_x & n_y & n_z \\
\dfrac{-n_x n_y \cos\phi - n_z \sin\phi}{\sqrt{n_x^2 + n_z^2}} & \sqrt{n_x^2 + n_z^2}\,\cos\phi & \dfrac{-n_y n_z \cos\phi + n_x \sin\phi}{\sqrt{n_x^2 + n_z^2}} \\
\dfrac{n_x n_y \sin\phi - n_z \cos\phi}{\sqrt{n_x^2 + n_z^2}} & -\sqrt{n_x^2 + n_z^2}\,\sin\phi & \dfrac{n_y n_z \sin\phi + n_x \cos\phi}{\sqrt{n_x^2 + n_z^2}}
\end{bmatrix}$$

$$(6.42)$$

Transformation of any one of the above vectors, parts of the total load or displacement vector, involves multiplication by the transformation matrix $[R]$. Since the load vector $\{F^e\}$ and the displacement vector $\{\delta^e\}$ consist of four 3-component vectors each, the transformation of the 12-component vectors $\{F^e\}$ and $\{\delta^e\}$ from the position of the element along the x axis (local) to the general space location (global), will be:

$$\{F^e_G\} = [R_G]\{F^e_G\} \tag{6.43}$$

$$\{\delta^e_G\} = [R_G]\{\delta^e_G\} \tag{6.44}$$

where:

$$[R_G] = \begin{bmatrix} [R] & 0 & 0 & 0 \\ 0 & [R] & 0 & 0 \\ 0 & 0 & [R] & 0 \\ 0 & 0 & 0 & [R] \end{bmatrix}$$

Equation (6.40) then relates loads and displacements in the local coordinate system. The vectors are referred to the global coordinate system upon multiplication by the matrix R. Then:

$$\{F^e\}[R_G] = [K^e]\{\delta^e\}[R_G] \tag{6.45}$$

Therefore, the matrix relating loads and displacements in the global coordinate system is $[R_G^{-1}][K^e][R_G]$. Inversion of matrix $[R_G]$ is not necessary, because due to the reciprocity of rotation the inverse of $[R_G]$ equals its transpose, thus $[K^e_G] = [R_G^T][K^e_L][R_G]$.

The stiffness matrices of the elements have to be combined to yield the system stiffness matrix relating the system nodal forces to the system nodal displacements.

The way to assemble the stiffness matrix of the system out of the element stiffness matrices has already been discussed and will be further exemplified later in this chapter.

6.5.6. Boundary Conditions

The most usual boundary conditions are restraints to some of the degrees of freedom at certain nodes. For example, completely fixed node i means that the 6 displacements associated with this node $\delta_{6(I-1)+j}$, $j = 1, 2, \ldots, 6$ are zero. There are many ways to take this into account in the system equations:

$$[K]\{\delta\} = \{F\} \tag{6.46}$$

The best way, of course, is to eliminate all the constrained coordinates from the system and reduce accordingly the number of equations.

It is much simpler however, to do a trick: the j component of the six coordinates of node i has coordinate number in the global system $p = (i - 1)6 + j$. Then the *p*-row of matrix [*K*] is eliminated and the diagonal term *k* is set = 1. The corresponding component of the force vector is set to zero if the coordinate is completely restrained (fixed), or to a specific value if an initial displacement is forced at this coordinate.

6.5.7. System Assembly and Solution

To assemble the element stiffness matrices into the system stiffness matrix *K*, we must follow the procedure indicated in Equation (6.35). To this end, we note that element *I* has been defined by its node numbers $N_1(I)$, $N_2(I)$ and the coordinates $X(J)$, $Y(J)$ $Z(J)$ of the node $J = N_1$ or N_2. To each node, a number of degrees of freedom are assigned, in this case six. Then the element stiffness matrix will have dimensions 12 × 12 in this case.

To node *j*, therefore, correspond six coordinates, $6(j − 1) + k$, where $k = 1$ . . . 6. Therefore, to nodes N_1, N_2 correspond coordinates $6(N_1 − 1) + k$ and $6(N_2 − 1)+ K$, respectively, $k = 1$ . . . 6. These coordinates are assigned numbers *I*1 to *I*6. Finally, each element of the 12 × 12 stiffness matrix *K* is added to the corresponding coordinates of the global stiffness matrix. The process is shown in Equation (6.35). The 12 × 12 elements of the element stiffness matrix are denoted with black dots.

The process is repeated with each one of the elements filling up the stiffness matrix. This matrix thus obtained has some noticeable properties:

1. It is symmetric (and of course real) due to the well known reciprocity property of elastic systems.
2. In most cases it is banded. All elements contribute to the main diagonal but also to a narrow band about the diagonal. This is not true for all problems, however.

These two properties make the solution considerably easier.

The solution of the system

$$[K]\{x\} = \{c\} \tag{6.47}$$

of linear equations is usually performed with two broad categories of methods: "exact" and iterative. Here an "exact" method will be used, the Gauss elimination. Take, for example, the 3 × 3 system of linear equations:

$$2x + 4y + 6z = 28$$
$$3x + y + z = 8 \tag{6.48}$$
$$x + 3y + 4z = 19$$

Multiply the first equation by $−3/2$ and add to the second, then multiply the first equation by $−1/2$ and add to the third equation. Then:

$$2x + 4y + 6z = 28$$
$$-5y - 8z = -34 \qquad (6.49)$$
$$y + z = 5$$

Multiply the second equation by $1/5$ and add to the third equation. The system becomes:

$$2x + 4y + 6z = 28$$
$$-5y - 8z = -34 \qquad (6.50)$$
$$-\left(\frac{3}{5}\right) z = \frac{-9}{5}$$

This process is called *elimination*. The solution follows immediately.

From the third equation, we find $z = 3$. Substituting in the second, we find $y = 2$, and substituting in the first we find $x = 1$ and the system is solved. This process is called *back substitution*.

We will not address ourselves at this point to the question of efficiency. In smaller computers, the main problem is usually memory and time for the element computation rather than solution time. This is not always the case, however.

Solution of the linear system will yield the nodal displacements. For design calculations, the element stresses must be computed. To this end, the nodal forces for each element corresponding to the nodal displacements must be computed. Therefore, the displacement vector has to be multiplied by the element stiffness matrix (Equation (6.37)). The resulting nodal forces are in the global coordinate system. Usually these forces are needed in the local system so that strength of materials formulas can be applied for the determination of stresses. For this reason, first the nodal displacements are transformed into the local system, and then the nodal forces in the local system are obtained, multiplying by the local stiffness matrix:

$$\{\delta\}^e_{\text{local}} = [R^T]\{\delta\}^e_{\text{global}} \qquad (6.51)$$

Therefore:

$$\{F\}^e_{\text{local}} = [K^e]_{\text{local}}[R^T]\{\delta\}^e_{\text{global}} = [H^e]\{\delta\}^e_{\text{global}} \qquad (6.52)$$

The stress matrix $[H^e]$ is computed by the same procedure as the element stiffness matrix.

Example 6.6   Calculate displacements and member forces for a double crane hook shown in Figure E6.6a (dimensions in cm) made of 4 cm-thick plate. Data: $F = 4$ tons, $t = 4$ cm, $b = 15$ cm, $E = 2.1 \times 10^{11}$ Pa, $\nu = 0.3$, material steel.

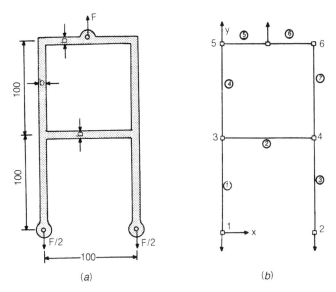

Figure E6.6a and b

DATA (From file FRAME42)

NODE COORDINATES

Node	x	y	z
1	50	0	0
2	50	99.99999	0
3	150	99.99999	0
4	150	0	0
5	50.00001	200	0
6	100	200	0
7	150	200	0

Mit RETURN to continue?

NODE RELEASES

Node	RX	RY	RZ	FX	FY	FZ
1	0	0	1	0	0	0
4	0	0	1	0	0	0

Hit RETURN to continue?

ELEMENT INCIDENCES

Member	Node 1	Node 2
1	1	2
2	2	3
3	3	4
4	4	5
5	5	6
6	6	7
7	7	8

Mit RETURN to continue?

Figure E6.6c

_____ ELEMENT PROPERTIES _____

No	Material	Area	Iy	Iz	Iyz	Roll angle
1	1	60	80	1125	1205	0
2	1	60	80	1125	1205	0
3	1	60	80	1125	1205	0
4	1	60	80	1125	1205	0
5	1	60	80	1125	1205	0
6	1	60	80	1125	1205	0
7	1	60	80	1125	1205	0

Hit RETURN to continue?

_____ LOADING _____

Node	Fx	Fy	Fz	Mx	My	Mz
6	0	5000	0	0	0	0

Hit RETURN to continue?

NODE DISPLACEMENTS

Element	ux	uy	uz	rotx	roty	rotz
1	0.00E+00	0.00E+00	0.00E+00	0.00E+00	0.00E+00	0.00E+00
2	2.60E−03	1.98E−02	0.00E+00	0.00E+00	0.00E+00	−9.41E−04
3	−2.60E−03	1.98E−02	0.00E+00	0.00E+00	0.00E+00	9.41E−04
4	0.00E+00	0.00E+00	0.00E+00	0.00E+00	0.00E+00	0.00E+00
5	−2.10E−03	3.97E−02	0.00E+00	0.00E+00	0.00E+00	4.77E−03
6	−1.61E−07	2.69E−01	0.00E+00	0.00E+00	0.00E+00	4.66E−10
7	2.10E−03	3.97E−02	0.00E+00	0.00E+00	0.00E+00	−4.77E−03

ELEMENT FORCES

Element	Fx	Fy	Fz	Mx	My	Mz
1	1.26E+02	−2.50E+03	0.00E+00	0.00E+00	0.00E+00	−4.08E+03
2	6.55E+02	1.53E−04	0.00E+00	0.00E+00	0.00E+00	−4.45E+03
3	1.26E+02	2.50E+03	0.00E+00	0.00E+00	0.00E+00	8.53E+03
4	−5.29E+02	−2.50E+03	0.00E+00	0.00E+00	0.00E+00	1.30E+04
5	−5.29E+02	−2.50E+03	0.00E+00	0.00E+00	0.00E+00	−4.00E+04
6	−5.29E+02	2.50E+03	0.00E+00	0.00E+00	0.00E+00	8.50E+04
7	−5.29E+02	2.50E+03	0.00E+00	0.00E+00	0.00E+00	−4.00E+04

Normal termination. Press any key.

Figure E6.6c (*Continued*).

Solution   Figure E6.6*b* shows a model, plane frame, that can be analyzed as a space frame in the *x–y* plane, setting all *z* coordinates of nodes equal to zero. The program FINFRAME of MELAB 2.0 will be used.

$$\text{Moments of inertia: } I_Y = \frac{bt^3}{12} = 80 \text{ cm}^4, \ I_Z = \frac{tb^3}{12} = 1125 \text{ cm}^4, \ I_p = I_Y + I_Z = 1205 \text{ cm}^4$$

$$\text{Section area } A = bt = 4 \times 15 = 60 \text{ cm}^2$$

(For numerical convenience ton, cm units will be used)
Young's modulus $E = 2100$ ton/cm^2
The model with node and element numbers is shown in Figure E6.6*b*.

Nodes 1 and 2 will be restrained: Node 1 in *x, y, z* and node 2 in *y, z*. This situation is statically equivalent to the given loading.

The results are shown on pages 441–442.

6.6. SURFACE MEMBERS

6.6.1. Membrane Stresses

Structural members that at every point have dimensions along one direction much smaller than dimensions in the other directions are called *surface structures*. Usual cases are plates and shells (Figure 6.36). An additional property that many times exists in machines is symmetry about an axis. Axisymmetric plates and shells are very common, such as rotating disks, pressure vessels, and toroidal shells.

Shells are capable of sustaining very high loads, mostly in the form of internal pressures. Therefore, many times they are very thin and their resistance to bending is very small. The loads are sustained by tensile stresses in the shell material. Such stresses are called *membrane stresses* (see Section 5.3.1).

6.6.2. Plane Stress–Strain Problems

Many machine components have plane form, such as the component shown in Figure 6.1. Analysis of complex plane geometric components with in-plane loading can easily be performed with the finite element method.

As in the case of the frame structure that was broken down to prismatic beam elements, for which a stiffness matrix relating nodal forces to nodal displacements could be found, the plane structure, such as the one in Figure 6.37, is divided in a number of triangles.

Vertices 1, 2, 3 will be used as nodes, and displacements u, v along coordinates x and y, respectively, will be used to describe the displacement within the triangle. This displacement will be a linear interpolation of the nodal displacement, which means that displacements will be linear functions of x, y.

Strains are derivatives of the displacements. Therefore, strains will be constant throughout the element. Consequently, the stresses within the element will be constant if linear elasticity will be assumed. This implies that the triangular element is small enough that the stresses in the vicinity of the element are nearly constant. In general, the stresses will be normal stresses σ_x, σ_y and shear stresses τ_{xy}. These stresses give forces on the edges of the triangle that will be assumed on the middle of each triangle side, as shown in Figure 6.10. These forces will be equally divided between the adjacent nodes. Take, for example, node 1. The nodal forces will be the sum of the contribution of stresses on edges 12 and 13:

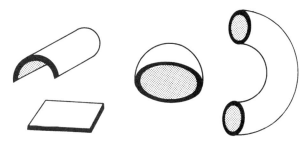

Figure 6.36. Surface structures.

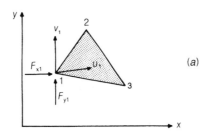

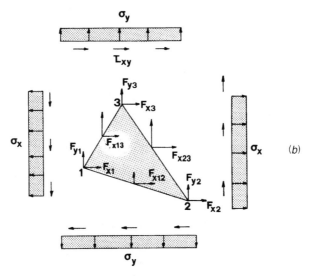

Figure 6.37. Free-body diagram of a plane triangular element.

$$F_{x13} = -\sigma_x(y_3 - y)t + \tau_{xy}(x_3 - x_1)t$$

$$F_{x12} = -\sigma_x(y_1 - y_2)t - \tau_{xy}(x_2 - x_1)t \tag{6.53}$$

$$F_{x1} = \frac{F_{x13} + F_{x23}}{2} = \frac{-\sigma_x(y_3 - y_2)t}{2} - \frac{\tau_{xy}(x_2 - x_3)t}{2}$$

$$F_{y1} = \frac{F_{y13} + F_{y23}}{2} = \frac{-\sigma_x(x_3 - x_2)t}{2} - \frac{\tau_{xy}(y_3 - y_2)t}{2}$$

In a similar way, the rest of the nodal forces can be related to the stresses. Writing these relations in matrix form, we obtain:

$$
\begin{bmatrix} F_1 \\ F_2 \\ F_3 \\ F_4 \\ F_5 \\ F_6 \end{bmatrix}
= -\frac{t}{2}
\begin{bmatrix}
y_3 - y_2 & 0 & x_2 - x_3 \\
0 & x_2 - x_3 & y_3 - y_2 \\
y_1 - y_3 & 0 & x_3 - x_1 \\
0 & x_3 - x_1 & y_1 - y_3 \\
y_2 - y_1 & 0 & x_1 - x_2 \\
0 & x_1 - x_2 & y_2 - y_1
\end{bmatrix}
\begin{bmatrix} \sigma_x \\ \sigma_y \\ \sigma_z \end{bmatrix}
\tag{6.54}
$$

or:

$$\{F^e\} = [Z]\{\sigma^e\} \tag{6.55}$$

As proposed above, the displacements within the elements are assumed to be linear functions of the position:

$$u = a_1 + a_2 x + a_3 y$$

$$v = a_4 + a_5 x + a_6 y \tag{6.56}$$

In matrix form:

$$
\begin{bmatrix}
1 & x & y & 0 & 0 & 0 \\
0 & 0 & 0 & 1 & x & y
\end{bmatrix}
\begin{bmatrix}
a_1 \\ a_2 \\ a_3 \\ a_4 \\ a_5 \\ a_6
\end{bmatrix}
= [f(x, y)]\{a\} \tag{6.57}
$$

Constants a_1, a_2, a_3, a_4, a_5, a_6 will be determined so that the displacements at nodes 1, 2, 3 are nodal displacements u_1, v_1, u_2, v_2, u_3, v_3. This yields:

$$\{a\} = [A]^{-1}\{\delta^e\} \tag{6.58}$$

where:

$$
[A] =
\begin{bmatrix}
1 & x_1 & y_1 & 0 & 0 & 0 \\
0 & 0 & 0 & 1 & x_1 & y_1 \\
1 & x_2 & y_2 & 0 & 0 & 0 \\
0 & 0 & 0 & 1 & x_2 & y_2 \\
1 & x_3 & y_3 & 0 & 0 & 0 \\
0 & 0 & 0 & 1 & x_3 & y_3
\end{bmatrix}
$$

The strains are now obtained from the displacements:

$$\varepsilon_x = \frac{\partial u}{\partial x} = a_2$$

$$\varepsilon_y = \frac{\partial v}{\partial y} = a_6 \tag{6.59}$$

$$\gamma_{xy} = \frac{\partial u}{\partial x} + \frac{\partial v}{\partial y} = a_3 + a_5$$

and in matrix form:

$$\{\varepsilon(x, y)\} = \begin{bmatrix} \varepsilon_x \\ \varepsilon_y \\ \gamma_{xy} \end{bmatrix} = \begin{bmatrix} 0 & 1 & 0 & 0 & 0 & 0 \\ 0 & 0 & 0 & 0 & 0 & 1 \\ 0 & 0 & 1 & 0 & 1 & 0 \end{bmatrix} \{a\} \tag{6.60}$$

$$= [C]\{a\} = [C][A^{-1}]\{\delta^e\}$$

The stresses are now obtained from the strains by the generalized Hooke's law:

$$\sigma_x = \frac{E}{1 - \nu^2} (\varepsilon_x + \nu\varepsilon_y)$$

$$\sigma_y = \frac{E}{1 - \nu^2} (\nu\varepsilon_x + \varepsilon_y) \tag{6.61}$$

$$\tau_{xy} = \frac{E}{1 - \nu^2} (1 - \nu)\gamma_{xy}$$

In matrix form,

$$\sigma(x, y) = \begin{bmatrix} \sigma_x \\ \sigma_y \\ \sigma_z \end{bmatrix} = \frac{E}{1 - \nu^2} \begin{bmatrix} 1 & \nu & 0 \\ \nu & 1 & 1 \\ 0 & 0 & (1 - \nu)/2 \end{bmatrix} \{\varepsilon(x, y)\} = [D]\{\varepsilon(x, y)\} \tag{6.62}$$

If the stresses are substituted into Equation (6.40), a matrix relation is obtained between nodal forces and nodal displacements, which yields directly the stiffness matrix of the element. Therefore, the stiffness matrix is:

$$[K^e] = [Z][D][C][A]^{-1} = [Z][D][B] \tag{6.63}$$

where $[B] = [C][A]^{-1}$. It can be shown that $[Z] = [B]^T$. Therefore:

$$\{K^e\} = [B^T][D][B] \tag{6.64}$$

The stiffness matrix can be written in explicit form (Weaver 1967):

$$[K^e] = \frac{t}{4\Delta}$$

$$\begin{bmatrix}
d_{11}(y_2 - y_3)^2 & d_{12}(x_3 - x_2)(y_2 - y_3) & d_{11}(y_2 - y_3)(y_3 - y_1) \\
\quad + d_{33}(x_3 - x_2)^2 & \quad + d_{33}(x_3 - x_2)(y_2 - y_3) & \quad + d_{33}(x_3 - x_2)(x_1 - x_3) \\
d_{21}(x_3 - x_2)(y_2 - y_3) & d_{22}(x_3 - x_2)^2 & d_{12}(y_3 - y_1)(x_3 - x_2) \\
\quad + d_{33}(x_3 - x_2)(y_2 - y_3) & \quad + d_{33}(y_2 - y_3)^2 & \quad + d_{33}(x_1 - x_3)(y_2 - y_3) \\
d_{11}(y_2 - y_3)(y_3 - y_1) & d_{12}(x_3 - x_2)(y_3 - y_1) & d_{11}(y_3 - y_1)^2 \\
\quad + d_{33}(x_1 - x_3)(x_3 - x_2) & \quad + d_{33}(x_1 - x_3)(y_2 - y_3) & \quad + d_{33}(x_1 - x_3)^2 \\
d_{21}(x_1 - x_3)(y_2 - y_3) & d_{22}(x_1 - x_3)(x_3 - x_2) & d_{12}(x_1 - x_3)(y_3 - y_1) \\
\quad + d_{33}(x_3 - x_2)(y_3 - y_1) & \quad + d_{33}(y_2 - y_3)(y_3 - y_1) & \quad + d_{33}(x_1 - x_3)(y_3 - y_1) \\
d_{11}(y_1 - y_2)(y_2 - y_3) & d_{12}(x_3 - x_2)(y_1 - y_2) & d_{11}(y_1 - y_2)(y_3 - y_1) \\
\quad + d_{33}(x_2 - x_1)(x_3 - x_2) & \quad + d_{33}(x_2 - x_1)(y_2 - y_3) & \quad + d_{33}(x_1 - x_3)(x_2 - x_1) \\
d_{21}(x_2 - x_1)(y_2 - y_3) & d_{22}(x_2 - x_1)(x_3 - x_2) & d_{12}(x_2 - x_1)(y_3 - y_1) \\
\quad + d_{33}(x_3 - x_2)(y_1 - y_2) & \quad + d_{33}(y_1 - y_2)(y_2 - y_3) & \quad + d_{33}(x_1 - x_3)(y_1 - y_2)
\end{bmatrix}$$

$$
\begin{bmatrix}
\begin{aligned}
&d_{12}(x_1 - x_3)(y_2 - y_3) \\
&\quad + d_{33}(x_3 - x_2)(y_3 - y_1) \\
&d_{22}(x_3 - x_2)(x_1 - x_3) \\
&\quad + d_{33}(y_2 - y_3)(y_3 - y_1) \\
&d_{12}(x_1 - x_3)(y_3 - y_1) \\
&\quad + d_{33}(x_1 - x_3)(y_3 - y_1) \\
&d_{22}(x_1 - x_3)^2 \\
&\quad + d_{33}(y_3 - y_1)^2 \\
&d_{12}(x_1 - x_3)(y_1 - y_2) \\
&\quad + d_{33}(x_2 - x_1)(y_3 - y_1) \\
&d_{22}(x_1 - x_3)(x_2 - x_1) \\
&\quad + d_{33}(y_1 - y_2)(y_3 - y_1)
\end{aligned}
&
\begin{aligned}
&d_{11}(y_1 - y_2)(y_2 - y_3) \\
&\quad + d_{33}(x_2 - x_1)(x_3 - x_2) \\
&d_{21}(x_3 - x_2)(y_1 - y_2) \\
&\quad + d_{33}(x_2 - x_1)(y_2 - y_3) \\
&d_{11}(y_1 - y_2)(y_3 - y_1) \\
&\quad + d_{33}(x_1 - x_3)(x_2 - x_1) \\
&d_{21}(x_1 - x_3)(y_1 - y_2) \\
&\quad + d_{33}(x_2 - x_1)(y_3 - y_1) \\
&d_{11}(y_1 - y_2)^2 \\
&\quad + d_{33}(x_2 - x_1)^2 \\
&d_{21}(x_2 - x_1)(y_1 - y_2) \\
&\quad + d_{33}(x_2 - x_1)(y_1 - y_2)
\end{aligned}
&
\begin{aligned}
&d_{12}(x_2 - x_1)(y_2 - y_3) \\
&\quad + d_{33}(x_3 - x_2)(y_3 - y_1) \\
&d_{22}(x_2 - x_1)(x_3 - x_2) \\
&\quad + d_{33}(y_1 - y_2)(y_2 - y_3) \\
&d_{12}(x_2 - x_1)(y_3 - y_1) \\
&\quad + d_{33}(x_1 - x_3)(y_1 - y_2) \\
&d_{22}(x_1 - x_3)(x_2 - x_1) \\
&\quad + d_{33}(y_1 - y_2)(y_3 - y_1) \\
&d_{12}(x_2 - x_1)(y_1 - y_2) \\
&\quad + d_{33}(x_2 - x_1)(y_1 - y_2) \\
&d_{22}(x_2 - x_1)^2 \\
&\quad + d_{33}(y_1 - y_2)^2
\end{aligned}
\end{bmatrix}
\tag{6.65}
$$

where 2Δ is the area of the triangular element and t is the thickness.

Of particular importance is the relation between stresses and displacements:

$$
\{\sigma^e\} = [H]\{\delta^e\} \tag{6.66}
$$

where:

$$
[H] = [D][B] \tag{6.67}
$$

which will eventually yield the stresses after the determination of displacements from the equations

$$
[K]\{\delta\} = \{F\} \tag{6.68}
$$

The system, or global, Equations (6.61) will be assembled from the element equations by composing the system stiffness matrix out of the element stiffness matrices.

In this element, triangular, with three nodes, for plane elasticity, the equations of equilibrium (6.51) were obtained directly due to the constant stresses assumed in the model. In other elements, this is no longer possible and other methods are available. Without going into much detail, we can state that energy considerations lead to the following equation for the element stiffness matrix, instead of Equation (6.63):

$$
[K^e] = \frac{1}{V} \int_V [B^T][D][B]\,dV \tag{6.69}
$$

where V is the element volume $= 2\Delta t$ and dV the differential volume $t\,dx\,dy$.

In this element, matrices $[B]$ and $[D]$ are constant within the element and the integration in Equation (6.62) is trivial, leading immediately to Equation (6.63).

The algorithm to implement the above procedure is essentially the same as for the space frame. Only the definitions of the nodal displacement and the nodal force vector and the element stiffness matrix procedure will change. This algorithm has been programmed in program FINSTRESS, which is presented in Appendix 6.III.

Example 6.7 A steel bracket as shown supports a load of 70,000 N ton uniformly distributed on the upper end of the bracket. Determine the stresses in the bracket, the maximum stress, and the stress concentration factor in the smallest section. Thickness $t = 100$ mm, $b_0 = 100$ mm, $b = 200$ mm, $h = 200$ mm.

Solution The program FINSTRESS will be used. To simplify the problem (although it is not necessary), we note that only one quarter needs to be considered (Figure 6.7*b*), with half load on the upper half surface and restraining the horizontal motion on the center line (*x* axis). One point will be fully supported to have a determinate system.

Next, an automatic mesh generation scheme will be employed, program AUTOMESH of MELAB 2.0. To this end, the area is divided by i planes y–z into $i - 1$ strips, which are divided by j lines into $j - 1$ quadrilaterals each. The latter are divided by a diagonal into 2 triangle elements.

The program asks for the number of divisions i and j and calculates node coordinates end element nodes. Material properties are uniform.

The program has two main options: **plotting** the finite element mesh or **computing.** If we select **computing** the program produces the results on the screen. **Postprocessing** gives a color map. The maximum stress found in the vertical direction is 8.087 MPa and the nominal stress is $70,000/(100 \times 100) = 7$ MPa. Therefore, $k_T = 8.087/7 = 1.15$.

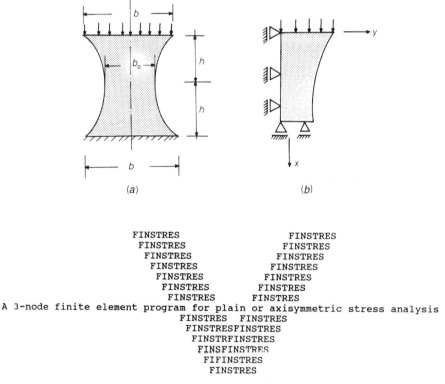

(a)   (b)

```
          FINSTRES                      FINSTRES
          FINSTRES                      FINSTRES
          FINSTRES                      FINSTRES
          FINSTRES                      FINSTRES
          FINSTRES                      FINSTRES
          FINSTRES                      FINSTRES
          FINSTRES                      FINSTRES
A 3-node finite element program for plain or axisymmetric stress analysis
          FINSTRES   FINSTRES
          FINSTRESFINSTRES
          FINSTRFINSTRES
          FINSFINSTRES
          FIFINSTRES
          FINSTRES
```

Figure E6.7*a* and *b*

TYPE OF PROBLEM

<1>Plane stress   <2>   Plane strain
<3> Axisymmetric, plain stress   <4> Axisymmetric, plain strain
              enter your selection...? 1

Enter: Modulus of Elasticity, Poisson Ratio, thickness ? 2.1e6,.3,100

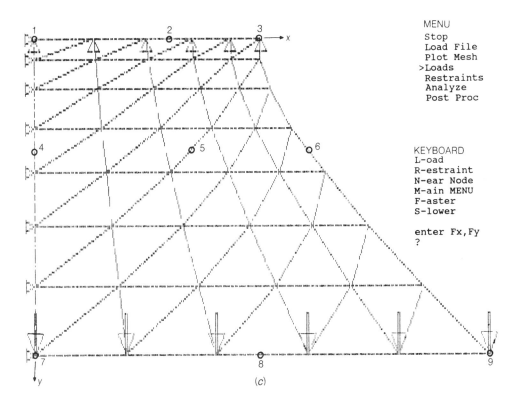

MENU
 Stop
 Load File
 Plot Mesh
>Loads
 Restraints
 Analyze
 Post Proc

KEYBOARD
L-oad
R-estraint
N-ear Node
M-ain MENU
F-aster
S-lower

enter Fx,Fy
?

(c)

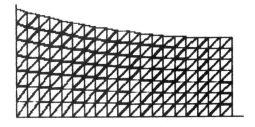

Screen dump of finite element mesh.

(d)

Figure E6.7c and d

6.6.3. Axisymmetric Solids

Stress analysis in axisymmetric solids can be performed in a similar way as in the case of the plane stress–strain problem (Section 6.6.2). Assuming the geometry of Figure 6.38, let y be the axis of symmetry and x the radial direction. The coordinate axes are now shown in Figure 6.12.

The displacement and force vectors are the same: $\{F\} = \{F_{r1}\ F_{z1}\ F_{r2}\ F_{z2}\ F_{r3}\ F_{z3}\}$, $\{\delta\} = \{u_1\ v_1\ u_2\ v_2\ u_3\ v_3\}$. In addition to the previously defined components of strain, there is a circumferential component $\varepsilon_t = u/r$, while the other terms are the same. Therefore, another row and column are introduced between rows and columns 3 and 4 in the matrix $[C]$, and

$$\{\varepsilon(r, z)\} = \begin{bmatrix} \varepsilon_x \\ \varepsilon_z \\ \varepsilon_t \\ \gamma_{rz} \end{bmatrix} = \begin{bmatrix} 0 & 1 & 0 & 0 & 0 & 0 \\ 0 & 0 & 0 & 0 & 0 & 0 \\ 1/r & z/r & 0 & 0 & 0 & 0 \\ 0 & 0 & 1 & 0 & 1 & 0 \end{bmatrix} \{a\} \tag{6.70}$$

Therefore, matrix $[B] = [C][A]^{-1}$ will be:

$$[B] = \frac{1}{2\Delta}$$

$$\begin{bmatrix} z_2 - z_3 & 0 & z_3 - z_1 & 0 & z_1 - z_2 & 0 \\ 0 & r_3 - r_2 & 0 & r_1 - r_3 & 0 & r_2 - r_1 \\ \dfrac{r_2 z_3 - r_3 z_2}{r} + & & \dfrac{r_3 z_1 - r_1 z_3}{r} + & & \dfrac{r_1 z_2 - r_2 z_1}{r} + & \\ + (z_2 - z_3) & 0 & + (z_3 - z_1) & 0 & + (z_1 - z_2) & 0 \\ + \dfrac{z}{r}(r_3 - r_2) & & + \dfrac{z}{r}(r_1 - r_3) & & + \dfrac{z}{r}(r_2 - r_1) & \\ r_3 - r_2 & z_2 - z_3 & r_1 - r_3 & z_3 - z_1 & r_2 - r_1 & z_1 - z_2 \end{bmatrix}$$

$$\tag{6.71}$$

where Δ is the area of the triangle:

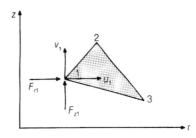

Figure 6.38. Axisymmetric triangular ring element.

$$2\Delta = \begin{vmatrix} 1 & r_1 & z_1 \\ 1 & r_2 & z_2 \\ 1 & r_3 & z_3 \end{vmatrix} \tag{6.72}$$

With the introduction of the circumferential stress, the stress–strain relation can be written in the form (now it is always plane strain):

$$\begin{bmatrix} \sigma_r \\ \sigma_z \\ \sigma_\theta \\ \tau_{rz} \end{bmatrix} = \frac{E(1-\nu)}{(1+\nu)(1-2\nu)} \begin{bmatrix} 1 & \dfrac{\nu}{1-\nu} & \dfrac{\nu}{1-\nu} & 0 \\[2mm] \dfrac{\nu}{1-\nu} & 1 & \dfrac{\nu}{1-\nu} & 0 \\[2mm] \dfrac{\nu}{1-\nu} & \dfrac{\nu}{1-\nu} & 1 & 0 \\[2mm] 0 & 0 & 0 & \dfrac{1-2\nu}{2(1-\nu)} \end{bmatrix} \tag{6.73}$$

$$\begin{bmatrix} \varepsilon_r \\ \varepsilon_z \\ \varepsilon_\theta \\ \gamma_{rz} \end{bmatrix} = [D\{_1(x, y)\}]$$

which also defines the new matrix D. Finally,

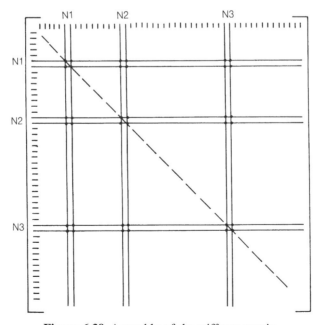

Figure 6.39 Assembly of the stiffness matrix.

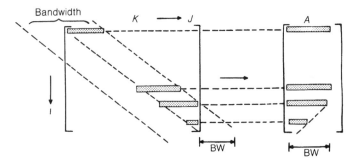

Figure 6.40 Banded matrix.

$$[K^e] = \frac{1}{V} \int_V [B^T][D][B] 2\pi r \, dr \, dz \tag{6.74}$$

Matrix B is a function of r, and the integration in Equation (6.72) has to be performed numerically. However, for small enough elements, it is possible to evaluate B at the centroid of the element:

$$r = \frac{(r_1 + r_2 + r_3)}{3}, \, z = \frac{(z_1 + z_2 + z_3)}{3} \tag{6.75}$$

with small effect on accuracy. Then, integration in Equation (6.67) is trivial and:

$$[K^e] = [B^T][D][B] \tag{6.76}$$

Matrix $[H^e]$ is again $[D][B]$.

The program FINSTRESS can be modified, changing the definition of matrices B and D and their smaller dimension from 3 to 6. In fact, this has been included as an option in the program.

6.6.4. System Assembly and Solution

To assemble the element stiffness matrices into the system stiffness matrix K, the procedure indicated in Equation (6.25) must be followed. To this end, we note that element I has been defined by its node numbers $NI(I)$, $NJ(I)$, $NM(I)$ and the coordinates $X(J)$, $Y(J)$ of node J. For element I, node J is $NI(I)$ or $NJ(J)$ or $NM(J)$. To each node, a number of degree of freedom are assigned. For example, in plane elasticity, two degree of freedom per node are assigned. Then the element stiffness matrix will have dimension 6×6 in this case.

To node J, therefore, correspond two coordinates, the $2J$-1 and the $2J$. Therefore, for nodes NI, NJ, NM correspond coordinates $2NI$-1, $2NI$, $2NJ$-1, $2NJ$, $2NM$-1, $2NM$. These coordinates are assigned numbers $I1$ to $I6$. Finally, each element Ip, Iq of the 6×6 stiffness matrix KE is added to the corresponding coordinates p, q of the global stiffness matrix. The process is shown in Figure 6.39. The 6×6

elements of the element stiffness matrix are denoted with black dots. The process is repeated with each one of the elements filling up the stiffness matrix.

This matrix thus obtained has some noticeable properties:

1. It is symmetric (and of course real) due to the well-known reciprocity property of elastic systems.
2. In most cases it is banded. All elements contribute to the main diagonal but also to a narrow band about the diagonal. This is not true for all problems, however.

These two properties make the solution considerably easier.

The solution of the system:

$$[K]\{\delta\} = \{F\} \tag{6.77}$$

will be done in the way explained in a previous section using the Gauss elimination method. To help with the memory problem, we shall use the two properties mentioned above, namely the symmetry and banded form of the matrix. We define a matrix A with a number of rows equal to the number of rows of matrix K but number of columns equal to roughly half of the bandwidth BW, including the diagonal (Figure 6.40). The required computer space ($N \times BW$) is much less than $N \times N$, usually on the order of 10%.

The relationship between the stiffness matrix K and the band storage matrix A is:

$$K(I, J) = A(K, L) \tag{6.78}$$

where

$$K = \min(I, J), L = ABS(I - J) + 1 \tag{6.79}$$

During the assembly process, only the upper part of matrix $[K]$ needs to be filled in if matrix $[A]$ is going to be used. In fact, matrix $[A]$ is filled directly without using $[K]$ at all.

6.6.5. Preprocessing and Postprocessing

The numerical methods of stress analysis widely employed today in machine design involve an enormous amount of input data and output results. This has prompted the development of preprocessing methods for automatic input generation and postprocessing to automatically sort the results for identification and presentation of the important results and to present them in an effective way.

The automatic mesh generation presented already in this chapter (Section 6.4) is used to prepare inputs for finite element analyses in the form of grids with triangular or rectangular elements.

Example 6.8 A steel spur gear is secured on the shaft by a shrink fit. It was found that the hoop stress σ_t at the outer surface was 65 N/mm^2 by strain gage measurements. Find the pressure between shaft and gear.

Solution We first use AUTOMESH to define the mess. We need to model only the right half of the hub. The y axis then coincides with the axis of symmetry. The x axis is on the top of the screen. From the menu we select **Automesh,** then we define nodes (50, 50), (50, 100), (150, 50), (150, 100), in this sequence and hit **A**. We enter 2 horizontal layers, 2 vertical layers, then 6 horizontal desired layers, 6 vertical desired layers. Then we hit **M** to return to the main menu. We save the mesh in the file MESHE6_8.DAT. Now we quit **AUTOMESH** to return to MELAB 2.0 and select **Structural, Finstress.** We select **Axisymmetric, Plane Strain** (selection 3). We enter $E = 21,000$ (N/mm^2) and Poisson's ratio $\nu = 0.3$. We select constraints. No boundary constraints exist here, but we need to enter the loads. The left-most nodes are the inner surface of the hub. On each one, we hit button 1 of the mouse and specify horizontal load 1 and vertical load 0.

The internal pressure was taken arbitrarily to, say, 1 Pa. The resulting hoop stress at the outer diameter will be computed, and the internal pressure will then be the ratio of the measured to the computed stress. Indeed, the maximum hoop stress in the outer nodes is found to be 1.66 MPa, and the true value is 65 MPa, thus the true inner pressure is $1 \times 65/1.66 = 39$ MPa. Figure E6.8c shows the element displacements.

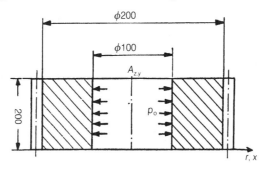

Figure E6.8a

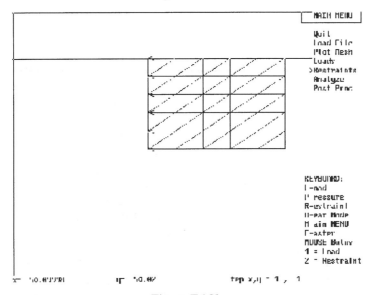

Figure E6.8b

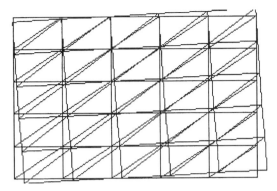

Figure E6.8c

REFERENCES

Faux, I. D., and M. J. Pratt. 1981. *Computational Geometry for Design and Manufacture.* Chichester: Ellis Horwood, and New York: Halsted Press.

Holzer, H. 1921. *Die Berechnung der Drehschwingungen und ihre Anwendung im Maschinenbau.* Berlin: J. Springer.

Rooney, J., and P. Steadman. 1987. *Principles of Computer-Aided Design.* London: Pitman.

Spiegel, M. R. 1968. *Mathematical Handbook of Formulas and Tables.* Schaum's Outline Series in Mathematics. New York: McGraw-Hill.

Weaver, W., Jr. 1967. *Computer Programs for Structural Analysis.* New York: Van Nostrand.

ADDITIONAL READING

Angel, I. O. *Advanced Graphics with the IBM PC.* London: Macmillan, 1985.

———. *A Practical Introduction to Computer Graphics.* London: Macmillan, 1981.

Dimarogonas, A. D. *Vibration for Engineers,* 2nd ed. Upper Saddle River, N.J.: Prentice-Hall, 1996.

Fadeeva, V. N. *Computation Methods in Linear Algebra.* New York: Dover, 1959.

Foley, J. D., and A. van Dam. *Fundamentals of Interactive Computer Graphics.* Reading, Mass.: Addison-Wesley, 1982.

Harrington, S. *Computer Graphics: A Programming Approach.* New York: McGraw-Hill, 1983.

Knox, C. S. *CAD/CAM Systems Planning and Implementation.* New York: Dekker, 1986.

Newman, W. M., and R. F. Sproul. *Principles of Interactive Computer Graphics,* 2nd. ed. New York: McGraw-Hill, 1979.

Pestel, E. C., and F. A. Leckie. *Matrix Methods in Elastomechanics.* New York: McGraw-Hill, 1966.

Rockey, K. C. *The Finite Element Method.* New York: Halsted Press, 1975.

Rogers, D. F., and I. A. Adams. *Mathematical Elements of Computer Graphics.* New York: McGraw-Hill, 1976.

Zienkiewicz, O. C. *The Finite Element Method.* London: McGraw-Hill, 1977.

PROBLEMS[2]

6.1–6.10. [D] Use a solid modeler of your choice to draw plane views of the machine components shown in Figures P6.1–6.10, respectively (all dimensions are in mm).

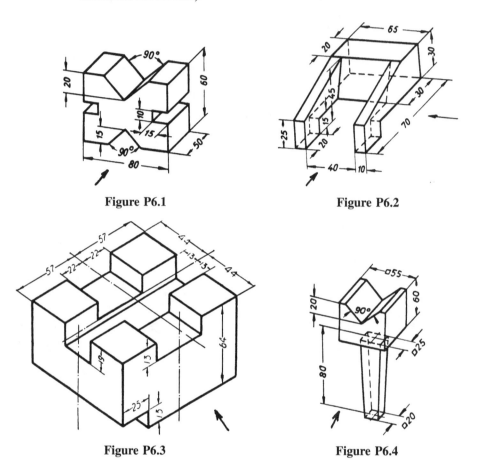

Figure P6.1

Figure P6.2

Figure P6.3

Figure P6.4

[2] [C] = certification, [D] = design, [N] = numerical, [T] = theoretical problem

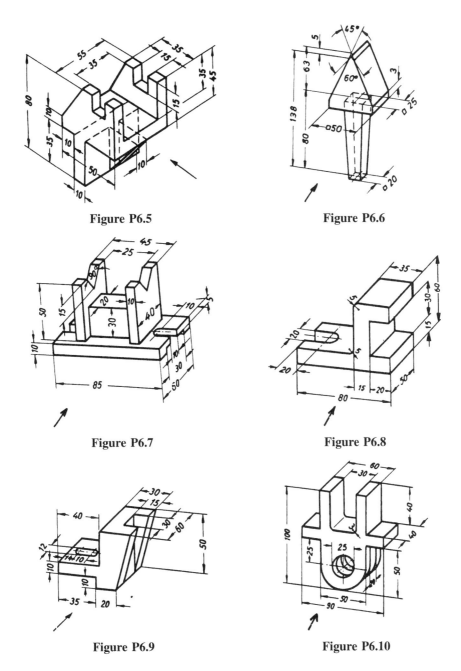

Figure P6.5

Figure P6.6

Figure P6.7

Figure P6.8

Figure P6.9

Figure P6.10

6.11. [T] A rectangle of dimensions 100×200 mm is placed with one corner at the origin, the long side along the x axis and the short side along the y axis.

1. Write the wireframe representation.

2. Write the wireframe representation of the rectangle that results with a parallel translation along the x axis by 50 mm and scaling by a factor 2.

6.12. [T] A rectangle of dimensions 100×200 mm is placed with its area center at the origin, the long side along the x axis and the short side along the y axis.

1. Write the wireframe representation.
2. Write the wireframe representation of the rectangle that results with a parallel translation along the y axis by 50 mm and rotation about the z axis by angle $45°$.

6.13. [T] A circle of diameter 200 mm is placed on the x–y plane with the center at the origin.

1. Write the wireframe representation approximation of the circle with a six-sided polygon.
2. Write the wireframe representation of the circle that results with a parallel translation along the x axis by 50 mm and scaling by a factor 2.

6.14. [T] A circle of diameter 200 mm is placed on the x–y plane with the center at point $(200, 200)$.

1. Write the wireframe representation approximation of the circle with a six-sided polygon.
2. Write the wireframe representation of the circle that results with a parallel translation along the x axis by 50 mm and rotation about the z axis by angle $45°$.

6.15. [T] A rectangle of dimensions 100×200 mm is placed with one corner at $x = 100$ mm, $y = 50$ mm, the long side along the x axis and the short side along the y axis.

1. Write the wireframe representation.
2. Write the wireframe representation of the rectangle that results with a parallel translation along the y axis by 50 mm and rotation about an axis parallel with the z axis passing through the corner further away from the origin by angle $45°$.

6.16–6.20. [T] Use AUTOMESH or any preprocessor available in your system to generate a triangular mesh as indicated in Figures P6.12–P6.20, respectively. The scale is 1 division = 10 mm. (Hint: for complex regions, dashed lines indicate division to curved or straight-line quadrilaterals first, then use of the function Add of AUTOMESH. The number of nodes is indicated on the sides of such quadrilateral regions. Coincident sides need to have the same number of nodes).

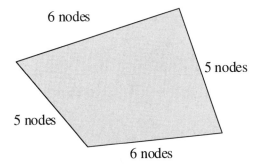

Figure P6.16

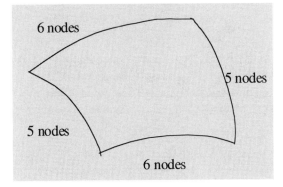

Figure P6.17

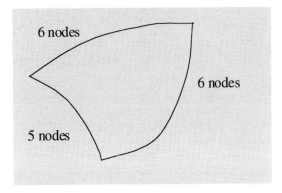

Figure P6.18

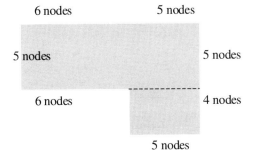

Figure P6.19

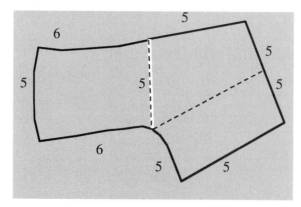

Figure P6.20

6.21. [T] Derive transfer matrices for torsional loading of a cylindrical shaft of uniform diameter between nodes. Torques are applied at the nodes and the shaft is clamped at one end, free at the other. Then write a program similar to TRANSFER MATRIX but with two-dimensional vectors (angle of twist, torque) to solve torsion of shafts problems.

6.22. [T] Do the same as in Problem 6.21, but for axial loading of shafts and beams. The state vector is (axial motion, axial force).

6.23. [T] For a beam where at some node there is change of direction of the center line by some angle, the slope and moment remain the same, but deflection and shear force change by rotation. Derive the relation of the state vector before and after the node and write in a matrix form, thus deriving a change of direction point matrix. You now have to include the axial force in all transfer matrices and state vector because on rotation it gives a shear component.

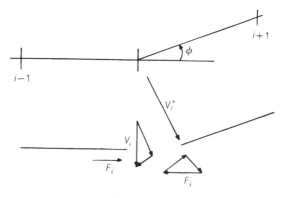

Figure P6.23

6.24. [T] Derive transfer matrices and state vectors to combine transverse and torsional loading of beams.

6.25. [T] Derive transfer matrices and state vectors to combine transverse and longitudinal loading of beams.

6.26. [N] The crankshaft of a reciprocating pump has maximum transmitted torque $T = 5000$ Nm. The material is steel, $E = 2.1 \times 10^{11}$ Pa, $\nu = 0.3$, $d = 30$ mm, $H = 120$ mm, $l = 100$ mm, $L = 2l$ mm, and the section of the transverse members is orthogonal of thickness 30 mm and width 60 mm. Determine the maximum stresses and the rotation of the one end relative to the other, using SPACEFRAME or any finite element program available to you.

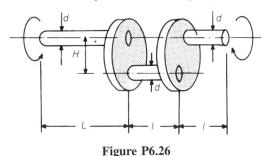

Figure P6.26

6.27. [N] A steel coil compaction press works in a frame of tubular steel, as shown in the rectangular cross-section 160×160 mm and thickness 10 mm. The 65-ton maximum force acts between points A and B. Using SPACEFRAME or any finite element program available to you, find the member forces and the change of the distance (AB) after loading. With proper boundary conditions, only 1/8 of the frame needs to be considered (dotted line).

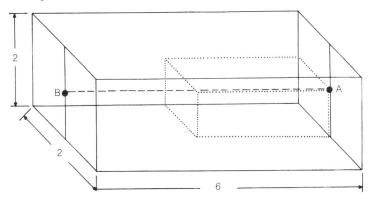

Figure P6.27

6.28. [N] A portal crane is loaded as shown. Determine the maximum load if the maximum tensile stress is not to exceed 1200 kp/cm² and the horizontal load at point A is 20% of the vertical. The section is tubular steel 200×200 mm with thickness 15 mm for the columns and beam and strip 100×20 mm for the diagonal members. Use SPACEFRAME or any finite element program available to you, with simple supports at the rails.

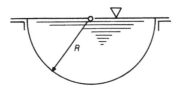

Figure P6.28

6.29. [N] A marine gear has a hub and rim connected by 4 rods of rectangular cross-section $b \times w$. At point A there are three forces, radial F, axial F, and tangential F. Assuming that hub and rim are absolutely rigid, determine the stresses in the rods using SPACEFRAME or any finite element program available to you. Data: $F_a = 1500$ N, $F_r = 2000$ N, $F_t = 3000$ N, $b = 100$ mm, $w = 200$ mm, $d = 400$ mm, $D = 1000$ mm, $E = 1.9 \times 10^{11}$ Pa, $v = 0.6$.

(Hint: Assume clamped connection to the hub (ground) and model the rim as a rectangle with beams of cross-section one order of magnitude greater than that of the rods).

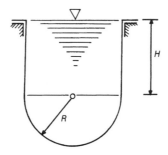

Figure P6.29

6.30. A car bumper is a beam connected to the body by way of two springs, in the simplest version. If we assume force P and a rigid car frame, the deflection at point A of the bumper due to P will be y. The energy of elastic deformation will be $Py/2$.

Suppose now that the car moves with a speed V and bumps on a solid object at point A. Since at maximum deflection, when the car stops, the elastic energy equals to kinetic energy $mV/2$ of the car, find the maximum speed V so that for a bumper that you can measure on some car, the maximum stress at point A will not exceed the yield strength (say 2200 kp/cm²).

(Hint: Analyze first with SPACEFRAME or any finite element program available to you for a force $P = 1000$ N and find the deflection and the maximum stress. Then calculate the maximum allowable P and from the elastic energy find V.)

6.31–6.34. [C] For the plane sections shown of thickness 1 cm and loading as shown, find maximum stresses. Data (wherever applicable): $M = 1000$ Nm, $R = 300$ mm, $r = 180$ mm, $a = 50$ mm, $b = B = 200$ mm, $d = 60$ mm, $D = 200$ mm, pressure $p = 100$ bar, $a = 200$ mm.

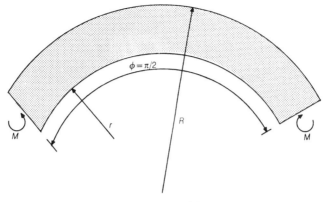

Figure P6.31

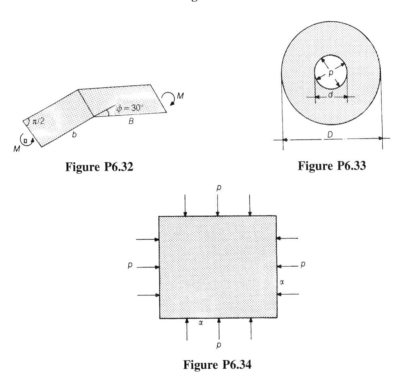

Figure P6.32

Figure P6.33

Figure P6.34

6.35. [N] A water dam with $w = 4$ m, $B = 16$ m, $H = 30$ m is made of concrete with density 2700 kg/ton. Find the stresses at the bearing surface of the dam.

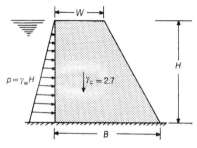

Figure P6.35

6.36. [N] An aluminum ring with $d = 60$ mm, $d = 130$ mm, thickness $b = 30$ mm is fitted into a steel bore considered as perfectly rigid. If the force of the punch is 6 tons, determine the resulting pressure between the ring and the rigid bore. The punch rings can also be considered rigid. For aluminum, $E = 0.62$ Mbar.

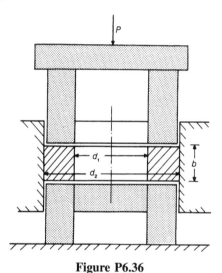

Figure P6.36

6.37. [N] In the previous problem, suppose that the steel ring of the bore is not rigid but has outer diameter $D = 180$ mm, width $B = 60$ mm, and the inner ring is symmetrically located. Find the resulting pressure.

6.38. [N] A turbine disc with $d = 300$ mm, $D = 900$ mm, $B = 100$ mm, $b = 60$ mm rotates at 3600 rpm. Use the axisymmetric FINSTRESS or any finite element program available to you, or an EXCEL spreadsheet, with node loads from inertia ("centrifugal") dividing equally on the three nodes the load of each element. Find the maximum stress and the increase of the inner diameter d. Neglect interference pressure (no restraints of the inner diameter).

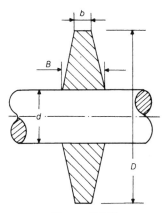

Figure P6.38

6.39. [N] The revolving head of a machine to atomize the insulation plastic which covers the lower part of large Coca-Cola bottles consists of a conical steel disk rotating at 12000 rpm. Assuming no interference between shaft and disk, determine the maximum stress, with FINSTRESS, or any finite element program available to you, or an EXCEL spreadsheet as in the previous problem. Data: $D = 300$ mm, $d = 100$ mm, $b = 10$ mm, $a = 30°$.

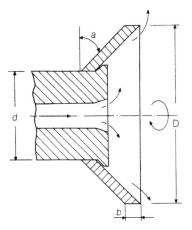

Figure P6.39

6.40. [N] The end of a high-pressure test tube is hemispherical, as shown. Using the axisymmetric FINSTRESS or any finite element program available to you, or an EXCEL spreadsheet, determine the maximum stresses. Take the straight part length equal to one diameter and boundary conditions on the left restraints of displacement along the y axis.

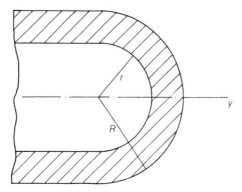

Figure P6.40

CHAPTER 7

MACHINE DESIGN SYNTHESIS

Electric generators were able to reach sizes of 1 million kW with the use of large-scale optimization.

OVERVIEW

In the foregoing chapters, we saw how we can design the machine as a system (Chapter 1), find the motion of each component (Chapter 2), find the forces that this component carries and the internal forces and moments that the different sections of the component carry (Chapter 3), select the appropriate material and manufacturing process (Chapter 4), and finally size the component by making capacity (material strength) equal to demand (working stress) (Chapter 5). In Chapter 6, we developed modeling tools to describe the operation of the machine as a system.

In the present chapter, we shall study how to use machine models to design the machine in its final form and make sure we have extracted the best possible performance out of the design selected.

7.1. MACHINE DESIGN FORMALIZATION

7.1.1. Machine Synthesis

In Chapter 6, we saw methods for the formalization of a machine by way of modeling its components. By *formalization* we mean the minimum possible detailization of the structure and operation of the machine to engineering accuracy requirements enabling design of the machine with the aid of mathematical methods. In Chapter 1, we performed a design decomposition as a means of initial conceptual formalization for a more effective conceptual design. In Chapters 2 and 3, we developed a system methodology to assign operation requirements to machine components so that they can be designed. In Chapters 4 and 5, we developed methods that can be used for the sizing of machined components for their operational requirements. All these make up phase A of the machine design process (Figure 7.1). Now all the parts need to be assembled together into a system so that the system operational requirements will be met. This is not always guaranteed, for many reasons, such as:

1. In performing the system analysis (for example, static, kinematic, dynamic analyses), we may have made some assumptions about some forms and sizes of parts that might not agree with the final ones resulting from the component design.

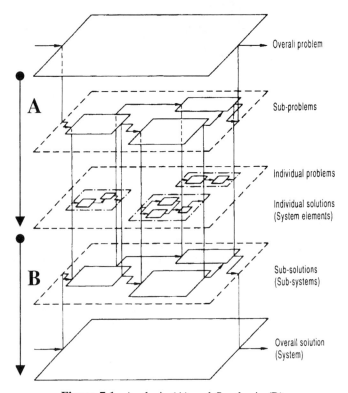

Figure 7.1. Analysis (A) and Synthesis (B).

2. The system may have to meet certain requirements that component integrity is not sufficient to achieve.

3. The solution found may not be the most economical one.

To address these concerns, a formalization of the process of combining the components into the system (phase B in Figure 7.1) and using mathematical analysis to achieve the most consistent and best possible performance with the lowest possible cost is the subject of *machine synthesis*.

7.1.2. The Machine Macromodel

In Chapter 6, we used *micromodels* in the form of differential equations describing the performance of a machine component in an infinitesimal area. In system dynamics and control, design synthesis is usually performed with a *metamodel* that operates on a certain transformed space, such as in the phase plane, to name one example.

In design synthesis, a system model, usually in the form of a system of algebraic equations, plays a cardinal role. Termed a *macromodel,* it relates the *system design parameters* to the system operational requirements by way of a system of algebraic equations. Every component is designed by way of a system of equations. For example, the stress equation *demand = capacity* relates the component design parameters, such as the component geometry, to the component operational requirements. Some components are assembled into a subsystem, a higher hierarchical level in the task decomposition scheme. Not all component design parameters are involved in the macromodel of the subsystem. Those design parameters necessary to describe the subsystem performance are termed *system* or *global design parameters,* while the rest are the *component* or *local design parameters.* It is true that every design parameter affects performance in the final analysis. It is of paramount importance, however, to keep the macromodel as simple as possible so that the designer has to make some choices in developing the macromodel and deciding which parameters are global or local.

The same distinction is made when the macromodel moves one level higher into the task decomposition scheme. Local design parameters are the ones necessary to describe the subsystem, while global design parameters are the ones necessary to describe the system performance.

We saw some macromodels already in Chapters 1, 2, and 3. For example, the equations of closure for a linkage are a kinematic macromodel. The link lengths are global design parameters for the linkage macromodel, while the detailed dimensions of each link are local design parameters for the design of the particular link. Moreover, in Chapter 1, Section 1.2.4, we developed subsystems and components (modules) that exchange matter, energy, and information. Schematics of the associated macromodels are, for example, Figures 1.16 and 1.17. For every module J that is enclosed in its boundary, there should be:

- A set of m global design parameters x_{j1}, x_{j2}, . . . , x_{jm} of the modules that affect their input–output relationships (performance)

- A set of n global design parameters of the system $y_1, y_2, \ldots, y_n$ that include the values describing the exchange of matter, energy, and information among modules
- A set of p operating requirements $o_1, o_2, \ldots, o_p$ of the system

The macromodel consists of a system of q equations and r inequalities that the system parameters should satisfy so that the system meets the operating requirements, including safety and reliability:

$$g_\eta(x_{ij}, y_k, o_l) = 0, \qquad i = 1 \text{ to } m, j = 1 \text{ to } s,$$

$$k = 1 \text{ to } n, l = 1 \text{ to } p, \eta = 1 \text{ to } q \qquad (7.1a)$$

$$h_\zeta(x_{ij}, y_k, o_l) > 0, \qquad i = 1 \text{ to } m, j = 1 \text{ to } s,$$

$$k = 1 \text{ to } n, l = 1 \text{ to } p, \zeta = 1 \text{ to } r \qquad (7.1b)$$

Equations and inequalities 7.1 are the mathematical description of the macromodel.

As we have mentioned already, the model has a weak dependence upon the local module parameters since we have identified as global parameters only the most important ones. Use of the model, therefore, needs to be iterative most of the time. The macromodel can be used for two main purposes:

1. At the *design stage,* to find the proper global design parameters and from them the local design parameters of the modules. Equations need to be solved for the determination of the unknown design parameters.
2. At the *commissioning stage.* The machine has been designed and perhaps even manufactured, and we need to evaluate the performance—to see, in other words, that the machine satisfies the operating requirements. In this case, we only need to see that all equations and inequalities are satisfied.

Example 7.1 For the design of an electric drill, it was decided during the conceptual design stage that the motor will be a 0.5 kW, 4000 rpm universal motor, the output speed at the drill will be 300 and 600 rpm, selectable, a first stage of speed reduction will be with a pair

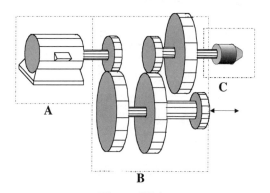

Figure E7.1a

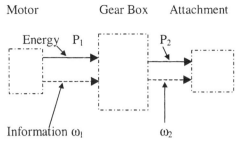

Figure E7.1b

of helical gears to 1000 rpm, and the final speed will be obtained with another stage with a double pair of spur gears with sliding pinion, as shown in Figure E7.1a. Develop a system macromodel for static and kinematic performance.

Solution   The system schematic is shown in Figure E7.1b. The boxes with interrupted line indicate boundaries of subsystems, the motor, gearbox, and tool adapter. The global design parameters are: voltage $V = 120$ V, power factor cos $\phi = 0.9$, current I, motor power $P_0 = 0.5$ kW, motor speed $\omega_o = 4000 \times 2\pi/60 = 419$ rad/s, $\omega_1 = \omega_o = 419$ rad/s, $\omega_2 = 300 \times 2\pi/60 = 31.4$ and $\omega_2 = 600 \times 2\pi/60 = 62.8$ rad/s, efficiency of the motor η_0, efficiency of the gearbox η_1, torque between motor and gearbox T_1, between gearbox and tool adapter $T_{2/300}$, $T_{2/600}$.

Local design parameters are, for the gearbox for example, the gear dimensions, the shaft and bearing dimensions, the size of the keys, etc. They will have no part in the macromodel.

The macromodel will consist of the power equations:

$$P_1 = P_0\eta_0, \; P_2 = P_1\eta_1$$

the torque equations:

$$T_1 = P_1/\omega_1, \; T_{2/300} = P_2/\omega_{2/300}, \; T_{2/600} = P_2/\omega_{2/600}$$

the electric power equation:

$$P_0 = IV \cos \phi$$

7.1.3. Kinematic Synthesis of Linkages

An interesting application of machine synthesis is in the kinematic synthesis of linkages, the process of determining the design of a linkage that will perform a kinematic task, such as:

- *Function generation.* The angles of rotation (motion, in general) of the input and output link should be in an approximate functional relationship within a certain range of the input link rotation. Examples are many linkages used in heating and cooling installations for primary and secondary air control, oil burners, and oil and air supply control.

- *Path generation.* One of the links, usually the coupler, takes prescribed positions for prescribed angles of rotation (motion, in general) of the input link. Examples are garage door openers and automobile hood linkage.
- *Motion generation.* The angle of rotation of the input link should be in an approximate functional relationship, within a certain range of the input link rotation, with the velocity of a point on the coupler or the angular velocity (motion, in general) of the output link.

On the other hand, selecting the type of linkage and selecting the number of links are the subjects of *type* and *number synthesis,* respectively, and they are part of the conceptual design discussed in Chapter 1. *Dimensional synthesis,* the determination of the geometry of the linkage to perform the desired kinematic task, is the subject of dimension synthesis, which will be discussed here.

Synthesis of linkages has been one of the earliest tasks in machine design, and a number of algebraic and graphical methods have been devised for this purpose. They are described in texts on kinematics of linkages. Computer aided synthesis of linkages is now widely used, even though graphical methods can help the designer in preliminary design and in getting an insight into the particular design.

The equation of closure of a mechanism is of the general form:

$$F(\phi, \psi, x_1, x_2, \ldots, x_n) = 0 \tag{7.2}$$

where ϕ, ψ are the input and output parameters (usually angles of rotation) and x_1, $x_2, \ldots, x_n$ are global design parameters, such as the link lengths and initial location of the links. For multiple loop linkages, more than one closure equation is necessary to describe the linkage. The desired performance of the linkage is usually described by a set of m prescribed values of the input and output ϕ_1, ψ_1, ϕ_2, ψ_2, $\ldots$, ϕ_m, ψ_m, termed *accuracy points,* and the synthesis task is to find the global design parameters $x_1, x_2, \ldots, x_n$.

In general, if $m = n$, application of the closure equation (7.2) n times will yield sufficient equations for the determination of the unknowns. Usually this is a difficult task due to the usually highly nonlinear form of the equations of closure. Thus, most of the time an approximation is sought to engineering accuracy.

For the solution of the closure equations, one of the numerical methods is usually employed, such as the Newton method. It is usually expedient, instead of trying to satisfy the equation of closure, to try to minimize the deviation of the performance of the linkage from the desired one, such as the sum of the squares of all such deviations at the accuracy points.

Linear equations are easily solved, and designers have been using linear approximations to nonlinear problems. Thus, instead of solving Equation (7.2) for the design of a four-bar linkage as a function generator, for example, one can rewrite the equation relating input–output 2.9a in the form of the *Freudenstein equation* (Erdman and Sandor 1984):

$$K_1 \cos \phi_2 + K_2 \cos \phi_4 + K3 = -\cos(\phi_2 - \phi_4) \tag{7.2a}$$

where:

$$K_1 = \frac{r_1}{r_4}$$

$$K_2 = \frac{-r_1}{r_2} \qquad (7.2b)$$

$$K_1 = \frac{r_3^2 - r_1^2 - r_2^2 - r_4^2}{2r_2 r_4}$$

r_1, r_2, r_3, r_4, the link lengths, base, crank, coupler, and follower, respectively.

One can assign three pairs of input–output angles (ϕ_2, ϕ_4) and apply Equation (7.2a). This will yield three linear algebraic equations in three unknowns K_1, K_2, K_3. Solving this system of equations, we can determine K_1, K_2, K_3, and assigning an arbitrary value to the length of the base link r_1, we can determine the link lengths from K_1, K_2, K_3:

$$r_4 = \frac{r_1}{K_1}$$

$$r_2 = \frac{-r_1}{K_2}$$

$$r_3^2 = 2r_2 r_4 K_1 + r_1^2 + r_2^2 + r_4^2$$

Example 7.2 Formalize the design of a four-bar linkage as a function generator so that the change of the output angle from the initial position $\Delta\psi$ will be related to the change of the input angle from the initial position $\Delta\phi$ as $\Delta\psi = \Delta\phi^2$ for five distinct values of $\Delta\phi = 0°$, 10°, 20°, 30°, 40°.

Solution The input and output angles are ϕ_2 and ϕ_4 and the link lengths will be labeled r_1, r_2, r_3, r_4, base, crank, coupler, and follower, respectively.

From Equations (2.9):

$$\phi_4 = \pi - \arccos \frac{r_1^2 + s^2 - r_2^2}{2r_1 s} \pm \arccos \frac{r_4^2 + s^2 - r_3^2}{2r_4 s}$$

where:

$$\gamma = \pm \arccos \frac{r_3^2 - s^2 + r_4^2}{2r_3 r_4}$$

$$s = (r_1^2 + r_2^2 - 2r_1 r_2 \cos \phi_2)^{1/2}$$

Let ϕ_{20} and ϕ_{40} be the input–output angles at the starting position. The desired input–output relationship is:

$$(\phi_4 - \phi_{40}) = (\phi_2 - \phi_{20})^2 = (\Delta\phi_2)^2$$

The sum of the squares of the deviation of the performance from the desired one is:

$$E = [(\phi_{41} - \phi_{40}) - 10^2]^2 + [(\phi_{42} - \phi_{40}) - 20^2]^2$$
$$+ [(\phi_{43} - \phi_{40}) - 30^2]^2 + [(\phi_{44} - \phi_{40}) - 40^2]^2 \qquad (a)$$

where:

$$\phi_{4i} = \left(1 - \arccos \frac{r_1^2 + s^2 - r_2^2}{2r_1 s_i} \pm \arccos \frac{r_4^2 + s^2 - r_3^2}{2r_4 s_i}\right) \times \frac{180}{\pi} \qquad (b)$$

where:

$$s_i = (r_1^2 + r_2^2 - 2r_1 r_2 \cos \phi_{2i})^{1/2}$$

$\phi_i = 10°, 20°, 30°, 40°$ for $i = 1, 2, 3, 4$, respectively

The Grashof criterion (Equation (2.4)) might be introduced, too. To this end, we rewrite the Grashof criterion (see Chapter 2 for the definition of the parameters used) in the form:

$$p + q + \ell + s - 2(\ell + s) > 0 \qquad (c)$$

The mathematical formalization of the synthesis problem is:

Find r_1, r_2, r_3, r_4, so that Equation (b) and Inequality (c) are to be always satisfied and function E in Equation (a) has the smallest possible value.

7.2. MACHINE DESIGN OPTIMIZATION

7.2.1. Introduction

In machine design, as in almost any design problem, the designer has to determine a number of design parameters, to unambiguously specify a machine or even the simplest machine element. Looking at any drawing, one can observe a number of dimensions that are not obvious for the designer and have to be selected in a rational way. In addition, materials, surface finish, heat treatment, and other manufacturing processes must be specified. On the other hand, the machine or component has to meet certain operational requirements so that it will perform its duty satisfactorily without failing. Finally, the part must be appealing to the end user, having low price or other features the user wants.

Take, for example, the case of the design of a simple rod to resist an extensional axial force F. A number of rectangular shapes of dimensions $a \times b$ are available, and a number of materials, each having a certain strength S and price c_m per unit mass. The surface finishing of the rod has also a cost per unit of surface c. The cost of the rod, if L is the desired length, is:

$$C = Lab\rho c_m + 2L(a + b)c_s \qquad (7.3)$$

where ρ is the material density.

The selection of parameters a, b, material (cost, density) is not altogether arbitrary, as the rod must sustain an axial load F without failing. This means that the following design equation must be satisfied:

$$S \geq \frac{F}{ab} \tag{7.4}$$

Common engineering sense dictates that the equal sign in the design equation will yield the minimum rod section and the minimum cost. Therefore, there is a relation between material, a, and b, which can, in this case, be introduced in cost equation (7.3):

$$b = \frac{F}{aS}, \quad C = \frac{LF\rho c_m}{S} + 2L \left(a + \frac{F}{aS} \right) C_s \tag{7.5}$$

The cost now depends on many parameters, which can be grouped in four categories:

1. Functional requirements (F)
2. Material properties (S, ρ, c_m)
3. Geometry (a, L)
4. Manufacturing processes (c_s)

It must be noted that some of the parameters might be specified beforehand, such as length and surface finish, due to additional functional requirements or availability that cannot be expressed mathematically, while other parameters are related, such as density, strength, and material cost. It seems, then, that the material and the rod width a are free for the designers to select. However, they want to minimize the cost. To this end, they plot the cost for the available materials I, II, III as a function of the width a, as in Figure 7.2. It seems that the lower cost is for material I and rod width a_0 and the resulting minimum cost is C_0.

It turns out that although in the beginning the solution of this design problem had some apparent ambiguity, this was cleared up by the process of obtaining minimum cost.

We can draw some more general conclusions for the analysis of this simple design problem.

In general, a design problem might have a number of design parameters much greater than the number of available design equations. Therefore, the design problems seem to have many solutions. In fact, if n is the number of design parameters and p is the number of design equations $(n > p)$, for every arbitrary choice of

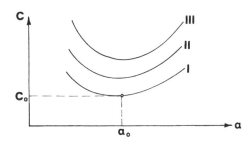

Figure 7.2. Cost of the rod for three different materials, I, II, III.

$n - p$ design parameters, the remaining p design parameters can be determined from the p design equations. Therefore, the design problem has at least a $(n - p)$ple infinity of solutions, not to account for multiple solutions of the design equations.

The designer, especially the student, in a situation like this has the feeling that his design is arbitrary, to the extent that his choice of the $n - p$ design parameters is arbitrary. All these solutions are formally "correct." Not all, however, are equally successful. This is apparent in the fact that although all the designers have at their disposal, in general, the same equations, the same manufacturing facilities, the same materials, and so on, they end up with designs that do not all have the same commercial success. The experience, ingenuity, and creativity of the designer lead to the better selection of the arbitrary design parameters. Now, what is a "better" design? What feature distinguishes a "good" from a "bad" design? There is no simple answer, beyond that it depends on the design. For example, for steel construction, the cost. For a ship, the operation cost. For a spacecraft component, the weight. For a function generator mechanism, the maximum error. Furthermore, in many cases the quality of the design depends on more than one parameter, sometimes difficult to combine quantitatively into one.

In most cases, the designer can identify a single criterion or a combination of criteria in the form of a function of the design parameters that better satisfies his objectives as it becomes smaller. Cost is a typical example. This function is called *objective function*. For every arbitrary combination of the design parameters, this function yields a single value, which the designer has to minimize. To the aid of the designer come modern methods, computer aided, that can enable an intelligent selection of the design parameters to yield a better or even the best possible design, called the *optimum design*. The corresponding procedure is called *optimization*.

A machine design class was asked once to design a beam from a list of materials with specific loading and length and one design equation, the strength of the beam. The outcome was interesting: beam cost ranging from \$320 to \$1500 among the several student solutions.

Sometimes there are no restrictions on the n design parameters and the problem can be stated as follows:

Minimize the objective function $f(x)$, where $x = \{x_1 \quad x_2 \quad \ldots \quad x_n\}$ and x_1, x_2, $\ldots$, x_n are yet unknown design parameters.

This is known as the *unconstrained optimization problem*. Such a problem is not usual in machine design. Its discussion, however, is important because more complex problems can be transformed to an unconstrained optimization problem.

In most cases, the design parameters are not independent:

1. There are restrictions on the range of the design parameters. For example, dimensions cannot be negative and some of them have upper limits due to space limitations.
2. There are equations or inequalities that the design parameters must satisfy. They are called, respectively, equality and inequality constraints. Equality constraints are, for example, speed, capacity, transmitted force, torque, and

moment of a machine or member. Inequality constraints are, for example, total weight limitations, limiting stresses and deflections, and space limitations.

The associated problem is called the *constrained optimization problem*. Many times, if there are only p equality constraints, it is possible to use them in order to reduce the number of design parameters by a number of p, by elimination. Then the problem is reduced to an unconstrained one with $(n - p)$ unknown design parameters, called *decision variables*. In fact, this was done in the example of the tension rod when the dimension b was eliminated by way of the application of an equality constraint, the limiting stress.

7.2.2. The General Optimization Problem Statement

In most design applications, the optimization problem can be stated as follows: Find an appropriate x that minimizes $f(x)$ (objective function)

Subject to

$$g_i(x) > 0, i = 1, 2, \ldots, m \tag{7.6}$$

$$h_j(x) = 0, j = 1, 2, \ldots, p \tag{7.7}$$

where:

$$x = \{x_1\ x_2 \ldots x_n\} \tag{7.8}$$

Functions g and h are termed, respectively, the *equality* and *inequality constraints*.

When functions f, g, h are linear, we call the problem linear programming. The definition of the nonlinear programming problem will follow logically.

Parameters $x_1, x_2, \ldots, x_n$ are the design parameters of the system and, in general, their range is restricted as follows:

$$x_{j\min} < x_j < x_{j\max}, j = 1, 2, \ldots, n \tag{7.9}$$

In general, there are two methods to attack the optimization problem:

1. In cases where we possess simple mathematical expressions for the objective function and the constraints, it is sometimes possible to obtain explicit expressions for the minima, even the global minimum.
2. In cases of lack of mathematical expressions or very complicated equations, one has to resort to numerical methods.

One special feature of the machine design optimization is that usually there are many objectives, which have to be combined in one objective function.

In many engineering applications, there is a common denominator, the cost. For example, capital cost and operating cost of a machine can be easily combined into one figure. There is no general rule for the selection of the objective function. It just depends on the problem. We shall return to the question of the selection of the objective function later on.

Example 7.3   An engineer wants to design a rectangular container with the following specifications:

$$\text{Contained volume} = V_0 = 1 \text{ m}^3, \text{ maximum base area} = A_0 = 1 \text{ m}^2$$

The design objective is to find the container dimensions for minimum cost of the container's material (thickness given).

Solution   The design parameters will be:

$$x_1 = \text{length } L, \ x_2 = \text{width } B, \ x_3 = \text{height } H$$

The objective function is the container surface, which must be minimized:

$$F = 2(LB + LH + BH) \tag{a}$$

The equality constraints (volume $= V_0$) are:

$$g = LBH - V_0 = 0 \tag{b}$$

The inequality constraints is on the base area:

$$h = A_0 - LB \geq = 0 \tag{c}$$

In order to meet the design objective, we must find a set of values L, B, H to give the minimum function f (total area) and satisfy the constraints of the problem. We must take into account some physical limitations on the design parameters. For example:

$$0 < L, B, H < M \tag{d}$$

In order to visualize the optimization process, we substitute the value of H from Equation (*b*) into Equation (*a*):

$$F = 2 \left[LB + \frac{(L + B)V_0}{LB} \right] \tag{e}$$

We eliminate one parameter using one equality constraint. This suggests that in general, we can eliminate some of the design parameters using an equal number of equality constraints. This is not always possible, or even wise, as we shall see later on.

The remaining parameters, the decision variables L and B, are now to be subject to optimization under the remaining inequality constraint.

Now we plot the function f in an $(L - B)$ Cartesian plane (Figure E.7.3a) in the form $f = C(f)$ for the given volume $V_0 = 1$.

For several values C we will have different contours. We note that from the $L - B$ plane, we can select only values (L, B) that are below hyperbola of Equation (*d*):

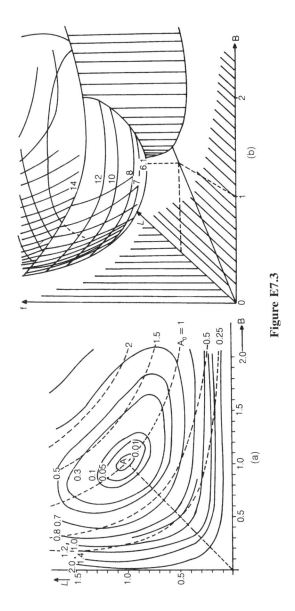

Figure E7.3

$$LB = A_0 \qquad (f)$$

These hyperbolas, for several values of A_0, have been plotted with dotted lines. We see that if we disregard the inequality constraint, there is a minimum of the function at point A (1, 1) and the minimum of the function is 6. Therefore, the container will have the shape of cube as one should expect from well known geometric rules. This type of minimum is called an unconstrained minimum. If we add constraint (d), we note that for values $A_0 \geq 1$ the minimum is valid and the inequality constraint is ineffective. However, for $A_0 < 1$ the unconstrained minimum is not acceptable anymore because it violates the inequality constraint.

The new minimum will be found as the point of contact of a contour line and a hyperbola. For example, for $A_0 = 0.25$ the minimum is at point B (0.5, 0.5) and the value of the function is about 9. This minimum is called *constrained minimum*. In this case, the inequality constraint appears as an equality.

Because of the small number of parameters involved and the form of the objective function, we have only one minimum in the permissible range of variation of the decision variables. In other cases, however, we might have several minima. We shall term them *local minima* and the *absolute minimum* will be called *global minimum*.

7.2.3. Analytical Methods

As stated above, the general optimization problem consists of the following:

Minimize $f(x)$, $x = \{x_1, x_2, \ldots, x_n\}$, subject to equality and inequality constraints.

Well-known methods for maxima and minima of functions of one variable can be used.

Case 1: Unconstrained Minimum of a Function

1. The conditions for relative minimum of a function $f(x)$ in n variables $x_1, x_2, \ldots, x_n$ are:

$$\frac{\partial f}{\partial x_1} = 0, \, i = 1, 2, \ldots, n \qquad (7.10)$$

or

$$\nabla f(x) = 0 \qquad (7.11)$$

2. For minimum, some additional conditions must be met because some of the solutions of Equation (7.5) might yield maxima or saddle points. There are mathematical criteria to distinguish the minima, but in most computer solutions it is more convenient just to compare the solutions for the relative minimum.

Conditions (7.10) provide us with n equations in n unknowns. Therefore, we can in principle solve this system and determine the design variables. This system does not necessarily have a unique solution. Every solution of the problem is a local extremum. As pointed out above, comparison will isolate the lowest minimum and no further consideration should be given to the other solutions. There is no general method to determine the lowest minimum, the global minimum.

Case 2: Equality Constraints. We want the minimum of a function with the equality constraints:

$$g_i(x) = 0, \ i = 1, 2, \ldots, m \tag{7.12}$$

We form the Lagrangian function:

$$P = f + \sum_{i=1}^{m} \lambda_i g_i(x) \tag{7.13}$$

In optimization theory, it is proven that extremum of this function corresponds to a minimum of the objective function subject to the equality constraints. In fact, at a minimum of the objective function, the Lagrangian function has a saddle point. This can be seen in Figure 7.3 for a function $f(x)$ of one variable subject to the constraint $g(x) = 0$.

Function $y = f(x)$ meets the x-axis at A and the minimum of $f(x)$ subject to $g(x) = 0$ is simply point B with a dashed line. The Lagrangian function $P = f(x) + \lambda g(x)$ is plotted. This function apparently has a saddle point at B.

The constants λ are called *Lagrange multipliers*. The necessary conditions for minima are:

$$\frac{\partial P(x, \lambda)}{\partial x_1} = 0, \ i = 1, 2, \ldots, n \tag{7.14}$$

We possess $m + n$ Equations (7.12) and (7.14) with $m + n$ unknowns $x_1, x_2, \ldots, x_n, \lambda_1, \lambda_2, \ldots, \lambda_m$. Therefore, the problem is formally solved.

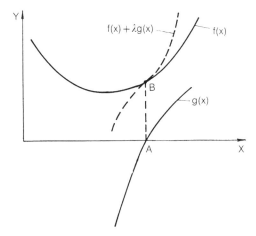

Figure 7.3. Geometric interpretation of a Lagrange multiplier.

Case 3: Inequality Constraints. This is the case when we want the minimum of a function subject to the following inequality constraints:

$$h_j(x) \geq 0, j = 1, 2, \ldots, p \tag{7.15}$$

We introduce at this point p new variables, called slack variables, to be determined later, $u = \{u_1, u_2, \ldots, u_p\}$ and we transform the inequalities to the following equalities:

$$h_j(x) - u_j = 0, j = 1, 2, \ldots, p \tag{7.16}$$

We have here a problem of equality constraints as before and p additional variables u. Then it is a case 2 and can be solved with the method of Lagrange multipliers. Therefore, the objective function will be a part of a Lagrangian function.

$$P(x, \lambda, u) = f(x) + \sum_{i=1}^{m} \lambda_j g_j(x) + \sum_{j=m+1}^{m+p} \lambda_j(h_j - u_j) \tag{7.17}$$

This is an unconstrained optimization problem in n x's, $p + m$ λ's, and p u's.

An important observation must be made at this point if we recall some results observed in Example 7.3. There it was shown that the inequality constraint was not always active. In other words, the minimum can be such that the inequality constraint is satisfied and therefore does not influence the value of the minimum. In other circumstances, the minimum violates the constraint, and therefore the acceptable minimum will be at the inequality limits, a hyperbola in the example, which corresponds to the inequality becoming equality. This corresponds, in the first case, in $\lambda = 0$ and u non-zero, and in the second case, in $u = 0$ and multiplier λ non-zero.

In analytical solutions, one has to try all possible combinations to be sure of the true minimum. In large problems, this is very difficult and we resort to numerical methods that automatically find the true minimum.

7.2.4. Numerical Methods

There are numerous difficulties in applying analytical methods for design optimization, except for cases when the design equations are fairly simple. The equality constraints are mostly nonlinear, and elimination of design parameters sometimes is impractical. Many times solutions are obtained with numerical methods and the constraints do not have analytical expressions or, even more, are not continuously differentiable. Therefore, one has to resort to numerical methods. Moreover, even if analytical formulation of the problem is feasible, solution of the system of equations for the determination of the design parameters, usually highly nonlinear, has to be done with numerical methods.

Most numerical methods start with the definition of a penalty function, which includes the objective function and the constraints and is subject to unconstrained optimization. It comes from the objective function $f(x)$ with proper development to include also equality and inequality constraints, as was discussed in Section 7.2.3. This function will be used in the form

$$P(x) = f(x) + K \sum_{i=1}^{m} [g_i(x)]^2 + L \sum_{i=1}^{p} \langle h_i(x) \rangle^2 \qquad (7.18)$$

where $\langle h(x) \rangle = h(x) < 0$ and zero otherwise.

The penalty function multipliers K and L (not to be confused with the Lagrange multipliers, although they are in some ways similar) have to be selected in a way that will ensure the proper contribution of the penalty terms and the objective function in the penalty function.

The optimization procedure starts with small values of K and L and is repeated from the last optimum point with progressively higher values of K and L.

Hard constraints, that is, constraints that must be kept absolutely, require large multipliers. "Soft" constraints, that is, constraints that accept small violation, need only small multipliers. In general, the multiplier is a weighting function expressing the relative importance of each penalty term in the penalty function. The squaring of the constraint functions is for smooth boundaries that, as will be apparent later, help for numerical convenience.

A geometric interpretation of the penalty function can be obtained again for one variable x and objective function $f(x)$ (Figure 7.4). There is an equality constraint $g(x) = 0$ that yields a constrained minimum at point B. Then the penalty function is plotted for three different values of parameter K. We can observe that, indeed, for $K = K_3$ and K_2, the miminum of the penalty function is at B. There are values of K, however, that can shift the minimum, to C in this case, $K = K_1$.

A similar geometric interpretation of the inequality constraints is shown in Figure 7.5. An inequality constraint $h(x) > 0$ is imposed. If the limit point A is on the right of the unconstrained minimum, which is at D (Figure 7.5a), the penalty function is plotted for three different values of L. It is observed that for $L = L_1$ or L_2, point B is, indeed, the constrained minimum. For a smaller L, however, and penalty function L_3, the minimum is shifted towards C at point B'.

If limit point A is on the left (Figure 5.4b) of the unconstrained minimum at D, then the minimum is still at C regardless of the value of L and the inequality constraint is inactive.

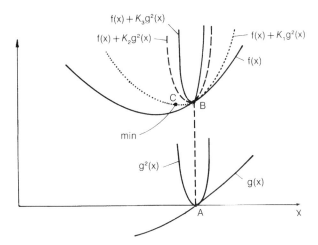

Figure 7.4. Penalty function, equality constraint.

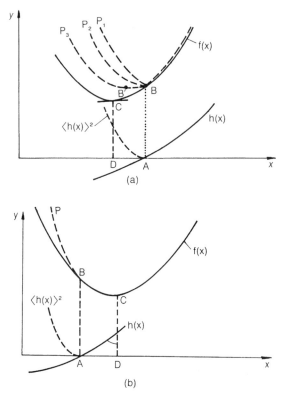

Figure 7.5. Penalty function, inequality constraint: (*a*) limit point A on right situation; (*b*) limit point on left situation.

Several versions of the method, some having advantages over others, exist in the literature. This method looks more complicated than elimination of parameters by way of the equality and, perhaps, inequality constraints. It has certain advantages, however:

1. Many times the equations have a form that does not allow for elimination.
2. Using inequality constraints as equalities for parameter elimination does not always lead to an optimum, because, as shown in Example 7.3, this is not always the case. One then has to obtain two solutions: one without the inequality constraint and the other using it as equality. This procedure is often tedious. On the other hand, the method of penalty function takes care of the inequalities automatically. For larger problems, the process of elimination is out of the question anyway.
3. An analytical method with the Lagrange multipliers is possible in principle but requires the solution of a large number of nonlinear algebraic equations, usually a tedious task.

7.2.5. Search Algorithms

7.2.5.1. First-Order Methods: Steepest Descent. In an optimization problem, we might look for either maxima or minima, depending on the nature of the problem. We do not lose generality if we assume from this point that we look for minima because, when a function $f(x)$ has a maximum, the function $-f(x)$ has a minimum.

Augustin Cauchy (1789–1857) was born to a royalist law-yer's family. When he was four years old, the family took refuge (this being the time of the French Revolution) in Ar-cueil, a village near Paris where Laplace lived at that time. In Laplace's house, a meeting place for royalists and scientists, he met Lagrange, who was impressed with young Cauchy's grasp of mathematics. In 1805, he entered the Ecole Poly-technique, and in 1807, he entered the École des Ponts et Chaussées, a famous engineering school. He practiced engi-neering while working on mathematics in his spare time. In 1816, he became a member of the Paris Academy, and sub-sequently he taught at the Sorbonne and the École Polytech-nique. He made fundamental contributions to mathematics and the theory of elasticity, where he introduced the idea of principal stresses.

Most of the methods for seeking minima with a step-by-step algorithm are really extensions of the steepest descent method, introduced by Cauchy in 1847. (The reader must be careful when using the many methods because there is a tendency to underestimate their limitations and to generalize their range of application.)

In order to visualize the search strategy, we start with a two-dimensional, un-constrained model (Figure 7.6). The penalty function (objective in this case) is $f(X_1, X_2)$, subject to optimization. Suppose that we have a guess at the decision variables $A(X_1, X_2)$. We want to change the values of X_1 and X_2 in order to obtain a better (smaller) value of the function $f(X_1, X_2)$. As a first approximation, we assume that the objective function is linear with respect to X_1 and X_2 about point A. In other words, we substitute surface $f(X_1, X_2) = C$ with a plane aAa tangent to it at point A. Then, for small variations ΔX_1 and ΔX_2, the value of the function will be, for a sufficiently small step, at point B:

$$f(X_1 + \Delta X_1, X_2 + \Delta X_2) \cong f(X_1, X_2) + \frac{\partial f}{\partial X_1} \Delta X_1 + \frac{\partial f}{\partial X_2} \Delta X_2 \qquad (7.19)$$

The length (AB) of the step will be:

$$\delta_1^2 = \Delta S^2 = \Delta X_1^2 + \Delta X_2^2 \qquad (7.20)$$

The change of the function BB', approximated by the tangent plane, will (from Equation (7.19)) be:

$$\Delta f = \frac{\partial f}{\partial X_1} \Delta X_1 \pm \frac{\partial f}{X_2} \Delta X_2 \qquad (7.21)$$

In order that this change be a maximum,

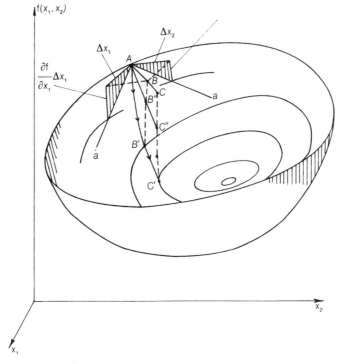

Figure 7.6. Steepest descent on a surface.

$$\frac{\partial}{\partial X_1} (\Delta f) = \frac{\partial}{\partial x_2} (\Delta F) = 0 \tag{7.22}$$

which yields, with Equation (7.20),

$$\frac{\Delta X_1}{\Delta X_2} = \pm \frac{\partial f / \partial X_1}{\partial f / \partial X_2} \tag{7.23}$$

Equations (7.20) and (7.23) give the lengths ΔX_1 and Δ_2. The new point $B'(X_1 + \Delta X_1, X_2 + \Delta X_2)$ can be written in an operational form:

$$X_B = X_A - \delta \Delta f_A(X) \tag{7.24}$$

where δ is the step length ($= \Delta S = AB$).

The derivatives $\partial f / \partial X_1$ and $\partial f / \partial X_2$ can be calculated explicitly or numerically, depending on the form of the objective function. We note that the vector $\Delta \mathbf{f}$ is the one that is perpendicular to the isocontour line that passes through point A (Figure 7.6).

We note that further steps along the same direction give smaller values of the objective function, up to some point C, which means that the step length was not the proper one. Therefore, Equation (7.24) has to be modified as follows:

$$\{X_C\} = \{X_A\} - \lambda_C\{\nabla f_A(X)\} \tag{7.25}$$

We can find point C and the value of $\lambda_C = AC$ by several methods. For example, we can proceed with successive steps until we observe an increase of the objective function. Then we can apply locally some numerical method to locate C. In Equations (7.24) and (7.25), from Equations (7.20) and (7.23),

$$\{\Delta f_A(X)\} = \left\{\frac{\partial f}{\partial X_1}\ \frac{\partial f}{\partial X_2}\right\}\left[\left(\frac{\partial f}{\partial X_1}\right)^2 + \left(\frac{\partial f}{\partial X_2}\right)^2\right]^{-1/2} \tag{7.26}$$

We can generalize the method for n decision variables $X_1, X_2, \ldots, X_n$ applying the same formula:

$$\{\Delta f_A(X)\} = \left\{\frac{\partial f}{\partial X_1}\ \frac{\partial f}{\partial X_2} \cdots \frac{\partial f}{\partial X_n}\right\}\left[\left(\frac{\partial f}{\partial X_n}\right)^2 + \left(\frac{\partial f}{\partial X_2}\right)^2 + \cdots + \left(\frac{\partial f}{\partial X_n}\right)^2\right]^{-1/2} \tag{7.27}$$

This method requires the computation of the function and its first derivative. It belongs to the class of first-order methods. In computer applications, however, the derivative is computed numerically using small increments of the design variables and computing the change of the function.

7.2.5.2. Zero-Order Methods: The Monte Carlo Method.

The previous method and similar ones converge very fast in some cases, but they are very sensitive to discontinuities and singularities, which very often exist in machine design. For example, in applications related to steam properties, problems are encountered when crossing the saturation line, which appears as a steep ridge.

A very simple and efficient method is the Monte Carlo method. As the name suggests, it is based on random selection of directions. In the two-dimensional cases, for example, we start from point O in Figure 7.7. We select a number of arbitrary directions, say three, and calculate the value of the function at points A, B, and C on a circle with center O and a radius r, the length of the step. To pick up an arbitrary direction, we take two random numbers r_1 and r_2 and form the vector $s(X_{01} + r_1, X_{02} + r_2)$ of length r where X_{01} and X_{02} are the coordinates of the point O.

We then compare the values of the objective function at points A, B, and C. Suppose that the value of the function at A is the smallest among the three points. Then we proceed along the direction of point A until we find point D, where the random search is repeated. We have to test at least two directions at every point. We can now generalize the method as follows.

We start from point O_0 $(X_1, X_2, \ldots, X_n)$ (Figure 7.7) and construct random directions:

$$\{S_j^*\} = \{r_{j1}, r_{j2}, \ldots, r_{jn}\} \tag{7.28}$$

where $j = 1, 2, \ldots, m \geq$ and r_{ji}, $i = 1, 2, \ldots, n$ are $m \times n$ arbitrary or random

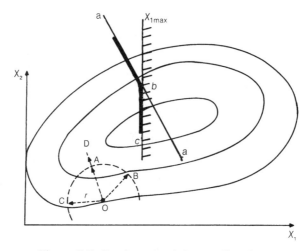

Figure 7.7. Random search in two directions.

numbers selected with a random-number generator. We normalize vector s to a desired length as follows:

$$\{S_j\} = [\{S^*_j\}[r^2_{j1} + r^2_{j2} + \cdots + r^2_{jn}]^{-1/2}\delta \tag{7.29}$$

Then we calculate the function at several points:

$$\{X_j\} = \{X_0\} + \{S_j\} \tag{7.30}$$

We then move along the direction s that gives the smallest value of the function to determine the value of λ_m.

This method gives very good results in the cases where the function $f(X)$ can be calculated without much effort.

For a two-dimensional problem, the two methods are shown in Figure 7.8.

7.2.5.3. Optimization in One Direction. In any numerical search for the minimum of the objective function, one always encounters the problem of optimization along the optimum direction. The lowest point along this direction has to be found

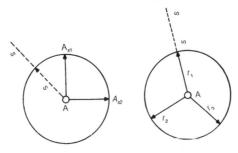

Figure 7.8. Steepest descent and random search strategies.

in order to start the new search. This seems easy, since the minimum in one variable can be found with a differentiation. This is not usually the case, however, because the directional derivative of the objective function is usually not available. Only values of the objective function can usually be computed. In this case, one proceeds with a preselected step until the objective function starts increasing again, for example, at the fourth step (Figure 7.9). Two strategies will be described here for the location of point C, the minimum along the direction s.

7.2.5.3.1. Lagrangian Interpolation. As the function is evaluated at successive steps in the s direction, it is checked to see when it starts increasing, thus having passed the minimum. The last three points are used in a three-point Lagrangian interpolation function (F):

$$F = \frac{(x - x^{j+1})(x - x^{j+2})}{(x^j - x^{j+1})(x^j - x^{j+2})} f_j + \frac{(x - x^j)(x - x^{j+2})}{(x^{j+1} - x^j)(x^{j+1} - x^{j+2})} f_{j+1}$$

$$+ \frac{(x - x^{j+1})(x - x^{j+2})}{(x^j - x^{j+1})(x^j - x^{j+2})} f_j + \frac{(x - x^j)(x - x^{j+1})}{(x^{j+2} - x^j)(x^{j+2} - x^{j+1})} f_{j+2}$$

(7.31)

where x is the coordinate along the direction s, f_j is the value of the objective function at point x_j, and $j = 1, 2, 3$, corresponding to the last three steps of the unidirectional search.

This is a second-order polynomial with respect to x. Upon differentiation, it yields the local minimum of this function. The new point is used with the nearest two points for further iteration with the same procedure.

Higher-order interpolation has also been used. The appropriate Lagrange interpolation function must be found, and a local minimum is obtained with one of the unidirectional optimization methods discussed below, such as the golden section method.

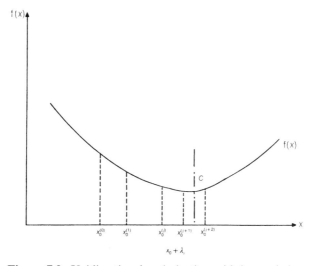

Figure 7.9. Unidirectional optimization with interpolation.

7.2.5.3.2. The Golden Section Method (Figure 7.10a). From the last three points, when the functions start increasing, the extreme two, *A* and *B*, are kept. Two new points, *C* and *D*, are calculated at distances $a(2s)$ from the two ends, where a is the golden section ratio 0.382 or some convenient fraction, usually around 0.3. The lower of the two points, say *C*, is found. The procedure is repeated between points *A* and *D*. This procedure converges rapidly to the required minimum. The Golden Section principle, known from ancient times, was demonstrated by Leonardo da Vinci in his studies of human proportions (Figure 7.10*a*). The method stops when a sufficient accuracy has been achieved or the number of steps exceed a predetermined number.

7.2.5.4. Constraints on the Design Variables. The constraints $X_{1j} < X_j < X_{uj}$ on the design variable can be simply observed during advance along a unidirectional search by setting the value of *X* to the limit that is violated. For example, consider a two-dimensional objective function with a search along line $a - a$. At point *b*, X_1 becomes greater than its higher limit $X_{1\max}$ (Figure 7.11). While the

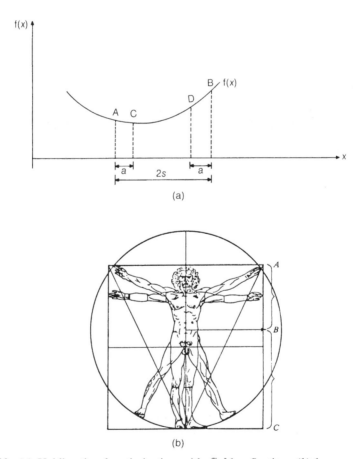

Figure 7.10. (*a*) Unidirectional optimization with Golden Section; (*b*) human proportions and Golden Section (after Leonardo da Vinci). BA/AC = the Golden Section ratio.

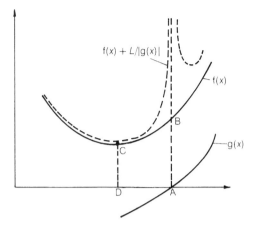

Figure 7.11. Interior point algorithms.

values of λ change normally, the values of $X_{1\max}$ are kept equal to $X_{1\max}$ until the power point c is detected. A new search must be conducted for an optimum direction at this point.

7.2.5.5. Interior Point Algorithms. The form of the penalty function of Equation (7.18) belongs to a broad category of algorithms called *exterior point algorithms* because the search can start from any initial point, even if the inequality constraints are violated. The penalty terms will force the search gradually to approach the feasible region, that is, the region where the inequality constraints are not violated.

If we have enough information for the behavior of our system to select an initial guess in the feasible region, there are penalty functions that will keep the search within the feasible region imposing very high values of the penalty terms when the boundaries of the region are approached. For example, functions of the form

$$f(x) = \ln[g(x)] \text{ or } \frac{1}{g(x)} \tag{7.32}$$

have very high values when $h(x)$ approaches 0. The associated algorithms are known as interior point ones. The form of the penalty function is indicated in Figure 7.11.

Some results of design optimization are shown in Figures 7.12–7.14. Further applications will be encountered in later chapters.

7.2.6. Design for Minimum Cost

In any society, especially in a free-enterprise economy, the engineer designs under cost control. For most machines and industrial products, the objective function is the profit, which must be maximized. Almost invariably, this is synonymous with minimizing the cost. Constraints are then imposed on certain design features such as stresses and deflections as a part of applicable design rules. Furthermore, addi-

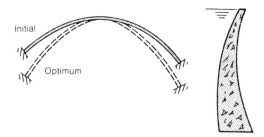

Figure 7.12. Optimum design of an arch and a dam (from Gottfried and Weisman 1973).

tional constraints can be imposed on such considerations as environmental impact and safety.

The cost of an industrial product consists of several components, which can be differentiated into two fundamental groups: *fixed costs* and *variable costs*.

Fixed costs are related not to the quantity of products manufactured, but to the cost of having the manufacturing facility in operation. Such costs are:

1. *Capital costs:* The manufacturing facility requires an investment that should be repaid within a certain schedule because the invested capital has a financial cost and the facility has a finite life during which the cost must be repaid (depreciation).
2. *Labor costs:* In some cases, the labor costs are independent of the production and have to be accounted as fixed costs, such as supervision.
3. *Maintenance costs:* These include materials and labor to maintain the facility.
4. *Fixed operating costs:* These include consumables, energy, taxes, etc., not related to the production volume.

Variable costs are directly related to the amount of products manufactured. Such costs are:

1. Materials needed for the product itself.
2. Materials for the facility operation that are related to the production volume.
3. Labor costs directly related to the production volume.

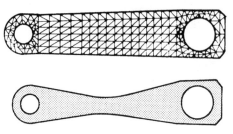

Figure 7.13. Optimum automotive arm design using the finite-element method (from Gottfried and Weisman 1973).

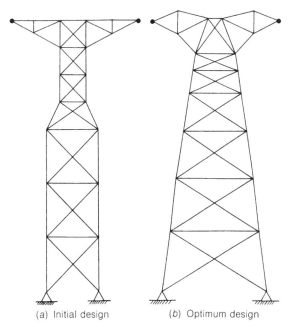

(a) Initial design   (b) Optimum design

Figure 7.14. Optimum truss design for overhead transmission lines (from Gottfried and Weisman 1973).

Usually the design engineer does not have to go through the optimization of the total production process when he designs an improved part or a new machine in the production process. With the planned production of the facility, *unit costs* are available for the estimation of the cost of machine or component, which include fixed and variable costs. Unit costs include material and machining costs:

1. *Material cost:* This is computed on the basis of the available size of the stock from which the final dimensions will be made and not the dimensions of the finished product. This cost varies between location and time. In a particular locality, at a certain time, some representative costs (in \$/kg) are:

Low-carbon steel   = 0.30
Alloy steel   = 0.5–2.0
Cast iron (product) = 0.20
Aluminum   = 2.0
Copper   = 3.0
Copper alloys   = 3.0–6.0

2. *Machining cost:* For some operations, this depends on the amount of machined volume (e.g., drilling), for others it depends on the amount of machined surface (e.g., grinding), while for others it depends on the length of machining (e.g., keyways). Some are on a preoperation basis (e.g., quenching). Some representative machining costs based on time and locality are:

Drilling ($/cm³)   = 0.1
Machining ($/cm²)   = 0.02
Grinding ($/cm²)   = 0.01
Welding, hand ($/cm/pass)   = 0.03
Welding, machine ($/cm/pass)   = 0.01
Cleaning, debarring ($/cm)   = 0.005
Painting ($/cm²)   = 0.005
Gear cutting ($/cm³)   = 0.20

Therefore, the cost function will have the form

$$C = c_m \rho V + \sum (\text{process cost} \times \text{process quantity}) \qquad (7.33)$$

where V is the volume of the material stock, c_m is the unit cost of the material per unit of weight, and ρ is its density.

The cost function will be supplemented by the constraints to yield the penalty function for optimization.

Example 7.4 Solve the problem of Example 7.3 using EXCEL with data: Volume $V_0 = 10$ m³, minimum base area $A_0 = 4$ m². Use the EXCEL module Solver.

Solution The design variables are H, B, L. The objective function is:

$$f(H, B, L) = 2(HB + BL + LH) \qquad (a)$$

The equality constraint is:

$$g(H, B, L) = HBL - V_0 = 0 \qquad (b)$$

And the inequality constraint is:

$$h(B, L) = A_0 - BL > 0 \qquad (c)$$

The objective function will be:

$$P(H, B, L) = 2(HB + BL + LH) \qquad (d)$$

Here we have a constrained optimization problem in three variables, H, B, and L. The module Solver of EXCEL solves this problem. In the spreadsheet, we use cells B8–B11 for the unknowns Height, Length, Width, and we give them initial values, say Height = 1, Length = 1, Width = 1. We put function g in cell G7, function h in cell G9, and function P in cell G12. We select **Tools, Solver.** We set Solver to minimize cell G12 by changing cells B8:B11. The results are shown in Table E7.4.

TABLE E7.4

Example 7.4: Optimum design of a water container							
DATA:							
Vol0=	10	m2					
Area0	4	m2					
				Constraints:			
Results:				Equality:	Volume-Vol0	2.11E-07	
Height =	2.5						
Length =	1.999902			Inequality:	Area0-Area	9.53E-09	
Width =	2.000098						
				Objective Function:		28	
Cell G7:	=Height*Width*Length-Vol0						
Cell G9:	=Area0-Width*Length						
Cell G12:	= 2*(Height*Width+Width*Length+Length*Height)						

Example 7.5 Formulate the problem of Example 7.3 by way of the penalty function with data: Volume $V_0 = 10$ m^3, minimum base area $A_0 = 4$ m^2. Use the EXCEL module Solver.

Solution The design variables are H, B, L. The objective function is:

$$f(H, B, L) = 2(HB + BL + LH) \tag{a}$$

The equality constraint is:

$$g(H, B, L) = HBL - V_0 = 0 \tag{b}$$

And the inequality constraint is:

$$h(B, L) = A_0 - BL > 0 \tag{c}$$

Therefore, the penalty function will be:

$$P(H, B, L) = 2(HB + BL + LH) + K(HBL - V_0)^2 + M \langle A_0 - BL \rangle^2 \tag{d}$$

Here we have an unconstrained optimization problem in three variables, H, B, and L. The module Solver of EXCEL solves just this problem. Let us use it now as a tool only to find the minimum of the penalty function P.

In the spreadsheet w, use cells $B8$–$B11$ for the unknowns Height, Length, Width, and we give them initial values, say Height = 1, Length = 1, Width = 1. We define the penalty function in cell $G14$. The function $\langle h(x) \rangle$ for the inequality constraints can be easily programmed in EXCEL as MIN(0, $h(x)$). We select **Tools, Solver.** We set Solver to minimize cell $G14$ by changing cells $B8$:$B11$. The results are shown in Table E7.5.

TABLE E7.5

Example 7.5: Optimum design of a water container						
Penalty function approach						
DATA:				**Weight Factors:**		
Vol0=	10	m2		Kfactor=	100	
Area0	4	m2		Mfactor=	100	
				Constraints:		
Results:				Equality:	Volume-Vol0	−0.00995
Height=	2.495886					
Length=	2.000651			Inequality:	Area0-Area	−0.00261
Width=	2.000651					
				Objective Function:		27.9788
Cell G7:	=Height*Width*Length-Vol0					
Cell G9:	=MIN(Area0-Width*Length)			**Penalty Function:**		27.98939
Cell G12:	=2*(Height*Width+Width*Length+Length*Height)					
Cell G14:	=G12+Kfactor*G7^2+Mfactor*G9^2					

Example 7.6 A torque arm is to be designed for an automobile application, such as in Figure E7.6. A vertical force P on the right produces a torque PL on the shaft on the left. Simple strength of materials formulas should be used.

1. The maximum stress at section $A - A$ should not exceed some value:

$$\sigma_{max} = \frac{M}{I} y_{max} = \frac{P(L - D/2)}{I} \frac{D}{2} \leq \frac{s_y}{N}, \; g(a, D) = 1 - \frac{P(2L - D)D}{4I} \frac{N}{s_y} = 0$$

$$I = 2a^2 \left(\frac{D - a}{2}\right)^2 + \frac{2a^4}{12} = \frac{a^2(D - a)^2}{2} + \frac{a^4}{6}$$

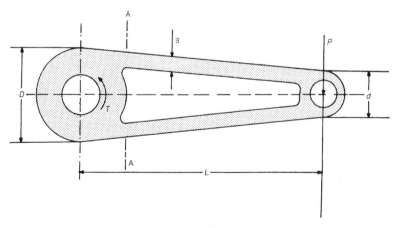

Figure E7.6

2. The axial load on the two connecting rods is approximately

$$F = \frac{PL}{D - a}$$

and it should not exceed the buckling load:

$$F \le \frac{4\pi^2 EI}{l^2}, \; h(a, D) = \frac{2\pi^2 Ea^2}{3(2L - D - d)} \frac{(D - a)}{PL} - 1 > 0$$

Determine the optimum D and a (thickness) for minimum volume of the arm.

Data: $E = 2.1 \times 10^{11}$ N/mm^2; $L = 200$ mm; $d = 30$ mm; $P = 5000$ N; $\dfrac{s_y}{N} = 100$ N/mm^2

Solution   The volume is:

$$V = \frac{\pi D^2}{4} a + 2 \left(L - \frac{D}{2} - \frac{d}{2} \right) a^2$$

The stress equation is used as an equality constraint:

$$g(a, D) = 1 - \frac{P(2L - D)}{4I} \frac{N}{s_y} = 0$$

The buckling equation is used as an inequality constraint:

$$h(a, D) = \frac{4\pi^2 Ea^4}{3(2L - D - d)} \frac{(D - a)}{P} - 1 > 0$$

Then the penalty function is assumed in the form

$$P(a, D) = V + K_1[g(a, D)]^2 + K_2 \langle h(a, D) \rangle^2$$

The program OPTIMUM was used with $K_1 = 10^6$ and $K_2 = 10^4$. The optimum is at $a = 17.8$ mm and $D = 46.8$ mm. At this point, $g = -0.047$ and $h = 13,704$. This means that the error in satisfaction of the stress equation is approximately 5% and the buckling equation is inactive. The first attempt was with $K_1 = K_2 = 10$. It yielded smaller volume but great value of the equality constraint, which means that the stress equality was not satisfied. Progressively higher values for K were used until the equality constraint was approximately satisfied.

Example 7.7   Design a four-bar linkage as a function generator so that the change of the output angle from the initial position $\Delta\psi$ will be related to the change of the input angle from the initial position $\Delta\phi$ as $\Delta\psi = \delta\phi^2$ for five distinct values of $\Delta\phi = 0, 10, 20, 30, 40°$.

Solution   The solution of the equations of closure will be programmed in an EXCEL spreadsheet using Solver.

The input and output angles ϕ_2 and ϕ_4 will be labeled Phi $= \phi$ and Psi $= \Delta\psi$, respectively. The initial input angle is Phi0, cell $B6$, and the link lengths will be labeled r_1, r_2, r_3, r_4, cells $B7$–$B10$, respectively. Cells $B12$–$B16$ have the input angles Phi in increments of 10°, or $10 \times \pi/180 = 0.17$ rad.

From Equations (2.9):

$$\phi_4 = \pi - \arccos\left(\frac{r_1^2 + s^2 - r_2^2}{2r_1 s}\right) \pm \arccos\frac{r_4^2 + s^2 - r_3^2}{2r_4 s}$$

where:

TABLE E7.7

	A	B	C	D	E	F	G	H	I
	Example 7.7: OPTIMIZATION OF A FOUR-BAR LINKAGE AS FUNCTION GENERATOR								
	Input angle phi $(=\phi_2)$, output angle psi$(=\phi_4)$								
	Link Lengths r_1, a_2, r_3, r_4 (Base link r_1=1)								
	We try initially r_2=1.5, r_3=2, r_4=2.5, Phi0=0,5 keeping r_1=1. We use Solver to adjust them for minimum error.								
	Design Parameters								
	Phi0=	0.774597			Psi_0=	2.313327			
	r_4=	2.461334		Longest Link =		2.461334			
	r_3=	1.761026		Shortest Link =		0.830422			
	r_2=	0.830422		Sum of Links =		6.052782			
	r_1=	1		Grashoff p+q+s+l-2(s+l)			−0.53073		
	We now use equations 2.9 to compute the mechanism output angle psi = ψ.								
	Point	input phi	sigma	gamma	output psi	Dphi-Mech	Dpsi-Mech	Dpsi-funct	Diff^2
	0	0.774597	0.708936	0.052969	2.313327	0	0	0	0
	1	1.036396	0.918527	0.28646	2.822565	0.261799	0.509238	0.511663	5.88E-06
	2	1.298196	1.114648	0.419598	3.040632	0.523599	0.727305	0.723601	1.37E-05
	3	1.559995	1.292927	0.528151	3.200679	0.785398	0.887352	0.886227	1.27E-06
	4	1.821794	1.449864	0.619642	3.336415	1.047198	1.023088	1.023327	5.71E-08
							Error:	Σ(Diff^2)	2.09E-05

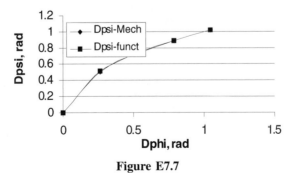

Figure E7.7

$$\gamma = \pm \arcos\left(r_3{}^2 - s^2 + \frac{r_4{}^2}{2r_3r_4}\right)$$

$$s = (r_1{}^2 + r_2{}^2 - 2r_1r_2 \cos \phi_2)^{1/2}$$

In cells $C12$–$C16$ are the lengths s. In cells $D12$–$D16$ are the angles γ. In cells $E12 - E16$ are the output angles $\psi = \phi_4$.

Phi0, the initial input angle, is a parameter to be found. The initial output angle Psi0 is determined when the values of the link lengths and the initial input angle are known. We put it in cell $E5$. The base link length r_1 will be arbitrarily set to 1 because the angular input–output relationship does not change with change in scale. Thus, the unknown link lengths are r_2, r_3, r_4. (See Table E7.7.)

In cells $F12$–$G16$ are the differences of the input and output angles minus the corresponding initial angles. Thus they are $\Delta\phi$ and $\Delta\psi$. The desired output angle $D_y = Df2$ is placed in cells $H12$–$H16$. Because we would like the values of the columns G and H to be identical, for an ideal mechanism, we compute their difference and place it in column I.

We now use Solver to find the values in cells $B6$–$B9$ so that cells $G13$–$G16$ will be zero. To this end, we first assign initial values to the unknown parameters, 0.5, 1., 1., 0.5 in cells $B6$–$B9$ and then we select Tools, Solver and set target cell $G13$ to 0 and, as constraints, cells $G14$, $G15$, $G16$ also to zero. We may also put additional constraints as $r\_2 > 0$, $r\_3 > 0$, $a4 > 0$. The Grashof criterion might be introduced, too, To this end, we rewrite the Grashof criterion (Equation (2.4)) in the form

$$p + q + \ell + s - 2(\ell + s) > 0$$

This is performed in cells D6:G10 and an additional constraint is used in **Solver**: $G10 \geq 0$.

We now hit **Solve**. Depending on the problem, Solver might find a solution or it might find the best it can. A plot of cells $H12$–$H16$ will show the mechanism input–output performance (Dpsi-mechanism) and the desired one (Dpsi-function $= D$phi^2) (Figure E7.7).

CASE STUDY 7.1: Optimization of a Solar Collector Orientation Linkage

Solar tracking by means of linkages is a simple, reliable, and economical solution to improve efficiency of solar collectors. An optimization method for this purpose was used in this project.

The apparent altitude and azimuth of the sun change with the hour of the day, the season, and the latitude of the station. A diagram for a specific date and locality is shown in Figure CS7.1. The azimuth is plotted against the hour of the day for January 21 at a latitude of 40° North. For the same locality, the altitude plotted against the hour of the day is also shown in Figure CS7.1 and the total surface insolation versus hour of the day and surface inclination for a south-facing surface is shown in Figure CS7.2.

An ideal tracking should provide for a space motion to track both altitude and azimuth. Though this can be achieved with spatial mechanisms, in this example only the azimuth tracking is considered. It is evident from Figure CS7.1 that one should approximate the azimuth functions in a range of angles–60° to 60°.

The corresponding ranges of the hours will be 8 a.m. to 4 p.m. The algebraic method of synthesis will be utilized with three accuracy points selected by means of the Tschebytschev polynomials in a way shown in Figure CS7.3. Finally, the

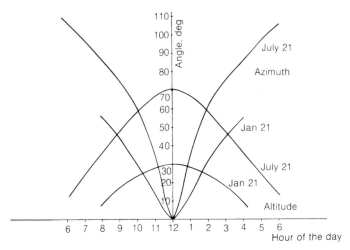

Figure CS7.1. Azimuth and altitude of sun versus hour of the day for latitude 40°N at January 21 and July 21.

following accuracy points were used in a form of couples of coordinate points for the system of Figure CS7.3 (January 21):

Hours	Degrees
$x_1 = 8.4$	$y_1 = -52$
$x_2 = 12$	$y_2 = 0$
$x_3 = 15.6$	$y_3 = 52$

Range $E = 100°$

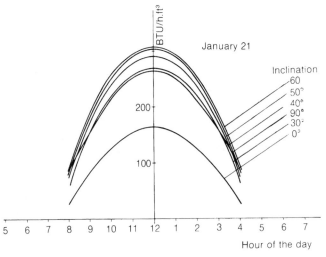

Figure CS7.2. Total surface insolation versus hour of the day for 40°N latitude at January 21 for different surface inclinations.

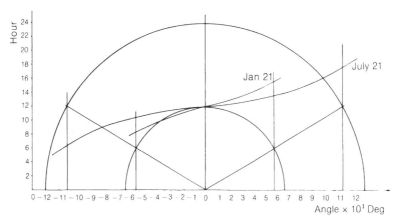

Figure CS7.3. Spacing of accuracy points (Tschebytschev) for the function to be generated, angle versus hour of the day.

The mechanism used is shown in Figure CS7.4. The input angle is provided by way of a geared motor delivering 1 revolution per 24 hours. The output angle ψ corresponds to the azimuth angle y. Because the range y is very wide, the range of ψ is reduced by way of a gear couple attached to the link 3, having a gear ratio π/ES so that the range of φ is 0 to π and the range of ψ is $-\pi/S$ to π/S, leaving S as an arbitrary parameter. Therefore, the transformation from (x, y) to (φ, π) will be:

$$\varphi = \varphi_1 + \frac{x - x_1}{24} 2\pi, \; \psi = \psi_1 + \frac{y - y_1}{2ES} \qquad (a)$$

It is apparent from Equation (a) that the choice of the initial angles φ_1 and ψ_1 is arbitrary. The selection of these angles will be optimized using as an objective function the ratio of the total energy per day (U) received by the collector to the total energy (U_α) available for the collector ideally corrected for azimuth only with a fixed angle to the horizontal at noon ω_0, while U_i will be

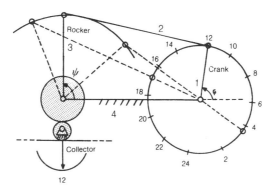

Figure CS7.4. Optimum mechanism configuration for January 21.

the daily total irradiation for ideal orientation, both for azimuth and altitude. The inverse function $f = U_\alpha/U_i$ will be minimized by appropriate selection of the gear ratio of the output and the initial angles φ_1 and ψ_1 using the steepest descent method. Additional constraints will be:

$$0.25 < \frac{\alpha_1}{\alpha_4}, \frac{\alpha_2}{\alpha_4}, \frac{\alpha_3}{\alpha_4} < 4$$

$$\alpha_1 + \alpha_2 + \alpha_3 > \alpha_4, \text{ etc.}$$

where the functions U_α, U_c are obtained by numerical integration of the expressions from sunrise time x_0 to sunset $x_0 + E_x$.

$$U_\alpha = \int_{x0}^{x_0+E_x} q_i \cos(\alpha - \omega_0)dx$$

$$U_c = \int_{x0}^{x_0+F_x} q_i \cos(\alpha - \omega_0)\cos(\psi - \psi^*)dx$$

where q_i is the instantaneous irradiation rate, ψ is the output angle delivered by the linkage, and ψ^* is the ideal output angle corresponding to the correct azimuth; in other words, $\psi - \psi^*$ is the absolute linkage error.

The optimization procedure used the program OPTIMUM and yielded a variety of mechanisms having a very low error in the energy function U_c owing to the stationary character of the cosine function about zero. For example, the following linkage was obtained.

Data:
January 21. Sunrise time = 8 a.m.; sunset time = 4 p.m.

$$U_\alpha = 2127.45 \text{ Btu/ft}^2; \ U_i = 2182 \text{ Btu/ft}^2$$

Results:

$$\alpha_1 = 0.3766; \ \alpha_2 = 1.057; \ \alpha_3 = 0.6007; \ \alpha_4 = 1 \text{ (link lengths)}$$

Input angle at 12 noon = 82.65°

Output gear ratio = 0.19

Surface daily total energy, $U_c = 2127.36 \text{ Btu/ft}^2$

It is observed that U_c is almost identical with the ideal value of $U_\alpha = 2127.45$ Btu/ft². We turn now to the accuracy of the output angle, which is most important for focusing collectors: in Figure CS7.5, the correct function and the linkage output angle are plotted against the hour of the day. A maximum error of 3.62% is observed, which is tolerable for most focusing collector applications. The mechanism is shown in Figure CS7.4.

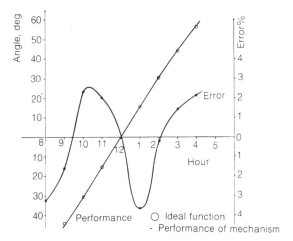

Figure CS7.5. Performance of solar tracking mechanism. Ideal and generated angle versus hour of the day. Error percent versus hour of the day.

REFERENCES

Erdman, A. G., and G. N. Sandor. 1984. *Mechanism Design,* 1st ed. Englewood Cliffs, N.J.: Prentice Hall.

Gottfried, S. B., and J. N. Weisman. 1973. *Introduction to Optimization Theory.* Englewood Cliffs, N.J.: Prentice-Hall.

ADDITIONAL READING

Dimarogonas, A. D. *Machine Design Optimization.* General Electric, Technical Information Series. Schenectady, N.Y.: General Electric, 1972.

Fiacco, A. V., and G. P. McCormick. *Nonlinear Programming: Sequential Unconstrained Optimization Techniques.* New York: John Wiley & Sons, 1968.

Johnson, R. C. *Optimum Design of Mechanical Elements.* New York: John Wiley & Sons, 1961.

Mischke, C. R. *An Introduction to Computer-Aided Design.* Englewood Cliffs, N.J.: Prentice-Hall.

Reclaitis, G. V., A. Ravindran, and K. M. Ragsdell. *Engineering Optimization: Methods and Application.* New York: John Wiley & Sons, 1983.

Siddal, J. N. *Optimal Engineering Design.* New York: Marcel Dekker, 1982.

PROBLEMS[1]

7.1. [D] Formulate Problem 3.18 as a synthesis problem. Modules are the lever, the pin, and the piston.

---

[1][C] = certification, [D] = design, [N] = numerical, [T] = theoretical problem.

7.2. [D] Formulate Problem 3.20 as a synthesis problem. Modules are the lever, the pad, and the drum.

7.3. [D] Formulate Problem 4.10 as a synthesis problem. Modules are the lever, the pin, and the valve.

7.4. [D] Formulate the design of a slider-crank linkage as a synthesis problem. Modules are the crank, the connecting rod (coupler), and the piston.

7.5. [D] Formulate the design of a quick-return linkage of the shaper machine of Example 2.4 as a synthesis problem. Modules are the crank, the rocking rod (coupler), and the oscillating slider.

7.6. [D] Design a four-bar linkage to produce the function $\Delta\psi = \cos(\Delta\psi)$. Formulate the problem as a synthesis problem.

7.7. [D] A coupler point of a four-bar linkage is to trace approximately a straight line segment. Formulate this as a synthesis problem. (Hint: Setting the second derivative of the function of the path of the coupler point equal to zero accomplishes a short almost straight line segment. Setting higher derivatives to zero extends the limits of the almost straight line segment. Use Solver to make the second derivative of the path equal to 0 with higher derivatives set to zero.)

7.8. [D] Design a four-bar linkage to produce within a short range of motion of the crank constant angular velocity of the follower of half value the angular velocity of the input crank. Formulate the problem as a synthesis problem.

7.9. [D] Formulate Problem 2.19 as a linkage synthesis problem. Modules are the piston, the crank, the follower, the coupler, and the track frame. The maximum slope of the track bed is 45°.

7.10. [D] Formulate Problem 2.20 as a linkage synthesis problem. Modules are the crank, the follower, and the coupler.

7.11. [D] A circular rod of diameter d will be machined to a rectangular shape to give a beam with maximum resistance moment in bending $I/c = W = bh^2/6$. Given d, find analytically the optimum of b. Then, for $d = 10$ cm, define the objective function and the equality constraint of the problem and define the penalty function.

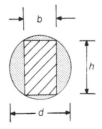

Figure P7.11

7.12. [D] A cylindrical container (20 m³) is filled with water. The stress near the end is pr/t, where p is the pressure, r is the radius $d/2$, and t is the thickness.

The material has allowable stress 100 N/mm^2. Analytically, find the diameter and height which gives minimum weight of the container. The thickness t is uniform.

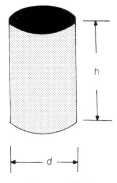

Figure P7.12

7.13. [D] A three-stage gear transmission has overall gear into 10 or $\omega_1/\omega_4 = 10$. On the motor at $\omega = \omega_1$, the equivalent moment of inertia for gear i that has $\omega = \omega_i$ is $I_{eq} = (\omega/\omega_1)^2 I_p$ where I_p is the polar moment of inertia of the particular gear. Assume that I_p is $1/2mr^2 = (\pi/2)\rho br^4$ where r is the radius, ρ is the density, and b is the constant width. Find analytically the three gear ratios i_1, i_2, i_3 to have:

1. Total gear ratio $i_1 \ i_2 \ i_3 = 10$
2. Minimum equivalent moment of inertia on the motor shaft, if $r_1 = r_2 = r_3$

(From Gottfried and Weisman 1973.)

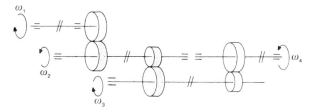

Figure P7.13

7.14. [D] A hook carrying a 10 ton load is supported from a roof through two circular rods AB and AC. If the load does not always have vertical direction, there is also a maximum horizontal force of 5 tons in either direction. Calculate analytically the distance BC (x) for the minimum diameter of the rods where $h = 1$ m if the allowable stress is 8 kN/cm^2. Do not consider buckling.

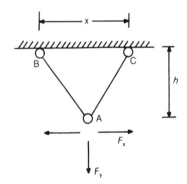

Figure P7.14

7.15. [D] A pressure vessel has internal pressure $p = 10$ bar and a volume $V = 4$ m^3. The tangential stress $pd/2t$ should not exceed $10kN/cm^2$. Find analytically the values of L and d for minimum weight. The thickness t is constant throughout the vessel.

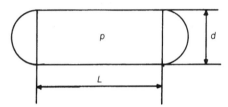

Figure P7.15

7.16. [D] The vessel shown is filled with water. The thickness t is calculated based on tensile strength at point A, which should not exceed $12kN/cm^2$. Find analytically the values of d and h for minimum vessel weight if the volume is 6 m^3 and the thickness is constant.

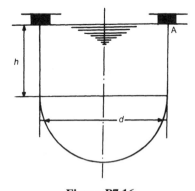

Figure P7.16

7.17. The shaft shown is made of steel with $E = 2.1 \times 10^5$ MPa, density 7800 kg/m^3, and allowable tensile stress 120 MPa. The shaft is loaded by its own weight. Find analytically the optimum values of ratios L_1/d, d_1/d for maximum L.

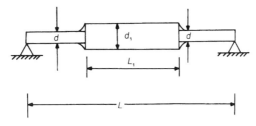

Figure P7.17

7.18. A tapered beam (Figure P7.18) loaded by its own weight is made of mild steel with $E = 2.1 \times 10^5$ MPa, density 7800 kg/m^3, and allowable tensile stress 120 MPa. Find the proper value of h/H to achieve maximum length L.

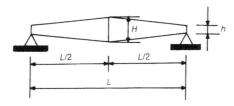

Figure P7.18

7.19. A rotating disk as shown is used for storage of kinetic energy. If $\omega = 300$ rad/s, the maximum allowable stress is 30 N/mm^2 and the density is 2200 kg/m^3, find t_1, t_2 for maximum stored kinetic energy per unit mass of the disk.

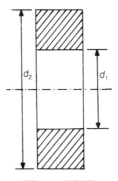

Figure P7.19

7.20–7.27. [D] Solve Problems 7.11–7.18 with the module Solver of EXCEL, forming a proper penalty function.

7.28. [D] Figure P7.28 shows a typical electromagnetic circuit. A coil C with copper of diameter d and n turns is fed with electric current to produce a certain magnetic flux in the core. The designer has the choice of using more copper winding or more iron core. To optimize this selection, formulate the optimization problem of minimizing the cost of the materials for the given flux, current, material cost, width α, and thickness w of the core at the lower section, gap g, and mean length of the magnetic path L. Develop the design equations, then write the appropriate objective function and the constraints.

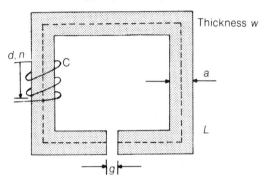

Figure P7.28

7.29. [D] An electric transmission line (length l, diameter d) is made of aluminum and carries electric power W at a constant voltage V. Owing to the current i, there is a power loss i^2R, where R is the ohmic resistance of the line. The designer has the choice of either greater cross-section with lower resistance, lower losses and greater capital cost, or smaller cross-section with greater resistance, greater losses, and smaller capital cost. To optimize this choice, for minimum cost per year, assume a utilization factor and year energy flow constant, cost of line per kg, cost per kWh, fixed percentage of the capital cost as interest and depreciation. Then formulate the objective functions and constraints.

7.30. [D] Discuss the optimization of the selection of the actual power to pump a constant flow of water to a certain distance and height. Tabulate the data you need and formulate the objective function and the boundary conditions.

CHAPTER 8

DESIGN FOR SHEAR STRENGTH: FASTENING AND JOINING

SS *Schenectady:* a liberty ship that had a weld failure. During World War II, liberty ships were made welded, for the first time, to speed up the supply to Europe. Many sank due to yet-unknown problems with weld fatigue.

OVERVIEW

In the foregoing chapters, we saw how we can design the machine as a system (Chapter 1), find the motion of each component (Chapter 2), find the forces that this component carries and the internal forces and moments that the different sections of the component carry (Chapter 3), select the appropriate material and manufacturing process (Chapter 4), and finally size the component by making capacity (material strength) equal to demand (working stress) (Chapter 5). In Chapter 6, we developed modeling tools to describe the operation of the machine as a system. In Chapter 7, we studied how to use machine models to design the machine in its final form and make sure we have extracted the best possible performance out of the design selected. Altogether, in Chapters 1–7, we studied the machine as a system, and we developed the methodology of sizing the machine components for strength in Chapter 5.

Sizing of machine components is not based only on strength; some mechanisms of failure are not directly related to strength, such as wear. Moreover, certain machine components are highly standardized and have special methods of sizing. For these reasons, in Chapters 8–14, we shall study the different methods of sizing elements of machines. In this chapter, we shall study the application of the methods developed in Chapter 5 for the design of specific machine elements that are designed mostly on the basis of strength and are used for fastening and joining, such as screws, rivets, and welds.

8.1. JOINTS OF MACHINE ELEMENTS

The several machine elements required to form a working machine frequently have to be connected together. In other cases it is advantageous to form a complicated machine element out of multiple components that have to be connected together to yield the final element. Some reasons, such as producibility, components made out of dissimilar materials, and serviceability with parts that have high wear rate, etc., were presented in previous chapters. Because in most cases joints have the purpose of carrying loads, we shall use them as examples of how to apply methods of design for strength.

Because joint elements are used in a variety of cases, they have been standardized extensively. The standardization of such elements must be examined first.

A joint element, such as a bolt, can be used in a variety of tasks. The same-size bolt can be used in an automobile, an aircraft engine, or a paper mill. On all occasions its purpose is to carry a certain load in a certain mode. We can get an idea of the frequency of use of joint elements if we consider the fact that an automobile might have more than 10,000 parts, a machine tool might have up to 20,000 parts, and a rolling mill might have up to 1,000,000 parts.

Joints may be permanent or removable. The selection is based on the purpose and on the economy of the joint. Permanent connections cost less in general, but they have the disadvantage that if the parts are disconnected, the joints must be destroyed and cannot be used again. On the other hand, permanent joints are safer, especially in parts that sustain dynamic loads. Therefore, for permanent joints the main consideration is strength. For separable joints permitting assembly and disassembly, additional security from accidental separation is necessary.

Joints are held together either by the forces of molecular cohesion, such as welded joints, or by mechanical means, such as riveted and bolted joints.

The main consideration in the design of joints is to make them function as much as possible as if all the connected elements were a single solid part and to make them comply with the condition of equal strength of the joint and connected elements. Otherwise the material of the connecting element will not be utilized to its full capacity and the cost will be unnecessarily high.

There are occasions when the joints are part of fluid-carrying vessels under pressure. In this case the joint has the additional condition of fluid tightness. For this purpose the contacting surfaces must be held together by a pressure exceeding the pressure of the fluid. If high accuracy under the application of the load is required, the joint must in addition have sufficient rigidity.

Several types of joints are categorized in Figure 8.1. *Permanent* joints are used to join parts that are not expected to be disconnected again during their lifetime. If this needs to be done, permanent damage will occur in the parts used. Such joints are welds, rivets, and interference fits. *Detachable* joints are used when the joint might need to be disassembled for maintenance or operational purposes and assembled again without damage to the parts. Such joints are bolts and pins.

8.2. RIVETS

8.2.1. Field of Application

A rivet is a short, round bar with heads at each end. One of the heads is made beforehand on the body of the rivet, and the second head, called the *closing head*, is formed during riveting (Figure 8.2). A riveted joint is made by inserting rivets into holes in the elements to be connected while holding the elements together in some way so that the holes coincide. Then the closing head is formed by riveting, which also expands the shank of the rivet, as is shown in Figure 8.2a, b, and c. Some applications are shown in Figure 8.3, and some types of rivets are shown in Figure 8.4. Rivets are perhaps the oldest known type of machine element. Today they are used in many applications:

1. When the material is such that welding might temper the heat-treated components or warp them and change their shape

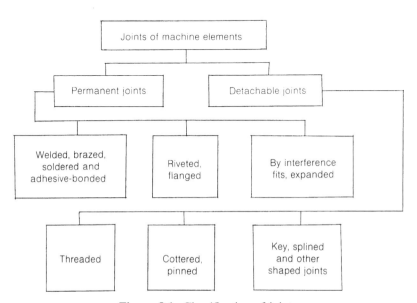

Figure 8.1. Classification of joints.

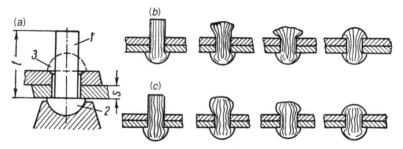

Figure 8.2. Formation of riveting.

2. In materials that are very dissimilar or nonweldable
3. In the case of very heavy repeated impact and vibrational loads
4. In cases where maximum safety is required, such as in aircraft structures

Standard dimensions of rivets are shown in Table 8.1, and the usual types of riveted connections are shown in Figure 8.5. The type of riveting selected depends on the loads that riveting will carry.

Materials used for machinery riveting are shown in Table 8.2, with allowable stresses for static and dynamic loads. For steel construction, local codes are mandatory and must be consulted.

For pressure-vessel and boiler riveting, the ASME or other applicable pressure vessel codes may be applied.

Riveting for light and aircraft structures needs special treatment and will be discussed later in this chapter (Section 8.2.4).

8.2.2. Shear-Loaded Rivet Joints

When a transverse force Q is acting on cross-sections of a bar and the other internal forces are equal to zero, this type of loading is called *shear*. In this case only shearing stresses arise at the section.

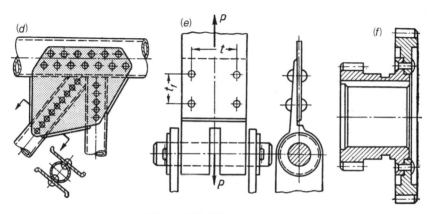

Figure 8.3. Riveted joints.

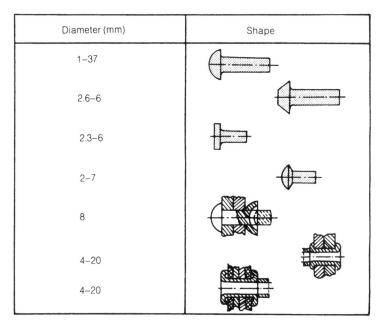

Diameter (mm)	Shape
1–37	
2.6–6	
2.3–6	
2–7	
8	
4–20	
4–20	

Figure 8.4. Types of rivets.

In most practical problems, a transverse force acts simultaneously with a bending moment and a longitudinal force so that normal stresses as well as shearing stresses usually act over the cross-sections. However, in cases where shearing stresses are considerably predominantly larger than normal stresses, only shearing strength analysis needs to be considered in the design. A typical example of such a simplified but, as experience shows, quite reliable analysis is the calculation of the shear strength of riveted, bolted, and welded joints.

William Fairbairn (1789–1874) was born in Kelso, Scotland, son of a poor farmer. He started helping his father in the farm but at 15 became an apprentice in a power station. In the evening, he studied mathematics and literature. He worked as a mechanic in London and Manchester. He became well known for the many improvements he made in cotton mills. His interest turned to iron properties and he started experimenting with its strength. In 1830, he became interested in iron ships and made extensive studies of riveted joints. He did further work on boilers and their riveting and participated in the design of the Conway and Britannia Bridges. In 1860, he became a fellow of the Royal Society and was awarded its gold medal. He wrote several books on machine design that were very popular in his time.

Figure 8.6a shows two plates joined by means of rivets (lap joint). If failure of each rivet occurs along a single shear plane, the rivet joint is referred to as single-shear (Figure 8.6a); if along two planes, the joint is referred to as double-shear (Figure 8.6b); and so on.

TABLE 8.1. Dimensions of Rivets

Rivet		10	12	(14)	16	(18)	20	22	24	27	30	(33)	36
	d	10	12	(14)	16	(18)	20	22	24	27	30	(33)	36
	d_1	11	13	15	17	19	21	23	25	28	31	34	37
Bolt (eq.)		M10	M12	—	M16	—	M20	—	M24	—	M30	—	M36
Pressure vessels	D	18	22	25	28	32	36	40	43	48	53	58	64
	k	7	9	10	11.5	13	14	16	17	17	21	23	25
	R	9.5	11	13	14.5	16.5	18.5	20.5	22	24.5	27	30	33
	r	1	1.6	1.6	2	2	2	2	2.5	2.5	3	3	4
Steel construction	D	16	19	22	25	28	32	36	40	43	48	53	58
	k	6.5	7.5	9	10	11.5	13	14	16	17	19	21	23
	R	8	9.5	11	13	14.5	16.5	18.5	20.5	22	24.5	27	30
	r	0.5	0.6	0.6	0.8	0.8	1	1	1.2	1.2	1.6	1.6	2
Sunken heads	D	14.5	18	21.5	26	30	31.5	34.5	38	42	42.5	46.5	51
	k	3	4	5	6.5	8	10	11	12	13.5	15	16.5	18
	r	27	41	58	85	113	124.5	75.5	91	111	114	136	164

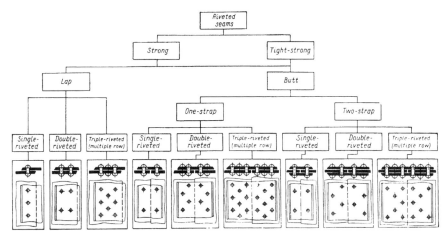

Figure 8.5. Classification of riveted joints.

TABLE 8.2. Allowable Stresses in Riveted Joints (MPa): (a) Plate Material; (b) Rivet Material

(a)

| | | ASTM | | | | | ASTM A159* | | ASTM A48* | |
| | | A284 | A284 | A570 | A572 | A572 | | | | |
	Loading	GC	GD	GD	G50	G55	G1800	G3000	G4000	Class 55
	static	120	140	160	180	220	35	65	100	135
Tension	pulsating	85	100	120	140	170	25	40	75	100
	alternating	70	85	95	110	130	20	35	50	70
	static	170	195	225	250	310	50	90	140	190
Bending	pulsating	95	110	130	155	185	28	45	80	110
	alternating	75	95	100	120	145	20	40	55	80
	static	240	280	320	360	410	65	130	200	270
Shear	pulsating	170	200	240	280	340	45	85	130	170
	alternating	140	170	190	220	260	35	65	100	130

(b)

| | Shear | | | | Bearing | | Tension | | |
| | A284 | A570 | A572 | A572 | A570 | A572 | A284 | A570 | A572 |
	GrC	GrD	Gr50	Gr60	GrD	Gr50	GrC	GrD	Gr50
static	140	180	225	280	360	440	70	90	110
pulsating	100	140	170	200	280	340	50	70	85
alternating	85	110	130	170	220	260	40	55	65

*Cast irons

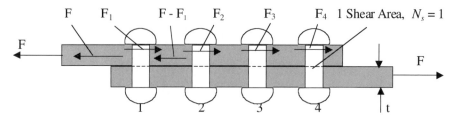

Figure 8.6. Shear distribution in riveted joints.

The results of the solution of this problem for a lap joint with the same cross-sectional area of the sheets being riveted are presented in Table 8.3. It can be seen that unequal loading conditions for rivets become more distinct with increase in number of rivets in a row. With six rivets the shearing forces in the outer rivets (first and sixth) are about 2.5 times larger than in the middle rivets (third and fourth).

In addition, the assumption is made that shearing stresses are distributed uniformly over the shear plane, though actually, as experimental observations show, their distribution is not uniform. However, a strict theoretical solution of this problem is rather involved, as there are clearances between rivets and sheets, frictional forces between sheets, and so on. Furthermore, rivets are usually made of the most ductile steels, and hence irregularity in the distribution of shearing stresses due to the occurrence of plastic deformations disappears by the moment of failure, by stress relaxation, as discussed above.

In cases where elements of different cross-sectional area are connected, unequal loading conditions for rivets are encountered. The most overstressed rivets are on the side of the sheet having the smaller cross-sectional area. Experiments show, however, that for ductile materials the rivets fail simultaneously under static loading. This is due to the fact that the forces in the rivets are equalized by the moment of failure as a result of ductility of the material and the clearances between the rivets and the sheets.

Under the action of a static load F, design is based on the average shear stress:

$$\tau_{av} = \frac{F}{nA} \tag{8.1}$$

where $A = \pi d^2/4$ is the cross-sectional area of a rivet with diameter d, F is the force acting on the joint, and n is the number of rivets.

In the case of impact and vibratory loading, unequal loading conditions for rivets must be taken into consideration in the form of stress concentration. Indeed, the end rivets carry more load than the center ones. For example, in the four-rivet lap joint shown in Figure 8.6a, the force transmitted is F. The first rivet carries some

TABLE 8.3. Riveting Stress Concentration Factor K_r

No. of Rivets	1	2	3	4	5	6
K_r	1	1	1.059	1.16	1.30	1.44

load F_1. The remaining force on the upper strip on the right of rivet 1 is $F - F_1$, between rivets 2 and 3 the force is $F - F_1 - F_2$, between rivets 3 and 4 the force is $F - F_1 - F_2 - F_4$. The system is redundant, and therefore to find the forces F_1, F_2, F_3, F_4, one needs to consider the rivet and the plate deformations. This analysis yields:

$$F_1 = \frac{1.16F}{4}$$

$$F_2 = \frac{0.84F}{4}$$

$$F_3 = \frac{0.84F}{4} \tag{8.2}$$

$$F_4 = \frac{1.16F}{4}$$

The maximum loads are at the end rivets, and the stress concentration factor is $K_r = 1.16$. For other numbers of rivets, the respective riveting stress concentration factors are shown in Table 8.3.

Therefore, for reversing shear loading, the maximum rivet shear stress is:

$$\tau_{max} \quad \frac{K_r F}{nA} \tag{8.3}$$

For static loading or for single and double row riveting, it always is:

$$K_r = 1 \tag{8.4}$$

8.2.3. Rivet Sizing

The design equation for the shearing strength of rivets is then, from Equation (8.3):

$$\tau_{av} = \frac{F}{nN_sA} = \frac{Q}{A} \leq \frac{S_{sy}}{N} \tag{8.5}$$

where N_s is the number of sections of the rivet that will fail in shear. For example, in a lap joint such as in Figure 8.6, $N_s = 1$. For a double lap joint (Figure 8.7), $N_s = 2$. Moreover, S_{sy}/N is the allowable shearing stress.

From Equation (8.5), the necessary number of single-shear rivets can readily be determined for static loading:

$$n = \frac{4NF}{\pi d^2 S_{sy} N_s} \tag{8.6}$$

For *reversed dynamic loading,* the effective fatigue strength $S_{se} \approx 0.577S_e$ needs

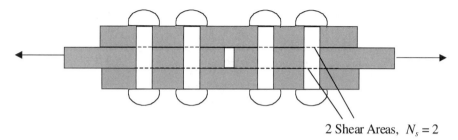

2 Shear Areas, $N_s = 2$

Figure 8.7. Number of shear planes.

to be used (see Chapter 5, and the load sharing equation (8.1b)). Equation (8.6) becomes:

$$n = \frac{K_r F}{\pi d^2 S_{se} N / 4N}$$ (8.7)

For combined static and dynamic loading, the Goodman criterion should be used (see Chapter 5).

The magnitude of allowable shearing static stresses is usually established experimentally to reveal the influence of irregularity in stress distribution on the strength of a joint, the influence of frictional forces, the influence of clearances, and so on. In the design of rivets, it is generally accepted that $S_{sy} = (0.6$ to $0.8)S_y$ (Table 8.2).

In addition to shear, bearing strength also has to be provided when rivet joints are designed. Bearing stresses are checked over the contact area between the rivets and sheets being connected. The bearing area of one rivet is assumed to be $A_b = td$, the projection of the bearing area on a plane that is perpendicular to the direction of the force and the riveted plate. The bearing stresses are considered to be uniformly distributed over the bearing area, due to the allowance of plastic deformation and stress redistribution, and the condition of bearing strength of the riveting is:

$$\sigma_b = \frac{F}{n' A_b} \leq \frac{S_b}{N}$$ (8.8)

where S_b is the allowable bearing stress, A_b is the bearing area, and n' is the number of rivets for sufficient bearing stress. For steels, tests have shown that:

For drilled or punched holes:

$$S_b \approx 1.1 S_u$$ (8.8a)

For drilled or punched and then reamed holes:

$$S_b = 2 S_u$$ (8.8b)

From Equation (8.8), the necessary number of rivets can be determined on the basis of the bearing strength:

$$n' = \frac{K_r F}{S_b A_b} \qquad (8.9)$$

Of the two quantities n and n', the greater is taken. Again, for static loading $K_r = 1$, for reversed dynamic loading K_r can be found for multiple row rivetings in Table 8.3.

Decker suggests the values given in Table 8.4 for pressure vessels. This table also gives some details for typical steel riveting.

A riveting might fail with a number of failure mechanisms (Figure 8.8) in addition to the one described above, shear and bearing failures, as of course might any pin-type joint such as bolting of spot welding. In Figure 8.8, each mode of failure diagram is accompanied by its design equation.

In general, the characteristic load for a riveting is the load per unit width. Depending on this figure, the proper type of riveting is selected that experience shows, will be near optimum.

(a) *By shear failure of the rivet* (Equation (8.5):

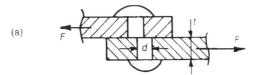

(a)

$$\tau_{av} = \frac{4 K_r F}{n N_s \pi d^2} \leq \frac{S_{sy}}{N} \qquad (8.10)$$

(b) *By tensile failure of the plate along the riveting:*

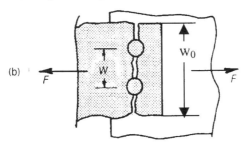

(b)

$$\sigma_{av} = \frac{K_r F}{t(w_0 - n_r d)} \leq \frac{S_y}{N} \qquad (8.11)$$

n_r is the number of rivets in the row, two in the case shown.

Figure 8.8. Modes of failure of riveted joints.

(c) *By crushing of the rivet or the plate owing to excess bearing pressure* (Equation (8.8)):

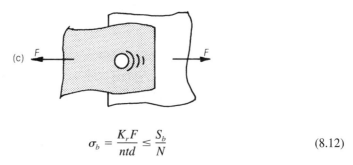

$$\sigma_b = \frac{K_r F}{ntd} \leq \frac{S_b}{N} \qquad (8.12)$$

For S_b, see Equation (8.8a).

(d) *By double shear of the plate:*

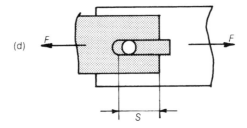

$$\tau = \frac{K_r F}{2t(s - d/2)} \leq \frac{S_{sy}}{N} \qquad (8.13)$$

(e) *By shear tear-out of the plate:*

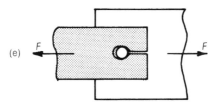

Approximately, the tear is assumed to be due to a tearing force F:

Figure 8.8. (*Continued*)

$$\sigma = \frac{K_r F}{t(s - d/2)} \leq \frac{S_y}{N} \qquad \textbf{(8.14)}$$

(This equation is approximately equivalent with the one of case d.)

Design Procedure 8.1: Design of a Shear-Loaded Riveting

Input: Shear load, material properties, safety factor. **Output:** Riveting dimensions

Step 1: Select type of riveting, number of rows (usually we start with one row), number of shear sections of the rivet N_s.

TABLE 8.4. Riveting Joints for Pressure Vessels[a]

No.	Figure	Rows	P_{1mm} (kP/mm)	V_{max}	d_1 (mm)	t (mm)	e	e_1	e_2	s_L
1	a	1	50	0.6	$\sqrt{50s} - 4$	$2d_1 + 8$	$1.5d_1$	—	—	—
2	a	2	40–95	0.7	$\sqrt{50s} - 4$	$2.6d_1 + 15$	$1.5d_1$	$0.6t$	—	—
3	a	3	70–135	0.75	$\sqrt{50s} - 4$	$3d_1 + 22$	$1.5d_1$	$0.5t$	—	—
4	b	1	35–95	0.65	$\sqrt{50s} - 5$	$2.6d_1 + 10$	$1.5d_1$	—	$1.35d_1$	$0.6 \ldots 0.7s$
5	c	2	85–160	0.8	$\sqrt{50s} - 6$	$5d_1 + 15$	$1.5d_1$	$0.4t$	$1.5d_1$	$0.8s$
6	d	2	85–160	0.8	$\sqrt{50s} - 6$	$5d_1 + 15$	$1.5d_1$	$0.4t$	$1.5d_1$	$0.8s$
7	b	2	85–135	0.75	$\sqrt{50s} - 6$	$3.5d_1 + 15$	$1.5d_1$	$0.5t$	$1.35d_1$	$0.6 \ldots 0.7s$
8	c	8	130–230	0.85	$\sqrt{50s} - 7$	$6d_1 + 20$	$1.5d_1$	$0.4t$	$1.5d_1$	$0.8s$
9	b	3	110–240	0.7	$\sqrt{50s} - 7$	$3d_1 + 10$	$1.5d_1$	$0.6t$	$1.5d_1$	$0.8s$
10	c	4	190–320	0.85	$\sqrt{50s} - 8$	$6d_1 + 20$	$1.5d_1$	$0.4t$	$1.5d_1$	$0.8s$
11	b	4	180–320	0.7	$\sqrt{50s} - 8$	$3d_1 + 10$	$1.5d_1$	$0.6t$	$1.5d_1$	$0.8s$

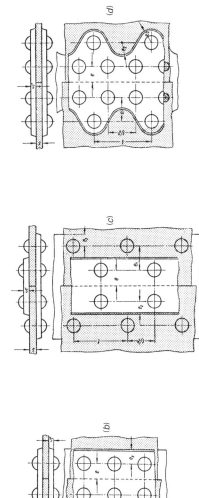

(a)　(b)　(c)　(d)

[a] After Decker, 1973.

Step 2: Use Equations (8.10) and (8.11) to determine the plate thickness and rivet diameter, if the plate width is known, or the plate width and the rivet diameter, if the plate thickness is known.

Step 3: Rivet diameter should be of the order of the plate thickness. If not, select a different type of riveting and go to step 1.

Step 4: Use Equation (8.12) to check the bearing strength. If not sufficient, select a different type of riveting and go to step 1.

Step 5: Use Equations (8.12) or (8.13) to determine the plate margin *s*.

Step 6: Check that there is sufficient space for the rivet heads ($w_0 - n_r(3d) > 0$). If not sufficient, select a different type of riveting and go to step 1.

Example 8.1 Design a riveted lap joint connecting two plates of the same thickness $t = 16$ mm having two cover plates (Figure 8.7) if $F = 60$ tons. The allowable stresses are:

$$\frac{S_y}{N} = 160 \text{ N/mm}^2, \frac{S_{sy}}{N} = 100 \text{ N/mm}^2, \frac{S_b}{N} = 320 \text{ MPa}$$

Solution In this case the rivets are in double shear because in order for a joint to fail there must be a failure in each rivet by shearing along two planes. Take $K_s = 1$ (static loading), $N_s = 2$ (double shear plane), the diameter of a rivet $d = 20$ mm. The necessary number of shear planes is found according to Equation (8.6):

$$n = \frac{K_r F}{\pi d^2 S_{se} N_s / 4N} = \frac{4 \times 600000}{2 \times 3.14 \times 20^2 \times 100} = 9.5$$

Consequently, it is necessary to take 10 rivets. The required number of rivets based on bearing strength is given by Equation (8.9) or (8.12).

$$n' = \frac{P}{tdS_b} = \frac{600,000}{16 \times 20 \times 320} = 5.85$$

The design against resistance to shearing has proven to be decisive. We take 10 rivets on each side of the joint in three rows with 3 rivets in a row. The section area A of the sheet based on the tension on the sheet:

$$A = \frac{F}{S_y/N} = \frac{600,000}{160} = 3750 \text{ mm}^2$$

Hence, for the thickness $t = 16$ mm we find the width of the sheet (Equation (8.11)):

$$w_0 = \frac{K_r FN}{tS_y} + n_r d = \frac{600,000}{2 \times 16 \times 100} + 3 \times 20 = 295 \text{ mm}$$

The length of the trailing edge will be found from Equation (8.13):

$$s = \frac{K_r FN}{2tS_y} + \frac{d}{2} = \frac{600,000}{2 \times 16 \times 100} + \frac{20}{2} = 210 \text{ mm}$$

8.2.4. Prestressed Rivet Joints

Bolts and rivets may, in certain load situations, be axially loaded, with the tensile loading being predominant. This is the case when rivets are loaded by thermal stresses developed during cooling, for example.

During cooling, steel rivets with hot riveting almost invariably reach thermal stresses higher than the material yield stress because initially the rivet is hot and the joined elements relatively cold. Therefore, after hot riveting there is generally remaining stress equal to the rivet material yield stress. In fluid-tight joints this is desirable because the axial force developed presses the two parts together and the force might also be sufficient to support the joint load by friction.

For cold riveting, heat is produced during the plastic deformation of the rivet head, which increases the rivet temperature. Upon cooling, thermal stresses are also developed, usually below the yield strength of the material. The situation is shown in Figure 8.9. At the beginning of the hemispherical head formation, the hammering force is P_b, the stress is S_y, and the diameter is d. At the end of the hammering, the force is P_e, corresponding to a diameter approximately $2d$ and a stress equal to tensile strength S_u. Therefore:

$$P_b = \frac{\pi d^2 S_y}{4}$$

$$P_e = \pi d^2 S_u$$

$$(8.15)$$

The change in height is computed for invariable volume, $\Delta h = 5d/3$, and assuming a linear law of variation of the hammering force, the mechanical work during the riveting process is:

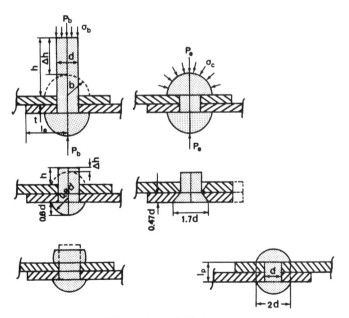

Figure 8.9. Cold riveting.

$$W = \frac{(P_b + P_e)\Delta h}{2} \tag{8.16}$$

Assuming that most of the produced heat enters the rivet, the rise of temperature will be:

$$\Delta T = \frac{W}{c\rho V} \tag{8.17}$$

where c is the specific heat, ρ is the density, and V is the volume of the rivet. Therefore:

$$\Delta T = \frac{5\pi}{6} d^3 \frac{S_y/4 + S_u}{c\rho V} \tag{8.18}$$

The thermal stress during cooling is:

$$\sigma = \alpha E \Delta T = \frac{5\pi}{6} \frac{\alpha E}{c\rho} \frac{d^3}{V} \left(\frac{S_y}{4} + S_u\right) \tag{8.19}$$

where α is the coefficient of thermal expansion. The friction force per rivet is:

$$Q = fF_{\text{rivet}} = \frac{fN_f \sigma \pi d^2}{4} \tag{8.20}$$

where f is the friction coefficient between rivet head and plate, usually $f > 0.3$, and N_f is the number of friction surfaces that transmit the shear force. For example, for the riveting of Figure 8.6, $N_f = 1$, while for the riveting of Figure 8.7, $N_f = 2$.

For hot riveting, the rivet is heated first above the point where the shear strength has a sharp decline (above 1250°C for most steels) and then the rivet head is formed by forging. Since the plate temperature does not increase appreciably, cooling of the rivet almost invariably causes the thermal stress to rise above the yield strength, and as the rivet cools it continues to yield. When the rivet is cold, the stress in the rivet is approximately equal with the yield strength S_y. Thus, the rivet force is $S_y \pi d^2/4$ and the friction force per rivet is $fN_f S_y \pi d^2/4$. All rivets can be considered as transmitting the same friction force. Therefore:

$$F = \frac{fN_f S_y \pi d^2}{4} \tag{8.21}$$

This equation needs to be satisfied in order for the direct shear force to be transmitted by friction instead of shear force on the rivet. This is desirable in some cases, such as when we need the riveting to be pressure tight, as in boilers and pressure vessels.

Design Procedure 8.2: Design of Friction-Riveted Joints

Input: Shear load, material properties, safety factor. **Output:** Riveting dimensions

Step 1: Select type of riveting, number of rows (usually we start with one row), number of friction surfaces per rivet N_f.

Step 2: Use Equations (8.9b) and (8.14b) to determine the plate thickness and rivet diameter, if the plate width is known, or the plate width and the rivet diameter, if the plate thickness is known.

Step 3: Rivet diameter should be of the order of the plate thickness. If not, select a different type of riveting and go to step 1.

Step 4: Use Equation (8.9c) to check the bearing strength. If not sufficient, select a different type of riveting and go to step 1.

Step 5: Use Equations (8.9c) or (8.9d) to determine the plate margins.

Step 6. Check that there is sufficient space for the rivet heads ($w_0 - n_r(3d) + 5$ (mm) > 0). If not sufficient, select a different type of riveting and go to step 1.

Example 8.2 An aluminum alloy rivet of diameter $d = 5$ mm, volume 300 mm³ is cold-formed. Determine the shear force that the rivet can transfer by friction alone. For the material, $S_u = 240$ MPa, $S_y = 130$ MPa, $\rho = 2700$ kg/m³, $c = 921$ J/kg°C, $E = 74$ GPa, $\alpha = 22 \times 10^{-6}$/°C. The friction coefficient between plates is $f = 0.3$.

Solution The residual stress after the formation of the rivet is (Equation (8.14)):

$$
\begin{aligned}
\sigma = \alpha E \Delta T &= \frac{5\pi}{6} \frac{\alpha E}{c\rho} \frac{d^3}{V} \left(\frac{S_y}{4} + S_u \right) \\
&= \frac{5\pi}{6} \frac{22 \times 10^{-6} \times 74 \times 10^9}{921 \times 2700} \frac{0.005^3}{300 \times 10^{-9}} \left(\frac{130 \times 10^6}{4} + 240 \times 10^6 \right) \\
&= 19.4 \times 10^7 \text{ N/m}^2
\end{aligned}
$$

The force imposed by the rivet on the plate is:

$$
F = \sigma A = 19.4 \times 10^7 \times \pi \frac{0.005^2}{4} = 3009 \text{ N}
$$

The friction force

$$
F_f = fF = 0.3 \times 3009 = 1.143 \text{ N}
$$

Example 8.3 A design engineer is seeking an alternative design using riveting, instead of welding, for welded pressure vessel shown in Figure E8.3. Calculate the wall thickness and the longitudinal riveting if the internal pressure is 3.2 MPa, the material has $S_y = 48$ N/mm², and the safety factor must be $N = 3$.

Solution The longitudinal seam transmits force per unit length, for $D = 1200$ mm, $p = 3.2$ MPa (N/mm²):

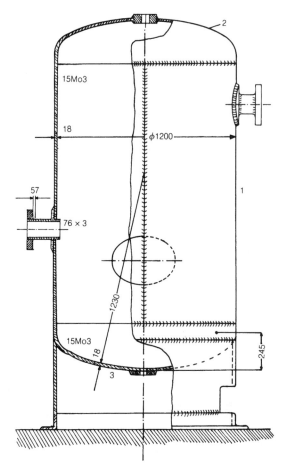

Figure E8.3. (After Decker 1973.)

$$P_{1mm} = \frac{pD}{2} = 1200 \frac{3.2}{2} = 1920 \text{ N/mm}$$

Riveting of type 9 is selected (Table 8.4), with weakening factor $v = 0.7$. Therefore, the wall thickness will be:

$$t = \frac{Dp}{2(S_y v/N)} = 17.5 \text{ mm}$$

Allowing 1 mm for corrosion, wall thickness 20 mm is selected. As for the rivets, from Table 8.4:

$$d_1 = (50s)^{1/2} - 7 = 24.6 = 25 \text{ mm}$$

The spacing is $w = 3d_1 + 10 = 85$ mm, $e = 1.5d_1 = 38$ mm, $e_1 = 0.6w = 51$ mm, $e_2 = 1.5d_1 = 38$ mm, and $s_L = 0.8t = 16$ mm.

The weakening factor, taken as 0.7, should be tested:

$$v = \frac{s - d_1}{s} = \frac{85 - 25}{85} = 0.706$$

which is acceptable.

8.2.5 Compound Rivet Joints

In many instances, a joint might have a compound loading. Such joints are, for example, riveting, bolting, and welding with any loading condition, in particular shear, bending, and torsion.

Take, for example, compound riveting (it could also be bolting or spot welding) (Figure 8.10). It is loaded by a direct shear force Q and moment M in the same plane. The direct shear force can be assumed equally distributed among the rivets. Each rivet is loaded by a shear force (termed *direct load*):

$$P_d = \frac{Q}{n} \tag{8.22}$$

Moreover, each rivet is loaded due to the moment M. The shear forces P_1, P_2, . . . , P_n owing to the moment M are not equal in magnitude or direction.

Assuming that the plate is rigid and the rivets elastic, the shear deformation of each rivet, and thus the shear force of each rivet owing to bending, termed *indirect load,* is proportional to its distance from the area centroid. This is:

$$\frac{P_1}{a_1} = \frac{P_2}{a_2} = \frac{P_3}{a_3} = \cdots = \frac{P_n}{a_n} \tag{8.23}$$

Balance of moments requires:

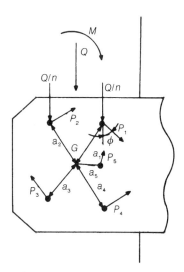

Figure 8.10. Compound joint.

$$M = P_1 a_1 + P_2 a_2 + \cdots + P_n a_n \tag{8.24}$$

Equations (8.15) and (8.16) yield:

$$P_i = M a_i \Big/ \sum_j a_j^2 \tag{8.25}$$

The total shear force on rivet i is then:

$$F_i = \left[P_i^2 + \left(\frac{Q}{n}\right)^2 + 2P_i \left(\frac{Q}{n}\right) \cos \phi_i \right]^{1/2} \tag{8.26}$$

where ϕ_i is the angle between the force P_i and the direct shear force.

The force P_i is perpendicular to the line connecting each rivet with the center of area of the riveting, which has coordinates:

$$x_G = \left(\sum_j x_i\right) \Big/ n, \; y_G = \left(\sum_i y_i\right) \Big/ n$$

if x is in the direction of the direct shear force and y is in a direction perpendicular with it. The Cartesian components of the force P_i are therefore:

$$F_{yi} = \frac{P_i}{(1 + \lambda_i^2)^{1/2}}, \; F_{xi} = \lambda_i F_{yi}$$

$$\lambda_i = \frac{y_i - y_G}{x_i - x_G} \tag{8.27}$$

Thus, an alternative expression for the resultant shear force per rivet, is:

$$F_i = \left[\left(F_{xi} + \frac{Q}{n}\right)^2 + F_{yi}^2 \right]^{1/2} \tag{8.28}$$

The rivet with the maximum resultant shear force F_{max} needs to be found. In most cases, this is apparent from the location of the rivet (the more distance from the geometric center of the riveting, the higher the resultant shear force) and the direction of the direct and indirect shear forces (the more they approach collinear position, the higher the resultant shear force). When the location of the rivet with the maximum resultant shear force is not obvious, all the rivet resultant shear forces need to be found and the maximum one sorted out. The design equation for the rivet with the maximum shear stress will then be:

$$\tau_{av} = \frac{F_{max}}{\pi d^2 / 4} \le \frac{S_{sy}}{N} \tag{8.29}$$

Program RIVETS, part of the MELAB 2.0 package, performs the design of compound riveted joints. This program can also be used for bolting and spot welding.

Example 8.3a Eccentrically Loaded Compound Riveted Joint

Figure E8.3a shows an eccentrically loaded compound rivet joint loaded as shown. If the plate thickness is 10 mm and the plate material has $S_y = 300$ MPa, find the rivet diameter for safety factor $N = 3$.

Solution   The direct shear force on each rivet (direct load) is (Equation 8.14a)):

$$P_d = \frac{Q}{n} = \frac{50,000}{6} = 8,333 \ N$$

The distance of the corner rivets from the center of the riveting (the origin shown) is $(75^2 + 50^2)^{1/2} = 90$ mm.

The force on the top-right rivet 1 due to the moment $M = 50,000 \times (0.150 + 0.050) = 10,000$ Nm (indirect load) will be found from Equation (8.17):

$$P_i = Ma_i / \sum_j a_j^2$$

$$P_1 = \frac{10,000 \times 10^3 \times 90}{4 \times 90^2 + 2 \times 50^2} = 24,064 \ N$$

It is apparent that this rivet will have the highest resultant force because it has the highest distance and the direction of the indirect force has a small angle with the direct force.

From the triangle $O12$ the angle $\angle 1O2 = \phi_1$, $\cos \phi_1 = 50/90 = 0.555$ and Equation (8.18) yields:

$$F_1 = \left[P_1^2 + \left(\frac{Q}{n} \right)^2 + 2P_1 \left(\frac{Q}{n} \right) \cos \phi_1 \right]^{1/2}$$

$$= [24,064^2 + 8,333^2 + 2 \times 24,064 \times 8,333 \times 0.555]^{1/2} = 29,514 \ N$$

Design Equation (8.18c) yields:

$$\tau_{av} = \frac{F_{max}}{\pi d^2 / 4} \le \frac{S_{sy}}{N}, \ d^2 = \frac{4NF_{max}}{S_{sy}\pi} = \frac{4 \times 3 \times 24,064}{300 \times 10^6 \times \pi} = 376 \times 10^{-6} \ m^2$$

or $d = 19.4$ mm. The next available diameter is $d = 20$ mm. For plate thickness $t = 10$

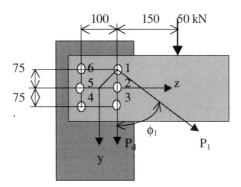

Figure E8.3a

mm, we get $d = t + 8 = 18$ mm, thus the diameter of the rivet found is close to the suggested value. Otherwise, more or fewer rivets or another layout should be selected and the computation repeated.

The spreadsheet of Example (8.6) can be used for this problem with all out-of-the-riveting-plane forces set equal to zero.

8.2.6. Stress Concentration

For the end rivets in multiple-row riveting, stress concentration occurs as discussed above. The associated theoretical stress concentration factors K_r can be found from Table 8.3.

Stress concentration also exists in the riveted part due to the hole, which, from the photoelastic studies of Heywood (1965), is curve (b) in Figure 8.11:

$$K_{tn} = \frac{\sigma_{max}}{F/(D - d)t} = 2 + \left(\frac{1 - d}{D}\right)^3 \tag{8.30}$$

where the subscript n indicates that the stress concentration factor was computed with the average stress over the net area. If the pressure of the rivet in the hole is taken into account, the stress concentration factor increases (curve (a) in Figure 8.11).

For ductile materials and static loads, only the reduction in the available section is accounted for owing to stress relieving. The section reduction factor is then:

$$v = \frac{A}{A_0} \tag{8.31}$$

where A_0 is the theoretical section between adjacent rivets Dt and A is the actual section $(D - d)t$. For pressure-vessel riveting, section reduction factors and design details are given in Table 8.4.

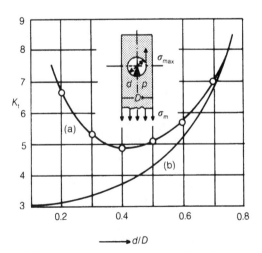

Figure 8.11. Stress concentration in a riveted connection.

Safety factors for most riveting applications, such as construction and pressure vessels, are dictated by local codes. Usually, for steel construction, factors of safety for riveting are taken to be between 1.5 and 2, for pressure vessels and steam boilers, up to 5.

Design Procedure 8.3: Design of Compound Riveted Joints—Forces in the Riveting Plane

Input: Shear load, material properties, safety factor. **Output:** Riveting dimensions

Step 1: Select riveting topology and the number of rivets.

Step 2: Find the direct shear force Q and the in-plane moment M applied to the riveting.

Step 3: Determine the direct shear stress in each rivet (Equation 8.14*a*).

Step 4: Determine the indirect shear force due to the moment on each rivet (Equation 8.17).

Step 5: Determine the resultant shear force on each rivet (Equations (8.18*a*), (8.18*b*)).

Step 6. Find the maximum resultant shear force and use Equation (8.18*c*) as a design equation to find the rivet diameter.

Step 7: Compare the rivet diameter you found with the plate thickness. If incompatible, select a different type of riveting and go to step 1.

(For out-of-plane loading, see Example 8.6.)

Example 8.4 A riveted joint for steel construction consists of two hot-rolled profiles ⌐ 60 × 60 × 6 mm and gusset plate. The joint transmits force $U_1 = 154$ kN. The two parallel ⌐ section members are joined with the 10 mm gusset plate with two 15 mm diameter rivets per member. Check the joint for riveting strength, if the safety factor is 1.1, and for the rivet, members, and gusset plate $S_{sy} = 120$ N/mm², $S_y = 160$ N/mm², $S_b = 300$ N/mm².

Solution For transmission of the force there are four rivets of 15 mm diameter each. The rivet average shear stress is:

$$\tau_{av} = \frac{U_1}{4\pi d^2/4} = \frac{154{,}000}{4\pi \times 15^2/4} = 217.6 \text{ mm}^2$$

The actual safety factor is:

$$N = \frac{S_{sy}}{\tau_a} = \frac{120}{217.6} = 0.55 < 1.1$$

The gusset plate has bearing stress

$$\sigma_1 = \frac{U_1}{wd_1 n} = \frac{154{,}000}{10 \times 15 \times 4} = 256.6 \text{ N/mm}^2$$

The corresponding safety factor is:

$$N = \frac{300}{256.6} = 1.17 > 1.1$$

The axial member is a double angle. It has a section of 691 mm² (see Appendix C.4). The section is reduced by $6 \times 15 = 90$ mm², owing to the rivet. Therefore, the net section is $A = 691 - 90 = 601$ mm². Therefore, because there are two loaded members, the average tensile stress is:

$$\sigma = \frac{U_1}{2A} = \frac{154,000}{2 \times 601} = 128 \ \text{N/mm}^2$$

The corresponding safety factor is:

$$N_s = \frac{S_y}{\sigma} = 1.25 > 1.1$$

The riveting will fail by rivet shear.

Example 8.5   The electromagnetic coupling shown in Figure E8.5 has the friction material riveted on the coupling plate by three rivets at 0°, 90°, 180°, 270° angles. The rivet material is ASTM A572 Grade 50 and the maximum transmitted torque is 500 Nm. Find the safety factor of the riveting. Rivet spacing is 90°.

Solution   The total transmitted peripheral force is:

$$P = \frac{T}{r} = \frac{500}{0.185} = 2710 \ \text{N}$$

This is divided into $3 \times 4 = 12$ rivets. Therefore, per rivet:

$$P_n = \frac{P}{n} = \frac{2710}{12} = 226 \ \text{N}$$

The rivet section is $A = 32.17$ mm². The average shear stress is:

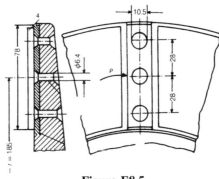

Figure E8.5

$$\tau_a = \frac{P_n}{mA} = \frac{226}{1 \times 32.17} = 7 \text{ N/mm}^2$$

For the rivet material,

$$S_y = 225 \text{ N/mm}^2$$

The yield strength in shear is, then, with the von Mises criterion:

$$S_{sy} = 0.5775_s = 113 \text{ N/mm}^2$$

Therefore, the safety factor is:

$$N_s = \frac{S_{sy}}{\tau_a} = \frac{113}{7} = 16$$

The bearing load is:

$$\sigma_b = \frac{P_n}{sd} = \frac{226}{4 \times 6.4} = 8.83 \text{ N/mm}^2$$

The bearing strength is:

$$S_c = 2S_y = 2 \times 230 = 460 \text{ N/mm}^2$$

The associated safety factor is:

$$N_c = S_c\sigma_c = 52$$

The smaller safety factor is to be applied, which is 16.

Example 8.6 Out-of-Plane Loading of a Riveting

The riveting shown in Figure E8.6, symmetric about the y- and z-axes, is loaded by forces $F_x = 40$ kN, $F_y = 20$ kN, $F_z = 20$ kN, $M_x = 30$ kNm, $M_y = -15$ kNm, $M_z = 15$ kNm.

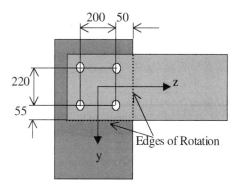

Figure E8.6

The rivet material is steel with $S_y = 220$ MPa and the safety factor is $N = 3.5$. Find the required rivet diameter assuming that there is no residual stress due to rivet formation and bending moments M_y (negative) and M_z (positive) produce elastic rotations about the edges shown.

Solution   We shall use EXCEL to find the solution under the most general loading. The spreadsheet is shown in Table E8.6. We used Solver to make cell F31 ($\sigma_{eq} - S_y/N$) equal to zero by changing cell $D10$ (the rivet diameter). The result is $d = 16.2$ mm, specifying initially any arbitrary value, say $d = 10$ mm. The normal stress will be tension due to the axial force F_x and the moments M_y and M_z:

$$\sigma_x = \frac{F_x}{A} + \frac{M_y}{I_y} c_y + \frac{M_z}{I_z} c_z$$

Shear stresses are direct shear stresses due to the forces in the plane of the riveting F_y and F_z and the torque M_z:

$$\tau_y = \frac{F_y}{A} - \frac{M_x r}{I_x} \frac{z}{r}, \ \tau_z = \frac{F_z}{A} + \frac{M_x r}{I_x} \frac{y}{r}$$

The shear and normal stresses are combined with the von Mises criterion of failure to yield the maximum equivalent stress at each rivet. The maximum of the rivet stresses is compared with the allowable stress to yield the design equation:

$$\sigma_{eq\text{-}max} = \left[\sqrt{\sigma_x^2 + 3(\tau_y^2 + \tau_z^2)} \right]_{max} = \frac{S_y}{N}$$

With EXCEL and Solver we modify the diameter to satisfy the design equation.

8.3. BOLTS

8.3.1. Kinematics and Force Equilibrium: Power Screws

Bolt or *screw joints* are separable joints held together by screw fastening, such as screws, bolts, studs, and nuts, or by thread cut on the parts to be joined (Figure 8.14). The thread is formed as a helical groove into the surface of a cylindrical bar or hold and has a cross-section complying with the corresponding thread profile. The term *screw* is used with any part with external thread, while a *nut* is a part with a hole that has an internal thread.

The *outer diameter* will be designated here by d_1 and equals the *nominal diameter d* of the thread, d_3 is the *inner diameter* of the thread, and d_2 is the *pitch diameter* of the thread. The distance of two adjacent threads is termed *pitch P*. For standard coarse metric threads the other dimensions determining the thread geometry are shown in Figure 8.14. Different types of threads are included in Appendix C.

Screw joints are used very extensively in machine design. Because of their standardization, their cost is relatively low and the parts joined can be disassembled very easily, usually without destroying the bolts.

TABLE E8.6

Example E8.6. Design of a Generally Loaded Rivet Joint

Step 1: Enter Data

Select Number and layout of riveting:

No of Rivets m = 4

Rivet Coordinates:

No	y	z		Loading:		Other Data	
1	0.11	0.1		$F_x =$	5000	$a_z =$	0.15
2	0.11	-0.1		$F_y =$	2000	$a_y =$	0.15
3	-0.11	-0.1		$F_z =$	2000	$S_y =$	2.20E+08
4	-0.11	0.1		$M_x =$	2000	Safety Factor N =	3.5
				$M_y =$	1000	Failure criterion $\alpha =$	3
				$M_z =$	1000	(3 for von Mises, 4 for Tresca)	

Step 2: Assume a rivet diameter (any guess will do)   Rivet Diameter d = 0.018912

Step 3: Find Geometric properties of riveting

Riveting cross-section = $m\pi^2/4$ = 0.237653191

The coordinates of the rivet centroid are:

$y_g = \Sigma y_i A_i / \Sigma A_i = A\Sigma y_i/m =$ 0

$z_g = \Sigma z_i A_i / \Sigma A_i = A\Sigma z_i/m =$ 0

The moments of inertia of the riveting section are:

$I_z = \Sigma(y_i\text{-}yg)^2 A_i =$ $\Sigma(y_i\text{-}yg)^2(\pi d_i^2/4) =$ 9.06383E-06

$I_y = \Sigma(z_i\text{-}zg)^2 A_i =$ $\Sigma(z_i\text{-}zg)^2(\pi_i^2/4) =$ 1.12361E-05 $I_x = I_y + I_z =$ 2.03E-05

TABLE E8.6. (Continued)

Step 4: Compute Stresses

Rivet Area $A = \pi^2/4 =$ 0.000280904

No.	c_y	c_z	r	σ_x	τ_y	τ_z	τ_{res}	σ_{eq}
1	0.11	0.1	0.148660687	55385000.18	1261742	11632199	1716121	62857143
2	0.11	-0.1	0.148660687	37585305.57	12617422	-8072260	1497867	45669874
3	-0.11	-0.1	0.148660687	13312994.73	-9057483	-8072260	1213257	24876374
4	-0.11	0.1	0.148660687	31112689.34	-9057483	11632199	1474266	40249691

Maximum equivalent stress $\sigma_{max} =$ 62857143

Step 5: Apply Design Equation and use Solver to determine the unknown diameter.

Design Equation: $\sigma_{max} - S_y/N =$ (must be zero)	-1.04308E-07

We use Solver to compute the rivet diameter (cell H12) that makes the left-hand of the design equation (cell E27) equal to zero.

536

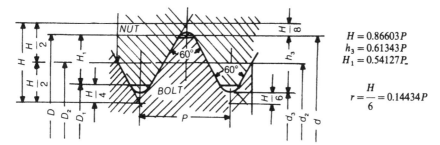

$$H = 0.86603P$$
$$h_3 = 0.61343P$$
$$H_1 = 0.54127P$$

$$r = \frac{H}{6} = 0.14434P$$

Figure 8.12. Standard bolt threads.

In steel structures, use of bolts considerably simplifies the manufacture and assembly.

According to their purpose, screw threads are classified as follows:

1. *Fastener threads* are usually very heavily loaded and can develop high fastening forces with relatively small torque owning to the wedge effect of their thread.

2. *Fastening and sealing threads* are for both fastening parts together and preventing the leakage of fluids at high pressure.

3. *Power threads for transmitting motion* reduce friction and are of a special shape, such as trapezoidal or orthogonal.

Joseph Whitworth (1803–1887), the son of a Congregational minister, was born in Stockport. At the age of fourteen he was apprenticed to a Derbyshire cotton-spinner. Whitworth studied the machinery in the factory and was critical of the poor standards of workmanship. This inspired him to become an engineer. From 1821 he worked as a mechanic. Over the next few years, he built a successful knitting machine (1835) and a horse-drawn mechanical roadsweeper (1842). He produced guns for the British and the French armies. By 1860, Whitworth's specifications for sizes of screw threads were generally accepted throughout Britain. He was an active philanthropist and paid for many engineering scholarships. He was knighted for his contributions to mechanical engineering.

Accordingly, the international standards[1] have specified different kinds of threads, some of which are shown in Appendix C.1.

A bolt is loaded, in general, by an axial force P that is transferred to the bolt from the nut (Figure 8.14). The contact surface of the bolt with the nut is not perpendicular to the bolt axis but forms an angle β' in respect to it and is deter-

---

[1] An Internet search through a substantial number of standards can be done at http://www.nssn.org/

Figure 8.13. Standard bolt.

mined by the angle of the helix α and another angle β in respect to the axis, which corresponds to the inclination of the thread (Figure 8.15).

To move the nut, we have to apply a peripheral force P_u. This force must balance the load and the friction force $P_n f$, where P_n is the normal force at the contact and f is the coefficient of friction between the bolt and the nut. The friction force has a positive or a negative sign depending on the direction of motion.

If ρ is the friction angle, $\tan \rho = f$, then, from Figure 8.14:

$$P_u = P \tan(\alpha \pm \rho) \tag{8.32a}$$

where the "+" sign refers to tightening and the "−" sign to untightening.

When, as usual, $\beta > 0$, the normal force changes and becomes $P/\cos \beta$. It is equivalent to the normal force for an orthogonal thread with $\beta = 0$ when the effective friction angle ρ' is introduced and then:

$$P_u = P \tan(\alpha \pm \rho') \tag{8.32b}$$

where:

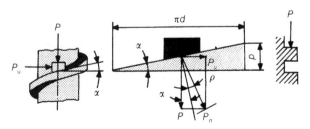

Figure 8.14. Force equilibrium on thread.

Figure 8.15. Forces on orthogonal and triangular threads.

$$\tan \rho' = \frac{\tan \rho}{1 + \cos^2 \alpha \tan^2 (\beta/2)} \cong \frac{\tan \rho}{\cos(\beta/2)}$$

The required torque at the nut is:

$$M_T = \frac{P_u d_2}{2} = \frac{P \tan(\alpha \pm \rho') d_2}{2} \qquad (8.33)$$

where d_2 is the pitch diameter of the thread.

Because of the friction, the work applied to the nut is not all transformed into the work pushing the nut upwards and the efficiency of the thread is the ratio of the useful work to the applied work, which is equal to the ratio of the moment without friction divided by the moment with friction:

$$\eta_t = \frac{\tan \alpha}{\tan(\alpha + \rho')} \qquad (8.34)$$

During the untightening, the efficiency will be:

$$\eta_u = \frac{\tan \alpha}{\tan(\alpha - \rho')} \qquad (8.35)$$

The above equation means that if $\alpha - \rho'$ is less than 0, the efficiency is negative and we have to apply moment in order to untighten the nut. If, however, the efficiency is positive, it means that the nut has the tendency to untighten itself freely. In order for this to happen, we must have:

$$\tan(\alpha - \rho') > 0 \qquad (8.36)$$

or in other words $\alpha > \rho'$. This suggests that the smaller the thread angle α, the more steady is the bolt. For this reason, for bolts which are loaded with dynamic load and there is a danger of self-untightening of the nut, we use special threads with small pitch and consequently small angle α. The screw efficiency is plotted in Figure 8.16. Negative friction angle corresponds to untightening efficiency.

When the nut is tightened against a stationary part, there is friction between the nut and the part during rotation of the nut. In this case, the applied torque has to overcome this friction as well. If we assume that the mean diameter of the appli-

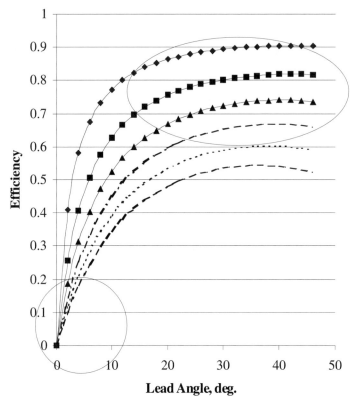

Figure 8.16. Efficiency of screws (friction angle ρ': ♦ = 0.05, ■ = 0.10, ▲ = 0.15, – – = 0.20, . . . = 0.25, –.–. = 0.30). The circle indicates the preferred area of operation for the fasteners, the ellipse, that for the power screws.

cation of the friction forces on the nut is d_A (generally the average of the two diameters of bolt and nut), and f_A is the coefficient of friction between the nut and the stationary part, there will be friction from the thread and the nut seating:

$$M = M_T + M_A = P \left[\frac{\tan(\alpha \pm \rho')d_2}{2} + \frac{f_A d_A}{2} \right] \tag{8.37}$$

In Table 8.5, coefficients of friction are tabulated for normal practices in bolt design and utilization.

Nut friction coefficient decreases the efficiency of the screw. This efficiency, adding the collar friction in the numerator of the fraction in Equation (8.34), is:

$$\eta = \frac{\tan \alpha}{\tan(\alpha \pm \rho') \pm f_A d_A / d_2} \tag{8.38}$$

Nut friction contributes towards a more stable bolt. In this case, the condition for stability of the bolt is:

TABLE 8.5. Friction Coefficients for Nuts and Bolts

Material		Lubrication		
Bolt	Nut	None	Oil	MoS$_2$
Surface Treatment				
Mn–Ph	None	0.14 . . . 0.18	0.14 . . . 0.15	0.10 . . . 0.11
Zn–Ph	None	0.14 . . . 0.21	0.14 . . . 0.17	0.10 . . . 0.12
8 μm Zn	None	0.125 . . . 0.18	0.125 . . . 0.17	
7 μm Cd	None	0.08 . . . 0.12	0.08 . . . 0.11	
8 μm Zn	5 μm Zn	0.125 . . . 0.17	0.14 . . . 0.19	
7 μm Cd	6 μm Cd	0.08 . . . 0.12	0.14 . . . 0.15	

$$\frac{\tan(\alpha - \rho')d_2}{2} < \frac{f_A d_A}{2} \tag{8.39}$$

To increase the friction between the nut and the station parts, thus making a more stable bolting, special safety parts are utilized (Figure 8.12).

Example 8.7   The tool carriage of a lathe is driven by a power screw that has a single-start ACME thread of diameter $d = 50$ mm, inner diameter $d_3 = 24$ mm, thread angle $\beta = 29°$, and pitch $l = 12$ mm. The maximum axial force is 2 kN at rotating speed 30 rpm. The thrust is supported by the end of the screw with a collar having inner diameter 5 mm and outer diameter 10 mm. If the friction coefficient is 0.15 for the thread and 0.10 for the collar, determine the power to drive the screw and the screw efficiency.

Solution   The screw thread has pitch diameter $d_2 = (d + d_3)/2 = (50 + 24)/2 = 37$ mm, the helix angle is $\tan \alpha = l/\pi d = 12/50\pi = 0.0764$, $\alpha = 4.37°$, $\beta/2 = 145°$, $\rho = a \tan(0.15) = 8.53°$. The collar mean diameter $d_A = (5 + 10)/2 = 7.5$ mm.
Equation (8.38) gives:

$$M = P\left[\frac{\tan(\alpha \pm \rho')d_2}{2} + \frac{f_A d_A}{2}\right]$$

$$= 2000\left[\tan\left(4.37 + \frac{8.53}{\cos(14.5)}\right)\frac{0.037}{2} + 0.10\frac{0.0075}{2}\right]$$

$$= 9.4 \text{ Nm}$$

The power required is:

$$W = M\omega = 9.4 \times 2\pi\frac{30}{60} = \underline{29.6 \text{ W}}$$

The efficiency is (Equation (8.38)):

$$\eta_t = \frac{\tan\alpha}{\tan(\alpha + \rho') + f_A d_A / d_2}$$

$$= \frac{\tan(4.37)}{\tan(4.37) + \tan(0.15)/\cos(14.5) + 0.10 \times 7.5/37}$$

$$= 0.300 \text{ or } \underline{30\%}$$

Figure 8.16 with $\alpha = 0.99$ rad, $\rho = 0.14$ rad gives approximate efficiency 0.4 (without the collar friction).

8.3.2. Design for Strength

In principle, bolts can be made from a variety of materials; in fact, for special applications they are indeed made of many different materials. Steel bolts are the most widely used and they have been standardized.

The ISO has specified twelve categories of bolt strengths, as shown in Table 8.6, where the number designation means approximately tensile strength in kg-f/mm^2.

The Society of Automotive Engineers uses the standard shown in Table 8.7, where in the left column is a sign that is pressed on the head of the bolt.

Table 8.7 has a column with a *proof load*. This is a material strength, slightly lower than the yield strength, especially in high-strength steels. The reason is that for most steels yield strength is determined on the basis of 0.2% yielding, which is not acceptable for bolts because maintaining the thread geometry is of utmost importance. In all strength considerations for bolts, proof strength (if specified) should be used instead of yield strength.

Bolts are also made from the following materials, among others:

1. Aluminum 2024-T4 + 2011-T3 for machine bolts

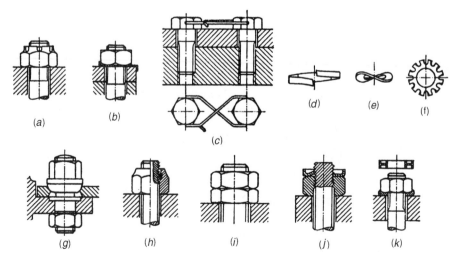

Figure 8.17. Different bolts, nuts, and securing washers (from Niemann 1965).

TABLE 8.6. Allowable Static and Dynamic Loads for Bolts (ISO)

Bolt		3.6	4.6	4.8	5.6	5.8	6.6	6.8	6.9	8.8	10.9	12.9	14.9
						SAE/(ISO Class)							
S_u	N/mm²	340	400	400	500	500	600	600	600	800	1000	1200	1400
S_y	N/mm²	200	240	320	300	400	360	480	—	—	—	—	—
$S_{y0.2}$	N/mm²	—	—	—	—	—	—	—	540	640	900	1080	1260
δ_u	%	25	25	14	20	10	16	8	12	12	9	8	7
Nut		4			5		6			8	10	12	14
S_u	N/mm²	400			500		600			800	1000	1200	1400

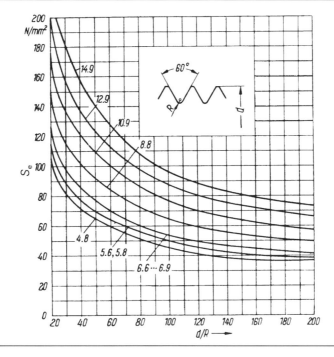

2. Copper, brass, bronze for general use, especially for chemical industries
3. Nickel alloys for extremely low and high temperatures and acid environment
4. Inconel nickel alloy for high temperatures and acid environments
5. High alloys for magnetic applications and where high hardness and high resistance to oxidation are demanded
6. Nylon for resistance to temperature, impact, vibration, chemicals and for electrical insulation
7. Polyvinyl chloride for acid environments
8. Teflon for resistance to oxidation and relatively high temperatures
9. Zinc, cadmium, nickel, and chromium coatings or plating for protection of the bolts in acid or humid environments

TABLE 8.7. SAE Standards of Materials and Strengths for Bolts

Identification Mark	Grade Designation	Fastener Description	Material	Is Mfr's Identification Symbol Required?	Nominal Size Range (in.)	Proof Load (psi)	Yield Strength (min. psi)	Tensile Strength (min. psi)	Brinell	Rockwell	Remarks
No mark	SAE grade 0	Bolts and screws	Steel	Yes	$\frac{1}{4}$ to $1\frac{1}{2}$	—	—	—	—	—	
No mark	SAE grade 1	Bolts and screws	Carbon steel	Yes	$\frac{1}{4}$ to $1\frac{1}{2}$	—	—	55,000	207 max.	B95 max.	Equivalent to ASTM A307, grade A
	GM 255-M	Bolts and screws	Carbon steel	Optional							
No mark	SAE grade 2	Bolts and screws	Carbon steel	Yes	$\frac{1}{4}$ to $\frac{1}{2}$	55,000	—	69,000	241 max.	B100 max.	
					over $\frac{1}{2}$ to $\frac{3}{4}$	52,000	—	64,000	241 max.	B100 max.	
	GM 260-M	Bolts and screws	Carbon steel	Optional	over $\frac{3}{4}$ to $1\frac{1}{2}$	28,000	—	55,000	207 max.	B95 max.	
![symbol]	SAE Grade 3	Bolts and screws	Medium carbon steel	Yes	$\frac{1}{4}$ to $\frac{1}{2}$	85,000	—	110,000	207/269	B95/104	
					over $\frac{1}{2}$ to $\frac{5}{8}$	80,000	—	100,000	207/269	B95/104	
![symbol]	SAE Grade 5	Bolts and screws	Medium carbon steel, heat treated	Yes	$\frac{1}{4}$ to $\frac{3}{4}$	85,000	—	120,000	241/302	C23/32	Equivalent to ASTM A449
					over $\frac{3}{4}$ to 1	78,000	81,000	115,000	235/302	C22/32	
	GM 280-M	Bolts and screws		Optional	over 1 to $1\frac{1}{2}$	74,000	77,000	105,000	223/285	C19/30	
![symbol]	SAE grade 5.1	Sems	Carbon steel, heat treated	Yes	$\frac{3}{8}$ and smaller	85,000	—	120,000	241/375	C23/40	
	GM 275-M	Sems		Optional							
![symbol]	SAE grade 6	Bolts and screws	Medium carbon steel, heat treated	Yes	$\frac{1}{4}$ to $\frac{5}{8}$	110,000	—	140,000	285/331	C30/36	
					over $\frac{5}{8}$ to $\frac{3}{4}$	105,000	—	133,000	269/331	C28/36	
![symbol]	SAE grade 7	Bolts and screws	Low alloy steel, heat treated	Yes	$\frac{1}{4}$ to $1\frac{1}{2}$	105,000	110,000	133,000	269/321	C28/34	Threads rolled after heat treatment
	GM 290-M	Bolts and screws		Optional							
![symbol]	SAE grade 8	Bolts and screws	Low alloy steel, heat treated	Yes	$\frac{1}{4}$ to $1\frac{1}{2}$	120,000	125,000	150,000	302/352	C32/38	Equivalent to ASTM A354, grade BD
	GM 300-M	Bolts and screws		Optional							
![symbol]	GM 455-M	Bolts and screws	Corrosion resistant steel	Optional	$\frac{1}{4}$ to $1\frac{1}{2}$	40,000	—	55,000	143 min.	B79 min.	

Source: Camcar

8.3.3. Stress Concentration and Safety Factors: Bolt Design

Bolts: Stress concentration can be present in bolts and screws for many reasons. Bolts have inherent stress concentration due to their geometry. The thread of the bolt is a succession of notches with substantial stress concentration. For fatigue loading, the effective fatigue stress concentration factor (K_f), for standard ISO thread bolts might be taken for three common bolt materials from Table 8.8. Lower values are for lower grades.

Nuts. Stress concentration appears also in the first threads, which are more heavily loaded than the distant ones, as one can see from Figure 8.18. This results in a nonuniform stress distribution (Figure 8.19) on the threads of both the bolt and the nut. Because of this, increase of the nut length beyond a point does not add to the strength of the thread. The associated stress concentration factor for bolt/nut threads $K_{b/n}$ is between 2 and 4 for cut and untreated threads and standard nuts; the lower value is for standard coarse-pitch metric threads and the higher value is for finer threads. This value must be applied for both static and dynamic loading because substantial yielding of the threads is not allowed for geometric reasons. Moreover, the method of manufacturing the thread and the form of the nut affect stress concentration in the threads. Therefore, the stress concentration factor *for the threads* is given the form:

$$K_{f-t} = \frac{K_{b/n}}{k_2 k_3} \tag{8.40}$$

where:

$K_{b/n}$ is the stress concentration factor along the nut discussed above, between 2 and 4.

k_2 is due to method of manufacturing, 1.0 for cut threads, 1.2 for rolled threads, 1.3 for rolled and heat-treated threads, 1.4 for quenched and grinded threads.

k_3 is due to the form of the nut, 1.0 for standard nuts, up to 1.2 for elastic nuts, such as the ones shown in Figure 8.20 (designs *b*–*d*).

Fatigue strength of bolts can be improved by mechanical working of threads (rolling, shot-peening) by 20–30%, surface hardening by 40%, and cold-rolled stock by 60%.

Stress concentration in bolt heads is also present. Special designs have been devised to allow for uniform transmission of the bolt load to the connected elements through the head and nut (Figure 8.20). Designs in Figure 8.20*c*, *d*, and *f* substantially reduce the stress concentration in the thread to almost 1.

TABLE 8.8. Fatigue Stress Concentration Factors K_f for Bolts

Bolt Material	Cut Threads	Rolled Threads
Carbon steels	3.0–4.0	2.4–3.2
Alloy steels	4.0–6.0	3.2–4.8
Titanium alloys	4.5–5.5	3.6–4.4

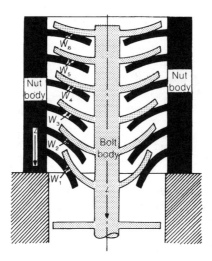

Figure 8.18. Demonstration of stress concentration in screw threads (from Horger 1965).

In the following, different modes of failure and corresponding design methods will be presented.

8.3.3.1. Mode of Failure: Failure of Bolt or Screw by Axial Load, Static OR Dynamic.

As the pitch diameter of the thread is used for power screw kinematic and efficiency determination, the inner diameter d_3 is used for strength calculations because it determines the minimum stress bearing area, termed *stress area*, $A = \pi d_3^2/4$. If the bolt is loaded with axial force F, the average stress needs to be compared with the allowable yield strength:

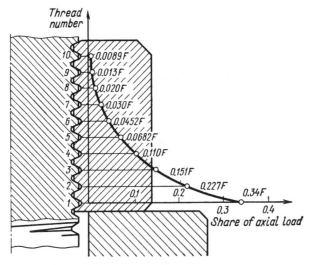

Figure 8.19. Stress concentration in a nut.

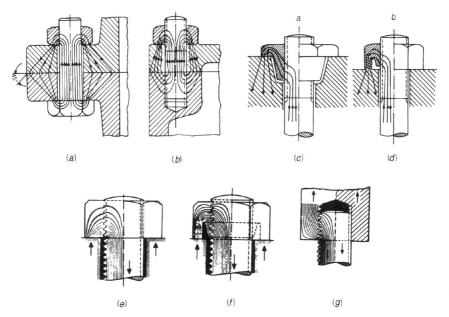

Figure 8.20. Stress flow lines between bolt and nut according to Niemann 1965.

$$\sigma_{\mathrm{av}} = \frac{F}{\pi d_3^2/4} \leq \frac{S_y}{N} \tag{8.41}$$

from which the inner diameter d_3 can be computed. If the axial loading is dynamic and fully reversing, the stress needs to be compared with the allowable fatigue strength:

$$\sigma_{\mathrm{av}} = \frac{K_f F_r}{\pi d_3^2/4} \leq \frac{S_e}{N} \tag{8.42}$$

8.3.3.2. Mode of Failure: Failure of Bolt or Screw by Combined Axial Load, Static AND Dynamic.

For combined static and dynamic loading, the Goodman theory will be used (see Chapter 5). The equivalent static stress is $\sigma_{\mathrm{eq}} = \sigma_m + \sigma_r(S_e/S_u)$ and the Goodman line and the yield line criteria of failure will yield, respectively:

$$\sigma_{\mathrm{eq}} = \frac{F_m + K_f F_r S_u/S_e}{\pi d_3^2/4} \leq \frac{S_u}{N} \tag{8.43}$$

$$\sigma_{\mathrm{eq}} = \frac{F_m + F_r}{\pi d_3^2/4} \leq \frac{S_y}{N} \tag{8.44}$$

where F_m and F_r are the static and dynamic components of the axial force, respectively. Equations (8.43) and (8.44) will yield, in general, different results. If they are used for certification, the smallest safety factor is to be accepted. If they are used for sizing, the largest bolt diameter needs to be used.

Table 8.6 gives the material properties for steel bolts according to ISA. For different materials and types of bolts, material properties need to be obtained from generic material properties, as explained in Chapter 5. For combined, static and dynamic, loads, the methods of Chapter 5 need to be applied.

The bolt–nut combination also needs to have sufficient strength for different modes of failure:

8.3.3.3. Mode of Failure: Failure of Threads by Bearing Stress.

The bearing area of the thread, as projected on a plane perpendicular with the axis of the bolt, is:

$$A_b = n \left(\pi \frac{d_1^2 - d_3^2}{4} \right)$$

where n is the number of threads in the nut. The bearing stress design equation is then:

$$\sigma_b = \frac{K_{f-t}F}{n\pi[(d_1^2 - d_3^2)/4]} \leq \frac{S_b}{N} \tag{8.45}$$

Since $n = L/P$, where L is the nut length and P is the thread pitch:

$$\sigma_b = \frac{K_{f-t}FP}{L\pi(d_1^2 - d_3^2)/4} \leq \frac{S_b}{N} \tag{8.46}$$

Stress concentration factors for threads can be found from Equation (8.40).

For dynamic or combined static and dynamic loads, Equation (8.46) is modified as was done in Equations (8.43) and (8.44).

8.3.3.4. Mode of Failure: Failure of Threads by Shearing.

This usually happens at the pitch diameter. The shear area of both the thread and the nut is $A_s = [\pi(d_1 + d_3)/2](L/2) = L\pi(d_1 + d_3)/4$:

$$\tau_{av} = \frac{K_{f-t}F}{L\pi(d_1 + d_3p)/4} \leq \frac{S_{sy}}{N} \tag{8.47}$$

For dynamic or combined static and dynamic loads, Equation (8.47) is modified as was done in Equations (8.43) and (8.44).

In both Equations (8.46) and (8.47), the strength of the weaker material will be used, usually the nut material. From these two equations two nut lengths L are computed. The larger is the one to be accepted. Stress concentration factors for threads can be found in Table 8.6.

Experience suggests the following guidelines for the length L of the nut or the depth of a threaded hole:

Steel bolt: steel nut 0.8*d* for the same grade of steel, 1.0–1.4*d* for dissimilar steels, cast iron nut 1.0–1.4*d*, aluminum nut 2.2*d*, aluminum alloy nut 1.0*d*–1.4*d*. Bolts made with other materials and all critical bolt joints need to be designed with the above guidelines and Equations (8.41)–(8.45).

Design Procedure 8.4: Design of Axially Loaded Bolts or Screws

Input: Axial, material properties, safety factor. **Output:** Bolt diameter, nut length.

Step 1: Use Equations (8.41) or (8.42) if the axial load is static or dynamic, respectively, to compute thread inner diameter d_3. Use yield strength for static loading, effective fatigue strength, and fatigue stress concentration factor for dynamic loading. Use Equations (8.43) and (8.44) for combined static and dynamic loading.

Step 2: From a table of standard bolts or screws, such as the tables in appendix C1, find the first inner diameter that is larger than the inner diameter found in step 1. Find the corresponding nominal diameter d, pitch diameter d_2, and thread pitch P.

Step 3: Use Equations (8.46) and (8.47) to compute independently the nut length. Select the largest of the two nut lengths. Compare with suggested values.

Step 4: Use Equations (8.37) and (8.39) to determine the required tightening torque and whether the bolt is self-locked or the nut needs to be locked, respectively.

Example 8.8 An air cylinder of bore $D = 150$ mm has a maximum pressure $p = 0.875$ MPa and the end flange is mounted by six bolts. The flange is sealed with an elastic gasket so that there is no need for appreciable prestressing of the bolts and all the load of the bolts is due to the air pressure. Find the bolt diameter for static strength and the required depth of the thread in the cylinder head for SAE class 4.8 carbon steel for both bolt and nut, cut threads, friction coefficient 0.15 for both the thread and the nut collar, safety factor $N = 6$.

Solution

Bolt: The flange load due to the cylinder pressure is:

$$F_{\text{tot}} = \frac{P \pi D^2}{4} = 875{,}000 \pi \times \frac{0.150^2}{4} = 15{,}462 \text{ N}$$

For each bolt, $F = 15{,}462/6 = 2577$ N. The yield strength of the bolt material (Table 8.6) is $S_y = 320$ MPa. Equation (8.41) gives:

$$\frac{F}{\pi d_3^2/4} = \frac{S_y}{N}, \ d_3^2 = \frac{4FN}{\pi S_y} = \frac{4 \times 2577 \times 6}{\pi \times 320 \times 10^6}, \ d_3 = 0.008178 \text{ m or } 7.8 \text{ mm}$$

From Table C.1 of Appendix C for standard threads, the next standard bolt is <u>M10 with d = 10 mm</u>, $d_3 = 8.16$ mm, $d_2 = 9.06$ mm, pitch $P = 1.5$ mm.

Threads: For the threads, the stress concentration factor is between 2 and 4, $K_{bn} = 4$ is selected, Equation (8.40) for $k_2 = k_3 = 1$, thus $K_f = 4$. The ultimate strength is $S_u = 400$ MPa (Table 8.6). The bearing strength is $S_b = 1.1 S_u = 1.1 \times 400 = 440$ MPa.

For the nut bearing strength, design equation (8.46) gives:

$$\sigma_b = \frac{K_f PF}{L\pi(d_1^2 - d_3^2)/4} \le \frac{S_b}{N}, \frac{4 \times 1.5 \times 10^{-3} \times 2{,}577}{L\pi(10^2 - 8.16^2) \times 10^{-6}/4} = \frac{440 \times 10^6}{6}$$

$$L = \frac{3.5 \times 1.5 \times 10^{-3} \times 2{,}577 \times 6}{\pi(10^2 - 8.16^2) \times 10^{-6} \times 440 \times 106/4} = 0.008 \text{ m or } L = 8 \text{ mm}$$

The yield strength in shear is (distortion energy–von Mises criterion of failure) $S_{sy} = 0.577 S_y$ = $0.577 \times 320 = 184.5$ MPa. Equation (8.47) gives:

$$\frac{K_f PF}{L\pi(d_1 + d_3)/4} \le \frac{S_{sy}}{N}, \frac{4 \times 1.5 \times 10^{-3} \times 2{,}577}{L\pi(10 + 8.16) \times 10^{-3}/4} = \frac{184.6 \times 10^6}{6}$$

$$L = \frac{1.8 \times 1.5 \times 10^{-3} \times 2{,}577 \times 6}{\pi(10 + 8.16) \times 10^{-3} \times 184.5 \times 10^6/4} = 0.0064 \text{ m or } 6.4 \text{ mm}$$

The largest length, $L = 8$ mm, is accepted.

Example 8.8a   Solve Example 8.8 for fatigue strength if the pressure is applied as a pulsating one from 0 to the maximum pressure.

Solution

Bolt: The maximum flange load is:

$$F_{tot} = p\pi D^2/4 = 875{,}000\pi \times 0.150^2/4 = 15{,}462 \text{ N}$$

For each bolt, $F_{max} = 15{,}4626 = 2{,}577$ N. Therefore:

$$F_m = F_{max}/2 = 2{,}577/2 = 1288 \text{ N}, \ F_r = F_{max}/2 = 2{,}577/2 = 1288 \text{ N}$$

The yield strength (Table 8.6) is $S_y = 320$ MPa and the ultimate strength is $S_u = 400$ MPa. The fatigue strength (Table 8.6) for $d/R = 40$, $S_e = 65$ MPa.
From Table 8.8 $K_f = 3.5$ is selected. Thus:

$$F_m + \frac{K_f F_r S_u}{S_e} = 1288 + \frac{3.5 \times 1288 \times 400}{65} = 29{,}030 \text{ N}$$

Equation (8.43) gives for the Goodman line:

$$\frac{F_m + K_f F_r S_u/S_e}{\pi d_3^2/4} = \frac{S_u}{N}, \ d_3^2 = \frac{4FN}{\pi S_y} = \frac{4 \times 29{,}030 \times 6}{\pi \times 400 \times 10^6}, \ d_3 = 0.024 \text{ m or } 24 \text{ mm}$$

Equation (8.44) gives for the yield line:

$$\frac{F_m + F_r}{\pi d_3^2/4} = \frac{S_y}{N}, \ d_3^2 = \frac{4(F_m + F_r)N}{\pi S_y} = \frac{4 \times 29{,}030 \times 6}{\pi \times 320 \times 10^6}, \ d_3 = 0.0081 \text{ m or } 8.1 \text{ mm}$$

$d_3 = 24$ mm is selected, the maximum one.

From Table C.1 of Appendix C for standard threads, the next standard bolt is $\underline{M30}$ with $d = 30$ mm, $d_3 = 25.72$ mm, $d_2 = 27.71$ mm, pitch $P = 35$ mm.

Threads: For the threads, the stress concentration factor is between 2 and 4, $K_f = 3.5$ is selected (Equation 8.40) for $k_2 = k_3 = 1$, thus $K_{b-n} = 3.5$. The bearing strength is $S_b = 1.1 S_u = 1.1 \times 400 = 440$ MPa. For the nut bearing strength, design equation (8.46) gives (since the Goodman line was dominant above we need not to check the yield line):

$$\sigma_b = \frac{F_m + K_f F_r S_u / S_e}{(L/P)\pi(d_2^2 - d_3^2)/4} \leq \frac{S_b}{N}, \frac{29{,}030 \times P}{L\pi(10^2 - 8.16^2) \times 10^{-6}/4} = \frac{440 \times 10^6}{6}$$

$$L = \frac{29{,}030 \times 3.5 \times 10^{-3} \times 6}{\pi(10^2 - 8.16^2) \times 10^{-6} \times 440 \times 10^6/4} = 0.049 \text{ m or } 49 \text{ mm}$$

The yield strength in shear is (distortion energy–von Mises criterion of failure) $S_{sy} = 0.577 S_y = 0.577 \times 320 = 184.5$ MPa. Equation (8.47) gives:

$$\frac{F_m + K_f F_r S_u / S_e}{L\pi(d_1 + d_3)/4} \leq \frac{S_{su}}{N}, \frac{3.5 \times 10^{-3} \times 27{,}048}{L\pi(10 + 8.16) \times 10^{-3}/4} = \frac{400 \times 10^6}{6}$$

$$L = \frac{29{,}030 \times 6}{\pi(10 + 8.16) \times 10^{-3} \times 400 \times 10^6/4} = 0.028 \text{ m or } 28 \text{ mm}$$

The largest length, $L = 49$ mm, is accepted.

8.3.4. Prestressing of Bolt Joints

For bolts, the axial force is known from the applied torque on the nut and Equations (8.31)–(8.33). For both bolts and rivets, applications exist where loads are expected in service in addition to the initial loads imposed with tightening the nut or riveting. Examples are the flanges of pressure vessels, engine heads, and so on (Figure 8.14). The head bolts are tightened initially to develop an axial force F_v (prestress), without internal pressure. When such pressure p is applied, a force is applied on the head, F_B.

During prestressing, the bolt is elongated by δ_s, while the flange is compressed by δ_f (subscript f denotes flange and s bolt). Upon application of pressure p, with a head force F_B per bolt, the bolt is further elongated while the compression of the flange decreases. Because the elements are elastic, there is a linear relationship between forces and deformations on the bolt and flange (Figure 8.21a,b). At equilibrium, without pressure, the forces on both bolt and flange are equal to the prestress force F_V. Approaching the two force-deformation diagrams, for common point V, we obtain the diagram of Figure 8.21c. There, the bolt and flange force-displacement relationships are described by the two straight lines. Geometry demands that the sum $\delta_s + \delta_f$ remain constant. External load F_B moves the equilibrium point to the right and the bolt load is F_s while the flange load is F_f. Obviously, always $F_B = F_s - F_f$. The increase of the bolt load is F_{diff}.

At point V, the prestress force is:

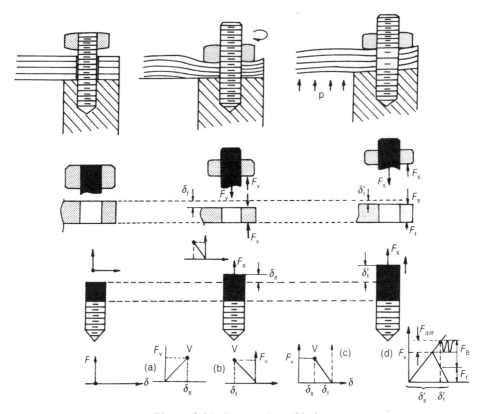

Figure 8.21. Prestressing of bolts.

$$F_V = \frac{A_f E_f \delta_f}{l} = \frac{A_s E_s \delta_s}{l}, \frac{\delta_s}{\delta_f} = \frac{A_f}{A_s} \cdot \frac{E_f}{E_s} \qquad (8.48)$$

because for a prismatic bar $\delta = Fl/(AE)$. Here A is the cross-section, l is the length, F is the force, and E is the Young's modulus. The geometry of Figure 8.23b yields:

$$\frac{F_{\text{diff}}}{F_B} = \frac{\delta_f}{\delta_f + \delta_s}, F_{\text{diff}} = F_B \frac{1}{1 + \delta_s/\delta_f}$$

$$= C_K F_B, C_K = \frac{1}{1 + \delta_s/\delta_f} \qquad (8.49)$$

$$F_v = (1 - C_K)F_B \qquad (8.50)$$

Factor C_K is computed from Equations (8.51) and (8.52). The former equation assumes a prismatic flange, which is not very realistic because only a portion of the flange is deformed, around the deformed bolt.

We rewrite the expression for factor C_K using the fact that $\delta = F/k$, where k is the stiffness of the elastic member (bolt or flange):

$$C_K = \frac{1}{1 + \delta_s/\delta_f} = \frac{1}{1 + k_f/k_s} \tag{8.51}$$

To compute the stiffness factor C_K, first we need to find the stiffnesses of the participating elements.

8.3.4.1. Bolt Stiffness k_s.
Unless the bolt has the special forms, such as in Figure 8.22, it is a prismatic bar of diameter d and length L. Therefore:

$$k_s = \frac{AE_s}{L} \tag{8.52}$$

where L is the bolt length, A is the stress area of the bolt $\pi d_3^2/4$, and E_s is the Young's modulus of elasticity of the bolt material.

8.3.4.2. Flange Stiffness k_f.
The flange stiffness may consist of several parts:

1. *A drilled-through plate*, the upper flange in Figure 8.24. Most authorities agree that the stiffness of that part can be approximated by the stiffness of a frustrum cone that has inner diameter d, outer diameter at the top $D = 1.5d$, for standard bolts. If a stiff washer is used, D is the outer diameter of that washer. There is no general agreement on the cone angle α. Recent investigations (Oswood 1979) suggest the value $\alpha = 25°–33°$. Unless experimental

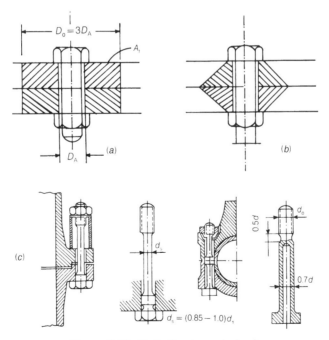

Figure 8.22. Bolts for dynamic loads.

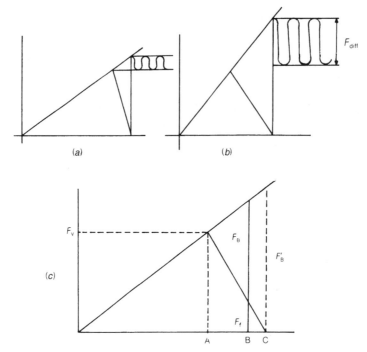

Figure 8.23. Prestressing for fluid-tight joints.

results for the particular application are available, the conservative value of $\alpha = 33°$ is recommended.

The outer diameter at distance x from the top of the frustrum cone is $D + 2 \tan \alpha$. The area of the cross-section of the frustrum cone at that point is $A_{fc} = \pi(D + 2 \tan \alpha)^2/4 - \pi d^2/4$. The deformation of the frustrum cone will be:

$$\delta_{fr} = \int_0^h \frac{4F\,dx}{E[\pi(D - \tan \alpha)^2 - \pi d^2]} \tag{8.53}$$

$$= \frac{F}{\pi E d \tan \alpha} \ln \frac{(2h \tan \alpha + D - d)(D + d)}{(2h \tan \alpha + D + d)(D - d)}$$

where h is the thickness of the plate. The frustrum cone stiffness is:

$$k_{fr} = \frac{F}{\delta_{fr}} \tag{8.54}$$

$$= \frac{\pi E d \tan \alpha}{\ln[(2h \tan \alpha + D - d)(D + d)]/[(2h \tan \alpha + D + d)(D - d)]}$$

Using Equation (8.52), the function $k_{fr}/k_s = 4k_{fr}/\pi E d(d/h)$ was plotted in Figure 8.24b as a function of $(h/d)\tan \alpha$. In terms of this function, C_K can be computed

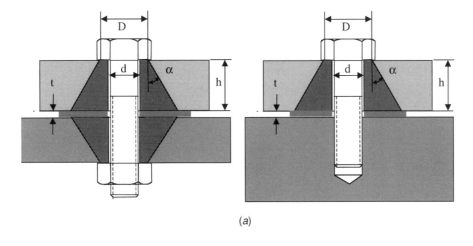

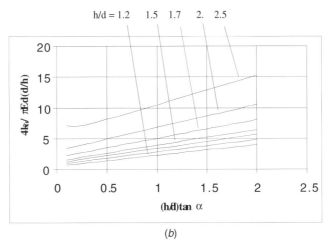

(b)

Figure 8.24. (a) Flange stiffness model; (b) frustrum cone stiffness function $k_{fr}/k_s = 4k_{fr}/\pi Ed(d/h)$.

by Equation (8.51) if the only elasticity of the flange is due to one frustrum cone, thus $k_f = k_{fr}$ and $C_k = 1/(1 + k_{fr}/k_s)$.

2. A *stud bolted directly in a large part* (Figure 8.24b). The participating flange is usually very stiff, as compared with the stiffness of the bolt or the upper plate, and can be neglected. Only the upper part is considered as in 1 above.

3. For one-piece flanges, C_K has very low values and the reversing force will be large, which might lead to fatigue failure. To overcome this, the flange can be made softer using a *soft gasket* between the two faces of the flange. The gasket is prismatic with area of cross-section A_g and thickness t. Therefore, its stiffness is:

$$k_g = \frac{A_g E_g}{t} \tag{8.55}$$

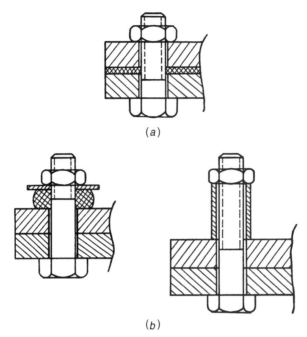

(a)

(b)

Figure 8.25. (a) Increasing elasticity of the flange; (b) Increasing elasticity of the bolt.

where E_g is the modulus of elasticity of the gasket material. Usually the gasket is much softer than the flange.

4. *A compound flange*, in general, has stiffness that consists of the stiffness of the upper plate k_u, of the lower plate k_l, if such plate exists (Figure 8.24a), and of the gasket k_g, if such gasket is used. k_u and k_l are computed with Equation (8.54), k_g with Equation (8.55). The total flange stiffness is found by noting that deformations δ are additive. Since stiffness is inversely proportional with the deformation:

$$\frac{1}{K_f} = \frac{1}{K_u} + \frac{1}{K_l} + \frac{1}{K_g} + \cdots \tag{8.56}$$

For time-varying pressure p, usually pulsating, the bolt might fail by fatigue. The bolt load fluctuates between F_V and F_s. The mean load is then $F_m = (F_V + F_s)/2$ while the load range is $F_r = (F_s - F_V)/2$. It is apparent that the dynamic load reduces when the slope of the force-deformation line, on the left, decreases. Therefore, for dynamic loaded joints, slender bolts are designed with as great a length as possible (Figure 8.23c).

The prestress force F_V must be high enough so that at full external load F_B, the flange, will not have negative compression, that is, $F_f > 0$. From Figure 8.23a, it is apparent that slender bolts and rigid flanges result in lower dynamic loads on the bolt but the same external load F_B results in much higher unloading of the flanges. Therefore, when fluid sealing is of prime importance, the bolt is made stiff and the flange soft (Figures 8.22c, 8.25a,b).

Equations (8.40) and (8.41) were found on the basis that, at maximum external load, the flange will have zero load. Then:

$$F_B = F_s - F_f = \frac{F_V}{1 - C_K} \tag{8.57}$$

If the joint is to be fluid-tight, as usually happens, at maximum external force the flange must be under compression. Therefore, it must be designed so that after the maximum external load F_B is applied, the flange is still under some compression. The situation is shown at point B in Figure 8.23c, where F_f is the remaining flange load. Denoting the ratio $AC/AB = N_p$, from a similar triangle analogy it is found that the preload F_V and the differential force F_{diff} are:

$$F_V = (1 - C_K)N_p F_B \tag{8.58}$$

$$F_{\text{diff}} = C_k N_p F_B \tag{8.59}$$

$$F_{\text{max}} = F_V + F_{\text{diff}} = N_p F_B \tag{8.60}$$

When $N_P = 1$, Equation (8.58) is equivalent to Equation (8.49). The value of N_P depends on the machining on the flange surfaces and the bolt spacing. It is suggested that $N_p = 1.1–1.2$ for static load and $N_p = 1.20–1.30$ for dynamic loading. Lower values are for permanent connections, higher ones for reused connections. The ratio N_P is called the *preload factor*. For the computation of the steady and alternating forces on the bolt, we note that:

$$F_m = F_V + \frac{F_{\text{diff}}}{2} = \left(1 - \frac{C_k}{2}\right) N_p F_B \tag{8.61}$$

$$F_r = \frac{F_{\text{diff}}}{2} = \frac{C_k N_p F_B}{2} \tag{8.62}$$

Design Procedure 8.5: Design of Preloaded Bolts

Input: Operational Load FB, material properties, safety factor. **Output:** Bolt diameter, nut length.

Step 1: Use Figure 8.24 or one of Equations (8.51)–(8.57) to determine the stiffness constant C_k. Select a preload factor N_p (1–4).

Step 2: Use Equations (8.48a) and (8.48b) to determine the static and alternating component of the bolt load.

Step 3: Use Procedure 8.4 to determine the bolt/nut dimensions.

Example 8.8b   A refrigeration compressor of bore $D = 60$ mm has a maximum pressure $p = 1.5$ MPa and the end flange is mounted by four bolts. Find the bolt if it will be made of SAE class 8.8 carbon steel for both bolt and nut, cut threads, for safety factor $N = 3$.

Solution   The flange load due to the cylinder pressure is $F_{\text{tot}} = p\pi D^2/4 = 1.5 \times 10^6 \pi \times 0.060^2/4 = 4{,}241$ N. For each bolt, $F = 4{,}241/4 = 1060$ N. The yield strength of the bolt material (Table 8.6) is $S_y = 640$ MPa, the ultimate strength is $S_u = 800$ MPa, the fatigue strength is $S_e = 92$ MPa, for $d/R = 50$.

From Figure 8.24b, assuming $h/d = 1.5$, $a = 33°$, $(h/d)\tan \alpha = 0.78$, $k_f/d_s = 4$, and then $C_k = 1(1 + 4) = 0.2$. A refrigeration compressor requires a fluid-tight head. It is not a critical application but dynamic loading has to be considered and we select $N_p = 1.2$. Equations (8.61) and (8.62) give:

$$F_m = \left(1 - \frac{C_k}{2}\right) N_p F_B = \left(1 - \frac{0.2}{2}\right) \times 1.2 \times 1060 = 1145 \text{ N}$$

$$F_r = \frac{F_{\text{diff}}}{2} = \frac{C_k N_p F_B}{2} = \frac{0.2 \times 1.2 \times 1060}{2} = 127 \text{ N}$$

From Table 8.8, $K_f = 3.5$ is selected. Thus:

$$F_m + \frac{K_f F_r S_u}{S_e} = 1145 + \frac{3.5 \times 127 \times 800}{92} = 3865 \text{ N}$$

Equation (8.43) gives for the Goodman line:

$$\frac{F_m + K_f F_r S_u / S_e}{\pi d_3^2 / 4} = \frac{S_u}{N}, \ d_3^2 = \frac{4FN}{\pi S_y} = \frac{4 \times 3865 \times 3}{\pi \times 800 \times 10^6}, \ d_3 = 0.049 \text{ m or } 4.9 \text{ mm}$$

Equation (8.44) gives for the yield line:

$$\frac{F_m + F_r}{\pi d_3^2 / 4} = \frac{S_y}{N}, \ d_3^2 = \frac{4(F_m + F_r)N}{\pi S_y} = \frac{4 \times (1145 + 127) \times 3}{\pi \times 600 \times 10^6}, \ d_3 = 0.0028 \text{ m or } 2.8 \text{ mm}$$

The larger of the two diameters, $d_3 = 4.9$ mm, is selected.

From Table C.1 of appendix C for standard threads, the next standard bolt is M8 with $d = 8$ mm, $d_3 = 6.3$ mm, $d_2 = 7.1$ mm, pitch $P = 1.5$ mm.

Nut: For the threads, the stress concentration factor is between 2 and 4, $K_f = 4$ is selected (Equation 8.40) for $k_2 = k_3 = 1$, thus $K_{f-t} = 4$. The bearing strength is $S_b = 1.1 S_u = 1.1 \times 800 = 880$ MPa.

Because the bolt operates on the Goodman line, the maximum force $F_{\max} = F_m + F_r = 1125 + 127 = 1152$ N will be used for the nut bearing strength. Design equation 8.44 gives:

$$\sigma_b = \frac{P(F_m + K_f F_r S_u / S_e)}{L \pi (d_1^2 - d_3^2)/4} \leq \frac{S_b}{N}, \ 1.5 \times 10^{-3} \times \frac{1152}{L \pi (8^2 - 6.3^2) \times 10^{-6}/4} = \frac{880 \times 10^6}{3}$$

$$L = \frac{4 \times 10^{-3} \times 1152 \times 3}{\pi (8^2 - 6.3^2) \times 10^{-6} \times 880 \times 10^6/4} = 0.0096 \text{ m or } 9.6 \text{ mm}$$

The yield strength in shear is (distortion energy–von Mises criterion of failure) $S_{sy} = 0.577 S_y = 0.577 \times 600 = 346.2$ MPa. Equation (8.45) gives:

$$\frac{F_m + K_f F_r S_u / S_e}{L\pi(d_1 + d_3)/4} \leq \frac{S_{sy}}{N}, \frac{3865}{L\pi(10 + 8.16) \times 10^{-3}/4} = \frac{346.2 \times 10^6}{3}$$

$$L = \frac{3865 \times 3}{\pi(10 + 8.16) \times 10^{-3} \times 346.2 \times 10^6/4} = 0.0034 \text{ m or } 3.4 \text{ mm}$$

The largest length, $L = 9.6$ mm, is accepted.

Example 8.9 Develop an EXCEL spreadsheet for design of bolts for general loading. As an application, design a bolt to carry a continuous axial load 10,000 N, a reversing axial load 5,000 N, a continuous shear load 10,000 N, a reversing shear load 5,000 N. The material is carbon steel with $S_u = 3.00E + 08$ MPa, $S_y = 2.20E + 08$ MPa, $S_e = 1.20E + 08$ MPa. The friction coefficient between bolt and nut is 0.2 and between nut and base is 0.2. The fatigue stress concentration factor for the thread is $K_f = 2$ and the safety factor should be at least $N = 2$.

Solution The spreadsheet is shown in Table E8.9.

8.4. WELDED JOINTS

8.4.1. Field of Application

Welded joints are permanent joints formed by localized heating with the addition of a welding element or by welding together of the two mating parts.

Welded joints are the most advanced type of permanent joints because the properties of the welded components and the weldment are closest to those of the solid member under certain conditions. Furthermore, very complex members can be fabricated by welding.

All structural steels, including high alloys, and also nonferrous metals and alloys and plastics can be efficiently welded. Low-carbon steels (<0.3% C) can be effectively welded. Higher-carbon steels (>0.3% C) and most alloy steels can be welded, but they tend to develop cracks and thus are subject to fatigue failure. Nonferrous alloys sometimes require special measures, such as preheating and welding under inert gas atmosphere.

The weld can be "shaved": the surface can be machined or grinded, and heat-treated to reduce stress concentration due to notches and cracks and to reduce internal stresses that are generated during welding and cooling of the weld.

Metal arc welding is performed with an electric arc that is formed between a metal electrode and the mating parts. The heat generated by the arc melts both the electrode and the material near the surface of the mating parts. This process can be automated—a process called *submerged arc welding*—and can give a very high quality of joints. It can be applied to a wide range of thicknesses of the welded parts and can join parts of ordinary structural steels as well as of high-strength alloys. *Robots* and *automatic welding machines* have been very effectively used in welding processes, reducing cost and improving quality.

TABLE E8.9

	B	C	D	E	F	G	H	I	J
1	**Example 8.9: DESIGN OF BOLTS FOR GENERAL LOADING (no prestress)**								
2	**Data (SI units):**								
3	Loading	Static	Reversing	Material Properties	Yield	$S_y =$	2.20E+08	in shear:	$S_{sy} =$	1.26E+08
4	Axial	10000	5000		Ultimate	$S_u =$	3.00E+08			
5					Fatigue	$S_e =$	1.20E+08			
6					Bearing	S_b	3.30E+08	($=1.1S_u$)		
7	Friction Coefficient					Bolt	Nut			
8		Bolt/nut	0.2		Stress Concentration $K_f =$	2	3			
9		Nut/base	0.2		Safety Factor N =	3	3			
10	**Results:**								
11	**a. Design of the bolt:**								
12	Friction angle (bolt/nut) =				0.1973956				
13	Equivalent force (von Mises):				35000				
14	Computed stress diameter				Goodman	0.024651			
15	(equation 8.39bb)				Yield line	0.016138			
16	Effective stress diameter:				0.024651				
17	We select now a standard bolt based on the stress area or stress diameter (Appendix C.1):								
18	Nominal (outer) diameter =				0.018				
19	Pitch diameter =				0.016				

TABLE E8.9 (*Continued*)

	B	C	D	E	F	G	H	I	J
20	Stress diameter =				0.015				
21	Pitch =				0.0035				
22	Pitch Angle =				0.040198				
23	Nut friction diameter =				0.036				
24	Efficiency (tightening) =				0.166082				
25	Efficiency (untightening) =				−0.253743				
26	Tightening Torque =				1682.12				
27	Untightening Torque =				−173.032				
28	The bolt is		**Self-locking**						
29	**b. Design of the nut:**								
30	Threads-shear								
31	Equation 8.39e yields the nut length								
32	L =	0.009454003							
33	Threads-bearing:								
34	Equation 8.39d yields the nut length								
35	L =	0.008880167							
36	Nut length =				0.009454				

561

To avoid oxidation due to the high temperatures, either a paste is used with the electrode (the paste melts at a relatively low temperature and subsequently covers the weld, preventing the oxygen from coming into contact with the hot metal) or an inert gas is supplied (which again prevents the oxygen from coming in contact with the metal when very hot).

Resistance welding is based on heating the conducting surfaces of the joint by passing an electric current through them and applying pressure on the seam. Resistance welding is mostly used in large lot and mass production.

Induction welding is based on heating the conducting surfaces of the joint by imposing a high-frequency electromagnetic field on the mating parts and applying pressure on the seam. The induction currents produce ohmic heating of the metal and subsequent local melting and welding. Induction welding can be highly automated and is also used in large lot and mass production.

Friction welding is performed by making use of the heat generated in the relative motion of the parts being joined.

There are a very great number of different methods of welding, which suggests the wide use of these methods of joining machine elements.

With respect to the mutual position of the welded components, welded joints can be classified into the following categories (see Figure 8.26):

1. *Butt joints*: One of the parts is a continuation of the other. The end faces are welded together. Up to 3 mm thickness welding is performed without edge preparation, up to 20 mm with slant machining of the edges and V-weld, up to 40 mm with slant machining of both sides of the edges and X-weld, beyond 40 mm thickness with U-form welding.

2. *Lap joints*: The side surfaces of the welded parts partly overlap. Welding is performed at one or both the edges over the side of the other part. This is also called *fillet weld*.

3. *Tee joints*: The joint parts are perpendicular to each other or less frequently at some other angle. The end face of one part is welded to the side of the other.

4. *Corner joints*: The joint parts are perpendicular to each other or at some other angle and are welded together along the adjacent edges.

Butt joints are closest in behavior to solid components and their application is gradually increasing at the expense of lap joints.

Because of their geometry, welded joints require less base metal for the components in comparison with riveted joints (usually 15–20% less).

Welded parts are much lighter than steel castings (by up to 50%) and cast iron castings (by up to 30%).

One disadvantage of welded joints is in welding performed by hand, where reliability depends very much on the skill of the welder. There are, however, methods to test the quality of the weld and improve reliability. Another disadvantage of welded joints is that due to the high temperatures in the area of the weld as compared with the base metal, very high thermal stresses are developed, which deform the welded parts and leave behind internal stresses.

Corrective measures and processing are available for reducing or preventing distortion from welding, such as the symmetry of the welds, reduction of amount

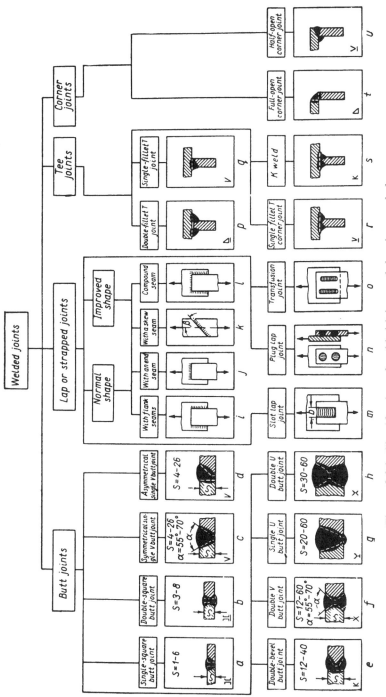

Figure 8.26. Classification of welded joints and drawing symbols.

of weld metal, and proper application of the welding sequence. There are also methods of relieving stresses by heating after the welding. The metal near the weld is annealed.

In particular, welded parts to be manufactured after welding to a high standard of accuracy should be tempered to prevent warping due to the residual stresses.

A special type of welded joint results when an intermediate element is used, an element with a strength much less than that of the basic materials. These joints are called *soldered*, *brazed*, or *adhesive joints.*

Soldering and *brazing* are processes for joining parts by means of a material with a low melting point compared to that of the basic material. These processes are mainly for joining metal parts. Soldering is known from ancient times and recently has been applied to much more critical joints than previously. Most widely used are lap joints. In comparison to welding, soldering and brazing have a number of advantageous features. The soldering or brazing temperature is low and causes no change in the properties of the metals being joined. In a lap joint, soldering or brazing is carried out along a surface, showing more uniform stress distribution than in welds and enabling shorter joints to be used. The influence of defects on soldered or brazed joints is relatively small. Further, because the surface carrying the load can be very large as compared with welding, soldered and brazed joints can reach the strength of welding.

For joining of various materials, *adhesives* based on synthetic resins have found extensive application in recent years. These adhesives have very high strength and have been used in critical machines and structures such as aircraft and bridges. They also have the advantage that they can be used for joining parts of very dissimilar materials. The main disadvantage of such adhesives is their low resistance to higher temperatures.

Adhesives applied in mechanical design are divided into two groups. Most widely used are the *adhesives based on organic polymeric resins*, with a heat resistance not over the range 300–350°C. The shear strength of such adhesives at normal temperatures reaches 18–20 N/mm^2 and at 200–300°C is reduced drastically. The second group includes *adhesives based on organic silicon compounds and inorganic polymers*, which have a heat resistance up to 1000°C but also have higher brittleness. The shear strength of such adhesives at normal temperature is between 7 and 8 N/mm^2.

Adhesives have been very efficiently used to strengthen the fastening between parts that are joined by other methods.

Some of the more usual welded joints are shown in Figure 8.26.

8.4.2. Sizing of Welded Joints

The stress distribution in a weld is very complicated. Attempts have been made to map the stress distribution in welds (Shigley and Mischke 1989, Norris 1945, Salakion and Glaussen 1937). In engineering practice, the allowable stresses, safety and stress concentration factors, and so on, are based on the convention that failure is due to shear at the narrowest section (termed *throat*) of the weld, the shear stress being the ratio of the shear force divided by the throat area of the weld. Therefore, this is the approach we will follow in the sequel.

Due to the catastrophic consequences of some early weld failures in pressure vessels and steel construction, design of welds is highly regulated. The designer is advised to seek the codes applicable to the design at hand and comply with the appropriate regulations.[2] The design methodology that follows is to be applied only for preliminary design or in cases not covered by the codes.

Usually the electrode used in a welding must be as close to the base metal as possible. For welding electrodes, the American Welding Society[3] (AWS) specifies the symbol *Eyyxx*, where *yy* is the tensile strength in ksi and *xx* is the type of welding (Table 8.9).

8.4.2.1. Mode of Failure: Stress Failure Due to an Axial Load in a Butt Weld.
Figure 8.26, cases *a* to *d* shows butt welds loaded by an axial load *P*. The weld is loaded in tension and the design equation is:

$$\sigma_{\text{av}} = \frac{K_f P}{tb} = \frac{S_y}{N} \tag{8.63}$$

where *t* is the plate thickness and *b* is the weld width. K_f is the fatigue stress concentration factor for dynamic loading, to be discussed in Section 8.4.4. For static loading, $K_f = 1$.

8.4.2.2. Mode of Failure: Shear Failure Due to a Symmetric Shear Load in a Fillet Weld.
The shear load passes through the welding area center and generates only direct shear on the weld. Figure 8.27 shows a lap joint of two sheets by means of transverse and side fillet welds. It is assumed that the dangerous section in the weld coincides with the plane passing through the bisector of the right angle *ABC* (Figure 8.27). Thus, for one transverse fillet weld the area of the dangerous section is $Lh/\sqrt{2} = L(0.707h)$, where *h* is the leg of the weld; in the case represented in Figure 8.27, the leg of the weld is equal to the thickness *h* of the upper sheet.

The shearing stresses are assumed to be uniformly distributed over the area of the dangerous section. Moreover, attempts to find the stress with analytical methods have not been successful due to the uncertainty of the weld geometry and stress distribution. Long design practice has established a methodology by which the

TABLE 8.9. AWS Standard Welding Electrodes

AWS Class	S_u (psi)	S_y (psi)
E60xx	62,000	50,000
E70xx	70,000	57,000
E80xx	80,000	67,000
E90xx	90,000	77,000
E100xx	100,000	87,000
E120xx	120,000	107,000

[2] For example, the ASME Boiler Code and the AISC Code (AISC, 400 N. Michigan Ave, Chicago, IL 60611).
[3] An Internet search through a substantial number of standards can be done at http://www.nssn.org/

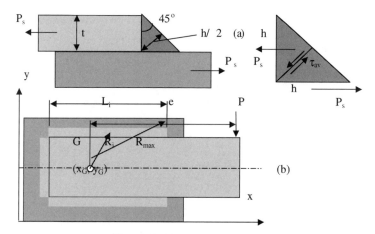

Figure 8.27. Weld geometry.

average shear stress is computed by division of the transmitted force by the throat area. This assumption is conservative, and because safety and derating factors are based on it, it is recommended for use in design instead of more detailed analytical or numerical methods.

Based on the above assumptions, the design equation for a transverse fillet weld of length L, shear load P_s, and throat width $0.707h$, is:

$$\tau_s = \frac{K_f P_s}{0.707hL} = \frac{vS_{sy}}{N} \tag{8.64}$$

In the case of complex welds, with line segments of lengths $L_1, L_2, \ldots$, strength considerations obviously require that the total allowable resistance of the welds not be less than the force acting on the joint:

$$\tau_s = \frac{K_f P_s}{0.707h\Sigma L_l} = \frac{S_{sy}}{N} \tag{8.65}$$

where $0.707h$ is the effective weld throat section width and L_i is the length of each segment.

By using this equation, it is possible, given the dimension h, to determine the required length of welds. Very rarely, the extent of the weld (lengths L_i) is known and the weld width h is sought. One can solve Equation (8.65) for h in this case. This is unusual, however, because k cannot be much different than the plate thickness, in general. If the plate is thicker, it does not become warm enough and the welding is prone to be defective (*cold welding*). On the other hand, if the plate is thinner than the welding, the plate is overheated and might melt locally (*hot welding*).

8.4.2.3. Mode of Failure: Shear Failure Due to an Eccentric, In-Plane, Shear Load Due to Torsion.

An eccentric shear load produces direct shear stresses

$$\tau_d = \frac{K_f P}{A} = \frac{K_f P}{\Sigma(A_i)} = \frac{K_f P}{0.707 h \Sigma(L_i)} \tag{8.66}$$

but also indirect shear stresses due to the twisting of the shear area by the moment Pe of the eccentric force P, where e is the moment arm, the distance of the area center of weld surface from the line of application of the force P (Figure 8.27b). Thus:

$$\tau_{\text{ind}} = \frac{K_f P e R_{\text{max}}}{0.707 h \, I_p} \tag{8.67}$$

where I_u is the polar moment of inertia of the weld of unit width:

$$I_u = \Sigma(L_i R_i^2) + \frac{L_i^3}{12} \tag{8.68}$$

where R_{max} is the maximum distance of any part of the weld from its area center, L_i is the length of the weld segment i, and R_i is the distance of the center of the segment i from the area center of the weld area. In Equation (8.69), the polar moment of inertia of the line segments $L_i^3/12$ can usually be neglected. The designer should always try to place the weld segments away from the area center of the weld so that the weld will be economical. Thus, in a well-designed weld, the direct polar moments of inertia of the weld segments are indeed negligible.

Considering the weld segment as a line rather than a rectangle does not result in appreciable error because the width of the weld is usually much smaller than the overall dimensions of the welded joint, that is, $h << R_i$. Thus, whether the distance R_i ends at the middle of the rectangular weld segment or the middle of the line representing it is not important.

As in the case of eccentric, in-plane shear load of rivet joints, the direction of direct and indirect shear stresses is not the same and at the same point they form angle ϕ_i. The resultant stress yields the design equation, by comparison of the maximum stress among all segments with the allowable shear stress:

$$\tau_{\text{max}} = K_f (\tau_d^2 + \tau_{\text{ind}}^2 + 2\tau_d \tau_{\text{ind}} \cos \phi_i)^{1/2} \leq \frac{S_{sy}}{N} \tag{8.69}$$

8.4.2.4. Mode of Failure: Bending Failure Due to an Eccentric, Out-of-Plane, General Load.

An eccentric general load (Figure 6.28a) produces:

Direct shear stresses in the direction of the loads F_x, F_y:

$$\tau_x = \frac{K_f F_x}{A} = \frac{K_f F_x}{\Sigma(A_i)} = \frac{K_f F_x}{0.707 h \Sigma(L_i)} \tag{8.70}$$

$$K_f \tau_y = \frac{K_f F_y}{A} = \frac{K_f F_y}{\Sigma(A_i)} = \frac{K_f F_y}{0.707 h \Sigma(L_i)} \tag{8.71}$$

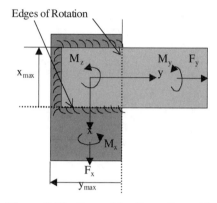

Figure 8.28. General loading of a weld.

Direct normal stresses in the direction of the load F_z:

$$\sigma_2 = \frac{K_f F_z}{A} = \frac{K_f F_z}{\Sigma(A_i)} = \frac{K_f F_z}{0.707h\Sigma(L_i)} \tag{8.72}$$

Indirect torsional shear stresses due to the twisting of the shear area by the moment M_z:

$$\tau_{\text{ind}} = \frac{K_f M_z R_{\max}}{I_z} = \frac{K_f M_z R_{\max}}{\Sigma(A_i R_i^2)} = \frac{K_f M_z R_{\max}}{0.707h I_u} \tag{8.73}$$

Indirect normal stresses due to the bending of the weld area by the moments M_x, M_y:

$$\sigma_{\text{ind-x}} = \frac{K_f M_x x_{\max}}{I_x} = \frac{K_f M_x x_{\max}}{\Sigma(A_i x_i^2)} = \frac{K_f M_x x_{\max}}{0.707h I_{uy}} \tag{8.74}$$

$$\sigma_{\text{ind-y}} = \frac{K_f M_x y_{\max}}{I_y} = \frac{K_f M_x y_{\max}}{\Sigma(A_i y_i^2)} = \frac{K_f M_y y_{\max}}{0.707h I_{ux}} \tag{8.75}$$

where:

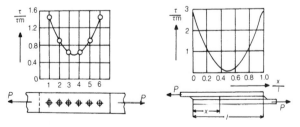

Figure 8.29. Stress concentration in long welds and rivet joints.

x_max, y_max are the maximum distances from the respective edges of possible rotation in bending. If the two plates are not forced together during welding, the edges pass through the area center of the weld.

$I_{uy} = \Sigma(L_I x_i^2) + L_{ix}^3/12$ is the area moment of inertia per unit width of the weld about the y-axis through the area center of the weld.

$I_{ux} = \Sigma(L_I y_i^2) + L_{iy}^3/12$ is the area moment of inertia per unit width of the weld about the x-axis through the area center of the weld.

L_{ix} and L_{iy} are the projections of the weld segment length L_i on axes x and y, respectively. $L_{ix}^3/12$ and $L_{iy}^3/12$ are usually small and can be neglected.

In general, there is a combined loading of normal and shear stresses, and a theory of failure needs to be applied. ISO, based on extensive experiments, suggested that owing to the complexity of the stress distribution, the test results suggest an equivalent stress of the form (Niemann 1975):

$$\sigma_\text{eq} = \sqrt{\sigma^2 + 1.8\tau^2} \tag{8.76}$$

Therefore, the design equation is:

$$\sigma_\text{max} = K_f \left\{ (\sigma_{d-z} + \sigma_{\text{ind-}x} + \sigma_{\text{ind-}y})^2 + 1.8 \left[\left(\tau_{d-x} - \frac{\tau_{\text{ind-}z} y}{R} \right)^2 \right. \right.$$
$$\left. \left. + \left(\tau_{d-y} + \frac{\tau_{\text{ind-}z} x}{R} \right)^2 \right] \right\}^{1/2} \le \frac{S_y}{N} \tag{8.77}$$

8.4.3. Spot Welds

Spot welds are used almost invariably to take direct shear load. Here we shall use Equation (8.66) with the appropriate expression for the area of the spot weld:

$$\tau_d = \frac{K_f P}{A} = \frac{K_f P}{\Sigma(A_i)} = \frac{K_f P}{n(\pi d^2/4)} \le \frac{S_{sy}}{N} \tag{8.78}$$

where d is the diameter of each spot weld and n is the number of spots.
From Equation (8.78), the number of spots required can be computed.

8.4.4. Stress Concentration and Safety Factors

Transverse welds show behavior similar to that of multiple riveting and the threads: the end parts of the weld are higher stressed, as shown in Figure 8.28. For ductile materials, yielding is acceptable and thus is not a problem for static loads. For dynamic loads and brittle materials, long transverse seams must be avoided, since the stress concentration is difficult to estimate and a long weld is in this case not economical. Moreover, for such cases, a fatigue stress concentration factor must be used in the design equations. For different types of welds, effective fatigue stress concentration factors K_f are shown in Figure 8.32. Figure 8.30 shows the trans-

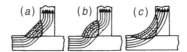

Figure 8.30. Stress flow lines in T-welds (from Decker 1973).

mission of stress between the welded elements, while Figure 8.31 shows stress concentrations across the weld.

Spot welds have stress concentration factors that depend on the spot diameter, which in turn depends on the sheet metal thickness. For steel sheets, fatigue stress concentration factors are given in Table 8.11.

Safety factors are suggested by the applicable codes. For structural applications, in the absence of specific code for the design at hand, the values of the safety factor suggested by the AISC (American Institute for Steel Construction) Code can be used: for tension in butt welds, $N = 1.7$, for compression $N = 1.1$, bending $N = 1.5–1.7$, shear $N = 1.45$.

For machine applications, the safety factor can be estimated[4] as:

$$N = N_1 N_2 N_3 \tag{8.79}$$

where:

- N_1 is the weld quality factor: 1–1.2 for inspected and heat-treated weld, free of voids and other defects, 2 for noninspected and non–heat-treated welds.
- N_2 is the load factor: 1.15 for design based on the main load without accounting for additional loads, 1 for accurately estimated loads.
- N_3 is the weld type factor, due to the fact that some welds are easier to make good quality and have more unequal distribution of stresses than others. For butt welds and lap-fillet welds in compression or tension or bending, $N_3 = 1.0$, in shear $N_3 = 1.2$, for corner welds and T welds, $N_3 = 1.2$.

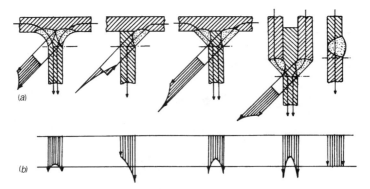

Figure 8.31. Stress concentration across the weld (from Decker 1973).

---

[4] By interpretation of DIN 4100 and DIN 15018.

TABLE 8.10. Spot Welds Fatigue Stress Concentration Factor

Thickness s (mm)

Parameter	1	1.5	2	3	5
Spot diameter d (mm)	4–8	6–10	8–10	10–12	10–14
K_f	1–2	1.5–2.5	2–2.5	2.5–2.9	2.5–4

The values of the safety factor from Equation (8.79) are suggested in the absence of applicable codes and standards. If they are available, the designer, for reasons of liability, is strongly advised to apply them, even if sometimes they appear to be overconservative.

For boilers and pressure vessels, higher safety factors need to be used due to the severe consequences of failure. In such designs, usually $N = 1.5$–2.0 or more, but the designer is strongly advised to use the applicable code. In the United States, this would be the ASME boiler code for land applications and the U.S. Register of Shipping or other applicable codes for ship boilers[5] and pressure vessels.

Design Procedure 8.6: Design of Welded Joints

Input: Load(s), material properties, safety factor. **Output:** Weld dimensions.

Step 1: Draw a schematic with dimensions and other data of the problem.

Step 2: Select a weld material, compatible with if not the same as the plate material.

Step 3: Select a type of weld (butt, fillet, spot, etc.), update the schematic, identify unknown dimensions. Find the area center of the weld.

Step 4: Observe the load, if it is dynamic, select a fatigue stress concentrated factor K_f (Section 8.4.4 and Figure 8.32 if seam weld or Table 8.11 if spot weld).

Step 5: Depending on the number and direction of loads, select a design equation (from 8.66–8.78).

Step 6: Use the design equation to (a) determine some unknown from the weld geometry, usually some length (sizing), or (b) solve for the safety factor (certification).

Example 8.10 Design a welded lap joint connecting two plates of the same thickness $t = 16$ mm having two cover plates (Figure 8.7) if $F = 60$ tons. The allowable shearing stress for the weld is $S_{sy}/N = 110$ N/mm^2. Assume static loading and inspected weld, thus $K_f = 1$ and $N = 1.5$.

Solution In order to leave a place for laying out side fillet welds, we take the width of the cover plate b_c to be somewhat narrower than the width of the sheet b_s, i.e., for thickness t, $b_s = b_s - 2t = 235 - 32 = 203$ mm. For equal strength, the cross-sectional area of the two cover plates must not be less than the cross-sectional area of the sheet, i.e., $2b_c t_c \geqslant F_s$. Hence the thickness of the cover plate is:

---

[5] Several accidents plagued the operation of the first steam ferryboats crossing the Mississippi River, taking a heavy toll in human lives.

Weld type	Design width, $t =$	Loading: Tension compression	Loading: Bending	Loading: Torsion
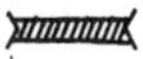	s	1	0.83	1.25
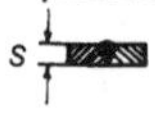	s	2	1.66	2.38
	s	1.43	1.19	1.78
	s	1.09	0.91	1.37
	s	1.43	1.19	1.78
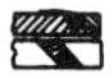	s	1.25	1.02	1.54
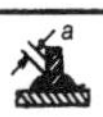	$2a$	3.12	1.45	3.12
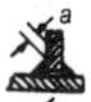	$2a$	2.85	1.43	2.86
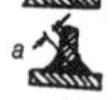	$2a$	2.44	1.15	2.44
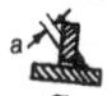	a	4.55	9.1	4.55
	s	1.59	1.25	2.0
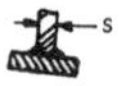	s	1.79	1.25	2.22
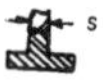	s	1.43	1.19	1.79
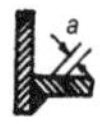	a	4.55	9.9	4.55
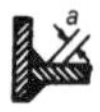	$2a$	3.33	1.67	3.33
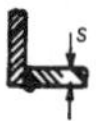	s	2.22	1.81	2.10
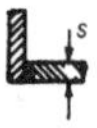	s	1.67	1.33	2.0
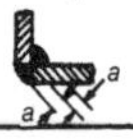	$2a$	2.87	1.43	2.86
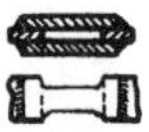	$2a$	4.55		
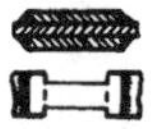	$2a$	4.0		
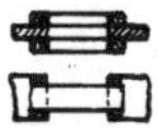	$2a$	4.0		
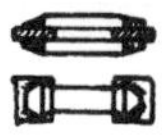	$2a$	2.08		

Figure 8.32. Fatigue stress concentration factor for different weld types (from Niemann 1965).

$$t_c \geq \frac{16 \times 235}{2 \times 203} = 9.2 \text{ mm}$$

Assume $t_c = 10$ mm. The necessary working length of the side fillet welds, L, can be determined from the condition $S_{sy}/N = P/(2 \times 0.707hL)$. Let the weld thickness be $h = 10$ mm. Two welds are carrying the load.

The length of the weld should be:

$$L = \frac{600,000}{2 \times 0.707 \times 40 \times 110} = 97.5 \text{ mm}$$

Example 8.11 The pressure vessel shown in Figure E8.3 is to be welded and made of a hot-rolled steel grade with $S_y = 480$ N/mm^2 at the operating temperature of 200°C. A local code requires that $N = 2$. Calculate the wall thickness of the vessel for an internal static pressure $p = 3.2$ MPa.

Solution

Main vessel: The peripheral (tangential or hoop) stress is:

$$\sigma_t = \frac{pr}{t} \leq \frac{S_y}{N}, t = \frac{pr}{S_y/N}$$

The thickness t will be computed on the basis of tangential stress only at the axial weld because at points of the peripheral seam near the tangential weld the axial stress is one half the peripheral one. Therefore:

$$t = \frac{3.2 \times 600}{480/2} = 8 \text{ mm}$$

The curved end elements have the longitudinal stress

$$\sigma_z = \frac{K_t pr}{2t} \leq \frac{S_y}{N}, t = \frac{K_t pr}{(S_y/N)}$$

Therefore:

$$t = \frac{3.2 \times 600}{2(480/2)} = 4 \text{ mm}$$

Because the thickness should be uniform, the maximum thickness of 8 mm is selected.

Example 8.12 The node point shown in a welded steel structure is loaded by axial loads $V = -30$ kN, $D = 50.75$ kN, $U_2 = 40$ kN, $U_1 = 85$ kN. The weld thickness is $a = 5$ mm. Check the weld for strength if the local building code requires allowable shear strength of the weld 105 N/mm^2. Assume static loading, thus $K_f = 1$.

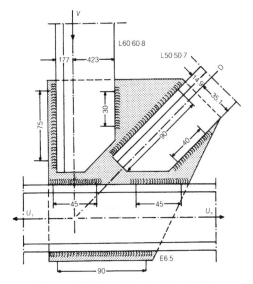

Figure E8.12 (After Decker 1973.)

Solution

1. For the force V:

$$\tau = \frac{V}{\Sigma \alpha L} = \frac{30{,}000}{5(75 + 30)} = 57.14 \text{ N/mm}^2$$

2. For the force D:

$$\tau = \frac{D}{\Sigma \alpha L} = \frac{50{,}750}{5(90 + 40)} = 78.1 \text{ N/mm}^2$$

3. For the force U: the flange transmits the difference

$$U = U_1 - U_2 = 85 - 40 = 35 \text{ kN}$$

Therefore:

$$\tau = \frac{U}{\Sigma \alpha L} = \frac{35{,}000}{5(90 + 45 + 45)} = 39 \text{ N/mm}^2$$

All welds are safe, but overdesigned. Weld thickness or length for every member can be reduced.

Example 8.13 For the compound welded bevel gear shown, transmitting a torque of 10 kNm and an axial force $F_a = 7500$ N, size the weld if the allowable shear stress is $S_{sy}/N = 120 \text{ N/mm}^2$. Assume static loading, thus $K_f = 1$, and disregard any radial forces.

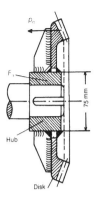

Figure E8.13 (After Decker 1973.)

Solution

1. The torque is transmitted mainly through the peripheral weld of the disk to the hub. The section more severely loaded is the one facing the disk. If the thickness there is *s*, the section modulus for small *s* is:

$$W = \frac{\pi^2 ds(d/2)^2}{(d/2)} = \frac{2\pi d^2 s}{2}$$

The shear stress is:

$$\tau = \frac{M}{W} = S_{sy}/N.$$

Therefore:

$$120 = \frac{10{,}000 \times 10^3 \times 2}{2 \times (\pi \times 75^2 \times s/2)}$$

$$s = \frac{10{,}000 \times 10^3 \times 2}{2(\pi \times 75^2 \times 120)} = 4.7 \text{ mm}$$

2. The axial force (F_a) will be loading primarily the reinforcing fins, and the force is transmitted through the four welds of the two fins at the hub. If *L* is the length at each weld and the same weld thickness is used, the shear stress will be, for $a = s/(2)^{1/2}$ = 3.5 mm:

$$\tau = \frac{F_a}{\Sigma \alpha L} = \frac{F_a}{4aL} = 120$$

Therefore:

$$L = \frac{F_a}{4 \times 120a} = \frac{7500}{4 \times 120 \times 3.5} = 4.46 \text{ mm}$$

8.5. INTERFERENCE FIT JOINTS

Interference fit joints (as has been seen already in Chapter 4) are formed by a fit where the shaft has a diameter larger than the hole of the hub. The joining is performed by applying pressure or heating or cooling one or another part so that the shaft will enter freely into the hole, and by the temperatures being equalized afterwards, the fit will be formed. During the contraction of the hub, pressure is developed between the shaft and the hole. Thus, mutual displacement of the joined parts is prevented by frictional forces at the contact surfaces.

Their chief advantage is easy manufacturing and assembly and the heavy loads they can carry. The disadvantage is that they present some difficulty in disassembly and during that process the surfaces might be damaged.

Cylindrical interference fit joints have very wide application for heavy loads and when there is no need for frequent assembly and disassembly. Typical examples of parts joined by interference fit are cranks, wheels, and bands for railway rolling stock, rims of gears and worm gears, turbine disks, rotors of electric motors, and ball and roller bearings (Figure 8.33).

The tightness of the joint is determined by the amount of interference, which depends on the category and quality of the fit established by the existing standard system of tolerances and fits, as presented in Chapter 4.

The most commonly used interference fits of grade 6, 7, and 8 tolerances are, in order of decreasing interference, $H7/u7$, $H7/s6$, $H7/r6$, $H7/p6$.

Press-fitted joints are used for lighter loads, and thermal expansion fit joints are used for heavier loads.

The limitations presented by the surface damage possible during disassembly can be overcome by tapered fit joints. These joints are obtained by a taper usually of 1:50. Critical joints are assembled by shrinkage or expansion or by hydraulic forcing. The same method is used for disassembly.

Design of interference fits for strength was discussed in Chapter 5.

8.6. OPTIMUM DESIGN FOR STRENGTH

8.6.1. Objective

The objective of any design for strength effort is usually the optimum distribution of the load-carrying capacity of a structure in order to achieve the smallest possible weight and the lowest cost, or usually both, since they are in close relationship. No general rules can be discussed, however, because the corresponding procedures are problem-oriented and the general ideas described in Chapter 7 apply, together with the methods of analysis discussed in this chapter and Chapter 4. Some pertinent problems will also be discussed here.

8.6.2. Optimum Design of Pin Joints

Multiple-pin joints, such as riveting and bolting, under shear loading, have an inherent deficiency in that the stress distribution within the elements is not uniform. For practical purposes, it is not technically sound to use fasteners of different size

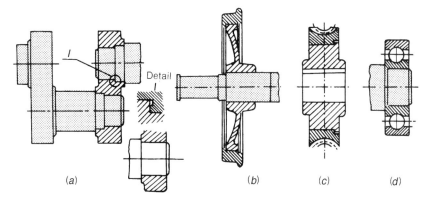

Figure 8.33. Shrink-fitted connections.

for the same joint, even for different joints of the same structure. Therefore, the design objective is to find the optimum distribution of the fasteners so that for a given load the minimum number will be used. The formulation of the optimization problem should be:

1. *Objective function:* Number of fasteners of given size (cost)
2. *Design variables:* Spacing of fasteners, i.e., their location
3. *Constraints*: Strength, space limitations

8.6.3. Optimization of Welds

Substantial savings in the amount of weld needed to carry a given load can be achieved by optimum distribution of the weld. The problem can be formulated as follows:

1. *Objective function:* Total weld length for given weld thickness
2. *Design variables:* The beginning and end points defining the several (perhaps) weld segments
3. *Constraints:* Strength, maximum available length

CASE STUDY 8.1: Dynasham Truck Co. Truck Suspension Bolt Failures[6]

Dynasham Truck Co, Ltd., in Chicago, Ill., has built heavy trucks since 1922. Project Engineer Lance O'Leary investigated the problem of persistent fractures of a truck suspension pivot bolt and its rubber bushing.

The suspension system is shown in Figure CS8.1. One of the two pivot bolts, shown in the bottom left corner of the system, pivots the suspension leaves on

---

[6] Based on Case Study ECL 44 of the Engineering Case Library. Originally prepared by J. A. Alic and H. O. Fuchs of Stanford University. By permission of the trustees of the Rose-Hulman Institute of Technology. All names and places in the case were disguised to preserve anonymity.

Figure CS8.1. Dynasham truck suspension system.

the bracket that is attached to the chassis of the truck. Both bolts are marked with a circle. The bolt and bushing assembly is shown in Figure CS8.2.

The bolt is 7/8 in. diameter, SAE grade 7, and is designed for a static load of 18,000 lb. However, the bolt takes the braking and accelerating loads, which are variable, and should be designed for fatigue.

In August 1958, the pivot bolt material specification was changed to SAE grade 8. This is the same material as grade 5 but quenched in oil and tempered at 800° F minimum. Minimum yield strength is 124,000 psi. Thus, because there are two shear surfaces:

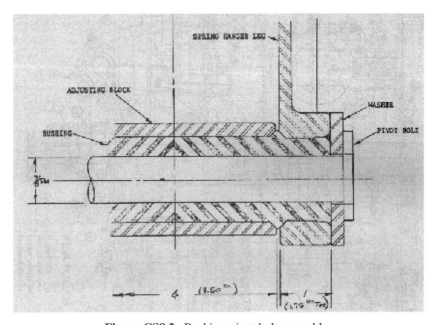

Figure CS8.2. Bushing-pivot bolt assembly.

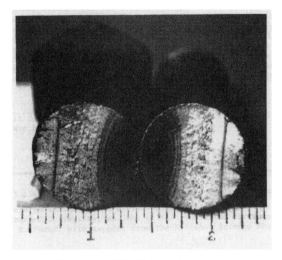

Figure CS8.3. Rule is in inches.

$$\tau_{av} = \frac{18,000}{2 \times p \times 0.875^2/4} = 14,974 = \frac{124,000}{N}, \, N = 8.28$$

The safety factor $N = 8.28$ seems adequate, but only for static loading. Some of these trucks weigh 127,000 lb, and for the braking load for a coefficient of friction 0.90, typical in truck design, there is an additional horizontal load 0.9 × 127,000 = 114,300 lb. This is divided among four rear-suspension pivot bolts, 114,300/4 = 28,575 lb per bolt. The resultant load is $(18,000^2 + 28,575^2)^{1/2} =$ 33,771 lb. Therefore:

$$\tau_{av} = \frac{33,771}{2 \times p \times 0.875^2/4} = 28,094 = \frac{124,000}{N}, \, N = 4.41$$

This is marginal enough to warrant a fatigue analysis because the horizontal load is dynamic and dominant.

In February 1959, 380 trucks had the Dynasham air suspension and 100 had bolt failures. A failed bolt is shown in Figure CS8.3. In October 1959, Dynasham changed the 7/8 in. to 1 in. grade 8 bolts. Moreover, Lance O'Leary suspected the shock-absorbing capacity of the bushings and ran experiments with different bushing materials. Failure with neoprene bushings occurred at 29,000 cycles and with natural rubber at 44,000 cycles.

Then he experimented with combinations of different materials until he found a particular combination of two short natural rubber bushings with a nylon spacer in between. No failures were reported thereafter.

What we can learn from this case? Lance O'Leary worked for about a year and a half and spent a lot of money in experimentation, mainly because he did not do a relatively simple fatigue analysis of the bolt on its elastic support, thus accounting for bending, too. Instead, he preferred the trial and error process, which eventually worked but at considerable expense of time and money. □

REFERENCES

Decker, K.-H. 1973. *Maschinenelemente.* Munich: K. Hanser Verlag.

Heywood. 1965. In O. J. Harger, ed., *ASME Handbook,* 2nd ed. New York: McGraw-Hill.

Horger, O. J., ed. 1965. *ASME Handbook,* 2nd ed. New York: McGraw-Hill.

Niemann, G. 1975. *Mashinenelemente,* 2nd rev. ed. Berlin: Springer-Verlag.

Norris, C. H. 1945. "Photoelastic Investigations of Stress Distribution in Transverse Fillet Welds." *Welding Journal* 24:557.

Oswood, C. C. 1979. "Saving Weight on Bolted Joints." *Machine Design* (Oct. 25).

Salakian, A. G., and G. E. Glaussen. "Stress Distribution in Fillet Welds: A Review of the Literature." *Welding Journal* 16 (May):1.

Shigley, J. E., and C. R. Mischke. *Mechanical Engineering Design.* New York: McGraw-Hill, 1989.

ADDITIONAL READING

Deutscher Normenausschus (DNA). *Schweisstechnische Normen.* DIN Taschenbuch 8. Berlin: DNA, 1971.

Duggan, T. V., and J. Byrne. *Fatigue as a Design Criterion.* London: Macmillan, 1977.

Hánchen, R., and K.-H. Decker. *Neue Festigkeitsberechnung für den Maschinenbau.* Munich: K. Hanser Verlag, 1967.

Hoyt, S. L., ed. *ASME Handbook: Metals Properties.* New York: McGraw-Hill, 1954.

Kverneland, K. O. *World Metric Standards in Engineering.* New York: Industrial Press, 1978.

Machine Design. 1980. Fastening and joining issue. Cleveland: Penton/IPC.

Niemann, G. *Maschinenelemente,* 2nd rev. ed. Berlin: Springer-Verlag, 1965.

Thum, A., and A. Erker. *Schweissen im Maschinenbau.* Berlin: VDI Verlag, 1943.

PROBLEMS[7]

8.1. [D] The riveted steel construction node shown (Figure P8.1) for a crane structure is loaded by a horizontal load $F = 47.5$ kN. The thickness of the gusset plate equals the thickness of the web of the horizontal member. The steel is of ASTM A572 grade 42 and the rivets are made of the same material. Compute the rivet diameter for the riveting of the horizontal beam, if the safety factor is 1.5.

---

[7][C] = certification, [D] = design, [N] = numerical, [T] = theoretical problem.

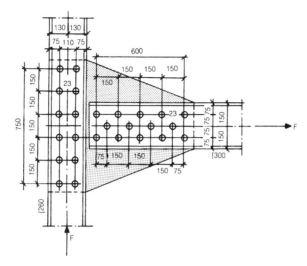

Figure P8.1 (After Decker, 1973).

8.2. [D] In the riveting of Problem 8.1, compute the diameter of the rivets for the riveting on the vertical column.

8.3. [D] A plate of 8 mm thickness is joined with a vertical steel beam of 10 mm thickness by way of two rivets and supports an eccentric weight $F = 12$ kN. The rivets are 16 mm in diameter. If the material for all components is construction hot-rolled steel ASTM A572 grade 42 and the rivets are made from the same material, determine which one of the two solutions has higher safety factor.

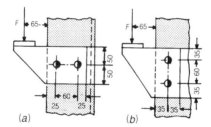

Figure P8.3

8.4. [D] Solve Problem 8.3, except that the riveting is as in Figure P8.4, the thickness of the column web is 12 mm, the force $F = 35$ kN, and the material is ASTM A570 grade D steel. Rivet diameter is 16 mm and plate thickness is 8 mm.

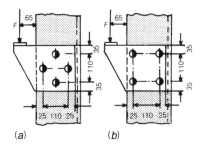

Figure P8.4

8.5. [D] The steel construction for the overhead crane support shown is made of ASTM A284 grade D structural steel and supports a force of $F = 72$ kN. The thickness of the web and the two plates is 14 mm. The eight rivets of design (a) have diameter 20 mm. Determine the rivet diameter of design (b) (12 rivets) to have the same strength as design (a).

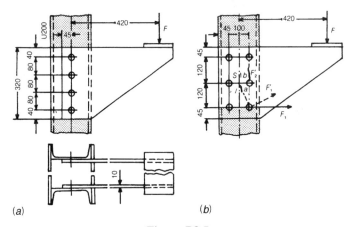

Figure P8.5

8.6. [D] The chain gear (sheave) shown is fixed on the plate of a toothed coupling with six equidistant 6.4 mm diameter rivets located on a circle of diameter 120 mm. It rotates at 2000 rpm and transmits power 43 kW. The material of the gear and the rivets is carbon steel SAE 1018. Determine the safety factor of the joint.

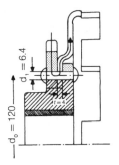

Figure P8.6

8.7. [D] The blade of a centrifugal blower is attached to the rotating disk by four rivets as shown in Figure P8.7. Their material is ASTM A284 grade D steel and the material of the blade and disk is ASTM A570 grade D steel. The blade weighs 0.3 kg and its mass center is at 175 mm radius. Determine the necessary rivet diameter if the safety factor must be $N = 1.8$ and $n = 3000$ rpm. Assume that only centrifugal forces are substantial and disregard aerodynamic forces.

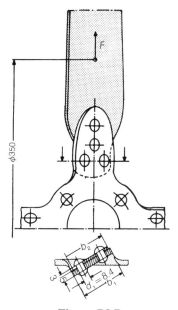

Figure P8.7

8.8. [D] For a disk clutch, the plate is fixed on the disk by eight rivets, located at equal distances on a circle of diameter 70 mm, of material carbon steel 1020 while the material of the disk and the plate is carbon steel 1025. Determine the necessary rivet diameter for a safety factor $N = 2$. The friction moment is generated by an axial force 80 kN at a concentric circle of diameter 130 mm with a coefficient of friction $f = 0.3$.

8.9. [D] The strip band of a band brake is transforming a force $F = 12$ kN. It is fixed on the pulling bracket by five rivets. The material of the bracket and the rivet is ASTM A284-D structural steel while the band is made of a spring steel with $S_u = 400$ N/mm^2 and $S_y = 320$ N/mm^2. Determine the width of the band and the rivet diameter for a safety factor $N = 4$.

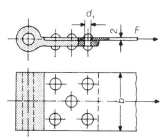

Figure P8.9

8.10. [D] A brake drum is fixed on a disk rim with eight rivets as shown. Assuming material steel SAE 1020 for rivets, rim and drum, brake moment 1000 Nm, safety factor $N = 1.7$, determine the rivet diameter.

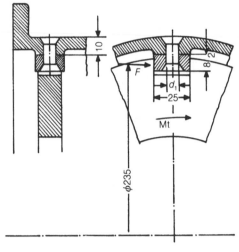

Figure P8.10

8.11. [D] In the node of Problem 8.1, there is a vertical force also on the horizontal member of 50 kN at a distance of 2000 mm to the right of the right side of the vertical column. Determine the rivet diameter.

8.12. [D] Solve Problem 8.7 if there is also a peripheral force, owing to the flow, applied at the center of mass on the plane of riveting, of magnitude 1000 N.

8.13. [D] Solve Problem 8.9 for fatigue strength of the rivet and the band.

8.14. [D, T] Solve Problem 8.10 if the drum is heated owing to abnormal braking to 150°C while the temperature of the rim remains at 60°C. Rivet length is 18 mm. Note that due to the thermal expansion of the drum, its diameter increases by $D\alpha\Delta T$, where D is the diameter of the drum, α is the coefficient of thermal expansion, and ΔT is the temperature difference $150 - 60$ degrees. This elongates the rivets by an equal amount and creates thermal stress in them.

8.15. [D] Two steel plates, riveted on two U-beams, support the shaft and pulley system of a hoist. The cable force is 6500 N. The plate has thickness 8 mm and is joined to the U-beam by three rivets of 16 mm diameter. The web thickness of the U-beam is 10 mm. The cable force varies a great number of times during the operation of the hoist, from the given maximum value to zero. Find a proper material for the rivets for fatigue strength to assure a safety factor $N = 2.5$. The beam and plate material is hot-rolled construction steel ASTM A572 grade 42.

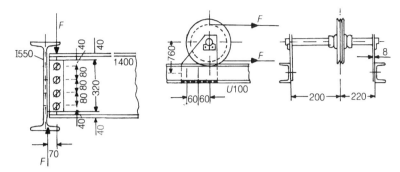

Figure P8.15

8.16. [D] The turn-buckle shown consists of two threaded studs with opposite threads connected with the nuts on one piece. When the common nut is rotated, the two threaded studs are pulled towards one another. Design a 2 ton pulling force turn-buckle of class 4.6 material. The turn-buckle is rotated under load and the safety factor is taken as $N = 3$. Assuming length $L = 100$ mm, design the section between nuts if the turning is performed by a 10 mm diameter rod inserted between the turn-buckle rods, which have circular cross-section and centers 40 mm apart.

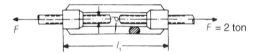

Figure P8.16

8.17. [D, C] A hook shown in Figure P8.17 is made of AISI 1020 steel and is rated at 100 kN. Find the necessary nut length and the safety factor for the bolt.

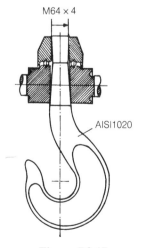

Figure P8.17

8.18. [C] In the steel construction shown, the load is 45 kN. Determine the smallest safety factor for either one of the two bolts. Assume that the center of rotation of the horizontal beam is point A_4 and that the bolt loads are proportional with the distance from point A_4.

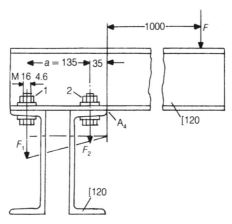

Figure P8.18

8.19. [D] A coupling is joined by bolts as shown. The coupling material is ASTM A159 grade 3000 and the bolts are category 8.8. The transmitted torque is 10 kNm and the desired safety factor is 2.2. The torque must be transmitted with friction, supplied with sufficient bolt preload. The coefficient of friction is 0.2 between coupling faces. For the case of loosening the bolts, their shank must be able to support the torque by shearing. Determine the necessary number of bolts and the tightening torque.

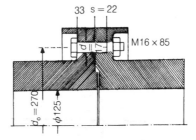

Figure P8.19

8.20. [D] In the friction coupling shown, the transmitted torque is 600 Nm and the shaft diameter is $D = 60$ mm. The coupling is held together by four bolts on each side and a total force 20,000 N. The safety factor must be at least 2.5. Determine:

1. The coefficient of friction between shaft and coupling
2. The bolt diameter of material class 5.6
3. The tightening torque

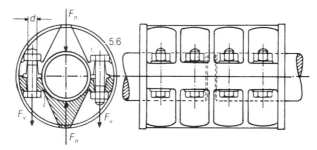

Figure P8.20

8.21. [D] The follower bearing of a reciprocating engine is made of two parts joined together by two bolts of ISO/SAE Class 8.6. Between the two parts is a copper gasket 3 mm thick. To maintain contact during the operation of the pulling force, $P = 30$ kN, the bolts are properly preloaded. Assuming an equivalent flange of outer diameter $2d$ and inner diameter $1.1d$, determine the bolt diameter, the preload force, and the tightening torque. The threads are cut and rolling-hardened. The safety factor must be 1.8 and the preload factor 1.

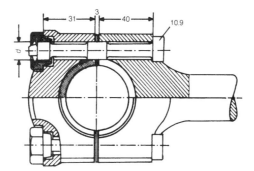

Figure P8.21

8.22. [D] The cast iron head flange of a small diesel engine is secured on the engine body by six bolts, category 8.8. The maximum pressure is 4 N/mm^2. The threads are cut without other treatment. The safety factor is $N = 2$, the flange thickness is 20 mm. Determine the bolt diameter and the tightening torque if the preload factor is 1.5.

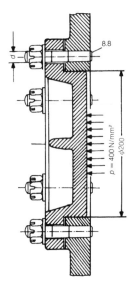

Figure P8.22 (From Decker, 1973.)

8.23. [D] The left end bearing of the worm for a worm gearbox is secured by a flange with four bolts. From experience it is known that $\delta_f / \delta_s = 0.6$, the safety factor is $N = 1.5$, and the preload factor is 1.1. Determine the bolt diameter for cut threads without further treatment. The bolt is class 8.8.

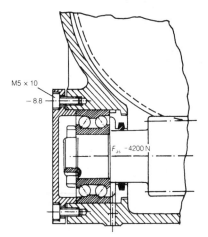

Figure P8.23

8.24. [D] A gear pump delivers pressure at 2 N/mm^2, assumed uniform over the flange. It is fixed by six bolts SAE class 5.6. The preload factor is 1.5. Determine:

1. The bolt diameter for rolled and heat-treated threads with safety factor $N = 1.8$
2. The tightening torque
3. The safety factor if a 3 mm thick copper gasket is used with the previously designed bolts

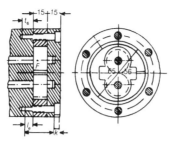

Figure P8.24

8.25. [D] A pinion gear is secured on a shaft with a tapered fit with a 10% taper. The torque transmitted is 500 Nm and the coefficient of friction between shaft and gear is 0.25. Determine:

1. The required axial force to force the gear onto the shaft

2. The relation of axial force versus radial movement versus axial movement

3. The bolt diameter if, during operation, there is an axial force on the gear to the left direction of pulsating type from 0–5 kN and the safety factor is $N = 2$

4. The nut length necessary to support the axial force if the thread is cut and the safety factor is 2

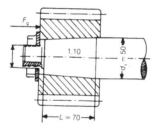

Figure P8.25

8.26. [D] A hand press for a 2 ton rated capacity is operated with a lever on each end of which the operator can apply a 200 N force. The screw has a standard ACME thread. Determine:

1. The bolt diameter and the applied torque if the nut is made of phosphor–bronze, with grease lubrication. The screw material is AISI 1020 steel and the safety factor is $N = 2.2$

2. The efficiency of the screw

3. Whether the screw is self-locked

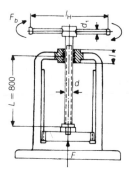

Figure P8.26

8.27. [D] A hub pulling mechanism consists of a screw and a holding frame as shown. It is rated at 5 kN load and the nut is directly formed on the cast iron flange. Determine:

1. The bolt diameter, the tightening torque, and the efficiency. The bolt material is carbon steel AISI 1020, the safety factor is $N = 2$, and the bolt is oil-lubricated

2. The maximum allowable travel L of the screw

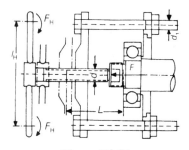

Figure P8.27

8.28. [D] A screw jack has an oil-lubricated phosphor–bronze nut and an AISI 1015 steel screw with standard square thread 36×6. It is rated at 2 ton load. Determine:

1. The torque required at maximum load

2. The efficiency of the screw

3. The safety factor

4. Whether the bolt is self-locked.

8.29. [D] The steam valve shown has a standard ISO screw and a phosphor–bronze nut. Assuming 500 N peripheral force at a 170 mm radius at the valve wheel:

1. Determine the screw diameter and efficiency, for a safety factor $N = 2$.

2. Check for self-locking.

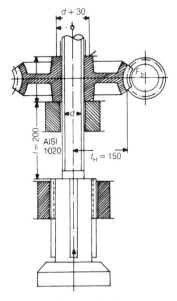

Figure P8.29

8.30. [D] A 200 kN press consists of a power screw with square threads, phosphor–bronze nut, grease-lubricated, operated by a worm gear system. Determine the screw diameter for a factor of safety $N = 2$, the efficiency, and the loading and unloading torque.

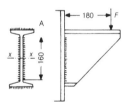

Figure P8.30

8.31. [C] The I-beam shown is welded on a vertical steel column by welds of 6 mm width with a E60 electrode. The beam is loaded as shown by a 6 ton force. Assuming that the shear force is uniformly distributed, determine the safety factor of the welded joint.

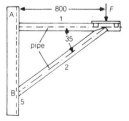

Figure P8.31

8.32. [C] A supporting structure is made of 80 mm diameter pipe with 3 mm thickness, welded on a vertical angle section. The welding has a width $a = 5$ mm and the load F is static. Assuming a weld around the pipe, determine, on the basis of the welds on the vertical column, the maximum permissible force F if the weld is made of quality E70 electrode.

8.33. [D] The bracket of Figure P8.33 is welded, as shown, on a horizontal shaft. If the load F is 60 kN and the thickness of the plates is 12 mm, determine the weld needed with E70 electrodes, for static load, with safety factor $N = 2$.

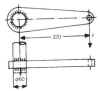

Figure P8.33

8.34. [D] The lever shown is welded with class E60 electrode and loaded as shown with a force $F1 = 350$ N on the left and appropriate force at the right, for equilibrium. The strips welded to the web have width 20 mm and thickness 10 mm. For a safety factor $N = 1.5$, determine the required weld width a.

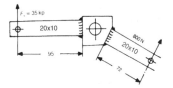

Figure P8.34

8.35. [D] The gear shown is welded on the hub and transmits a 500 Nm constant torque. Determine the proper electrode quality for a weld width $a = 5$ mm on both sides. The safety factor is $N = 1.6$.

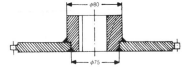

Figure P8.35

8.36. [D] The pulley shown transmits power 11 kW at 100 rpm. Determine the required weld width h for electrode quality E70, safety factor $N = 2$.

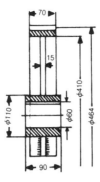

Figure P8.36

8.37. [D] The wheel assembly shown is loaded by force $F = 6$ kN. Determine the required weld width h for electrode quality E70, safety factor $N = 2$. Assume only axial force at the weld.

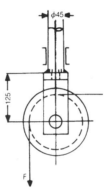

Figure P8.37

8.38. [D] The crane beam shown is loaded with weight $Q = 50$ kN. Determine the required weld width h for electrode quality E70, safety factor $N = 2$.

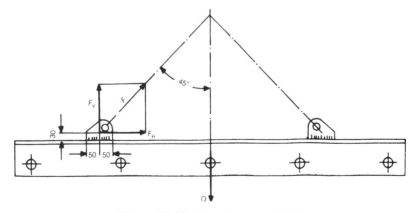

Figure P8.38 (From Decker, 1973.)

8.39. [D] The weld of the wheel assembly shown is loaded eccentrically with force $F = 3$ kN. Determine the required weld width h for electrode quality E70, safety factor $N = 2$.

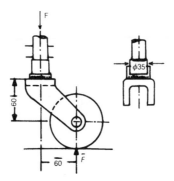

Figure P8.39

8.40. [D] The weld of the cable pulley assembly shown is loaded eccentrically with force $F = 5$ kN. Determine the required weld width h for electrode quality E70, safety factor $N = 2$.

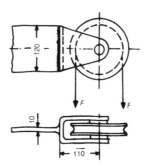

Figure P8.40

8.41–8.45. [D] Solve Problems 8.36 to 8.40 if the applied load is pulsating, that is, varies between zero and the given maximum value, for fatigue strength.

8.46. [C] The node shown of a steel construction is made with spot welding and loaded as shown. Find the safety factor if the material is ASTM A284, grade D. 1 kp $\approx$ 10 N.

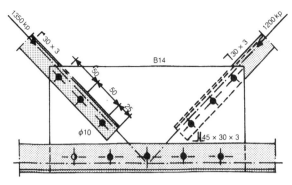

Figure P8.46

8.47. [C] For the light-steel construction detail shown, with material steel ASTM A284, grade D, find the maximum of the loading force F if the safety factor must be $N = 2$.

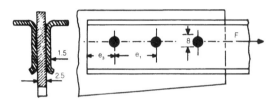

Figure P8.47

8.48. [D] The V-belt pulley shown is made of steel ASTM A284, grade D, spot-welded. The belt is 180° around the pulley, transmits 60 Nm torque, and has total tension 2 kN (the sum of forces at both branches) with a wedge angle 30°.

The spot welds are loaded with half of the transmitted torque. Disregard the wedge effect. Determine:

1. The number of spot welds of 6 mm diameter needed to keep the halves of the pulley together

2. The number of spot welds between the left half of the pulley and the hub, of 6 mm diameter.

The material is steel ASTM A284, grade D, and the safety factor is $N = 2.5$.

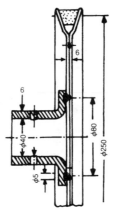

Figure P8.48

8.49. [D] A control device has a strip of thickness 2 mm fixed on a disk with four spot welds. The material disk and strip is carbon steel AISI 1020. The motion of the disk is intermitted in cycles of 0.3 sec as shown. With a safety factor $N = 2.2$, determine the diameter of the spot welds.

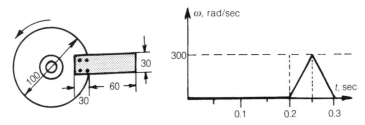

Figure P8.49

8.50. [D] A disk clutch consists of a hub and a disk fixed by spot welds of 5 mm diameter on the hub. If the maximum transmitted torque is 50 Nm, the disk and hub material is steel ASTM A572, grade 50, and the safety factor is 1.5, determine the number of spot welds required.

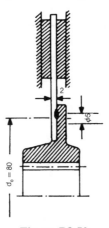

Figure P8.50

8.51. [C] A pinion gear with inner diameter = 50 mm, outer diameter = 80 mm, width = 60 mm, and material carbon steel AISI 1030 is press-fitted on a shaft from carbon steel AISI 1015 with fit *H7/t*6. Determine, for precision-machined surfaces:

1. The maximum transferred torque
2. The safety factor against material yielding (the shaft and bore surfaces are precision-machined)
3. The preheating temperature required for assembly

8.52. [D] A control arm has a maximum activating force 150 N in the position shown. It is press-fitted on a 15 mm diameter shaft and the hub has 37 mm outer diameter and 22 mm width. The material of the shaft and hub is carbon steel AISI 4140. Determine the required fit for a safety factor against slipping 1.5 and against material failure 2. The fit surfaces are ground to a roughness of 10 μm.

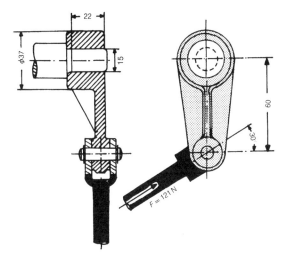

Figure P8.52

8.53. [C] A disk and a hub of ASTM A284, grade D steel, is shrink-fitted on a shaft of material AISI 1015 steel. The shaft diameter is 80 mm and the fit is *H7/u*6. The surfaces are normally machined. Determine:

1. The maximum torque transferred with a safety factor against slipping 2.2
2. The safety factor against hub failure

Assume that the disk and the hub act independently, transferring a certain torque each.

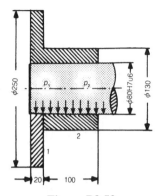

Figure P8.53

8.54. [D] The crank arm shown is press-fitted on a shaft. The shaft material is ASTM A572 grade 65 steel and the crank is of grade 50 steel. Determine the required fit for safety factors against yielding 1.5 and against slipping 2.5. If the crank is to be press-fitted, determine the required axial force. Shaft diameter 30 mm, hub width 60 mm, length 40 mm.

8.55. [D] The flywheel of an internal combustion engine is shrink-fitted on the crankshaft as shown. The moment of inertia of the flywheel is 4 kg/m² and the maximum expected angular acceleration is 100 rad/s². Determine the required fit if the factor of safety against slipping is 1.8, against yielding 2. The material of the shaft is AISI 1030 steel, that of the flywheel is cast iron ASTM A 159, grade G3000, and the surfaces are ground.

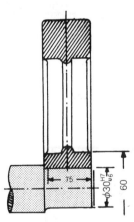

Figure P8.55

CHAPTER 9

DESIGN FOR RIGIDITY

Railroad car suspension systems are designed for low rigidity.

OVERVIEW

Starting with Chapter 8, we reached the second part of the book, devoted to design methods for machine elements. Sizing of machine components is not based only on strength, as in Chapter 8. There are mechanisms of failure that are not directly related to strength, such as wear.

Machine elements fail many times because of excessive sudden deformation due to loss of their elastic stability. Most of the time, this occurs at stresses far below the allowable stresses for the material. Respective methods for design of machine elements will be discussed in this chapter.

9.1. RIGIDITY REQUIREMENTS IN MACHINE DESIGN

Rigidity is the capacity of machines to sustain loads without appreciable change of their geometry. The deformation due to loading might not be high enough, many times, to geometrically disturb the operation of the machine. However, it might cause other effects, such as instability, friction, and vibration, which in turn might influence proper machine operation.

Some requirements imposed on the rigidity of machine parts are:

1. Strength during dynamic instabilities or shock loads
2. Performance of the parts in connection with mating parts
3. Easy manufacturing
4. Satisfactory operation of the machine as a whole

The conditions, based on the performance of mating parts, are of prime importance. For example, the proper rigidity of shafts determines the satisfactory behavior of bearings, gears, worm drives, and other drives whose components are mounted on the shafts. The forces exerted by drives bend shafts and lead to interference in bearings and gears. Permissible deflection and angles of inclination are determined on the basis of rigidity requirements.

Requirements for rigidity on the basis of manufacturing processes are of prime importance for various parts, especially with mass production. Rigidity requirements based on conditions of satisfactory operation of the machine as a whole are usually quite specific. Here, design proceeds for the working loads and the accuracy specifications of the workpiece.

The rigidity of machine components is determined by their inherent rigidity when they are dealt with as beams, plates, or shells with idealized supports and by their contact rigidity, that is, the rigidity of the surface layers at the areas of contact.

Of prime importance for most parts subject to considerable loads is their inherent deformation. But in precision machinery at relatively low loads, the contact deformations in unsecured joints (between mutually movable mating parts) play an essential and even predominant role in the balance of displacements.

The contact of parts may be under conditions of:

1. Initial contact in a point or line, as in pressing two balls or cylinders together
2. A large nominal area of contact

Stability, one form of which is *buckling*, is a criterion determining the dimensions of:

1. Long, thin parts in compression
2. Thin plates subjected to compression acting on opposite edges
3. Shells subjected to external pressure
4. Thin-walled hollow shafts

The most common parts checked for buckling are the screws of jacks, lead screws, piston rods, and compression springs. Many components are checked for stability in the steel structures of machinery.

Sometimes low rigidity in the form of high flexibility is wanted in many machine members, such as springs. This property permits constant loads to be maintained in pressure-tight components and mechanisms, vibration isolation, and so on.

Because the dynamic characteristics of mechanical systems depend on rigidity (or flexibility), this property is many times carefully controlled in order to maintain proper dynamic properties of the machine. More on that point will be discussed in Chapter 14.

9.2. RIGIDITY OF MACHINE COMPONENTS

9.2.1. Rigidity Material Indices

High strength of a material is not necessarily associated with high rigidity. One can assert the opposite: Improvement of strength properties (S_u, S_y, etc.) has little effect on elastic properties (E, G, v). Because deflection, in general, is influenced by loads, geometry, and elastic properties, improvement of material strength results in smaller sections for the same load and thus higher deflections and smaller rigidity. Therefore, a representative index expressing material rigidity would be, for example, E/S_y. For machines where a major part of the load is determined by the mass of the machine, the material density (or specific weight) γ is equally important. The associated index would be $E\gamma/S_y$ in this case.

For the tension rod, rigidity is defined as the ratio of axial load to resulting axial deformation, also called *spring constant:*

$$k = \frac{P}{\delta L} = \frac{EA}{L} \tag{9.1}$$

where A is the cross-section, L is the length, E is the Young's modulus. For maintaining constant rigidity k, the weight of the part γLA, where γ is the specific weight, is minimized when the weight of the parts:

$$W = A\gamma L = \frac{L^2 k}{E/\gamma} \tag{9.2}$$

is minimized. For given length and rigidity, the material of minimum weight is the one with maximum E/γ index.

For maximum rigidity at given material strength S_u, load P, and length L:

$$k = \frac{EA}{L} = \frac{P}{L}\frac{E}{S_u} \tag{9.3}$$

and the index E/S_u must be maximum, or E/S_y, depending on the design requirements.

Similarly, minimum weight, maximum strength, and maximum rigidity for minimum weight are achieved for given geometry A, L, and load P when:

$$\frac{1}{W}\, kN = \frac{1}{AL\gamma} \times \frac{EA}{L} \times \frac{S_y}{P/A} = \left(\frac{A}{LP_{max}}\right)\left(\frac{ES_y}{\gamma}\right) \qquad (9.4)$$

is maximized. The general index ES_y/γ applies, therefore, generally for cases where tensile stresses are the main material load. For flexure and torsion, similar results can be obtained.

For the basic material categories, rigidity indices are shown in Table 9.1. Let us compare the rigidity, strength and weight of structures when changing over from castings of pig iron to castings of aluminum and carbon steels without altering the parts' shape. Assuming the rigidity, strength, and weight of gray cast iron structures as equal to unity, we obtain:

Parameter	Iron	Aluminum Alloys	Low-Carbon Steel	High-Carbon Steel
Rigidity (E)	1	0.9	2.6	2.6
Strength (S_y)	1	0.72	2.3	4.6
Specific weight (γ)	1	0.39	1.1	1.1
Index (ES_y/γ)	1	1.66	5.4	10.8

Thus, the transfer to aluminum alloy castings hardly affects rigidity, somewhat lowers strength (approximately 30%), and significantly reduces the weight of the structure (2.5 times). Changing to low-carbon steel increases rigidity and strength approximately 2.5 times and the weight is practically the same.

The above-listed relations are based on the fact that Young's modulus has a constant value for steel and only slightly depends on the presence (in normal amounts) of alloying elements, heat treatment, and strength characteristics of alloys of the given metal. For example, in respect to steels, beginning from low-carbon and up to high-alloy ones, Young's modulus varies from 190,000–220,000 MPa and the shear modulus G varies from 79,000–82,000 MPa. With regard to aluminum alloys, $E = 70,000$–75,000 MPa and $G = 24,000$–27,000 MPa. Consequently, for the manufacture of identically shaped components when rigidity is the first requirement and the stress level is low, the use of low-cost materials is recommended (low-carbon steels). This will not affect the rigidity of the construction but will enable the cost to be reduced.

The above does not hold when the strength of the construction is as important as the rigidity. Thus, for instance, identically shaped structures, one made from low-carbon steel and the other from alloy steel, will have the same rigidity, but the load capacity of the first structure will be as many times less as the tensile strength of carbon steel is less than that of alloy steel.

For equally rigid parts, the greatest strength advantage is with materials having the highest S_y/E ratio (high-strength steels, titanium alloys, and wrought aluminum alloys). In terms of weight advantage (in this case, proportional to the γ/E factor), the above-listed materials possess approximately equivalent values. Worse weight

TABLE 9.1. Strength and Rigidity Characteristics of Structural Materials

Material	Specific Weight, γ, kgf/dm³	Ultimate Tensile Strength, S_u, kgf/mm²	Yield Limit, $S_{0.2}$, kgf/mm²	Modulus of Elasticity, E, kgf/mm²	Modulus of Shear, G, kgf/mm²	Rigidity Characteristics			Generalized Factor, $\dfrac{S_{0.2}E}{\gamma}\cdot10^{-5}$
						$\dfrac{E}{\gamma}\cdot10^{-3}$	$\dfrac{E}{S_u}\cdot10^{-2}$	$\dfrac{E}{S_{0.2}}\cdot10^{-2}$	
Carbon steels	7.85	35–80	21–48	21000	8000	2.67	2.6	1.56	1.3
Alloy steels		100–180	80–145				1.17	0.94	3.8
High-strength steels		250–350	225–315				0.6	0.54	8.4
Grey cast irons	7.2	20–35	14–25	8000	4500	1.1	2.3	1.6	0.3
High-strength cast irons	7.4	45–80	32–56	15000	7000	2	1.9	1.3	1.1
Aluminum alloys cast	2.8	18–25	13–17.5	7200	2500	2.67	2.9	2.04	0.45
Aluminum alloys wrought		40–60	28–42				1.2	0.83	1.1
Magnesium alloys cast	1.8	12–20	8–13	4500	1500	2.3	2.1	1.35	0.32
Magnesium alloys wrought		25–30	18–21				1.4	1.2	0.52
Structural bronzes	8.8	40–60	32–48	11000	4200	1.25	1.85	2.3	0.6
Titanium alloys	4.5	80–150	70–135	12000	4200	2.66	0.8	0.72	3.6

characteristics among metals for machine applications are held by bronzes and gray cast irons.

In practice, the choice of material is determined not only by the strength–rigidity characteristics, but also by some other properties. That is why preference is given to those design features that enable strength and rigidity to be enhanced even when materials of low strength and rigidity are used.

9.2.2. Design Rules for High Rigidity

Some design methods to increase rigidity are:

1. Avoidance of bending that is weak from the viewpoint of rigidity and strength, changing the force-carrying mechanism to compression and tension
2. For parts working in bending, rational support positioning
3. Rational increase of section inertia moments, reinforcing joints and the parts transferring forces from one section to another
4. For thin-walled box-shaped parts, using optimum shapes

9.2.3. Rational Sections

It is important that the increase in rigidity not be accomplished by increasing weight of the machine. In general, the solution of the problem involves strengthening the sections that under the given loads are subjected to the highest stresses and removing weight from the unloaded or slightly loaded areas. In flexure, the stressed sections are those farthest from the neutral axis. In torsion, the external fibers are mostly stressed; moving radially and inwards, the stresses become weaker and at the center are zero. Consequently, for these cases it is more rational to concentrate material at the periphery and remove it from the center.

Generally, the greatest rigidity and strength characteristics with smallest weight are found in components with thin-walled sections—parts such as box sections, tubes, and shells.

Table 9.2 gives rigidity and strength (load-carrying capacity) comparisons for differently shaped sections. The base of the comparison depends upon similar weight conditions of parts, expressed as similar cross-sectional areas. The strength and rigidity improvements are obtained by proper material distribution in the regions of the highest acting stresses.

For cylindrical sections, the moment of inertia I and the section modulus W of a solid round section are taken as the basis of comparison—with respect to the other parts, a solid square-shaped section.

Figure 9.1 shows reinforcements of beams by transversal ribs and stiffening boxes: (a) by stiffening partitions; (b) by stiffening boxes; (c) by semicircular stiffening elements.

Diagonal stringers in the form of webs strongly improve rigidity (Figure 9.2a,b) and also load section stiffening (Figure 9.3). Thus, constructions with longitudinally formed stiffening rib angles at the transition points where vertical walls change to horizontal ones (Figure 9.3b) have greater rigidity than the original construction (Figure 9.3a) in spite of formal reduction in inertia moment. The rigidity

TABLE 9.2. Rigidity/Strength of Sections of Same Weight

Section	Ratios		Inertia I/I_0	W/W_0
		—	1	1
	$\dfrac{d}{D}$	0.6	2.1	1.7
		0.8	4.5	2.7
		0.9	10	4.1
		—	1	1
	$\dfrac{h}{h_0}$			
		1.5	3.5	2.2
		2.5	9	3.7
		3.0	18	5.5
		—	1	1
	$\dfrac{h}{h_0}$			
		1.5	4.3	2.7
		2.5	11.5	4.5
		3.0	21.5	7.0

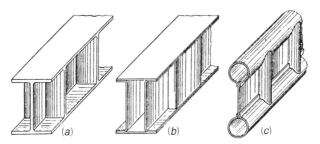

Figure 9.1. Beams with rigidity: (*a*) by stiffening partitions; (*b*) by stiffening boxes; (*c*) by semicircular stiffening elements.

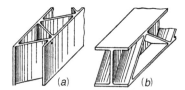

Figure 9.2. Reinforced beams.

parameter also increases when the longitudinal rib has transverse stiffening ribs spaced over the part length (Figure 9.3*c,d*).

Table 9.3 illustrates how longitudinally arranged webs affect the rigidity of profiles during flexure and torsion. One diagonal stringer will suffice; another stringer will enhance the rigidity but only to a small degree.

Ribbing is used to improve rigidity, particularly of cast housing-type components (Figures 9.4 and 9.5).

9.3. STABILITY OF MACHINE ELEMENTS

9.3.1. Buckling

Stability, loosely defined, is the property of a system to return to its original state after it has been displayed from the position of equilibrium. Otherwise, the system is said to be *unstable*. A system that has lost stability may behave in different ways. Usually a transition to a new position of equilibrium occurs, commonly accompanied by large displacements, the development of large plastic strains, or complete collapse. In some cases, a structure continues to operate properly and perform its basic functions after loss of stability, as, for example, in thin-walled panels in aircraft structures. Finally, a system that has lost stability may set into continuous oscillatory motion.

The most common case of instability is the buckling of a centrally compressed bar (Figure 9.6). At a sufficiently large force, the bar cannot maintain the straight-line configuration and deflects laterally.

A thin-walled tube (Figure 9.7) under external pressure may lose stability. The circular shape of the section then changes into an elliptical one and the tube is completely flattened, though the stresses in the walls are far from reaching the yield point at the moment of buckling.

The same tube may lose stability under axial compression (Figure 9.8). A similar phenomenon takes place when a tube is subjected to torsion (Figure 9.9).

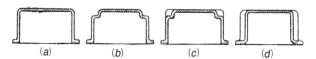

Figure 9.3. Local section stiffening.

TABLE 9.3. Section of Increased Rigidity by the Use of Longitudinal Webs

Profile	Factors				
	I_{flex}	I_{tors}	G	$\dfrac{I_{flex}}{G}$	$\dfrac{I_{tors}}{G}$
	1	1	1	1	1
	1.17	2.16	1.38	0.85	1.56
	1.55	3	1.26	1.23	2.4
	1.78	3.7	1.5	1.2	2.45

It is apparent that instability is most pronounced in light thin-walled structures, such as slender rods, thin shells, and thin walls. Therefore, in the design of such elements stability analysis of both separate components and the system is required in addition to the strength analysis.

For the analysis of stability it is assumed that the system is ideal, that is, if it is a compression member, its axis is perfectly straight, the material is homogeneous, and the forces are applied centrally.

The simplest elastic element is a bar compressed by central forces P (Figure 9.10a). This problem was first formulated and solved by Euler in the middle of the 18th century. Hence, the expressions *Euler problem* or *stability of an Euler bar* are often used when reference is made to the stability of a centrally compressed bar.

Suppose that the bar has deflected slightly for some reason. This always happens because a perfectly straight bar or a perfectly central force is not possible in an engineering structure. The coordinates of the elastic curve of the bar are denoted by z and y (Figure 9.10a). For small deflections, bending theory of beams states

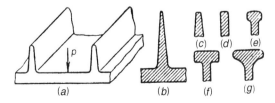

Figure 9.4. Ribbed sections (in order of increasing strength).

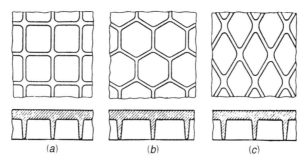

Figure 9.5. Rib.

that curvature (y'') is proportional to the bending moment at the particular point (Beer and Johnston 1991):

$$EIy'' = M \tag{9.5}$$

The bending of the bar occurs in the plane of minimum rigidity and so the quantity I is understood to be the minimum transverse moment of inertia of the section.

The bending moment M at a section located at any point z is equal to $-Py$.

$$EIy'' = -Py \tag{9.6}$$

Let

$$\frac{P}{EI} = k^2 \tag{9.7}$$

Equation (9.6) then becomes:

$$y'' + k^2y = 0 \tag{9.8}$$

whence the solution is a harmonic function

$$y = C_1 \sin kz + C_2 \cos kz \tag{9.9}$$

Figure 9.6. Buckling of a column.

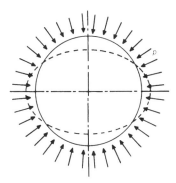

Figure 9.7. Buckling of a ring or pipe under external pressure.

Leonhard Euler (1707–1783) was born in Basel, Switzerland. He studied mathematics there under Johann Bernoulli. At the age of 20 he was appointed to the Chair of Mathematics at the new St. Petersburg Academy, formed by Peter the Great, where he wrote his *Mechanica sine motus scientia analytice exposita,* introducing analytical methods in the study of mechanics. Fourteen years later he became the head of the Prussian Academy and moved to Berlin, invited by Frederick the Great. After 25 years, he returned to St. Petersburg for the remainder of his life, at the invitation of the Empress Catherine II. Euler was the most prolific writer in the history of mathematics. Forty years after his death, the Russian Academy was still publishing his papers.

Figure 9.8. Collapse of a cylindrical shell under axial compressive force.

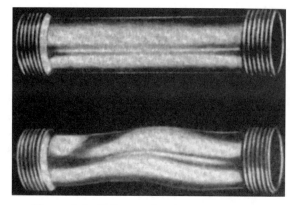

Figure 9.9. Collapse of a pipe under torsion.

Constants C_1 and C_2 must be chosen so as to satisfy the boundary conditions: when $z = 0$, $y = 0$, and $z = l$, $y = 0$. From the first condition, it follows that $C_2 = 0$, and from the second condition:

$$C_1 \sin kl = 0 \tag{9.10}$$

This equation has two possible solutions: either $C_1 = 0$ or $\sin kl = 0$.

In the first case, the displacements y (Equation (9.9)) are identically zero for $C_1 = C_2 = 0$, and so the bar maintains the straight-line configuration. This case is of no interest to us. In the second case, $kl = \pi n$, where n is an arbitrary integer. Taking into account expression (9.7), we obtain $P = \pi^2 n^2 EI/l^2$. This means that for the bar to maintain a curvilinear configuration, the force P must take a definite value. This minimum force P, when $n = 1$, is:

$$P_{cr} = \frac{\pi^2 EI}{l^2} \tag{9.11}$$

This force is termed the *critical load* or *Euler's load*. When $n = 1$, $kl = \pi$ and the elastic curve (Equation (9.9)) becomes:

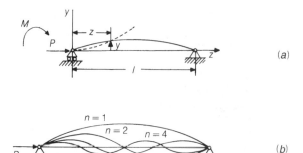

Figure 9.10. Modes of buckling of compressed columns.

$$y = \frac{C_1 \sin \pi z}{l} \tag{9.12}$$

The bar buckles in the half-wave of a sine curve with a maximum deflection C_1.
For any whole-number value of n:

$$y = \frac{C_1 \sin \pi n z}{l} \tag{9.13}$$

and the elastic curve of the bar is represented by a curve in the form of n half-waves (Figure 9.10b).

Our solution, however, does not say what happens $P > P_{cr}$, $kl \neq \pi$. Then it follows from Equation (9.10) that $C_1 = C_2 = 0$ since $\sin kl \neq 0$. This means that function y (Equation 9.9)) is identically zero and the bar remains straight. Thus, at $P = P_{cr}$ the bar assumes a curvilinear shape and becomes straight again at a value somewhat greater than P_{cr}, thus contradicting the accepted physical concepts of mechanics of buckling.

These predicaments can readily be overcome if we take into account that the differential equation (9.6) is approximate and applicable only for small deflections. If this equation is written accurately (Timoshenko and Gere 1961), we obtain:

$$EJ \frac{1}{\rho} = \frac{EIy''}{(1 + y^2)^{3/2}} = -Py \tag{9.14}$$

where ρ is the radius of curvature. With a force P greater than critical, the displacements grow so rapidly that the quantity y^2 in the denominator cannot be neglected.

Solution of the nonlinear differential equation (9.14) can be obtained numerically. This solution will yield the maximum deflection at mid-span as a function of the applied load (Figure 9.11, curve a).

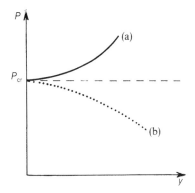

Figure 9.11. Post-buckling behavior of (a) beams and (b) shells.

AISC recommendations: $\mu = 2.1$ $\mu = 0.65$ $\mu = 0.8$

Nonzero maximum deflection occurs for $P > P_{cr}$, and the critical load P_{cr} coincides with the one calculated with linear analysis (Equation (9.7)). Such a procedure is called *post-buckling analysis*. It can be seen from Figure 9.11 that post-buckling behavior of the rod is relatively smooth, although for small increases of the load above the critical value, very high deflections occur, which might lead to the yield point being exceeded and unwanted permanent damage.

There are instability cases, however, in which the situation is worse. For example, in Figure 9.11*b*, the dotted line indicates the post-buckling behavior of certain curved plates. When the critical load is reached, the plate moves to another equilibrium point with a sudden jump.

Equation (9.7) was obtained for a bar with guided hinges (simple supports) at the two ends. Other end conditions will yield similar results. In general:

$$P_{cr} = \frac{\pi^2 EI}{(\mu l)^2} \tag{9.15}$$

where the *column end factor* μ takes the values indicated in Figure 9.12.

The end factors have some uncertainty because, for example, perfect fixing of one end is nearly impossible. Thus, the end factor for a fixed end should be taken between fixed and simple supported. For this reason, the American Institute of Steel Construction (AISC 1980) recommends $\mu = 2.1$ for the fixed–free column (second case in Figure 9.12), $\mu = 0.65$ for the fixed–fixed column (seventh case), and $\mu = 0.8$ for the fixed–simply supported column (last case).

The compressive stress in the column is:

$$\sigma_{cr} = \frac{P_{cr}}{A} = \frac{\pi^2 E}{(\mu \lambda)^2} \tag{9.16}$$

where $r^2 = I/A$ is the *radius of gyration* of the section and $\lambda = l/r$ is the *slenderness ratio*. This relationship is plotted as curve BCD in Figure 9.13.

If we define a *safety factor for buckling* N_b, the design equation for buckling will be:

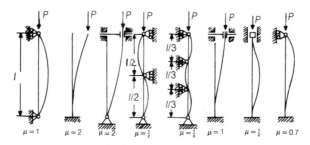

Figure 9.12. Buckling of columns with different boundary conditions.

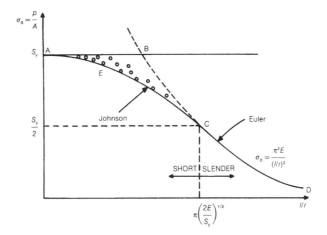

Figure 9.13. Buckling near the yield strength of the material.

$$P_{cr} = \frac{P}{N_b} \tag{9.17}$$

where P is the operational compressive load.

Buckling is not limited to compressed columns. Analysis of each of these cases is beyond the scope of this text. However, some useful results for critical buckling loads of miscellaneous cases are shown in Table 9.4.

In the preceding analysis, *small* deflections were assumed; that is, small in respect to the column length. For relatively short columns, the Euler critical load deviates from experimental results (open circles in Figure 9.13). The Euler critical load is in satisfactory agreement with experiments up to a compressive stress equal to one half of the yield strength.

The column might undergo considerable plastic deformation before becoming unstable. In machine design, this is usually not acceptable, because geometric limitations are violated by the resulting permanent deformation. Therefore, the mean compressive stress should not exceed the yield strength, point A. To connect points A and C and approximate the experimental results, many empirical relations have been used. One of the proposed approximations, the Johnson formula, is the very convenient quadratic function:

$$\sigma_0 = \frac{P_{cr}}{A} = S_y \left[1 - \frac{S_y(l/r)^2}{4\pi^2 E} \right] \tag{9.18}$$

It is apparent that for $\sigma_{cr} > S_y/2$, which corresponds to slenderness ratio $\lambda_0^2 = 2\pi^2 E/S_y$, the *Euler solution* (Equation (9.16)) is valid; otherwise the *Johnson solution* (Equation (9.18)) should be used. In the first case, the column is termed *slender*. In the second case it is termed *short*.

Design Procedure 9.1: Design of a Stable Column in Compression

Input: Compressive load, material properties, length, supports.
Output: column cross-section dimensions

TABLE 9.4. Design Equations for Stability of Structural Members

Geometry	Description	Critical Load
	Buckling of a ring or tube of thickness t under external pressure p_{cr}	$p_{cr} = 3EI/R^2$ $I = t^3/12$
	Lateral buckling of a thin beam under a moment on its plane $I_p = I_x + I_z$, Length L	Hinged ends: $M_{cr} = \dfrac{\pi}{L}(EI_zGI_p)^{1/2}$ Fixed ends: $M_{cr} = \dfrac{2\pi}{L}(EI_zGI_p)^{1/2}$
	Torsional buckling of a thin rod in torsion	$T_{cr} = 2\pi EI_p/L$
	Arch or shell clamped at both ends	$p_{cr} = EI(k^2 - 1)/R^3$ $k\tan\alpha\cot k\alpha = 1$
	Column on elastic foundation constant $\beta = \bar{\beta}L^4/\pi^4 EI$ $\bar{\beta}$: soil modulus (Table 9.6)	B.C. $P_{cr} =$ pinned–pinned $2(\beta EI)^{1/2}$ free–free $(\beta EI)^{1/2}$
	Circular plate force P per unit arc length	Boundary $P_{cr} =$ free $2.88t^2E/R^2$ fixed $9.79t^3E/R^2(1 - \nu^2)$
	Long tube under axial thrust pressure p	$p_{cr} = \dfrac{2tE}{R[3(1 - \nu^2)]^{1/2}} \cdot \dfrac{t}{R}$
	Rectangular plate of thickness t under uniform force per unit length P	$P_{cr} = \dfrac{\pi^2Eh^2}{3(1 - \nu^2)}\left(\dfrac{1}{a^2} + \dfrac{1}{b^2}\right)$

Step 1: Sketch the column and the boundary conditions.

Step 2: Select cross-section form.

Step 3: Use the Euler equation (9.15) and Figure 9.12 to determine the smallest cross-section area moment of inertia *I*.

Step 4: Find the unknown dimension, assuming that the moment of inertia found in step 2 is the minimum one, in case of sections with varying moment of inertia with orientation of the respective axis.

Step 5: Use Equation (9.16) to determine the critical compressive stress. If less than $S_y/2$, the solution found is valid; exit this procedure. Otherwise, go to the next step.

Step 6: Use Johnson's equation (9.18) to determine the slenderness ratio *r*.

Step 7: From the slenderness ratio and the section geometry, find the unknown section dimension.

Example 9.1 Critical Load of a Column

Determine the critical load for an aluminum alloy pipe with inner diameter 30 mm, outer diameter 40 mm. Young's modulus $E = 71,000$ N/mm², yield strength $S_y = 180$ N/mm², and length 1.2 m, fixed at one end and hinged on the other, safety factor in buckling $N_b = 1$.

Solution The moment of inertia and the cross-section area are:

$$I = \frac{\pi(D^4 - d^4)}{64} = \frac{\pi(256 - 81)10^4}{64} = 8.59 \times 10^4 \text{ mm}^4$$

$$A = \frac{\pi(D^2 - d^2)}{4} = \frac{\pi(1600 - 900)}{4} = 549 \text{ mm}^2$$

The radius of gyration $r = (I/A)^{1/2} = (85900/549)^{1/2} = 12.5$ mm. For the given boundary conditions, $\mu = 0.7$ (Figure 9.12). The Euler critical load is:

$$P_{cr} = \frac{\pi^2 E}{(\mu L/r)^2} = \pi^2 \times \frac{71,000}{(0.7 \times 1200/12.5)^2} = 26,226 \text{ N}$$

The Euler critical stress is:

$$\sigma_{cr} = \frac{P_{cr}}{A} = \frac{26226}{549} = 47.77 \text{ N/mm}^2$$

Hence

$$\sigma_{cr} = 47.77 < \frac{S_y}{2} = \frac{180}{2} = 90 \text{ N/mm}^2$$

Thus, the Euler critical load is valid. Thus:

$$P_{max} = \frac{P_{cr}}{N_b} = \frac{26,226}{1} = 26,226 \text{ N}$$

Example 9.2   Design of a Column of Circular Cross-Section

Using EXCEL, design a solid round bar in compression. Test data are axial load 10,000 N, length 0.5 m, Young's modulus $E = 210,000$ MPa, yield strength 220 MPa, boundary conditions factor $\mu = 1$.

Solution   The spreadsheet is shown in Table E9.2.

TABLE E9.2

1	**EXAMPLE 9.2**					
2	**Design of a column of circular cross-section for axial load**					
3						
4	**DATA:**					
5	Length $L =$		0.5			
6	Axial Load $F =$		10,000			
7	B.C. $\mu = :$		1			
8	Young's Modulus $E =$		2.10E + 11			
9	Yield Strength $S_y =$		2.20E + 08			
10						
11	**RESULTS:**					
12	Diameter $d =$		0.010528228			
13	$I =$		1.2062E − 09			
14	Radius of gyration r =		1.38554E − 05			
15	Euler load =		10000	$= \pi^2 EI/(\mu L)^2$		
16	Johnson load =		−2.38842E + 13	Equation 9.18		
17	Compressive stress =		114868138.1	$= F/A$		
18	Slenderness ratio =		36086.88987			
19	Applicable load =		10000			
20	Difference $F - F_{appl} =$		−5.30959E − 07			
21	Difference $\sigma - S_y =$		−105131861.9			
22	We now use Solver to make the target, difference [critical load (cell C19) − the axial load (cell C6)],					
23	cell C20, equal to zero and the compressive stress (cell C17) below S_y (cell C21 < 0 − constraint)					
24	by changing the diameter (cell C12).					
25	Result: Diameter =		0.010528228		Euler Column	

9.3.2. The Beam-Column

The buckling analysis in section 9.3.1 was done on the assumption of a perfectly central load, one that applies on the centroid of the section. This is not always the case. In jib cranes, for example, there is a column-compressing load at a substantial distance from the centroid.

The bending moment at any point $A(z, y)$ is now $M = -(y + e)P$ (Figure 9.14). Therefore, Equation (9.6) will have the form:

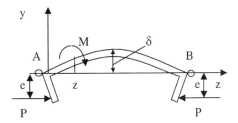

Figure 9.14. Eccentrically loaded column.

$$EIy'' = -(y + e)P \tag{9.19}$$

or, for $k^2 = P/EI$:

$$y'' + k^2y = -K^2e \tag{9.20}$$

This is a nonhomogeneous harmonic equation and its general solution is (Timoshenko and Gere 1961):

$$y(z;t) = C_1 \sin(kz) + C_2 \cos(kz) + e \tag{9.21}$$

where C_1 and C_2 are undetermined constants. The solution is found using the same boundary conditions as in Section 9.3.1:

$$y(0, t) = 0, \quad C_2 = -e, \quad y(l;t) = 0, \quad C_1 = \frac{e \cos(kl)}{\sin(kl)} = 0 \tag{9.22}$$

Therefore:

$$y(z;t) = e[\cotan(kl) \sin(kz) + 1 - \cos(kz)] \tag{9.23}$$

The maximum deflection of the beam-column is at $z = 1/2$:

$$z_{\max} = \delta = e\sec\left(\frac{1}{2}\sqrt{\frac{P}{EI}}\right) \tag{9.24}$$

At that point, the maximum bending moment is:

$$M_{\max} = -P(e + \delta) = -Pe \sec\left(\frac{1}{2}\sqrt{\frac{P}{EI}}\right) \tag{9.25}$$

The maximum compressive stress at this point is the sum of the maximum bending stress and the compressive stress due to the axial load:

$$\sigma_{\mathrm{c}} = \frac{P}{A} - \frac{Mc}{I} \tag{9.26}$$

$$\sigma_{\mathrm{c}} = \frac{P}{A}\left[1 + \frac{ec}{r^2}\sec\frac{1}{2r}\sqrt{\frac{P}{EA}}\right] \tag{9.27}$$

where $r^2 = I/A$ is the radius of gyration of the section. Since this stress should not exceed the compressive yield stress, $\sigma_{\mathrm{c}} < S_{\mathrm{yc}}$, this yields the critical load:

$$P_{\mathrm{cr}} = \frac{AS_{\mathrm{yc}}}{1 + (ec/r^2)\sec[(l/2r)\sqrt{P/AE}]} \tag{9.28}$$

This equation is termed the *secant formula* and relates the critical load for an eccentrically loaded column to both the material strength and the Young's modulus. For short compression members, called *struts*, we note that if $l \to 0$, the secant in the denominator of Equation (9.28) tends to 1 (because $1/\sin x \to 1$ if $x \to 0$) and Equation (9.28) becomes, for struts:

$$P_{\mathrm{cr}} = \frac{AS_{\mathrm{yc}}}{1 + (ec/r^2)} \tag{9.29}$$

Example 9.3   A jib crane (Figure E9.3) has vertical column of height 3.0 m, hollow square cross-section of inner width 100 mm, and outer width 120 mm. The horizontal beam has length $e = 2.0$ m, and the load can move at distances between 0.3 m and 2 m from the vertical column. The material of the column is low-carbon steel with $S_{\mathrm{yc}} = 220$ MPa and Young's modulus $E = 2.1 \times 10^{11}$ Pa. Find the critical load P_{cr} for $e = 2$ m and $e = 0.3$ m.

Solution   The moment of inertia is:

$$I = \frac{a_2^4 - a_1^4}{12} = \frac{0.120^4 - 0.100^4}{12} = 8.95 \times 10^{-6}\ \mathrm{m}^4$$

$$\text{Area } A = (a_2^2 - a_1^2) = (0.120^2 - 0.100^2) = 0.0044\ \mathrm{m}^2$$

$$\text{Radius of gyration } r^2 = \frac{I}{A} = \frac{8.95 \times 10^{-6}}{0.0044}, r = 0.045\ \mathrm{m}$$

The secant formula is:

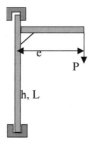

Figure E9.3

TABLE E9.3.

$I = 8.94667E - 06$	$e = 2$
$S_{yc} = 2.00E + 0.8$	$c = 0.06$
$A = 0.0044$	$L = 3$
$r = 0.045092498$	
$E = 2.10E + 11$	
$P = 1.47E + 04$	(Give it initially any value)
$P_{cr} = 1.47E + 04$	B3/(1+(E2*E3/B5^2)/COS((E4/
	(2*B5)*SQRT(F7/(B4*B6)))))

Use Solver to make the difference below zero by changing P.

$P - P_{cr} = 0.00E + 00$

$$P_{cr} = \frac{AS_{yc}}{1 + (ec/r^2)\ \sec[(l/2r)\sqrt{P/AE]}}$$

It cannot be solved explicitly for P_{cr} and can be solved by iteration. The EXCEL solution using Solver is shown in Table E9.3: The solution is $P_{cr} = 14,700$ N

We repeat the procedure for $e = 0.3$ to find critical load $P_{cr} = 89,300$ N.

If only bending is considered, the maximum bending stress will equal the yield strength, $S_y = Mc/I = Pec/I$. For $e = 2$ m:

$$P_{cr} = \frac{IS_y}{(ec)} = \frac{8.94667 \times 10^{-6} \times 2.1 \times 10^8}{(2 \times 0.06)} = 14,911 \text{ N}$$

(compare with 14,700 N of the full eccentric column solution).

For $e = 0.3$ m:

$$P_{cr} = \frac{IS_y}{(ec)} = \frac{8.94667 \times 10^{-6} \times 2.1 \times 10^8}{(0.3 \times 0.06)} = 99,407 \text{ N}$$

(compare with 89,300 N of the full eccentric column solution).

We note that the effect of buckling is more substantial as the load moves closer to the column.

9.3.3. Computer Aided Column Stability Analysis

With the computer algorithms and programs presented in Chapter 6, stability questions cannot be answered because the associated mechanisms have not been included. For example, in the transfer matrix and stiffness matrix analysis with straight beam elements, axial force and the resulting moment (as in Equation (9.6)) do not appear. This indicates the proper course of action.

Recall the beam equations that are used to obtain transfer stiffness matrices:

$$\left.\begin{array}{r} EIy'' = M \\ M'' = q(x) \end{array}\right\} \tag{9.30}$$

In view of the additional moment due to the axial force P, Equations (9.30) take the form:

$$EIy'' = M - Py \left.\vphantom{\begin{matrix}1\\1\end{matrix}}\right\} \tag{9.31}$$
$$M'' = q(x)$$

Combining in one equation:

$$EIy^{IV} + Py'' = q(x) \tag{9.32}$$

The general solution of this equation is not a harmonic function, as in Equation (9.9), but (for $k = (P/EI)^{1/2}$):

$$y(x) = C_1 \cos hkx + C_2 \sin hkx + C_3 \cos kx + C_4 \sin kx + \frac{1}{EI} \int \int \int \int q dx \tag{9.33}$$

For stability analysis we assume, at the moment, $q(x) = 0$.

The transfer matrix for the beam element with axial force P will be obtained with application of the same boundary conditions (see Chapter 4). Then:

$$L = \begin{bmatrix} 1 & \sin kl/k & (1 - \cos kl)/EIk^2 & (kl - \sin kl)EIk^3 & 0 \\ 0 & \cos kl & \sin kl/EIk & (1 - \cos kl)EIk^2 & 0 \\ 0 & (P \sin kl)/k & \cos kl & \sin kl/k & 0 \\ 0 & 0 & 0 & 1 & 1 \end{bmatrix} \tag{9.34}$$

This transfer matrix can be utilized to yield the solution with the boundary conditions at the end of the beam. Buckling means a non-zero solution, which is possible only if the determinant of the coefficient matrix is zero.

$$\det [A(P)] = 0 \tag{9.35}$$

Because the elements of matrix A are functions of axial force P, the proper design approach is to seek the lowest value of P that satisfies Equation (9.35).

Sometimes, due to numerical inaccuracies, this process might be slow or even impossible. For this reason we usually seek with the methods of Chapter 7 (program OPTIMUM) or with the Solver function of EXCEL the minimum of the function $\det [A(P)]$ and test if it is close enough to zero.

Another method is to assume a small lateral load and obtain some response quantity, such as deflection at mid-span. With a gradual increase of the axial load P, when the critical load is reached, the response quantity will increase rapidly. Exactly the same approach is used with the stiffness matrix formulation.

One word of caution: transfer matrix (9.21) must be used very carefully and only when stability analysis is required, although it appears that it could replace the matrix in Equation (9.34), making static analysis programs versatile and usable also for stability analysis. This is not the case, because some of the elements of the transfer matrix in Equation (9.34) are differences of two almost equal members, for example $1 - \cos kL$. Small errors in the evaluation of the $\cos kL$ for small P will have great effect on the results. For this reason, Equation (9.34) must be used only when necessary and with caution. For example, very short beams will make kL very small, and this will amplify the above unwanted effects.

The program TMSTABIL is part of the MELAB 2.0 package.

Example 9.4   Using the program TMSTABIL, find the limiting axial load of the shaft shown of diameter, uniform, $d = 50$ mm, material carbon steel with $E = 2.1E5$ N/mm^2.

Solution   The program TMSTABIL is used with the data indicated in the output and Figure E9.6a. The bar is uniform to compare with known values of the buckling load. This is a beam-column, and thus we do not anticipate a definitive buckling load. The moment of inertia is:

$$I = \frac{\pi d^4}{64} = 3.07E - 7, \quad EI = 2.1E11 \times (3.07E - 7) = 64,427 \text{ m}^4$$

The Euler critical load is:

$$P_{cr} = \frac{\pi^2 EI}{L^2} = \frac{\pi^2 \times 64,427}{1.1^2} = 525,513 \text{ N}$$

for pinned ends.

Successive runs are performed with increasing axial load and one of the deflection parameters (say θ_4, the slope at the right end) is recorded. At 525 kN, there is a rapid increase, indicating instability. As expected, this coincides with the Euler critical load of 524 kN. At that load, the deflections and slopes determine the buckling mode, plotted in Figure 9.6b.

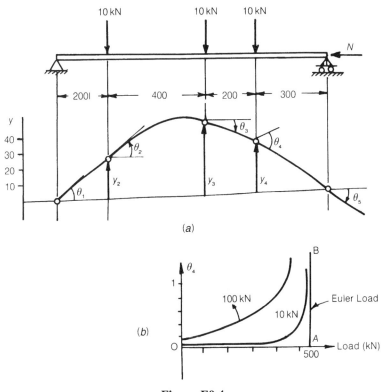

Figure E9.4

As expected, it has a sinusoidal shape. The lateral loads cause the deviation from the theoretical Euler deflection curve *OAB*. At 100 kN lateral loads, this deviation is more noticeable.

9.3.4. Design of Machine Frames

The criteria for performance and reliability of housing-type components are strength, rigidity, and durability.

Strength is the basic criterion for components subject to heavy loads, mainly impact (shock) and variable loads.

Rigidity is the main criterion of performance for the great majority of housing-type components. Increased elastic deflections in such components usually lead to faulty operation of the mechanisms, poor accuracy, and the initiation of vibrations. Wear life is important for components with sliding friction.

When high rigidity is desirable, the components are made of materials with a high modulus of elasticity, such as cast iron and steel used without heat treatment.

Housing-type components of transportation machinery, such as engine crankcases, as well as components subject to heavy inertia forces, are frequently made of light alloys. Such alloys possess increased strength per unit mass.

Most housing-type components are of cast iron. This is based on the feasibility of obtaining intricate geometric shapes and the comparatively low cost of lot production, in which the cost of the die is distributed over a large number of castings.

Housing-type components of welded design are employed to reduce the mass and overall dimensions and, in job and small-lot production, also to lower production costs and shorten the lead time (time required to begin production).

Welded components are made of:

1. Elements of simple shapes in poorly equipped and job and small-job production
2. Fabricated elements in sufficiently well-equipped lot production
3. Press-worked elements of refined, streamlined shapes in large-lot and mass production

Weldments made up of cast elements considerably simplify the required castings.

Housing-type components that should be of minimum mass, but are not subject to appreciable loads and do not require dimensional stability, can be efficiently made of plastics. Such components include the housings of portable and hand-held machines and tools, instruments, covers, and hoods.

The best practice is to design housing-type components subject to bending and torsion with thin walls whose thickness is usually determined from the condition of castability. Components subject to torsion should, wherever possible, be of closed, box section, and those subject to bending should have the main part of the material placed as far as possible from the neutral axis. If openings or hatches are required, to enable the inner cavity of the casting to be utilized, they should not coincide in location in opposite walls of the casting. The weakening of the casting by such openings should be compensated for by the provision of flanges or rigid covers. The most effective method of economizing on materials in machine manufacture is, as a rule, the reduction of the wall thicknesses. The required rigidity of the walls can be provided for by the proper ribbing.

Recommended wall thicknesses for iron castings are listed in Table 9V on the basis of the so-called equivalent overall size of the casting. The equivalent overall size, that is, the size of a box-like casting of cubic shape that is equivalent, with respect to boundary conditions, to the casing being designed, can be estimated as:

$$N = \frac{2L + B + H}{4} \tag{9.36}$$

where L, B, and H are the length, width, and height of the casting. Inner walls and ribs cool more slowly than do outer ones, and therefore, to ensure simultaneous cooling, they should be 80% the thickness of the outer walls and ribs.

The height of ribs or fins should not be more than five times their thickness. To comply with conditions of castability, walls of steel castings are designed from 20–40% thicker than those of iron castings. Nonferrous casting alloys allow substantially thinner walls than iron castings. Thick-walled castings are used when the overall size of the component is strictly limited.

The walls and sections should, as far as possible, be of the same thickness. If a constant thickness cannot be maintained, there should be a gradual blending of the thinner into the thicker sections (Figure 9.15).

Castings should be designed so that in cooling the members are free to shrink without developing excessive residual stresses. For this reason, the ribs of round cast plates are designed with a curved shape (Figure 9.16b), a special network of ribbing is provided, and so on.

To prevent an accumulation of nonmetallic inclusions and the formation of blowholes, it is best to avoid large flat surfaces if, according to the molding conditions, they are to be poured in the horizontal position (Figure 9.16c,d).

To simplify pattern making, the geometric shapes of elements of castings should be ones easy to machine; they should be bounded by planes and cylindrical and conical surfaces.

Special efforts should be made to design castings to that they are easy to mold. It should be feasible to mold simple casings in one half-mold (Figure 9.17b) or

TABLE 9.5. Recommended Wall Thickness for Iron Castings

Equivalent overall size, m	Thickness of outer walls, mm	Thickness of inner walls, mm	Equivalent overall size, m	Thickness of outer walls, mm	Thickness of inner walls, mm
0.4	6	5	2.0	16	12
0.75	8	6	2.5	18	14
1.0	10	8	3.0	20	16
1.5	12	10	3.5	22	18
1.8	14	12	4.5	25	20

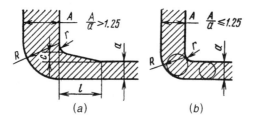

Figure 9.15. Casting details.

with a single flat parting line (Figure 9.17*d*). To facilitate withdrawing of the pattern from the mold, sidewalls should be designed with a slight inclination from the vertical (Figure 9.17*f*); otherwise it will be necessary to provide foundry draft on both the outside and inside surfaces.

Wherever possible, the shapes of casings should be designed so that no cores or loose pieces on the patterns are required (Figure 9.17*h*). The possibility of withdrawing the pattern from the mold (without employing loose pieces or cores) is checked as follows. An imaginary stream of rays, perpendicular to the parting plane, should not produce shadowed portions (Figure 9.17*g*). If cores are an absolute necessity, holes should be provided (for production purposes) to reliably anchor the covers (Figure 9.17*j*). Cores should be unified wherever possible.

Provision should be made in job-lot production for the sweep-molding of large casings (Figure 9.17) using a strickleboard or sweep template.

9.4. DESIGN OF ELASTIC ELEMENTS: SPRINGS

9.4.1. Field of Application

In Chapter 8, we saw several types of joint elements with the main purpose of connecting as rigidly as possible other, generally larger, elements. This is not al-

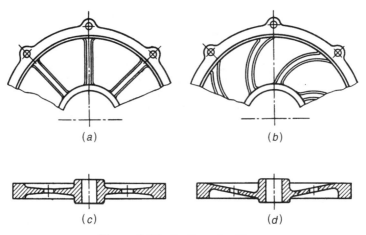

Figure 9.16. Casting of pulleys.

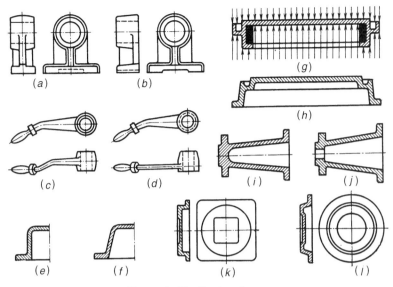

Figure 9.17. Casting forms.

ways the case. Many times joining two machine elements needs to be done in a flexible way. Such joints are used when we need the following:

1. To produce constant forces, even if the two joining elements will move in respect to one another to some extent
2. To adjust machine members when clearances exist or can be developed
3. To accumulate elastic strain energy, which subsequently will be used to drive an element to some motion
4. For vibration isolation when we expect dynamic loads and want to minimize maximum stresses on the machine
5. To absorb impact energy when impact loads are expected in a machine, in order to keep the peak impact stresses as low as possible

Linear elements are loaded by forces in one direction and are deformed in the same direction. In most cases, there is a linear relationship between force and displacement:

$$F = k\delta \tag{9.37}$$

where k is a constant depending on the material and geometry of the spring, called the *spring constant*. Such springs include helical springs (Figure 9.18).

Torsion springs are loaded with a torsional moment M_T, and there is a relative twist between the two ends of the spring, which are again related with a linear relationship:

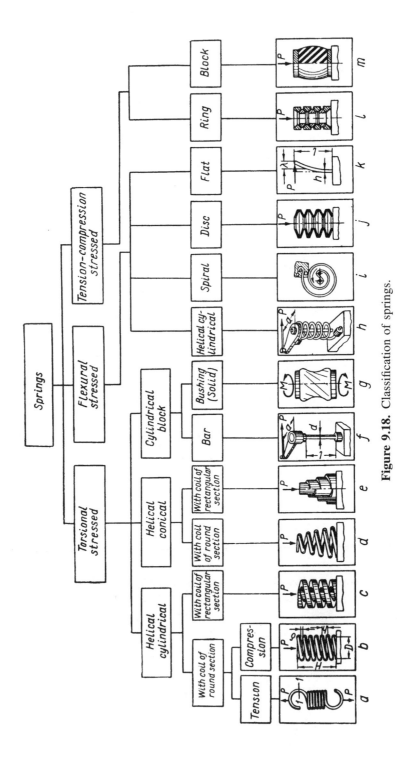

Figure 9.18. Classification of springs.

$$M_T = k_T \, \Delta\varphi \tag{9.38}$$

where k_T is the torsional spring constant. Such springs are shown later in this section.

It is apparent that linear and torsional springs can be used interchangeably by the use of proper load transmission mechanisms. Usually springs are made out of hard materials, mostly metals. However, many applications of springs are made out of plastics or rubber, used primarily for machine mountings. Such springs are shown in Figure 9.26. For even more elasticity, springs are made sometimes using the compression of a large volume of air (Figure 9.27).

Materials for springs must have high elastic properties and high strength and must be stable over time. Because springs made out of weaker materials must have large dimensions, for dynamic loadings, which are very usual in spring applications, there are high inertia forces that must be avoided. Because the modulus of elasticity does not change considerably among various types of steel, high-strength steels are mostly used, as presented in Chapter 4.

The main steel alloys used include high-carbon steels, manganese steel, chromium steel, chromium vanadium steel. However, the most widely used are carbon steels because they have high strength and low price.

Manganese, silicon, and chromium manganese steels have higher strength and can be hardened, which enables them to be used for springs of small cross-sections. Chromium vanadium steel has high mechanical strength and also high endurance limit, heat resistance, and good mechanical processing properties, and therefore is used for critically loaded springs and in particular for repeated loading, such as in valve springs for engines. Springs operating in corrosive environments are made of nonferrous alloys such as several types of bronzes. Steel springs in the same environments can resist oxidization when coated with cadmium and other coatings. For critical applications, springs are processed to increase their strength with methods such as *shot-peening* or *plastic free stress*. The later is achieved by way of application of a load greater than design load to produce plastic strains. Upon release, the remaining stresses act in an opposite way to applied stresses, enabling the spring to take sometimes 25% higher loads.

9.4.2. Design of Torsion Bars

A torsion bar (Figure 9.19) has a well-known relation between torque and angle of twist:

$$\Delta\varphi = \frac{M_T L}{G I_p} \tag{9.39}$$

which considers the torsion bar as a perfect cylinder under the action of the torque. By definition, the spring constant is then

$$k_T = \frac{G I_p}{L} \tag{9.40}$$

Maximum stresses at the outer fiber due to torsion are:

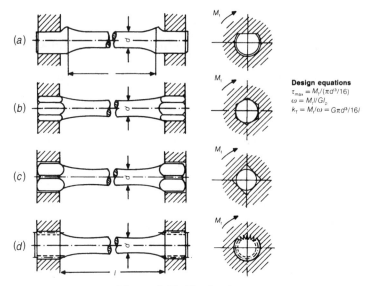

Figure 9.19. Torsion bars.

$$\tau_{\max} = \frac{M_{\mathrm{T}}}{I_p} R \tag{9.41}$$

if pure torsion is assumed and bending and shear are not present.

As in the case of buckling, that is, stability of thin rods under axial compressive loads, long twisted rods can undergo torsional buckling, assuming a helical form. According to the theory of elastic stability (Timoshenko and Gere 1961), there is a critical torsion

$$T_{\mathrm{cr}} = \frac{2\pi EI}{L} = \frac{\pi^2 E d^4}{32L} \tag{9.42}$$

For designing a torsional spring in the form of a torsion bar, the maximum torque T and a specific value (or a maximum value or an allowable range) of the spring constant are given. Depending on other circumstances, such as environment and temperature, a material is selected and then its strength properties are known. Design parameters are the length and the diameter, which are calculated by way of the two Equations (9.40) and (9.41), and a stability check is made by way of Equation (9.42). In other cases, the length is specified and then, by way of the same equations the diameter and the material strength will be determined, from which the proper material will be selected, if the design is feasible. For dynamic loads, equivalent stresses must be computed with the method presented previously. Design equations for torsion bars are summarized in Figure 9.19.

9.4.3. Axially Loaded Helical Springs

In such a spring the wire is at any section twisted by a torque equal to $PD/2$, where P is the axial force acting on the centerline of the coil and D is the mean coil diameter. If we consider a small part of the wire as a torsion bar, for the angle of twist, Equation (9.39) applies. This twist gives deflection in the center of the spring equal to $\Delta\varphi D/2$. Therefore, if the total wire length of one turn is πD, the total deflection will be:

$$\delta = \frac{8N_a PD^3}{Gd^4} \tag{9.43}$$

where N_a is the number of active turns of the spring, D is the mean diameter of the spring, d is the wire diameter, G is the shear modulus of the spring material (Figure 9.20). Therefore, the spring constant will be:

$$k_T = \frac{P}{\delta} = \frac{Gd^4}{8N_a D^3} \tag{9.44}$$

The same twisting torque of the wire produces maximum shear stresses at the outer fiber due to torsion:

$$\tau_{max} = \frac{T(d/2)}{J_p} = \frac{PR(d/2)}{\pi d^4/32} \tag{9.45}$$

Because $R = D/2$:

$$\tau_{max} = \frac{8PD}{\pi d^3} K_s \tag{9.46}$$

where $C = (2R/d)$ and K_s is a stress concentration and correction factor (Wahl 1963):

$$K_s = \frac{4C-1}{4C-4} + \frac{0.615}{c} \tag{9.47}$$

The first fraction on the righthand side of Equation (9.47) is a stress concentration

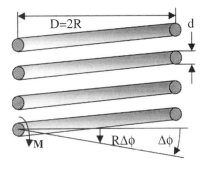

Figure 9.20. Geometry of a helical spring.

factor due to the curvature of the beam, while Equation (8.45) is correct only for straight cylindrical rods.

Helical springs in compression are also subject to buckling, which in reality is torsional buckling of the wire but appears as column buckling. Applying Equation (9.42) for the torsional buckling of the wire, we can use the Euler equation with an equivalent moment of inertia (Niemann 1965):

$$I_{eq} = \frac{Ld^4}{64N_a D(1 + \nu/2)} \tag{9.48}$$

where ν is the Poisson ratio. Then the Euler equation can be applied:

$$P_{cr} = \frac{\pi^2 E I_{eq}}{(\mu L^2)} \leq P_{max} \tag{9.49}$$

The coefficient μ, as previously, characterizes the boundary conditions at the point of support of the spring. The same values apply as for the buckling of columns (Figure 9.12).

In the design of a helical spring in axial loading, the maximum load and the spring constant are usually given. A material is selected and its mechanical properties determined. Design parameters are the mean diameter, the wire diameter, and the number of acting turns of the coil. In principle, the three design equations can yield the unknown design parameters. Because there are other design considerations, it is more rational to use the computer to calculate for a number of active turns, say from 5 to 50 in steps of 5, the diameters d and D by the two nonlinear algebraic equations:

$$f_1(d, D) = \frac{Gd^4}{8N_a D^3} - k = 0 \tag{9.50}$$

$$f_2(d, D) = \frac{8PD}{\pi d^3} \left(\frac{C - 0.5}{C - 1} \right) - \frac{S_{sy}}{N} = 0 \tag{9.51}$$

The yield strength in shear S_{sy} needs to be replaced by the fatigue strength in shear S_{se} if the loading is alternating. For combined dynamic loads, one has to use the methods of Chapter 5; the Goodman method is suggested.

The solution of these equations is performed with a nonlinear equation solver, such as Solver of EXCEL or the Newton–Raphson method, and each solution is checked for buckling. Finally, the feasible solutions are tabulated and the designer must pick the most suitable solution based on some other requirements. One obvious requirement might be, for example, that the spring must be allowed to be compressed by a certain distance before the subsequent turns of the coil come in contact. If L_0 is the free length, the compressed length L at compressive load P is:

$$L = L_0 - \frac{P}{k} \tag{9.52}$$

But the fully compressed length is $N_a d$. Therefore:

$$L = L_0 - \frac{P}{k} \leq N_a d \tag{9.53}$$

Moreover, the angle of inclination of the helix of the spring should not be excessive. Most designers limit it to $10°$ because, for one thing, the assumptions on which Equations (9.50) and (9.51) are based might not be valid. Therefore, at the maximum extensional load length L (or at the unloaded condition if load is compressive, length L_0):

$$\tan(\theta) = \frac{L}{\pi D N_a} \leq \tan(10°) = 0.176 \tag{9.54}$$

Also, for practical purposes, the number of active turns cannot be very small. Other times there are space limitations for the external diameter of the coil. In cases where it is necessary that the requirement of stability be waived, an external guide to the spring with proper lubrication is provided to minimize friction and wear and support the spring to avoid lateral buckling. In this case care must then be exercised for high-speed engines because heating of the spring or pipe and rapid deterioration might occur.

For relatively thick wire, the shear stress due to the actual force must be also included in the calculation. In occasions, rather rare, where the coil is made out of ductile material and for static loading, the correction factor due to curvature can be neglected due to the redistribution of stresses because of local yielding. For brittle materials, and especially for dynamic loading, this factor always has to be taken into account.

Design Procedure 9.2: Design of an Axially Loaded Cylindrical Helical Spring

Input: Maximum load, material properties, length, supports, stiffness. **Output**: spring dimensions.

Step 1: Sketch the spring and the boundary conditions.

Step 2: Out of the three unknowns, wire diameter d, helix diameter D, number of turns N_a, select one arbitrarily or by using available design constraints or engineering judgment.

Step 3: Use Equations (9.50) and (9.51) to determine the other two unknown design parameters.

Step 4: Use the Euler equation to check the wire stability (Equation (9.49)).

Step 5: Check for acceptable free length (Equation (9.53)).

Step 6: Check for acceptable slope of the spring helix (Equation (9.54)).

Step 7: If the applied force is alternating at a high frequency, check with methods of vibration analysis for surging.

9.4.4. Optimum Design of Helical Springs

It appears that at least two parameters in a helical spring design are arbitrary. This naturally leads to optimization. Depending on the problem at hand, other limitations might be imposed, such as space and maximum inclination of the spiral, while the most obvious objective function is the weight. The program HELICAL SPRINGS (MELAB 2.0 package) uses the exhaustive search method to tabulate the results for a parameter study with given limits in the material and the number of turns. A typical computation follows the listing of the program.

Because the number of design parameters may be larger than the design equations, spring design can be optimized. For helical springs, for example, the formulation of the optimization problem follows:

Data:	Free length L, travel δ, load P, spring constant k, material S, E, G
Design variables:	N, D, d
Objective function:	$f(N, D, d) = \pi^2 D d^2 N / 4$, the volume of the spring

Constraints:
 Equality constraints:
  (a) Maximum stress:
  $$g_1(N, D, d) = S_{sy} - (16PR/\pi d^3)(C - 0.25)/(C - 1)$$
  $$= 0, \quad C = d/D, \quad R = D/2$$
  (b) Spring constant:
  $$g_2(N, D, d) = k_T - Gd^4/64NR^3 = 0$$
 Inequality constaints:
  (a) Buckling load:
  $$h_1(N, D, d) = \pi^2 EI\mu/L^2 - P > 0$$
  (b) Travel:
  $$h_2(N, D, d) = L - Nd - \delta > 0$$
  (c) Helic angle 10°:
  $$h_3(N, D, d) = 0.176 - L/\pi ND > 0,$$
  because $\tan \alpha = L/\pi ND$, $\tan 10° = 0.176$
  (d) Maximum diameter:
  $$D_{max} > D, \quad h_4(N, D, d) = D_{max} - D > 0$$
Penalty function:

$$P(N, D, d) = f(N, D, d) + K(g_1^2 + g_2^2) + L(\langle h_1 \rangle^2 + \langle h_2 \rangle^2 + \langle h_3 \rangle^2 + \langle h_4 \rangle^2)$$

Example 9.5 The suspension spring for a car is of the form of Figure 9.21, made of AISI 9255 steel with $S_y = 1130$ N/mm², has strip width 50 mm, thickness 5 mm, length 1200

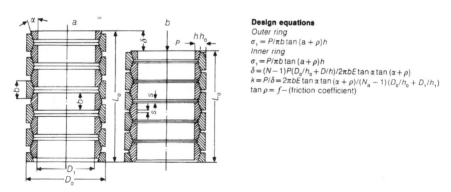

Design equations
Outer ring
$\sigma_1 = P/\pi b\, t \tan{(a + \rho)}h$
Inner ring
$\sigma_1 = P/\pi b\, t \tan{(a + \rho)}h$
$\delta = (N - 1)P(D_o/h_o + D/h)/2\pi bE \tan{\alpha} \tan{(\alpha + \rho)}$
$k = P/\delta = 2\pi bE \tan{\alpha} \tan{(\alpha + \rho)}/(N_a - 1)(D_o/h_o + D_1/h_1)$
$\tan \rho = f-$ (friction coefficient)

Figure 9.21. Design equations for ring springs (from Decker 1973).

mm, and consists of seven strips. The total weight of the car is $G = 10$ kN. The load is distributed 45% front and 55% rear. Determine:

1. The safety factor for the spring strength
2. The spring constant
3. The natural frequency of the car on these springs

Solution Each strip behaves as a beam independent from the others supporting an equal portion (1/7th) of the load. The section modulus is:

$$W = \frac{bh^2}{6} = \frac{50 \times 5^2}{6} = 208.3 \text{ mm}^3$$

1. The stresses in the front spring are:

$$\sigma_F = \frac{(0.45G/4)L}{nW} = \frac{(0.45 \times 1000/4) \times 600}{7 \times 208.3} = 463 \text{N/mm}^2$$

The stresses in the rear spring are:

$$\sigma_R = \frac{(0.55G/4)L}{nW} = \frac{(0.55 \times 1000/4) \times 600}{7 \times 208.3} = 566 \text{N/mm}^2$$

The corresponding safety factors are:

$$N_F = \frac{1130}{463} = 2.44, \ N_R = \frac{1130}{566} = 2.00$$

2. The spring constant is, for half-spring of length 600 mm (Figure 9.20), with layers:

$$k = \frac{P}{\delta} = \frac{3nEI}{K_t L^3}$$

The stress concentration factor $K_t = 1.36$, the section moment of inertia:

$$I = \frac{bh^3}{12} = \frac{50 \times 5^3}{12} = 520.8 \text{ mm}$$

Therefore:

$$k = \frac{3 \times 7 \times 21,000 \times 520.8}{1.36 \times 600^3} = 7.82 \text{ N/mm}$$

3. The natural frequency is, for eight half-springs supporting the car:

$$\omega_n = \left(\frac{k}{m}\right)^{1/2} = \left(\frac{8 \times 7.8 \times 1000}{1000}\right)^{1/2} = 7.9 \text{ rad/s}$$

Example 9.6 Develop a spreadsheet for the design of a helical spring. For an application, design a helical spring with Density $\rho = 7800$ kg/m^3, spring constant $K = 3.60 \times 10^5$ N/m, maximum static load $P_m = 1.23 \times 10^4$, max. dynamic load $P_r = 0$, Young's modulus $E = 2.1 \times 10^{11}$ Pa, shear modulus $G = 1.05 \times 10^{11}$ Pa, Poisson's ratio $\nu = 0.3$, shear fatigue strength $S_{se} = 2.00 \times 10^8$ Pa, shear strength $S_{sy} = 3.00 \times 10^8$ Pa, free length $L = 0.1$ m.

Solution The spreadsheet is shown in Table E9.6.

9.4.5. Design of Other Springs

It was found in the case of helical springs that the most important design equations are the ones relating the external load to deformation (stiffness) and stress (strength). In some types of springs, stability is to be ensured. In the following (Figures 9.21–9.27), we list a number of different springs together with their design equations. The derivation of these equations is left to the reader.

9.5. DESIGN OF MACHINE MOUNTS AND FOUNDATIONS

Machines that are expected to transmit substantial static or dynamic forces through their pedestals are installed on foundations. A typical arrangement of this kind is shown in Figure 9.27. The machine, of mass m mounted on a massive foundation of mass M, rests directly on soil or some other elastic material such as cork, rubber, or springs. It can therefore be represented as a one-degree-of-freedom system. The spring constant k can be determined from the dimensions and properties of the elastic material. If the mass M rests on soil, the spring constant will be:

$$k = Ak_s \tag{9.55}$$

where A is the footing surface and k_s is a constant called the coefficient of compression of the soil (lb/in^3 or similar). Typical properties of relevant materials and soils are given in Table 9.6.

Because foundation directly affects the machine operation and the environmental effects of machine vibration and noise, its design is a part of the machine design effort. Usually the machine rests on a heavy base that is resiliently supported by some elastic substance, or directly on soil.

The purpose of this construction is to keep at a minimum the force transmitted through the foundation to the surroundings. Therefore, we are interested in the force transmitted. This is the force carried through the springs, which is, accounting also for damping:

$$f = kx + c\dot{x} \tag{9.56}$$

where x is the displacement, which can be computed from the imposed force $F \sin(\omega t)$ and the properties of the system $(m + M)$ and k. The amplitude of this force will be:

TABLE E9.6

DESIGN OF HELICAL SPRINGS	
Example 9.6	
π	3.14159
gravity constant g	9.81
DATA:	
Density $\rho =$	7800
Spring Constant $K =$	$3.60E + 05$
Maximum Static Load $P_m =$	$1.23E + 04$
Max. Dynamic Load $P_r =$	$0.00E + 00$
Young's modulus $E =$	$2.10E + 11$
Shear modulus $G =$	$1.05E + 11$
Poisson's ratio $\nu =$	$3.00E - 01$
Shear fatigue strength $S_{se} =$	$2.00E + 08$
Shear strength $S_{sy} =$	$3.00E + 08$
Free length $L =$	$1.00E - 01$
RESULTS:	
No. of Turns $N =$	7.90
Helix Diameter $D =$	$1.11E - 01$
Wire Diameter $d =$	$2.34E - 02$
CALCULATIONS:	
Equivalent moment of inertia (Equation (9.48)) I_{eq}	$4.63E - 10$
$C = D/d =$	$4.76E + 00$
$k_s = (1 + 2C)/(2C)$	1.104973215
Static shear stress (Equation (9.45)) =	300,000,000
Dynamic shear stress (Equation 9.45)) =	0
Equivalent shear stress $= \sigma_m + (S_y/S_{se})\sigma_r =$	300,000,000
Euler buckling load $\pi^2 EI/L^2 =$	95995.41353
Equiv. shear stress $-S_{sy}$ (must be ≤ 0) =	0
Calc. spring const. $-K$ (must be $= 0$) =	0
Buckling load $-$ maximum load (must be > 0) =	83733.41353
$\tan(a) - 0.176$ (must be <0) = $L/\pi ND - 0.176 =$	-0.13984129
Material Volume (must be minimum) = $\pi DN(\pi d^2/4) =$	0.001189872
Surge natural frequency Hz	4348.515818
HELP: DESIGN CASES:	
CASE 1: Find wire d and helix D with given G and shear strength S_{sy}. Make B30 and B31 zero.	
CASE 2: Find d, D, no. of turns N for given G and S_{sy} and minimum volume.	

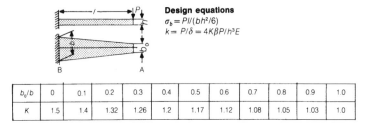

Design equations

$\sigma_b = Pl/(bh^2/6)$

$k = P/\delta = 4K\beta P/h^3 E$

b_o/b	0	0.1	0.2	0.3	0.4	0.5	0.6	0.7	0.8	0.9	1.0
K	1.5	1.4	1.32	1.26	1.2	1.17	1.12	1.08	1.05	1.03	1.0

Figure 9.22. Design equations for tapered beam springs.

$$F_T = kX + ic\omega X = \frac{F_0[1 + (2J\omega/\omega_n)^2]^{1/2}}{\{[1 - (\omega/\omega_n)^2]^2 + (2\zeta\omega/\omega_n)^2\}^{1/2}} \qquad (9.57)$$

where ω_n is the natural frequency $(k/m)^{1/2}$ and ζ is the fraction of critical damping $c/[4(km)^{1/2}]$, being 2.7–3.5% for concrete frames, 0.5–2.5% for steel frames, 5–6% for heavy mass resting on soil, 0.5–1.0% for steel spring support, 1–2% for cork or rubber support. Therefore, the ratio of transmitted to imposed force (T_R) is:

$$T_R = \frac{F_T}{F_0} = \left\{ \left[1 - \left(\frac{\omega}{\omega_n} \right)^2 \right]^2 + \left(\frac{\zeta\omega}{\omega_n} \right)^2 \right\}^{-1/2} \qquad (9.58)$$

This ratio is called *transmissibility* and is given in Figure 9.28. For quick calculations, we sometimes use the static deflection δ_{st} of the system due to its own weight. This can readily be measured by loading the foundation with a static weight w and multiplying the resulting static deflection by $(M + m)g/w$. The natural frequency is:

$$\omega_n = \left[\frac{kg}{(m + M)g} \right]^{1/2} = \frac{(g)^{1/2}}{(\delta_{st})^{1/2}} \qquad (9.59)$$

The following design formula is popular for the natural frequency:

$$f_n = 188(\delta_{st})^{-1/2} \qquad (9.60)$$

where n is in counts per minute (cpm) and static deflection is in inches. In terms of the frequency $f = \omega \times 60/2\pi$ (cpm), for $\zeta = 0$ we obtain:

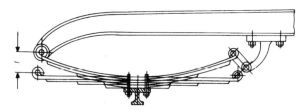

Figure 9.23. Leaf springs.

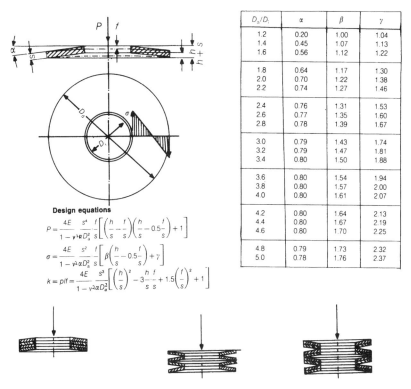

D_a/D_i	α	β	γ
1.2	0.20	1.00	1.04
1.4	0.45	1.07	1.13
1.6	0.56	1.12	1.22
1.8	0.64	1.17	1.30
2.0	0.70	1.22	1.38
2.2	0.74	1.27	1.46
2.4	0.76	1.31	1.53
2.6	0.77	1.35	1.60
2.8	0.78	1.39	1.67
3.0	0.79	1.43	1.74
3.2	0.79	1.47	1.81
3.4	0.80	1.50	1.88
3.6	0.80	1.54	1.94
3.8	0.80	1.57	2.00
4.0	0.80	1.61	2.07
4.2	0.80	1.64	2.13
4.4	0.80	1.67	2.19
4.6	0.80	1.70	2.25
4.8	0.79	1.73	2.32
5.0	0.78	1.76	2.37

Design equations

$$P = \frac{4E}{1-\nu^2\alpha D_a^2}\frac{s^4}{s}\frac{f}{s}\left[\left(\frac{h}{s}-\frac{f}{s}\right)\left(\frac{h}{s}-0.5\frac{f}{s}\right)+1\right]$$

$$\sigma = \frac{4E}{1-\nu^2\alpha D_a^2}\frac{s^2}{s}\frac{f}{s}\left[\beta\left(\frac{h}{s}-0.5\frac{f}{s}\right)+\gamma\right]$$

$$k = p/f = \frac{4E}{1-\nu^2\alpha D_a^2}\frac{s^3}{s}\left[\left(\frac{h}{s}\right)^2-3\frac{h}{s}\frac{f}{s}+1.5\left(\frac{f}{s}\right)^2+1\right]$$

Figure 9.24. Design equations for Belleville springs.

$$T_R = \left(\frac{(2\pi f)^2\delta_{st}}{g}-1\right)^{-1} \tag{9.61}$$

From Equation (9.58), we observe that for $\zeta = 0$ we have effective isolation ($T_R < 1$) only if $(\omega/\omega_n)^2 - 1 > 1$ or $\omega > \omega_n(2)^{1/2}$. From Figure 9.30, however, we observe that for $\omega/\omega_n = 3.0$ the transmissibility is about 0.1 and decreases very slowly after that. The value $\omega = 3\omega_n$ is very often used for foundation design.

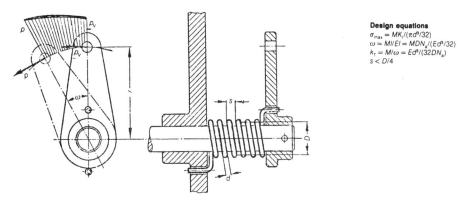

Design equations
$\sigma_{max} = MK_i/(\pi d^3/32)$
$\omega = Ml/EI = MDN_a/(Ed^4/32)$
$k_T = M/\omega = Ed^4/(32DN_a)$
$s < D/4$

Figure 9.25. Design equations for torsional helical springs (from Decker 1973).

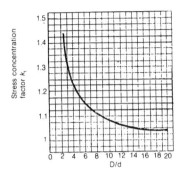

Figure 9.26a. Design equations for torsional helical springs (from Decker 1973).

In Figure 9.30, we observe also that damping has no effect on the point where effective isolation ($T_R < 1$) starts. This point is $\omega/\omega_n = (2)^{1/2} = 1.41$. For frequencies $\omega < 1.41\omega_n$, however, damping reduces the transmissibility ratio, and thus the transmitted force. For frequencies $\omega > 1.41$, however, damping increases the transmissibility ratio and the transmitted force. This observation shows that damping is not always a desirable feature in engineering systems.

The spectrum of Figure 9.30 can also be used to study the transmissibility of motion in the case of base excitation. Thus, if the base has a motion $y = Y \cos \omega t$, the mass will have a motion $x = X \cos(\omega t - \varphi)$. The ratio of resulting to imposed motion (output/input) is $X/Y = H(\omega)$, where $H(\omega)$ is exactly the function plotted in Figure 9.30.

Physically, the situation might represent the isolation of an instrument, for example, mounted on a vibrating floor. The engineer very often has to design a flexible mounting of the instrument in order to minimize its vibration if the floor or the base vibrates. Here the motion transmissibility ratio equals the force transmissibility ratio in the previous case of a harmonic force imposed on the mass. They are thus both plotted in Figure 9.30. From this figure, we again observe that effective motion isolation is incurred for $\omega > 1.41\omega_n$ only, regardless of damping. For higher exciting frequencies, damping has an adverse effect. As a general rule,

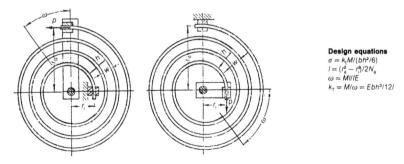

Figure 9.26b. Design equations for torsional helical springs (from Decker 1973).

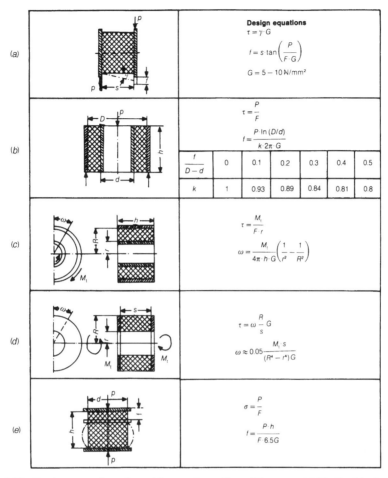

Figure 9.27. Design equations for rubber springs (from Niemann 1965). In this figure, F is the cross-sectional area, f is the deformation, ω is the angle of rotation, G is the shear modulus. Niemann suggests $G = 5$–10 N/mm^2. However, elastomers with G outside this range are also available.

therefore, for a given ω, one must try to provide a flexible mounting (low k, thus high ω/ω_n) with light damping.

Example 9.7 A refrigeration compressor has mass 250 kg and rotates at 1750 rpm. Find the dimensions of the four cylindrical mounts such as the ones in Figure 9.26*e* for height h equal to diameter d that deliver damping factor $\zeta = 0.1$ and have shear modulus $G = 15$ MPa, so that the transmissibility ratio will be 0.1 or less.

Solution Let m be the mass of the compressor and k the stiffness of the four mounts. The rotating frequency is $\omega = \text{rpm} \times \pi/60 = 1750 \times 2\pi/60 = 183$ rad/s. The natural frequency is $\omega_n^2 = k/m$ and the transmissibility ratio is determined from Equation (9.58), which can be written as follows:

TABLE 9.6. Elastic Properties of Soils

		$k_s = \overline{\beta}$	
		Vertical	Horizontal
	Permissible Load	lb/in.³	lb/in.³
Soil Type	psi/(N/mm²)	N/mm³(× 2 × 10⁵)	N/mm³
Grey plastic silty clay with sand and organic silt	15/10	75/200	110/300
Brown, saturated silty clay with sand	22/15	110/300	170/460
Dense silty clay with sand	75/50	190/520	315/850
Medium moist sand	30/20	110/300	200/550
Dry sand with gravel	30/20	110/300	200/550
Fine saturated sand	35/25	150/400	340/925
Medium sand	35/25	150/400	340/925
Loesial, natural moisture	45/30	170/460	370/1000
Gravel	90/62	375/1000	1000/2700
Sandstone	150/10	350/950	1000/2700
Limestone	165/115	400/1100	1200/3250
Granite, partly decomposed	600/415	1000/2700	3500/9500
Granite, sound	850/590	1050/2850	3600/9800

$$T_R^2 = \left[1 - \left(\frac{\omega}{\omega_n} \right)^2 \right]^2 + \left(\frac{J\omega}{\omega_n} \right)^2$$

$$\left(\frac{\omega}{\omega_n} \right)^4 - (2 - J)\left(\frac{\omega}{\omega_n} \right)^2 - T_R^2 + 1 - 0$$

The positive solution of this bi-quadratic equation is:

$$\left(\frac{\omega}{\omega_n} \right)^2 = \frac{(2 - J) + \sqrt{(2 - J)^2 + 4(-T_R^2 + 1)}}{2}$$

$$= \frac{(2 - 0.1) + \sqrt{(2 - 0.1)^2 + 4(-0.1^2 + 1)}}{2} = 2.32$$

Therefore, $\omega/\omega_n = 1.52$, $\omega_n = 1.52\omega = 1.52 \times 183 = 278.16$ rad/s. The total mount stiffness is:

$$k = \omega_n^2 m = 278.16^2 \times 250 = 19.3 \times 10^6 \text{ N/m}$$

Because the four mounts are assembled in parallel, each mount should have stiffness

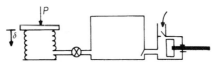

Figure 9.28. Air springs.

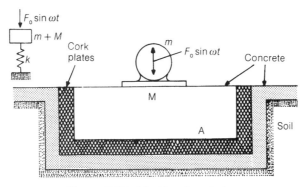

Figure 9.29. Machine foundation.

$$k_m = \frac{k}{4} = \frac{19.3 \times 10^6}{4} = 4.8 \times 10^6 \text{ N/m}$$

From Figure 9.26e, the stiffness $k_m = P/f = 6.5\ GF/h$, where f is the deformation, F is the cross-sectional area $\pi d^2/4$ and $h = d$ (by the problem statement) the height of the cylinder. Therefore:

$$k_m = \frac{6.5G(\pi d^2/4)}{d}$$

$$d = \frac{4K_m}{6.5G\pi} = \frac{4 \times 4.8 \times 10^6}{6.5\pi \times 15 \times 10^6} = 0.0626 \text{ m or } 63 \text{ mm}$$

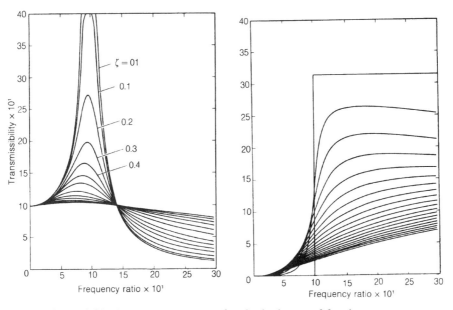

Figure 9.30. Response spectrum of a single-degree-of-freedom system.

CASE STUDY 9.1

The supporting column for a power-plant installation of length $L = 6$ m is loaded with a central, compressive load of 350 kN and is fabricated with two channel sections with welded cross-plates. The column material is structural steel ASTM A284 grade D with $S_y = 225$ N/mm², $E = 210,000$ N/mm². The safety factor against buckling must be at least 1.4. Select the proper channel section and find the spacing of the plates and the distance of the channel sections.

Solution For buckling in a direction perpendicular to the figure plane, the plates do not contribute and the channel sections take all the buckling load. Therefore, by the Euler formula:

$$2I_z = \frac{P(\mu L)^2}{\pi E} = \frac{350,000(6,000)^2}{\pi^2 10,000} = 6.1 \times 10^6 \text{ mm}^4$$

The channel section will then have $I_z = 3.05 \times 10^6$ mm⁴.
   The section width 120 mm is selected (Appendix D.2) with:

$$I_z = 304E6 \text{ mm}^4, \quad I_x = 0.31E6 \text{ mm}^4, \quad A = 1330 \text{ mm}^2$$

In the y-direction, the section must be also stable. The moment of inertia of the section in the y-direction is, if a_0 is the distance of the centroid, equal to 15.4 mm:

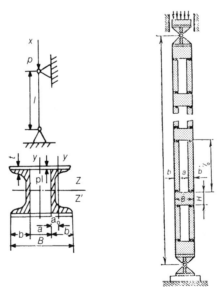

Figure CS9.1

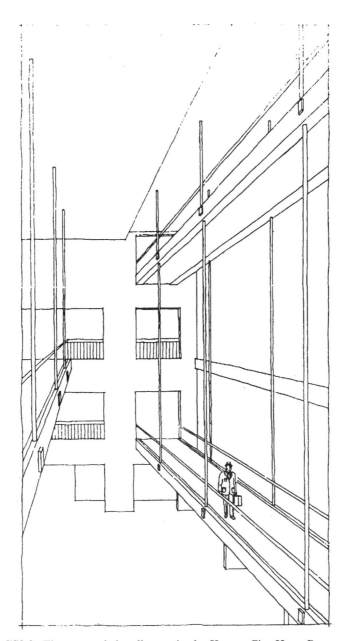

Figure CS9.2. The suspended walkways in the Kansas City Hyatt Regency Hotel.

$$I'_y = 2I_y + 2a \left[\left(\frac{a}{2} \right) + a_0 \right]^2 = 2 \times 0.31E6 + 2 \times 1330 \left(\frac{a}{2} + 15.4 \right)^2$$

This moment of inertia must be equal to $2I_z$ to have equal buckling resistance in both directions. Therefore:

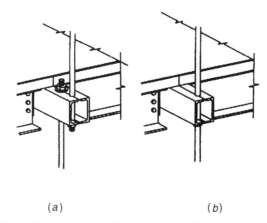

(a)   (b)

Figure CS9.3. Connection detail: (a) as built, (b) as originally designed.

$$\left(\frac{a}{2} + 15.4\right)^2 = \frac{6.1 \times 10^6 - 2 \times 0.31 \times 10^6}{2 \times 1330} = 2060.2$$

and $a = 60$ mm.

Each section between two adjacent plates at distance l must resist local buckling. In the z-direction, this condition is obviously satisfied. In the x-direction, it must be, from the Euler formula:

$$I_x = 0.31 \times 10^6 > \frac{(P/2)(\mu l)^2}{\pi^2 E}$$

Therefore, for fixed-fixed mounting:

Figure CS9.4. The failed connection.

$$l_0^2 = \frac{0.31E6 \times \pi^2 E}{\mu^2 P/2} = \frac{0.31E6 \times \pi^2 \times 210,000}{2^2 \times 350,000/2}$$

$$l_0 = 1698 \text{ mm}$$

□

CASE STUDY 9.2

In 1981, the Kansas City Hyatt Regency Hotel had its opening night. The architects had designed several suspended walkways overlooking the main lobby (Figure CS9.2). Suddenly, one side of the suspended walkways collapsed, killing more than 100 people. An investigation was commenced into the causes of the accident.

Investigation   In a search of the original architectural drawings and the detailed drawings of the construction of the building, it was found that the original design suspended the walkways from the roof through vertical bolts that emanated from the roof and went down all the way to the lower walkway. At the level of each walkway, a nut and washer supported the horizontal structure.

During construction, it was decided that using these very long bolts was difficult and expensive, and thus it was decided to support the top walkway from the ceiling, the walkway below from the top one, and so on, as shown in Figure CS9.3. Thus, every walkway support supported not only the weight of the walkway, but also the weight of all the walkways below. Such a connection failed, as shown in Figure CS9.4.

The failure was by local buckling of the channel section, which was loaded as an eccentric column. The professional engineers who signed the drawings were found responsible and were severely disciplined by having their professional engineering licenses revoked.   □

REFERENCES

Bccr, F. P., and E. R. Johnston. 1981. *Mechanics and Materials.* New York: McGraw-Hill.

Decker, K.-H. 1973. *Maschinenelemente,* 6th ed. Munich: C. Hanser Verlag.

Niemann, G. 1965. *Maschinenelemente.* Berlin: Springer-Verlag.

Timoshenko, S., and J. M. Gere. *Elastic Stability.* New York: McGraw-Hill, 1961.

Wahl, A. M. 1963. *Mechanical Springs.* New York: McGraw-Hill, 1963.

ADDITIONAL READING

American Institute of Steel Construction (AISC), *Manual of Steel Construction.* Chicago AISC, 1980.

Decker, K.-H., *Maschinenelemente,* 1st ed. Leipzig: VEB, 1962.

Dimarogonas, A. D. *Vibration for Engineers,* 3rd ed. Upper Saddle River, N.J.: Prentice-Hall, 1996.

Johnson, R. C. *Optimum Design of Mechanical Elements.* New York: John Wiley & Sons, 1961.

Orlov, P. *Fundamentals of Machine Design.* Moscow: Mir, 1976.

Roark, R. J., and W. C. Young. *Formulas for Stresses and Strain.* New York: McGraw-Hill, 1975.

Simitsis, G. *Elastic Stability of Structures.* Englewood Cliffs, N.J.: Prentice-Hall, 1976.

PROBLEMS[1]

9.1. [C]. The connecting rod of a steam engine has circular cross-section with diameter $d = 30$ mm, is made of SAE 1020 steel, has length $L = 350$ mm, and can be considered as simply supported at the two ends. Find the maximum load that the rod can take without buckling.

9.2. [C]. The vertical column supporting a water tank is a hollow tube with inner diameter $d_1 = 300$ mm, outer diameter $d_1 = 310$ mm, is made of SAE 1020 steel, has length $L = 14$ m, and can be considered as fixed at the lower end and free at the upper end. Find the maximum weight of the tank that the column can take without buckling.

9.3. [C]. The connecting rod of an indexing mechanism has rectangular cross-section 25×40 mm, is made of SAE 1020 steel, has length $L = 500$ mm, and can be considered as simply supported at the two ends. Find the maximum load that the rod can take without buckling.

9.4. [C]. The screw of a screw press has circular cross-section with minor diameter $d_3 = 40$ mm, is made of SAE 1040 steel, has length $L = 600$ mm, and can be considered as simply supported at the lower end and fixed at the upper end. Find the maximum load that the press can take without buckling.

9.5. [C]. A girder in a steel construction is a U-section 50×25 mm, thickness 5 mm (see Appendix D.2), is made of SAE 1015 steel, has length $L = 4$ m, and can be considered as fixed at both ends. Find the maximum compressive axial load that the girder can take without buckling.

9.6. [D]. Redesign the screw of the hand press of Problem 8.26, checking the buckling load of the screw and assuming pinned ends.

9.7. [D]. For Problem 8.27, determine the buckling load of the screw, assuming pinned end conditions for the screw.

9.8. [D]. In Problem 8.28, find the maximum allowable loaded length of the screw, assuming pinned end conditions and a safety factor for buckling of at least 3.

9.9. [D]. Solve Problem 8.29, checking screw buckling for fixed-end conditions.

9.10. [D]. Solve Problem 8.30, checking screw buckling for fixed-pinned end conditions.

---

[1] [C] = certification, [D] = design, [N] = numerical, [T] = theoretical problem.

9.11. [C]. The load-carrying end effector of a ship crane consists of two rods of rectangular cross-section in triangular arrangement (Figure P9.11) of width 60 mm (on the plane of the triangle) and thickness 30 mm. The smaller angle of the isosceles triangle is 30° and the length of the equal sides is 1 m. The material of the rods is SAE 1030 steel. If the two ends of each rod are assumed pinned, find the maximum load that the crane can carry, with safety factor 3, for both stress and buckling.

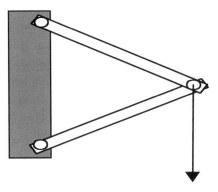

Figure P9.11

9.12. [C]. An acrobat is performing his act on a 500 mm diameter circular platform fixed on the top of a 6 m-high steel column of hollow circular cross-section of inner diameter 50 mm and thickness 3 mm (Figure P9.12). The column is fixed at the lower end and free at the upper end, where the platform is affixed. The column material is SAE 1030 steel. Determine the minimum safety factor if the 800 N weight of the acrobat is located anywhere on the platform.

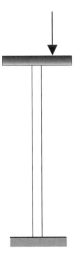

Figure P9.12

9.13. [C]. The connecting rod of an indexing mechanism has rectangular cross-section 25 × 40 mm, is made of SAE 1020 steel, has length $L = 500$ mm, and can be considered as simply supported at the two ends. The load can be applied anywhere within the cross-section of the rod. Find the maximum load that the rod can take without buckling or failing.

9.14. [C]. The screw of a screw press has circular cross-section with minor diameter $d_3 = 40$ mm, is made of SAE 1040 steel, has length $L = 600$ mm, and can be considered as simply supported at the lower end and fixed at the upper end. The load can be applied anywhere within the cross-section of the screw. Find the maximum load that the press can take without buckling.

9.15. [C]. In an electrowelded pipe leak testing facility, the pipe has length 12 m, inner diameter 56 mm, outer diameter 60 mm, and is tested under a compressive force P at both ends through two sealing blocks (Figure P9.15). The pipe is made of SAE 1020 carbon steel. The sealing blocks can apply the compressive force anywhere on the periphery of the pipe, due to manufacturing inaccuracies on the plane ends of the pipe. Find the maximum compressive force that the pipe can take. Assume both ends pinned.

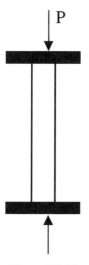

Figure P9.15

9.16. [C]. A helical spring (Figure 9.20) is made of spring steel and has dimensions $D = 30$ mm, $d = 2$ mm. It is loaded with an axial force $F = 500$ N. Find the spring constant for axial loading and the maximum stress.

9.17. [C]. A cantilever (Figure 9.20) has $b = 100$ mm, $b_0 = 40$ mm, $l = 300$ mm, $h = 5$ mm, is made of spring steel, and is loaded by a lateral force at the free end $F = 300$ N. Find the spring constant and the maximum stress.

9.18. [C]. A leaf spring (Figure 9.23) consists of four leaves of thickness $h = 5$ mm, width $b = 50$ mm, length $= 300$ mm, is made of spring steel, and is loaded by a lateral force at mid-span $F = 3$ kN. Find the spring constant and the maximum stress.

9.19. [C]. A Belleville washer spring (Figure 9.24) consists of six washers of thickness $s = 1$ mm, inner diameter $D_1 = 10$ mm, outer diameter $D_2 = 25$

mm, height h = 4 mm, is made of spring steel, and is loaded by an axial force F = 1000 N. Find the spring constant and the maximum stress.

9.20. [C]. A cylindrical torsional helical spring (Figure 9.25) consists of six turns, has wire diameter d = 1 mm, mean spring diameter D = 30 mm, is made of spring steel, and is loaded by a torque at one end T = 50 Nm. Find the torsional spring constant and the maximum stress.

9.21. [C]. A cylindrical torsional flat helical spring (Figure 9.25) consists of 3.5 turns, has wire diameter d = 1 mm, width b = 50 mm, length = 300 mm, is made of spring steel, and is loaded by a torque about the axis of the spring T = 20 Nm. Find the torsional spring constant and the maximum stress.

9.22. [C]. A cylindrical torsion bar spring (Figure 9.19), diameter d = 0 mm, length = 300 mm, is made of spring steel, and is loaded by a torque T = 500 Nm. Find the torsional spring constant and the maximum stress.

9.23. [C]. A cylindrical rubber spring (Figure 9.27) has diameter D = 30 mm, height = 25 mm, is made of rubber with Young's modulus E = 10 MPa and shear modulus 5 MPa. Find the constants for axial and torque loading.

9.24. [C]. A cylindrical rubber bellows air spring (Figure 9.28) has bellows diameter D = 150 mm, height = 125 mm, and is filled with air with sufficient pressure p to support a dead load P = 1000 N. Find the constants for additional axial loading assuming isentropic compression of the air.

9.25. [C]. A cylindrical helical spring (Figure 9.20) consists of 10 turns, has wire diameter d = 2 mm, mean spring diameter D = 30 mm, is made of spring steel, and is loaded by an axial force F = 50 N. Find the axial spring constant and the maximum stress.

9.26. [D]. The mechanism shown is a safety device for an elevator. If the cable brakes, then the two springs push the rods upwards to the horizontal position in order to force the pads on the wall and stop the elevator. At the operation position, the rods are at an angle 20° with the horizontal. They must reach the horizontal position, if the cable breaks, in 0.5 sec. The rods are of rectangular cross-section 30 mm wide and 60 mm high. The distance l = 300 mm and the material for the rods is steel. Determine the spring dimensions, number of turns, and force length if at the compressed position it is fully compressed and, when the rods are horizontal, it is halfway released. The spring must be made of spring steel AISI 9260.

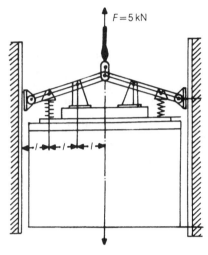

Figure P9.26

9.27. [D]. A railroad car has a 40 ton weight, loaded. It has a bumper consisting of two helical springs and must be in the fully compressed position if the car comes to a stop against the end wall at a maneuvering speed of 5 MPH. Design a proper spring.

9.28. [D]. The return of seating actions for an intake valve of an automobile engine are secured by a coil spring. The maximum travel of the valve is 8 mm, and at the closed valve position it must deliver a force of 150 N. Five to 8 turns are suggested. Design the spring if the maximum force, when the valve opens, must not exceed 250 N.

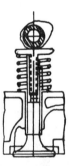

Figure P9.28

9.29. [C]. The pressure relief valve shown uses a spring made of AISI 9260 spring steel. If at the position shown the spring was compressed to 1/2 of the maximum compression, determine:

1. The pressure at which the relief valve opens
2. The pressure required to force the spring at the fully compressed position

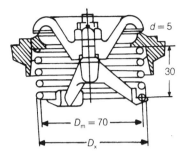

Figure P9.29

9.30. [D]. The return spring of a control arm must pull the arm with a 20 N force at the position shown and 25 N force at maximum extension, which is 12 mm. Design a spring of AISI 9260 spring steel for fatigue strength (large number of operations) with a safety factor of $N = 2$. The total length of the spring $L + 2L_H$ must not exceed 50 mm.

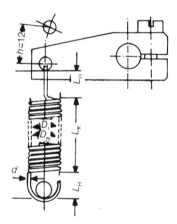

Figure P9.30

9.31. [D]. A torque wrench consists of a torsion rod of diameter d and length l. A pointer fixed on the lower end indicates at the upper end the total angle of twist, proportional to the applied moment. The wrench is rated at 100 Nm and the angle of twist at that torque is 20°. Design the torsion rod of AISI 1035 carbon steel with a factor of safety $N = 1.5$.

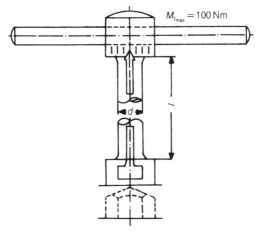

Figure P9.31

9.32. [D]. A torsion bar for an automobile supports a 3 kN load (approximately 1/4 of the automobile mass) as shown. The travel of point A from the dead load, 2 kN, to the full load of 3 kN must not exceed 50 mm. The natural frequency with the dead load only must not exceed 1 Hz. Is the design feasible under a safety factor for strength $N = 1.6$?

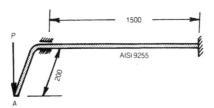

Figure P9.32

9.33. [D]. The ratchet mechanism for a toothed wheel consists of a helical spring around a pin of 20 mm diameter that forces the ratchet arm on the wheel as shown. The spring is presented to deliver 20 N force at the position shown. Design the spring so that at the twisted position the point of contact with the wheel moves to the tip of the tooth, which has a height of 5 mm, the over-stress should not exceed 10%. The material is spring steel AISI 9260 and the design must be for fatigue strength with a safety factor $N = 2$ for strength.

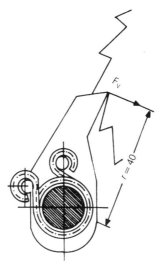

Figure P9.33

9.34. [C]. Twelve Belleville washers as shown form a spring for a rolling mill. If the washer material is carbon steel AISI 1035 and the maximum travel is 15 mm, determine the maximum static load F_1 and the maximum pulsating load F_2 for safety factor $N = 1.5$. There is no initial prestress of the spring.

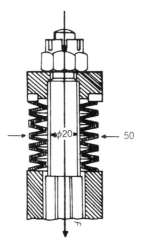

Figure P9.34

9.35. [D]. Twelve Belleville washers as shown form a support spring for vibration isolation of a heavy machine. The machine weight is 6 tons, equally divided among four springs. The natural frequency for vertical vibration should not exceed 20 Hz. Determine:

1. The required spring constant
2. The washer dimension for fatigue strength if the dynamic load is 15% of the static load, the material is AISI 1035 steel, and the safety factor is 1.5.

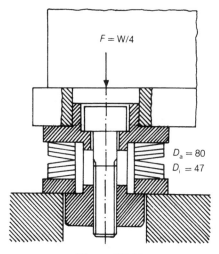

Figure P9.35

9.36. [D]. Design the spring of Problem 9.35 with a rubber spring.

9.37. [D]. A rubber spring is used to reduce impact in a highly elastic machine. The machine has weight 1000 N, equally divided among four rubber springs. It is expected that the rubber spring will take impact of the above weight at speed 1 m/s. During impact, the rubber spring must not be compressed more than 20%. Determine the diameter d. $S_u = 5$ MPa.

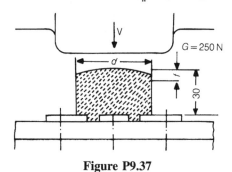

Figure P9.37

9.38. [D]. A sensitive mechanism aboard a destroyer ship weighs 100 N and is mounted on four rubber springs of the type of Figure 9.27. The dominant vibration frequency in the vicinity of the mechanism is 30 Hz. Design the rubber spring for effective vibration isolation, that is, with natural frequency of the mechanism on the spring no greater than 10 Hz.

9.39. [D]. An automobile engine weighs 2300 N and is mounted on two rubber springs of the type of Figure 9.27. The modulus of elasticity of the particular rubber is 30 N/mm². The rubber spring has the shape of a rectangular pad 60 × 60 mm and thickness 30 mm. Determine the natural frequency of the engine on the springs.

9.40. [D]. The torsion bar of Problem 9.32 is to be replaced with two rubber springs, at the support points, of the type in Figure 9.27. Design the rubber springs.

9.41. [C]. A printing press has a length of 15 m, weighs 60 ton, and operates at 2000 rpm. It must be isolated to avoid transferring its vertical vibration and noise to the surroundings. To this end, a concrete base plate is made of thickness 1 m and density of 2700 kg/m³. This plate rests on gravel. Determine the necessary base plate dimensions for effective isolation. Determine the transmissibility ratio at 2,000, 4,000, 10,000 rpm rotating speed of the press for $\omega/\omega_n = 6$.

9.42. [D]. A reciprocating refrigeration compressor operates at 3000 rpm and consequently vibrates in the vertical and horizontal directions. It weighs 10 kN and is installed on a concrete base with cork support as in Figure 9.29. Determine the dimensions of the concrete base and the required cork plate thickness if cork has $G = 20$ N/mm², $E = 40$ N/mm², $S = 1$ N/mm².

9.43. [D]. An air compressor operating at 1500 rpm was installed with an arrangement such as in Figure 9.29. The concrete base was 1×1 m, 300 mm high, weighing 8.1 kN, while the compressor weight was 5 kN. The cork plate had $E = 40$ N/mm². Find the cork plate thickness if the transmissibility ratio is required to be below 0.001, assuming fraction of critical damping $\zeta = 0.05$.

CHAPTER 10

DESIGN OF DRY FRICTION ELEMENTS

International Harvester tractor. Farm and earth-moving equipment have low speed and substantial power and thus sustain high loads and torques. They must have powerful brakes and clutches.

OVERVIEW

In Chapters 8 and 9, we developed design methods based mostly on material strength and flexibility, respectively. The reason was that the respective mechanisms of failure were material failure and large displacements. There are other mechanisms of failure in machines, including sliding friction and wear of machine parts. For example, an automobile brake pad never breaks but has to be replaced periodically because of excessive wear of the friction material. In this chapter, we shall study design methods based on failures due to dry friction, with applications to clutches and brakes.

10.1. DRY FRICTION

10.1.1. Introduction

Because motion is inherent in machines, so is friction (a) between components that have relative motion and (b) where that force is transmitted through their point of contact. If at the point of contact the respective points of each element are moving with different linear velocities, there is *sliding friction.* If at the point of contact the respective points of each element are moving instantaneously with the same linear velocity, there is *rolling friction.*

Friction is an often unwanted but often necessary condition. It is unwanted because the friction force is usually in an opposite direction to the velocity of the friction point and thus mechanical work is transformed into heat flow. In addition, the friction surfaces are gradually damaged because of *wear* at the points of contact. Special methods are used to reduce friction and wear, and we will discuss the corresponding machine design methodology in Chapters 11 and 12.

There are, however, applications where the friction force is used to transfer power from one element to another and thus the friction force is performing a useful task. In this chapter, we shall discuss methods of machine design based on transmission of force by sliding friction. An important application is the transmission of rotary motion from one rotating shaft to another under the control of the machine operator (a person or a controller). If both shafts are moving upon engagement of the friction force between them, the associated element is termed *clutch.* If one shaft is fixed and the purpose of the frictional engagement is to make the other shaft stop, the associated element is termed *brake.*

Clutches and brakes can also be used in rectilinear or general motion.

10.1.2. Sliding Friction

It was mentioned in Chapter 4 that real surfaces are always irregular. If amplified, they would look as in Figure 10.1. Curve (*b*) is the actual geometry and curve (*a*) shows the surface with amplification in the vertical direction much greater than that in the horizontal direction. It appears that the surface profile consists of two wave forms, one with a large wavelength, called *waviness,* and another with a much smaller wavelength, called *roughness.*

When two solids with real surfaces, as in Figure 10.1, approach each other under a normal force, the tips of the surface irregularities come in contact first. They deform elastically first and then plastically for most engineering situations. The

Figure 10.1. Geometry of machined surfaces.

approach of the two surfaces stops when the interaction forces between surface irregularities sum up to the normal load.

The real area of contact A_r is the sum of the contact surfaces of the surface asperities, usually much smaller than the apparent area of contact A_a (Figure 10.2).

At contact points, there is almost always plastic deformation, accompanied by elastic deformation of the neighboring areas (Figure 10.3).

Hertz studied contact mechanics macroscopically and proved that when a sphere is pressed against a plane, the apparent area of contact will be a circle (Figure 10.4), with diameter

$$d = 1.75 \left[Wr \left(\frac{1}{E_1} + \frac{1}{E_2} \right) \right] \tag{10.1}$$

The stress distribution has an elliptic shape and the maximum interface pressure is

$$p_{\max} = 0.42 W^{1/3} r \left[\left(\frac{1}{E_1} + \frac{1}{E_2} \right) \right]^{-2/3} \tag{10.2}$$

where E is the elasticity modulus of solids 1 and 2.

Plastic deformation of the surface asperities is so sudden that the heat of the plastic deformation raises the interface temperatures and local bonds may be created. The force necessary to break these bonds during sliding is the main contribution of the friction force in most engineering cases. Molecular forces, microcutting, and elastic hysteresis are some of the other mechanisms contributing to the friction force.

In metal contacts, the prevailing mechanism for dry friction is usually bond formation, and this helps in the estimation of the friction coefficient.

Ernst and Merchant suggested that the real area of the contact has magnitude determined by the assumption that the external load N is balanced by general yielding of the softer material:

$$N = A_r S_u \tag{10.3}$$

The friction force F needed to break the bonds of surface A_r with ultimate shear strength S_{su} must be:

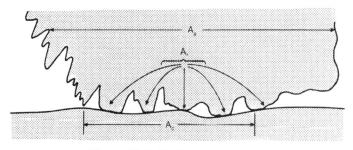

Figure 10.2. Real area of contact.

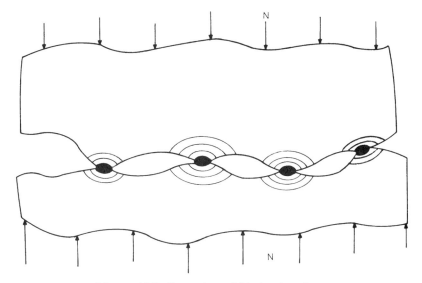

Figure 10.3. Formation of friction junctions.

$$F = A_r \, S_{su} \tag{10.4}$$

Dividing through by Equation (10.3) results in

$$f = \frac{F}{N} = \frac{S_{su}}{S_u} \tag{10.5}$$

The idea of solid lubricants, such as graphite and MoS_2 and surface coatings is to reduce the shear strength without reducing the bulk strength S_u of the supporting material. This drastically reduces the coefficient of friction, as Equation (10.5) indicates. Ice-skating is based on the same equation.

Because the coefficient of friction is reduced if the mating materials are very different physico-chemically, bond formation is difficult. For this reason, sliding pairs in machines are always made from dissimilar materials. Because, for reasons

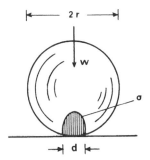

Figure 10.4. Hertz contacts.

of strength, one of the two materials is usually steel, the other one must be not steel but, for example, cast iron, bronze, and babbitt.

Some materials have inherently good antifriction properties. Cast iron, for example, has graphite in its structure, which acts as an antifriction coating.

The resistance to the sliding of one surface over the other—the frictional force—is due to a variety of mechanisms, such as the resistance offered by the interaction of the projections on both surfaces. During the process of sliding, very high local temperatures will be generated. If the surfaces are only lightly pressed against each other, the tendency will be for the projections to be smeared over, and this will be facilitated by the local high temperatures and possible local melting of the metals.

The process of smearing over the projections of both surfaces is considerably facilitated if there is a lubricant in the space between them. Local melting of the projections is followed by rapid cooling, and many times the ultimate result is a very highly polished, extremely hard amorphous layer on the surface. The process of forming such layer is known as *running-in*.

Another important factor is that when metal comes into contact with a lubricant, an absorbed film of lubricant will be formed on its surface. There are indications that the nature of this film, formed with the first oil to contact the surface, is of importance. The process of running-in can be facilitated by the use of solid lubricants such as graphite and molybdenum bisulphide.

10.1.3. Temperatures at Sliding Contacts

When two bodies come into sliding contact, nearly all the energy dissipated by friction appears as heat, which is distributed between the two bodies and raises their temperature appreciably at the area of the sliding contact. The knowledge of these temperatures at the sliding interface is of fundamental importance to the tribological behavior of the materials and has immediate application in the fields of lubrication, metal cutting, grinding, and the design of, for example, gears, bearings, forming tools, mechanical seals, and stationary or moving electric contacts.

Steady rubbing over a given area or point contact tracing of a closed path at high speed results in high temperatures over the contact surfaces of the stationary and the moving body.

Heat is transferred through the cooled surfaces of sliding bodies by convection. The average steady surface temperature rise due to a long engagement depends on the rate of heat produced, the product of friction force and sliding speed, and the resistance of the two elements to the flow of heat. This has been expressed in the form:

$$\Delta T_0 = \frac{\dot{Q}}{l_1 k_1 H_1 + l_2 k_2 H_2} \tag{10.6}$$

where $\dot{Q} = UF$ is the rate of heat generation, U is the sliding speed, F is the frictional force, H_i is the dimensionless resistance of one of the mating parts i to the heat flow (Table 10.1), k is the thermal conductivity of the material (Table 5.1), l is a characteristic length indicated in Table 10.1 and 1, 2 are indices for the two mating parts. This yields the average surface temperature as a function of the total

TABLE 10.1. Heat Resistance Coefficients (h Is the Heat Transfer Coefficient)

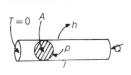	Cylinder, heat input at one end, the other at constant temperature	$H = \dfrac{A}{l^2}\dfrac{N}{\tan h(N)}$, $N = l\left(\dfrac{hp}{kA}\right)^{1/2}$ $H = l\left(\dfrac{hp}{kA}\right)^{1/2}$ for $h = 0$
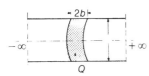	Rotating cylinder of infinite length, radius l, friction on cylindrical surface band of width $2b$	$H = \dfrac{2}{S_1}\pi^2\beta^2$, $\beta = \dfrac{b}{l}$ with $S_1 = \displaystyle\int_0^\infty \dfrac{\sin^2(\beta\gamma)\,d\gamma}{\gamma^2\left[\gamma\dfrac{I_1(\gamma)}{I_0(\gamma)} + \dfrac{lh}{k}\right]}$
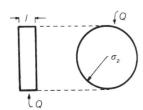	Full cylindrical ring of radius a_2 and thickness l. Friction on the cylindrical surface of width l side surface cooled with heat transfer coefficient h	$H = 2\pi m a_2\dfrac{I_1(ma_2)}{I_0(ma_2)}$, $m^2 = \dfrac{2h}{kl}$ $I(x)$ modified Bessel Functions of the first kind
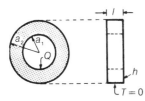	Circular ring with inner radius a_1, outer radius a_2, thickness l. Friction on the inner surface. Outer surface insulated. Side surface at heat transfer coefficient h	$H = 2\pi a_1 m\psi$ with $m^2 = \dfrac{2h}{kl}$ and $\psi = \dfrac{I_1(ma_2)K_1(ma_1) - I_1(ma_1)K_1(ma_2)}{I_0(ma_1)K_1(ma_2) + I_1(ma_2)K_0(ma_1)}$ $K(x)$ modified Bessel Functions of the second kind
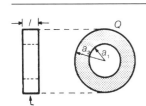	Circular ring with inner radius a_1, outer radius a_2, thickness l. Friction on the outer surface, inner surface insulated. Side surface at heat transfer coefficient h	$H = 2\pi a_2 mg$ with $g = \dfrac{I_1(ma_1)K_1(ma_2) - I_1(ma_2)K_1(ma_1)}{I_0(ma_2)K_1(ma_1) + I_1(ma_1)K_0(ma_2)}$
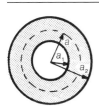	Circular ring with inner radius a_1, outer radius a_2. Friction on the side surface at radius a. Inner and outer surfaces insulated	$H = 2\pi a_2 m(X + Y)$ with $X = \dfrac{I_1(ma_1)K_1(ma) - I_1(ma)K_1(ma_1)}{I_0(ma)K_1(ma_1) + I_0(ma_1)K_0(ma)}$ $Y = \dfrac{I_1(ma)K_1(ma_2) - I_1(ma_2)K_1(ma)}{I_0(ma)K_1(ma_2) - I_1(ma_2)K_0(ma)}$

frictional heat generated and the properties of the sliding bodies. Although it is a macroscopic average temperature, it can be used for all practical design purposes as maximum contact temperature. This temperature is used in selecting the friction, brake, and seal materials and calculating the capacity of these elements to carry the pressure loads. High temperatures due to friction often cause the yielding of the seal teeth due to the pressure difference between the two faces of the tooth, cracking of brake and clutch disks and drums, among other reasons.

The average temperature rise of the disk or cylindrical drum of the clutch or brake for a short application of the braking force can be easily calculated, assuming that the friction pad has high resistance to the flow of heat and all friction heat goes into the metal disk or drum, which has volume V, density ρ, and heat capacity c_p (see Appendix E for values for the most common metals). Observing that the heat generated $Q = FU\Delta t = F\Delta s$ raises the temperature of the clutch or brake by ΔT:

$$\Delta T = \frac{Q}{\rho V c_p} \tag{10.6a}$$

Equation 10.6a is often used in the design of friction elements, under the assumption that all generated heat flows into one of the sliding parts and raises its temperature. This is a conservative assumption, but for many design applications is a realistic one. Most of the time, one of the friction parts (usually the stationary one) is made of a nonmetallic material with low thermal conductivity and thus most of the heat flows into the moving part, usually made of a metal with high thermal conductivity and substantial heat-entering surface.

Example 10.1   A disk brake shown in Figure E10.1 consists of a disk and two pads rubbing on the disk. It has dimensions $d_1 = 30$ mm, $d_2 = 250$ mm, thickness $l = 10$ mm, and the brake pads can be assumed to operate at radius $a = 90$ mm. The disk is made of carbon steel. The coefficient of friction was measured as $f = 0.3$. The brake force applied by a hydraulic cylinder on the pads is 1000 N.

1. Find the maximum interface temperature for long application of the brake, assuming that practically all the heat goes into the disk because of the low thermal conductivity of the pad, and that at the surface of the disk the convection heat transfer coefficient is $h = 40$ W/m^2 °C h. The speed of rotation is 100 rpm.
2. Find the average temperature rise of the disk if the brake is applied for 10 sec and the volume of the disk is 300 cm^3.
3. Find the average temperature rise of the disk if the brake is applied to bring the car of mass $m = 1000$ kg from a speed $u = 80$ km/h to a complete stop.

Solution

1. For carbon steel, thermal conductivity $k = 190$ kJ/m h°C $= 52.7$ J/m s°C, $c_p = 0.4$ kJ/kg °C (Appendix E). Therefore, from Table 10.1, case 6, $m = (2h/kl)^{1/2} = (2 \times 40/52.7 \times 0.01)^{1/2} = 12.3$. From tables of Bessel functions (see Spiegel 1968), we obtain the values of X and Y (Table 10.1), and then:

$$H = 2\pi a_2 m(X + Y) = 2\pi \times 0.125 \times 4.1 \times 0.23 = 0.74$$

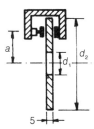

Figure E10.1

From Equation (10.15), if all the heat is entering the disk:

$$\Delta T_o = \frac{Q}{lkH} = \frac{f F \omega \alpha}{akH} = \frac{f P (2\pi n/60)}{kH} = \frac{0.3 \times 1000 \times 2\pi \times 100/60}{52.7 \times 0.74} = \underline{80°C}$$

2. The mean temperature rise for $\Delta t = 10$ sec, $U = \omega R = (2\pi n/60)(0.030 + 0.250)/2 = 1.46$ m/s, is:

$$\Delta T = \frac{FU\Delta t}{\rho V c_p} = \frac{1000 \times 1.46 \times 10}{7800 \times 300 \times 10^{-6} \times 400} = \underline{15.5°C}$$

3. The total heat entering the four disks will equal the kinetic energy of the car:

$$Q = \frac{mv^2}{2} = \frac{1000 \times (80,000/3600)^2}{2} = 247,000 \text{ J}$$

The mean temperature rise for four disks with volume $V = 4 \times 300 \text{ cm}^3$ is:

$$\Delta T = \frac{Q}{\rho V c_p} = \frac{247,000}{7800 \times 4 \times 300 \times 10^{-6} \times 400} = \underline{66°C}$$

10.1.4. Wear Due to Sliding Friction

Wear is not a process with a single mechanism. Major contributors to sliding wear are the interaction of asperities, fatigue, oxide removal, and molecular interaction. Most severe wear mechanism, where it exists, is the tearing in depth caused by work hardening due to plastic deformation. During the approach of the surfaces, plastic deformation of asperities might lead to their hardening. When, during sliding, the bond breaks, it happens away from the bond interface, at some depth where the material is softer. This results in removal of large debris and quick surface deterioration. For this reason, softer material should never be steel, which in most forms is subject to work hardening.

Wear will be apparent after two surfaces have been rubbing together for a considerable period. A good design ensures that the maximum amount of wear is confined to those parts that can be readily replaced. For example, journals that

TABLE 10.2. Wear Rates

Material	Hardness $(10^{12}\ N/m^2)$	$K \times 10^{-4}$
Soft steel on soft steel	18.6	70
Hard steel on:		
Brass 60/40	9.5	6
Teflon	0.5	0.25
Brass 70/3	6.8	1.7
Plexiglass	2.0	0.07
Hardened steel	85	1.3
Stainless steel	25	0.47
Polyethylene	0.17	0.0043

cannot readily be replaced are made of hard materials, while mating bearing "brasses" or shells are made of soft material so that most of the wear takes place in the softer bearing material.

If good fluid film lubrication could always be maintained—that is, if the mating surfaces could always be separated by a film of lubricant—extremely little, if any, cutting wear would result, and would only take place during the starting and stopping periods, when full fluid film lubrication is not possible.

Abrasive wear results from the intrusion of some abrasive such as grit. This can be reduced to a minimum by the use of filters in lubrication systems (Figure 10.5a).

Wear due to scuffing can only be reduced by improvement in lubrication conditions.

Corrosion wear results from chemical action, which is often due to acidic deterioration products in the lubricating oil, though it could arise from naturally occurring free acids in fatty oils or residual acid from bad refining.

The following definitions of terms used in connection with wear were agreed upon at the Conference of Lubrication and Wear arranged jointly by ASME and the Institution of Mechanical Engineers in October 1957.

Wear: Progressive loss of substance from the surface of a body brought about by mechanical action. (Usually it reduces the serviceability of the body, but it can be beneficial in its initial stages in running-in.)

Abrasion: Wear caused by fine solid particles.

Embedding: Inclusion of solid particles, abraded or foreign matter, in the surfaces of wearing parts.

Fretting Corrosion: Destruction of metal surfaces by a combination of sliding and corrosive (usually oxidation) actions when there is a small relative movement at the contacting surfaces.

Pitting: Local wear characterized by removal of material to a depth comparable with surface damage.

Scoring: Scratches across rubbing surfaces produced without modification of the general form.

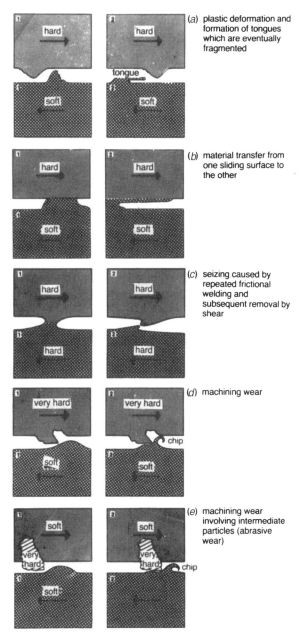

(a) plastic deformation and formation of tongues which are eventually fragmented

(b) material transfer from one sliding surface to the other

(c) seizing caused by repeated frictional welding and subsequent removal by shear

(d) machining wear

(e) machining wear involving intermediate particles (abrasive wear)

Figure 10.5. Wear mechanisms (from Engel and Klingele 1981).

Scuffing: Cross-damage characterized by the formation of local welds between sliding surfaces.

Seizure: The stopping of a mechanism as the result of interfacial friction.

Spalling: Wear, commonly associated with rolling bearings, which involves the separation of flakes of metal from the surface.

The wear rate can be expressed as volume wear:

$$\frac{V}{L} = \frac{KF_N}{S_b} \qquad (10.7)$$

where: V = the volume of wear debris
  L = the sliding distance
  F_N = the normal force
  S_b = the yield bearing pressure of the material, approximately the material yield strength if test results are not available
  K = a wear coefficient having the values in Table 10.2

Friction materials must have a high coefficient of friction with low wear rate. With metallic materials this combination is very difficult. For this reason, while one of the mating parts is usually steel, for strength purposes, the other is usually nonmetallic, such as asbestos, leather, or plastics.

Design Procedure 10.1: Design of Pair of Sliding Elements for Friction and Wear

Input: Maximum load, material properties, sliding speed, expected life. **Output:** Dimensions of friction surfaces.

Step 1: Sketch the two parts and place on the sketch loads and motion data.

Step 2: From dimensions, speed, and expected life, find the total sliding distance L.

Step 3: From Table 10.2 find the wear coefficient K.

Step 4: Apply Equation (10.7) to find the wear volume V.

Step 5: Find the friction area required so that the product of it with the allowed depth of wear will equal V.

Step 6: For the particular application, estimate the friction coefficient (for example, from Table 10.3).

Step 7: From Equation (10.6), determine the maximum temperature of the friction surfaces and check whether the friction material can withstand it.

Example 10.2 The collector rings of an electric generator are made of hard steel and the brushes of 70/30 brass. The brushes, of section 40×40 mm, are pressed against the rings by a force of 10 N and have a length of 40 mm. The useful length is 30 mm. The yield strength of the brass is $S_y = 80$ N/mm². Determine the life of the brushes, in hours, if the machine rotates at 3600 rpm.

Solution From Table 10.2, the wear coefficient is $k = 1.7 \times 10^{-4}$. Equation (10.7) yields the sliding distance:

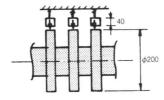

Figure E10.2

$$L = VS_y/kN = \frac{40 \times 40 \times 30 \times 80}{1.7 \times 10^{-4} \times 10}$$

$$= 2.25 \times 10^9 \text{ mm} = 2250 \text{ km}$$

The time is:

$$t = \frac{L/\pi d}{n/60} = \frac{2.25 \times 10^9/\pi \times 200}{3600/60}$$

$$= 59{,}948 \text{ sec} = 16.7 \text{ hr}$$

10.2. CLUTCHES AND BRAKES

10.2.1. Classification

Couplings are elements connecting two moving elements of a machine, usually two rotating shafts with the same axis of rotation. Couplings with controlled engagements, meaning those that during operation can be activated to connect two rotating members and the opposite, are called *clutches*. *Brakes* are clutches that connect one rotating member to nonrotating ones, usually the machine frame. Further, clutches and brakes are classified according to the means of effecting the engagement:

> *Disk clutches* and *brakes* use a concentric disk connected to one rotating element and one or more pads pressed on the disk and connected to the other element.
>
> *Drum* or *rim clutches* and *brakes* use a concentric cylinder connected to one rotating element and one or more pads pressed on the inner or outer surface of the cylinder and connected to the other element.
>
> *Band clutches* and *brakes* use a concentric cylinder connected to one rotating element and one flat band wrapped about part of the cylinder and connected to the other element.

Clutches and brakes have very similar operating characteristics and are studied together. Almost invariably, torque is transmitted via sliding friction between two mating surfaces.

Mechanical strength is not the main issue in the design of a clutch or brake. The main design considerations are:

1. In order to generate sufficient friction force to carry the operating force or torque, appropriate *actuating force* needs to be applied.
2. Because sliding friction is associated with heat generation, the *temperature rise* needs to be limited to temperatures that materials can tolerate and at which the associated *energy loss* will not be excessive.

George Westinghouse, Jr. (1846–1914) was born in Central Bridge, New York, the son of a man who made farm machinery in New York. Westinghouse had a creative mind, and his father's shop was just the place to try out new ideas. At age 19 he received his first patent: a design for a rotary engine. In 1866, perhaps the year that changed his life, George was riding a train that was suddenly brought to a halt to avoid colliding with a wrecked train on the rails ahead. Inspecting the site, he mused that there must be a safer way to stop a heavy train. Existing braking systems were inadequate. Working with compressed air—the idea used to power rock drills in tunneling—Westinghouse began to experiment with a new type of braking system for trains. At 22 years of age, he developed the air brake, a device that stopped trains using compressed air. During his career, he formed about 70 companies, first among them the Westinghouse Electric Co. With a total of 361 patents to his credit, the last being granted four years after his death, he was one of the most creative engineers of all times.

10.2.2. Friction Bands

Perhaps the simplest brake/clutch consists of a flexible band around a cylindrical drum (Figure 10.6, see page 674). The drum is fixed to the one rotating member while the band is fixed to the other (clutch) or to the stationary frame (brake). One end is fixed to a pivot point A and the other to a mechanical or hydraulic forcing device. During operation, the two ends of the band are stressed by forces P and $P + dP$ at the two ends, normal force dN and frictional force $f dN$ from the drum, where f is the coefficient of friction. The friction forces between drum and band have total moment about the center of rotation equal to the shaft torque. Along the contact zone the band tension P is variable, function of the angle θ.

An elementary part of the band corresponding to an angle $d\theta$ will be acted upon by forces P and $P + dP$ at the two ends, normal force dN and friction force $f dN$ from the drum. Summing up forces in the radial direction, since $\sin dx = dx$:

$$(P + dP) \sin \frac{d\theta}{2} + P \sin \frac{d\theta}{2} - dN = 0 \qquad (10.8)$$

$$dN - P d\theta \qquad (10.8a)$$

In the tangential direction, force equilibrium leads to:

$$(P + dP) \cos \frac{d\theta}{2} - P \cos \frac{d\theta}{2} - fdN = 0 \qquad (10.8b)$$

$$dP \pm fdN = 0 \qquad (10.8c)$$

(upper sign for clockwise, lower sign for counterclockwise rotation). Therefore, eliminating dN between (10.8a) and (10.8c):

$$\frac{dP}{P} = \pm fd\theta \qquad (10.9)$$

Assuming constant coefficient of friction and integrating along the contact arc θ:

$$\int_{P_1}^{P_2} \frac{dP}{P} = \pm f \int_0^\theta d\theta, \quad \frac{P_1}{P_2} = e^{\pm f\theta} \qquad (10.10)$$

This relation was obtained by Euler. The friction torque will be:

$$T = (P_1 - P_2)R = P_2(e^{\pm f\theta} - 1)R \qquad (10.11)$$

Therefore, for given torque T, the required actuating force will be:

$$P_2 = \frac{T}{(e^{f\theta} - 1)R} \qquad (10.12)$$

For pressure p between band and drum at angle θ:

$$dN = pbRd\theta \qquad (10.13)$$

Substituting dN from Equation (10.8a):

$$p = \frac{P}{bR} \qquad (10.14)$$

and the maximum value is for $\theta = \theta_0$, $P = P_1$:

$$p_{max} = \frac{P_1}{bR} \qquad (10.15)$$

$$p_{max} = \frac{T}{bR^2} \frac{e^{\pm f\theta_0}}{e^{\pm f\theta_0} - 1} \qquad (10.16)$$

Design Procedure 10.2: Design of a Friction Coupling or Brake

Input: Maximum operational torque, material properties, sliding speed, expected life. **Output:** Dimensions of friction surfaces, actuation force.

Step 1: Sketch the two parts and place on the sketch loads and motion data.

Step 2: Select a friction material, hard surface material and find the coefficient of friction f and allowable bearing pressure p_a.

Step 3: From free-body analysis, the friction coefficient and the operation torque find the actuation forces (for example, Equations (10.10) and (10.11)).

Step 4: Find the normal force on the friction surface.

Step 5: Find the dimensions of the friction surface to obtain maximum bearing pressure equal to the allowable bearing pressure pa, for example, Equation (10.16).

Design is based on the required transmitted torque and the allowable maximum pressure of the lining material. Control parameters are the arc θ_0, radius R, width b, while the friction coefficient f depends on the selected materials and environment.

Example 10.3 The band brake for a construction lifting crane is hand-operated by a lever arm as shown in Figure 10.5a. The operator force is 300 N and the lever has $l_2 = 100$ mm, $l_1 = 400$ mm. The band is perpendicular to the lever at the position shown and the contact arc is $\theta = 120°$. The diameter of the drum is 250 mm, lined with asbestos friction material. Determine: (a) the braking moment for both directions of rotation; (b) the required steel bandwidth.

Solution From Table 10.3, for asbestos lining on steel, the coefficient of friction is 0.3 and the allowable pressure is 0.3 N/mm². For *clockwise* rotation, the force P_2 is, from the lever arm equilibrium:

$$P_2 = \frac{Fl_1}{l_2} = \frac{300 \times 400}{100} = 1200 \text{ N}$$

Equation (10.32) yields the braking torque:

$$T = P_2 R(e^{f\theta} - 1) = 1200 \times 0.125[e^{0.3 \times 2\pi/3} - 1] = 131 \text{ Nm}$$

For *counterclockwise* rotation:

$$T = P_2 R(e^{f\theta} - 1) = 1200 \times 0.125[e^{-0.3 \times 2\pi/3} - 1] = 70 \text{ Nm}$$

For *clockwise* rotation:

$$P_2 = 1200 \text{ N}$$

$$P_1 = P_2 e^{f\theta} = 1200 e^{0.3 \times 2\pi/3} = 2249 \text{ N}$$

The maximum pressure is given by Equation (10.14). The bandwidth is then:

$$b = \frac{P_1}{p_{max}R} = \frac{2249}{0.3} \times 125 = 60 \text{ mm}$$

For *counterclockwise* rotation:

$$b = \frac{P_2}{p_{max}R} = \frac{1200}{0.3} \times 125 = 32 \text{ mm}$$

because $P_1 < P_2$.

10.2.3. Friction Materials

For adequate clutch or brake performance, the following requirements are imposed on friction materials:

1. High coefficient of friction and its stability, i.e., small variation in the coefficient upon changes in velocity, pressure, and temperature
2. Wear resistance, including resistance to seizing and the tendency to grab
3. Heat resistance, including resistance to thermal fatigue, i.e., the capacity to withstand elevated temperatures without failure, retaining the necessary properties of the material for a prolonged period

Dry clutches and brakes usually have friction pairs consisting of steel or cast iron on a lining of some asbestos-based friction material.

Thermosetting (phenol–cresol–formaldehyde) resins, natural or synthetic rubbers, or both resin and rubber together are used as the bond for the friction material. Friction linings may be:

1. A fabric woven of asbestos and cotton fibers and metal wire, molded at high temperature.
2. Molded in press molds from short-fibered asbestos. Band stock for shoe-type clutches and brakes is rolled from the same materials.

Due to environmental concerns about asbestos, other lining materials are now in use.

At working temperatures up to 300°C, for asbestos friction materials, the allowable pressure is up to 0.6–1.0 N/mm². It is better practice to glue on friction linings than to rivet them. Gluing allows wear to a greater depth and increases the effective area. This doubles the service life.

Heavy-duty applications require high surface strength and high heat conduction coefficient for pad linings. In such cases, sintered metals can be used, copper and /or iron particles with friction modifiers, manufactured by fusion at high pressure and temperature.

In cluthes manufactured in large lots, it proves expedient and environmentally sound to use disks with *cermet* (metal–ceramic) linings made by sintering. Up-to-date cermet friction materials contain the following components: copper or iron, which constitutes the base and provides for heat disposal, graphite and lead, which serve as the lubricant, and asbestos and quartz sand, which increase the friction. Cermet friction materials have higher wear resistance and thermal conductivity than ordinary asbestos materials. Their properties change less when they are heated. The cermet materials are applied on the steel disks or shoes and are joined by sintering

TABLE 10.3. Properties of Lining Materials for Clutches[a]

Mating Materials		Friction Coefficient			Maximum Temperature, °C		Maximum pressure p_a[b] (kPa)
Material	Against	Dry	Light lubrication	Full lubrication	Intermittent	Continuous	
Woven fabric	Steel, cast	0.65–0.4		0.2–0.1	150	100	50–1200
Woven asbestos	iron, cast	0.5–0.3	0.35–0.15	0.2–0.1	300	200	50–2000
Molded asbestos	steel	0.4–0.2		0.15–0.1	500	250	50–8000
Sintered metal		0.65–0.45		0.15–0.1	300	250	50–8000
Molded paper	Cast iron	0.35–0.2	0.2	0.1	160		
Molded paper	Steel	0.55–0.25	0.25–0.15		160		
Leather	Metal	0.6–0.3	0.35	0.15			
Cork	Metal	0.35	0.3				
Bronze	Bronze, cast iron	—	0.15	0.1–0.02			
Cast iron	Cast iron		0.1–0.05				
Steel band	Cast iron		0.1–0.15	0.1–0.03			
Hard steel	Hard steel						700–3000
Ceramic/metal[c]	Cast iron	0.1–0.4	0.05–0.1	0.05–0.1	540		1000
Ceramic/metal[c]	Hard steel	0.1–0.3	0.05–0.1	0.05–0.1	540		2100

[a]From Niemann 1954.

[b]Higher values for smooth operation, lower ones for rough operation. Higher values result in shorter life.

[c]From Shigley and Mischke 1989.

672

under pressure. First the steel surfaces are copper-plated. The thickness of a disk with a cermet coating is 30–40% less than that of the glued-on friction lining. This means that the axial overall size of the clutch is correspondingly reduced.

The friction elements of clutches that are to run in oil are made of steel with subsequent hardening or sulphocyaniding, as well as of combinations of materials: hardened steel on a friction plastic, or hardened steel on a cermet in the form of a lining (for large-lot production).

10.2.4. Disk Clutches and Brakes

Such elements are extensively employed. They have working surfaces of the simplest shape. Even with a small overall size, they can have a large friction surface. The force required for engagement is not very large because it consecutively applies pressure on the friction surfaces and is not distributed among them.

Clutches may be single-disk (with two friction surfaces) or multiple-disk. The disk is keyed to one shaft and is compressed for engagement between two flanges keyed to the other shaft. These clutches are widely used in automobiles for which clear-cut disengagement is a desirable feature (Figure 10.6).

A multiple-disk clutch consists of a housing, a sleeve, a set of disks linked to the housing, a set of disks linked to the sleeve, and a pressure mechanism.

The valuable features of multiple-disk clutches are:

1. High load capacity in conjunction with small overall size, especially diametral size, which is of prime importance for high-speed drives
2. Smooth engagement
3. The possibility of varying the number of disks, which is an essential advantage in limiting the number of components of different sizes in standard clutches

One drawback of multiple-disk clutches is poor disengagement, especially for clutches mounted on vertical shafts.

Hardened steel or metal–ceramic disks are commonly used for clutches operating with lubrication. In dry clutches, one set of disks (usually the outside disks) has friction linings.

Disks are linked to the housing and sleeve by means of parallel-sided or involute spline joints. It should be noted that the joint between the disks and the sleeve, and even the housing, is subject to very high stresses. Frequently, the disks wear grooves on the sides of the splines, which prevent smooth engagement by interfering with axial motion of the disks. For this reason, the surfaces of the splines must be properly hardened.

As a rule, not more than eight to twelve disks are used. The use of more disks leads to nonuniform pressure between them, owing to friction on the splines and the poor disengagement.

The following clearances are to be provided between the disks when the clutch is released:

1. For metal disks: from 0.5–1 mm for single- and two-disk clutches and 0.2–0.5 mm for multiple-disk clutches

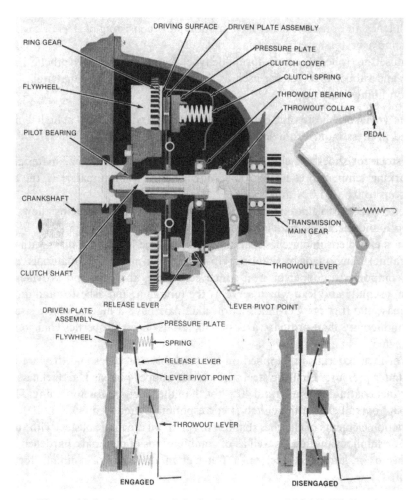

Figure 10.6. Automotive disk clutch (courtesy of Mobil Oil Corp.)

2. For nonmetallic disks: from 0.8–1.5 mm for single- and two-disk clutches and 0.5–1 mm for multiple-disk clutches

Disk disengagement can be improved by providing spreading springs on one-half of the disks (for instance, the inside disks). These are not flat in the free state (such disks are tapered or wavy and are also called sine disks). Such disks behave like springs themselves.

It is more difficult to ensure proper disk disengagement on vertical shafts. Disks are sometimes made with different outside diameters for such installations, providing steps on which they rest. When such a clutch is disengaged, each disk drops by gravity onto its own step.

Multiple-disk normally disengaged clutches, used in speed gearboxes, are frequently of twin design. This enables two gears to be alternatively engaged to the shaft to change speeds and to reverse rotation.

The torque that can be transmitted by a disk clutch can be found by writing the equation of equilibrium of torques on the disk (Figure 10.7). The normal pressure

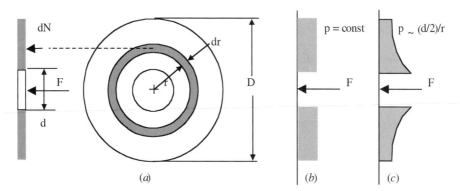

Figure 10.7. (*a*) Pressure distribution on a clutch disk; (*b*) uniform pressure; (*c*) uniform wear.

on a differential ring of inner radius r and thickness dr results in a differential normal force $dN = p(2\pi rdr)$. The friction force $2fp\pi rdr$ gives torque about the axis of rotation $dT = 2fp\pi r^2dr$, and the total friction can be found by integration over the surface of the disk:

$$T = 2\int_{d/2}^{D/2} fp\pi r^2dr \qquad (10.16a)$$

To perform the integration, we need the distibution of the pressure over the surface of the clutch. This is known, in general, but some assumptions can be made depending on the design of the plate that pressures the disk on the friction plate. Moreover, the friction torque must be higher than the operation torque by a safety factor $N_F > 1$, the *margin of frictional engagement* or *factor of safety against slipping,* which is very similar to the safety factor used for strength equations. For clutches, the design values of N_F need to be between 1.25 and 1.5.

Uniform pressure. If the pressure plate is flexible enough to allow a uniform transmission of the force, the pressure is nearly uniform and we can set $p = p_a$, the maximum value that the material can sustain without excessive wear. Then:

$$T = \frac{fp_a\pi n_s(D^3 - d^3)}{12N_F} \qquad (10.16b)$$

where n_s is the number of friction surfaces of the clutch, in the case of a multiple surface clutch.

The actuating force F is simply the product of pressure and area or the integral:

$$F = \int_{d/2}^{D/2} pdA = \int_{d/2}^{D/2} p2\pi rdr = p_a2\pi \int_{d/2}^{D/2} rdr = \frac{p_a2\pi(D^2 - d^2)}{8} \qquad (10.16c)$$

$$F = \frac{p_a\pi(D^2 - d^2)}{4} \qquad (10.16d)$$

where D and d are outside and inside diameters of the annular friction surfaces (see Figure 10.7), the ratio d/D usually ranges from 0.5–0.7, f is the coefficient of friction, p_a is the allowable pressure on disk lining.

Equations (10.16b) and (10.16d) are the design equations for clutches of uniform pressure.

Uniform wear. If the pressure plate is rigid enough, the thickness of the disk will be constant and thus there will be uniform wear, the pressure is not uniform. Experiments show that the pressure has the maximum value p_a at the inner circle $d/2$. Elsewhere it is to a good approximation:

$$p = \frac{p_a(d/2)}{r} \tag{10.16e}$$

Equation (10.16a) then yields:

$$T = \frac{n_s l}{N_F} \int_{d/2}^{D/2} f p_a d\pi r dr = \frac{f p_a \pi (D^2 - d^2)}{8 N_F} \tag{10.17}$$

The actuating force F is simply the product of pressure and area or the integral:

$$F = \int_{d/2}^{D/2} p dA = \int_{d/2}^{D/2} p 2\pi r dr = p_a d\pi \int_{d/2}^{D/2} dr = \frac{p_a \pi d(D - d)}{2} \tag{10.18}$$

$$F = \frac{p_a \pi (D - d)}{2} \tag{10.19}$$

Equations (10.17) and (10.19) are the design equations for clutches of uniform wear.

An especially critical mechanism of friction clutches is the control device. Lever-cam pressure mechanisms are quite extensively used for controlling clutches for transmitting low and medium torques if automatic controls are not required.

Motion is transmitted from the manual control lever to a collar that is shifted along the axis of the shaft. Most pressure mechanisms have operating levers. When the collar is shifted, its beveled surface engages one end of the operating levers, whose other end applies pressure to the friction components of the clutch (Figure 10.6).

Pressure mechanisms must have a considerable mechanical advantage. This is obtained by properly selecting the lengths of the arms of the operating levers, the angle of the bevel on the collar, and so on.

Design Procedure 10.3: Design of a Disk Clutch

Input: Maximum operational torque, material properties, sliding speed, expected life. **Output:** Dimensions of friction surfaces, actuation force.

Step 1: Sketch the two parts and place on the sketch loads and motion data.

Step 2: Select a friction material, hard surface material and find the coefficient of friction f and allowable bearing pressure p_a. Select number of friction surfaces n_s.

Step 3: Use Equations (10.16*b*) and (10.16*d*) to find the unknown diameters *d* and *D*, in the case of a uniform-pressure design.

Step 4: Use Equations (10.17) and (10.19) to find the unknown diameters *d* and *D*, in the case of a uniform-wear design.

Alternatively, if the inner diameter d is known and the number of friction surfaces is to be determined:

Step 3: Use Equations (10.16*b*) and (10.16*d*) to find the unknown outer diameter *D* and the number of friction surfaces n_s, in the case of a uniform-pressure design.

Step 4: Use Equations (10.17) and (10.19) to find the unknown outer diameter *D* and the number of friction surfaces n_s, in the case of a uniform-wear design.

Example 10.4 A single-disk clutch for an automobile transmits maximum torque at 4100 rpm, where it delivers 92 kW power. The two friction surfaces are lined with asbestos friction material rubbing against the cast iron flywheel. The external diameter of the disk lining is 300 mm and the internal diameter is 100 mm. The friction coefficient is $f = 0.3$ and maximum allowable pressure is 0.3 N/mm². Assuming uniform wear, determine:

1. The factor of safety against clutch slipping
2. The required actuating force

Solution The maximum torque is:

$$T_{max} = 9550 \frac{P}{n} = 9550 \frac{92}{4100} = 214 \text{ Nm}$$

The area (*A*) of the lining is:

$$A = \frac{\pi(d_2^2 - d_1^2)}{4} = \frac{\pi(300^2 - 200^2)}{4} = 39,270 \text{ mm}^2$$

The allowable actuating force is:

$$F = Ap_{max} = 39,250 \times 0.3 = 11,781 \text{ N}$$

The maximum capacity of transmitting torque is, from Equation (10.17):

$$T_{max} = \frac{fp_a \pi(D^2 - d^2)}{8N_F}$$

Thus, the margin of safety against slipping is:

$$N_F = \frac{fp_a \pi(D^2 - d^2)}{8T_{max}} = 0.3 \times 0.3 \times \pi(300^2 - 100^2)/(8 \times 214) = 13.21$$

10.2.5. Conical Clutches and Brakes

One of the members of an ordinary cone clutch has an internal conical working surface and the other member has a mating external cone. The clutch is engaged and disengaged by axially shifting one of the members. The conical friction surfaces enable considerable normal pressure and friction forces to be produced with a relatively small engaging force (principle of a wedge mechanism).

To avoid self-engagement and to facilitate disengagement, the cone angle (angle between an element of the conical surface and the shaft axis) is taken from greater than the angle of static friction. For metallic friction surfaces, the cone angle is taken from 8–10° or more, while for asbestos-based linings it is taken from 12–15° and more.

Advantages of cone clutches are their ready release and comparatively simple design. But they have essential shortcomings as well: their considerable radial overall size and the strict requirements regarding the coaxiality of the shafts being joined. For these reasons, cone clutches, previously extensively employed in the engineering industries, have only restricted application today. The torque that can be transmitted by a cone clutch with a mean radius R_m of the friction surfaces with width b is:

$$T = \frac{2}{N_f} \pi R_m^2 b p_a f \tag{10.20}$$

Hence, the width b of the friction surface for a selected value of R is:

$$b = \frac{N_F T}{2\pi R_m^2 p_a f} \tag{10.21}$$

Usually $\psi = b/R_m = 0.3$–0.5. Another procedure in design is first to assign the value of ψ. Then:

$$R_n = \left[\frac{N_F T}{2\pi \psi p_a f}\right]^{1/3} \tag{10.22}$$

The force required to engage the clutch is:

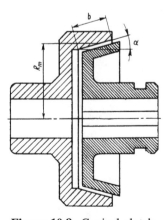

Figure 10.8. Conical clutch.

$$F = \frac{N_F T \sin a}{R_m f} \tag{10.23}$$

where α is the angle between an element of the cone surface and the shaft axis (cone angle).

In deriving this last equation, we assume that the force of friction on the working surfaces (in accordance with the direction of the sliding velocity) acts in the peripheral direction and does not impede clutch engagement.

Example 10.5 For safety reasons, to guard against control malfunction, a typical motor drive for an overhead crane consists of a conical rotor pressed axially against a conical brake by a helical spring. When electrical voltage is applied, the rotor moves to the left, the brake is released, and rotation starts. The motor is designed for a maximum torque of 400 Nm. Design the conical brake if the friction surfaces are asbestos lining against cast iron. The inner diameter of the cone must be at least 100 mm, the cone angle must be $a = 10°$, $f = 0.3$, $p_{max} = 0.3$ N/mm^2, and the safety factor is $N_F = 2$ against slipping.

Solution If b is the unknown width of the friction surface, the outer diameter d_2 is (if $d_1 = 150$ mm, the inner diameter):

$$d_2 = d_1 + 2b \tan \alpha$$

and the mean radius is:

$$R_m = \frac{d_1 + d_2}{4} = \frac{d_1 + b \tan \alpha}{2}$$

Equation (10.20) yields:

$$N_s T = 2\pi R_m^2 b p_{max} f = 2\pi \left(\frac{d_1 + b \tan \alpha}{2} \right)^2 b p_{max}$$

$$(100 + b \tan 10°)^2 b = \frac{N_s T}{2\pi p_{max} f} = 2 \times \frac{400,000}{2\pi \times 0.3 \times 0.3} = 1.415 \times 10^6$$

Iterating on $b = 1.41 \times 10^6 / (100 + b \tan 10)^2$ or using Solver of EXCEL, we find $\underline{b = 101.45}$ mm. Then:

$$R_m = \frac{(d_1 + b \tan \alpha)}{2} = 100 + 101.45 \times \tan 10 = 117.8 \text{ mm}$$

The required axial force that the helical springs must provide will be:

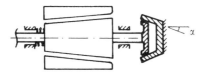

Figure E10.5

$$F = \frac{TNs}{R_m} \tan \alpha = \frac{400{,}000 \times 2 \times \tan 10°}{117.8} = 1195 \text{ N}$$

10.2.6. Radial Air Clutches

In these clutches, friction is developed between the shoes of a rubber tire, secured to one clutch member, and the cylindrical rim of the second member (drum). To engage the clutch, air under pressure is admitted into the tire, which expands so that the shoes are pressed uniformly against the drum.

The tire shown in Figure 10.9 transmits a peripheral. It consists of: (a) an elastic rubber inner tube that holds the air, (b) multiple-ply load-carrying lining of tough rubberized fabric (cord), and (c) an outer tread of rubber. The tire is manufactured by hot vulcanization and constitutes an integral unit.

The shoes are secured to the tire by plain pins that pass through holes in the tread and are kept from falling out by a wire threaded through holes in the pins. The tire is heat-insulated from the shoes with a lining that also protects it against the products of wear.

The shoes are coated by a friction lining held by glue. This lining is usually made of an asbestos fabric band impregnated with phenolic resin.

The merits of radial tire-type air clutches are:

1. Convenient controls.
2. The possibility of regulating the limiting torque and the rate of engagement so that the clutch can be a safety device.
3. Compensation for axial, radial, and angular displacements of the shafts being jointed without subjecting them to appreciable radial or axial loads; in practice, radial displacements up to 2 or 3 mm can be observed, and even considerably larger axial displacements.
4. Self-compensation for wear so that periodic adjustment is unnecessary.
5. Noise attenuation, shock cushioning, and damping of torsional vibrations.

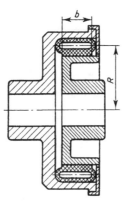

Figure 10.9. Radial pneumatic clutch, outwards expanding (from Decker 1973).

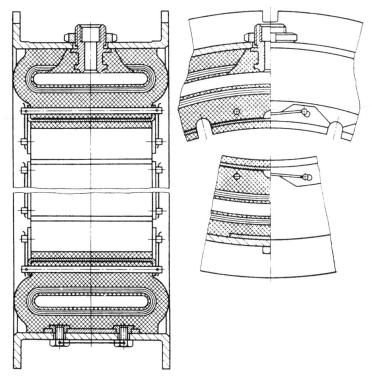

Figure 10.10. Radial pneumatic clutch, inwards expanding (from Decker 1973).

The drawbacks of these clutches include high cost of the tire, aging of the rubber, and the liability to damage by oil, alkalis, or acids getting on the rubber.

These clutches operate well in the temperature range from -20 to $+50°C$.

Tire-type air clutches are employed mainly in heavy engineering: in oil-well winches, in drives from a marine engine to the propeller, in excavators and other similar machinery.

10.2.7. Cylindrical Clutches and Brakes

Such elements consist of a drum and one or more friction pads pressed against the inner or outer surface of the drum by a variety of force mechanisms. On the basis of friction interface on the inner or outer surface, they are classified as inner or outer clutches or brakes. On the basis of forcing mechanism they are classified as:

1. *Centrifugal clutches*: The friction pads are on the driving shaft, and when the speed exceeds some value, the centrifugal force is enough to provide the normal force required to drive the drum and the driven shaft.
2. *Pneumatic clutches:* The friction material is glued on the outer (or inner) surface of a rubber tire that, when inflated, creates enough normal pressure between the mating surfaces to drive the drum.

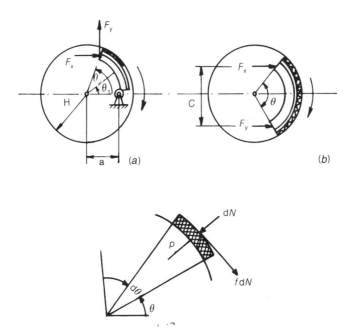

Figure 10.11. Force transmission in drum clutches/brakes.

3. *Mechanical clutches*: A mechanism provides the force on the friction pad necessary to drive the drum.
4. *Hydraulic clutches*: The force is supplied by one or two hydraulic actuators.

For a given torque and maximum permissible pressure of the friction material, the dimensions and the actuating force need to be calculated.

Except for pneumatic clutches, the pressure distribution between the rubbing surfaces is not exactly uniform. In the case of the pivoted pads, the pressure depends on the distance from the pivot point, and the utilization of the friction material is poor. For this reason, the tendency is to use two forces upon each pad to ensure almost uniform pressure between the mating surfaces. The length of the pads is chosen as a fraction of the circle periphery for the same reason. Assuming constant pressure $p_{\max}$, constant coefficient of friction f, radius of drum R, width of pads b, transmitted torque T, the total actuating pad force will be (Figure 10.11), for each of the n pads:

$$F = \frac{T}{(fRn)} \tag{10.24}$$

The resulting pressure will be:

$$p = \frac{F}{(R\varphi b)} = p_{\max} \tag{10.25}$$

For a pivoted pad (Figure 10.11*a*), a realistic assumption for the pressure distribution is that it is proportional at each point to the distance of the point from the line connecting the center of the drum to the pivot. That is:

$$\frac{p}{\sin \theta} = \frac{p_a}{\sin \theta_a} \tag{10.26}$$

where θ_a is the angle of maximum pressure. It equals $\pi/2$ if $\theta_2 > \pi/2$ and θ_2 when $\theta_2 < \pi/2$. The maximum pressure p_{max} obviously is assumed to be equal to the maximum allowable pressure of the friction material. Then:

$$p(\theta) = \frac{p_a \sin \theta}{\sin \theta_a} \tag{10.27}$$

$$dN = pbR \, d\theta = \frac{p_a bR}{\sin \theta_a} \sin \theta \, d\theta \tag{10.28}$$

The friction forces fdN give moment about the pivot point:

$$M_f = \int_{\theta_1}^{\theta_2} fdN(R - a \cos \theta)$$

$$= \frac{fp_a bR}{\sin \theta_a} \int_{\theta_1}^{\theta_2} \sin \theta \, (R - a \cos \theta)d\theta$$

$$= \frac{fp_a bR[a(\cos 2\theta_2 - \cos 2\theta_1)/4 - R(\cos \theta_2 - \cos \theta_1)]}{\sin \theta_a} \tag{10.29}$$

The normal forces dN give moment about the pivot point, similarly:

$$M_n = \int_{\theta_1}^{\theta_2} dNa \sin \theta$$

$$= \frac{p_a bRa}{\sin \theta_a} \int_{\theta_1}^{\theta_2} \sin^2 \theta d\theta \tag{10.30}$$

$$= \frac{p_a bRa}{4 \sin \theta_a} [2(\theta_2 - \theta_1) - \sin 2\theta_2 + \sin 2\theta_1]$$

The actuating force must supply the balance moment. This yields:

$$F = \frac{M_n \pm M_f}{c} \tag{10.31}$$

The $\pm$ sign stands for counterclockwise rotation of the drum and the pad from the pivot on in the direction of rotation. For a two-pad design, Equation (10.31) needs to be applied for each pad separately. However, the force is usually applied by a hydraulic cylinder equally on both pads. In this case, each of the two pads does not produce the same torque.

The proper choice of the design parameters might be $M_n = M_f$ and then $F = 0$. This means that the coupling is self-actuating and does not depend upon the magnitude of the actuating force. This feature, however, is used only for safety devices because such coupling is rather sudden and uncontrollable and is usually accompanied by high impact forces.

The coupling torque per pad will be, accounting for the safety against slipping:

$$T = \frac{1}{N_F} \int_{\theta_1}^{\theta_2} f R dN = \frac{f p_a b R^2}{N_F \sin \theta_a} \int_{\theta_1}^{\theta_2} \sin \theta d\theta = \frac{f p_a b R^2}{\sin \theta_a} (\cos \theta_1 - \cos \theta_2) \quad (10.32)$$

Example 10.6   A drum brake is shown in Figure E10.6 with two centrally pivoted brake pads. The pad angle $\theta = \theta_2 - \theta_1 = \pi/2$, the diameter of the drum is $d = 220$ mm, and the lever geometry is determined by $l_2 = 200$, $l_1 = 200$, $e = 60$ mm, $L = 160$ mm. Asbestos friction material will be used in a cast iron drum, $f = 0.3$, $p_a = 0.3$ N/mm². Determine, for maximum braking torque 180 Nm and under the assumption of uniform pressure:

1. The required actuating force F
2. The width of the pads
3. The required actuating force and pad width if the pad pivots are fixed on the vertical levers

Solution

1. Because the pads do not have a fixed pivot, the pressure is nearly uniform and for force F_1 on the pivot, the friction force will be $f F_1$ and the torque transmitted will be:

$$T = f F_1 d, \quad 180 = 0.3 \times F_1 \times 0.220, \quad F_1 = 2727 \text{ N}$$

The actuating force will be, by equilibrium of moments about the pivot:

$$F = \frac{(l_1 + l_2) F_1}{l_2} = \frac{0.200 \times 2727}{0.200 + 0.200}, \quad \underline{F = 1363 \text{ N}}$$

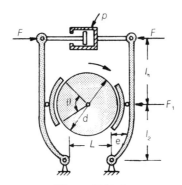

Figure E10.6

2. Assuming uniform pressure distribution:

$$p_a = \frac{F_1}{b\pi d/4}$$

or

$$b = \frac{4F_1}{p_a\pi d} = \frac{2727}{0.3 \times \pi \times 220}, \quad \underline{b = 13.1 \text{ mm}}$$

3. Now the pivot is at the lower point of the arm, at angle from the vertical:

$$\theta_0 = a \tan \frac{(L/2)}{l_1} = a \tan \frac{60}{200} = 0.29 \text{ rad or } 16.7°$$

From the geometry of Figure E10.6:

$$\theta_1 = \frac{\pi}{2} - \theta_0 - \frac{\pi}{4} = \frac{\pi}{4} - 0.29 = 0.495 \text{ rad}$$

and

$$\theta_2 = \theta_1 + \frac{\pi}{2} = 0.495 + \frac{\pi}{2} = 2.195 \text{ rad}, \quad \theta_\alpha = \theta_2 = 2.195 \text{ rad}$$

The results will be obtained using Equations (10.29)–(10.32). Table E10.6 was developed in EXCEL.

Design critique: We note that both the actuating force and the pad width are substantially larger in the fixed pad pivot design, due to the nonuniform pressure distribution.

10.2.8. Safety and Automatic Clutches

Self-acting (self-controlled) clutches release automatically under certain operating conditions. The operating conditions of certain classes of machinery, such as crushers, excavators, cultivators, and machines for the press-forming of metals, include systematic overloads. In machinery not subject to impacts and in the machining of homogeneous workpieces, such as machine tools, overloads may be caused by, for example, selection of excessively high machining variables (speeds, feeds, or depths of cut) or dulling or breakage of the cutting tools. Overloads occurring in the operation of mechanisms may be caused by, for example, lubrication failures, seizing, or jamming.

As to its action, an overload may build up gradually (for instance, in the dulling of a tool), or it may be of the impact type.

The function of a safety link can be performed not only by special safety clutches, by other elements of the drive that permit slipping upon overloads: friction clutches, hydraulic clutches, and hydraulic and pneumatic drives. In hydraulic drives, for instance, overloads are prevented by safety values. In pneumatic drives, any appreciable overload is impossible owing to the elasticity of the working medium—air.

TABLE E10.6.

Example 10.6						
Design of a pivoted drum brake/clutch with two symmetric pads						
DATA:						
Torque	180	Nm	**RESULTS:**		units	
Max pressure p_a	0.3	MPa	**Pad width** (eq. 10.32)	**40.1315**	**mm**	
Friction coefficient	0.3		Friction moment M_f (eq. 10.29)	7188999	Nmm	
Drum Diameter	220	mm	Pressure moment M_n (eq. 10.30)	386943.2	Nmm	
Entry angle θ_1	0.495	rad	Actuating force F_1 (eq. 10.31) 18939.85	Nmm		
Exit angle θ_2	2.295	rad	Actuating force F_2 (eq. 10.31)	-17005.1	N	
Max angle θ_a	2.295	rad	**Total actuating force**	**1934.72**	**N**	
Actuating force distance c	400	mm				
Pivot distance a	80	mm				

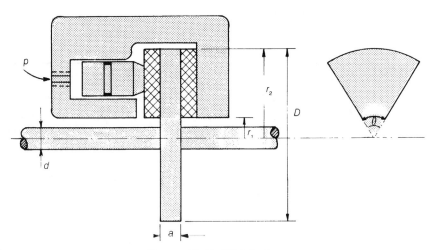

Figure 10.12. Disk brake.

10.2.9. CAD-Optimum Design of Clutches and Brakes

The general idea behind computer aided design and optimization of clutches and brakes is to achieve optimum performance with minimum dimensions and cost. The design objectives depend on the particular application, and no general formulation can be made at this point. By way of example, we will set out some aspects of this class of problems (Figure 10.12).

An automobile brake is to be designed of the disk type. The brake consists of a disk of diameter D and two friction pads between radii r_1 and $r_2 = D/2$ and angle θ. The car has design speed v and mass m per wheel. There is a linear relationship between car speed v and angular velocity ω of the shaft:

$$v = L\omega \tag{10.33}$$

where L is a constant with dimensions of length. The pad area is $\pi(r_2^2 - r_1^2)(\theta/2\pi)$, or:

$$A = \frac{\theta(r_2^2 - r_1^2)}{2} \tag{10.34}$$

The average pad pressure, equal to the maximum allowed pressure p_a, is:

$$p = p_a = \frac{P}{A} = \frac{2P}{\theta(r_2^2 - r_1^2)} \tag{10.35a}$$

$$P = \frac{p_a\theta(r_2^2 - r_1^2)}{2} \tag{10.35b}$$

where P is the braking force on the pad. The moment of the friction forces on both sides of the disk is approximately:

$$M = \frac{Pf(r_1 + r_2)}{2} = \frac{p_a f \theta(r_1 + r_2)(r_2^2 - r_1^2)}{4} \tag{10.36}$$

To find a moment of inertia J equivalent to the mass per wheel m, the kinetic energy is used:

$$\frac{mv^2}{2} = \frac{J\omega^2}{2} \text{ or } J = mL^2 \tag{10.37}$$

Newton's law gives:

$$M = J\dot{\omega} = \frac{mL^2\gamma}{L} = mL\gamma, \ \gamma = \frac{M}{mL} \tag{10.38}$$

where g is the required deceleration. The time to stop the car from a speed v is:

$$t = \frac{v}{\gamma} = \frac{umL}{M} = \left(\frac{umL}{4p_a f \theta}\right)\left[\frac{1}{(r_1 + r_2)(r_2^2 - r_1^2)}\right] \qquad (10.39)$$

The kinetic energy of the car $mv^2/2$ will be transformed to heat, which will increase the temperature of the disk by (Equation (8.16a) $\Delta T = Q/\rho V c_p$, or:

$$\Delta T = \frac{mv^2/2}{\rho(a\pi r_2^2/2)c_p} \qquad (10.40)$$

Equation (10.39) gives the objective function $t(r_1, r_2, a)$, which must be minimized under certain constraints:

1. The maximum disk temperature (Equation (10.40)), should not exceed the temperature allowed by the brake material T_{max}.
2. The radius r_1 has a minimum and r_2 has a maximum allowed by the space available.

Constraint (1) can be waived because it can be used for the determination of a after the radii have been optimized. In this case, the design variables are the two radii r_1 and r_2.

10.3. FRICTION BELTS

10.3.1. Overview

A belt drive consists of a driving and a driven pulley and a belt that is mounted around the pulleys, with a certain amount of preload. It transmits peripheral force from one pulley to the other.

Some arrangements of belt drives are shown in Figure 10.13. More than one driven pulley can be used, and also in limited applications the pulleys can be in different planes.

The usual form of belts is indicated in Figure 10.14. Flat belts have a narrow rectangular cross-section (Figure 10.14a). The V-belts have trapezoidal cross-section (Figure 10.14b) and timing belts or toothed belts providing transmission without slip (Figure 10.14d). Round belts are also used also, and they have circular cross-section.

In many applications, instead of one large belt being used, several belts of smaller thickness are utilized, to reduce bending stresses and fatigue loading, as will be shown later (Section 10.3.3).

Belts have certain advantages that make their use very wide:

1. They are capable of transmitting motion at considerable distances.
2. Their operation is smooth and they can absorb impact loads on one drive and not transmit them to the other drive.
3. They can slip, preventing overloading of motors during start-up of the machine with high inertial loads.

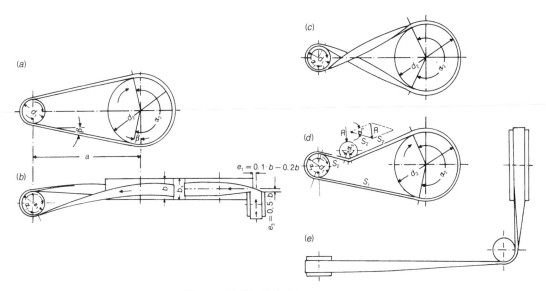

Figure 10.13. Belt drive geometry.

4. They operate at very high speeds.
5. Their cost is usually low.

Disadvantages of belt drives are:

1. Considerable overall size, much greater than other types of drives, such as gear drives.
2. Because of slipping, which is not constant, the relation of the two speeds of rotation of driven and driving pulley is not constant in time.
3. Frequent maintenance is required because the belt needs frequent tensioning and its service life is usually short.

The classical material for flat belts is leather, which comes in three types, standard, flexible, and very flexible. It has the highest coefficient of friction among all belt materials; in fact, the coefficient of friction increases for high peripheral velocities.

Natural fabrics are made with cotton, linen, silk, and animal hair. They can be manufactured to longer lengths and in closed loops and thus have more quiet operation. Different thicknesses can be obtained with multiple layers.

Balata belts have very good properties. They are made from the natural rubber of the tropical apple tree. They have substantially higher strength than leather belts.

Figure 10.14. Belt sections: (*a*) flat; (*b*) V-belts; (*c*) round; (*d*) toothed-timing belts.

Though sensitive to higher temperatures, they tolerate dust and humidity well. Balata rubber can be used with several plies of fabric or steel matrix.

Natural or synthetic rubber is widely used with fabric reinforcements. They tolerate higher temperatures, dust, and humidity well but have higher density and are not appropriate for high speeds.

Plastic materials, such as polyamide, nylon, and perlon, are used in combination. They are stiff and have little extension and creep, so they are manufactured in closed loop and do not need substantial adjustment in time. They have to be used with small thickness because of the high elastic modulus, but they can be used at very high speeds and they tolerate harsh environments well. Due to their high strength, they can be used for high power transmission.

10.3.2. Mechanics of Belt Operation

It is assumed that when the drive does not operate or is idling, no power is transmitted and there is an initial tension in the belt. Obviously the tension in each of the two branches of the belt around the pulley is the same, S_0 (Figure 10.15a). When the drive transmits power, the driving end of the belt is pulled with a greater force than the follower end, $S_1 > S_2$ (Figure 10.15b, d). These two forces are related with the Euler equation, developed in Section 10.2.2 for band brakes and clutches. However, in developing Equation (10.10), it was assumed that the band was stationary. The belt moves, and thus in the radial force equilibrium there is a d'Alembert (centrifugal) force dmu^2/R, where $dm = qd\theta$, the mass of the elementary part of the belt of angle $d\theta$, and q is the mass per unit length. Following the same procedure as in Section 10.2.2, we obtain:

Force equilibrium in the radial direction:

$$Pd\theta = dN + qv^2d\theta \tag{10.41}$$

Force equilibrium in the tangential direction:

$$dP = fdN \tag{10.42}$$

Eliminating dN and integrating:

$$(P - qv^2)d\theta = \frac{1}{f} dP \tag{10.43}$$

$$\int_0^\theta fd\theta = \int_{S_1}^{S_2} \frac{dP}{P - qv^2} \tag{10.44}$$

$$\frac{S_1 - qv^2}{S_2 - qv^2} = e^{f\theta} = m \tag{10.45}$$

$$S_1 - S_2 = \frac{m - 1}{m}(S_1 - qv^2) = U = \frac{H}{v} \tag{10.46}$$

Therefore, the transmitted power is:

$$H = \frac{m-1}{m} v(S_1 - qv^2) \qquad (10.47)$$

Equations (10.45) and (10.46) can be used to determine the forces of the two branches of the belt S_1 and S_2 if the coefficient of friction f, the belt arc θ, the power H, and the belt velocity v are known:

$$S_1 = \frac{mH/v + (m+1)qv^2}{m-1} \qquad (10.48)$$

$$S_2 = \frac{H/v + (m+1)qv^2}{(m-1)} \qquad (10.49)$$

Equation (10.45) was obtained on the basis of the derivation by Euler in 1775 and assumes that the belt is not extended under the influence of the force. This of course is not true, and therefore the above relation must be considered as approximate, and because the coefficient of friction f is measured experimentally on operating drives, the Euler equation is calibrated for practical applications and is the basic design equation for belt drives.

10.3.3. Belt Kinematics

In Figure 10.15a, a belt is shown around a pulley in idle operation. As mentioned above, both branches of the belt have the same tension. Let us draw lines perpendicular to the belt in equal distances, say at unit lengths. Their distances during rotation remain constant and equal to the unit length.

Suppose now that the belt transmits a certain power N. This requires that the tension in the two branches of the belt be different, their difference producing the power. Let these two forces be S_1 and S_2. It is apparent that $S_1 > S_0$ and $S_2 < S_0$. That means that branch 1 of the belt is elongated and branch 2 of the belt is contracted. Therefore, the lines previously traced will have greater distances in branch 1 and smaller distances in branch 2.

Let q be the mass per unit length of the belt. If we assume that the deformation of the belt does not change its volume, then, because the mass rate entering the pulley and the one leaving the pulley must be equal, the *principle of conservation of mass* requires that q must be constant, the belt mass per unit length. Let v be the velocity of the belt in each position. It is obvious now that in the loaded belt the mass per unit length in the two branches is different and is $q_2 > q_0 > q_1$. Therefore, it must be $v_1 > v_0 > v_2$.

If q_0 refers to belt with initial tension S_0 and q refers to an unloaded belt, and if e is the strain during the initial tension, then we expect that $S_1 > S_0$ and $S_2 < S_0$. Also as the strain increases, the mass per unit length decreases so that $q(1 + e) = q_0 = $ constant. Therefore, since the volume per unit time $qv = $ constant:

$$\frac{v}{1+e} = \text{constant} \qquad (10.50)$$

The transmitted force around the pulley also changes in the loaded belt from S_1 to

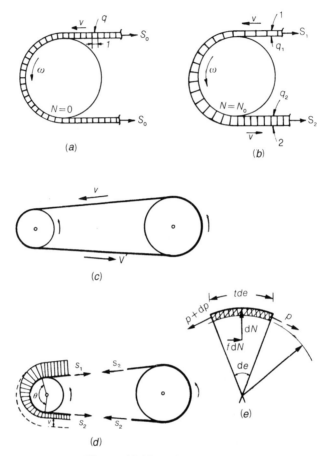

Figure 10.15. Belt kinematics.

S_2. Therefore, the strain e changes accordingly and so does the velocity according to Equation (10.50). This phenomenon is called *elastic creep*, and it is responsible for deviations from the theoretical Euler equation because this creep is associated with sliding of the belt around the pulley, making the coefficient of friction variable, as it depends on the friction velocity.

From the above discussion, it is apparent that the two branches of the belt have different strains e_1 and e_2. The difference is called the *belt slip:*

$$s = e_1 - e_2 \tag{10.51}$$

Because of Equation (10.50), we obtain:

$$\frac{v_1}{1 + e_1} = \frac{v_2}{1 + e_2} \tag{10.52a}$$

$$\frac{v_1}{v_2} \approx 1 + (e_1 - e_2) = 1 + s \tag{10.52b}$$

Returning to Figure 10.15b, we note that the periphery of the pulley has of course the same velocity everywhere. Because the velocity of the belt changes, there must be a point where the velocity of the belt and the velocity of the pulley coincide. This is a point A where the belt is leaving the pulley and there has the smaller pulling force. From point A to the driving of belts branch is a gradual slip. With this assumption, which is experimentally substantiated, if ω_1 and ω_2 are the angular velocities of the driving and the driven pulleys respectively and d_1 and d_2 are the respective diameters, we obtain:

$$\left.\begin{array}{l} v_1 = \dfrac{\omega_1 d_1}{2}, \quad v_2 = \dfrac{\omega_2 d_2}{2} \\[2mm] i = \dfrac{\omega_1}{\omega_2} = \dfrac{(1 + s)d_2}{d_1} \end{array}\right\} \tag{10.53}$$

Equation (10.53) gives the transmission ratio i, which is related not only to the diameters of the pulleys but also to the relative belt slip, which depends on the operating conditions and is not constant. This is a serious shortcoming of friction belts, and they cannot be used in applications where the two pulleys are required to be in constant speed relation with one another.

The transmission ratio for flat belts is up to 5, with a tension pulley it is up to 10, and with V-belt drives with transmission, a ratio up to 15 can be effectively used.

As mentioned above, the belt at idle operation has a preload force S_0 and assumes forces S_1 and S_2 during operation. The corresponding strains will be respectively e_0, e_1, and e_2. If the volume is constant, the contraction of one branch must equal the elongation of the other. If L is the length of each branch of the belt, it must be:

$$\left.\begin{array}{l} (e_1 - e_0)L = (e_0 - e_2)L \\[2mm] e_1 + e_2 = 2e_0 \end{array}\right\} \tag{10.54}$$

Because the strain $e = S/(EA)$, Equation (10.54) yields the Ponçelet equation:

$$S_1 + S_2 = 2S_0 \tag{10.55}$$

Using Equations (10.48) and (10.49) with (10.55), we obtain (since $U = H/v$):

$$S_0 = \frac{(1 + m)U + 2(m + 1)qv^2}{2(m - 1)} \tag{10.56}$$

If we neglect the centrifugal forces, the ratio

$$d = \frac{U}{2S_0} = \frac{m - 1}{m + 1} \tag{10.57}$$

is called the *pull ratio* and it should be, for efficient operation, for flat belts between 0.4 and 0.6 and for V-belts between 0.7 and 0.9.

10.3.4. Belt Stress Analysis

The pull ratio is an important factor for belt operation. The slip factor and the efficiency of the belt are functions of the pull factor, as is shown in Figure 10.16. Efficiencies η of belts are usually high, around 95%. As shown in Figure 10.16 for a typical belt, near the point where there is a rapid increase in the slip factor, it is approximately the maximum of the efficiency. Before that point, at low pull factor, the belt is heavy and many losses occur that reduce efficiency. After that optimum point, sliding friction increases rapidly and so do the losses, lowering the efficiency.

Experimental observations show that the optimum value of the pull factor is 0.6 for elastic and leather belts, 0.5 for cotton belts, 0.4 for wool belts, and 0.5 for plastic belts.

When the belt operates at high speeds, there are considerable inertial forces due to the circular motion (centrifugal forces). Assuming an elementary part of the belt (Figure 10.17), on which the tension force due to the centrifugal force is V while the elementary centrifugal force is dV, we observe that the centrifugal force is:

$$dV = \frac{v^2 dm}{r} = qv^2 d\theta \tag{10.58}$$

Equilibrium of the elementary part of the belt (Figure 10.17) requires that $dV = 2V \sin(d\theta/2) \approx V d\theta$. With equation (10.58):

$$V = qv^2 \tag{10.59}$$

The belt is stressed due to the following:

1. *Static stresses* due to the maximum tension S_1 over the cross-sectional area $A = bt$:

$$\sigma_1 = \frac{S_1}{A} = \frac{Um}{A(m-1)} \tag{10.60}$$

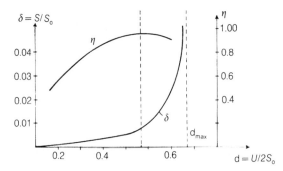

Figure 10.16. Performance of belts.

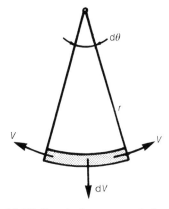

Figure 10.17. Inertia forces on a belt element.

2. *Tensile stresses* due to the centrifugal forces, if ρ is the material density:

$$\sigma_u = \frac{V}{A} = \frac{qv^2}{A} = v^2\rho \tag{10.61}$$

3. *Bending stresses* due to the curvature of the belt wrapped around the pulleys:

$$\sigma_b = \frac{Et}{d_{1,2}} \tag{10.62}$$

where there is a different bending stress at every pulley, if the diameters are different. The total maximum stress will be the sum of the above stresses (1–3), all tensile:

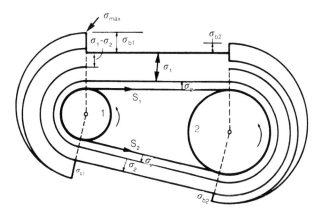

Figure 10.18. Stresses on belts.

$$\sigma_{\max} = \sigma_1 + \sigma_u + \sigma_b = \frac{Um}{A(m-1)} + v^2\rho + \frac{Et}{d_{1,2}} \tag{10.63}$$

All stresses except the bending stress are static stresses, while the bending stress is pulsating. Therefore, for fatigue analysis the mean and the range of stress are:

$$\sigma_m = \frac{Um}{A(m-1)} + v^2\rho + \frac{Et/d_{\min}}{2} \tag{10.64}$$

$$\sigma_r = \frac{Et/d_{\min}}{2} \tag{10.65}$$

The distribution of stresses around a belt is shown in Figure 10.18, where it is indicated that the maximum stress occurs at the beginning of the arc of the driving pulley, at the entering branch.

For all the above calculations, some geometric factors need to be known, such as the length of the belt, the arc of contact, and the distance of the pulleys a. If the two diameters and the pulley distance are known, the other geometric parameters can be computed as shown in Figure 10.19, following simple geometric rules.

It is apparent that the force- and power-transmitting capacity of the system depends on the arc of contact of the smaller pulley. For this reason, to increase capacity of belt drives, especially for high transmission ratios where the arc of contact in the smaller pulley is small, auxiliary pulleys are used to increase the arc of contact. Such a system can also provide for the initial tension. This and other systems for pretensioning belts are shown in Figure 10.20.

10.3.5. Flat Belt Design

For most belting applications, the process of designing is merely a process of engineering application, selecting belts according to manufacturing catalogs. Each manufacturer gives also a suggested design procedure for his belts. This of course is associated with the material properties that are given by such manufacturers, in a way that is related to a method of design. Many manufacturers, however, supply information about the basic material properties for their belts. In this case, it is suggested to proceed with a rational design method to check and substantiate the manufacturer's application rules.

In most design applications, the power transmitted H and the speeds of the two pulleys, n_1 and n_2, are given and in some cases the distance between the two pulleys. Design should begin from the determination of the diameter of the driving pulley which is the more critical. Some empirical relations, however, are based entirely on the idea of optimum selection of the driving pulley diameter. This is associated with the centrifugal stresses on the belt, which, if the diameter of the driving pulley is large enough, lead to such values that produce tension on the belt greater than the initial tension S_0, which means that there is no normal force between the belt and the pulley and the transmitted power is 0. To quantify this, let us return to Equation (10.47):

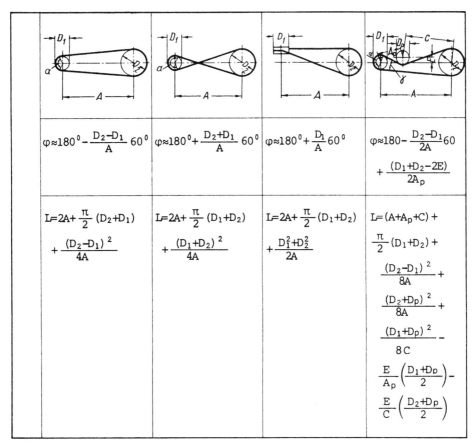

Figure 10.19. Geometric relations in belt drives.

$$H = \frac{m-1}{m} \, v(S_1 - qv^2) \qquad (10.66)$$

Equation (10.66) implies that the power the belt can transmit depends on the velocity. For zero velocity we have transmitted power zero and for some value of the velocity

$$v = \left(\frac{S_1}{q}\right)^{1/2} \qquad (10.67)$$

the quantity in parentheses in Equation (10.66) becomes zero and therefore the belt does not transmit power. In between those two points is a third-degree curve that has a maximum when:

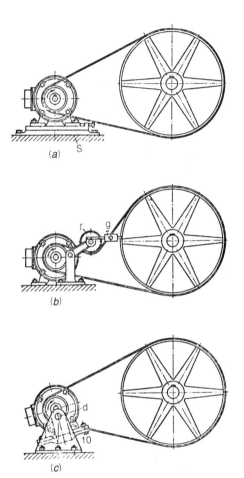

Figure 10.20. Belt tension methods.

$$\frac{d}{dv}[v(S_1 - qv^2)] = S_1 - 3qv^2 = 0$$

$$v_{opt}^2 = \frac{S_1}{3q} = \frac{S_y/N}{3A_b\rho} = v^2 = \left(\frac{\pi n d_1}{60}\right)^2 \tag{10.68}$$

where A_b is the cross-section area of the belt and ρ is the belt material density. This determines the diameter of the small pulley:

$$d_1 = \frac{60}{\pi n_1}\left(\frac{S_y/N}{3\rho}\right)^{1/2} \tag{10.69}$$

This equation can be used only qualitatively because it does not take into account many other factors influencing optimum design, such as the cost of the pulley itself. In general, it yields high values for the pulley diameters for the best use of the belt materials, but the pulley and the whole system become larger and heavier. Setting

$S_y/N = S_c$, the power transmitted is $H = FU = S_c bt(\pi nd/60)$ and can be introduced in Equation (10.69). Niemann (1965) suggests the following form of Equation (10.69):

$$d_1 = y_1 \sqrt{\left(\frac{d_1}{s}\right)_{\min}} \sqrt[3]{\frac{HC}{S_c n_1}} \qquad (10.70)$$

where d_1 is the diameter of the smaller pulley in m, H is the power transmitted in W, S_c is the allowable stress in the belt, n_1 in rpm, d_1/s is the minimum allowed diameter to avoid premature failure due to high bending stresses, which are pulsating and cause fatigue, and y_1 is a constant between 80 and 100 and takes the higher values for wider belts. Values of $(d_1/s)_{\min}$ are given in Table 10.4, on the basis of which the diameter of the smaller pulley from Equation (10.70) can be checked.

The determination of the diameter of the driven pulley is made with Equation (10.53) in the form:

$$d_2 = \frac{d_1 i}{1 + s} \approx 0.985 d_1 i \qquad (10.71)$$

At this point a thickness of the belt must be selected, based on available sizes or suggested diameter to belt thickness ratio, where they are available. Then the width of the belt can be computed from the relation

$$b = \frac{HC}{H_0} \qquad (10.72)$$

where C is the service factor, H is the power to be transmitted, and H_0 is the power capacity of the belt per unit width. This capacity is given in most manufacturers' catalogs and, if not available, can be computed from Equations (10.48) and (10.57), with $A = bt$, in the form

$$H_0 = \frac{H}{b} = \frac{(S_c - \rho v^2 - Et/d)vt(m - 1)}{m} \qquad (10.73)$$

In this relation, S_c is the service strength, much lower than the tensile strength obtained with static tests. This is because at a point far below the static strength, the creep increases rapidly and the belts lose their tension. Repeated retensioning can be done, but it requires continuous attention and service and the operation will be less reliable. For this reason, experimental values of the creep strength S_c are used, given in Table 10.4 for some belt materials, instead of the static strength S_n given also for reference purposes only.

A similar situation exists for the elasticity modulus E used for the computation of the bending stresses. Repeated bending results in local creep in the tensile stress area, which is apparent when we cut a belt and let it free on a plane surface. It will take a curved form owing to unequal creep along the thickness. For this reason, an apparent value of the elasticity modulus E is used that corresponds to correct bending stresses when Equation (10.69) is used. Such values are given also in Table 10.4. Therefore, wherever strength and elasticity modulus appear in the design equations, S_c and E_c from Table 10.4 must be used.

Again, H_0 is the power capacity of the belt per unit length and for service factor equal to 1. The service factor C depends on the particular application and can be broken down to five subfactors:

$$C = C_1 C_2 C_3 C_4 C_5 \tag{10.74}$$

Values of the five subfactors are given in Table 10.5. For V-belts, two more factors are used, C_6 and C_7, given also in Table 10.5.

Equations (1.72) and (1.73) are the basic design equations for flat belts. Manufacturers suggest different empirical methods based on extensive tables for H_0 for their particular types of belts, instead of using Equation (10.72). The designer needs to consult with those tables also because the strength values given in Table 10.4 are average values among different manufacturers. Moreover, because failure of the belt is gradual by structural deterioration, the strength values of Table 10.4 are only nominal. Exceeding the allowable stresses will simply result in shorter life of the belt. Up to twice the values given for strength in Table 10.4 can be used, provided that the calculated life of the belt is acceptable. Most designers prefer to do this because it leads to smaller belts and pulleys and smaller forces on shafts and bearings, while on the other hand belts have to be changed more frequently.

To find the life of the belt, we must now check the latter for fatigue. Fatigue tests on friction belts have not revealed specific endurance limits. The point where the deterioration of belt starts has great uncertainty, and it seems that after a number of reversals of the stress, deterioration starts independently of the magnitude of stress range. Therefore, the endurance limit must be considered as a statistical quantity with a great deviation. In that respect, service life calculation based on fatigue strength is based on an equation (using the expression for the fatigue curve):

$$\sigma_{max}^m N = \text{constant} \tag{10.75}$$

where N is the number of cycles and σ_{max} is the maximum stress. If the fatigue strength is assumed at a number of cycles N_n, usually taken to be 10^6, and this strength S_n, then:

$$\sigma_{max}^m N = S_n^m N_n \tag{10.76}$$

The number of cycles to failure of the belt is then:

$$N = N_n \left(\frac{S_n}{\sigma_{max}^m} \right) = \frac{3600 L_h z \upsilon}{L} \tag{10.77}$$

where L_h is the life in hours, z is the number of bends per one full cycle of the belt (equal to the number of pulleys), and υ/L is the number of full revolutions of the belt per second. On this basis, the service life of the belt is:

TABLE 10.4 Allowables for Flat-Belt Design[a]

Type	s, mm	u_{max}, m/s	E_b, MPa	S_c, MPa	$(s/d_1)_{max}$	T_{max}, °C	ρ, g/cm³	f (dry)
Leather	3–20	30	70	4	0.033	35	1.0	$0.22 + 0.012u$ (u, m/s)
Leather, flexible	3–20	40	60	4.5	0.04	35	0.95	
Leather, very flexible	3–20	50	50	5.5	0.05	45	0.9	
Rubber, balata, cotton	3–8	40	50	4.4	0.035	45	1.2	0.5
Rubber, cotton	3–7	40	50	4	0.033	70	1.25	0.5
Balata, cotton	3–8	40	50	4.4	0.04	40	1.25	0.5
Balata, steel	4, 5	40	30	5.2	0.05	40	1.25	0.5
Artificial silk	2–18	50	40	4.2	0.04	70	1.0	0.35
Artificial cotton	2–10	50	40	4	0.04	70	1.1	0.8
Cotton	2–12	50	40	3.7	0.05	70	1.3	.3
Camel hair	3–6	50	40	4	0.05	70	1.1	0.3
Silk, linen-silk	0.4–12	60	40	9	0.06	70	0.95	0.3
Nylon, Perlon	0.4–5	65	250	19	0.04	75	1.1	0.15
Composites, leather lining	1–4	80–100	250	19	0.01	60	1.15	$0.22 + 0.012u$
Composites, rubber lining	1–4	80–100	250	19	0.01	75	1.15	0.75

[a]Adopted from Niemann 1965, with modifications and updates.

TABLE 10.5. Design Factors for Belts

C_1, overload factor,   1.0–1.5
C_2, environment factor,   1.0–1.3
C_3, continuous operation factor, if fatigue test not performed,

Hours per Day	C_3
3–4	1.45
8–10	1.5
16–18	1.9
24	2.0

C_4, contact angle factor ($= 1$ if equation (10.73) used)

θ	80°	100°	140°	180°
C_4	1.5	1.35	1.1	1

C_5, tension factor,
    $C_5 = 1$, tension with bolts
        $= 1.2$, with belt shortening (flat belts)
        $= 0.8$, self-tensioned system
C_6, pulley selection factor (V-belts)
    $C_6 = 1$, if d_1 computed by equation (10.69)
        $= d_1/d_0$, d_0 the value from equation (10.69)
C_7 multiple belt factor (V-belts)
    $C_7 = 1$, 1 V-belt
        $= 1.25$, multiple V-belts
f, friction coefficient, 0.25 to 0.40

$$L_h = \frac{N_n(S_n/\sigma_{max})^m}{3600zv/L} \tag{10.78}$$

In the above equations, σ_{max} is supposed to be the range of the alternating stress. Most experiments of that type to determine the endurance limit are not made on a regular material testing machine because the belt cannot be tested very easily in compression. For that reason, belts are tested in an asymmetrical cycle, similar to real operation in service. Therefore, unless detailed values are given for the endurance limit under similar conditions as for metals, the range σ_{max} in Equation (10.78) must be taken equal to the maximum stress that appears on the belt, given by Equation (10.57). The values in Table 10.4 for endurance limits are given on that basis. Finally, the pull factor and the slip coefficient have to be computed and checked against experience.

10.3.6. Design of V-Belts

V-belts contact the pulley in a groove, as shown in Figure 10.21. The angle of the groove is usually 35–39°. Its purpose is to increase the normal force between belt and pulley by the wedge effect and thus increase the friction force and the trans-

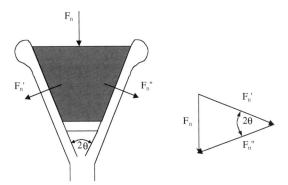

Figure 10.21. V-belt wedge effect.

mitted power. V-belts are highly standardized, and dimensions of standard belts are given in Table C.4 of Appendix C.

V-belts are relatively inexpensive because of standardization and mass production, easily replaceable, and relatively quiet. On the negative side, they have limited life, do not have exact transmission ratio because of slippage, and cannot be used at very high speeds. Moreover, they come only in standard lengths, and because they creep in operation, tension mechanisms must be provided.

V-belts can be designed exactly the same way as flat belts, except that the size of the belt is selected and the number of belts is computed. To this end, we first note that the total friction force is:

$$F = f(F_n' + F_n'') = \frac{f}{\sin\theta} F_n \tag{10.79}$$

which is equivalent to a flat belt with a friction coefficient $f' = f/\sin\theta$. The wedge angle 2θ is usually between 35–39°, thus approximately $f' \approx 3f$. From that point on, the calculations proceed as in the case of flat belts, with a modified coefficient of friction f'. Eventually, instead of Equation (10.72), a similar equation can be used giving the *number* of V-belts needed to transmit the given power:

$$z = \frac{H_p C}{H_0} \tag{10.80}$$

Again C is the service factor and H_0 is the power transmitted for the given conditions by one V-belt.

Design Procedure 10.4: Design of a Flat Belt Drive

Input: Maximum operational torque, material properties, rotating speeds, expected life. **Output:** Belt dimensions, preload.

Step 1: Sketch the drive and place on the sketch loads and motion data.

Step 2: Select a belt material and find its properties from Table 10.4.

Step 3: Use Equation (10.70) to find the smaller pulley diameter d_1.

Step 4: Use Equation (10.71) to find the diameter of pulley d_2.

Step 5: Find the peripheral velocity, (Equation (10.45)), and test against the maximum value in Table 10.4 for the material selected. If not satisfactory, change the material.

Step 6: Find the contact angle a from the appropriate formula in Table 10.15.

Step 7: Use Equation (10.50) to find the belt forces and the preload.

Step 8: Select a belt thickness and find the specific belt capacity H_0 using Equation (10.73) or nomograms from manufacturers catalogs.

Step 9: Find the service factor (Table 10.5).

Step 10: Use Equation (10.72) to find the width of the belt b.

Step 11: Find the maximum stress (Equation (10.57)).

Step 12: Find the service life (Equation (10.78)).

Step 13: Review the results. If unacceptable, change assumptions and repeat the design.

The unit power transmitted H_0 is usually given in manufacturers' catalogs or in the absence of this can be computed by way of Equation (10.57), in the form of Equation (10.81), because here the section area A is known, with the values of thickness t and width b given in Table C.4 of Appendix C.

$$H_0 = \frac{(S_c - \rho v^2 - Et/d)Av(m - 1)}{m} \qquad (10.81)$$

Conventionally, we use for the area of the cross-section $A = bt$ because the strength values given in Table 10.4 were tabulated with this understanding. For belts in Table C.4 that do not appear in Table 10.4, material properties of belts with similar area of cross-section may be used. If more than one V-belt is used, their rating has to be reduced (C_7 factor) because after a certain operation the tensions of the belts are not equal and therefore there is no equal distribution of the load.

Equations (10.80) and (10.81) are the basic design equations for V-belts. In addition, Equation (10.78) will yield the life of the belt, an important design consideration. Manufacturers suggest different empirical methods based on extensive tables for H_0 for their particular types of belts, instead of using Equation (10.81). The designer needs to consult with those tables also because the strength values given in Table 10.4 are average values among different manufacturers.

The programs FLATBELTS and V-BELTS for the preliminary design of flat and V-belts, respectively, are included in the MELAB 2.0 package.

Design Procedure 10.5: Design of a V-belt Drive

Input: Maximum operational torque, material properties, rotating speeds, expected life. **Output:** Belt dimensions, preload.

Step 1: Sketch the drive and place on the sketch loads and motion data.

Step 2: Select a belt size and find its properties from Table 10.5a.

Step 3: Use Equation (10.70) to find the smaller pulley diameter $d1$.

Step 4: Use Equation (10.71) to find the diameter of pulley $d2$.

Step 5: Find the peripheral velocity (Equation (10.45)) and test against the maximum value in Table 10.4a for the material selected. If not satisfactory, change the material.

Step 6: Find the contact angle a from the appropriate formula in Table 10.15 and the effective friction coefficient (Equation (10.79)).

Step 7: Use Equation 10.50 to find the belt forces and the preload.

Step 8: Find the specific belt capacity H_0 using Equation (10.81) or nomograms from manufacturers' catalogs.

Step 9: Find the service factor (Table 10.5).

Step 10: Use Equation (10.80) to find the number of belts z.

Step 11: Find the maximum stress (Equation (10.57)).

Step 12: Find the service life (Equation 10.78)).

Step 13: Review the results. If unacceptable, change assumptions and repeat the design.

Example 10.7 A 30 kW, 1450 rpm electric motor drives an ammonia compressor at 500 rpm. Design a flat belt for this transmission if the distance of the two pulleys is 1500 mm.

Solution The solution is shown in the spreadsheet shown in Table E10.7. A plastic belt 5 mm thick was initially selected.

Example 10.8 Design a V-belt drive system for the auxiliary power system of an automotive engine. The crankshaft pulley 1 drives the alternator 2, the water pump 3, and the air-conditioning compressor 4, rated at n_1 = 4000 rpm, n_2 = 5000 rpm, H_2 = 0.5 kW, n_3 = 3500 rpm, H_3 = 0.4 kW, n_4 = 4000 rpm, N_4 = 1 kW. The operation is 3 hours per day, the friction coefficient is f = 0.3, and the wedge angle is 40°.

Solution The apparent friction coefficient (Equation (10.79)) is:

$$f' = \frac{f}{\sin(40)} = 0.877$$

The contact angle for the driving pulley is θ = 75°, from Figure E10.8. The total power that the driving pulley transmits is H − 0.5 + 0.4 + 1 = 1.9 kW. The rest of the computations are organized in a spreadsheet shown in Table E10.8.

The result n = 4.5 (5 belts) is unacceptable due to space limitations. We note, however, that the service life of 163 million hours is excessive. Thus, we can use a higher value for S_c, as indicated in Table 10.4. Indeed, entering S_c = 4 N/mm^2, we obtain from the spreadsheet n = 1.85 and L_h = 163,000 hours. The design is then acceptable, with two V-belts type A.

CASE STUDY 10.1: The Roverson Crane Brake Failure[1]

On December 11, 1964, a model HO-8A, Serial Number 9HS 3512 crane, manufactured by the Hans Roverson Crane Company of Atlanta, Georgia, in 1958,

---

[1] Based on Case Study ECL 184A of the Engineering Case Library. Originally prepared by R. Ganeriwal and H. O. Fuchs of Stanford University under an NSF grant. By permission of the trustees of the Rose-Hulman Institute of Technology.

TABLE E10.7. Design of Flat Belts

DATA:	Source	Symbol	Value	Units	
Modulus of elasticity	Table 10.4	E_c	60	MPa	Service Factors:
Weight density	"	γ	11000	N/m³	$C_1 =$ 1.5
Allowable stress	"	S_c	6.1	MPa	$C_2 =$ 1
Fatigue strength	"	S_n	6	MPa	$C_3 =$ 1
Friction coefficient	"	f	0.3		$C_4 =$ 1.06
Max. peripheral velocity	"	v_max	60	m/sec	$C_5 =$ 1.2
Power	Problem	H	30	kW	
Shieve rotating speed	"	n_1	1450	rpm	$C\_fac =$ 1.908
Pulley rotating speed	"	n_2	500	rpm	
Center distance	"	a	1500	mm	
Selected belt thickness	Selection	t	5	mm	$=C\_1*C_{-2}*C_{-3}*C_{-4}*C_{-5}$
No of bends per turn	Problem	z	2		
Fatigue exponent	Table 10.4	m_exp	6		
RESULTS:					
Shieve diameter	Eq. 10.70	d_1	329.4	mm	Equation used
Pulley diameter	Eq. 10.71	d_2	941.02	mm	$= 1200*(H/n\_1)^\wedge(1/3)$
Peripheral velocity	Eq. 10.45	v	25.011	m/sec	$= 0.985*d\_1*(n\_1/n\_2)$
Contact angle	Geometry	$theta$	2.7309	rad	$= (2*PI()*n\_1/60)*(d\_1/2)/1000$
Pull factor	**Eq. 10.50**	**m**	**2.2689**		$= 2*ACOS((d\_2-d\_1)/(2*a))$
Peripheral force	Eq. 10.48	U	1199.4	N	$= EXP(f*theta)$
Belt tension	Eq. 10.50	S_2	945.28	N	$= H*1000/v$
	"	S_1	2144.7	N	$= U/(m-1)$
	"	S_0	1545.0	N	$= U*m/(m-1)$
					$= U*(m+1)/(2*(m-1))$
Pull ratio	Eq. 10.57	d	0.3881		$= (m-1)/(m+1)$
Specific Belt Capacity	Eq. 10.73	H_0	3876	W/m	$= (S\_c*10^\wedge6-:\gamma\gamma*v^\wedge2/9.81-E\_c*10^\wedge6*t/d\_1)*v*(t/1000)*(m-1)/m$
Belt Width	**Eq. 10.72**	**b**	**0.181**	**m**	$= H*1000*C\_fac/H\_0$
Belt length	Fig. 10.19	L	5057.9	mm	$= 2*a+PI()*(d\_1+d\_2)/2+(d\_2-d\_1)^\wedge2/(4*a)$
Maximum stress	Eq. 10.57	s_max	4.4879	MPa	$= (S\_c*10^\wedge6-:\gamma\gamma *v^\wedge2/9.81-E\_c*10^\wedge6*t/d\_1)/10^\wedge6$
Service life	**Eq. 10.78**	**L_h**	**160.38**	**Hours**	$=10^\wedge6*(S\_n/s\_max)^\wedge m\_exp/(3600*z*v/(L/1000))$

TABLE E10.8. Design of V-belts

DATA:	Source	Symbol	Value	Units		
					Service Factors:	
Modulus of elasticity	Table 10.4	E_c	12	MPa	$C\_1 =$	1.5
Weight density	"	g	11000	N/m³	$C\_2 =$	1
Allowable stress	"	S_c	2.5	MPa	$C\_3 =$	1.45
Fatigue strength	"	S_n	2	MPa	$C\_4 =$	1.06
Friction coefficient (app)	"	f	0.877		$C\_5 =$	1.2
Max. peripheral t velocity	"	v_max	90	m/sec		
Power	Problem	H	1.9	kW	C_6	1
Shieve (1) rotating speed	"	n_1	4000	rpm	C_7	1.25
					$C\_fac =$	3.45825
Pulley (2) rotating speed	"	n_2	5000	rpm		
Belt thickness	Table 10.4	t	8	mm		
Belt width	"	b	13			
No of bends per turn	Problem	z	4			
Fatigue exponent	Table 10.4	m_exp	8			
Belt length		1500	mm			
RESULTS:					**Equations used**	
Shieve diameter	Eq. 10.70	d_1	93.629445	mm	$= 1200*(H/n\_1)\wedge(1/3)$	
Pulley diameter	Eq. 10.71	d_2	73.7800027	mm	$= 0.985*d\_1*(n\_1/n\_2)$	
Peripheral velocity	**Eq. 10.45**	**v**	**19.609705**	m/sec	$= (2*PI()*n\_1/60)*(d\_1/2)/1000$	
Contact angle	Geometry	theta	1.30899694	rad	$= EXP(f*theta)$	75°
Pull factor	Eq. 10.50	m	3.15185231		$= H*1000/v$	
Peripheral force	Eq. 10.48	U	96.8907992	N	$= U/(m-1)$	
Belt tension	Eq. 10.50	S_2	45.0266957	N	$= U*m/(m-1)$	
	"	S_1	141.917495	N	$= U*(m+1)/(2*(m-1))$	
	"	S_0	93.4720953	N	$= (m-1)/(m+1)$	
Pull ratio	**Eq. 10.57**	**d**	**0.5182873**		$= (S\_c*10\wedge6-g*v\wedge2/9.81-E\_c*10\wedge6*t/d\_1)*v*(t/1000)*(m-1)/m$	
Specific Belt Capacity	Eq. 10.73	H_0	1452.91805	W		
No of belts required	**Eq. 10.80**	**n**	**4.52239892**		$= H*1000*C\_fac/N\_0$	
Maximum stress	Eq. 10.57	s_max	1.04349438	MPa	$= (S\_c*10\wedge6-g*v\wedge2/9.81-E\_c*10\wedge6*t/d\_1)/10\wedge6$	
Service life	**Eq. 10.78**	**L_h**	**1.63E+08**	Hours	$= 10\wedge6*(S\_n/s\_max)\wedge m\_exp/ (3600*z*v/(L/1000))$	

707

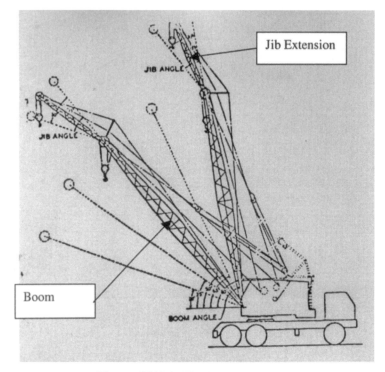

Figure CS10.1. The Roverson crane.

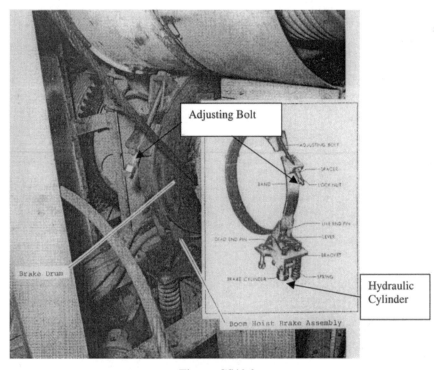

Figure CS10.2

was delivering buckets filled with concrete at a construction site of the bridge at Poway Road and Highway 395 in San Diego, California. Suddenly, the concrete-filled bucket fell and injured Mr. Arnold Ware, a construction worker.

Investigation

The crane was rated as having a 40 ton lifting capacity at 12 ft radius. Its delivery system consists of 90 ft main boom and a 30 ft boom extension (Figure CS10.1). Inspection of the crane after the accident revealed that:

1. The accident was caused by a failure of the boom brake that secures the boom in its desired location and does not allow rotation about its pivot. Moreover, the brake is of the band-brake type and the slack of the band can be adjusted by an adjusting screw that was found broken (Figure CS10.2). Therefore, the band was open and the brake could not stop the boom from falling.
2. There was a ratchet system to secure the boom in place (as required by California law), thus providing additional safety, but the operator did not engage it.
3. At the time of the accident, the length of the boom + jib assembly was 90 + 30 = 120 ft instead of the maximum 100 ft suggested by the manufacturer.
4. The spring that operates the band was found enhanced by another spring inserted in it.
5. The band adjusting bolt failed due to fatigue, as revealed by breach marks observed on the fractured surface.

Based on the above results of the inspection, points 3 and 4 suggest that the crane was regularly overloaded. This is underlined by point 5, the fatigue failure of the bolt. Thus, one should not look at the loading conditions at the time of the accident but rather at the loading patterns.

The band is spring-operated, as shown in the insert of Figure CS10.2. The spring keeps the band under tension, providing constant braking. When the operator wants to change the angle of the boom, the hydraulic cylinder is engaged, which overcomes the spring force and relieves the pressure from the brake. The preload of the spring, and thus the tension on the brake band, is determined by the adjusting bolt. Adding another spring meant that the preload of the spring, and thus the bolt load and the braking torque, were increased by more than 100% because the added spring had to have a smaller helix diameter than the original one. Moreover, the operation of the hydraulic cylinder was intermittent, thus the load on the bolt was pulsating. This explains the bolt and brake failure.   □

REFERENCES

Decker, K.-H. 1973. *Maschinenelemente,* 6th ed. Munich: C. Hanser Verlag.

Engel, L., and H. Klingele. 1981. *An Atlas of Metal Damage.* Munich: C. Hanser Verlag.

Niemann, G. 1954. In *Hutte des Ingenieurs Taschenbuch,* vol. 1. Berlin: Wilhem von Ernst & Sohn.

———. 1965. *Maschinenelemente.* Berlin: Springer-Verlag.

Shigley, J. E. 1972. *Mechanical Engineering Design,* 2nd ed. New York: McGraw-Hill.

Shigley, J. E., and C. R. Mischke. 1989. *Mechanical Engineering Design.* New York: Mc-Graw-Hill.

Spiegel, M. R. 1968. *Mathematical Handbook of Formulas and Tables.* Schaum's Outline Series in Mathematics. New York: McGraw-Hill.

ADDITIONAL READING

Dimarogonas, A. D., and D. Michalopoulos. "A Compilation of Heat Distribution Parameters at Sliding Contacts." *Tribology* (1981): 225.

Halling, J. *Introduction to Tribology.* London: Wykeham, 1976.

Kragelskii, I. V. *Friction and Wear.* London: Butterworths, 1965.

Machine Design 53:6 (1981). Mechanical Drives Reference Issue.

Oliver, L. R., C. O. Johnson, W. F. Breig. "V-belt Life Prediction and Power Rating." *Trans. ASME J. Eng. for Industry* (1976): 340.

Ummen, A., ed. *Keilriemen.* Essen: Verlag Ernst Heyer, 1965.

PROBLEMS[2]

10.1. [C] For the disk of Example 10.1, find the average temperature in the disk for a short application of the brake to bring the car from a 60 MPH speed to complete stop. The mass of the car is 1200 kg and the heat convection over the surface of the disk can be neglected. All heat can be assumed to enter the disk because of the low thermal conductivity of the lining material.

10.2. [C] The drum of Example 10.6 is shown in Figure P10.2. Determine the drum temperature, assumed uniform, if the brake is applied for 15 rev. Neglect heat losses of the drum and assume all heat enters the drum.

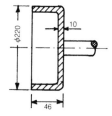

Figure P10.2

10.3. [C] A cam-and-follower mechanism consists of a nearly circular cam 1 of diameter d_1 =120 mm and a follower disk 2 of diameter d_2 = 40 mm, pressed against the cam by a force F = 100 N by a lever and a helical spring. The width of the disks is 15 mm and the coefficient of rolling friction was found by measurement to be f = 0.05. Assuming convection heat-

---

[2][C] = certification, [D] = design, [N] = numerical, [T] = theoretical problem.

transfer coefficient $h = 50 \text{ W/m}^2 \text{ °C } h$, determine the interface temperature of the rolling surfaces if the cam rotates at 3000 rpm.

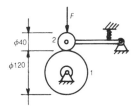

Figure P10.3

10.4. [C] In a friction welding operation of two circular carbon steel shafts of diameter $d = 100$ mm, one piece is stationary while the other revolves with a constant speed n. At the same time, an axial force F is applied. If the coefficient of friction is $f = 0.35$, determine the required speed of rotation to reach a temperature of 1300°C required. The axial force is $F = 200$ kN and the surface heat transfer coefficient for the stationary shaft is 30 W/m^2 °C h and for the rotating one 6 W/m^2 °C h. Assume infinite length.

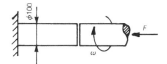

Figure P10.4

10.5. [C] Solve Problem 10.4 for the friction welding of two carbon steel disks, one circular, the other annular with a slight taper of angle $a = 5°$. The axial force is 10 kN and the heat transfer coefficient is as in Problem 10.4.

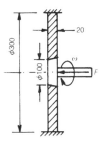

Figure P10.5

10.6. [C] A steel shaft of diameter $d = 40$ mm rotates at a constant speed of 125 rpm loaded by a 1500 N force. It is supported by a Teflon bearing of width 30 mm. Determine after 20 hours of operation the bearing wear and the vertical position of the center of the shaft in respect to the initial one.

Figure P10.6

10.7. [C] An indexing mechanism consists of a toothed wheel of diameter $d =$ 60 mm of hard steel and a follower of 60/40 brass pressed against the wheel by a force $F = 180$ N. The sliding section of the follower is 20×20 mm and the operation is without lubrication. If the limit of wear for the follower is 1 mm, determine the follower life.

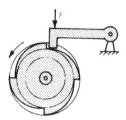

Figure P10.7

10.8. [C] A disk brake such as the one in Example 10.1 was tested for wear. The asbestos pads had friction area 2000 mm² and were pressed against the rotating disk for 10 sec every minute. The speed of rotation was 4000 rpm and the braking force was 1000 N. After 100 hours of testing, the wear of the pads was 3 mm. The yield strength of the asbestos lining was measured $S_y = 2$ N/mm² by a static test. Determine the coefficient K for the asbestos lining.

10.9. [C] A revolving pole of an advertisement sign that weighs 5 kN is supported by two 50 mm diameter polyethylene bearings (such as the one in Figure P10.6) with yield strength $S_y = 40$ MPa. If the pole rotates at 20 rpm, determine the life of the bearing if the linear wear allowed, by geometry, is 10 mm.

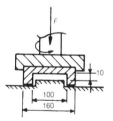

Figure P10.9

10.10. [C] A differential expansion detector for a steam turbine rotor consists of a brass 70/30 slider and a helical spring pre-stressed to a force $F = 150$ N applied on the slider. When the rotor expands to the right, it contacts the slider and the 150 N force is then applied between rotor and slider. The sliding surface is 10×10 mm². During an overhaul, the slider was inspected and 8 mm wear was measured. Determine the total duration of the differential expansions of the rotor, revolving at 3600 rpm. If, between overhauls, the turbine made 150 start-ups, determine the average duration of the differential expansion, given that such expansion is expected during every start-up.

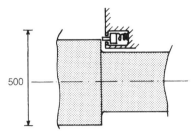

Figure P10.10

10.11. [D] On a differential band brake, the force applied is $P = 300$ N and the braking torque is $T = 1000$ Nm. The drum diameter is 280 mm and the contact angle is 230°. One end of the band is hinged at a distance of 60 mm from the lever pivot, as shown. Determine the proper position for the hinge of the other end to support the braking torque. Both ends of the band are perpendicular to the lever. The lining is asbestos friction material and the drum is made of cast iron. Then design the band of carbon steel SAE 1025 with a safety factor $N = 3$.

Figure P10.11

10.12. [D] The band brake shown is activated by force $F = 250$ N. The lining is asbestos friction material and the drum is cast iron. Determine the braking torque and the lining width for operation of the brake with both directions of rotation.

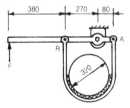

Figure P10.12

10.13. [D] A double-drum hoisting machine has a band brake with common band, as shown. If the braking torque required for each drum is 800 Nm, determine the required activating force P and the width of the asbestos lining. The drums are made of cast iron.

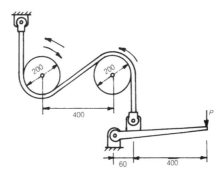

Figure P10.13

10.14. [C] For the brake of Problem 10.12, determine the braking torque for both directions of rotation if the left end of the band is hinged with the other end at point *B*.

10.15. [C] For the system of the Problem 10.13, determine the activating force if the direction of the drum on the left is reversed. Determine also the tension of each part of the band.

10.16. [D] A single-disk automotive clutch is designed for 80 kW rated power at 3600 rpm. Both sides of the disk have lining material with minimum diameter 140 mm. The lining is asbestos friction material. For a factor of safety against slipping $N = 3$, determine the external lining diameter and the actuating force.

10.17. [D] An engine for a bulldozer is rated at 550 hp at 3000 rpm. It uses a multiple-disk clutch with asbestos lining. The friction surface is between diameters $d_1 = 200$ mm and $d_2 = 300$ mm. If the factor of safety against slipping is 2.5, determine the required number of clutch disks.

10.18. [D] An automobile weighs 1100 kg and the four brakes must bring it to a complete stop from a speed of 90 mph at a 60 m distance. The braking force is distributed 60% in the front and 40% in the rear wheels and is assumed constant during braking. Design a disk brake for the front wheels with asbestos lining with one disk and two pads. The distance of the area center of pads from the axis of rotation should be approximately 120 mm, while the wheel diameter is 500 mm.

10.19. [D] In Problem 10.18, design the rear wheel brakes as drum brakes with two 120° pads each and internal drum diameter 360 mm. The pads are forced at both ends by hydraulic cylinders. (Shigley 1972.)

10.20. [D] Solve Problem 10.19 for pad activation with one cylinder and the other pad end hinged to the opposite one. (Shigley 1972.)

10.21. [D] The inflated rubber clutch shown in Figure 10.7 is rated at 40 kW, 3600 rpm. The coefficient of friction between rubber and steel drum is $f = 0.35$ and the shear strength of the rubber is 6N/mm². The available air pressure is 1 N/mm². Determine the main dimensions of the clutch, trying to keep the ratio b/R between 0.3 and 0.8.

10.22. [D] A centrifugal clutch consists of four pads free to move outwards against the inner cylindrical surface of the drum. Coupling is achieved by frictional forces between pads and drum. The pads are forced to rotate with the inner rotor. Such a clutch was rated at 10 hp, 1000 rpm. The pads had asbestos lining. The inner diameter of the drum was 300 mm. Determine the size and mass of the pads. Find pad size.

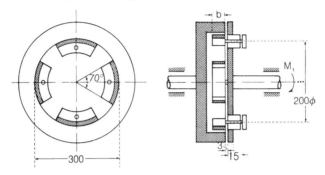

Figure P10.22

10.23. [C] An automobile has a 1200 kg mass with a center of mass as shown, with $l_1 = 1000$, $l_2 = 1200$, $h = 900$ mm. In braking, without slipping, frictional forces develop, equal to deceleration times mass. Determine the distribution of the load G to the front and rear wheels:

1. At constant speed
2. At deceleration 0.6 g

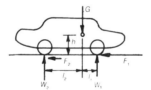

Figure P10.23

10.24–10.29. [D] On a hoisting crane, a load of 1 ton is lowered with a cable on a 200 mm diameter drum. For the brakes shown in Figures P10.24–P10.30, design the brake so that the load will come to a complete stop from a 1 m/s speed in 1 sec.

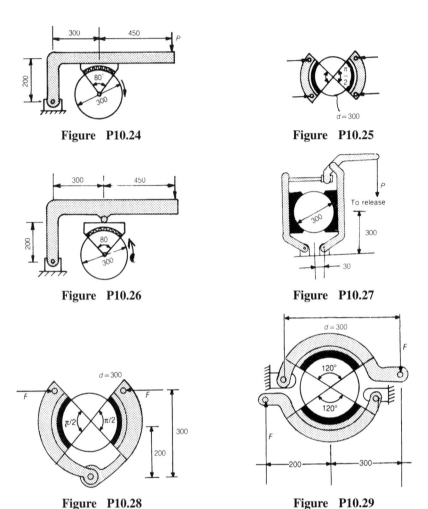

Figure P10.24   Figure P10.25

Figure P10.26   Figure P10.27

Figure P10.28   Figure P10.29

10.30. [C] The main shaft of a milling machine is powered by an electric motor at $N = 20$ kW at 1750 rpm with a 160 mm pulley, while the driven pulley has $d_2 = 300$ mm. The distance of the two shafts is 1400 mm. The selected leather belt is 3 mm thick and 160 mm wide. Check the belt for strength and determine the fatigue life.

10.31. [D] In Problem 10.30, design a proper rubber belt and determine the fatigue life.

10.32. [D] A 10 hp, 750 rpm motor drives a mixing machine at 150 rpm. A plastic flat belt is used. Determine:

1. The belt thickness and width and the pulley diameters
2. The service life
3. The proper tightening weight

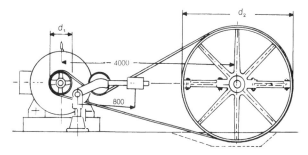

Figure P10.32

10.33. [C] A 10 kW motor drives a press with 350 rpm, while the diameter of the motor pulley and the tightening pulley is 125 mm and of the driven pulley is 1000 mm. The tightening weight $G = 60$ N. A rubber belt is selected of 4 mm thickness and 140 mm width. Check the strength of the belt and determine the fatigue life and the forces on the shaft owing to the belt tension.

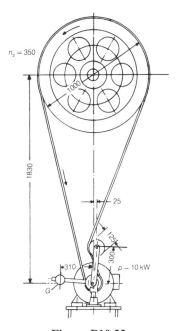

Figure P10.33

10.34. [D] In Problem 10.33, design a plastic belt and determine the service life.

10.35. [D] A paper machine is driven by an 8 kW, 1700 rpm motor at equal speed. Design the proper V-belting and determine the service life.

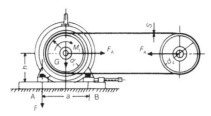

Figure P10.35

10.36. [D] An air compressor is driven by an electric motor of 3 kW, 2800 rpm to a speed of 710 rpm. Design the proper V-belting and determine the service life and the forces on the shafts.

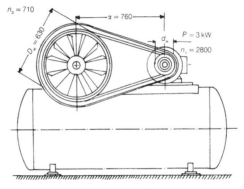

Figure P10.36

10.37. [D] The crankshaft of an automobile engine powers the water pump and ventilator of 6 hp and the electric generator of 0.3 hp. The crankshaft speed is 1000 rpm. Determine the V-belts required.

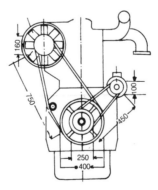

Figure P10.37

CHAPTER 11

LUBRICATION AND BEARING DESIGN

SS *Imperator*, 1913, 52,000 tons. The need for very high thrust for the propulsion of ships prompted the early studies in thrust bearing design.

OVERVIEW

In Chapters 8 and 9, we developed design methods based mostly on material strength and flexibility, respectively. The reason was that the respective mechanisms of failure were material failure and large displacements. There are other mechanisms of failure in machines, among them sliding friction and wear of machine parts. In Chapter 10, we studied design methods based on failures due to dry friction, with applications to clutches and brakes.

In this chapter, we shall study machine element design methods based on the prevention of failures related to hydrostatic and hydrodynamic lubrication, the use of a flowing fluid to prevent rubbing surfaces from touching one another during sliding friction.

11.1. BEARINGS AND LUBRICATION

11.1.1. Classification

Machines almost invariably have moving components that are loaded by forces. These forces need to be transmitted to the machine base, and this will generate friction and wear at the points of contact of the stationary with the moving components. Sometimes, as in the cases discussed in Chapter 10, friction is a desired feature. In general, however, friction is associated with energy loss to heat and wear of the sliding surfaces. Since sliding is a situation that we cannot avoid altogether, techniques have been developed to reduce the sliding friction through antifriction elements termed *bearings*. Figure 11.1 shows one scheme of classification.

Bearings may be classified on the basis of the principle employed. In *sliding bearings,* the two sliding surfaces are separated only by a layer of a fluid or solid that reduces friction and wear. *Rolling* (or *antifriction*) *bearings* employ the principle of rolling by using rolling elements between the sliding surfaces to reduce friction and wear (see Chapter 12).

Bearings are also classified by the direction of the applied force in relation to the axis of rotation. *Journal bearings* support loads perpendicular to the axis of rotation (lateral). *Thrust bearings* support loads parallel to the axis of rotation (axial).

The use of a liquid or solid substance of low resistance to shear is termed *lubrication*. According to the lubrication principle used, bearings may be classified as *hydrodynamic, hydrostatic,* or using *boundary* or *thin film lubrication.*

11.1.2. Bearing Materials

A bearing material must support the load imposed upon it while the loaded surfaces have relative motion. The coefficient of friction between moving and stationary parts must be as low as possible. This can be achieved by a film of lubricant between shaft and journal, but this is not always possible. When a continuous film of lubricant does not exist, a number of factors assume importance. Some metals have a natural *lubricity* or tendency to slide readily over each other, and this is generally more marked when dissimilar surfaces rub on each other. Others weld together more readily, a property that leads to scuffing, pick-up, and seizure. Many surfaces are able to absorb lubricants because of their chemical nature or the porosity of physical constitution of their surfaces. Fatty oils are more readily absorbed by surfaces than mineral oils and also have far better surface-wetting properties than mineral oils. Even the adventitious presence of microscopic amounts of water or mild oxidation products can have a similar effect to that of a lubricant.

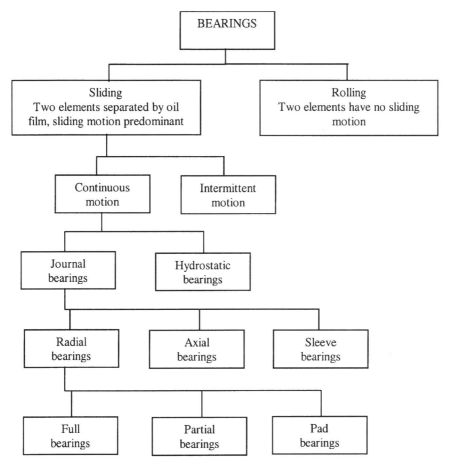

Figure 11.1. Classification of bearings.

Other considerations are that the alloy should have high thermal conductivity so that the heat generated by friction is readily conducted away; that it should be durable, that is, it should have a comparatively long life in service; that it should be easily replaced; and that it should have appreciable resistance to corrosion. Bearing metals that are to be cast should flow freely when molten and have minimum contraction on cooling.

11.1.2.1. Copper Alloys. Phosphor-bronze is a popular bearing metal. The phosphorus is added to make the alloy more molten for casting purposes, to give a sounder casting by thus removing the oxygen, and to form some phosphide of copper the presence of which improves wear-resisting properties of the alloy.

The so-called aluminum-bronzes fall into two groups. Up to 7.5% aluminum gives *alpha bronze,* while from 8–14% aluminum gives an alloy with a duplex structure.

Bronzes are used for average bearing pressures up to 30 MPa.

11.1.2.2. Lead. Lead has a low coefficient of friction, provides a soft "bedding" surface, and yields sufficiently to take up small misalignment. It will not dissolve in copper and hence does not form a solid solution but disperses as very small droplets throughout the metal.

Lead can be added to the normal bronzes and brasses to improve machineability, or in larger amounts to provide a sound bearing material. It can also be added to copper in proportions of up to 30% to give "leaded bronzes." This material is used in copper–lead big-end bearings for aircraft and other heavy duty internal combustion engines.

11.1.2.3. White Metal Bearing Alloys—Babbitts. Popularly known as the white metals are alloys of lead, tin, antimony, and a little copper. They are divided into two groups: those in which the major constituent is tin (referred to as tin-based) and those in which lead is the principal constituent (lead-based). All are relatively soft and are used as bearing linings, that is, as surface coatings of harder materials forming the structural component of the step or "brass."

Tin-based alloys are often called babbitt metals after their inventor, but Isaac Babbitt's original metal consisted of a mixture of 24 lb tin, 8 lb antimony, 4 lb copper, to every pound of which 2 lb of tin was added. This gives a percentage composition of tin 88.9%, antimony 7.4%, copper 3.7%. In addition to these constituents, present-day tin-based alloys may contain some lead, nickel, or cadmium. Small amounts of copper combining with tin tend to form a fibrous network that prevents segregation of the hard particles that may form, thus giving a more uniform structure. Lead tends to increase resistance to deformation, antimony to increase hardness, and cadmium to increase hardness and tensile strength.

Lead-based alloys, in general, are less expensive than tin-based alloys, but they will not stand up in performance. Antimony increases the tensile strength of the alloy, but in excess of 15% it can cause brittleness. Tin increases hardness and tensile strength.

The thickness of the babbitt layer is determined from the maximum allowable bearing wear during running-in and normal operation. It is usually 1% of the bearing diameter. In large diameters, the babbitt is cast in place with the liner and then machined. For mass production and smaller diameters, the antifriction material is bonded on strip stock and pressed to the final shape. In this case, the strip stock is 1.5–2.5 mm thick and the antifriction layer is 0.2 to 0.3 mm thick.

Babbitts are used for pressures up to 20 MPa and temperatures up to 110°C.

11.1.2.4. Cadmium-Based Bearing Alloys. These are heavy duty bearing alloys that are much superior in mechanical strength to tin-based white metals. Nickel, copper, and silver may be present in small proportions. Their drawback is susceptibility to corrosion.

11.1.2.5. Sintered Bearings. Porous bearings, mainly bushes, can be produced by the methods of powder metallurgy, and having a porous structure, they can be impregnated with oil. Such sintered bearings generally contain, in powdered form, the same materials as ordinary bearings with the additional of powdered graphite of the order of $4\frac{1}{2}$–$5\frac{1}{2}\%$.

11.1.2.6. "Dry" Bearings. Ordinary sleeve-type bearings require a continuous supply of lubricant. A possible alternative is the oil-impregnated porous bearing. Such a bearing contains a limited amount of lubricant, and its life may be short if it is subjected to heat or the action of solvents.

Where the complete absence of any lubricant is highly desirable or it is necessary to eliminate periodical servicing, "dry" bearings may be used. These are made of a material that inherently possesses the property of being able to slide smoothly over other materials with minimum frictional resistance. Such a material is polytetrafluorethylene, generally known as PTFE. It is, however, mechanically weak and dimensionally unstable and has low thermal conductivity. A very good bearing material can be made by impregnating sintered bronze with PTFE and adding a small amount of lead. Alternatively, PTFE may be used as a very thin shell bonded to a steel backing.

Other unusual bearing materials are carbon, which has a rather specialized application in the case of clutch thrust washers, and nylon, which is meeting with some success when used in inaccessible positions. Nylon is tough, hard-wearing, and easy to mold and machine, and any lubrication difficulties can be overcome by the use of molybdenum bisulphide powder.

11.2. THEORY OF HYDRODYNAMIC LUBRICATION

11.2.1. Fluid Viscosity

The viscosity of a liquid can be simply defined as its resistance to flow and is a measure of the friction between the liquid molecules when moving past each other. In his *Law of viscous flow,* Newton first stated the relation between the mechanical forces involved as follows: "The internal friction (i.e., viscosity) of a fluid is constant with respect to the rate of shear."

Absolute or dynamic viscosity can be defined in terms of the simple model shown in Figure 11.2, with two parallel flat plates, distance h apart, the upper one moving and the lower one stationary, separated by a film of oil. In order to move the upper plate, of area A, at a constant velocity U across the surface of the oil and cause adjacent layers to flow past each other, a tangential force F must be applied. Since oil will adhere to the two surfaces, the lowest layer of molecules

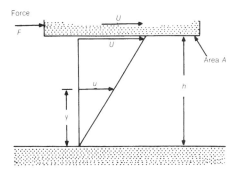

Figure 11.2. Lubricated plate in parallel motion.

will remain stationary, the uppermost layer will move with the velocity of the upper plate, and each intermediate layer will move with a velocity directly proportional to its distance y from the stationary plate.

The shear stress on the oil causing relative movement of the layers is equal to F/A. The rate of shear R of a particular layer, sometimes called the *velocity gradient,* is defined as the ratio of its velocity u to its perpendicular distance from the stationary surface y, and is constant for each layer, i.e., $R = u/y = U/h$. Newton deduced that the force F required to maintain a constant velocity U of the upper plate was proportional to the area and to the rate of shear U/h, or:

$$F = \frac{\mu A U}{h} \tag{11.1}$$

where μ is the coefficient of viscosity or absolute viscosity.

Absolute viscosity is thus defined by:

$$\mu = \frac{\text{Shear stress}}{\text{Rate of stress}} = \frac{F/A}{U/h} \tag{11.2}$$

Since the dimensions of shear stress are $[MLT^{-2} L^2]$ and of rate of shear $[T^{-1}]$, it can be seen that the units of absolute viscosity are mass divided by length times time, i.e., $[M/LT]$, or force seconds divided by length squared.

Two units of absolute viscosity have been used, the poise and the reyn, the former based on the centimeter-gram-second (cgs) system and the latter on the inch-pound-second system of units. Both fundamental units of viscosity are too large for practical problems or presentation of data, and they are therefore subdivided into the centipoise (1 cP = 0.01 poise) and microreyn (1 $\mu R = 10^{-6}$ reyn). In the SI system, the unit of absolute viscosity is Nsec/m^2 or Pa $\cdot$ s.

For conversion of absolute viscosity units:

$$1 \text{ reyn} = \frac{1 \text{ lb fs}}{\text{in.}^2} = \frac{454 \times 981}{2.54^2} \frac{\text{dyn sec}}{\text{cm}^2} = 6.903 \times 10^4 \text{ P} = 6.903 \times 10^6 \text{ cP}$$

or approximately:

$$1 \ \mu R = 7 \text{ cP}$$

$$1 \text{ cP} = 1 \text{ mPa} \cdot \text{s}$$

The most traditional methods of measuring viscosity were based on measuring the time needed for a certain oil quantity to flow through a small pipe with gravity, a principle known for measurement of time since ancient Egypt. For this reason, the kinematic viscosity was introduced:

$$v = \frac{\mu}{\rho} \tag{11.2a}$$

with units L^2/T where ρ is the oil density. In the cgs system, the unit was cm^2/s, called stokes (St), with the most widely used unit being the centristoke (cSt). The SI unit is then m²/s equal to 10^4 St or 100 cSt. Therefore, 1 sSt is 10^{-2}(m²/s). Oil density depends on oil type and is 850–930 kg/m³ for most mineral oils with an average value of 890.

A nomogram for viscosity units conversion is shown in Figure 11.3, including some of the empirical scales used in the United States and Europe. An increase in temperature or a decrease in pressure weakens the intermolecular bonds in a fluid and leads to a reduction in viscosity. To be precise, therefore, viscosity should always be quoted at a specified temperature and pressure; if the pressure is omitted, it is understood to be atmospheric. Some oils are less sensitive to changes in viscosity with temperature than others; these are said to have a high viscosity index. Figure 11.4 shows this dependence for some oils.

Pressure affects much less viscosity. For hydrodynamic lubrication, this effect is usually negligible. An approximate expression for the dynamic viscosity μ at pressure p as function of the viscosity μ_0 at atmospheric pressure is:

$$\mu = \mu_0 \exp(ap) \tag{11.3}$$

where $a = (1.3 \text{ to } 3.5)10^{-4}$ mm²/N for most mineral oils.

Since the SAE crankcase oil specification is based on viscosity requirements at two temperatures only (0°F and 210°F), it becomes possible by suitable formulation

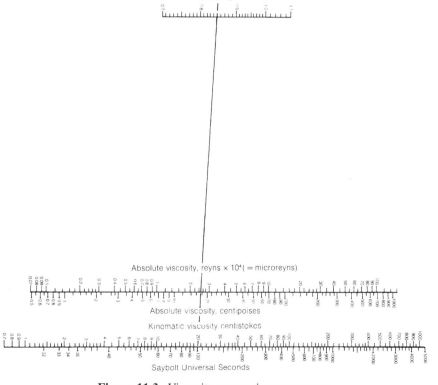

Figure 11.3. Viscosity conversion nomogram.

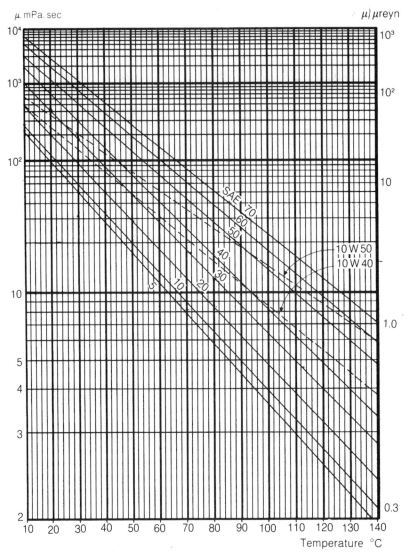

Figure 11.4. Viscosity of SAE oil grades.

to make an oil meeting the requirements of two SAE grades simultaneously. Such an oil is described as a multigrade oil. For example, crankcase oils SAE 10W and SAE40 would be classified as SAE 10W/40. The advantage of multigrade oils is that they allow operation over a considerably wider range of temperatures than would be possible with a single-grade oil.

It must be pointed out that the values of the viscosity given are mean values, and substantial deviations from these values for different oil qualities are common. For this reason, standards organizations specify usually the mean and limit values.

ISO uses the designation ISO VG x, where x is the mid-point kinematic viscosity in cSt (mm^2/s) at 40°C. Deviations are ±10%. For industrial lubricants, ISO specifies grades 2, 3, 5, 7, 10, 15, 22, 32, 46, 68, 100, 220, 320, 460, 680, 1000, 1500. The midpoint viscosity of the first four is exactly 2.2, 3.2, 4.6, 6.8, respectively. For the rest, it equals the grade number.

For computer applications, it is convenient to express the viscosity–temperature function by way of an empirical relation:

$$\text{Log}_{10}(\text{Log}_{10}\mu) = c_1 T + c_2 \tag{11.4}$$

Application of this equation for two temperatures, say 40° and 100°C, gives the constants c_1 and c_2 for each lubricant grade. This temperature is in degrees Centigrade and then μ is in mPa.s. For the usual SAE grades used for sleeve bearings the constants c_1 and c_2 are:

Grade →	SAE10	SAE20	SAE30	SAE40	SAE50	SAE60	SAE70
c_1	0,0067968	−0,00624	−0,00611	−0,00612	−0,00542	−0,00513	−0,00515
c_2	0,4593269	0,493615	0,538256	0,579809	0,595444	0,619336	0,642911

Figure 11.4 shows the temperature dependence of commonly used SAE grades.

Example 11.1 An automobile manufacturer specifies SAE 30 oil for an engine. The design temperature of the oil during normal engine operation is 80°C. Determine the dynamic and kinematic viscosities of the oil at that temperature, if it is known that the density of the oil is 880 kg/m³.

Solution Figure 11.4 shows dynamic viscosity at 80°C for SAE 30 oil:

$$\mu = 12.2 \text{ MPa.s} = 12.2 \text{ cP} = 1.77 \text{ } \mu\text{reyn}$$

The kinematic viscosity is:

$$\upsilon = \frac{\mu}{\rho} - \frac{12.2 \times 10^{-3}}{880} = 1.38 \times 10^{-5} \text{ m}^2/\text{s} = 13.8 \text{ cSt}$$

From Figure 11.3, 13.8 cSt corresponds to 73 Saybolt Universal Seconds.

11.2.2. Viscous Flow in a Concentric Bearing

Petroff applied Newton's law of viscous flow to calculate the frictional torque and power loss in certain types of journal bearings that tend to run concentrically because they carry little or no transverse loads. Figure 11.5 shows a vertical guide bearing of this type. Oil completely fills the clearance space, which is small compared with the shaft diameter; it is also assumed that negligible axial flow of oil takes place. The resisting torque is the moment about the axis of rotation of all shear stresses (shear stress × area × radius):

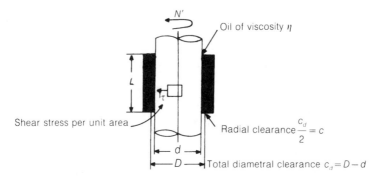

Figure 11.5. Lubricated concentric cylinders.

$$T = \tau(\pi DL)R \tag{11.5a}$$

where the shear stress is:

$$\tau = \frac{\mu u}{c} = \frac{\mu \pi DN}{c} \tag{11.5b}$$

and D is the bearing diameter, N^1 is the frequency of rotation (rps), L is the bearing length, c is the radial clearance (radius of bearing–radius of journal).

Then the Petroff equation is obtained:

$$T = \pi^2 \frac{\mu NLD^3}{2c} \tag{11.5c}$$

The power loss is:

$$\dot{W} = \omega T = \frac{\pi^3 N^2 LD^3 \mu}{c} \tag{11.5d}$$

We can define a friction coefficient:

$$f = \frac{T}{WR} = \frac{T}{p_m LDR} = 2\pi^2 \frac{R}{c} \frac{\mu N}{p_m} = 2\pi^2 \frac{c}{R} S \tag{11.5e}$$

$$\frac{R}{c} f = 2\pi^2 S \tag{11.5f}$$

where W is the bearing load, p_m is the bearing mean pressure W/LD, and S is the

---

[1] It should not be confused with the safety factor N and the rotating speed n (rpm).

Sommerfeld number[2] $S = (\mu N/p_m)(R/c)^2$. This number plays an important role in bearing design, and we will discuss it in more detail later.

It is apparent that the coefficient of friction depends on two groups of parameters:

$$\text{Geometry: } \frac{R}{c}$$

$$\text{Operating conditions: } \frac{\mu N}{p}$$

Petroff's equation is of particular value in estimating the power losses in journal bearings that are lightly loaded and operate nearly concentric at high rotational speeds, such as turbine bearings. Even in the case of moderately loaded bearings running with a considerable degree of eccentricity, Petroff's equations may be applied as a first approximation to assess the minimum losses to be expected. Experiments of Guembel (Cameron 1976, Niemann 1943) suggest that for high speeds ($S > 1/2\pi$) the Petroff equation predicts with adequate accuracy the bearing friction, since the bearing runs nearly concentric. For high loads ($S < 1/2\pi$), Guembel suggested:

$$f\frac{R}{c} = 4\pi^2 \sqrt{S} \tag{11.5g}$$

Note that the power loss and hence the frictional heat developed become of increasing significance in large diameter bearings since T is proportional to D^3.

Arnold Sommerfeld (1868–1951) attended the Gymnasium and the University in Königsberg. In 1891, he was awarded his doctorate from Königsberg. In 1893, he left Königsberg for Göttingen, where he became Klein's assistant. Sommerfeld's first work was on the mathematical theory of diffraction. From 1897, he taught at Clausthal, where he became professor of mathematics at the mining academy. In 1900, he became professor of mechanics at the Technical University of Aachen, where he developed the first solution of the lubrication equation. In 1906, he became professor of theoretical physics at Munich and worked on atomic spectra. Sommerfeld had built up a very famous school of theoretical physics at Munich but its 30 years of fame ended with the Nazi rise to power. In 1940, the school closed, but by that time Sommerfeld was 71 years old. He survived World War II, only to die in a street accident in Munich.

---

[2] It must be pointed out that some authors define the Sommerfeld number differently. A notable notation is the one used by most German authors, $S_0 = 2\pi/S$.

Example 11.2 On a viscosity-measuring instrument, a cylinder of diameter $D = 30$ mm and length $L = 30$ mm rotates with $N = 12$ rps in a cylindrical cavity with a radial clearance of $c = 1$ mm. The torque on the stationary cylinder is measured $T = 0.5$ Nmm. Find the absolute oil viscosity.

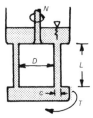

Figure E11.2

Solution The Petroff equation (11.5c) yields:

$$\mu = \frac{2cT}{\pi^3 NLD^3} = \frac{2 \times 0.001 \times 0.5 \times 10^{-3}}{\pi^2 \times 12 \times 0.003 \times 0.003^3} = 0.0104 \text{ Pa} \cdot \text{s}$$

or

$$\mu = 10.4 \text{ mPa} \cdot \text{s} = 10.4 \text{ cP} = 1.51 \ \mu\text{reyn.}$$

Example 11.3 A lightly loaded sleeve bearing is split and the upper half has a wide groove as shown in Figure E11.3. The radial clearance is 0.05 mm, the journal rotates at 3600 rpm, and oil SAE 10 is used. The bearing temperature is 60°C. Find the power loss, assuming that the journal rotates in a concentric position.

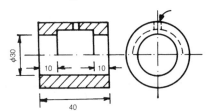

Figure E11.3

Solution Petroff's equation will be used. To this end, the lower half is one-half of a full bearing and the upper lands of width 10 mm are one-half of a full bearing each 10 mm long. The viscosity is, from Figure 11.4, $\mu = 12.5$ MPa.s. Therefore, the total power loss will be, for $N = 3600/60$ rps:

$$W = W_1 + 2W_2 = \frac{\pi^3 \times 60^2 \times 0.04 \times 0.03^3 \times 12.5 \times 10^{-3}}{5 \times 10^{-5}} \times 0.5$$

$$+ \frac{\pi^3 \times 60^2 \times 0.01 \times 0.03^3 \times 12.5 \times 10^{-3}}{5 \times 10^{-5}} \times 0.5 \times 2 = 15.06 + 7.53 = 22.6 \text{ W}$$

11.2.3. Viscous Flow in a Pipe

When a liquid flows in a parallel-sided pipe, it loses energy in friction between adjacent layers of the liquid and also between boundary layers of the liquid and the pipe walls. Provided that rate of flow does not exceed a certain critical velocity, the mode of flow is described as viscous or laminar flow and is governed by a fundamental relationship known as *Poiseuille's law*. The chief characteristics of viscous flow are that the velocity is independent of the internal roughness of the pipe and the velocity distribution across a diameter of the pipe is parabolic in form, with maximum velocity along the pipe axis. At flow velocities above the critical, the flow pattern becomes turbulent. The transition region can be defined by a non-dimensional criterion known as the *Reynolds number,* Re.

$$\mathrm{Re} = \frac{Vd}{v} \tag{11.6a}$$

where V is the fluid velocity, d is the pipe inner diameter, v is the kinematic viscosity (μ/ρ), and ρ the density of the oil.

If, under the specified conditions of flow, the value of Re is found to be less than 2000, it can be reasonably assumed that the mode of flow is laminar. For values of Re between 2000 and 4000, the mode of flow is indeterminate, but at Re above 4000, flow is almost certainly of a turbulent nature. Turbulent lubrication can be found only in very large bearing sizes or very high speeds.

Fox and McDonald showed that below the critical velocity the volume of liquid flowing through a narrow, parallel-sided channel in time t is given by:

$$q = \frac{\Delta p h^3}{12 \mu L} \tag{11.6b}$$

where q is volume of the liquid per unit time, Δp is the difference in the pressure causing the flow, h is the channel height, μ is the absolute viscosity, and L is the length of channel that has unit width.

11.2.4. Viscous Flow between Moving Parallel Plates

A flat plate, moving parallel to a stationary plate with a velocity U (Figure 11.2), is accompanied by a flow

$$q = \frac{wUh}{2} \tag{11.7}$$

assuming a linear variation of velocity across the plate distance h, constant along the width w.

11.2.5. Viscous Flow in a Wedge: The Reynolds Equation

Consider now a flat surface stationary and a curved surface, nearly parallel to the stationary one and moving in respect to it in the x-direction with velocity u. It also moves away from the stationary surface with a velocity $V = \dot{h}$, where $h(x, y; t)$ is the distance between the two surfaces at location x, y in the z-direction. The pressure is, in general, a function of x, y, and t.

The flow in a control volume of sides Δx and Δy and height h (variable) (Figure 11.6a) is considered constant, for incompressible liquid.

Flow in the x-direction:
Due to pressure difference (Figure 11.6b)

$$Q_x = \frac{h^3 \partial p}{12\mu \partial x} \Delta y \qquad (11.8a)$$

from Equation (11.6).

Owing to motion of the plate (Figure 11.6c):

$$Q_u = \frac{uh}{2} \Delta y \qquad (11.8b)$$

from Equation (11.7).

Flow in the y-direction:

$$Q_x = \frac{h^3 \partial p}{12\mu \partial y} \Delta x \qquad (11.8c)$$

from Equation (11.6), Figure 11.6b.
The total flow will be:

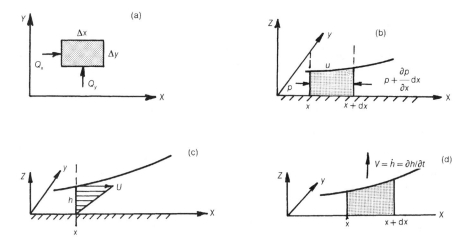

Figure 11.6. Elementary lubricated area.

$$Q = \frac{Uh}{2} \Delta y - \frac{h^3}{12\mu} \frac{\partial p}{\partial x} \Delta y - \frac{h^3}{12\mu} \frac{\partial p}{\partial y} \Delta x \tag{11.8d}$$

Because of continuity, the derivative of the flow will equal the change of volume due to the upwards motion of the moving plate with velocity $h = V$.

$$\Delta Q = \frac{\partial Q}{\partial x} \Delta x + \frac{\partial Q}{\partial y} \Delta y \tag{11.9}$$

$$\Delta Q = \frac{u}{2} \frac{\partial h}{\partial x} \Delta x \Delta y - \frac{\partial}{\partial x} \left(\frac{h^3}{12\mu} \frac{\partial p}{\partial x} \right) \Delta x \Delta y \tag{11.10}$$

$$- \frac{\partial}{\partial y} \left(\frac{h^3}{12\mu} \frac{\partial p}{\partial y} \right) \Delta x \Delta y = h \Delta x \Delta y$$

This yields the general form of the Reynolds equation (Reynolds 1886):

$$\frac{\partial}{\partial x} \left(\frac{h^3}{12\mu} \frac{\partial p}{\partial x} \right) + \frac{\partial}{\partial y} \left(\frac{h^3}{12\mu} \frac{\partial p}{\partial y} \right) = \frac{u}{2} \frac{\partial h}{\partial x} - h \tag{11.11}$$

Osborne Reynolds (1842–1912) was a mathematics graduate of Cambridge in 1867. He became the first professor of engineering in Manchester in 1868 and retired in 1905. He became a Fellow of the Royal Society in 1877 and, eleven years later, won their Royal Medal. His early work was on magnetism and electricity, but he soon concentrated on hydraulics and hydrodynamics. In 1886, he formulated a theory of lubrication, and three years later he developed the standard mathematical framework used in the study of turbulence. The "Reynolds number" used in modeling fluid flow is named after him.

From the foregoing discussion, it is apparent that the Reynolds equation is based on the following assumptions:

1. Laminar flow of lubricant within the bearing clearance.
2. Constant viscosity; temperature rise due to shearing of the oil and effect of pressure on viscosity are neglected.
3. Film thickness is much less than bearing dimensions so that pressure can be assumed constant with respect to depth.

Under certain simplifying assumptions, the Reynolds equation can be solved analytically (see Example 11.4). In its more general form, it can be solved numerically.

Example 11.4 A fixed Michell pad bearing for the main shaft of a ship consists of six pads as in Figure E11.4 of angle $\theta_0 = 50°$ between inner radius $a = 200$ mm and outer radius $b = 300$ mm. The oil has viscosity 50 cP and the minimum film thickness must not

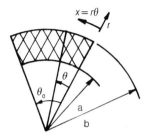

Figure E11.4

be less than 0.1 mm where the film thickness at the trailing edge is 0.5 mm greater than the one at the leading edge owing to the pad inclination. Determine the load-carrying capacity for speed of rotation 300 rpm assuming short bearing in the tangential direction, thus negligible derivatives in the tangential direction.

Solution   Neglecting $\partial/\partial x$ (or $\partial/\partial\theta$) terms owing to short bearing approximation, the Reynolds equation (11.11) becomes, for $y = r$, $x = r\theta$:

$$\frac{\partial}{\partial r}\left(h^3\frac{dp}{dr}\right) = 6\mu u\frac{dh}{dx} = k \tag{a}$$

where k is a constant. But $h = h_0 + (h_1 - h_0)\theta/\theta_0$, $dh/dx = dh/d(r) = h_1 - h_0/r\theta_0$, and $u = 2\pi/Nr$; therefore:

$$k = \frac{12\mu N\mu(h_1 - h_0)\theta}{\theta_0}$$

Integrating Equation (a) twice yields:

$$p = \frac{k}{2h^3}r^2 + \frac{c_1 r}{h^3} + c_2 \tag{b}$$

where for $r = a$ and b the pressure $p = 0$. Equation (b) for $r = a$, $p = 0$, and $r = b$, $p = 0$ yields:

$$c_1 = -\frac{k(b - a)}{2}, \quad c_2 = \frac{kab}{2h^3}$$

$$p = \frac{k}{2h^3}[r^2 - (b + a)r + ab]$$

The load is found by integration over the pad surface

$$W = \int_a^b \int_0^{\theta_0} \frac{k}{2h^3}[r^2 - (b + a)r + ab]r\,dr\,d\theta = \int_a^b [r^3 - (a + b)r^2 + abr]dr \int_0^{\theta_0} \frac{k}{2h^3}d\theta$$

The first integral has the value

$$I_1 = \left[\frac{r^4}{4} - (b + a)\frac{r^3}{3} + \frac{abr^2}{2}\right]\Big|_a^b = \frac{b^4 - a^4}{4} - \frac{b + a}{3}(b^3 - a^3) + \frac{ab}{2}(b^2 - a^2)$$

The second integral yields

$$I_2 = \int_0^{\theta_0} \frac{k}{2h^3}\,d\theta = \frac{\theta_0 k}{2(h_1 - h_0)}\int_{h_0}^{h_1} \frac{dh}{h^3} = \frac{\theta_0 k(h_1 + h_0)}{2h_0^2 h_1^2}$$

Therefore:

$$W = I_1 I_2$$

The numerical values are:

$$h_0 = 0.1 \times 10^{-6}\text{ m}, \quad h_1 = 0.6 \times 10^{-3}\text{ m}, \quad \theta_0 = 50° = 0.87\text{ rad},$$

$$\mu = 50 \times 10^{-3}\text{ Pa.s}, \quad N = 5\text{ rps}$$

$$k = 12 \times 5 \times 50 \times 10^{-3}\frac{0.6 \times 10^{-3} - 0.1 \times 10^{-3}}{0.87} = 1.72 \times 10^{-3}$$

$$I_1 = \frac{0.3^2 - 0.2^2}{4} - \frac{0.3 + 0.2}{3}(0.3^3 - 0.2^3) + \frac{0.2 \times 0.3}{2}(0.3^2 - 0.2^2) = 0.0108$$

$$I_2 = \frac{0.87 \times 5.17 \times 10^{-3} \times (0.1 \times 10^{-3} + 0.6 \times 10^{-3})}{2 \times (0.1 \times 10^{-3} \times 0.6 \times 10^{-3})^2} = 4.37 \times 10^8$$

Then the load-carrying capacity will be, per pad:

$$W = I_1 I_2 = 4.72 \times 10^6\text{ N}$$

For six pads, $W_0 = 28.32 \times 10^6$ N

11.3. DESIGN OF HYDRODYNAMIC BEARINGS

11.3.1. Slider Bearings

Slider bearings consist of two plane-surfaced members, separated by a wedge-shaped film of oil. The moving member, called the *slider* or *runner*, moves in the direction of convergence of the oil film; the other, stationary member is called the *pad* or *shoe*.

The simplest type of slider bearing is shown diagrammatically in Figure 11.7. A rectangular pad of dimensions B in the direction of motion and L at right angles to it is maintained at a small fixed angle of inclination to the slider, which moves with velocity U. The film thickness at the leading edge of the pad is h_0, at the trailing edge is h_1, and, at any distance x from the trailing edge, is h. Expressing h in terms of x and substituting in the Reynolds equation, it can be shown that for a pad in which $L >> B$, $\partial p/\partial y = 0$ (i.e., a pad of "infinite" width), the pressure

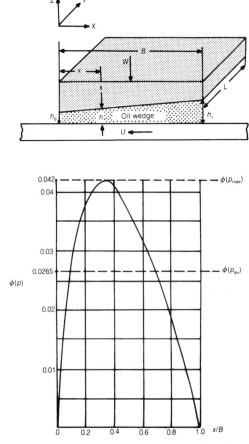

Figure 11.7. Pressure distribution in an oil wedge (from BP 1969).

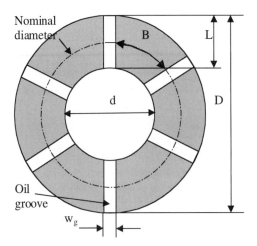

Figure 11.7a. Geometry of a thrust bearing.

distribution in the direction of motion is given by integrating Equation (11.11) twice (see also Example 11.4):

$$p = \frac{6\mu uB}{h_0^2} \cdot \phi \qquad (11.12)$$

where ϕ is a non-dimensional pressure function whose value depends on the angle of tilt of the pad. Evaluation of ϕ enables a pressure profile to be plotted across the face of the pad as shown in Figure 11.7. The average pressure across the pad p_{av} is found by integration to be:

$$p_{av} = \frac{6\mu UB}{h_0^2} \cdot \phi_{av} \qquad (11.13)$$

From this equation, the minimum film thickness at the trailing edge is given by:

$$h_0 = \left(\frac{6\mu UB}{p_{av}} \cdot \phi_{av} \right)^{1/2} \qquad (11.14)$$

A typical basis for design is to incline the pad so that the film thickness at the leading edge is twice that of the trailing edge (i.e., $h_1/h_0 = 2$). In this case, the pressure function ϕ as plotted in Figure 11.7 has a maximum value of about 0.042 and an average value of 0.0265. Calculations show that ϕ_{av} is not greatly affected by small changes in pad inclination provided h_1/h_0 lies within the range 1.5–3. For design purposes, therefore, it is sufficiently accurate to take ϕ_{av} as equal to 0.026.

The load-carrying capacity W of the simple rectangular pad is equal to the product of the average pressure and the pad area ($B \times L$):

$$W = \frac{6\mu uN^2L}{h_0^2} \cdot \phi_{av} \cong 0.026 \frac{6\mu uN^2L}{h_0^2} \qquad (11.15)$$

It appears that by the clearance being progressively reduced, the load capacity could be steadily increased and the coefficient of friction decreased. However, practical problems, such as the accurate machining of truly convergent planes of good surface finish and the mechanical and thermal distortion of the surfaces under working conditions, limit the extent to which this objective can be pursued. A modified arrangement of the slider, the *pivoted pad* bearing, developed by Michell (1905) in Australia and independently by Kingsbury (1898) in the United States, eliminates this problem.

Albert Kingsbury (1863–1943) was born in Morris, Illinois, son of a superintendent of the Stoneware Manufacturing Co. in Akron, Ohio. He entered Ohio State University in 1884, and in 1887 he moved to Cornell University, where he graduated in 1889 with a Mechanical Engineering degree. During his studies he did experimental work on bearing performance. He took a teaching job with New Hampshire College as professor of Mechanical Engineering in 1896. In 1898, Kingsbury built the first centrally pivoted pad bearing. In 1899, he moved to Worcester Polytechnic Institute and was also appointed General Engineer at Westinghouse. In 1924, he formed the Kingsbury Machine Works. His dispute with Michell for the tilting pad bearing patent became legendary. Eventually Kingsbury prevailed in the courts.

In the pivoted-pad bearing illustrated in Figure 11.8 the pad rests on a pivot through which the bearing load is transmitted. The pivoted pad is inherently stable but at the same time free to adjust its inclination to varying operating conditions. If the pivot is at the center of the pad, it is possible to run the slider in either direction without seriously affecting the performance of the bearing. For unidirectional running, the optimum position of the pivot is slightly beyond the middle of the pad towards the trailing edge. Five-ninths of the way along is often chosen in the design of such a bearing.

The equations developed for the fixed-pad bearing apply equally well to the pivoted-pad bearing, provided the pivot is located at the center of pressure of the pad. Given the bearing load, the minimum film thickness and the pad angle are found.

A number of pivoted pads can be assembled circumferentially to form a complete bearing capable of sustaining an axial load between rotating surfaces. The pattern of oil flow and the pressure distribution in the form of pressure contours are superimposed on the plan view of one of the sector-shaped pads. It will be observed that the flow of oil in a radial direction (i.e., the side leakage) is appreciable when the length of a pad is approximately equal to its width (Figure 11.9).

The foregoing equations were derived on the assumption that there is no side leakage of oil and consequently the pressure remains constant in that direction. In

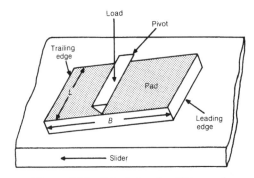

Figure 11.8. Principle of a tilting pad.

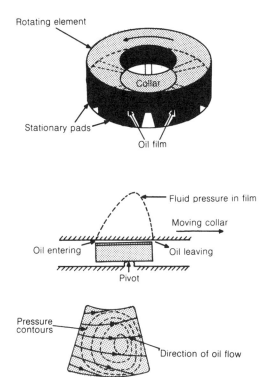

Figure 11.9. Thrust bearing.

fact, this could be true only for bearing pads having infinite length. However, the equations may be applied to real bearings with reasonable accuracy provided the ratio of L/B is greater than about 4. If L/B is less than 4, Kingsbury and Needs (Pugh 1970) have derived a series of *leakage factors* (K_L) that when introduced into the above equations for infinitely long bearing pads enable satisfactory solutions to be obtained for pads with L/B ratios of less than 4 (Figure 11.10). To a good approximation:

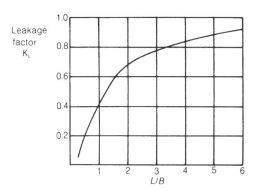

Figure 11.10. Leakage factors for bearings of finite width.

$$K_L \approx 0{,}539111 - 0{,}109706 \left(\frac{L}{B}\right) + 0{,}007724 \left(\frac{L}{B}\right)^2 - 0{,}039628 \left(\frac{L}{B}\right)^3 \quad (11.15a)$$

and $\phi_{av} \approx 0.026$. Then:

$$W = \frac{6\mu U B^2 L}{h_0^2} \cdot \phi_{av} \cdot K_L \quad (11.16)$$

$$h_0 = \left[\frac{6\mu 6 U B}{p_{av}} \cdot \phi_{av} \cdot K_L\right]^{1/2} \quad (11.17)$$

The most important criterion in design is the minimum film thickness, which should not be less than a certain critical amount. This figure is governed by such factors as the efficiency of filtration in the oil system, the surface finishes of the bearing parts, and the surface speed. Typically, h_0 is of the order of 0.1–0.4 mm and the pad dimensions are chosen so that the average oil film pressure is between 2 and 3 N/mm², depending on the application. Integrating the shear stress over the slider surface, we obtain the friction force:

$$F = \frac{\mu U L B}{h_0} \phi \quad (11.18)$$

where ϕ is a nondimensional function. Calculations show that the value of ϕ is relatively insensitive to changes in pad inclination and, for bearings in which h_1/h_0 is in the region of 2, may be taken as equal to 0.77.

The analysis for the plane slider can be applied to thrust bearings, which support axial load on a shaft, and the corresponding dimensions are shown in Figure 11.7a. All bearing parameters are measured on the nominal diameter, $(D + d)/2$, and the bearing pad is approximated by a rectangle of dimensions $L \times B$, as shown.

The power loss is equal to the product of the frictional drag F and the velocity U:

$$\text{Power loss} = \frac{\mu U^2}{h_0} \text{LB} \cdot \phi \cong 0.77 \frac{\mu U^2}{h_0} \text{LB} \quad (11.19)$$

Design Procedure 11.1: Design of a Thrust Bearing

Input: Maximum operational load, oil properties, rotating speed. **Output:** Dimensions of friction surfaces, actuation force.

Step 1: Sketch the bearing and place on the sketch loads and motion data.

Step 2: Assume an inner and outer bearing diameter d_1 and d_2.

Step 3: Calculate the peripheral speed at the outer diameter $U = \omega d_2/2$.

Step 4: Select a number of pads n_p and find the bearing length $L = (d_2 - d_1)/2$ and the average bearing width $B = \pi(d_2 + d_1)/n_p - w_g$. Assume a groove width w_g.

Step 5: From Equation (11.15a), find K_L. Take $\phi = 0.77$ and $\phi_{av} = 0.026$.

Step 6: Use Equation (11.16) to calculate the load-bearing capacity. If it is acceptably close the operational load, the design is accepted. If not, different bearing dimensions in step 2 should be selected and another iteration should be performed. This can also be done with Solver of EXCEL (see Example 11.6).

Step 7: Use Equations (11.17) and (11.19) to calculate the minimum oil film, h_0, thickness ratio, and power loss.

Example 11.5 A 6000 kW steam turbine moves a ship propeller at 180 rpm through a main shaft. The propeller has efficiency of 80% and the maximum speed of the ship is 50 km/h.

A thrust bearing is used on the main shaft to take the thrust load. It consists of 50° segments as shown in Figure 11.9, between radii and $a = 200$ and $b = 520$ mm. The oil used is SAE 40 maintained at 70°C and surface roughness, manufacturing tolerances, and oil filtration do not allow for minimum film thickness less than 80 μm. Determine the number of six-pad thrust bearings of such type required to sustain the load, assuming a rectangular pad and a Kingsbury and Needs solution. Determine then the power loss.

Solution The pad geometry is:

$$L = b - a = 520 - 200 = 320 \text{ mm}$$

$$B = R\theta_0 = \frac{a + b}{2} \theta_0$$

$$= \frac{200 + 520}{2} \times 50 \times \frac{3.14}{180} = 314 \text{ mm}$$

The mean peripheral velocity is:

$$U = \omega R = \frac{2\pi N(a + b)}{2} = \frac{2\pi(180/60)(0.200 + 0.520)}{2} = 6.78 \text{ m/s}$$

From Figure 11.4, the oil viscosity is $\mu = 25$ mPa.s.

The leakage factor (Figure 11.10), for $L/B \cong 1$, is $K_L = 0.4$. Therefore, the load-carrying capacity of one pad is (Equation (11.16)):

$$W = \frac{6\mu UB^2 L\phi_{\text{av}} K_L}{h_0^2} = \frac{6 \times 25 \times 10^{-3} \times 6.78 \times 0.314^2 \times 0.320 \times 0.026 \times 0.4}{(80 \times 10^{-6})^2}$$

$$= 52{,}142 \text{ N}$$

Each complete six-pad bearing can take load $6 \times 52{,}142 = 3.13 \times 10^5$ N. The thrust F of the ship is found from the power and speed:

$$P = 6 \times 10^6 W = FU = \frac{F \times 50{,}000}{3600}$$

$$F = \frac{6 \times 10^6 \times 3600}{50{,}000} = 0.42 \times 10^6 \text{ N}$$

The required number of thrust bearing stages of six pads each is:

$$0.42 \times 10^6 \times 3.12 \times 10^5 = 1.38$$

Two stages are sufficient.

The power loss, from Equation (11.19), assuming $\phi = 0.77$, is:

$$P_L = \frac{25 \times 10^{-6} \times 6.8^2 \times 0.32 \times 0.314 \times 0.77}{80 \times 10^{-6}} = 1118 \ W = 1.12 \text{ kW}$$

for each pad. For the two stages, twelve pads:

$$P_{\text{Tot}} = 12 \times P_L = 12 \times 1.12 = 13.40 \text{ kW}$$

A more accurate estimate would be obtained taking into account that two stages (instead of 1.4) were selected, therefore the minimum oil film thickness will be somewhat greater than 80 m. Equation (11.17) will give this value, which must be then used in Equation (11.19).

Example 11.6   Develop a spreadsheet solution, using EXCEL and Solver, for the solution of Example 11.5. One pad bearing is allowed, and the external bearing diameter needs to be computed (see Design Procedure 11.1).

Solution   The spreadsheet is shown in Table E11.6.

A larger-diameter single thrust bearing is found than in the results of Example 11.5 (1.5 m instead of 1.040 m outer diameter) and substantially greater losses (35 kW vs. 6.7 kW) due to the larger diameter.

Example 11.7   Optimize the design of Example 11.5 for minimum power loss.

Solution   The power loss will be minimized by changing the outer diameter and the oil viscosity. Design Procedure 11.1 was modified to implement this change. The power loss (cell C22) was targeted for minimization by changing cells B8 and B9 (note that we have moved the viscosity to be adjacent to the diameter). The equation of the bearing load with the load-carrying capacity of the bearing (cell E20 set to 0) was set as a constraint.

The results are interesting. Not only was the power loss cut from 35 kW to 27 kW (33% reduction), but the outer diameter was also reduced, from 1.5 m to 0.67 m, or by 55%. The oil film thickness is 190 μm, larger than the minimum 80 μm allowed. The required viscosity found is 469 mPa.s, which corresponds approximately to SAE 70 oil at 45°C. The spreadsheet is in Table E11.7 (page 744).

TABLE E11.6

	A	B	C	D
1	**Example 11.6. Design of a Slider Thrust Bearing**			
2	**Input:**			
3	Maximum operational load W (N)	3,12E+05	Oil groove width W_g	0,005
4	Oil Dynamic Viscosity M_0 (mPas)	25	No of Pads N_p =	6
5	Rotating speed (RPM)	180	Minimum Film Thickness h_0 =	0,00008
6	**Step 2:**			
7	Outer Diameter d_2 =	1,541465084		
8	Inner Diameter d_1 =	0,4		
9	**Step 3:**			
10	Peripheral speed U =	9,148938667	m / s	
11	**Step 4:**			
12	Pad Length L_p =	0,570732542	Pad Width W_p =	0,513274
13	Average Pressure p_av =	177509,3072	N/m^2	
14	L / B Ratio L_B =	1,111944361		
15	**Step 5:**			
16	K_L =	0,372192357		
17	**Step 6:**			
18	Load Capacity W_cap =	312000	W-W_cap=	−1,61E-07
19	**Use SOLVER to change d_2 so that W-W_cap = 0 (cell D18)**			
20	**Step 7:**			
21	Power Loss =	35401,00932	Watts	

11.3.2. Journal Bearings

A journal bearing consists of a shaft completely or partially surrounded by a bearing surface. Usually the shaft is rotating and the bearing remains stationary. Oil enters the clearance space through a hole or groove and moves in a circumferential direction, leaving the bearing by leakage at its ends.

Figure 11.11 illustrates the generation of a hydrodynamic oil film under various conditions of operation. For the sake of clarification, the clearance between sleeve and journal has been greatly exaggerated. At starting and stopping of the shaft (Figure 11.11a), boundary or thin-film conditions cause relatively high frictional resistance, but as soon as a sufficient speed of rotation is exceeded, journal and bearing surfaces separate and hydrodynamic conditions prevail, greatly reducing the frictional resistance (Figure 11.11b). If the load is raised further or the speed reduced, the eccentricity increases and the "attitude" angle, ϕ, decreases (ϕ is defined as the angle between the direction of the load and the line joining journal and bearing centers, measured in the direction of rotation). This reduces the minimum film thickness (Figure 11.11c). Increasing load or decreasing rotational speed will cause complete or partial breakdown of hydrodynamic conditions.

TABLE E11.7

1	B	C	D	E
2	**Example 11.7. Optimum Design of a Slider Thrust Bearing**			
3	**Input:**			
4	Maximum operational load W (N)	3,12E+05	Oil groove width W_g	0,005
5	Rotating speed (RPM)	180	Minimum oil film thickness h_0 =	0,00008
6	**Step 2:**			
7	Oil Dynamic Viscosity M_0 (mPas)	469,1037308	No of Pads N_p =	6
8	Outer Diameter d_2 =	0,672809379		
9	Inner Diameter d_1 =	0,4		
10	**Step 3:**			
11	Peripheral speed U =	5,055495097	m / s	
12	**Step 4:**			
13	Pad Length L_p =	0,13640469	Pad Width W_p =	0,2858608
14	Average Pressure p_av =	1333580,935	N/m^2	
15	L/B Ratio L_B =	0,477171656		
16	**Step 5:**			
17	K_L =	0,484215581		
18	**Step 6:**			
19	Load Capacity W_cap =	312000	W-W_cap=	−7,86E-07
20	Use SOLVER to change M_0 and d_2 so that B22 will be minimized (target) and add the constraint W-W_cap = 0 (cell D19 set equal to zero)			
21	**Step 7:**			
22	Power Loss =	26998,03949	Watts	
23	Computer oil thickness	0,000195959	m	
24				

At very light loads and high rotational speeds, on the other hand, the journal will rotate almost concentrically within the bearing, i.e., eccentricity $e = 0$ (Figure 11.16d). In this case, frictional torque and power losses due to viscous shear of the oil film can be readily estimated by application of the Petroff equation. Figure 11.12 illustrates how the state of lubrication and frictional resistance in a journal bearing are influenced by changes in the three main parameters that govern its performance, as was found in Petroff's equation.

The short section AB of the curve indicates a region of operation in which a state of boundary lubrication prevails. The fact that AB is nearly horizontal confirms that in this region frictional drag is almost uninfluenced by changes in speed, load, or viscosity. Section BC indicates a rapid fall in friction as $\mu N/p$ is increased, for example by a significant increase in the rotational speed; in this region, thin-film or mixed conditions of lubrication exist. At C, transition from thin-film to full

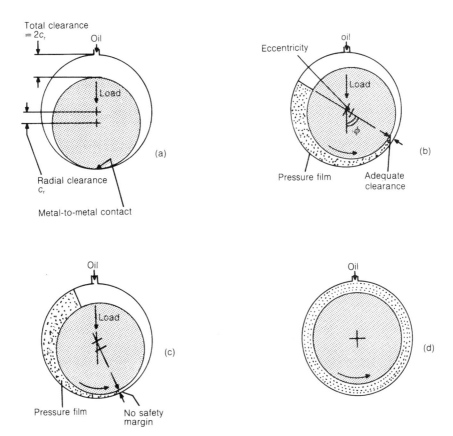

Figure 11.11. Unloading of a cylindrical bearing.

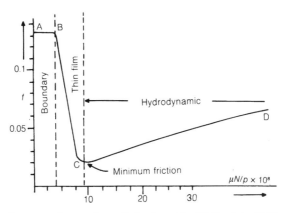

Figure 11.12. Bearing operating regimes (introduced by Stribeck in 1902 and M. D. Hershey in 1914).

hydrodynamic lubrication occurs, further increase in $\mu N/p$ causing a gradual rise in friction owing to the additional viscous shear.

11.3.3. Analytical Approximations—Long Bearings

11.3.3.1. The Infinitely Long Bearing. Figure 11.13 shows a cross-section of a 360° journal bearing with exaggerated clearance in which the geometrical relationships between the bearing and journal are illustrated. Taking as a datum $O'O$ the line joining bearing and journal centers, the film thickness h at an angular position θ in the direction of rotation is given by:

$$h = c_r(1 + \varepsilon \cos \theta) \tag{11.19a}$$

where c_r is the radial clearance (i.e., half the total clearance) and the eccentricity ratio is defined by $\varepsilon = e/c_r$.

The journal is capable of supporting a load equal to the resultant of the oil pressures acting on its surface. In order to predict the actual load-carrying capacity, we must first determine the circumferential pressure distribution from beginning to end of the load-bearing oil film. The Reynolds equation for pressure distribution in a hydrodynamic bearing (Equation 11.11) may be rewritten in polar form by putting $\partial x = r\partial \theta$. Then, if U is surface velocity (neglecting $\partial/\partial y$ terms and h), integrating once:

$$\frac{dp}{d\theta} = \frac{6\mu Ur(h - h')}{h^3} \tag{11.20}$$

By neglecting $\partial p/\partial y$ terms, a long bearing was assumed (in theory, "infinitely" long); then the above equation will correctly express the circumferential pressure gradient in any plane perpendicular to the bearing axis, since leakage of oil and

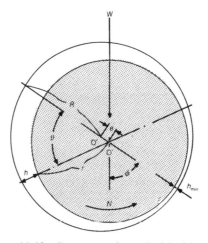

Figure 11.13. Geometry of a cylindrical bearing.

change of pressure in an axial or y-direction can be neglected. Substituting for h the relation for film thickness gives:

$$\frac{dp}{d\theta} = 6\mu UR \left[\frac{c_r(1 + \varepsilon \cos \theta) - c_r(1 + \varepsilon \cos \theta')}{c_r^3(1 + \varepsilon \cos \theta)} \right] \qquad (11.21)$$

where θ' is the position at which $dp/d\theta = 0$ or $h' = c_r(1 + \varepsilon \cos \theta')$. θ' is a constant to be determined by the boundary conditions that define the beginning and end of the generated pressure film. The pressure film can reasonably be expected to commence at the thickest part of the film (i.e., $p = 0$ when $\theta = 0$) and continue for slightly more than half the journal circumference (i.e., $p = 0$ when $\theta = +$ some angle θ' and $dp/d\theta = 0$). These boundary conditions, illustrated in the top diagrams of Figure 11.14, are physically realistic and generally confirmed by experimental evidence, but their exact adoption adds greatly to the complication of further mathematical analysis. Reynolds was only partially successful in solving the above equation for pressure distribution, which was subsequently integrated by Sommerfeld in 1904.

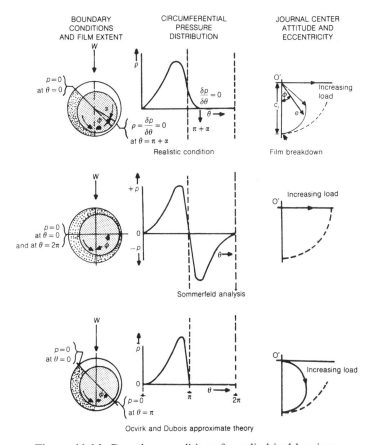

Figure 11.14. Boundary conditions for cylindrical bearings.

In his solution, Sommerfeld suggested that the pressure again falls to zero when $\theta = 2\pi$, with the implication that a continuous pressure film from $0–2\pi$ extends round the whole of the bearing circumference. In this case, the oil-film pressure p at any angular position θ is given by:

$$p = \frac{6\mu Ure}{c_r^2} \frac{(2 + \varepsilon \cos \theta)\sin \theta}{(2 + \varepsilon^2)(1 + \varepsilon \cos \theta)^2} \text{ and } \varphi = \cos^{-1}\left(\frac{-3}{2 + \varepsilon^2}\right) \quad (11.22)$$

In Figure 11.14, the pressure distribution is shown, as well as the bearing locus for three representative sets of boundary conditions. Realistic is the first, where the film breaks down when negative pressures are encountered (cavitation). The original Sommerfeld analysis for a full bearing accepts negative pressures. Ocvirk and Dubois boundary conditions are a very good approximation, as experiments have shown. This approximation facilitates analytical and numerical treatment of the problem.

Figure 11.15 shows the forces acting on a journal bearing under load. At equilibrium, the load, acting through the center of the bearing, must be equal and opposite to the resultant of the fluid pressures in the film surrounding the shaft. By resolving and equating these, Sommerfeld showed that the load capacity W of a full, infinitely long journal bearing is given by:

$$W = \frac{\mu UL(r/c_r)^2 12\pi\varepsilon}{[(2 + \varepsilon^2)(1 - \varepsilon^2)]^{1/2}} \quad (11.23)$$

where μ is the oil viscosity, U is the peripheral speed of the journal, L is the axial length of the bearing, c_r is the radial clearance, and c_r/r is the clearance ratio.

It is convenient to tabulate journal bearing performance in terms of a nondimensional criterion. Such a criterion is the Sommerfeld number, S, which relates

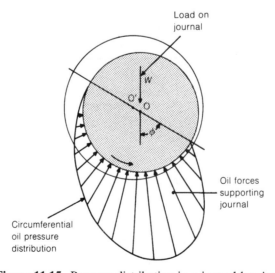

Figure 11.15. Pressure distribution in a journal bearing.

the clearance ratio to the other parameters of viscosity, speed, and load (Sommerfeld 1904):

$$S = \frac{\mu N}{p} \left(\frac{r}{c_r}\right)^2 \tag{11.24}$$

where μ is the absolute viscosity, N is the speed (rev/s), p is load per unit projected bearing area $= W/(LD)$, r is the journal radius, and c is the radial clearance. Equation (11.23) assumes a dimensionless form:

$$S = \frac{(2 + \varepsilon^2)(1 - \varepsilon^2)^{1/2}}{12\pi^2\varepsilon}, \quad \varphi = \frac{\pi}{2} \tag{11.25}$$

A full 360° journal bearing for which Equation (11.25) holds encounters problems of cavitation and stability. Partial bearings have been used. The 180° bearing (the bottom half only) is very widely used. Solution of the Reynolds equation for an infinitely long 180° bearing yields:

$$S_{180°} = \frac{(2 + \varepsilon^2)(1 - 6\varepsilon^2)^{1/2}}{12\pi^2\varepsilon \left[\frac{1}{4} + \varepsilon^2/\pi^2(1 - \varepsilon^2)\right]^{1/2}} \tan \varphi = \frac{-\pi(1 - \varepsilon^2)^{1/2}}{2\varepsilon} \tag{11.26}$$

For finite-length full 360° bearings, numerical solutions yield the pressure distribution for finite bearings (Figure 11.16). Upon integration of the pressures, the load can be related to the bearing eccentricity as in Figure 11.17, giving the eccentricity as function of the Sommerfeld number.

11.3.4. Analytical Approximations—Short Bearings

The present trend in industrial practice is to utilize narrow bearings in which the length-to-diameter ratio is less than 1. This reduced L/D ratio emphasizes the practical consequences of side flow of oil and introduces a factor of unreality into design procedures based on Sommerfeld's solution of the Reynolds equation.

An alternative solution that includes the axial oil flow and also retains a considerable part of the circumferential flow was proposed by Michell (1929) and Ocvirk

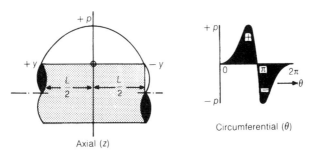

Figure 11.16. Axial pressure distribution.

and Dubois (1952) as a basis of a chart design procedure for narrow bearings. This latter analysis supplements the Sommerfeld analysis for narrow bearings, and the two may be considered as two halves of the same type of solution.

In Michell's solution, the pressure distribution is obtained neglecting the first term in Reynolds equation as much smaller than the second. Integration then yields:

$$p = \frac{\mu U}{rc_r^2}\left(\frac{L^2}{4} - y^2\right)\frac{3\varepsilon \sin \theta}{(1 - \varepsilon \cos \theta)^3} \tag{11.27}$$

A plot of this expression (Figure 11.16) shows that the pressure distribution is parabolic in the axial y-direction, falling to zero at the edges of the bearings; in a circumferential direction it follows the Sommerfeld pattern. If the "negative pressure" region from $\theta = \pi$ to $\theta = 2\pi$ is omitted, the overall pressure distribution is in good agreement with experimental results. Integration of the pressure gives:

$$\frac{1}{(L/D)^2 S} = \frac{4\pi^2\varepsilon}{(1 - \varepsilon)^{3/2}} \tag{11.28}$$

Solution of the 180° short bearing yields:

$$\frac{1}{(L/D)^2 S} = \frac{2\pi^2\varepsilon}{(1 - \varepsilon)^{3/2}} \tag{11.28a}$$

The *short bearing approximation* is the approximate method used for design of bearings having $L/D \leq 1/4$. The infinitely long bearing approximation is adequate for bearings having $L/D > 4$. If for a particular type of bearing S_∞ is the Sommerfeld number for the infinitely long bearing (Equation (11.25) or (11.26)) and S_0 the one for the short bearing approximations (Equation (11.28) or (11.28a)), a good approximation for a finite bearing $(4 > L/D > 0.25)$ and for preliminary design calculations is:

$$S_{\text{finite}} = S_\infty + S_0 \frac{1.4}{(L/D)^{0.22}} \tag{11.28b}$$

The error is less than 5% for most values of ε and L/D, but at extreme values of ε it might reach somewhat higher errors. This approximation also is very convenient for CAD applications.

For intermediate values of L/D, the bearings are termed *finite* and numerical methods are also used, or tabulated results (usually obtained with numerical methods) such as Figure 11.17 for the full cylindrical bearing. The curve labeled "Max W" designates the combination of ε and S that gives the maximum load-carrying capacity for good bearing operation, while the curve labeled "Min f" designates the combination of the same parameters that yields the minimum friction coefficient and thus the most efficient bearing. The shaded area between these two curves is the recommended operating range for the full 360° journal bearing.

It must be pointed out that the numerical results shown in Figures 11.17–11.19 are to be used only for the full 360° journal bearing. Because such bearings have

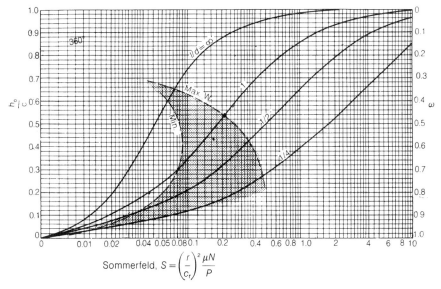

Figure 11.17. Minimum oil film thickness for a full (360°) cylindrical bearing. (After Raimondi and Boyd 1958; courtesy ASLE.)

problems such as cavitation and dynamic instability at high speeds (oil whirl), in critical applications partial bearings are used, for which the above numerical results are not applicable. In such case, the designer should not use these graphs and should either seek solutions applicable to the particular geometry or use the analytical approximations, if applicable.

11.3.5. Frictional Torque

For many practical purposes, it suffices, as a first approximation, to calculate the friction loss from the Petroff equation assuming a parallel-sided oil film, that is, as if the bearing were concentric.

For a loaded bearing running eccentrically, it is the larger viscous drag torque on the rotating member, T_J, that must be calculated in order to predict the magnitude of the heat generated. As a general rule, T_J can be expressed in the form

$$T_J = T_p \phi(\varepsilon) \tag{11.29}$$

where T_p is the torque predicted by the Petroff equation and $\phi(\varepsilon)$ is a friction factor involving the degree of eccentricity (at concentricity, $\phi(0) = 1$). As the eccentricity increases, the calculated value of $\phi(\varepsilon)$ rises gradually but still remains near unity until eccentricity ratios of the order of 0.5–0.6 are reached. However, at higher eccentricities, $\phi(\varepsilon)$ increases more rapidly and for very high eccentricities approaches a value of 3.

For the infinitely long and the short bearings, function $\phi(\varepsilon)$ can be readily computed. To this end, the power loss per unit time for the portion of the film at an angle d will be:

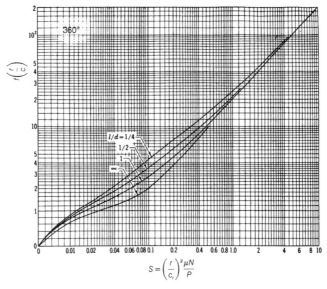

$$S = \left(\frac{r}{c_r}\right)^2 \frac{\mu N}{P}$$

Figure 11.18. Friction factor for a full (360°) cylindrical bearing. (After Raimondi and Boyd 1958; courtesy ASLE.)

$$d\dot{W} = Qdp = q\,\frac{\partial p}{\partial \theta}\,d\theta + Q\,\frac{\partial p}{\partial y}\,dy \tag{11.30}$$

Integrating

$$\dot{W} = \int_{\theta_1}^{\theta_2} \int_0^L Q(\theta, y) \left(\frac{\partial p}{\partial \theta}\,d\theta + \frac{\partial p}{\partial y}\,dy\right) = T_j\omega = T_p\omega\phi(\varepsilon) \tag{11.31}$$

where the flow Q is given by Equation (11.8). Integration under the assumptions of infinitely long and short, and 180° and 360°, bearings, yields the following results for the function $\phi(\varepsilon)$:

1. 360° bearing, infinitely long:

$$\phi_1 = \frac{2(1 + 2\varepsilon)^2}{(2 + \varepsilon^2)/(1 - \varepsilon^2)^{1/2}} \tag{11.32}$$

2. 180° bearing, infinitely long:

$$\phi_2 = \frac{2(1 + 2\varepsilon)^2}{(2 + \varepsilon^2)/(1 - \varepsilon^2)^{1/2}} \tag{11.33}$$

3. 360° bearing, short:

$$\phi_3 = \frac{[1 + \varepsilon^2/2(1 - \varepsilon^2)](L/D)^2}{(1 - \varepsilon^2)^{1/2}} \tag{11.34}$$

4. 180° bearing, short:

$$\phi_4 = \frac{\phi_3}{2} \tag{11.35}$$

Numerical solutions for finite bearings yielded friction coefficient functions given in Figure 11.18.

11.3.6. Heat Balance

In order for the actual oil temperature and hence the working viscosity to be estimated, a balance must be established between the frictional heat generated within the bearing clearance and the heat dissipated from the bearing itself. At thermal equilibrium, the frictional heat is partly carried away in the oil flowing out from the ends of the bearing and partly transmitted by conduction through the body of the bearing and along the shaft. The solution of the heat balance equation requires evaluation of the heat transfer along these three paths.

For forced-feed bearings, experiments show that almost all the frictional heat is removed by the oil flow. In the case of self-contained bearings, transfer of heat is then localized to the region of minimum film thickness and may give rise to substantial temperature gradients both circumferentially and axially. Because of the variety of designs in these bearings, it is possible to apply only rough, empirical rules.

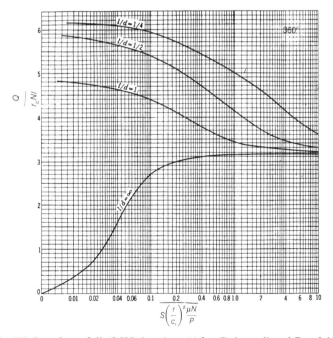

Figure 11.19. Oil flow for a full (360°) bearing. (After Raimondi and Boyd 1958; courtesy ASLE.)

11.3.7. Oil Flow

At any angular position θ where the film thickness is h, the circumferential flow Q_c is:

$$Q_c = \frac{ULh}{2} - \frac{h^3 L}{12\mu r}\left(\frac{\partial p}{\partial y}\right) - \frac{h^3}{12\mu}\frac{\partial p}{\partial y} \qquad (11.36)$$

Equation (11.36) may be now integrated with the known expressions for pressure for the infinitely long and the short bearings. This yields:

$$\text{Infinitely long: } Q_L = \frac{\pi c_r DLN(1 - \varepsilon^2)}{2(2\varepsilon^2)} \qquad (11.37)$$

$$\text{Short: } Q = \frac{cDLN(5 - 1 + 24)}{16} \qquad (11.38)$$

Oil flow for the 360° finite bearing has been obtained numerically and is given in Figure 11.19 in the form of a flow factor $Q/r_c NL$.

11.3.8. Journal Bearing Design

From practical experience in bearing design, a number of design rules have emerged that provide general guidance in choosing bearing dimensions and acceptable operating conditions.

11.3.8.1. Heat Generation. The product pU of the average pressure and the peripheral velocity is proportional to the generated heat FU, where F is the friction force, in proportion to p. Because the capacity to absorb and transmit the generated heat depends on the overall design and operating conditions, it is suggested for automotive engines that pU is up to 25–35 Nm/mm², high-speed internal combustion engines up to 30–50, rolling mill bearings 40–200, and steam turbines up to 100 Nm/mm². This experience has been included in Figure 11.20a, b.

For thermal calculations, the heat transfer coefficient at the bearing surface can be taken as $h = 10(1 + u^{1/2})$ W/m² °C, where u (m/s) is the velocity of the air flow around the bearing housing. If no forced flow is available, natural convection can be assumed with $h = 14$ W/m² °C.

For bearings without oil circulation, all generated heat has to be transmitted to the environment by the bearing enclosure surface A. The equilibrium temperature will be at the point where the generated heat, given by Equation (11.30), equals the heat removed by convection over the surface A:

$$\dot{W} = hA(T_b - T_e) \qquad (11.39)$$

where T_b is the bearing temperature and T_e is the environment temperature.

T_b must not exceed, for usual oils, 70°C for normal conditions and 90°C for heavily loaded bearings. If T_b is substantially higher than the environment temperature, it affects viscosity and the bearing calculations have to be repeated until the difference between assumed viscosity and the viscosity at the resulting bearing temperature are close enough.

If the temperature is above safe limits and changing design parameters within practical limits cannot resolve the problem, forced circulation and external oil cooling can be provided. The oil temperature rise will then depend on oil flow.

First the heat generation within the oil film has to be estimated. Then energy balance demands that:

$$\dot{W} = c\rho Q(T - T_0) \tag{11.40}$$

where T_0 is the oil supply temperature, T is an average oil temperature, Q is the oil supply to the bearing, and c is the specific heat.

The oil supply is usually greater than the oil circulation Q in the bearing in order to keep exit temperatures low. Usual design-temperature differences ΔT between oil supply and return are in the range 8–15°C and seldom exceed 20°C. The external oil supply will be:

$$Q_e = \frac{\dot{W}}{c\rho \Delta T} \tag{11.41}$$

For mineral oils, ρc can be taken as 1.36 N/mm^2 °C or 110 lb/in.^{2} °F at normal temperatures.

Self-contained bearings without external oil circulation have been successfully operated in the following range:

Diameter (mm)	Maximum Speed (rpm)
75	3600
200	1000
600	200

11.3.8.2. Clearance Ratio (c_r/r). The bearing performance is particularly sensitive to changes in this ratio. Clearance ratios are usually within the range of 0.001–0.0005. The lower figure applies to slow-speed bearings and it is usually necessary to increase the ratio with increasing speed to avoid undue heating. A diametral clearance of 0.001 in. per in. of shaft diameter is commonly accepted as standard practice, i.e., $c_r/r = 0.001$. Vogelpohl (1958) suggested:

$$\frac{r}{c_r} = 1000 + 0.4\sqrt{U} \tag{11.41a}$$

with U in m/s.

It is suggested that calculations be performed for a range of clearance ratios and the results tabulated. Then the optimum value can be selected by an overall evaluation of the results obtained.

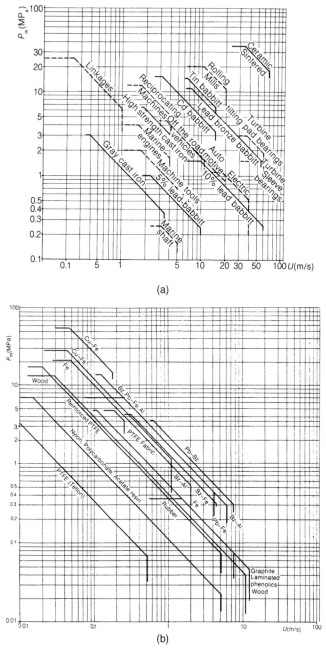

Figure 11.20. Design experience with fluid bearings: (*a*) based on applications (*b*) based on materials.

11.3.8.3. Minimum Oil Film Thickness (h_{min}). Production tolerances, surface finish of journal and bearing, and the effects of possible contaminants in the oil supply must all be considered. For small bearings (less than 50 mm diameter) having high-quality surface finishes and running at slow to medium speeds, h_{min} is usually designed to be not less than about 2.5 μm. Trumpler (1966) suggested minimum film thickness

$$h_{min} \ (\mu m) = 5 + 0.4D \tag{11.41b}$$

where D is in mm. Limiting factors for the minimum film thickness are the journal and bearing surface roughnesses, R_j and R_b, respectively. The minimum film thickness should also account for the maximum expected size of solid particles in the oil d_f, usually dictated by the filter type. For usual industrial quality filters, this size is 50 μm, while with special filters particle size can be limited to 5–10 m. Finally:

$$h_{min} > S(R_j + R_b) + d_f \tag{11.41c}$$

where S should be at least 2. This should be the minimum film thickness for the running fit selected.

11.3.8.4. Radial Clearance (c_r). Allowances may have to be made for wear and for possible shaft misalignment or deflection. For example, if the minimum clearance ratio desired is 0.001, allowance for wear could be 0.0005 and production tolerance 0.0005. Calculations of load capacity and performance must be made at both possible extremes resulting from these allowances.

11.3.8.5. Length-to-Diameter Ratio (L/D). There is no obvious advantage in making the bearing much longer than the shaft diameter, since reduced loading per unit projected area of bearing may be offset by increased frictional torque; the chances of edge loading owing to misalignment are also increased. If no special restrictions are imposed, an L/D ratio of 1.0 is a reasonable basis to adopt.

In general, length L is computed for the selected oil, clearance, and minimum film thickness.

11.3.8.6. Oil Grooves. The purpose of oil grooves is to distribute the lubricant in the bearing so that an effective pressure film may be formed as soon after the point of oil entry as possible (Figure 11.21). Grooves that are incorrectly placed, for example within or close to the loaded area, reduce the load-carrying capacity.

Design Procedure 11.2: Design of a Journal Bearing

Input: Maximum operational load W, oil viscosity μ, rotating speed ω. **Output:** Dimensions of bearing, operating parameters.

Step 1: Sketch the bearing and place on the sketch loads and motion data.

Step 2: Assume a bearing diameter D.

Step 3: Calculate the peripheral speed at the bearing diameter, $U = \omega D/2$.

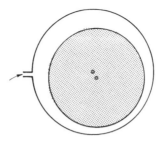

Figure 11.21. Geometry of a cylindrical journal bearing.

Step 4: Select a bearing material. From Figures 11.20a, b, from U find the maximum allowed average pressure $p_{max} = W/LD$. Find L.

Step 5: From Vogelpohl equation (11.41a), find the r/cr ratio. Find cr ($r = D/2$).

Step 6: Find Sommerfeld number (Equation (11.24)). From Figure 11.17 (for a full 360° finite bearing) or from Equation (11.25) or (11.26) or (11.28) or (11.28a) (depending on the bearing angle 360° or 180° and the L/D ratio), determine the eccentricity ratio?

Step 7: Find the eccentricity $e = \hat{\varepsilon}c$ and the minimum oil thickness $h_0 = c - e$. If the eccentricity ratio $\hat{\varepsilon} > 0.85$ or the minimum oil film thickness $h_0 > h_{min}$, use Equation (11.41b) or (11.41c). If these two tests fail, go to step 2 or 4 and assume a larger diameter and/or larger length, until the eccentricity and film thickness are acceptable.

Step 8: Find friction factor (Equation (11.5) or Figure 11.18) and power loss (Equation (11.4) or (11.29)) with ϕ from one of Equations (11.32)–(11.35).

Step 9: Find the bearing oil flow from Figure 11.19 or Equation (11.37) or (11.38) and oil supply for cooling from Equation (11.41).

Example 11.8   A turbine journal bearing has diameter 400 mm, rotates at $n = 3000$ rpm, uses oil with $\mu = 25$ mPa · s at the operating temperature, and the bearing radial load is 10 kN. Design the bearing and find the operating eccentricity, power loss, and oil flow.

Solution   The solution, based on Design Procedure 11.2, is implemented in the spreadsheet in Table E11.8.

Example 11.9   An electric generator rotor weighs 600,000 N and rotates at 3600 rpm. It is supported by two sleeve bearings. The journal diameter is 400 mm and the oil used is light oil SAE 30. The oil filter allows particles 60 μm and the surface roughnesses are 5μm. Determine:

1. The required bearing width
2. The operating parameters for normal operation

Assume bearing of infinite length.

Solution   The load per bearing is 600,000/2 = 300,000 N. The clearance ratio is selected $c_r/r = 0.001$. Therefore, the clearance is $c_r = 0.001r = 0.001 \times 200 = 0.2$ mm. The minimum oil film thickness:

TABLE E11.8

1	B	C	D	E
2	**PROCEDURE #11.2. Design of a Cylindrical Slider Bearing (Example 11.8)**			
3	**Input:(SI Units: N, m, s, except if otherwise stated)**		*Shaded shells are for user input*	
4	Maximum operational load W (N)	1,00E+04		
5	Oil Dynamic Viscosity M_0 (mPas)	6		
6	Rotating speed (RPM)	3000	Minimum Oil Film Thickness h_0 =	0,00008
7	**Step 2:** Assume a bearing diameter			
8	Bearing Diameter d_1 =	0,4		
9	**Step 3:**			
10	Peripheral speed U =	62,83185307	Rev per sec N_s =	50
11	**Step 4:** *Select bearing material: Pb Bronze babbit (fig 11.20a)*			
12	Average Pressure p_av =	200000	N/m^2	
13	Bearing Length L_b =	0,125	L/D ratio =	0,3125
14	**Step 5:** Find clearance			
15	r/c ratio r/c =	1003,170662	Radial clearance c_r	0,000199
16	**Step 6:**			
17	Operation Sommerfeld Number S =	1,509527065		
18	Eccentricity ratio ε =	0,733282014	Enter a guess first	
19	Computer Sommerfeld number S_c	0,019867785	Infinite, 360 deg	
20		0,032757857	Infinite, 180 deg	
21		1,775831081	Short, 360 deg	
22		0,222372708	Short, 180 deg	
23		1,509526256	Finite Brg, 360 deg	
24		0,228043307	Finite Brg, 180 deg	
25	Operational S - computed S_c	8,09439E-07	= S-C23 or other cell	
26	*Use **Solver** to find eccentricity ratio that gives operational Sommerfeld number S - computed above for one of the bearing types above = cell C25 = 0, enter the appropriate cell reference C19 to C24 to the formula in the equation in cell C25*			
27	**Step 7:**			
28	Eccentricity ε =	0,000146193	Min oil film thickness	5,3175E-5
29	*If minimum oil film thickness is less than the one in cell E6 (given) or equations 11.41b,c, go to step 2, increase diameter and repeat steps 3-7*			
30	**Step 8:**			
	Petroff Power Loss =	18662,75226	Watts	Power Loss
31	Function $\phi(\varepsilon)$, equations 11.32-35	7,052016274	Infinite, 360 deg	131610,03
32		3,526008137	Infinite, 180 deg	65805,016
33		0,227155617	Short, 360 deg	4239,3490
34		0,113577809	Short, 180 deg	2119,6745
35	**Step 9:**			
36	Oil Flow Q: Infinitely long bearing	0,000142625	m³/s	
37	Short bearing	0,000124393		

$$h_{min} = 2 \times (5 + 5) + 60 = 80 \ \mu m$$

$$\frac{h_{min}}{c} = \frac{80}{200} = 0.4$$

$$\varepsilon = \frac{1 - h_{min}}{c} = 1 - 0.4 = 0.6$$

A bearing temperature will be selected, and then a corresponding viscosity from Figure 11.4. It will be verified later.

1. The Sommerfeld number yields the bearing length:

$$L = \frac{SW}{\mu ND(r/c)^2}$$

The Sommerfeld number is obtained from Equation (11.25) for infinitely long, 180° journal bearing:

$$S = \frac{(2 + \varepsilon^2)(1 - \varepsilon^2)^{1/2}}{12\pi^2 \varepsilon} = \frac{(2 + 0.6^2)(1 - 0.6^2)^{1/2}}{12\pi^2 0.6} = 0.300$$

For a temperature of 60°C, $\mu = 12$ mPa.s. Then:

$$L = \frac{SW}{\mu ND(r/c)^2} = \frac{0.3 \times 300,000}{0.012 \times 60 \times 0.400 \times (1000)^2} = 0.578 \ m$$

and

$$\frac{L}{D} = \frac{0.578}{0.400} = 1.445$$

The Petroff torque is:

$$T_p = \frac{\pi^2 \ 12 \times 10^{-3} \times 60 \times 0.578 \times 0.4^3}{2 \times 0.0002} = 657 \ Nm$$

The correction $\phi(\varepsilon)$ is (Equation (11.33)):

$$\phi(0.6) = \frac{2 \times (1 + 2 \times 0.6^2)}{(2 + 0.6^2)(1 - 0.6^2)^{1/2}} = 0.91$$

Therefore, the frictional torque will be:

$$T_J = T_P = 0.91 \times 657.7 = 598 \ Nm$$

a little less than the Petroff torque. The heat generation is:

$$\dot{W} = T_J \omega = T_J \ 2\pi N = 2\pi \times 60 \times 1506 = 667,749 \ W$$

The oil flow for the infinitely long bearing is (Equation 11.37)):

$$Q_L = \frac{\pi c_r DLN(1 - \varepsilon^2)}{(2 + \varepsilon^2)} = 2.36 \times 10^{-3} \text{ m}^3/\text{s}$$

This concludes the bearing design.

Advanced Design

To account for the change in viscosity due to the bearing temperature, the heat balance requires (Equation (11.39))

$$\dot{W} = \rho c Q_L \Delta T$$

For $\rho c = 1.36$ Nm/m³°C:

$$\Delta T = \frac{\dot{W}}{\rho c Q_L} = 76°\text{C}$$

For an environmental temperature $T_0 = 20°$C, the oil temperature is 96°C. But it was assumed initially to be 60°C. Therefore, we repeat the calculation for several temperatures. The results follow:

$T_0(°\text{C})$	$\mu(\text{mPa} \cdot \text{s})$	ΔT	$T_e + \Delta T$
60	13	76	96
70	9	52.6	72.6
80	6.6	38.6	58.6

For $T_0 = 70°$C, a reasonable agreement is obtained. Repeating the calculation for $T_0 = 70°$C, $\mu = 9 \times 10^{-3}$ Pa.s:

$$L = 0.836 \quad \dot{W} = 225 \text{ kW}, \quad Q_L = 3.42 \times 10^{-3} \text{ m}^3/\text{s}$$

2. For the operating load of 300 kN per bearing, the computation is repeated to yield, with the length $L = 836$ mm found previously, operating temperature = 52°C and eccentricity ratio = 0.05, power loss 260 kW. Unloading the bearing resulted in very small eccentricity, which might present instability problems. This indicates that using a high safety factor for the load in bearings does not necessarily result in a safer bearing—the load-carrying capacity will be sufficient but problems of another nature might be encountered.

Example 11.10 In a circumferential groove bearing, the oil is supplied through a circumferential groove of enough depth to ensure uniform pressure around the journal, and the flow is along the journal axis.

For such a bearing of diameter $D = 100$ mm, length of the bearing lands $L = 50$ mm, carrying a load of 30 kN at 3000 rpm, determine the operating parameters, eccentricity ratio, oil flow, exit oil temperature, and power loss. The SAE 20 oil is supplied at 43°C and pressure 20 N/mm², and the clearance ratio c_r/r is 0.001.

Consider the bearing that consists of two equal short bearings of 360° oil film.

Solution Figure 11.4 gives, for an initial estimate of viscosity at 43°C for the SAE 20 oil, $\mu = 43$ mPa.s.

The Sommerfeld number is, for each bearing half:

$$S = \frac{\mu NLD}{W}\left(\frac{r_c}{c}\right)^2 = \frac{43 \times 10^{-3} \times 50 \times 0.050 \times 0.100}{30,000} \, 1000^2 = 0.358$$

The ratio $L/D = 50/100 = 0.5$.

For the infinitely long bearing:

$$\left(\frac{L}{D}\right)^2 S = \frac{(1 - \varepsilon^2)^{3/2}}{2\pi^2\varepsilon}$$

Solving for ε by iteration, $\varepsilon = (1 - \varepsilon^2)^{3/2}/[2\pi^2(L/D)^2 \, S]$, we obtain $\varepsilon = 0.422$.

The axial oil flow is independent of the bearing rotation, provided that cavitation does not take place because the Reynolds equation is linear and the principle of superposition holds. The oil flow (Equation (11.37)) for infinitely long bearings is:

$$Q_L = \frac{\pi c_r DLN(1 - \varepsilon^2)}{2(2\varepsilon^2)}$$

Therefore, for $c = 0.001r = 0.050$ mm:

$$Q_L = \frac{\pi \times 0.050 \times 10^{-3} \times 0.050 \times 50 \times (1 - 0.422^2)}{2(2 \times 0.422^2)} = 1.38 \times 10^{-4} \text{ m}^3/\text{s}$$

The Petroff torque (Equation (11.3)):

$$T_P = \frac{\pi^2\mu NLD^3}{2c} = \frac{\pi^2 \times 43 \times 10^{-3} \times 50 \times 0.050 \times 0.100^3}{2 \times 0.050 \times 10^{-3}} = 10.6 \text{ Nm}$$

Heat generation (Equation (11.40)) is:

$$\dot{W} = T_p\omega\phi(\varepsilon) = \frac{T_p2\pi N[1 + \varepsilon^2/2(1 - \varepsilon^2)](L/D)^2}{(1 - \varepsilon^2)^{1/2}} = 1019 \text{ W}$$

Temperature rise (Equation (4.40)) for $\rho c = 1.36E6$:

$$\frac{T - T_0\dot{W}}{\rho cQ} = 23°\text{C}$$

The exit temperature is:

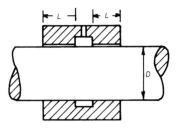

Figure E11.10

$$T = T_0 + 23 = 40 + 23 = 63°C$$

To assess the effect of oil temperature, we proceed as follows.

Repeating the calculation for oil temperature 59°C and so on until the exit temperature matches the assumed one, this leads to an exit temperature of 55°C with the following operating parameters: $S = 0.460$, $Q = 4.51 \times 10^{-4}$ m^3/s, $\dot{W} = 6173$ W, $T = 50°C$.

Example 11.11   The rotor of a turbo-compressor weighs 100 kN and is supported by two tilting pad bearings with four pads each. The light turbine oil is equivalent to SAE 5 oil and is extremely cooled to 50°C and fed to the bearings by axial grooves. Design the proper bearing and determine the operating characteristics. The rotating speed is 3600 rpm. Select clearance ratio is 0.002.

Solution   To start, an L/D ratio 1 is selected. From engineering experience (Figure 11.20a), a value of average pressure 2 MPa is selected. Each half carries a 50 kN load. Therefore:

$$P_m = \frac{W}{LD}, \ 2E6 = \frac{50E3}{LD}, \ LD = 0.025$$

Because $L/D = 1$, $D^2 = 0.025$, $D > 158$ mm. The clearance ratio is selected as 0.002.

For diameters 160, 170, 180, . . . , mm, the load-carrying capacity will be computed. Then the one that will be equal to the given bearing load will be selected.

We assume an average oil film temperature 70°C and Figure 11.4 gives $\mu = 8.1$ mPa.s. The clearance is $c_r = (c_r/r)r = 0.002r$.

The minimum film thickness $h_{min} = 5 + 0.04D$ and the eccentricity ratio is then $\varepsilon = 1 - h_{min}c_r$.

Figure 11.26, for the given ε, yields the Sommerfeld number S. Then the load is:

$$W = \frac{\varepsilon \mu NLD}{S} \left(\frac{r}{c_r}\right)^2$$

The procedure was programmed, with the following results: At $D = 185$ mm, the load-carrying capacity is $W = 51{,}979$ N, at $\varepsilon = 0.993$ and $S = 0.08$. The clearance is $c_r = 0.002 \times 0.105 = 0.370$ mm.

To find the thermal equilibrium, we must estimate the power loss. For $\varepsilon = 0.935$, Figure 11.17 gives the equivalent Sommerfeld number for the full sleeve bearing $S = 0.14$. From Figure 11.18, the friction parameter $f(f/c) = 0.9$. Therefore:

$$f = 0.9 \times 2 \times 10^{-3} = 1.8 \times 10^{-3}$$

The power loss for the two bearings is:

$$\dot{W} = \frac{2(fW \times \pi ND)}{2} = 1.8 \times 10^{-3} \times 50{,}000 \times 3.14 \times 60 \times 0.180 = 3216 \text{ W}$$

The oil flow parameter (from Figure 11.19) is $Q/RcNL = 4.8$. Therefore:

$$Q = 4.8 \times 0.100 \times 0.0004 \times 60 \times 0.200 = 9.2 \times 10^{-4} \text{ m}^3/\text{s}$$

The temperature rise is:

$$T - T_0 = \frac{\dot{W}}{\rho c Q} = \frac{3261}{1.36 \times 10^6 \times 2.3 \times 10^{-3}} = 2.65°C$$

The oil temperature is then 52.6°C and not 70°C as assumed. The computation is repeated until agreement is obtained. The final results are:

$$D = 160 \text{ mm}, L = 160 \text{ mm, load capacity, } W = 57{,}600 \text{ N,}$$

$$\dot{W} = 3125 \text{ W}, Q = 5.83 \times 10^{-4} \text{ m}^3/\text{s}, T = 54°C$$

11.3.9. Stability of Journal Bearings

Hydrodynamically lubricated journal bearings can exhibit a form of instability known as "oil-film whirl." This most readily occurs in high-speed, lightly loaded bearings such as those of some turbines and gives rise to the transmission of vibrations throughout the equipment, which is most undesirable because of the possible fatigue effects on components.

There are two types of oil-film whirl. In *half-speed* or *half-frequency* whirl, the journal performs an orbital path in the bearing. The locus of the journal center usually encloses the bearing center; its direction is in that of journal rotation and its speed is half the speed of rotation of the journal.

In *resonant* whirl, or *oil whip,* the journal center describes a similar path, but its frequency is that of the critical speed for mechanical whirling of the shaft alone. Resonant whirl usually occurs only at speeds above twice the critical whirling speed. In resonant whirl, the stiffness of the rotating shaft is therefore seen to be a controlling factor; the larger the diameter of the shaft relative to its length, the less likely is instability in the bearing.

The unstable interaction of rotating shaft with hydrodynamic oil film is to be avoided in the design of bearings. The following factors, some of which cannot be taken into account by the bearing designer because they are not under his control, all reduce the likelihood of oil-film whirl:

1. Reduced bearing clearance
2. Increased oil supply pressure
3. Lower running temperature
4. Higher oil viscosity
5. Lower running speed
6. Lower L/D ratio of journal and shaft
7. Higher bearing load
8. Absence of vibrations or shocks, including load fluctuations, that could cause an initial displacement of the journal from its equilibrium position.

Oil grooves in the bearing give greater stability, as does an "anti-whirl" bearing design such as a slightly elliptical bore (Figures 11.21 and 11.22).

Tilting pad bearings usually resolve oil-film whirl problems. They follow the motion of the shaft without generation of lateral forces that could initiate an unstable shaft motion.

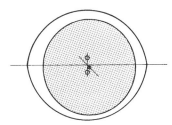

Figure 11.22. Geometry of an elliptical journal bearing.

11.3.10. Design of Tilting Pad Bearings

For axial bearings used for ship propulsion a problem of wedge formation existed that was solved by Michell, dividing the thrust surface into a number of individual pads with an inclination in respect to the direction of rotation, such as in Figure 11.23*a*. It was further found that improved operation could be achieved by non-fixed pads, free to turn about a pivot (Figure 11.23*b*). The pads were so pivoted that they could take up the angle of approach necessary for the formation of the wedge of lubricant. The complete thrust ring for such a thrust block is shown in Figure 11.24. The same idea has been used for journal bearings (Figure 11.25).

Michell journal bearings are also available. In these, tilting pads replace the ordinary fixed bearing and are suitably mounted to tilt freely to the angle necessary to form the wedge of lubricant, but they cannot move circumferentially with the shaft. Much higher bearing pressures then become possible, and better vibration attenuation. Load-carrying capacity for the usual types of pad bearing can be found from Figure 11.26. Power loss and oil flow can be taken as equal to the ones for the full journal bearing with the same eccentricity ratio, if appropriate design data are not available.

11.4. EXTERNALLY PRESSURIZED (HYDROSTATIC) BEARINGS

To keep two solid surfaces separated, fluid pressure must be applied at the interface. Consider a block of area A subjected to a load W and separated from a plane by

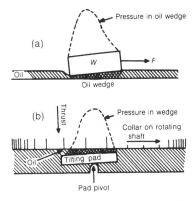

Figure 11.23. Pressure distribution on a tilting pad.

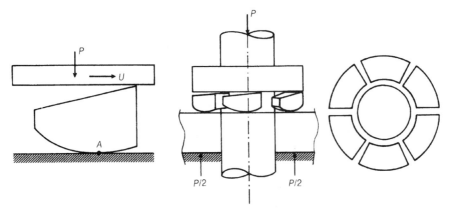

Figure 11.24. Geometry of an axial tilting thrust pad bearing.

a film of fluid held at a constant pressure p_s, as in Figure 11.27. The film can be sustained provided that:

$$W = p_s A \tag{11.42}$$

If p_s is not constant at all points, so that each element dA carries a load dW:

$$dW = p\,dA \tag{11.43}$$

For the total load:

$$W = \int_A p\,dA \tag{11.44}$$

where the integral is taken over the total area.

For flow along a uniform pipe of diameter d (Figure 11.28), the Poiseuille flow (Equation (11.6b)) can be written as:

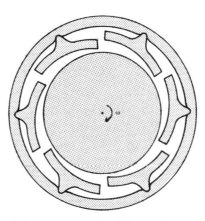

Figure 11.25. Geometry of a tilting pad journal bearing.

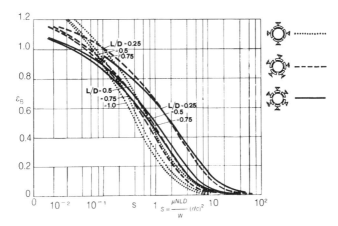

Figure 11.26. Eccentricity ratio for tilting pad journal bearings.

$$q = \frac{\pi d^4}{128\mu}\left(\frac{dp}{dx}\right) = \frac{\pi d^4}{128\mu}\left(\frac{p_2 - p_1}{L}\right) \tag{11.45}$$

For flow between parallel surfaces (Figure 11.29), Equation (11.6b) can be written as:

$$dq = \frac{h^3 dz}{12\mu}\left(\frac{dp}{dx}\right) = \frac{h^3 dz}{12\mu}\left(\frac{p_2 - p_1}{L}\right) \tag{11.46}$$

The pressure drop along the large section is very small in relation to the other section, and for such large sections we can therefore assume constant pressure (Figure 11.30).

For the geometry shown in Figure 11.31, for constant flow:

$$\frac{dp}{dx} = -\frac{12\mu dq}{h^3 dz} = \frac{k}{h^3} \tag{11.47}$$

where k is a constant.

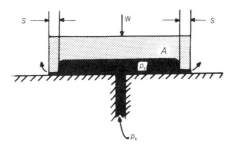

Figure 11.27. Externally pressurized (hydrostatic) bearing.

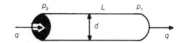

Figure 11.28. Pipe flow.

11.4.1. The Simple Pad Hydrostatic Bearing

A simple hydrostatic pad is supplied by constant pressure p_s along the central line and the x-direction. Flow is then assumed along the y-direction only. The load per unit width in the z-direction is:

$$W = 2 \left(\frac{p_s - 0}{2} \right) \frac{l}{2} = p_s \frac{l}{2} \tag{11.48}$$

If we now increase the load, we obviously have to increase p_s to maintain equilibrium.

There are two possible solutions. We may use not a constant pressure but a constant flow supply, or we may use a constant pressure supply together with a "compensating" element.

11.4.1.1. Constant Flow Supply. Suppose a pump supplies fluid at a constant flow rate q. Neglecting flow in the z-direction, the flow from the bearing in the y-direction is from the center to the two edges. Applying equation (11.46) for parallel-walled channels to one-half of the bearings gives (Figure 11.31):

$$\frac{1}{2} q = \frac{h^3}{12} \left(\frac{p_s - 0}{l/2} \right) \tag{11.49}$$

$$p_s = \frac{3q\mu l}{h^3} \tag{11.50}$$

and

$$W = \frac{p_s l}{2} = \frac{3q\mu l^2}{2h^3} \tag{11.51}$$

Since q is constant, any change in W is accommodated by a change in the film

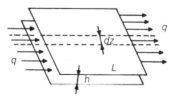

Figure 11.29. Plate flow.

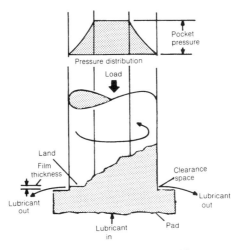

Figure 11.30. Hydrostatic bearing, uniform pressure.

thickness h, and the supply pressure p_s now varies to ensure a constant flow rate q. An improvement in the bearing design will significantly increase its load capacity (Figure 11.32). Since the film thickness in the recess is relatively large with respect to the film thickness at the lands, the pressure in the recess is essentially constant. The load capacity of such a bearing is:

$$W = p_s b + \left(\frac{p_s}{2}\right) l \qquad (11.52)$$

Considering the constant flow supply through the lands given by equation (11.49) as before, the load capacity is now:

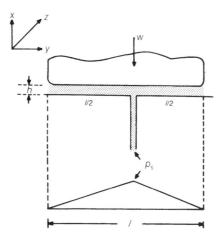

Figure 11.31. Hydrostatic bearing, uniform oil film thickness.

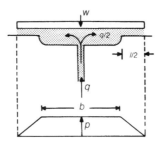

Figure 11.32. Hydrostatic bearing, mixed.

$$W = \frac{3q\mu l^2}{2h^3}\left(1 + \frac{2b}{l}\right) \tag{11.53}$$

This is an increased load capacity by a factor $1 + (2b/l)$, and this type of geometry is that most commonly employed in externally pressurized bearing designs. However, the use of a constant flow type of pump implies that every bearing would require its own separate pump, a very expensive arrangement. The advantage of the other alternative, a constant pressure pump, is that the same pump may supply a whole series of bearings, provided that its flow capacity can meet their total requirements.

11.4.1.2. Constant Pressure Supply. We have seen that a direct connection of a constant pressure supply to the bearing is not practical, since we would have to adjust the supply pressure to accommodate changes in load. Automatic adjustment can be achieved by a restricting element. The simplest compensating element is a length of capillary tube. Consider the system shown in Figure 11.33a, where a constant supply pressure p_s drops a pressure p_r at the bearing supply point. From the flow through a tube, we note, from Equation (11.44), that:

$$q = \frac{\pi d^4}{128\mu}\left(\frac{p_s - p_r}{l_c}\right) \tag{11.54}$$

The linear pressure drop along each half of the bearing (through which flows half of the fluid) must give flow

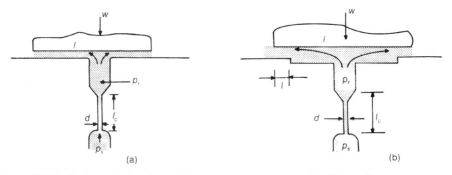

Figure 11.33. Hydrostatic bearings: (*a*) constant pressure supply; (*b*) capillary compensated.

$$\frac{q}{2} = \frac{h^3 w}{12\mu} \frac{p_r}{l/2} \tag{11.55}$$

where w is the width of the bearing in the z-direction. Thus, eliminating q, one obtains:

$$p_r = \frac{p_s}{1 + (h^3/k)} \tag{11.56}$$

where:

$$k = \frac{3\pi d^4 l}{128 l_c w}$$

The load carried is then:

$$W = \frac{p_r l w}{2} = \frac{p_s l w}{2[1 + (h^3/k)]} \tag{11.57}$$

This equation relates the load to the film thickness for a given geometry of bearing. The flow through such a bearing is then found from Equations (11.54) and (11.55):

$$q = \frac{k(p_s l w - 2w)}{1 + (h^3/k)} \tag{11.58}$$

Thus, although the supply pressure is constant, changes in load cause a change in the film thickness and the flow rate. For a constant p_s, any increase in load increases p_r and reduces flow rate. The operation is stable. A more useful geometry, in practice, is to have a large central recess together with narrow lands (Figure 11.33b). The load capacity for such a bearing then becomes:

$$W = \frac{p_s(b + l/2)w}{1 + (h^3/k)} \tag{11.59}$$

11.4.2. Externally Pressurized Journal Bearings

Figure 11.34 shows a a single-pad and a multi-pad arrangement for such a bearing. Each pad is supplied separately from a constant pressure supply via appropriate compensating elements. The oil then flows over the lands into the drainage channels and the calculations of Section 11.4.1 apply.

Such bearings are used in slow-speed applications where the speed is not high enough to sustain a hydrodynamic film. They are also used in heavy machinery, such as large steam turbines for start-up until they reach a speed high enough to assure an adequate hydrodynamic oil film thickness. The flow for each pad is:

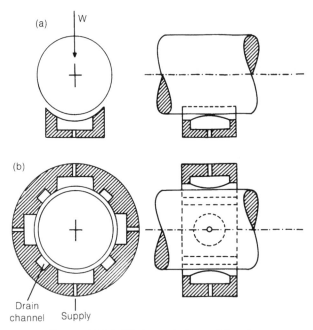

Figure 11.34. Hydrostatic journal bearing.

$$q = \frac{k[p_s(b + l/2)w - W]}{3\mu l(b + l/2)} \qquad (11.60)$$

For cases with oil flow in the z-direction or with more complex geometries, the basic behavior is similar.

Example 11.12 The hydroelectric generator shown has a thrust load of 800 kN. The thrust bearing is of the hydrostatic type, with a capillary compensator and a supply oil pressure 1000 MPa. Assuming a compensator of diameter $d_r = 1$ mm, $L = 200$ mm, oil SAE 5

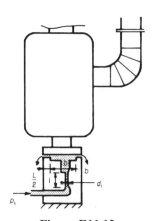

Figure E11.12

supplied at 50°C, and a land over recess diameter ratio $L/b = 0.2$, design the bearing for a minimum film thickness of 60 μm.

Solution The width of the bearing is the periphery of the land (Figure 11.33*b*):

$$2w = \pi(b + l)$$

For 50°C, oil SAE has viscosity 7 mPa.s. Then, from Equation (11.59), the load-carrying capacity is:

$$W = \frac{P_s(b + l/2)w}{1 + h^3 \cdot 128l_c w/(3\pi d^4 l)}$$

For $l = 0.2b$ the bearing recess b will be computed as follows:

$$W = \frac{p_s b^2 \times 1.1\pi \times 1.2/2}{1 + h^3 \times 128 \times l_c \times 1.2b/(2 \times 3 \times d^4 \times 0.2b)} = \frac{1.32\pi p_s b^2/2}{1 + 128l_c h^3/d^4}$$

Therefore:

$$b^2 = \frac{2W(1 + 256l_c h^3/d_r^4)}{1.31\pi p_s} = \frac{2 \times 800E3(1 + 128 \times 0.2 \times 60^3 E - 18/0.001^4)}{1.31\pi \times 1000 \times 10^6}$$

Finally, $b = 50$ mm. The other design parameters are:

$$D = 50 + 2 \times 0.2 \times 50 = 70 \text{ mm}, \quad l = 0.2b = 10 \text{ mm}, \quad w = \frac{1.2b}{2} = 30 \text{ mm}$$

The restriction coefficient is:

$$k = \frac{3\pi d_r^4 l}{128l_c w} = \frac{3\pi \times 0.001^4 \times 0.01}{128 \times 0.200 \times 1.22 \times 0.050/2} = 3.9 \times 10^{-14}$$

The intermediate pressure is:

$$p_r = \frac{p_s}{1 + h^3/k} = \frac{1000 \times 10^{-6}}{1 + (60 \times 10^{-6})^3/7.75 \times 10^{-14}} = 153 \text{ MPa}$$

The oil flow is:

$$q = \frac{\pi d_r^4(p_s - p_r)}{128l_c} = \frac{\pi \times 0.001^4(1000 - 13) \times 10^6}{128 \times 0.2} = 1.04 \times 10^{-4} \text{ m}^3/\text{s}$$

The pump power, for a pump efficiency $\eta = 75\%$, is:

$$W = \frac{p\dot{q}}{\eta} = \frac{10 \times 10^6 \times 1.28 \times 10^{-4}}{0.75} = 1719 \text{ W}$$

11.5. COMPUTER METHODS

The hydrodynamic lubrication theory has a strong analytical tool in the form of the Reynolds equation to predict bearing behavior. Solutions of this equation in analytical form have been obtained only in special cases, such as the infinitely long and the short bearing. In the ranges $0.5 < L/D < 2$ and $0.6 < \varepsilon < 0.9$, which are the most important in applications, such solutions deviate considerably from reality. For this reason, numerical methods have been employed, implemented in digital computers.

The finite-element method, introduced in Chapter 4, is widely used today for bearing analysis and design. The direct method used in Chapter 4 is not applicable here. One has to work directly with the Reynolds equation.

CASE STUDY 11.1: The Carter Automotive Fuel Pump

The journal bearing of the inner gear of the geared rotor (gerotor) fuel pump manufactured by Carter Automotive was showing premature seizing wear at the bearings during repeated start-up tests. The gerotor revolves about a 6.3 mm diameter journal. The bearing is lubricated with the gasoline that the system pumps. The task for the design engineer was to find the source of the problem and redesign the bearing to correct it.

Investigation

In the gerotor pump under consideration, the inlet and outlet regions are assumed to be under constant pressures, the inlet and outlet pressures, respectively. In the flow region, on the left of Figure CS11.1, the fluid pressure is assumed to be linearly varying with the polar angle. In the sealing region, on

Figure CS11.1. Gerotor geometry and assumed pressure distribution.

the right, the exact location of the sealing contact point cannot be determined exactly because it changes within a certain range with the rotation. Using the animation technique, which will be explained later, it was observed that an average location of the sealing point was at about 56° from the y-axis (12 oc) clockwise. Above this point the pressure is assumed equal to the outlet pressure; below this point, equal to the inlet pressure.

The exact pressure distribution is not known. It could be computed with elaborate numerical methods, but it is believed that the linear variation is a sufficiently good assumption, in view of the fact that the respective contributions on the inner bearing that eventually load the journal are opposite in direction and the bearing load comes mainly from the outlet region.

The pressure on the inner gear is integrated around the gear, and the resultant forces along two perpendicular directions are computed, namely along the line of centers and the pitch point (x) and perpendicular to it (y).

Two modes of operation are assumed:

Mode I: The resultant force in the x-direction is not transferred to the bearing but to the outer gear and the casing through the gear contact. This will happen in the case where the clearance between the inner gear and the journal plus the operating fluid film thickness between gear teeth are larger than the outer gear-casing clearance.

Mode II: It is more likely that the journal-bearing clearance is sufficiently small for the lateral force to be transmitted to the journal.

At a given speed and pressure, the bearing load is expressed as a dimensionless parameter, the Sommerfeld number:

$$S = \frac{\mu NLD(R/c)^2}{W}$$

where μ is the fluid dynamic viscosity, N is the speed of rotation (cps), L, D, R, c are the length, diameter, radius, radial clearance, respectively, and W is the bearing load. This number is related to the relative eccentricity ratio $\varepsilon = e/c$, where e is the operating eccentricity $= c - h$, where h is the minimum fluid film thickness—the critical design quantity. h should exceed the sum of the surface roughnesses plus the size of the expected particles that go through the filter.

The relation $S = f(\varepsilon)$ for a full 360° bearing can be found in this chapter.

The above procedure has been implemented in a computer program, GERO-PUMP, developed for this purpose. The program has the following inputs. Their numerical values are shown in Figure CS11.2.

Journal diameter (m)
Minimum bearing fit (diametral) clearance (m)
Maximum bearing fit (diametral) clearance (m)
Bearing length (m)
Fluid viscosity (Nsec/m^2)

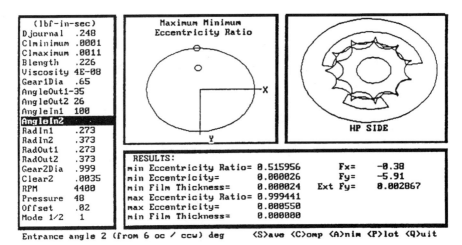

(lbf-in-sec)	
Djournal	.248
Clminimum	.0001
Clmaximum	.0011
Blength	.226
Viscosity	4E-08
Gear1Dia	.65
AngleOut1	-35
AngleOut2	26
AngleIn1	100
AngleIn2	.
RadIn1	.273
RadIn2	.373
RadOut1	.273
RadOut2	.373
Gear2Dia	.999
Clear2	.0035
RPM	4400
Pressure	48
Offset	.02
Mode 1/2	1

Maximum Minimum Eccentricity Ratio

HP SIDE

RESULTS:
min Eccentricity Ratio= 0.515956   Fx=   -0.38
min Eccentricity=   0.000026   Fy=   -5.91
min Film Thickness=   0.000024   Ext Fy=   0.002867
max Eccentricity Ratio= 0.999441
max Eccentricity=   0.000550
min Film Thickness=   0.000000

Entrance angle 2 (from 6 oc / ccw) deg   <S>ave <C>omp <A>nim <P>lot <Q>uit

Figure CS11.2. GEROPUMP output.

Outer diameter of the inner gear (m)
Exit angle 1 (from 12 oc/cw) deg
Exit angle 2 (from 12 oc/cw) deg
Entrance angle 1 (from 12 oc/cw) deg
Entrance angle 2 (from 12 oc/cw) deg
Inner radius at entrance packet (m)
Outer radius at entrance packet (m)
Inner radius at exit packet (m)
Inner radius at exit packet (m)
Outer gear diameter (m)
Minimum outer gear (diametral) clearance (m)
Rotational speed (rpm)
Exit pressure (N/m^2)
Offset of the outer packet arc center
Mode I or II

The units indicated are in the SI system. In fact, upon initiation of the program, the user is asked to select one from among three systems of units:

1. N, m, sec
2. N, mm, sec
3. lb-force, in, sec

The program produces the following results:

1. For the two extreme values of the journal-bearing clearance, the operating parameters, bearing loads F_x and F_y, eccentricity e, and minimum film thickness h are computed.

2. The pump geometry is plotted.
3. The gerotor rotation is animated to observe the gear interaction.

The designer can change any number of design parameters and repeat the computations.

For the data provided, it is clear that the bearing is marginally designed for hydrodynamic operation for the given fluid. The typical bearing design operation parameters are well within engineering practice. In fact, the typical average pressure is about 4 MPa and surface velocity 1.3 m/s. In the design charts of Figures 11.20a and b, we can observe that the operating point is not extreme. The only problem is that due to the small dimensions, very small clearances cannot be used.

The design is marginal in the sense that:

1. The range of clearances is such that some units operate under very small minimum fluid film thickness.
2. For the vast majority of the units, hydrodynamic lubrication is maintained, but at a minimum fluid film thickness, which is very small compared with the size of the unfiltered particles. For automotive class gasoline filters, this size is 5 μ or 200 μin.

In selecting clearances, it must be emphasized that large operating clearances do not necessarily mean large operating fluid film thicknesses, which is the main design criterion.

In Figure CS11.3, the fluid film thickness is plotted as a function of the operating clearance. There is an optimum clearance that yields the maximum fluid film thickness and consequently the better design. Above and below this clearance the operating fluid film thickness becomes smaller.

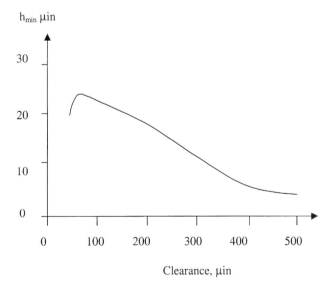

Figure CS11.3. Minimum fluid film thickness vs. operating clearance.

Most of the units are expected to survive, since the distribution of dimensions between the extreme values is stochastic. There is room, therefore, for design changes that will improve the design from a hydrodynamic point of view. For example, a quality 4 for the journal and the gear inner diameter and fit 6.3H4/g4 (mm) will yield clearances in the range 100–300 μin., which will ensure hydrodynamic lubrication. It corresponds, however, to minimum oil film thickness in the range 12–22 μin (see Figure CS11.3), much too small to accommodate fluid filtered with 200 μin. filters. The roughness for this quality is in the range of 1–16 μin., sufficient for this design.

Quality 4 can be avoided with proper sorting during production. The costs involved should be the guide for the extent of sorting and proper quality control procedures.

Redesign

In view of the very small minimum oil film thickness, much smaller than available automotive grade gas filters, it was suggested that this design cannot be improved further. It was decided that the gerotor design be replaced by a turborotor design that can operate with much larger clearances.

REFERENCES

British Petroleum (BP). 1969. *Lubrication Theory and Its Application*. London: BP.

Cameron, A. 1976. *Basic Lubrication Theory*. Chichester: E Horwood, and New York: Halsted Press.

Michell, A. G. M., 1929. "Progress in Fluid Film Lubrication." *Transactions of the ASME* 51 (1929): 153–163.

Niemann, G. *Maschinenelemente,* 1943. Berlin: Springer.

Ocvirck, F. W., and G. B. Du Bois. 1953. *Analytical Derivation and Short Bearing Approximation for Full Journal Bearings*. NASA Report, 1157.

Pugh, B. 1970. *Practical Lubrication*. London: Butterworths.

Raimondi, A. A., and J. Boyd. 1958. "A Solution for the Finite Journal Bearing and Its Application Analysis and Design I, II and III," *Trans. ASLE* 1, no. 1: 159–209.

Reynolds, O. 1886. "Theory of Lubrication, Part I." *Phil. Trans. Royal Soc.—London.*

Sommerfeld, A. 1904. "Zur Hydrodynamischen Theorie der Schmiermittel-Reibung." *Zeit. Math. Phys.* 50: 95–155.

Trumpler, P. R. 1966. *Design of Film Bearings*. New York: Macmillan.

Vogelpohl, G. 1958. *Betriebssichere Gleitlager.* Berlin: Springer.

ADDITIONAL READING

Cameron, A. *The Principles of Lubrication*. New York: John Wiley & Sons, 1976.

Dimarogonas, A. D. *Vibration for Engineers,* 2nd ed. Upper Saddle River, N.J.: Prentice Hall, 1996.

———. *History of Technology*. University of Patras, 1979 [in Greek].

Dimarogonas, A. D., and S. A. Paipetis. *Analytical Methods in Rotor Dynamics*. London: Elsevier, 1983.

Fox, R., and A. McDonald. *Introduction to Fluid Mechanics*. New York: John Wiley & Sons, 1978.

Huebner, K. H. *The Finite Element Method for Engineering*. New York: John Wiley & Sons, 1975.

Pinkus, O., and B. Sternlicht. *Theory of Hydrodynamic Lubrication*. New York: McGraw-Hill, 1961.

Stribeck, R. "Characteristics of Plain and Roller Bearings." *VDI Zeitschrift* 46 (1902).

Wills, J. G. *Lubrication Fundamentals*. New York: Marcel Dekker, 1980.

PROBLEMS[3]

11.1. [C] An oil sample was tested for viscosity. At 100°C a kinematic viscosity of 100 Saybolt Universal Seconds and a specific weight of 0.895 were measured. Find the kinematic and absolute viscosity in SI and English units and the nearest SAE grade.

11.2. [C] The viscosity of an oil sample was measured at 80°C. It had the value 8.8 mPa.s. Find the nearest SAE grade and the viscosity at 0°C.

11.3. [C] SAE 15W oil is defined as having a viscosity of 500 cP at $-18°C$ (0°F). Draw on Figure 11.4 the viscosity–temperature curve for the SAE 15W oil, by linear interpolation of the parameters c_1 and c_2 in the approximation $\log \mu = c_1 \exp(B/T) + c_2$.

11.4. [C] Determine the viscosity of the ISO VG 32 oil at a temperature of 100°C.

11.5. Determine the viscosity of the 10W-60 oil at a temperature of 50°C.

11.6. [D] The wheel bearing for an overhead crane is rated at 200 kN load and it will be made porous for self-lubrication. Select the proper material and lubricant on the basis of allowable *PV* values, keeping the ratio $L/D = 1$.

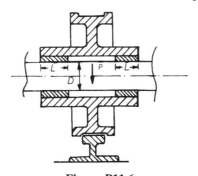

Figure P11.6

11.7. [D] The operation of a hydraulic system is controlled by a crank-and-

---

[3][C] = certification, [D] = design, [N] = numerical, [T] = theoretical problem.

follower mechanism. The maximum force transmitted through the crank journal is 2000 N. Design a proper porous bearing, keeping $L/D = 1$, for $\omega = 50$ rad/s.

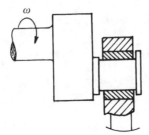

Figure P11.7

11.8. [D] A corner crane is supporting its vertical column by a combination journal and thrust bearing. The vertical force $F = 50$ kN and the horizontal force $F = 30$ kN. Design a proper porous journal and the thrust bearing. The thrust bearing must have an internal recess for lubricant accumulation on the order of one-half its external diameter.

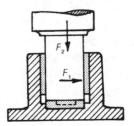

Figure P11.8

11.9. [D] Design a porous bush for the footstep bearing shown in Figure P11.8 if the journal must have a diameter of 60 mm, for strength reasons, and the horizontal load is 2000 N. The speed of rotation is very low.

11.10. [D] A swinging garage door has dimensions width 4 m and height 2.5 m, weighs 100 N, and the distance of the symmetrically placed hinges is 1.60 m. Design porous bushes for the hinges, assuming that the door is uniform and that it might sometimes be swinging at a rotating speed of $N = 2$ rps.

11.11a. [D] A fixed Michell pad bearing rotating at 500 rpm has six pads bounded by an outer circle of diameter $D = 600$ mm and an inner circle of diameter $d = 150$ mm (see Figure 11.7a). Between adjacent pads there is a 10 mm space. Lubrication is with SAE 10 oil at 70°C, and the h_1/h_0 ratio is 2. Determine:

1. The pad dimensions
2. If the minimum oil film thickness allowed is 100 μM, the load-carrying capacity for an infinite-width bearing approximation

11.11b. [D] A fixed Michell pad bearing rotating at 1000 rpm has six pads bounded by an outer circle of diameter $D = 300$ mm and an inner circle of diameter

$d = 100$ mm (see Figure 11.7a). Between adjacent pads there is a 10 mm space. Lubrication is with SAE 10 oil at 60°C, and the h_1/h_0 ratio is 2.5. Determine:

1. The pad dimensions

2. The minimum oil film thickness for a thrust load $W = 5$ kN, for an infinite-width bearing approximation

3. The power loss

11.12. [D] A tilting Michell–Kingsbury pad bearing rotating at 250 rpm has eight pads bounded by an outer circle of diameter $D = 600$ mm and an inner circle of diameter $d = 250$ mm (see Figure 11.7a). Between adjacent pads there is a 10 mm space. Lubrication is with SAE 10 oil at 60°C, and the h_1/h_0 ratio is 2.5. Determine:

1. The pad dimensions

2. The thrust load-carrying capacity if the minimum oil film thickness is 50 μM, using the infinite-width bearing approximation

3. The power loss

11.13. [D] A fixed Michell pad bearing rotating at 1000 rpm has six pads bounded by an outer circle of diameter $D = 300$ mm and an inner circle of diameter $d = 100$ mm (see Figure 11.7a). Between adjacent pads there is a 10 mm space. Lubrication is with SAE 10 oil at 60°C, and the h_1/h_0 ratio is 2.5. Determine:

1. The pad dimensions

2. The minimum oil film thickness for a thrust load $W = 5$ kN, for a Kingsbury and Needs finite-width bearing approximation

3. The power loss

11.14. [D] A tilting Michell–Kingsbury pad bearing rotating at 250 rpm has eight pads bounded by an outer circle of diameter $D = 600$ mm and an inner circle of diameter $d = 250$ mm (see Figure 11.7a). Between adjacent pads there is a 10 mm space. Lubrication is with SAE 10 oil at 60°C, and the h_1/h_0 ratio is 2.5. Determine:

1. The pad dimensions

2. The thrust load-carrying capacity if the minimum oil film thickness is 50 μM, using the Kingsbury and Needs finite-width bearing approximation

3. The power loss

11.15. [C] The main thrust bearing of a ship is a tilting Michell–Kingsbury pad bearing rotating at 250 rpm and has eight pads bounded by an outer circle of diameter $D = 600$ mm and an inner circle of diameter $d = 250$ mm, (see Figure 11.7a). Between adjacent pads there is a 10 mm space. Lubrication is with SAE 10 oil at 60°C, and the h_1/h_0 ratio is 2.5. During an emergency, it was decided that to maintain the ship speed the ship engine should run overloaded at 10% higher speed of rotation than rated. It is known to the ship engineer that change of the propeller speed by 10% will cause a change in thrust of about 20%. If the minimum oil film thickness at rated conditions is 50 mm, determine the minimum oil film thickness at 10% overspeed.

11.16. [C] Determine the Petroff power losses for the bearing shown for oil SAE 10 entering at 40°C and speed of rotation = 3200 rpm.

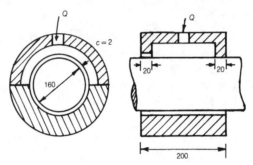

Figure P11.16

11.17. [C] The tilting pad bearing shown is supplied by oil from four equally spaced axial grooves. The diameter is 140 mm, the length L is 180 mm, the radial clearance is 0.2 mm, the oil is SAE 5, the oil temperature is 60°C, and the speed of rotation is 3000 rpm. Determine the Petroff power losses.

Figure P11.17

11.18. [T] Determine the expression of the Petroff equation for the partial bearing of angle ϕ.

Figure P11.18

11.19. [D] The bearing shown has two supply circumferential grooves, and the central circumferential groove is for return, together with the side leakage. If the diameter is 120 mm, the length of each land is $L = 30$ mm, the oil is SAE 20 at 50°C, the clearance ratio is $c_r = 0.001$, and the speed of rotation is $N = 45$ rps, determine the power loss for concentric bearing.

Then modify the length and diameter, keeping the average bearing pressure constant, for minimum power loss.

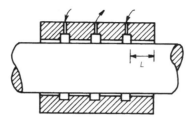

Figure P11.19

11.20. [C] The ring fed bearing shown is oiled by a ring that is dipped into oil and by the rotation it carries oil into the bearing. Determine the power loss for D = 140 mm, L = 200 mm, l = 600 mm, c_r/r = 0.001, oil SAE 40 at 60°C, with the Petroff equation for n = 1000 rpm.

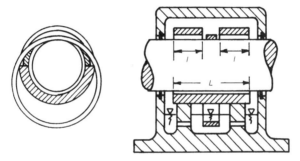

Figure P11.20

11.21. [D] The thrust bearing for a high-pressure turbine rotor is mounted on a shaft of 360 mm diameter and must support a thrust load of 300 kN at a speed of 3600 rpm. Turbine oil, equivalent to SAE 5, is supplied at 60°C. The bearing is of the Michell fixed pad type with a taper of 0.5% and a minimum film thickness according to Trumpler's equation (Section 11.3.8.3) must be maintained. Design a six-pad bearing for this purpose. Then determine the bearing operating parameters. Use an infinite-length solution and assume that the space between pads is 6% of the annular area.

11.22. [D] A turbocompressor operates at 6000 rpm, and the 60 kN thrust is taken by a five-fixed-pad Michell-type pad bearing of inner diameter D_1 = 100 mm, outer diameter D_2 = 200 mm, oil supply of 10 SAE at 50°C. Determine the minimum film thickness, oil film temperature, and oil flow for a taper of the pads 0.8% assuming infinitely long bearing and that 5% of the annular space is space between pads.

11.23. [D] The large Hoover Dam water turbine-generator rotors each weigh 1,800,000 lb. Since the shaft is vertical, the entire weight must be supported by the thrust bearing. The speed is 150 rpm. The bearing is immersed in an oil bath that contains the necessary water cooling coils. SAE 10 oil is circulated by the action of the rotating collar and is supplied to the bearing pads at 120°F. Assuming that the inside diameter of the pads may not be less than 28 in., the number of pads is eight, and 5% of the annular area is space between pads, calculate the proportions of this bearing and the horse-power loss. Show that all the criteria for satisfactory operation are met by your design and make a sketch to scale showing all important dimensions of the bearing surfaces (Trumpler 1966).

11.24. [D] The transatlantic passenger liner SS *Maasdam* (Holland-America Line), built in 1952, had a single screw driven by a propeller shaft 75 cm in diameter. Using full-turbine power of 8500 hp (shaft horsepower, or shp) and a shaft speed of 80 rpm, the ship speed is approximately 400 miles in 24 hours. The effective or two-rope horsepower (ehp) that establishes the propulsive thrust is related to shp by a "propulsive coefficient" (PC), which is defined as the ratio ehp/shp. Assume PC to be 0.70. Propose a design of tilting-pad thrust bearing for the propeller shaft, making an engineering sketch of the main features and presenting calculations to verify that your design meets the criteria of fatisfactory operation. Use SAE 20 oil delivered to the bearings at 90. Consider forward shift speed only, ignoring the problem of "reverse" (Trumpler 1966).

11.25. [D] The pad shown is known as stepped pad bearing. Determine the load-carrying capacity of such bearing for given geometry, speed, and oil properties, assuming infinite pad width.

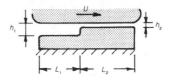

Figure P11.25

11.26. [D] The spindle shaft bearing of a heavy lathe has extreme operating conditions, load 6000 N at 450 rpm and 2800 N at 3200 rpm. It is lubricated by circulating SAE 30 oil supplied at 45°C, and the bearing has an L/D ratio of 0.5. Determine the bearing dimensions for the worst loading case, selecting appropriate minimum film thickness 20 μm and clearance ratio $c_r/r = 0.002$, by the short bearing approximation. Also determine the rest of the operating parameters.

11.27. [D] The bearing Problem 11.19 can be considered as three short bearings. With a short bearing approximation, determine the load-carrying capacity, oil film temperature, supply oil flow at atmospheric pressure, and power losses for a minimum film thickness of 25 m and clearance ratio according to the Vogelpohl equation.

11.28. [D] Solve Problem 11.27 for oil supplied at a pressure of 0.5 MPa.

11.29. [D] A four-axial groove bearing of diameter $D = 60$ mm, length 20 mm, clearance ratio $c_r/r = 0.001$, supplied by SAE 20 oil at 45°C is carrying a load of 3000 N at 3000 rpm. Using the short bearing approximation, determine the operating characteristics.

11.30. [D] The last pass of a rolling mill has a sleeve bearing loaded by 40 kN and an L/D ratio of 0.5. The clearance ratio is $c_r/r = 0.001$, and the minimum film thickness should be at least 10 μm. The speed of rotation is 4400 rpm. Determine the dimensions and operating characteristics of the bearing based on the short bearing approximation.

11.31. [D] Solve Problem 11.30 for an L/D ratio of 2 with the infinitely long bearing approximation.

11.32. [D] Solve Problem 11.20 for a 60 kN load with the infinitely long bearing approximation. If the outer surface of the bearing enclosure is 0.5 m^2 with natural cooling, determine the equilibrium temperature of the bearing.

11.33. [D] For the bearing of Problem 11.16, determine the operating conditions assuming a long 180° bearing approximation for a load 80 kN.

11.34. [D] Solve Problem 11.26 for full finite bearing. If available, compare results with Problem 11.26.

11.35. [D] Refer to Problem 11.36 and solve Problem 11.30 for full finite bearing. Compare with result of Problem 11.26.

11.36. [D] Refer to Problem 11.36 and solve Problems 11.20 and 11.32 for full finite bearing. Compare results with Problem 11.20.

11.37. [D]Refer to Problem 11.36 and solve Problems 11.16 and 11.33 for full finite bearing. Compare the results with Problem 11.16 for a 5 kN load.

11.38. [D] Refer to Problem 11.36 and solve Problem 11.31 for full finite bearing. Compare results with Problem 11.26.

11.39. [D] Design a hydrostatic pad bearing for the thrust bearing of Problem 11.21 with a constant pressure supply of 10 MPa and the same oil film thickness. Compare results with Problem 11.21, if available.

11.40. [D] Design a hydrostatic pad bearing for the thrust bearing of Problem 11.22 with a constant pressure supply of 10 MPa and the same oil film thickness. Compare results with Problem 11.21, if available.

11.41. [D] Design a hydrostatic pad bearing for the thrust bearing of Problem 11.23 with a constant pressure supply of 10 MPa and the same oil film thickness. Compare results with Problem 11.21, if available.

11.42. [D] Design a hydrostatic pad bearing for the thrust bearing of Problem 11.24 with a constant pressure supply of 10 MPa and the same oil film thickness. Compare results with Problem 11.21, if available.

11.43. [D] Design an externally pressurized pad bearing for Problem 11.8 with air supplied at 1.5 MPa pressure.

CHAPTER 12

DESIGN OF CONTACT ELEMENTS

A 5.61 ton, 1.2 m outer diameter four-row tapered roller bearing made by FAG, Schweinfurt, Germany.

OVERVIEW

In Chapters 8 and 9, we developed design methods based mostly on material strength and flexibility, respectively. The reason was that the respective mechanisms of failure were material failure and large displacements. There are other mechanisms of failure in machines, among them sliding friction and wear of machine parts. In Chapter 10, we studied design methods based on failures due to dry friction, with applications to clutches and brakes.

In Chapter 11, we studied machine element design methods based on the prevention of failures related to hydrostatic and hydrodynamic lubrication, the use of a flowing fluid to prevent rubbing surfaces from touching one another during sliding friction.

In this chapter, we will continue the design of lubricated elements, but with rolling friction.

12.1. DRY CONTACTS

The stress distribution in highly loaded contacts, lubricated or dry, closely follows the results of the theory developed for contacts between cylindrical and spherical surfaces. This is a rather complicated topic in the general theory of elasticity, and some useful results will be presented here.

Consider the contact for two spheres of diameters d_1 and d_2 under a compressive load P. Initially, the contact is a point, but due to elastic deformation, there is a circular area of contact with diameter $2a$, depending on the load. Hertz studied this problem and proved that the normal stresses over the contact area have an elliptical distribution (Figure 12.1) and that the radius of the contact circle is:

$$\alpha = \left[(3P/4) \, \frac{(1 - \nu_1^2)/E_1 + (1 - \nu^2)/E_2}{(1/d_1) + (1/d_2)} \right]^{1/3} \tag{12.1}$$

Heinrich Hertz (1857–1894) was born in Hamburg, the son of a lawyer. From a young age he was inclined towards mechanical engineering. He entered the Polytechnic Institute of Munich in 1877. He soon switched to Berlin to study physics with Helmholtz and Kirchhoff. He turned to experimental work in electrodynamics and was awarded a Ph.D. in 1880. In 1881, he presented his famous paper on contact of elastic bodies to the Berlin Academy, with a detailed analytical and experimental investigation. Next he became interested in hardness measurements and other problems of elasticity. In 1885, he became professor at Karlsruhe Polytechnic Institute. He died at 37.

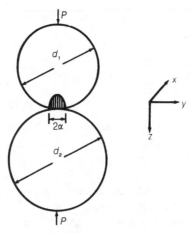

Figure 12.1. Contact of two spheres.

The maximum normal stress (p) is at the center of the contact circle and has a value of

$$p = \frac{3P}{(2\pi a^2)} = \frac{3p_m}{2} \tag{12.2}$$

when p_m is the average pressure $P/(\pi a^2)$.

The distribution of stresses along the center-line is shown in Figure 12.2 for a typical application. The maximum equivalent stress according to the maximum shear stress criterion appears below the contact surface along the center-line and has a value of approximately $0.3p_{max}$.

Solutions for different contact situations, spheres and cylinders, are given in Figure 12.3, where the maximum pressure p_0, the radius of the contact circle a (for spheres), or the half-width of the contact area (for cylinders) and the approach δ of the centers are given. Cylinder loads are assumed per unit length.

Rolling contacts are associated with repeated loading during each rolling pass. Fatigue failures usually occur in the form of surface cracks, which result in pitting, spalling and similar forms of wear. For this reason, the fatigue surface strength must now be defined.

If fatigue strength is not available from experiments, we can use the approximation that the maximum equivalent shear stress is about $0.3p_0$. For contact of cylinders, we obtain for maximum pressure:

$$p_0 = \frac{2P}{\pi a} \tag{12.3}$$

where a is the width of the contact area from Figure 12.3 and P is the external compressive force per unit length of the cylinder. If the maximum stress for fatigue failure is called *fatigue surface strength* S_{He} and the fatigue shear strength (using the Tresca criterion of failure) is $S_e/2$, then:

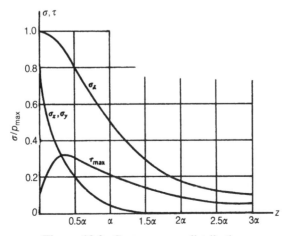

Figure 12.2. Contact stress distribution.

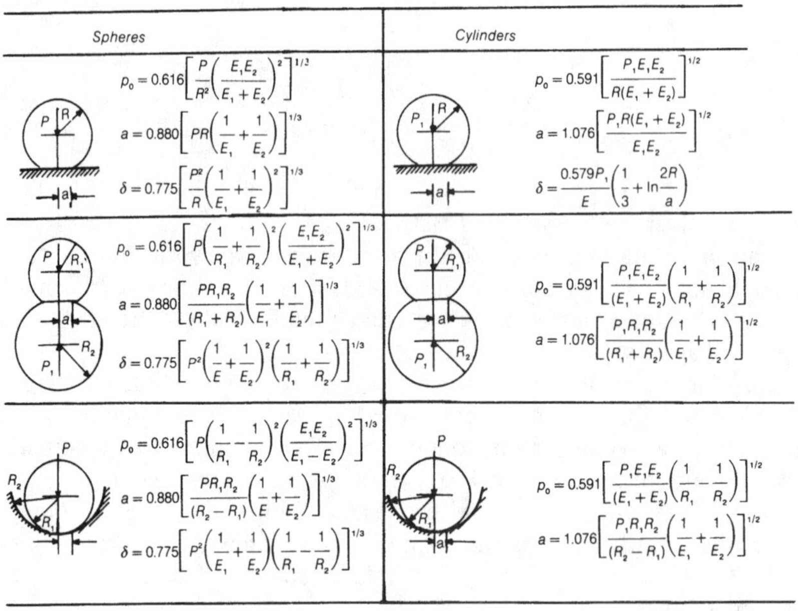

Spheres	Cylinders
$p_0 = 0.616\left[\dfrac{P}{R^2}\left(\dfrac{E_1 E_2}{E_1+E_2}\right)^2\right]^{1/3}$ $a = 0.880\left[PR\left(\dfrac{1}{E_1}+\dfrac{1}{E_2}\right)\right]^{1/3}$ $\delta = 0.775\left[\dfrac{P^2}{R}\left(\dfrac{1}{E_1}+\dfrac{1}{E_2}\right)^2\right]^{1/3}$	$p_0 = 0.591\left[\dfrac{P_1 E_1 E_2}{R(E_1+E_2)}\right]^{1/2}$ $a = 1.076\left[\dfrac{P_1 R(E_1+E_2)}{E_1 E_2}\right]^{1/2}$ $\delta = \dfrac{0.579 P_1}{E}\left(\dfrac{1}{3}+\ln\dfrac{2R}{a}\right)$
$p_0 = 0.616\left[P\left(\dfrac{1}{R_1}+\dfrac{1}{R_2}\right)^2\left(\dfrac{E_1 E_2}{E_1+E_2}\right)^2\right]^{1/3}$ $a = 0.880\left[\dfrac{PR_1 R_2}{(R_1+R_2)}\left(\dfrac{1}{E_1}+\dfrac{1}{E_2}\right)\right]^{1/3}$ $\delta = 0.775\left[P^2\left(\dfrac{1}{E_1}+\dfrac{1}{E_2}\right)^2\left(\dfrac{1}{R_1}+\dfrac{1}{R_2}\right)\right]^{1/3}$	$p_0 = 0.591\left[\dfrac{P_1 E_1 E_2}{(E_1+E_2)}\left(\dfrac{1}{R_1}+\dfrac{1}{R_2}\right)\right]^{1/2}$ $a = 1.076\left[\dfrac{P_1 R_1 R_2}{(R_1+R_2)}\left(\dfrac{1}{E_1}+\dfrac{1}{E_2}\right)\right]^{1/2}$
$p_0 = 0.616\left[P\left(\dfrac{1}{R_1}-\dfrac{1}{R_2}\right)^2\left(\dfrac{E_1 E_2}{E_1-E_2}\right)^2\right]^{1/3}$ $a = 0.880\left[\dfrac{PR_1 R_2}{(R_2-R_1)}\left(\dfrac{1}{E_1}+\dfrac{1}{E_2}\right)\right]^{1/3}$ $\delta = 0.775\left[P^2\left(\dfrac{1}{E_1}+\dfrac{1}{E_2}\right)\left(\dfrac{1}{R_1}-\dfrac{1}{R_2}\right)\right]^{1/3}$	$p_0 = 0.591\left[\dfrac{P_1 E_1 E_2}{(E_1+E_2)}\left(\dfrac{1}{R_1}-\dfrac{1}{R_2}\right)\right]^{1/2}$ $a = 1.076\left[\dfrac{P_1 R_1 R_2}{(R_2-R_1)}\left(\dfrac{1}{E_1}+\dfrac{1}{E_2}\right)\right]^{1/2}$

Figure 12.3. Contact stress and displacements for $\nu = 0.3$. For cylinders, the force P_1 is external force per unit length.

$$\tau_{\max} = 0.3 p_0 = 0.3 S_{He} = \frac{S_e}{2}, \qquad S_{He} = 1.66 S_e \tag{12.4}$$

Therefore, a fatigue surface strength of 66% greater than the fatigue strength is predicted. Because of friction, in dry contacts, experiments show that surface fatigue strength is at least $2S_e$ but for lubricated contacts 1.66 should not be exceeded.

Surface strength, however, is influenced by the material strength near the surface, which in turn is highly influenced by surface treatments. Buckingham, based on experiments, has suggested for steels the formula (Shigley 1972)

$$S_{He} = 400 H_B - 10,000 \tag{12.5}$$

where S_{He} is given in lb/in.^{2} and H_B is the Brinell hardness number.

12.2. ELASTOHYDRODYNAMIC LUBRICATION

The conditions of contact in such applications as gear teeth, cams, and roller bearings frequently result in such high contact pressures that hydrodynamic lubrication would appear to be impossible.

High contact pressures can cause appreciable local elastic deformation of the contacting surfaces with consequent change in pressure distribution. A further complication is that the viscosity of the lubricant will also be affected by the high temperatures and pressures in the oil film, as well as by the time of contact, that is, the time the oil film is subjected to load.

Figure 12.4 shows the film shape and pressure distribution between two disks in pure rolling, as in Figure 12.5, as predicted by elastohydrodynamic analysis. It will be noted that the high load elastically deforms the contacting area to localized restriction near the outlet. The pressure distribution closely follows that expected for dry elastic solids in contact over most of the region, but has a sharp rise at the exit side.

Sir William Hardy (1864–1934) was born in Erdington, Warwickshire. He was educated at Cambridge and graduated with a degree in biology. He was appointed lecturer in physiology at Cambridge in 1913, where he taught successive generations of medical students. His first work was in histology, and he was elected fellow of the Royal Society in 1902 as biologist. After World War I, he studied friction of dry and lubricated surfaces as a continuation of his previous work on colloids and surface tension. He introduced the term *boundary lubrication* and discovered the monomolecular lubricant films that adhere to the friction surfaces. He was knighted in 1927.

Considering the disks to be rigid solids and the lubricant to be incompressible and of constant viscosity, Martin (Pugh, 1970) derived the following expression for the minimum film thickness by solving the Reynolds equation presented in Chapter 11:

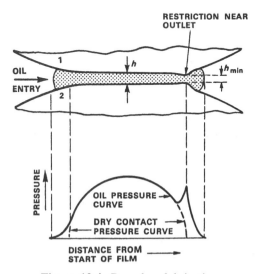

Figure 12.4. Boundary lubrication.

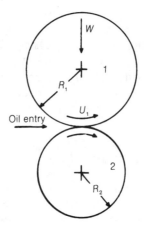

Figure 12.5. Boundary lubrication of two rotating cylinders.

$$h_{min} = \frac{k\mu(U_1 + U_2)R}{P} \tag{12.6}$$

where μ is the absolute viscosity, U_1 is the peripheral speed of surface 1, U_2 is the peripheral speed of surface 2, R is the mutual radius of curvature of roller pair $= R_1R_2/(R_1 + R_2)$, P is the load per unit width, and k is a constant.

This equation suggests an oil film thickness less than the height of the asperities of the surface roughness, indicating boundary lubrication. Therefore, a different mechanism must be investigated.

Experiments show that the load can be carried for long periods without surface failure or appreciably rapid wear. This suggests that the mechanism of lubrication must be of a hydrodynamic nature, that is, the mating surfaces must normally be separated by a continuous oil film undisturbed by metal surface irregularities.

Thus, the development of the oil film in this type of lubrication conditions cannot be explained by lubrication theory alone and the oil film thickness predicted by the Martin equation above agrees with experimental results for very light loads only. At higher loads, the elastic deformation of the disks renders the oil film thickness a function not only of the geometry but also of the pressure that deforms the material of the disk. Moreover, even moderate loads can result in high pressures, on the order of 1000 MPa. As discussed in Chapter 11, viscosity at such pressures increases drastically, changing the behavior of the oil film. Thus, the oil film thickness depends on three groups of parameters. Expressing the above influences in the semiempirical equation, modifying Equation (12.6):

$$\frac{h_{min}}{R} = 2.6 \left(\frac{\mu_0 U}{E'R}\right)^{0.7} (\alpha E')^{0.6} \left(\frac{W}{RE'}\right)^{-0.13} \tag{12.7}$$

The first term expresses the hydrodynamic influence, as also discussed in Chapter 11. The second term represents the change of viscosity with pressure with an exponential law $\mu = \mu_0 \exp \alpha(p - p_0)$, where μ_0 is the viscosity at pressure p_0,

α is a lubricant property of typical value 13–35 mm²/N for common lubricant oils. The third term refers to the elastic deformation, where E' is the apparent modulus of elasticity, which depends on the elastic properties of the materials of the two disks:

$$\frac{1}{E'} = \frac{[(1 - \nu_1^2)/E_1] + [(1 - \nu_2^2)/E_2]}{2} \tag{12.8}$$

where E_1 and E_2 are the elasticity moduli and ν_1 and ν_2 are the Poisson ratios of the two materials. Similarly, R is a geometric parameter combining the two tadii $1/R = (1/R_1) + (1/R_2)$. This type of lubrication is called *elastohydrodynamic* or *thin film* lubrication.

Example 12.1 An overall crane is rolling on four wheels of diameter D = 400 mm on a rail of width w = 60 mm. The rail is made of carbon steel AISI 1020 and the wheel cast ASTM A530/GR 60–40–15. The latter has fatigue strength 170 MPa. Determine the maximum load that each wheel can carry for continuous (fatigue) strength.

Solution If F is the unknown force, the maximum pressure at the contact zone will be (Figure 12.3), for $E = 2.1 \times 10^{11}$ N/m²:

$$p = 0.591 \left(\frac{FE}{2Rw} \right)^{1/2}$$

Solving for F_2:

$$F = \left(\frac{p}{0.591} \right)^2 \frac{2Rw}{E}$$

The surface strength, for dry constant, is $2S_e$. From the material tables in Chapter 6, the fatigue strengths of AISI 1020 carbon steel and the ASTM A530/GR 60–40–15 cast steel are 190 and 170 N/mm², respectively. Therefore, the value of 170 N/mm² will be used. The design fatigue strength will be:

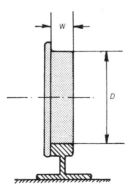

Figure E12.1

$$S_e = \frac{C_F C_S C_R C_H S_n'}{K_f} = 0.8 \times 0.7 \times 0.83 \times 1 \times 170/1 = 79 \text{ N/mm}^2$$

for 99% reliability. The surface strength is then $S_{He} = 2S_e = 158 \text{ N/mm}^2$. The allowable load is:

$$F = \left(\frac{158}{0.591}\right)^2 \frac{2 \times 200 \times 60}{2.1 \times 10^5} = 8168 \text{ N}$$

Example 12.2 Two disks of equal diameter $D = 200$ mm, as in Figure 12.5, are used in a rotating drum, carrying load of 50 kN and rotating at 800 rpm. Their material is hardened steel and the oil used is SAE 30 at 80°C. For this oil, $\alpha = 0.02 \text{ mm}^2/\text{N}$. Because of surface roughness and oil filtering solution, the minimum oil film thickness should not be less than 1 mm. Compute the width of the disks.

Solution For steel, $E = 2.1 \times 10^5$ MPa. Therefore, $E' = 2.31 \times 10^5$ MPa. The equivalent radius $R = R_1 R_2/(R_1 = R_2) = 100/2 = 50$ mm. The oil viscosity of SAE 30 at 80°C is $\mu = 12 \times 10^{-3}$ N · s/m². If the load is $P = 50,000$ N, then the load W per unit width is $W = P/w$ where w is the unknown width. Solving Equation (12.7) for w, because $U_1 = U_2 = 2\pi R n/60 = 2 \times 100 \times 800/60 = 8378$ mm/s $= U$, then substitution yields $w = 13$ mm.

12.3. ROLLING BEARINGS

A typical application of lubricated contacts under elastohydrodynamic lubrication is a type of bearing where rolling members are introduced between two rubbing surfaces, making use of the fact that the resistance to motion in this case is much less than in pure sliding. This concept is known from antiquity. Figure 12.6 shows the transportation of a colossus on rollers.

Rolling bearings have been known since Roman times. Figure 12.7 shows such a bearing of Nero's time, found in Italy, where a wheel of a chariot revolves around the axle through a roller bearing. Such bearings, described by Leonardo da Vinci, were reinvented at the turn of the century and have been very widely used ever since.

12.3.1. Classification

Figure 12.8 shows the classification of antifriction bearings. Each type of bearing has characteristic features that make it particularly suitable for certain applications. However, it is not possible to lay down rules for the selection of bearing types, since several factors must be considered and assessed relative to each other.

Deep-groove (*Conrad*) ball bearings are normally selected for small-diameter shafts and light loads. Cylindrical roller bearings and spherical roller bearings are considered for shafts of large diameter and higher loads.

If radial space is limited, then bearings with small sectional height must be selected, such as needle roller and cage assemblies, needle roller bearings without or with inner ring, certain series of deep groove ball bearings, and spherical roller bearings.

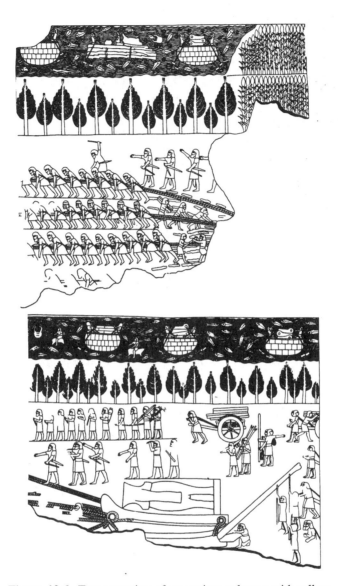

Figure 12.6. Transportation of an ancient colossus with rollers.

Where axial space is limited and particularly narrow bearings are required, then some series of single-row cylindrical roller bearings and deep-groove ball bearings can be used for radial and combined loads. For axial loads, needle roller and cage thrust assemblies, needle roller thrust bearings, and some series of thrust ball bearings may be used.

In Figure 12.8a, b, the capacity of antifriction bearings to support loads of radial and axial direction is indicated relative to similar deep-groove ball bearings.

Generally, roller bearings can carry greater loads than ball bearings of the same external dimensions. Ball bearings are mostly used to carry light and medium loads, while roller bearings are often the only choice for heavy loads and large-diameter shafts.

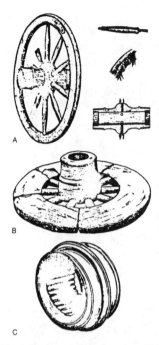

Figure 12.7. A Roman roller bearing (from Singer).

Thrust ball bearings are suitable only for moderate, purely axial loads. Single-direction thrust ball bearings can carry axial loads in one direction, and double-direction thrust ball bearings can carry axial loads in both directions. Cylindrical roller and needle roller thrust bearings (with or without washers) can accommodate heavy axial loads in one direction. Spherical roller thrust bearings, in addition to very heavy axial loads, can also carry a certain amount of simultaneously acting radial load.

The most important feature affecting the ability of a bearing to carry an axial load is its angle of contact, α. The greater this angle, the more suitable the bearings are for axial loading.

Single- and double-row angular contact ball bearings and taper roller bearings are mainly used for combined loads. Deep-groove ball bearings and spherical roller bearings are also used. Self-aligning ball bearings and cylindrical roller bearings can also be used to a limited extent for carrying combined loads. Four-point contact ball bearings and spherical roller thrust bearings should only be considered where axial loads predominate.

Single-row angular contact ball bearings, taper roller bearings, cylindrical roller bearings, and spherical roller thrust bearings can carry axial loads in one direction only. Where the direction of load varies, two such bearings arranged to carry axial loads in opposite directions must be used.

Where the axial component of the combined load is large, a separate thrust bearing can be provided for carrying the axial load independently of the radial load. Suitable radial bearings may also be used to carry axial loads only, such as deep-groove or four-point contact ball bearings. To ensure that these bearings are

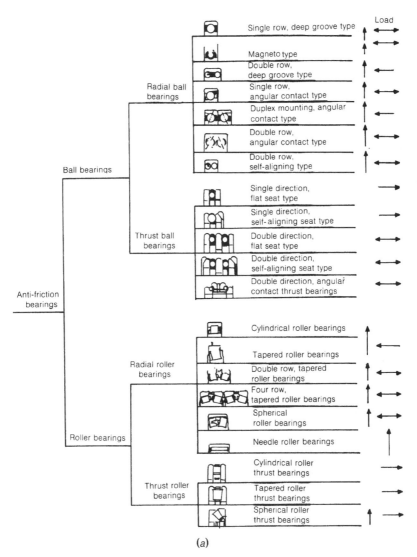

Figure 12.8. Classification of Rolling Bearings.

subjected only to axial loading, the outer rings must have radial clearance in their housings.

Where the shaft can be misaligned relative to the housing, bearings capable of accommodating such misalingment are required, namely self-aligning ball bearings, spherical roller bearings, and spherical roller thrust bearings. Misalignment can, for example, be caused by shaft deflection under load, when the bearings are fitted in housings positioned on separate bases and at a large distance from one another or when it is impossible to machine the housing seatings at one setting.

In Figure 12.8b, c, bearing classification specified by ISO and ANSI is indicated. ISO specifies for antifriction bearings a designation defined as follows (ISO #300):

1. The first four digits refer to the bore diameter in millimeters.

Sketch	Direction of applied load	Type	ISO/R 15 ANSI B3.14	International	U_{max} (m/s)
		Radial Deep groove Single-row	BC10 BC02 BC03	60 62 63 64	10–30
		Double-row self-aligning	BS02 BS22 BS03	12 13 22 23	10–20
		Radial-thrust Single-row	BT02	32	10–20
			ET03	33	
		Double-row Thrust	BE32 BE33	72 73	10–20
		Single-thrust	TA11	511, 512 513, 514	5–10
				522 523 562	

(b)

Figure 12.8. (*Continued*)

2. The following three letters indicate bearing type. For example, B = radial ball bearings, R = cylindrical roller bearings, T = thrust ball and roller bearing.

3. Two digits then designate the geometric proportions in the form of dimension series.

In the interests of both users and manufacturers of rolling bearings and for reasons of price, quality, and interchangeability, a relatively limited number of bearing sizes is desirable. Accordingly, ISO has laid down Dimension Plans for the boundary dimensions of metric rolling bearings (ISO/R 15, ISO/R 355, and ISO/R 104).

Sketch	Direction of applied load	Type	ISO/R 15 ANSI B3.14	International	U_{max} (m/s)
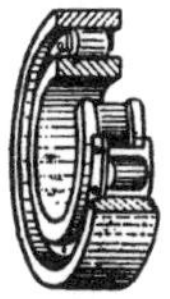	R 1.7	Radial Without lips on outer race	RN02 RN03	N	10–20
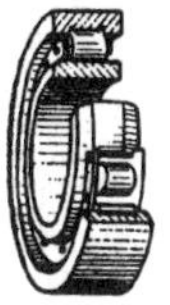	R 1.7	Without lips on inner race		NU	10–20
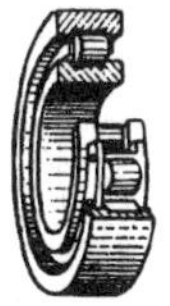	R 1.7 A ← → A	With one lip on inner race		NJ	10–20
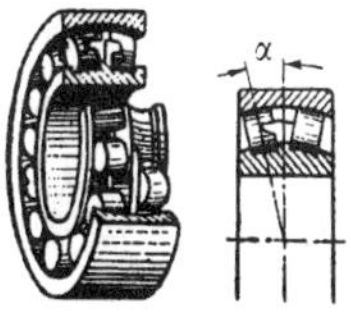	α R 2 A 0.2 A	Double-row self-aligning	SD22 SC22 SL24 Also 30, 31	213 222 223 230	10–20
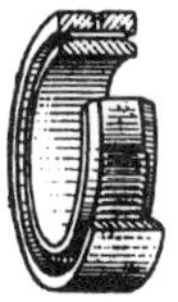	R	Needle with two lips on outer race			5–10
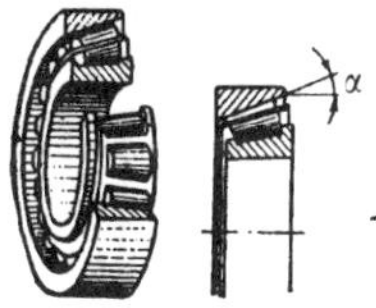	α R 1.9 A 0.7	Radial-thrust Single-row tapered	2FB 3CC 3DC 4DB 5DD	302, 303 313 322 323 329	5–15

(*c*)

Figure 12.8. (*Continued*)

In the ISO *Dimension Plans* for every standard bearing bore size *d*, a progressive series of standardized outside diameters *D* is specified (Diameter Series 8, 9, 0, 1, 2, 3, and 4, in ascending size order). For each Diameter Series, a series of bearing widths *B* is also specified (Width Series 0, 1, 2, 3, 4, 5, and 6, in ascending order of bearing width). The Width Series for radial bearings corresponds to the Height Series for thrust bearings.

The combination of a Diameter Series with a Width or Height Series is called a Dimension Series. Each Dimension Series has a number consisting of two figures:

the first figure indicates the Width (or Height) Series and the second the Diameter Series.

National or international standards have not yet been drawn up for all types of needle roller bearings, but certain dimensions have been established by usage, and interchangeability of needle roller bearings is thus largely assured. The same applies to spherical plain bearings.

Relative proportions for width and diameter series are shown in Figure 12.9.

The dimensional, form, and running accuracy of rolling bearings has been standardized by ISO. In addition to normal tolerances, ISO recommendations specify, for the fit between the rolling elements and the pathways, closer tolerances for higher-precision bearings (classes C1 and C2), for the normal-quality bearings (class 3) and larger than normal tolerances (classes 4 and 5). External dimension tolerances have been standardized as quality 1 (ABEC1) for normal bearings to quality 9 (ABEC9) for high-precision bearings.

ISO designation has been adapted by ANSI and the Anti-Friction Bearing Manufacturers Association (AFBMA) through its Annular Bearing Engineers Committee (ABEC). Antifriction bearing manufacturers, however, are still using different designations, the most universally accepted being those of the Swedish–German manufacturers, which are now converting rapidly to ISO designations. Therefore, the ISO/ABEC designation will be used here.

It must be stressed that not all the dimension series are available. That is, not all combinations of diameter and width series are in common use. The most popular, and readily available, are indicated in Figure 12.8, with both the ISO/AFBMA and the most common commercial designation. In general, diameter series 0, 1, 2, 3 and width series 0, 1, 2, 3 also are the most used. The older SAE designation specified:

Series	SAE	ISO/AFBMA
Extra light	100	01
Light	200	02
Medium	300	03
Heavy	400	04

The commercial designations are followed by a number corresponding to the diameter. For most commercial designations, this number is the bore diameter in

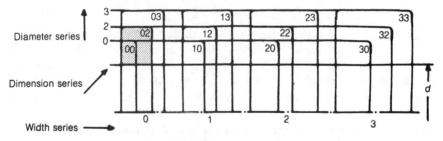

Figure 12.9. Dimension series of rolling bearings.

millimeters divided by 5, for diameters over 20 mm. For example, a deep-groove ball bearing eight series of bore diameter 100 mm would have the designations:

ISO/AFBMA:	100 BC02
SKF:	6220
New Departure-Hyatt:	3220

Many manufacturers provide conversion tables in their catalogs.

Manufacturers are also using prefixes and suffixes to indicate other features of the bearing design, such as sealing and mounting.

Antifriction bearings have, by their nature, very close tolerances. ABEC has specified five basic grades of precision, 1, 3, 5, 7, 9, with higher grade indicating increasing precision. Grade 3 is common, while higher grades, used for precision machinery, have progressively smaller tolerances. Tables for such tolerances can be found in manufacturers' catalogs or in AFBMA publications.

Henry Timken, the father of rolling bearings, was a German immigrant to United States. He started as an apprentice in a carriage shop and soon established his own successful carriage manufacturing business. In 1898, he was awarded the first patent for taper roller bearings, and in the same year he founded the Timken Roller Bearing Axle Company in St. Louis, Missouri. In 1902, he moved the company to Canton, Ohio, with a division in Detroit. The first applications of the Timken bearings were in horse-driven vehicles, but soon they found their way into the motor-driven ones. After World War I, motor driven vehicles came into general use, and ball and roller bearings became the standard bearings for vehicle axles.

12.3.2. Antifriction Bearing Design Database

A database can be established for standard bearing dimensions from antifriction bearing manufacturers and ISO catalogs in the SHAFTDES program, which automatically selects antifriction bearings from the database. It is part of MELAB 2.0. Only a small number of bearing types have been included as of this time.

At the time this book was written, on-line help was available on antifriction bearings from the Internet:

1. The design information group at the University of Bristol has elements of a machine design handbook online at http://www.dig.bris.ac.uk/hbook/. At this time, only antifriction bearing selection can be done on-line, free of charge.
2. SKF, one of the leading antifriction bearing manufacturers, has its catalog and engineering documentation on-line at http://www.skf.com. Though difficult to use and not yet complete, this is the most comprehensive database

on antifriction bearings available on-line today. Before use, free registration is mandatory.

12.3.3. Fatigue Load Rating

Fatigue life of materials in the high cycle fatigue range is described by a power law (see Chapter 5):

$$\sigma^m N = constant \tag{12.9}$$

where σ is the applied stress, N is the number of cycles to failure, and m is a constant.

Antifriction bearings are rated on the basis of the dynamic loading that they can sustain, purely radial or thrust for the respective bearings at 10^6 revolutions, according to ISO recommendations. If a bearing is designed for a different life, a different stress range must be applied. For all situations, since stress is proportional to the load:

$$P^p N = P_0^p N_0 \tag{12.10}$$

where $P_0 = C$ is the load-carrying capacity at $N_0 = 10^6$ revolutions. Therefore, the life of the bearing, in N revolutions, for applied load P is:

$$\frac{N}{N_0} = \left(\frac{C}{P}\right)^p \tag{12.11}$$

Here the exponent p is evaluated from experiments. Most bearing manufacturers specify $p = 3$ for ball and $p = 10/3$ for roller bearings. If n rpm is the speed of rotation, the nominal bearing life[1] $L_{10} = N/(60n)$ in hours will be, for a probability of failure less than 10%:

$$L_{10} = \frac{10^6}{60n} (C/P)^p \tag{12.12}$$

Standard bearing ratings C are given by bearings manufacturers for 10% probability of failure.

SKF suggests, (Table 12.1) machinery lives, if such specification for the designed bearing is not available.

The fatigue load rating C in kg-force was established based on elastohydrodynamic equations, such as (12.6), in the form

---

[1] In older literature, the 10% failure rate bearing life L_{10} is designated as L_{B10} or $B10$.

TABLE 12.1. Working Life of Rolling Bearings (SKF General Catalogue 1977)

Class of Machine	Life in Working Hours (L_h)
Instruments and apparatus that are used only seldom: Demonstration apparatus; mechanisms for operating sliding doors	500
Machines used for short periods or intermittently and whose breakdown would not have serious consequences: Handtools; lifting tackle in workshops; hand-operated machines generally; agricultural machines; cranes in erecting shops; domestic machines	4,000–8,000
Machines working intermittently and whose breakdown would have serious consequences: Auxiliary machinery in power stations; conveyor plant for flow production; lifts; cranes for piece goods; machine tools used infrequently	8,000–12,000
Machines for use 8 hours per day and not always fully utilized: Stationary electric motors; general-purpose gear units	12,000–20,000
Machines for use 8 hours per day and fully utilized: Machines for the engineering industry generally; cranes for bulk goods; ventilating fans; countershafts	20,000–30,000
Machines for continuous use 24 hours per day: Separators; compressors; pumps; mine hoists; stationary electric machines; machines in continuous operation on board naval vessels	40,000–60,000
Machines required to work with a high degree of reliability 24 hours per day: Pulp and papermaking machinery; public power plants; mine pumps; waterworks; machines in continuous operation on board merchant ships	100,000–200,000

For ball bearings:

$$C = 10\lambda(i \cos \alpha)^{0.7}\, z^{2/3}\, d^{1/8} \qquad \text{for } d_b < 25.4 \text{ mm} \qquad (12.13)$$

$$C = 36.47\lambda(i \cos \alpha)^{0.7}\, z^{2/3}\, d^{1.4} \quad \text{for } d_b > 25.4 \text{ mm} \qquad (12.14)$$

For roller bearings:

$$C = 56.2\lambda(il_r \cos \alpha)^{7/9}\, z^{3/4}\, d_r^{29/27} \qquad\qquad (12.15)$$

where i is the number of rows of rolling elements, α is the angle of contact defined in Figure 12.8, z is the number of rolling elements, d_b is the diameter of rolling balls (cm), d_r is the diameter of rollers (cm), l_r is the length of contact of rollers (cm), and λ is a constant that depends on the factor $\xi = d_b \cos \alpha/d_m$ for balls, and is $d_r \cos \alpha / d_m$ for rollers, where d_m is the diameter of the circle of the ball or roller centers.

The constant λ is given in Table 12.2 for different types of bearings.

At the application stage of design, fatigue load rating C is determined from manufacturers' catalogs. However, when the bearing geometry is known, C can be computed from Equations (12.13)–(12.15). Sometimes this is particularly useful for optimization and computer aided design studies.

The bearing load rating C is determined under prescribed conditions and needs adjustment for particular applications. To this end, the following revised life equation has been established by ISO:

$$L_h = a_1 a_2 a_3 L_{10} \tag{12.16}$$

where L_h is the adjusted bearing life in hours, n is the speed of rotation (rpm), a_1 is the life adjustment factor for reliability, a_2 is the life adjustment factor for material, and a_3 is the life adjustment factor for operating conditions. Finally, Equations (12.12) and (12.16) yield:

$$C = P \left(\frac{60 n L_h}{10^6 \, a_1 a_2 a_3} \right)^{1/3} \tag{12.16a}$$

This is the main design equation for antifriction bearings. A calculation of the adjusted rating life presupposes that the operating conditions are well defined and that the bearing loads can be accurately calculated, that is, the calculation should consider the load spectrum, shaft deflection, and so on.

For the generally accepted reliability of 90% and for conventional materials and normal operating conditions, $a_1 = a_2 = a_3 = 1$ and the two life equations become identical.

The factor a_1 for reliability is used to determine lives other than the L_{10} life, that is, lives that are attained or exceeded with a probability greater than 90%. Values of a_1 are given in Table 12.3.

It must be noted that the values of a_1 given by AFBMA (Table 12.3) are not the same as those in Chapter 6, because they are based not on the standard distribution but on the Weibull distribution.

Conventional material is considered to be the steel used for rolling bearings at the time the original fatigue rating equation was established by ISO. The improvements made to the standard steels used by different manufacturers have resulted in a value of the a_2 factor that is greater than 1. Even higher values of factor a_2 can

TABLE 12.2. Constant λ

Bearing	λ
Single row	$0.476 + 2.5a - 13a^2 + 17.5a^3$
Double row	$0.451 + 2.5a - 13a^2 + 17.5a^3$
Self-aligned	$0.176 + 1.5a - 2a^2 - a^3$
	$(a = \zeta - 0.05)$

TABLE 12.3. Reliability Factor

Reliability (%)	a_1	L_n
90	1	L_{10}
95	0.62	L_5
96	0.53	L_4
97	0.44	L_3
98	0.33	L_2
99	0.21	L_1

be applied when special steels, such electro-slag refined (ESR) or vacuum arc remelted (VAR) steels, are used.

Values of a_2 are given in manufacturers' catalogs. It is 1 for standard steels and 2 to 3 for special steels. The AFBMA suggests the value 3 for bearings of good quality.

Values of a_3 are related to bearing lubrication, provided bearing operating temperatures are not excessive. The efficacy of lubrication is primarily determined by the degree of surface separation in the rolling contacts of the bearing. Under the cleanliness conditions normally prevailing in an adequately sealed bearing arrangement, the a_3 factor is based on the viscosity ratio k.

$$k = \frac{\nu}{\nu_1} \tag{12.17}$$

This is defined as the ratio of the actual lubricant viscosity ν to the viscosity ν_1 required for adequate lubrication, both values being determined at the operating temperature. For the determination of ν_1, reference should be made to Figure 12.10. (from the SKF catalog) or similar recommendations of other manufacturers.

Because factors a_2 and a_3 are interdependent, bearing manufacturers usually replace them in Equation (12.13) with a combined factor a_{23} for material and lubrication. Provided cleanliness is normal, values of a can be obtained from Figure 12.11 (from the SKF catalog).

Finally, higher temperatures cause a reduction of the material strength. Therefore, at higher temperatures, the dynamic rating C must be reduced by 10% between 150–200°C, 25% between 250–300°C, and 40% over 300°C.

12.3.4. Bearing Loads

The loads acting on a bearing can be determined using normal engineering principles provided the external forces, such as weights, forces arising from the power transmitted or work done, and inertia forces, are known or can be accurately calculated.

Those loads arising from component weights, such as the inherent weight of shafts and parts mounted on them, vehicle weights, and inertia forces, are either generally known or can be calculated. However, for determining the loads arising from the working loads (roll forces, cutting forces in machine tools), shock loads,

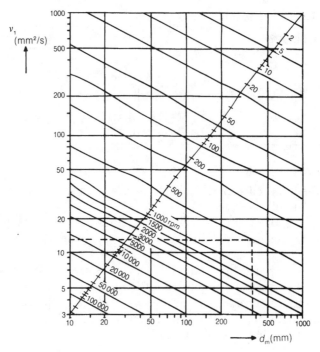

Figure 12.10. Speed coefficient ν_1, SKF bearings.

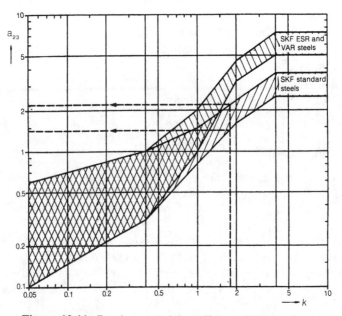

Figure 12.11. Bearing material coefficient, SKF bearings.

or additional dynamic loads, such as a result of unbalance, it is often necessary to estimate the load from experience gained with earlier machines and bearing arrangements.

For example, the theoretical tooth forces on a gear can be calculated if the power transmitted and the design characteristics of gear teeth are known. However, for calculating the bearing loads, additional forces arising externally and from within the gearing combination must be taken into account. The magnitude of these additional forces depends on the accuracy of the teeth and on the operating conditions imposed by the machines connected to the gearing. The effective tooth load, P_{eff}, used in bearing calculations is found by means of the equation

$$P_{eff} = f_k f_d P \qquad (12.18)$$

where P is the theoretical tooth load, f_k is the a factor for the additional forces arising in the gearing (see Table 12.4), and f_d is the a factor for the additional forces caused by the machines connected to the gearing (see Table 12.5).

As a further example, belt drives may be considered. Here the loads acting on the bearings are derived from the effective belt pull, which is dependent on the power being transmitted. To calculate the bearing loads, we must multiply the effective belt pull by a factor f. The factor f depends on the type of belt used, its initial tension, and the additional dynamic forces. Appropriate values of factor f are:

V-belts:	2–2.5
Plain belts with tension pulley:	2.5–3
Plain belts:	4–5

The larger values are suitable when the distance between the shafts is short, the peripheral speeds are slow, or the operating conditions are not known with high precision.

12.3.5. Combined Loads

Rated loads of antifriction bearings are either purely radial or thrust. Experimental investigations have shown that most bearings, as indicated in Figure 12.8, can take both radial and thrust loads. The equivalent bearing load, to compare with the load rating C, is the maximum of

$$P = \max\{VF_r, YF_a + VXF\} \qquad (12.19)$$

where P is the equivalent dynamic bearing load, F_r is the actual radial bearing

TABLE 12.4. Geometric Error Factor: Values for Factor f_k

Accuracy of Gears	f_k
Precision gears (pitch and form errors <0.02 mm)	1.05 to 1.1
Normal gears (pitch and form errors 0.02 to 0.1 mm)	1.1 to 1.3

TABLE 12.5. Factor for Operating Conditions: Values for Factor f_d

Operating Conditions, Type of Machine	f_d
Machines working without shock loads, e.g., electrical machines, turbines	1.0 to 1.2
Reciprocating engines, depending on degree of unbalance	1.2 to 1.5
Pronounced shock loads, e.g., rolling mills	1.5 to 3

load, F_a is the actual axial bearing load, X is the radial load factor for the bearing, Y is the axial load factor for the bearing, and $V = 1$ if the inner ring rotates, 1, 2 if the outer ring rotates.

Data required for the calculation of the equivalent dynamic bearing load are given in the bearing tables. The bearing types are too numerous to allow a complete presentation of bearing tables here. For reference only, Table 12.6 gives the combined load factors X and Y for several types of bearings, and Table 12.7 gives static and dynamic capacity of several types of bearings for a particular dimension series.

In the case of single-row radial bearings an additional axial load does not influence the equivalent bearing load P until its magnitude is such that the ratio F_a/F_r exceeds a specified value e given in bearing tables.

12.3.6. Fluctuating Loads

If loads of different magnitudes $P_1, P_2, \ldots, P_n$ act at respective numbers of revolutions $N_1, N_2, \ldots, N_n$ (Equation (12.10)), the approximate mean load for the remaining bearing life should not exceed, according to the Palmgren–Miner rule (see Chapter 5),

$$P = \left(\frac{P_1^p N_1 + P_2^p N_2 + \cdots}{N_1 + N_2 + \cdots} \right)^{1/p} \tag{12.20}$$

12.3.7. Static Loads

When a bearing under load stands still, makes slow oscillating movements, or operates at very slow speeds, the load-carrying capacity is determined not by the fatigue of the material but by the permanent deformation at the points of contact between the rolling elements and the raceways. This also applies to rotating bearings that are subjected to heavy shock loads during a fraction of a revolution. Generally, loads equivalent to the basic static load rating C_0 can be accommodated without detriment to the running characteristics of the bearing.

Loads made up of radial and axial components must be converted into an equivalent static bearing load. The equivalent static bearing load is defined as that radial load (for thrust bearings the axial load) that if applied would cause the same permanent deformation in the bearing as the actual loads. It is obtained by means of the general equation

$$P_0 = \max\{X_0 F_r + Y_0 F_a), F_r\} \tag{12.21}$$

where P_0 is the equivalent static bearing load, F_r is the actual radial load, F_a is the

TABLE 12.6. Combined Loads Factors

Bearing No.	$\dfrac{F_a}{F_r} \leq e$		$\dfrac{F_a}{F_r} > e$		
	X	Y	X	Y	e
Deep Grove Ball Bearings					
Series EL, R, 160, 60, 62, 63, 64, EE, RLS, RMS					
$\dfrac{F_a}{C_0} = 0.025$				2	0.22
= 0.04				1.8	0.24
= 0.07	1	0	0.56	1.6	0.27
= 0.13				1.4	0.31
= 0.25				1.2	0.37
= 0.5				1	1.44
Angular Contact Ball Bearings					
Series 72 B, 73 B	1	0	0.35	0.57	1.14
Series 72 BG, 73 BG					
A pair of bearings mounted in tandem	1	0	0.35	0.57	1.14
A pair of bearings mounted back-to-back or face-to-face	1	0.55	0.57	0.93	1.14
Series ALS, AMS	1	0	0.39	0.76	0.80
Series 32 A, 33 A	1	0.73	0.62	1.17	0.86
Self-aligning Ball Bearings					
135, 126, 127, 108, 129	1	1.8	0.65	2.8	0.34
1200–1203		2		3.1	0.31
04– 05		2.3		3.6	0.27
06– 07		2.7		4.2	0.23
08– 09	1	2.9	0.65	4.5	0.21
10– 12		3.4		5.2	0.19
13– 22		3.6		5.6	0.17
24– 30		3.3		5	0.2
2200–2204		1.3		2	0.5
05– 07		1.7		2.6	0.37
08– 09	1	2	0.65	3.1	0.31
10– 13		2.3		3.5	0.28
14– 20		2.4		3.8	0.26
21– 22		2.3		3.5	0.28
1300–1303		1.8		2.8	0.34
04– 05	1	2.2	0.65	3.4	0.29
06– 09		2.5		3.9	0.25
10– 22		2.8		4.3	0.23
2301		1		1.6	0.63
2302–2304	1	1.2	0.65	1.9	0.52
05– 10		1.5		2.3	0.43
11– 18		1.6		2.5	0.39

TABLE 12.6. (*Continued*)

Bearing No.	$\dfrac{F_a}{F_r} \le e$		$\dfrac{F_a}{F_r} > e$		
	X	Y	X	Y	e
Self-aligning Ball Bearings					
RL 4–RL 6		2.1		3.3	0.29
7– 8		2.3		3.6	0.27
9– 11		2.7		4.2	0.23
12– 14	1	2.9	0.65	4.5	0.21
15– 18		3.4		5.2	0.19
20– 36		3.6		5.6	0.17
38– 48		4.2		6.5	0.15
RM 3–RM 6		1.8		2.8	0.34
7– 10		2.1		3.3	0.29
11– 14	1	2.4	0.65	3.8	0.26
15– 18		2.7		4.2	0.23
20– 48		2.9		4.5	0.21
Spherical Roller Bearings					
23944–239/670	1	3.7	0.67	5.5	0.18
239/710–239/950		4		6	0.17
23024C–23068CA	1	2.9	0.67	4.4	0.23
23072CA–230/500CA		3.3		4.9	0.21
24024C–24080CA	1	2.3	0.67	3.5	0.29
24084CA–240/500CA		2.4		3.6	0.28
23120C–23128C	1	2.4	0.67	3.6	0.28
23130C–231/500CA		2.3		3.5	0.29
24122C–24128C		1.9		2.9	0.35
24130C–24172CA	1	1.8	0.67	2.7	0.37
24176CA–241/500CA		1.9		2.9	0.35
22205C–22207C		2.1		3.1	0.32
08C– 09C		2.5		3.7	0.27
10C– 20C	1	2.9	0.67	4.4	0.23
22C– 44C		2.6		3.9	0.26
48 – 64		2.4		3.6	0.28
23218C–23220C		2.2		3.3	0.31
21304–21305		2.8		4.2	0.24
06– 10		3.2		4.8	0.21
11– 19	1	3.4	0.67	5.	0.2
20– 22		3.7		5.5	0.18
22308C–22310C		1.8		2.7	0.37
11C– 15C		1.9		2.9	0.35
16C– 40C	1	2	0.67	3	0.34
44 – 56		1.9		2.9	0.35

TABLE 12.6. (*Continued*)

Bearing No.	$\dfrac{F_a}{F_r} \le e$		$\dfrac{F_a}{F_r} > e$		e
	X	Y	X	Y	
Taper Roller Bearings					
30203–30204				1.75	0.34
05– 08				1.6	0.37
09– 22	1	0	0.4	1.45	0.41
24– 30				1.35	0.44
32206–32208				1.6	0.37
09– 22	1	0	0.4	1.45	0.41
24– 30				1.35	0.44
30302–30303				2.1	0.28
04– 07	1	0	0.4	1.95	0.31
08– 24				1.75	0.34
31305–31318	1	0	0.4	0.73	0.82
32303				2.1	0.28
32304–32307	1	0	0.4	1.95	0.31
08– 24				1.75	0.34

actual axial load, X_0 is the radial load factor for the bearing, and Y_0 is the axial load factor for the bearing.

All data necessary for the calculation of the equivalent static bearing load are given in bearing tables, such as Tables 12.6 and 12.7.

The required static load rating C of a bearing can be determined by means of the equation

$$C_0 = s_0 P_0 \tag{12.22}$$

where C_0 is the basic static load rating, P_0 is the equivalent static bearing load, and s_0 is the static loading safety factor. At elevated temperatures, reduced hardness of the bearing material influences the static load-carrying capacity. In this case, the bearing manufacturer must be consulted.

The values of s_0 given below for a few typical applications can be used as a guide in determining the requisite basic static load rating of bearings that make occasional oscillating movements: variable-pitch propeller blades for aircraft, $s_0 > 0.5$; weir and sluice gate installations, $s_0 > 1$; moving bridges, $s_0 > 1.5$; crane hooks for large cranes without significant additional dynamic forces, small cranes for bulk goods with fairly large additional dynamic forces, $s_0 > 1.6$; generally for spherical roller thrust bearings, $s_0 > 2$.

Where there are large fluctuations in the applied load, and particularly when heavy shock loads occur during part of a revolution, it is essential to establish that the static load rating of the bearing is adequate. Severe shock loads can cause large unevenly distributed indentations around the raceways, which will seriously affect the running of the bearing. In addition, shock loads cannot usually be accurately calculated. Deformation of the housing resulting in an unfavorable load distribution in the bearing may also arise.

TABLE 12.7. Dynamic Capacity of Rolling Bearings

Deep groove ball bearing

Boundary dimensions			Basic load ratings dynamic	static
d	D	B	C	C_0
mm			N(1 N = 0.225 lbf)	
17	26	5	1320	915
	35	8	4650	2800
	35	10	4650	2800
	40	12	7350	4500
	47	14	10400	6550
	62	17	17600	11800
20	32	7	2040	1400
	41	8	5400	3400
	42	12	7200	4500
	47	14	9800	6200
	52	15	12200	7800
	72	19	23600	16600
25	37	7	2280	1700
	47	8	5350	4000
	47	12	8650	5600
	52	15	10300	6950
	62	17	17300	11400
	80	21	27500	19600
30	42	7	2280	1800
	55	9	8650	5850
	55	13	10200	6800
	62	16	15000	10000
	72	19	21000	14600
	90	23	33500	24000

Boundary dimensions			Basic load ratings dynamic	static
d	D	B	C	C_0
mm			N(1 N = 0.225 lbf)	
15	35	11	7650	3150
17	40	12	9650	3800
	40	16	9650	3900
	47	14	13700	5400
20	52	15	9300	4000
	52	18	9650	4150
	62	17	13700	5850
	62	24	18600	7500
25	52	15	12000	5600
	52	18	11800	5500
	62	17	16300	7500
	72	19	24000	10000
35	72	17	12000	6300
	72	23	16600	7800
	80	21	19300	9500
	80	31	30000	12900
40	80	18	14600	8000
	80	23	17300	9000
	90	23	22800	11800
	90	33	34500	15600
45	85	19	16600	9000
	85	23	17600	10000
	100	25	29000	15300
	100	36	41500	19300

Boundary dimensions			Basic load ratings dynamic	static
d	D	B	C	C_0
mm			N(1 N = 0.225 lbf)	
15	35	11	8150	4250
17	40	12	9800	5200
	40	16	14000	8150
	47	14	15600	8650
20	47	15	13400	7350
	47	18	18300	10800
	52	15	20400	11600
25	52	15	15300	8800
	52	18	20800	12900
	62	17	26000	15000
	62	24	38000	24500
30	55	13	14300	8500
	62	16	20400	12000
	62	20	29000	19000
	72	19	34000	20000
	72	27	45500	29000
	90	23	55000	34000
35	62	14	19000	11600
	72	17	29000	17600
	72	23	43000	29000
	80	21	43000	27000
	80	31	57000	38000
	100	25	68000	44000

Boundary dimensions			Basic load ratings dynamic	static
d	D	T	C	C_0
mm			N(1 N = 0.225 lbf)	
20	42	15	20800	15600
	47	15.25	23600	16600
25	52	16.25	29000	20000
	52	22.25	37500	28500
22	44	15	21600	16300
25	47	15	23200	18300
	52	16.25	26500	19300
	52	22	40000	32500
28	62	18.25	38000	26500
	62	18.25	32500	23200
	62	25.25	51000	39000
28	52	16	27000	21600
30	55	17	30500	24500
	62	17.25	34500	25500
	62	21.25	43000	34000
	62	25	55000	45500
30	72	20.75	48000	34000
	72	20.75	40500	29000
	72	28.75	65500	52000
32	58	17	31500	26000
35	62	18	36500	30500
	72	18.25	44000	32500
	72	24.25	56000	45000
	72	28	72000	62000

Boundary dimensions			Basic load ratings dynamic	static
d	D	H	C	C_0
mm			N(1 N = 0.225 lbf)	
20	35	10	9800	16000
	40	14	15300	25000
25	42	11	12200	22800
	47	15	19300	34000
	52	18	26000	45000
	60	24	40000	63000
30	47	11	12900	26500
	52	16	19600	37500
	60	21	31000	57000
	70	28	52000	90000
35	52	12	13400	30000
	62	18	27000	53000
	68	24	38000	695000
	80	32	63000	112000
40	60	13	18000	40000
	68	19	31000	64000
	78	26	50000	98000
	90	36	80000	146000
45	65	14	18600	45000
	73	20	31500	68000
	85	28	55000	110000
	100	39	93000	173000
50	70	14	19600	50000
	78	22	36000	61500

35	47	7	2360	2000
	62	9	9500	6950
	62	14	12200	8500
	72	17	19600	13700
	80	21	25500	18000
	100	25	42500	31000
40	52	7	2450	2200
	68	9	10200	7800
	68	15	12900	9300
	80	18	23600	16600
	90	23	31500	22400
	110	27	49000	36500
45	58	7	4650	3800
	75	10	12000	9300
	75	16	16300	12200
	65	19	25500	18600
	100	25	40500	30000
	120	29	58500	45500
50	65	7	4600	4250
	80	10	12500	10000
	80	16	16600	13200
	90	20	27000	19600
	110	27	47500	30000
	130	31	67000	52000
55	72	9	6400	5600
	90	11	15000	12200
	90	18	21600	17000
	100	21	33500	25000
	120	29	55000	41500
	140	33	76500	63000
60	78	10	6700	6100
	95	11	15300	13200
	95	18	22800	18300
	110	22	36500	28000
	130	31	63000	48000
	150	35	83000	69500
65	85	10	9000	8300
	100	11	16300	14600
	100	18	23600	19600

50	90	20	17300	10000
	90	23	17600	10600
	110	27	33500	17000
	110	40	49000	23600
55	100	21	20400	12500
	100	25	20400	12500
	120	29	39000	21600
	120	43	57000	28000
60	110	22	23200	14300
	110	28	26000	15600
	130	31	44000	25500
	130	46	67000	32500
65	120	23	23600	15600
	120	31	33500	20000
	140	33	47500	28000
	140	48	73500	38000
70	125	24	26500	17300
	125	31	34000	21200
	150	35	57000	34000
	150	51	83000	44000
75	130	25	29000	19600
	130	31	34000	22000
	160	37	61000	36500
	160	55	93000	51000
80	140	26	30500	21600
	140	33	37500	24500
	170	39	68500	40500
	170	68	104000	57000
85	150	28	37500	26000
	150	36	45000	29000
	180	41	75000	46500
	180	60	108000	61000
90	160	30	43000	29000
	160	40	54000	35500
	190	43	90000	54000
	190	64	116000	68000
95	170	32	49000	34000
	170	43	64000	42500

40	68	15	21200	13400
	80	18	38000	24000
	80	23	51000	34500
	90	23	51000	32500
	90	33	73500	51000
	110	27	88000	57000
45	75	16	26500	17600
	85	19	40000	25500
	85	23	54000	37500
	100	25	65500	41500
	100	36	95000	67000
	120	29	104000	69500
50	80	16	26500	17600
	90	20	42500	27500
	90	23	56000	40500
	110	27	80000	52000
	110	40	110000	80000
	130	31	127000	86500
55	90	18	31000	21200
	100	21	51000	34000
	100	25	67000	48000
	120	29	100000	67000
	120	43	134000	98000
	140	33	129000	86500
60	95	18	32000	22400
	110	22	62000	43000
	110	28	68000	68000
	130	31	112000	76500
	130	46	153000	114000
	150	35	153000	106000
65	100	18	32000	22800
	120	23	72000	51000
	120	31	106000	81500
	140	33	125000	85000
	140	43	173000	129000
	160	37	180000	127000
70	110	20	48000	34000
	125	24	72000	51000
	125	31	106000	81500

	80	22.75	62000	45500
	80	22.75	52000	38000
	80	32.75	81500	65500
	68	19	45000	40000
	80	19.75	51000	38000
	80	24.75	64000	50000
	80	32	88000	78000
	90	25.25	73500	56000
	90	25.25	63000	47500
	90	35.25	100000	83000
45	75	20	50000	44000
	80	26	72000	64000
	85	20.75	57000	44000
	85	24.75	68000	56000
	85	32	91500	81500
	100	27.25	91500	72000
	100	27.25	78000	60000
	100	38.25	120000	102000
50	80	20	52000	48000
	80	24	58500	56000
	90	21.75	64000	52000
	90	24.75	69500	57000
	90	32	98000	90000
	110	29.25	108000	83000
	110	29.25	91500	69500
	110	42.25	146000	127000
55	90	23	69500	64000
	95	30	95000	86500
	100	22.75	76500	61000
	100	26.75	90000	75000
	100	35	118000	106000
	120	31.5	122000	96500
	120	31.5	104000	80000
	120	45.5	170000	146000
60	95	23	71000	67000
	110	23.75	83000	65500
	110	29.75	108000	91500
	110	38	143000	134000

	95	31	67000	137000
	110	43	106000	204000
55	78	16	23600	62000
	90	25	49000	110000
	105	35	86500	180000
	120	48	137000	265000
60	85	17	27500	71000
	95	26	50000	118000
	110	35	90000	196000
	130	51	143000	265000
65	90	18	28500	78000
	110	27	51000	127000
	115	36	90000	196000
	140	56	166000	360000
70	95	18	32500	88000
	105	27	51000	127000
	125	40	102000	232000
	150	60	180000	400000
75	100	19	34000	98000
	110	27	52000	134000
	135	444	118000	270000
	160	65	208000	500000
80	105	19	34500	102000
	115	28	57000	153000
	140	44	122000	290000
	170	68	224000	550000
85	110	19	34500	106000
	125	31	72000	190000
	150	49	140000	340000
	180	72	236000	610000
90	120	22	39000	120000
	135	35	91500	240000
	155	50	153000	400000
	190	77	250000	670000
105	135	25	57000	173000
	150	38	102000	280000
	170	55	183000	480000
	210	85	305000	865000

TABLE 12.7. (Continued)

Block 1

Boundary dimensions			Basic load ratings dynamic	static
d	D	B	C	C_0
mm			N(1 N = 0.225 lbf)	
	120	23	43000	34000
	140	33	71000	56000
	160	37	91500	78000
70	90	10	9300	9150
	110	13	21600	19000
	110	20	29000	24500
	125	24	47500	37500
	150	35	80000	63000
	180	42	110000	104000
75	95	10	9650	9800
	115	13	22000	20000
	115	20	305000	26000
	130	25	51000	40500
	160	37	865000	72000
	190	45	118000	114000
80	100	10	9500	9800
	125	14	25500	23600
	125	22	36500	31500
	140	26	54000	45000
	170	39	95000	80000
	200	48	125000	125000
85	110	13	146000	15000
	130	14	26000	25000
	130	22	38000	33500

Block 2

Boundary dimensions			Basic load ratings dynamic	static
d	D	B	C	C_0
mm			N(1 N = 0.225 lbf)	
100	180	34	53000	35500
	180	46	75000	50000
	215	47	110000	59500
110	200	32	67000	49000
	200	53	96500	63000
	240	50	127000	66500
110	140	16	21600	23600
	170	19	44000	42500
	170	28	63000	57000
	200	38	112000	100000
	240	50	156000	166000
120	150	16	22400	25000
	180	19	47500	47500
	180	28	65500	61000
	215	40	112000	100000
	260	55	160000	166000
130	165	18	29000	32500
	200	22	61000	61000
	200	33	81500	78000
	230	40	120000	112000
	280	58	176000	193000
140	175	18	30500	35500
	210	22	62000	64000
	210	33	85000	83000

Block 3

Boundary dimensions			Basic load ratings dynamic	static
d	D	B	C	C_0
mm			N(1 N = 0.225 lbf)	
75	150	35	14600	102000
	150	51	204000	160000
	180	42	224000	163000
	115	20	49000	36000
	130	25	88000	63000
	130	31	120000	95000
	160	37	176000	125000
	160	55	250000	200000
	190	45	240000	173000
80	125	22	60000	44000
	140	26	96500	68000
	140	33	134000	106000
	170	39	176000	125000
	170	58	250000	200000
	200	48	275000	200000
85	130	22	62000	46500
	150	28	110000	78000
	150	36	153000	122000
	180	41	204000	146000
	180	60	285000	228000
	210	52	310000	223000
90	140	24	73500	56000
	160	30	134000	100000
	160	40	186000	150000

Block 4

Boundary dimensions			Basic load ratings dynamic	static
d	D	T	C	C_0
mm			N(1 N = 0.225 lbf)	
65	130	33.5	143000	116000
	130	33.5	122000	96500
	130	48.5	196000	173000
	100	23	71000	68000
	100	27	83000	83000
	110	34	12000	116000
	120	24.75	98000	78000
	120	32.75	129000	112000
	120	41	166000	153000
70	140	36	166000	134000
	140	36	140000	112000
	140	51	224000	200000
	110	25	86500	83000
	110	31	110000	109000
	125	26.25	108000	88000
	125	33.25	134000	118000
	125	41	173000	163000
75	150	38	190000	153000
	150	38	160000	127000
	150	54	250000	228000
	115	25	90000	88000
	115	31	112000	118000
	125	37	150000	146000
	130	27.35	120000	100000

Block 5

Boundary dimensions			Basic load ratings dynamic	static
d	D	H	C	C_0
mm			N(1 N = 0.225 lbf)	
110	145	25	58500	190000
	160	38	118000	365000
	190	63	204000	570000
120	155	25	68000	250000
	170	39	118000	380000
	210	70	240000	710000
	250	102	345000	1080000
130	170	30	80000	285000
	190	45	156000	500000
	225	75	255000	765000
	270	110	430000	1430000
140	180	31	81500	305000
	200	46	160000	530000
	240	80	285000	915000
150	190	31	83000	325000
	215	50	173000	585000
	250	80	290000	980000
160	200	31	86600	345000
	225	51	176000	610000
	270	87	340000	1180000
170	215	34	104000	415000
	240	56	208000	720000
	280	95	345000	1270000

Panel 1

Group				
90	150	28	64000	53000
	180	41	102000	90000
	210	52	134000	134000
95	115	13	15000	15600
	140	16	32000	29000
	140	24	45000	39000
	160	30	73500	62000
	190	30	110000	98000
	225	54	140000	146000
100	145	16	32500	31500
	145	24	46500	41500
	170	32	83000	69500
	200	45	118000	110000
105	125	13	15300	17000
	150	16	34000	32500
	150	24	46500	41500
	180	34	95000	78000
	215	47	134000	132000
	160	18	40000	38000
	160	26	56000	51000
	190	36	102000	143000
	225	49	140000	143000

Panel 2

Group				
150	250	40	127000	122000
	300	58	196000	224000
160	190	18	37500	43000
	225	22	71000	73500
	225	33	96500	96500
	270	42	134000	137000
	320	62	212000	250000
	200	20	38000	45500
	240	24	76500	80000
170	240	35	110000	112000
	290	45	143000	146000
	340	65	212000	250000
	215	20	47500	56000
	260	25	91500	96500
	260	38	129000	134000
	310	48	163000	180000
		22		
		28		
		42		
		52		

Panel 3

Group				
95	190	43	220000	160000
	190	64	300000	240000
	225	54	345000	260000
100	145	24	75000	58500
	170	32	150000	112000
	170	43	208000	170000
	200	45	250000	190000
	200	67	360000	300000
	240	55	375000	280000
	150	24	76500	60000
	180	34	166000	126000
	180	46	236000	193000
	215	47	290000	220000
	215	73	425000	355000
	250	58	415000	320000
105	160	26	90000	71000
	190	36	183000	137000
	225	49	335000	255000
	260	60	455000	345000

Panel 4

Group				
80	130	33.25	137000	120000
	130	41	176000	170000
	160	40	208000	173000
	160	40	176000	143000
	160	58	285000	265000
	125	29	116000	116000
	125	36	143000	153000
	130	37	153000	153000
	140	28.25	127000	104000
	140	35.25	160000	137000
	140	46	212000	208000
	170	42.5	232000	190000
	170	42.5	193000	153000
	170	61.5	320000	290000
180	225	34	104000	430000
	250	56	212000	750000
	300	95	375000	1370000

If the heaviest load to which the bearing is subjected acts during several revolutions of the bearing, the raceways will be evenly deformed and damaging indentations will not be produced.

It follows that, depending on the operating conditions, the heaviest load to act on a bearing should never exceed a certain value determined by the static safety factor s_0. Generally, the following minimum values can be used for s_0: applications where smooth, vibration-free running is assured, $s_0 = 0.5$; average working conditions with normal demands on quiet running, $s_0 = 1.0$; pronounced shock loads, $s_0 = 1.5$–2; high demands on quiet running, $s_0 = 2$.

Generally, for spherical roller thrust bearings, $s_0 > 2$.

For bearings that rotate very slowly and where the life is short, in terms of number of revolutions, the static load rating must also be taken into account. In such cases, the life equation can be misleading in giving an apparently permissible load that far exceeds the basic static rating.

12.3.8. Lubrication of Antifriction Bearings

In rolling bearings, the lubricant has a number of functions to perform. First, it must minimize the damage that results from the small degree of sliding that inevitably occurs. Second, the lubricant must protect the highly polished bearing surfaces from attack by the ambient environment, particularly if moisture is present. A corrosion-inhibiting additive is therefore required. Third, the lubricant should prevent dust and other contaminants from entering the bearing and causing premature failure. Finally, the lubricant may be required to act as a coolant for bearings operating at high speeds and under high loads.

Greases were the earliest and are still the most widely used lubricants for rolling bearings. As shown in Figure 12.12, the primary function of a grease pack is as a storehouse for the very small amount of oil required for lubrication. Among the advantages of grease as a lubricant are: It is an inexpensive and reliable system requiring a minimum of maintenance over a long period; the bearing is self-contained with a minimum of sealing problems; grease very successfully prevents the ingress of foreign matter.

Under carefully controlled conditions, grease lubrication has been successfully used up to $d_m n$ values of 0.6×10^6 ($d_m n$ = pitch diameter in mm $\times$ rotational

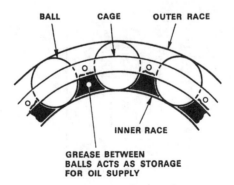

Figure 12.12. Grease lubrication of rolling bearings (from Timken 1983).

speed in rpm). At higher $d_m n$ values, the coolant property of the lubricant becomes important and some method of external oil feed must be provided. $d_m n$ is related to peripheral velocity (Figure 12.8) as $U = \pi d_m n / 60$.

Figure 12.13 shows the basic components of ball and roller bearings: these are the outer and inner races or rings, the rolling elements (balls or rollers) and the cage or separator, which locates and separates the rolling elements. During relative motion between the outer and inner ring, two distinct modes of contact occur, rolling contact between the rolling elements and the races and sliding contact between the cage pockets and the rolling elements. In ball bearings designed to have a high degree of osculation, such as deep-groove ball bearings, and in bearings functioning under severe conditions of load and speed, some slip must also occur between the rolling elements and the raceways of the rings. The primary function of the lubricant must be to reduce metallic interaction in these regions of contact in order to minimize wear and limit the frictional moment of the complete bearing.

In general, the frictional moment of a complete bearing will be composed of rolling resistance and hydrodynamic losses in the load-bearing components, sliding resistance and hydrodynamic losses in the cage system, and bulk churning of the lubricant within and without the bearing assembly. Frictional torque follows the trend shown in Figure 12.14.

At low rotational speeds, the frictional torque is constant, suggesting boundary lubrication. At higher speeds and/or viscosities, involving values of μN greater than about 3×10^3 (cSt × rev/min), a change occurs and the torque then becomes a function of both viscosity and speed, suggesting the predominance of hydrodynamic lubrication. For these conditions the frictional torque shows a certain similarity to the Petroff equation (see Chapter 11) and can be expressed in the form

$$M_0 = f_0 d^3 (\mu n)^{2/3} \tag{12.23}$$

where M_0 is the unloaded moment of frictional resistance, f_0 is a function of the type of bearing, d is the pitch diameter of the bearing, μ is the lubricant viscosity, and n is the speed of rotation.

The magnitude of the unloaded frictional moment is small in relation to the frictional moment caused by the normal bearing load. For a heavily loaded, low-speed bearing, the frictional resistance is found to be independent of viscosity and follows the equation

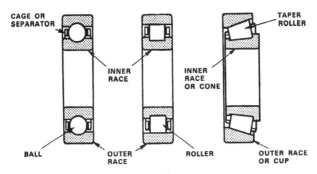

Figure 12.13. Details of roller bearings (from Timken 1983).

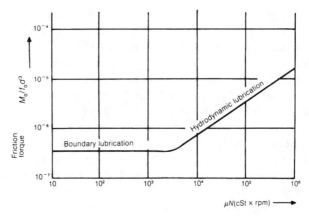

Figure 12.14. Friction torque of rolling bearings (from Timken 1983).

$$M_1 = f_1 W d \left(\frac{W}{C_0}\right)^c \qquad (12.24)$$

where f_1 and c depend upon the type of bearing, W is applied load in kg, and C_0 is the static load capacity of the bearing in kg. The magnitude of the total resistance to motion of a loaded bearing is given by the sum of the no-load torque M_0 and the load-dependent torque M_1.

For rough estimates, friction coefficients are usually given in manufacturers' catalogs, such as in Table 12.8 (from the SKF catalog).

12.3.9. Load Distribution within the Bearing

The bearing load is not equally distributed among the rolling elements of the bearing. It is obvious that the lower element carries heavier load and the side and upper

TABLE 12.8. Friction Coefficient for Rolling Bearings

Bearing type	f
Deep groove ball bearings	0.0015
Self-aligning ball bearings	0.0010
Angular contact ball bearings,	
single row	0.0020
double row	0.0024
Cylindrical roller bearings	0.0011
Needle roller bearings	0.0025
Spherical roller bearings	0.0018
Taper roller bearings	0.0018
Thrust ball bearings	0.0013
Cylindrical roller thrust bearings	0.0040
Needle roller thrust bearings	0.0040
Spherical roller thrust bearings	0.0018

ones carry virtually no load at all. Let P_0 be the load of the lower element and P_1, $P_2, \ldots$, the loads on the side elements. From Figure 12.3, the approach between two curved surfaces is $cP_1^{2/3}$, where c is a constant. The individual deflections therefore will be $\delta_0 = cP^{2/3}$, $\delta_1 = cP^{2/3}$, etc. Therefore, because $\delta_i = \delta_0 \cos i\theta$:

$$\left(\frac{P_i}{P_0}\right)^{2/3} = \frac{\delta_i}{\delta_0} = \cos i\theta \qquad (12.25)$$

The resulting vertical force, equal to the load W, is the sum of the vertical components

$$W = P_0 + 2P_1 \cos \theta + 2P_2 \cos 2\theta + \cdots \qquad (12.26)$$

Using Equation (12.25):

$$W = P_0 \left[1 + 2 \sum_{i=1}^{n} (\cos i\theta)^{5/2} \right] \qquad (12.27)$$

where n is the number of balls up to the horizontal position of one side only. If all the balls carried equal load, the load would be $P_0 n$. Therefore, there is a load concentration factor, with a similar meaning to stress concentration factor (Figure 12.15).

$$K_t = z \left(1 + 2 \sum_{i=1}^{n} \cos^{5/2} i\theta \right) / n \approx 4.36 \qquad (12.28)$$

where K_t is a function of the number of rolling elements z, with limiting values 1 for one element only and 4.36 for infinite number of elements. For most bearings it is about 4. If $K_t > 1$, the capacity of the rolling elements is not fully utilized. For special applications where a high capacity is desired for a small-volume bearing, a recess such as the one in Figure 12.16 can increase the capacity of the bearing by a better distribution of the load.

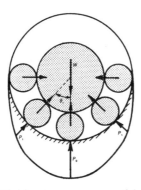

Figure 12.15. Rigid outer race, unequal load distribution.

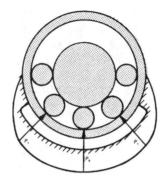

Figure 12.16. Flexible outer race for optimum load distribution.

Design Procedure 12.1: Selection of an Antifriction Bearing

Input: Maximum operational axial and radial loads, rotating speed. **Output:** Antifriction bearing type and dimensions.

Step 1: Sketch the bearing and place on the sketch loads and motion data.

Step 2: From the shaft strength calculation, find the nominal diameter D. Select a lubricant and find its kinematic viscosity ν (Equation (11.2a) and Figure 11.4).

Step 3: Select a working life L_n (Table 12.1), reliability factor α_1 (Table 12.3), life adjustment factor α_{23}; from the kinematic viscosity and Figure 12.10, find the material coefficient ν/ν_1 and use Figure 12.11.

Step 4: Use Equation (12.17) to find the life in millions of revolutions.

Step 5: Use a free-body diagram to determine the axial and radial loads on the bearing. If the values of the loads are theoretical, use Equation (12.18) with Tables 12.4 and 12.5, if applicable, to determine the true loads.

Step 6: Select a type of bearing from Table 12.7 or manufacturers' catalogs. From Table 12.6, find the load constants X and Y.

Step 7: Use Equation (12.19) to calculate the equivalent bearing load P.

Step 8: Use Equation (12.16) to determine the required bearing capacity C.

Step 9: With the nominal diameter d selected and the bearing capacity C, enter Table 12.7 or a manufacturers' catalog to find a bearing with capacity equal to or higher than C. If this is not possible, go to a higher diameter.

Step 10: For the bearing found, record the dimensions and the static load rating C_0. Check if it is greater than or equal to the static load C_0 from Equation 12.22. If not, go to step 6 and select a heavier bearing.

Example 12.3   The bearing on the upper left of Figure 12.17 (page 825) is used on a construction crane. Analysis of the system yielded, for this bearing, horizontal load 3500 N, vertical load 6500 N, and axial load 2000 N. The rotating speed is 2000 rpm and the desired bearing life is 6000 hours. Select a deep-groove ball bearing of common steel grade with a reliability of 99%. In the load computation, shock loads have been accounted for. Strength calculations for the shaft require that its diameter is at least 60 mm. SAE 30 oil is used, and it is expected to operate at 80°C.

Solution The total radial load is $F_r = (3500^2 + 6500^2)^{1/2} = 7380$ N. The axial load is $F_a = 2,000$ N. The oil viscosity of SAE 30 oil at 80°C is 12 mN · s/mm². For a density of 890 kg/m³, the kinematic viscosity is $\nu = \mu/\rho = 1.2 \times 10^{-2}/890 = 1.35 \times 10^{-5}$ m²/s = 13.5 mm²/s. The required kinematic viscosity, from Figure 12.19, is $\nu_1 = 12$ mm²/s. Thus, from Equation (12.17a), $k = \nu/\nu_1 = 13.5/12 = 1.125$. From Figure 12.11, $a_{23} = 1.2$, a midrange value for common steel grade.

If we start with deep-groove ball bearings, $F_a/F_r = 2,000/7,380 = 0.27$. From Table 12.6 and deep-groove ball bearings, we start with the assumption $F_a/F_r > e$ and then $X = 0.56$, $Y = 2$. Shaft rotates, thus $V = 1$. Equation (12.19) gives:

$$P = \max\{VF_r, YF_a + VXF_r\}$$

$$= \max\{1 \times 7,380, 2 \times 2,000 + 1 \times 0.56 \times 7380\} = 8,132 \text{ N}$$

Since additional loads have been accounted for, $f_k = 1$, $f_d = 1$, and $P_{\text{eff}} = P$ in Equation (12.18).

For radial deep groove bearings, the equivalent dynamic load will be, from Equation (12.16a):

$$C = P \left(\frac{60nL_h}{10^6 \, a_1 a_2 a_3} \right)^{1/3} = 8132 \left(\frac{60 \times 2000 \times 6000}{10^6 \times 0.21 \times 1.20} \right)^{1/3} = 115,403 \text{ N}$$

Therefore, using Table 12.7, first column, the bearing with $d = 100$, $D = 215$ mm, width $B = 47$ mm is selected. It has capacity $134,000 > 115,403$ needed. The static capacity is $C_0 = 132,000$ N and $F_a/C_0 = 2,000/132,000 = 0.015$, and our initial selection of X and Y in Table 12.6 was correct.

The static bearing capacity must now be checked. Equation (12.18) must be used. For small cranes, $S_0 = 1.6$. The equivalent static load is then:

$$P_0 = 0.6F_r + 0.5F_a = 5428 \text{ N}$$

and the required bearing static capacity should be:

$$C_0 = P_0 s_0 = 5428 \times 1.6 = 8685 \text{ N}$$

which is much lower than the selected bearing static capacity of 132,000 N.

The design is complete.

The solution was also obtained in the spreadsheet of Table E12.3 using Design Procedure 12.1.

Example 12.4 The wheel of a crane carries a maximum load of 50,000 N and revolves at 40 rpm when the crane is in motion. It is expected that only 30% of its operating life of 10,000 hours will it carry the maximum load. Another 30% of the time it will carry a 60% load, and the rest of the time it will carry the dead weight of the crane (18,000 N). Bending of the shaft requires that the diameter be at least 50 mm. Grease lubrication will be used at normal temperature, and 90% reliability will suffice. Select appropriate rolling bearings.

Solution For the diameter of 50 mm, thc 50 kN load looks heavy and roller bearings are initially selected. Each bearing carries half load. Because load is not constant, an equivalent

TABLE E12.3

Example 12.3. PROCEDURE 12.1. SELECTION OF AN ANTIFRICTION BEARING

Input: (shaded cells)				
Static Safety Factor		1.6		
Rotating Speed (RPM)	Reliability %	2000	99	
Step 2:				
Lubricant Kinematic Viscosity	Nominal Diameter D, mm	1.35E-05	60	
Step 3:				
Working Life Lh (table 12.1) (hours)		6000		
Reliability factor α_1 (table 12.3)	Speed coefficient $\nu_1 =$ (figure 12.10)	0.21	1.20E-05	
Step 4:				
Life (million revs) Lna	Material coefficient k =	720	1.125	
Life adj. Factor $\alpha_{23} =$ (Fig. 12.11)	Exponent p = 3 for ball or 10/3 for roller	1.2	3	
Step 5:				
Axial Load Fa =	Radial Load Fr =	2000	7380	
Load Constant X	Load Constant Y =	0.56	2	
Constant V (= 1 for shaft rotating, 1.2 for housing rotating)		1		
Step 7:				
Equivalent bearing load P		8132.8		
Step 8:				
Required **Bearing Dynamic Load C**		115403.08		
Step 9:				
Select Bearing: (Table 12.17)-Deep grove ball BRG	D= 215	B= 47		
Dynamic capacity of selecting bearing:	Static capacity of selected bearing	134000	132000	
Static constant X_0 =	Static Constant Y_0 =	1	1	
Step 10:				
Static load P_0 = (Table 12.17)	*Required* static capacity	9380	15008	

822

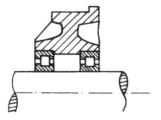

Figure E12.4

load will be computed with Equation (12.20). Because the number of revolutions is proportional to the percentage of operating time:

$$P = \left(\frac{50,000^p \times 0.30 + 30,000^p \times 0.30 + 18,000^p \times 0.40}{0.3 + 0.3 + 0.4} \right)^{1/p}$$

For $p = 10/3$, $P = 37,042$ N. Operating factors are $a_1 = 1$ (90% reliability specified), $a_{23} = 1$ (normal steel and adequate lubrication in conjunction with low speed). Therefore, for $f_k = 1$ (no gears), $f_d = 1$ (no machines connected), the effective load is 37,042 N. The required dynamic capacity of the bearing will be, by Equation (12.16):

$$C = 37,042 \left(\frac{60 \times 40 \times 10,000}{10^6} \right)^{3/10} = 96,107 \text{ N}$$

The roller bearing (column 3 in Table 12.7) with $d = 50$, $D = 110$, $B = 40$ mm has a dynamic capacity $C = 110,000$ N and a static capacity $C_0 = 80,000$ N. The dynamic capacity is adequate. The static capacity for $s_0 = 1.6$.

$$C_0 = P_0 s_0 = 37,042 \times 1.6 = 59,267 \text{ N}$$

which is less than the bearing capacity of 80,000 N. Therefore, the bearing selected is adequate.

Example 12.5 A cylindrical roller bearing SKF NU 318 made of standard steel is to rotate at $n = 500$ rpm under a constant radial load of 25 kN. The kinematic viscosity of the oil to be used at the expected temperature is 38 mm²/s. What is the expected life for reliability 90%? From bearing tables, $d = 90$, $D = 190$, $B = 43$ mm, $C = 220,000$ N, $C_0 = 160,000$ N.

Solution Because the load acts only in the radial direction, $P = F_r = 25,000$ N. The mean diameter is:

$$d_m = \frac{d + D}{2} = \frac{90 + 190}{2} = 140 \text{ mm}$$

From Figure 12.10, $\nu_1 = 21$ mm²/s. Therefore:

$$k = \frac{\nu}{\nu_1} = \frac{38}{21} = 1.8$$

The life adjustment factor is then $a_{23} = 1.4$–2.2. Therefore:

For reliability of 90%, $a_1 = 1$:

$$L = 1 \times (1.4\text{–}2.2)\left(\frac{220,000}{25,000}\right)^{10/3} = 1970\text{–}3095 \times 10^6 \text{ revolutions}$$

For reliability of 98%, $a_1 = 0.33$:

$$L = 0.33(1.4\text{–}2.2)\left(\frac{22,000}{25,000}\right)^{10/3} = 650\text{–}1021 \times 10^6 \text{ revolutions}$$

12.4. APPLICATION OF ROLLING BEARINGS

12.4.1. Locating Rolling Bearings

A rotating shaft generally requires two bearings to support and locate it radially and axially relative to the stationary part of the machine, such as the housing. Normally, only one of the bearings (the locating bearing) is used to fix the position of the shaft axially, while the other bearing (the nonlocating bearing) is free to move axially.

Axial location of the shaft is necessary in both directions, and the locating bearings must be axially secured on the shaft and in the housing to limit lateral movement. In addition to locating the shaft axially, the locating bearing is generally also required to provide radial support, and bearings that are able to carry combined loads are then necessary, such as deep-groove ball bearings, spherical roller bearings, and double-row or paired single-row angular contact ball bearings. A combined bearing arrangement, with radial and axial location provided by separate bearings, can also be used, such as a cylindrical roller bearing mounted alongside a four-point contact ball bearing or a thrust bearing having radial freedom in the housing.

To avoid cross-location of the bearings, the nonlocating bearing, which provides only radial support, must be capable of accommodating the axial displacements that arise from the differential thermal expansion of the shaft and housings. The axial displacements must be compensated for either within the bearing itself—as in cylindrical roller bearings and needle roller bearings—or between the bearing and its seating on the shaft or in the housing.

The bearings can also be arranged on the basis that axial location is provided by each bearing in one direction only. This arrangement is mainly used for short shafts supported by deep-groove ball bearings, angular contact ball bearings, taper roller bearings, or cylindrical roller bearings.

Some locating arrangements arc shown in Figure 12.17. In some cases, such as conical roller bearings, both bearings are locating, within certain limits, because of the nature of such bearings. Wear of the bearings makes an adjustment necessary to maintain close clearances.

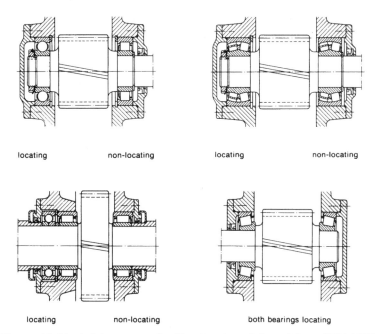

locating   non-locating   locating   non-locating

locating   non-locating   both bearings locating

Figure 12.17. Axial support of rolling bearings (adapted from SKF 1977).

Interference fits in general only provide sufficient resistance to axial movement of a bearing on its seating when no axial forces are to be transmitted and the only requirement is that lateral movement of the ring be prevented. Positive axial location or locking is necessary in all other cases. To prevent axial movement in either direction of a locating bearing, it must be located at both sides. When nonseparable bearings are nonlocating bearings, only one ring, that having the tighter fit, is axially located; the other ring must be free to move axially in relation to the shaft or housing.

Where the bearings are arranged so that axial locating of the shaft is given by each bearing in one direction only it is sufficient for the rings to be located at one side only.

Bearings having interference fits are generally mounted against a shoulder on the shaft or in the housing. The inner ring is normally secured in place by means of a lock nut and locking washer or by an end plate attached by set screws to the shaft end; the outer ring is normally retained by the housing end cover, but the threaded ring screwed into the housing bore is sometimes used.

Instead of shaft or housing abutment shoulders, it is frequently convenient to use spacing sleeves or collars between the bearing rings or a bearing ring and the adjacent component, such as a gear. On shafts, location can also be achieved using a split collar that sits in a groove in the shaft and is retained either by a solid outer ring that can be slid over it or by the inner ring of the bearing itself.

Axial location of rolling bearings by means of snap rings can save space, assist rapid mounting and dismounting, and simplify machining of shafts and housings. The dimensions of components adjacent to the bearings (shafts and housing shoulders, spacing collars, etc.) must be such that the surface against which the face of the bearing rings about is sufficiently large. It is also necessary to ensure that rotating parts of the bearing not come into contact with stationary components. The bearing tables give suitable dimensions for these features.

At corner points, suitable fillets must be provided, conforming to the appropriate bearing dimensions. The better the blending of the cylindrical shaft with the shoulder, the more favorable is the stress distribution. For very heavily loaded shafts, a larger fillet or more gradual change in diameter may be required. In this case, a spacing collar should be provided between the bearing and the shoulder to ensure a sufficiently large abutment surface for the inner ring. The spacing collar must be relieved adjacent to the shaft shoulder so that it does not contact the shaft fillet.

In order to facilitate dismounting of bearing rings it is sometimes desirable that slots be machined in the shaft or housing shoulders to accommodate the claws of bearing extractor tools.

12.4.2. Sealing

Bearings must be protected by suitable seals against the entry of moisture and other contaminants and to prevent the loss of lubricant. The effectiveness of the sealing can have a decisive effect on the life of a bearing.

In the design of a rolling bearing application, sealing must be carefully considered. Manufacturers' catalogs also contain information on available types of seals. In general, seals are classified as rubbing and nonrubbing.

Nonrubbing seals depend for their effectiveness on the sealing efficiency of narrow gaps (Figure 12.18), which may be arranged axially or radially or combined to form a labyrinth. This type of seal has negligible friction and wear and is not easily damaged. It is particularly suitable for high speeds and temperatures, where leaks can be a fire hazard.

The simple gap-type seal, which is sufficient for machines in a dry, dust-free atmosphere, is made up of a small radial gap formed between the shaft and housing. Its effectiveness can be improved by providing one or more grooves in the bore of the housing cover. The grease emerging through the gap fills the grooves and helps to prevent the entry of contaminants. With oil lubrication and horizontal shafts, right- or left-hand helical grooves can be provided in the shaft or the seal bore.

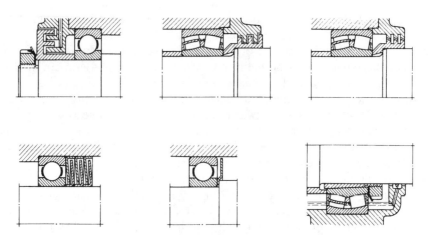

Figure 12.18. Labyrinth sealing (adapted from SKF 1977).

These serve to return any oil that may tend to leak from the housing. However, with this arrangement it is essential that the direction of rotation not vary.

Rubbing seals (Figure 12.19) rely for their effectiveness on the elasticity of the material exerting and maintaining a certain pressure at the sealing surface. The choice of seal and the required quality of the sealing surface depend on the peripheral speed.

Felt washers are mainly used with grease lubrication, such as in plummet-blocks. They provide a simple seal suitable for peripheral speeds up to 4 m/s and temperatures of about 100°C. The effectiveness of the seal is considerably improved if the felt washer is supplemented by a simple labyrinth ring. The felt washers or strips should be soaked in oil at about 80°C before assembly.

Where greater demands are made on the effectiveness of the rubbing seal, particularly for oil-lubricated bearings, lip seals are often used in preference to felt seals. A wide range of proprietary lip-type seals is available in the form of ready-

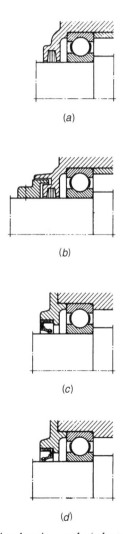

(a)

(b)

(c)

(d)

Figure 12.19. Rolling bearing seals (adapted from SKF 1977).

to-install units made up of a seal of synthetic rubber or plastic material normally enclosed in a sheet metal casing. They are suitable for higher peripheral speeds than felt washers. As a general guide, at peripheral speeds of over 4 m/s the sealing surface should be ground, and above 8 m/s, hardened or hard-chrome-plated and fine-ground or polished if possible. Referring to Figure 12.19, if the main requirement is to prevent leakage of lubricant from the bearing, then the lip should face inwards (*c*); if the main purpose is to prevent the entry of dirt, then the lip should face outwards (*d*).

Simple space-saving arrangements can be achieved by using bearings incorporating seals or shields at one or both sides. Bearings sealed or shielded at both sides are supplied, lubricated with the correct quantity of grease. Relubrication is not normally required, and they are primarily intended for application where sealing is otherwise inadequate or cannot be provided for reasons of space.

12.4.3. Lubricant Application

Under normal conditions, grease lubrication is usually applied. Grease is more easily retained than oil and also prevents the entrance of moisture and dust. Typical grease application is shown in Figure 12.20.

At higher speeds of rotation, oil lubrication is preferable (Figure 12.21). Additional reasons sometimes might be the necessity to remove heat generated at high rate or the existence of other oil-lubricated elements, such as gears, near the bearing. Manufacturers usually give application ranges for grease or oil lubrication.

CASE STUDY 12.1: Redesign of a Task Corp. Motor Ball Bearing[2]

In 1962, the Task Corporation in Anaheim, California, won a contract to design and manufacture 300 electric motors for the Geyser Pump Company. The pump

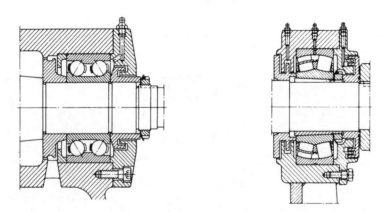

Figure 12.20. Oil lubrication of rolling bearings (adapted from SKF 1977).

---

[2] Based on Case Study ECL 14 of the Engineering Case Library. Originally prepared by Karl H. Vesper of Stanford University under an NSF grant. By permission of the trustees of the Rose-Hulman Institute of Technology.

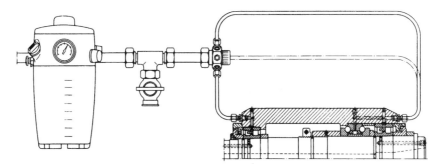

Figure 12.21. Forced oil lubrication (adapted from SKF 1977).

was one of two pumps required for the Thunder 99 aircraft hydraulic systems, under the following specifications:

Life time between scheduled maintenance: 2,500 h
Motor cooling with through-flow of the hydraulic fluid at the suction of the pump
Hydraulic fluid: Skydrol 500A
Fluid temperature: 65–180°F
Fluid pressure at the motor: 45–50 psi
Flow rate: 1–6 gallons/min
Power source: 400 Hz AC, 200 V
Rated power: 11.75 HP
Motor speed: 6,000 rpm
Motor efficiency > 65%
Motor weight < 18.75 lb
Maximum motor case temperature: 225°F.

In September 1963, the first motor was tested, and the ball bearings used in the design failed at 1800 hours. Engineer Jack Wireman of the Task Corp. was assigned to investigate the failure and redesign the motor bearings.

Investigation

A cross-section of the motor-pump assembly is shown in Figure CS12.1. The motor rotor is supported by two deep-groove (*Conrad*) ball bearings. The one towards the pump takes the thrust of the pump. The radial loading is the weight of the motor and impeller, 5.3 lb plus the dynamic unbalance, 5 lb, and the thrust load applied by a spring was estimated by the Geyser engineers to be between 50 and 70 lb and is supported by the front bearing. The Barden precision deep-groove bearing No. 204SST5 was selected with rated dynamic capacity 1400 lb. In the Task Corp. literature, it is noted that "Although no detailed life calculations were made on the bearings, catalog data and design experience indicated to Task engineers that they would last well beyond the 2500 hours" Calculations performed later estimated the bearing life to 6500 hours.

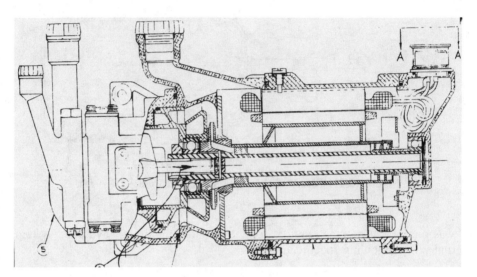

Figure CS12.1

Bearing Redesign

Without performing force and other measurements and calculations, the designers decided that the problem would be solved if they used a bearing with greater dynamic capacity. Thus, Barden bearing 204HJB1519 was used. Jack Wireman calculated that the life of this bearing would be 19,000 hours. The new bearing was installed and a test was performed. The new bearing failed in only 700 hours.

It became apparent that the load capacity was not a problem. It was suggested by one consultant that the problem was additional loading of the bearing due to thermal expansion or misalignment. A higher-precision bearing was one suggested solution, at a cost of $1000 per bearing (as compared with $10 for the Barden bearing) for a motor price of $800. An alternative was tested, a $3 New Departure 030204 bearing. Because the Air Force needed the pumps, it was decided to proceed with the new design and closely inspect the motors. It was found that at 1000 hours of operation none had failed. The final decision was to reduce the inspection time from 2500 hours to 1000 hours until more operation experience dictated otherwise.

The difference between the Barden and the New Departure bearings was in their quality class. The Barden 204SST5 and 204HJB1519 were of quality class C7 (high precision) while the New Departure 030204 bearing was of normal quality class C3; thus it had larger clearances and could better tolerate misalignment and thermal expansion while the high precision of the Barden bearings was not really needed because the motor rotor–stator clearance is several orders of magnitude greater than the bearing clearance.

REFERENCES

Pugh, B. 1970. *Practical Lubrication*. London: Butterworths.

Shigley, J. E. 1972. *Mechanical Engineering Design*, 2nd ed. New York: McGraw-Hill.

Singer, C. J. *A History of Technology.*

SKF. 1977. *General Catalog.* King of Prussia, Pa.: SKF.

Timken Co. 1983. *Bearing Selection Handbook.* Detroit: Timken.

ADDITIONAL READING

ANSI B3.14. *Standard.* [Can be obtained through the Anti-Friction Bearings Manufacturers Association, AFBMA #20 Standard.]

British Petroleum. *Lubrication Theory and Application.* London: BP Trading Co, 1969.

Dimarogonas, A. D. *History of Technology.* (in Greek). University of Patras, 1979 [in Greek].

Harris, T. *Rolling Bearing Analysis.* New York: John Wiley & Sons, 1966.

ISO Technical Committee 4 (ISO/TC4). *Standards.*

Mechanical Drives (1978). Machine Design Reference Issue.

Niemann, G. *Maschinenelemente.* Munich: Springer, 1965.

Palmgren, A. *Ball and Roller Bearing Engineering.* Philadelphia: SKF Industries, 1959.

Reshetov, D. N. *Machine Design.* Moscow: Mir, 1978.

Wills, J. G. *Lubrication Fundamentals.* New York: Marcel Dekker, 1980.

PROBLEMS[3]

12.1. [D] A large railroad car is designed for transporting large electric generators. The total weight of the car and generator is 500 tons. The wheels have diameter 600 mm, and the width of the wheel and the rail is 50 mm. The fatigue strength of the material for both is 200 N/mm². Each axle has two wheels. Determine the required number of axles for that car. The car will be moving very slowly and no dynamic loads should be considered.

12.2. [D] A cement kiln drum is rotated by way of several rolling rings of diameter $D = 2$ m. Each of them rolls on two rollers of diameter $d = 0.40$ m. The width of the rollers is 120 mm, and the material of the ring and the rollers is cast steel ASTM A 148/GF 80-50. Determine for each such station along the kiln drum how much load it can take.

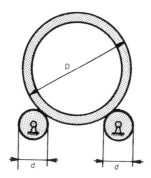

Figure P12.2

---

[3][C] = certification, [D] = design, [N] = numerical, [T] = theoretical problem.

12.3.   [D] A generator rotor weighs 120 tons and is turned on a lathe supported on two pairs of rollers at the two ends. The rotor diameter at the rolling station is 1 m, and the rollers have diameter 250 mm. The rotor material is Cr–Mo–V, which has much higher fatigue strength than the carbon steel material of the rollers, which is 280 N/mm². Determine the required width of the rollers.

12.4.   [C] A cam and follower mechanism, as shown, operates at 3200 rpm. The diameter of the follower is 40 mm, and the radius of the curvature of the cam is 100 mm at the point of maximum acceleration, which yields a maximum transmitted force of 6000 N. The lubricant used is SAE 10, and the temperature is kept at below 50°C. The width of the follower is 20 mm. Determine the minimum oil film thickness.

Figure P12.4

12.5.   [C] In Problem 12.4, if the material of the cam and follower is carbon steel AISI 1020, determine the safety factor for fatigue failure.

12.6.   [C] A deep-groove ball bearing (first column of Table 12.7) of size $d = 17$ mm, $D = 26$ mm, $B = 5$ mm is used in a gearing application with moderate shocks. The lubricant is SAE 80 at temperature 80°C. The axial load is 600 N and the radial load is 1200 N. For reliability 90%, $s_0 = 1$ and $a_{23} = 1.2$, determine the bearing life in hours. The speed of rotation is 3000 rpm.

12.7.   [D] A taper roller bearing will be used for the main shaft of a lathe, which rotates at a maximum speed of 4000 rpm. Maximum radial loads is 20,000 N, axial, 12,000 N. The oil used is SAE 30 at temperatures which do not exceed 40°C and $s_0 = 1$. The shaft diameter is at least 80 mm. Select the bearing for a reliability of 90% and 6000 hours of life.

12.8.   [D] A deep-groove ball bearing is used to support a radial load of 12,000 N, axial load of 5,000 N. No limit is set on the shaft diameter. If the operating life is 5000 hours, the desired reliability is 90%, $s_0 = 1$ and $a_{23} = 1.2$, select the proper bearing assuming adequate lubrication for $n = 1000$ rpm.

12.9.   [D] A taper roller bearing used in a piece of construction equipment has two modes of operation: in the forward mode, $E_a = 6000$ N, $F_r = 6000$ N, while in the backward mode, $F_a = 0$, $F_r = 900$ N. The speed of rotation is low and adequate lubrication is provided. Standard reliability of 90% is adequate, $s_0 = 1$ and $a_{23} = 1.2$. Select a proper bearing for a 10,000 hours life for the worst mode, $n = 33$ rpm.

12.10. [D] The thrust load in an axial fan is loading an axial ball bearing with 68,000 N. The diameter has no limitation, there is adequate lubrication, the reliability required is 98% and $s_0 = 1$. The bearing life is to be 10,000 hours. Select the proper bearing for $n = 100$ rpm.

12.11. [C] The bearing in Problem 12.6 was overloaded for 1000 hours by a factor of 1.5. Determine the remaining life under the design load.

12.12. [C] The bearing in Problem 12.7 was overloaded for 1000 hours by a factor of 1.5. Determine the remaining life under the design load.

12.13. [C] The bearing in Problem 12.8 was overloaded for 1000 hours by a factor of 1.5. Determine the remaining life under the design load.

12.14. [D] The bearing in Problem 12.9 operates for 80% of its life in the forward mode and 20% in the backward mode. Select the proper bearing.

12.15. [C] The bearing of Problem 12.10, owing to a malfunction, operates for 500 hours with erratic shock load. It was estimated that the shock factor was at least 2. Determine the life remaining after this incident, assuming quiet operation.

12.16. [D] An industrial centrifuge is loaded as shown in Figure P12.16. The main shaft is supported by two antifriction bearings as indicated. The oil used is SAE 30 at temperatures that do not exceed 40°C.
1. Make a reasonable assumption for the bearing that will take the thrust load.
2. Select the proper bearings assuming 95% reliability, bearings of standard SKF steel, $s_0 = 1.2$, $f_k = f_d = 1$.
3. Make a sketch showing the proper mounting, lubrication, and sealing.

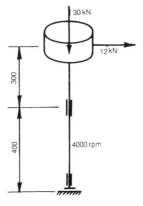

Figure P12.16

12.17. [D] An industrial plastics extruder is loaded as shown in Figure P12.17. The shaft is supported by two antifriction bearings as indicated. The oil used is SAE 30 at temperatures that do not exceed 40°C.

1. Make a reasonable assumption for the bearing that will take the thrust load.
2. Select the proper bearings assuming 95% reliability, bearings of standard SKF steel, $s_0 = 1.2, f_k = f_d = 1$.
3. Make a sketch showing the proper mounting, lubrication, and sealing.

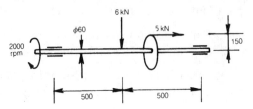

Figure P12.17

12.18. [D] An intermediate shaft of a crane is loaded as shown in Figure P12.18. The shaft is supported by two antifriction bearings as indicated. The oil used is SAE 30 at temperatures that do not exceed 40°C.

1. Select the proper bearings assuming 95% reliability, bearings of standard SKF steel, $s_0 = 1.2, f_k = f_d = 1$.
2. Make a sketch showing the proper mounting, lubrication, and sealing.

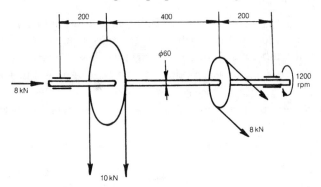

Figure P12.18

12.19. [D] The main shaft of a hoist is loaded as shown in Figure P12.19. The shaft is supported by two antifriction bearings as indicated. The oil used is SAE 30 at temperatures that do not exceed 40°C.

1. Select the proper bearings assuming 95% reliability, bearings of standard SKF steel, $s_0 = 1.2, f_k = f_d = 1$.
2. Make a sketch showing the proper mounting, lubrication, and sealing.

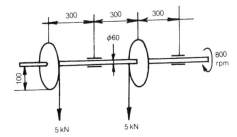

Figure P12.19

12.20. [D] An intermediate shaft is loaded as shown in Figure P12.20 by belt loads and a dead load. The shaft is supported by two antifriction bearings as indicated. The oil used is SAE 30 at temperatures that do not exceed 40°C.

    1. Select the proper bearings assuming 95% reliability, bearings of standard SKF steel, $s_0 = 1.2$, $f_k = f_d = 1$.

    2. Make a sketch showing the proper mounting, lubrication, and sealing.

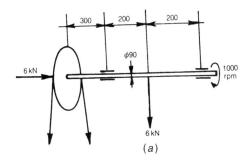

(a)

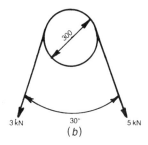

(b)

Figure P12.20

CHAPTER 13

DESIGN OF FIXED SPEED DRIVES

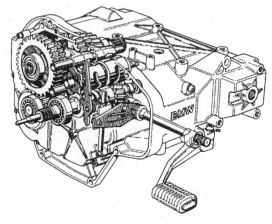

A five-speed manual gearbox for a motorcycle. (From the catalog of FAG, KgaA, Schwein-furt, Germany, reproduced by permission.)

OVERVIEW

In Chapters 8 and 9, we developed design methods based mostly on material strength and flexibility, respectively. The reason was that the respective mechanisms of failure were material failure and large displacements. There are other mechanisms of failure in machines, among them sliding friction and wear of machine parts. In Chapter 10, we studied design methods based on failures due to dry friction, with applications to clutches and brakes. In Chapter 11, we studied machine element design methods based on the prevention of failures related to hydrostatic and hydrodynamic lubrication, the use of a flowing fluid to prevent the rubbing surfaces from touching one another during sliding friction. In Chapter 12, we studied the design of lubricated elements with rolling friction.

There is a class of elements where all the above design considerations apply: the design of gearing. This will be the subject of the present chapter.

13.1. GENERAL DESIGN CONSIDERATIONS

13.1.1. Mechanisms of Failure

Tooth breakage is the most dangerous kind of gear failure. It is the result of high overloads of either impact or static action, repeated overloads causing low cycle fatigue, or multiple repeated loads leading to fatigue of the material.

Cracks are usually formed at the root of the teeth on the side of the stretched fibers where the highest tensile stresses occur together with local stresses due to the shape of the teeth. Fracture occurs mainly at a cross-section through the root of the tooth.

Fatigue pitting of the surface layers of the gear teeth (Figure 13.1a) is the most serious and widespread kind of tooth damage that may happen to gears that are enclosed, well lubricated, and protected against dirt.

As a result of the meshing of the teeth, the contact stresses at each point of the working surface of the teeth vary in a zero-plus cycle (with stresses of the same sign), while the stresses in the surface layers vary according to an alternating cycle, which, however, is unsymmetrical. Fatigue cracks usually originate at the surface, where stress concentration occurs due to micro-irregularities. If the hardened layer is comparatively thin, and also at high contact stresses, the cracks may originate deeper in the material.

Abrasive wear (Figure 13.1b) is the principal reason for the failure of open gearing and the closed gearing of machinery operating in environments polluted by abrasive materials.

Wear is nonuniform along the tooth profile because of the unequal sliding (rubbing) velocities and contact stresses. Wear increases dynamic loads and noise, weakens the teeth, and finally leads to tooth breakage.

Seizing of the teeth consists in localized molecular cohesion of the contacting surfaces due to the action of the high pressure and the rupture of the film of lubrication between them. As a contact is broken, particles of one of the surfaces are torn out by the other surface. These particles then score the rubbing surfaces of the teeth.

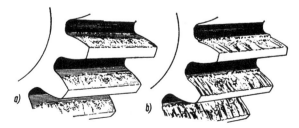

Figure 13.1. Gear teeth wear. (From Decker 1973.)

13.1.2. Materials and Manufacture

Requirements of materials for gears are to some extent similar to that for antifriction bearings. This is due to the high contact stresses inherent in gearing as in antifriction bearings. In addition, owing to the beam-like shape of the teeth and the repeated character of the load, considerable fatigue stresses are encountered. These stresses, however, are usually small compared to the contact stresses and the respective material strengths. Therefore, homogeneous materials yield gears with low bending stresses if contact stresses are near the limit, and thus surface hardening can increase the load-carrying capacity without exceeding the fatigue strength of the bulk material. For this reason, the principal materials used for load-carrying gears are steels, which can be heat-treated and surface-hardened. Cast irons are also used for gears, mostly for low speed and high sizes, because cast iron has considerable hardness and by nature has much more contact than fatigue strength in tension. Plastics and bronzes are used to a lesser extent.

Gear teeth can be cut either by direct or generating processes.

Direct cutting is performed with shape cutters (Figure 13.2a, b), which must have the form of the space between teeth. Because this space is different for different numbers of teeth, even for the same pitch, each cutter is assigned a range of modules and cutting is therefore approximate. In addition, the process is not productive and is used for low-accuracy gears. The advantage is that it can be performed in usual milling machines in a very simple way.

Generating processes (Figures 13.2c, d) use screw-type cutters with profiles identical to the basic rack corresponding to the pitch and module of the gear. The process is continuous and has definite advantages:

1. It is faster and more productive.
2. It is more accurate because the correct involute is generated regardless of the number of teeth.
3. It requires only one tool for every module.

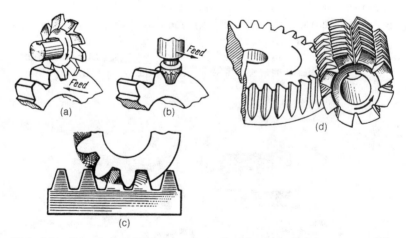

Figure 13.2. Manufacturing methods for gear teeth: (*a*) with a disk-type gear milling cutter; (*b*) with an end-mill type gear milling cutter; (*c*) with a rack-type gear shaping cutter; (*d*) with a gear hob.

For the above reasons, generating processes are used exclusively for mass production of gears.

13.2. DESIGN OF SPUR GEARS

13.2.1. Design of Spur Gears for Bending Strength

As the gears rotate, the point of contact between mating teeth moves along the profile, as one can observe in Figure 2.24. If the friction force is neglected, it can be assumed that the interaction force is perpendicular to the tooth profile at all times. It will be assumed at this point that only one tooth at a time carries the load, owing to cutting errors and high tooth stiffness. Multiple contact will be discussed later.

Under the single-contact assumption, the worst case is when the load is applied at the top of the tooth (Figure 13.4).

The normal force can be analyzed in tangential and radial components:

$$F_t = F_n \cos \varphi \tag{13.1}$$

$$F_r = F_n \sin \varphi = F_t \tan \varphi \tag{13.2}$$

where ϕ is the pressure angle (see Section 2.5) The tangential force is known from the system parameters, power H, angular velocity ω, pinion diameter d_1:

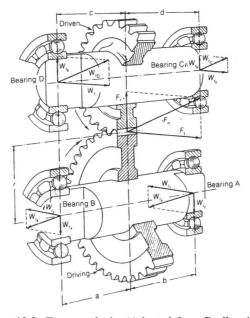

Figure 13.3. Force analysis. (Adapted from Dudley 1962.)

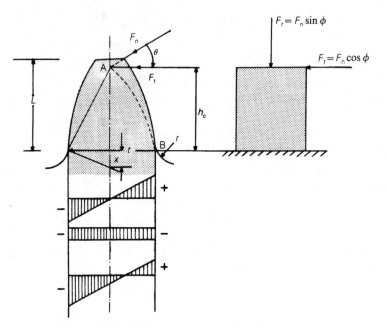

Figure 13.4. Beam strength analysis of spur gears.

$$F_t = \frac{2H}{(d_1 \omega)} \tag{13.3}$$

The tangential force results in bending stress, and the radial force results in compressive stress. The dangerous cross-section is at the root of the tooth in the zone of maximum stress concentration. Lewis (1897) proposed that this is the point B where a parabola from point A is tangent to the tooth profile (Figure 13.4).

Because fatigue cracks and failure begin on the side of the tooth subjected to tensile stress, the strength is determined for this side. The normal stress in the dangerous section due to bending is:

$$\sigma = \frac{F_t h_c}{bt^2/6} \tag{13.4}$$

where t is the thickness of the tooth in the dangerous section, b is the tooth width, and h_c is the moment arm of the force, the height of the tooth.

This can be rewritten as the *Lewis equation*:

$$\sigma = \frac{F_t}{(mbY)} \tag{13.5}$$

where:

$$Y = \frac{t^2}{6mh_c} \qquad (13.6)$$

is the *Lewis form factor*.

The maximum fillet stress is $\sigma_{max} = K_t F_t / (mYb)$, where K_t is the stress concentration factor, which has been determined by photoelastic methods (Figure 13.5). This factor accounts also for the normal force, which is omitted in the Lewis equation.

The stress concentration factor depends on the radius of the fillet and, consequently, is a function of the number of teeth and also of the heat treatment:

1. Hardened gears for which the effective stress concentration factor K_f is usually taken as equal to the theoretical factor K_f

2. Structurally improved and normalized gears for which $K_f = 0.9 K_t$, this being taken conditionally into consideration by a corresponding reduction of the safety factor.

The values of the rated load and the allowable material strength must be adjusted to the realistic service conditions. The American Gear Manufacturers Association (AGMA) suggested a modification of the Lewis equation in the form

$$\sigma = \frac{F_t K_s K_a K_m}{K_v mbJ} \qquad (13.7)$$

where the Lewis form factor Y has been modified to a *bending strength geometry factor J* to include the stress concentration factor. Furthermore, K_v is the *dynamic* or *velocity* factor, owing to additional dynamic loads at tooth engagement as a result of the cutting inaccuracies. It depends on peripheral velocity v and manu-

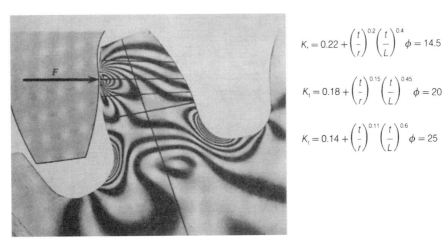

$$K_t = 0.22 + \left(\frac{t}{r}\right)^{0.2}\left(\frac{t}{L}\right)^{0.4} \quad \phi = 14.5$$

$$K_t = 0.18 + \left(\frac{t}{r}\right)^{0.15}\left(\frac{t}{L}\right)^{0.45} \quad \phi = 20$$

$$K_t = 0.14 + \left(\frac{t}{r}\right)^{0.11}\left(\frac{t}{L}\right)^{0.6} \quad \phi = 25$$

Figure 13.5. Stress concentration at the root of the tooth. (Adapted from Dolan and Broghamer 1942.)

facturing errors. AGMA defined a number of quality grades Q_v = 3–7 for commercial quality gears and 8–12 for precision gears. The following values are recommended (see also Figure 13.6):

$$K_v = \left[\frac{A}{A + \sqrt{20B}}\right]^B, A = 50 + 56(1 - B), B = \frac{(12 - Q_v)^{2/3}}{4} \quad (13.7a)$$

K_a is the application factor, depending on the uniformity of the load. If simulation or experimental results are not available, the overload factor can be taken as in Table 13.1.

K_m is the mounting factor, which can be taken from Table 13.2, depending on the mounting, the accuracy, and the rigidity of the shaft.

K_s is a size factor. In the absence of specific AGMA recommendations, one should use the value 1 or the recommendations for the same factor in Chapter 5, if uniformity of the material is not certain.

The endurance strength (S_e) is adjusted to the service conditions with the derating equation

$$S_e = \frac{S_n' K_L}{K_T K_R} \quad (13.8)$$

where:

K_T is the temperature factor. For temperatures below 70°, K_T = 1; for higher temperatures, for T (°C):

$$K_T = \frac{273 + T}{344} \quad (13.9)$$

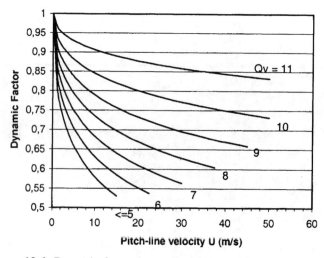

Figure 13.6. Dynamic factor kv. (Adapted from AGMA note 218.01.)

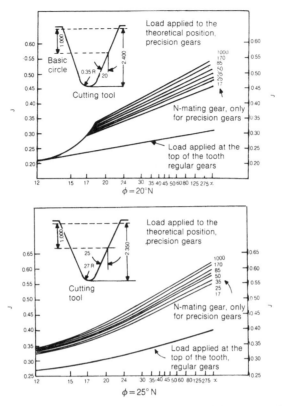

Figure 13.7. Bending strength geometry factor J. (Adapted from the AGMA note 218.01.)

TABLE 13.1. Application Factor K_a for Gears

Parameter	Formula
Center-to-center distance a	$a = \dfrac{(N_1 + N_2)m}{2} = 0.5(N_1 + N_2)m$
Whole depth of teeth h	$h = 2.25m$
Addendum h_a	$h_a = m$
Radial clearance c	$c = 0.25m$
Pitch diameter d	$d_1 = mN_1$
	$d_2 = mN_2$
Rolling circle diameter d	d_1 and d_2
Outside (addendum circle) diameter d_a	$d_{a1} = d_1 + 2m$
	$d_{a2} = d_2 + 2m$
Foot (dedendum circle) diameter d_f	$d_{f1} = d_1 - 2m - 2c$
	$d_{f2} = d_2 - 2m - 2c$

TABLE 13.2. Mounting Factor K_m for Gears

Source of Power	Driven Machine		
	Uniform	Moderate Shock	Heavy Shock
Uniform	1.00	1.25	1.75
Light shock	1.25	1.50	2.00
Medium shock	1.50	1.75	2.25

K_L is a *fatigue strength life modification factor*, defined by AGMA for less than 3 million load applications N_c as function of the tooth surface hardness BHN:

$K_L = 2.3194 N_c^{-0.0538}$, for 400 BHN
$K_L = 4.9404 N_c^{-0.1045}$, for case-carburized teeth
$K_L = 6.1514 N_c^{-0.1192}$, for 250 BHN
$K_L = 9.4518 N_c^{-0.1480}$, for 160 BHN

Above 3 million load applications, $K_L = 1.3558 N_c^{-0.0178}$ to $K_L = 1.6831 N_c^{-0.0323}$, depending on the service conditions.

K_R, the reliability factor, is t, the reliability factor of Chapter 4, Figure 4.29, or can be obtained from Table 13.3. Therefore, the design equation will be:

$$\frac{F_t K_s K_a K_m}{K_v m b J} \leq \frac{S_n' K_L}{K_T K_R} \tag{13.10}$$

Because the unknown is usually the module $m = 1/P$ and some of the parameters depend on it (such as J), this equation must be solved by iteration.

The geometry factor J incorporates into the Lewis form factor the stress concentration, a concept not well known in Lewis's time. Recommended values of AGMA for the geometry factor J are given in Figure 13.13. There, for gears of regular accuracy, the load is assumed on the top of the tooth, as the most conservative location, and one tooth in contact is assumed.

For gears of high accuracy, simultaneous contact of more than one tooth can be assured, and the geometry factor also depends on the number of teeth of the gear.

For computational purposes and for preliminary design calculations, the following expressions approximate the values of Figure 13.13:

TABLE 13.3. Reliability Factor K_R and C_R

Mounting	b/d_1					
	0.2	0.4	0.8	1.2	1.6	2.0
Symmetrically between bearings	1.0	1.05	1.10	1.25	1.40	(1.55)
Near one end very rigid shaft	1.00	1.08	1.20	1.40	(1.70)	—
Near one end flexible shaft	1.10	1.20	1.40	1.70	(2.0)	—
Cantilever pinion	1.25	1.55	(20)	—	—	—

$$\varphi = 20°, \text{ regular gears, } J = 0.32 \left(1 - \frac{1.14}{N_1^{0.546}} \right)$$

$\varphi = 20°$ high-accuracy gears,

$$J = 0.56 \left(1 - \frac{0.38}{\sqrt{N_2}} - \frac{0.88}{N_2} \right)\left(1 - \frac{0.26}{\sqrt{N_1}} - \frac{5.5}{N_1} \right) \qquad (13.11)$$

$$\varphi = 25°, \text{ regular gears, } J = 0.39 \left(1 - \frac{1.14}{N_1^{0.546}} \right)$$

$\varphi = 25°$, high-accuracy gears,

$$J = 0.63 \left(1 - \frac{0.38}{\sqrt{N_1}} - \frac{0.88}{N_1} \right)\left(1 - \frac{0.26}{\sqrt{N_2}} - \frac{5.5}{N_2} \right)$$

For certification, determination of the maximum power that a drive can transmit for given other operating requirements, we first solve Equation (3.10) for F_t. Then the maximum power that the drive can transmit on the basis of tooth bending strength is:

$$H_b = M_1\omega_1 = F_t \frac{\pi d_1 n_1}{60} = \frac{\pi d_1 n_1}{60} \frac{S'_n K_L K_v mbJ}{K_s K_d K_m K_T K_R} \qquad (13.11a)$$

Example 13.1 A pair of standard 20° involute spur gears has pinion speed $n_1 = 1785$ rpm and gear speed 714 rpm, $N_1 = 17$, $N_2 = 43$, $d_1 = 102$ mm, $d_2 = 258.06$ mm, center distance $a = 180.3$ mm, gear width $b = 63.1$ mm. Ninety percent reliability will suffice, and no shocks are expected in the motor with light shocks in the blower. The gears are to last for at least 1 million revolutions at a temperature only slightly exceeding that of the environment. The gears will be made of AISI 1015 steel surface-hardened to 270 BHN, and they will be machined but not ground, quality $Q_v = 6$. Find the maximum power that the gears can transmit without exceeding the allowable bending stress and the contact ratio.

Solution From Chapter 4, AISI 1015 carbon steel has a fatigue strength of $S_n' = 170$ MPa. For 270 BHN, the contact fatigue strength (Table 13.4) for both gears is:

TABLE 13.4. Surface Strengths of Gear Teeth

Heat Treatment	Surface Hardness	Group of Materials	S_H (N/mm²)
Structural improvement, normalization	BHN < 350	Carbon and alloy steels	2.8(BHN) − 70
Through hardening	38–50R_c		18(R_c) + 150
Surface hardening	40–50R_c		17(R_c) + 200
Carburization and hardening	56–65R_c	Alloy steels	23(R_c)
Nitriding	DPH 550–750		1050
No heat treatment	—	Cast iron	2(BHN)

$$S_H = 2.8 \times 270 - 70 = 686 \text{ MPa}$$

Also, for both gears:

$$E_1 = E_2 = 2.1 \times 10^{11} \text{ Pa}$$

The peripheral velocity is:

$$u = \frac{\pi n_1 d_1}{60} = \frac{\pi \times 1500 \times 0.102}{60} = 8.01 \text{ m/s}$$

From Figure 13.5, $K_v = 0.72$.
The gear ratio is:

$$R = \frac{43}{17} = 2.53$$

We select $C_s = 1$, $K_L = C_L = 6.1514 N_c^{-0.1192} = 6.1514 \times 1^{-0.1192} = 6.151$, $K_T = C_T = 1$, $K_R = C_R = 0.85$ (Table 13.3), $C_m = K_m = 1.55$ (Table 13.2), $K_a = C_a = 1.25$ (Table 13.1), $K_F = C_F = 1$. From Figure 13.6, for $\phi = 20°$, $N_1 = 17$, regular gear teeth, $J = 0.24$. The power will be computed using Equation (13.11a)):

$$H_b = \frac{\pi d_1 n_1}{60} \frac{S'_n K_L K_v m b J}{K_a K_m K_T K_R}$$

$$= \frac{\pi \times 0.102 \times 1785}{60} \times \frac{170 \times 10^6 \times 6.151 \times 0.72 \times 0.006 \times 0.0631 \times 0.24}{1.25 \times 1.55 \times 1 \times 0.85}$$

$$H_b = 360,000 \text{ W or } 360 \text{ kW}$$

The contact ratio is (Equation 2.69):

$$m_c = \frac{\begin{array}{c} [(N_2 + 2)^2 - N_2^2 \cos^2 \varphi]^{1/2} - N_2 \sin \varphi + \\ [(N_1 + 2)^2 - N_1^2 \cos^2 \varphi]^{1/2} - N_1 \sin \varphi \end{array}}{2\pi}$$

$$= \frac{\begin{array}{c} [(43 + 2)^2 - 43^2 \cos^2 20°]^{1/2} - 43 \sin 20° + \\ [(17 + 2)^2 - 17^2 \cos^2 20°]^{1/2} - 17 \sin 20° \end{array}}{2\pi}$$

$$= 1.523$$

13.2.2. Design of Spur Gears for Surface Strength

Contact between two perfect cylindrical surfaces is, in general, along a line. If the contact carries a normal force, zero area of contact would result in infinite stress. Thus, the two surfaces are deformed and a finite area of contact develops. The stresses there are very high and, since the application of the stress is repeated every rotation, fatigue cracks tend to form in this area, which are further amplified by the lubricating oil pressure as the oil is trapped into the crack. Thus, the crack propagates at an angle to the surface, and eventually a small piece of metal is removed. We call this *pitting*. It is observed in service that pitting usually originates at pitch point. Therefore, this is the place to expect surface fatigue failure (Figure 13.7). Maximum contact stresses in pressing together two cylinders that are in contact along elements of the cylindrical surfaces are given by the Hertz formula:

$$\sigma_H = \left[\frac{EF_n}{2\pi(1 - \nu)bR_{eq}} \right]^{1/2} \tag{13.12}$$

where F_n is the load normal to the surface, $E = 2E_1E_2/(E_1 + E_2)$ is the equivalent modulus of elasticity of the material, b is the gear tooth width, E_1 and E_2 are the Young's moduli of elasticity of the pinion and gear materials (if they are of the same material, $E = E_1 = E_2$), ν is Poisson's ratio, and R_{eq} is the equivalent radius of curvature $= R_1R_2/(R_1 + R_2)$.

In the equation for equivalent radius of curvature, the plus sign refers to external and the minus to internal gearing. For rack and pinion gearing, $R = R_1$ and $R_2 = \infty$.

From basic involute geometry, the radii of curvature applicable in the Hertz formula are the lengths of the generating tangent NN between the involute and the point of contact of NN with the basic circle (Figures 13.9 and 13.10), $R_1 = d_1 \sin \varphi/2$, $R_2 = d_2 \sin \varphi/2$.

For $\nu = 0.3$, equating to the surface fatigue strength S_H and solving for F_n:

$$F_n = d_1bQK \tag{13.13}$$

where b is the gear width:

$$Q = \frac{2d_2}{d_1 + d_2} = \frac{2N_2}{N_1 + N_2} \tag{13.13a}$$

a parameter that depends on the number of teeth only:

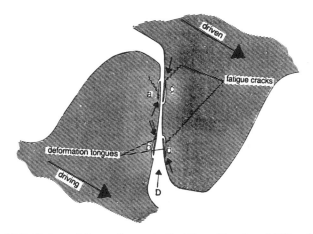

Figure 13.8. Fatigue failure of gear teeth. (From Engel and Klingele 1981.)

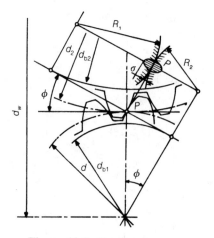

Figure 13.9. Contact stresses.

$$K = \frac{S_H^2 \sin \varphi(1/E_1 + 1/E_2)}{1.4} \tag{13.13b}$$

is a parameter that depends on the materials and the pressure angle φ. Equation (13.13b) is associated with the name of Earl Buckingham, who derived it.

Going back to the Hertz equation, AGMA suggested adjusting the load to the service conditions, as in the case of the beam strength of the teeth:

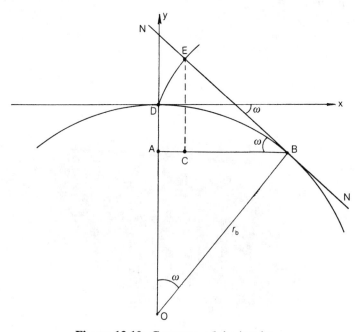

Figure 13.10. Curvature of the involute.

$$\sigma_H = C_p \left[\frac{F_t C_a C_s C_m C_f}{C_v b d_1 I} \right]^{1/2} \tag{13.14}$$

where:

C_a = the application factor, same as K_a
C_m = mounting factor, same as K_m
C_f = surface finish factor; in the absence of AGMA recommendations, one can use the surface factor defined in Chapter 5
C_v = dynamic or velocity factor, same as K_v
C_s = size factor:

$$1.0 \text{ for } m < 0.2(P > 5) \text{ and } 0.85 \text{ for } m > 0.2(P < 5) \tag{13.15}$$

C_H = gear hardness adjustment factor, 1 for the pinion

$$C_H = 1 + 8.98 \times 10^{-3}(\text{BHN}_1/\text{BHN}_2)(\text{R} - 1) \tag{13.16}$$

for the mating gear.

$$C_p = \sqrt{\frac{1}{\pi \left(\dfrac{1 - \nu_1^2}{E_1} + \dfrac{1 - \nu_2^2}{E_2} \right)}} \tag{13.17}$$

is a material factor, E is the modulus of elasticity, ν the Poisson's ratio, and

$$I = \frac{(\sin 2\varphi)(N_2/N_1)}{4(1 + N_2/N_1)} \tag{13.18}$$

is a *surface strength geometry factor*.

In this approach, recommended by AGMA, the maximum contact stress must be compared with the contact strength, adjusted for the service conditions:

$$\sigma_H < \frac{S_H C_L C_H}{C_T C_R} = S_{He} \tag{13.19}$$

where:

C_L = *surface strength life modification factor*, defined by AGMA for less than 10 million load applications N_c as $K_L = 2.466 N_c^{-0.056}$ and between $1.4488 N_c^{-0.023}$ and $2.466 N_c^{-0.056}$ otherwise.
C_T = temperature factor (same as K_T in Equation (13.9)).
C_R = reliability factor $C_R = 0.7 - 0.15 \log(1 - \text{R})$, for $0.9 \leq \text{R} < 0.99$ and $C_R = 0.5 - 0.25 \log(1 - \text{R})$, for $0.99 \leq \text{R} < 0.9999$.
S_H = contact fatigue strength. If test results are not available, contact fatigue strength can be estimated on the basis of surface hardness (Table 13.4).

For certification, determination of the maximum power that a drive can transmit for given other operating requirements, we solve Equation (13.14) and (13.19) for F_t, and then the maximum power on the basis of surface strength is:

$$H_s = M_1 \omega_1 = \frac{\pi d_1 n_1}{60} F_t = \frac{\pi d_1 n_1}{60} \frac{C_v b d_1 I}{C_a C_s C_m C_f} \left[\frac{S_H C_L C_H}{C_p C_T C_R} \right]^2 \tag{13.19a}$$

The maximum power that the drive can transmit is then the smallest of the powers computed on the basis of bending strength or surface strength (Equations (13.11a) and (13.19a)).

In Chapter 2, Section 2.5, it was pointed out that the number of teeth of the gear should be above a minimum that depends on the pressure angle as shown in Table 13.5.

Example 13.2   A pair of standard 20° involute spur gears has pinion speed $n_1 = 1785$ rpm and gear speed 714 rpm, $N_1 = 17$, $N_2 = 43$, $d_1 = 102$ mm, $d_2 = 258.06$ mm, center distance $a = 180.3$ mm, gear width $b = 63.1$ mm. Ninety percent reliability will suffice, and no shocks are expected in the motor, with light shocks in the blower. The gears are to last for at least 1 million revolutions at a temperature only slightly exceeding that of the environment. The gears will be made of AISI 1015 steel surface hardened to 270 BHN and will be machined but not ground, quality $Q_v = 6$. Find the maximum power that the gears can transmit without exceeding the allowable surface stress.

Solution   From Chapter 4, AISI 1015 carbon steel has a fatigue strength of $S_n' = 170$ MPa. For a 270 BHN, the contact fatigue strength is (Table 13.4), for both gears:

$$S_H = 2.8 \times 270 - 70 = 686 \text{ MPa}$$

Also for both gears, $E_1 = E_2 = 2.1 \times 10^{11}$ Pa

The peripheral velocity $u = \pi n_1 d_1/60 = \pi \times 1500 \times 0.102/60 = 8.01$ m/s, from Figure 13.5, $K_v = C_v = 0.72$.

The gear ratio is $R = 43/17 = 2.53$.

From Equations (13.17) and (13.18):

$$C_p = 0.591 \left(\frac{E_1 E_2}{E_1 + E_2} \right)^{1/2} = 0.59 \left(\frac{2.1 \times 10^{11}}{2} \right)^{1/2} = 1.9 \times 10^5$$

$$I = \frac{(\sin 2\varphi)(N_2/N_1)}{4(1 + N_2/N_1)} = \frac{\sin(2 \times 20°) \times 2.5}{4 \times (1 + 2.5)} = 0.115$$

From Equation (13.16), $C_H = 1 + 8.98 \times 10^{-3}[1/1](2.5 - 1) = 1.013$.

TABLE 13.5. Minimum Number of Gear Teeth

Pressure angle ϕ	Minimum Number of Teeth N_{min} gear-rack, $N_2 = \infty$ Equation (2.70b)	Minimum Number of Teeth N_{min} equal gears, $N_2 = N_1$ Equation (2.70b)
14.5°	32	23
20°	17	12
25°	12	9

The other constants:

$$C_s = 1$$
$$C_L = 2.466 N_c^{-0.056} = 2.466 \times 1^{-0.056} = 2.466$$
$$K_T = C_T = 1$$
$$C_R = 0.7 - 0.15 \log(1 - 0.9) = 0.85$$
$$C_m = K_m = 1.55 \text{ (Table 13.2)}$$
$$K_a = C_a = 1.25 \text{ (Table 13.1)}$$
$$K_F = C_F = 1$$

The power will be computed using Equation (13.19a):

$$H_s = \frac{\pi d_1 n_1}{60} \frac{C_v b d_1 I}{C_a C_s C_m C_f} \left[\frac{S_H C_s C_L C_H}{C_p C_T C_R} \right]^2$$

$$= \frac{\pi \times 0.102 \times 1785}{60} \frac{0.72 \times 0.0631 \times 0.102 \times 0.115}{1.25 \times 1 \times 1.55 \times 1}$$

$$\left[\frac{686 \times 10^6 \times 2.466 \times 1.013}{1.91 \times 10^5 \times 1 \times 0.85} \right]^2$$

$$= 292{,}000 \text{ W or } 292 \text{ kW}$$

13.2.3. Computer Aided Graphics of Spur Gears

In the design of standard spur gears, conventional techniques were used for the representation of the teeth because drawing them was a tedious effort if done by hand. With the advent of CAD, this is no longer a problem and there is a tendency to draw the actual form of toothed gearing. To this end, the involute has to be expressed by way of a plane function in a convenient coordinate system.

In Figure 13.10, a basic circle of center O and radius $r_b = r \cos \phi$ is used to draw an involute from point D by way of rolling the line NN. If after some rolling angle ω (not to be confused with the angular velocity) the point of contact of the rolling line is at B, the length (BE) will equal the arc $(BD) = r_b \omega$. In reference to coordinate system x, y:

$$x = (AB) - (BC) = r_b \sin \omega - \omega r_b \cos \omega$$

$$= r \cos \varphi \cos \omega (\tan \omega - \omega) \tag{13.20}$$

$$y = (EC) - (DA) = (EC) - (OD) + (OA)$$

$$= r \cos \varphi \cos \omega \left[\omega(\tan \omega + 1) - \frac{1}{\cos \omega} \right] \tag{13.21}$$

The limits of the involute will be imposed by the addendum and dedendum circles, that is, between points E_1 and E_2 where $OE_1 = r - 1.25m$, $OE_2 = r + m$, for standard tooth profiles. This gives the limits of the rolling angle:

$$\omega_{max} : r_b(1 + \omega_{max}^2)^{1/2} = r + m; \; \omega_{max} = \left[\frac{(1 + 2N)^2}{\cos^2 \varphi} - 1\right]^{1/2} \tag{13.22}$$

$$\omega_{min} : [1 + \omega_{min}^2]^{1/2} = r - 1.25m; \; \omega_{min} = \left[\frac{(1 - 2.5/N)^2}{\cos^2 \phi} - 1\right]^{1/2} \tag{13.23}$$

There is an obvious limitation, that $\omega > 0$. This means that the dedendum circle cannot be smaller than the basic circle. In case of smaller number of teeth than necessary to avoid interference, the involute is continued as a straight line along the radius DO. This happens when the quantity in the root becomes negative, which happens when $N < 2.5/(1 - \cos \phi)$. The procedure to draw a complete spur gear profile was programmed in the GEARPLOT module of MELAB 2.0.

13.2.4. Design Procedure for Spur Gears

For the design of spur gears, two design equations are available.

The AGMA/Lewis equation:

$$\frac{F_t K_a K_m}{K_v mJb} \le \frac{S_n' K_L}{K_R K_T} \tag{13.24}$$

The AGMA/Buckingham equation:

$$C_p \left(\frac{F_t C_0 C_s C_f}{\cos \varphi \; C_v bd_1 I}\right)^{1/2} \le \frac{S_H C_S C_L C_H}{C_T C_R} \tag{13.25}$$

Design data are usually the speed of the two gears, the power transmitted, and sometimes the center-to-center distance $a = (d_1 + d_2)/2$ is given if the design of the associated parts or space limitations demand it.

It is evident that the Buckingham equation does not include the module, owing to the fact that Hertz stresses depend on the radius of curvature and not the size of the tooth. Substituting the moment

$$M_1 = \frac{F_t d_1}{2} = \frac{P}{\omega_1} \tag{13.26}$$

where H is the power and ω_1 is the pinion angular velocity, and noting that $a = (d_1 + d_2)/2$, the gear ratio $R = N_2/N_1 = d_2/d_1$, thus:

$$d_1 = \frac{2a}{1 + R} \tag{13.27}$$

where R is the given gear ratio, the Buckingham/AGMA equation can be solved for a to yield:

$$a^2 = \frac{M_1(1 + R)^2}{2bIK} \qquad (13.27)$$

where:

$$K = \left(\frac{S_H C_S C_L C_H}{C_T C_R C_p}\right)^2 \frac{\cos \varphi \; C_v}{C_0 C_m C_f}$$

The constant K depends only on the operating conditions and general design features, except for the velocity constant K_v, which has a weak dependence on a because $v = \omega_1 d_1/2 = \omega_1 a/(1 + R)$. Therefore, Equation (13.23) has to be solved by iteration, starting with an average value for K_v. To find I (Equation (13.14)), we need only the gear ratio. At this point, the selection of b seems arbitrary. There are two ways to assign a value to it:

1. From design experience, it is recommended that the width factors b/a are between 0.3 and 0.4 for structurally improved gears and 0.25 and 0.315 for hardened gears, asymmetrically located on the shaft. If the gears are symmetrically located, which means better contact, the width factor b/a can be as high as 0.5. In terms of this factor, Equation (13.10) can be written as:

$$a^2 = \frac{M_1(1 + R)^2}{2(b/a)IK} \qquad (13.28)$$

Using Equation (13.22):

$$d_1^3 = \frac{4M_1}{(1 + R)(b/a)IK} \qquad (13.28a)$$

2. In cases where the center distance a is given, the material has to be selected properly to give values of b within the recommendations for the width ratio.

For fatigue strength of the tooth, the AGMA/Lewis equation is used. In terms of known parameters, Equation (13.10) can be written as:

$$m = \frac{4M_1 K_0 K_m K_R K_T}{K_v(b/a)d_1^2(1 + R)JK_L S_n''} \qquad (13.28b)$$

This equation gives the module of the gearing. To find the geometry factor J, we need to know the number of teeth. Design starts with the minimum number of teeth of the pinion $N_{1-\min}$ to avoid interference, given by Table 13.5. The number of gear teeth is then $N_2 = RN_1$.

Following the determination of m by Equation (13.28b), the resulting number of teeth is $N_1 = d_1/m$. The design number of the pinion teeth will then be $\max(N_{1-\min}, N_1)$.

Only the smallest gear is used for fatigue strength calculations, if the two gears are made of the same material. Otherwise, both gears must be used and the larger module, if different, should be adopted.

In the definition of the form factor J, distinction is made between common and precision gears. This is because only in precision gears may one assume that more than one tooth at one time is simultaneous contact. In common gears, even if kinematically the contact ratio is greater than 1, as always happens, owing to inaccuracies and the high rigidity of teeth, load sharing among teeth cannot be warranted.

In precision gears, even if the contact ratio is not an integer, it is accounted for because fatigue is a cumulative process and unloading of the teeth for a fraction of their loading time contributes to the fatigue life. Only in slow gearing, where a small total number of cycles is expected, must the design for strength be based on yield strength. In this case, the contact ratio should be always taken as 1 and, in principle, for precision gearing the integer part of the contact ratio, though precision gears are not used for slow applications.

For gearing optimization, the whole system has to be used for objective function, including shafts, bearings, and couplings. As a first approximation, however, the volume of the gear blanks should be minimized. To this end, the objective function is set up in the form

$$P(m, a, b) = \frac{\pi(d_2^2 + d_2^2)b}{4}$$

subject to the following constraints:

1. The Lewis/AGMA equation
2. The Buckingham/AGMA equation
3. The limits on minimum number of teeth
4. The limits on the width ratio b/a

A free parameter is the material. Several feasible designs from different materials must be found first and their total cost compared.

The module GEARDES of MELAB 2.0 computes the main dimensions for a spur gear drive. Design Procedure 13.1 can also be used.

Design Procedure 13.1: Design of Spur Gear Drives

Input: Gear ratio R, transmitted power P, rotating speed n_1. **Output:** Gear design.

Step 1: Sketch the gear drive and place on the sketch loads and motion data.

Step 2: Select bearing materials, find fatigue strength S_n' and S_H (Table 13.4) and Young's moduli for the materials E_1 and E_2 (1 = pinion).

Step 3: Determine factor K_v (Equation (13.26a)). Use initially $K_v = 1$.

Step 4: Select a pressure angle, normally 20°. From Table 13.1b, find the minimum pinion number of teeth N_1. From the gear ratio R, find $N_2 = RN_1$.

Step 5: From Equations (13.17) and (13.18) find C_p and I and from Equation (13.28) find K. Find the derating factors C_H (Equation (13.16)), C_L, K_L, C_T (Equation (13.9)), C_R, K_R (Table 13.3), C_m, K_m (Table 13.2), C_a, K_a (Table 13.1), C_F, K_F, C_s, K_s (Chapter 5), C_T, K_T (Equation (13.9)). Find the torque $M_1 = 60H/2\pi n_1$.

Step 6: Select the (width/center distance) b/a ratio in the range 0.25–0.4.

Step 7: Use Equation (13.27) to find the center distance a. Find the bearing diameters and width. Use Equations (13.27a) to find d_1 and d_2.

Step 8: Find the geometry factor J (Figure 13.6).

Step 9: From Equation (13.28), find the module m. If the number of teeth d_1/m is greater than the minimum selected in step 4, go to step 4 and assume the new number of teeth N_1 and repeat steps 5–9.

Step 10: Find the contact ratio from Equation (2.69).

Example 13.3 An electric motor of rated power 12 kW at 1785 rpm runs an air blower at 714 rpm with one stage of standard 20° involute spur gearing. Ninety percent reliability will suffice, and no shocks are expected in the motor, with light shocks in the blower. The gears are to last for at least 1 million revolutions at a temperature only slightly exceeding that of the environment. The gears will be made of AISI 1015 steel surface-hardened to 270 BHN and will be machined but not ground, quality $Q_v = 6$. Design the proper gearing.

Solution

Step 1: Sketch the gear drive and place on the sketch loads and motion data.

See Figure E13.3.

Step 2: Select bearing materials, find fatigue strength S_n' and S_H (Table 13.4) and Young's moduli for the materials E_1 and E_2 (1 = pinion).

From Chapter 4, AISI 1015 carbon steel has a fatigue strength of $S_n' = 170$ MPa. For a 270 BHN, the contact fatigue strength is (Table 13.4), for both gears:

$$S_H = 2.8 \times 270 - 70 = 686 \text{ MPa}$$

Also for both gears, $E_1 = E_2 = 2.1 \times 10^{11}$ Pa.

Step 3: Determine factor K_v (Equation (13.26a)). The range for $Q_v = 6$ (Figure 13.5) is 0.55–1. We select $K_v = C_v = 0.7$.

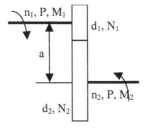

Figure E13.3

Step 4: Select a pressure angle, normally 20°, from Table 13.5 find the minimum pinion number of teeth N_1. From the gear ratio R find $N_2 = RN_1$.

The gear ratio is $R = 1785/714 = 2.5$. From Table 13.5, $N_1 = 17$, $N_2 = RN_1 = 2.5 \times 17 = 43$. The exact value of the gear ratio is $R = 43/17 = 2.53$.

Step 5: From Equations (13.17) and (13.18), find C_p and I and from Equation (13.28) find K. Find the derating factors C_H (Equation (13.16)), C_L, K_L, C_T (Equation (13.9)), C_R, K_R (Table 13.3), C_m K_m (Table 13.2), C_a, K_a (Table 13.1), C_F, K_F, C_s, K_s (Chapter 5), C_T, K_T, (Equation (13.9)). Find torque $M_1 = 60H/2\pi n_1$.

From Equations (13.17) and (13.18):

$$C_p = 0.591 \left(\frac{E_1 E_2}{E_1 + E_2} \right)^{1/2} = 0.59 \left(\frac{2.1 \times 10^{11}}{2} \right)^{1/2} = 1.9 \times 10^5$$

$$I = \frac{(\sin 2\varphi)(N_2/N_1)}{4(1 + N_2/N_1)} = \frac{\sin(2 \times 20°) \times 2.5}{4 \times (1 + 2.5)} = 0.115$$

From Equation (13.16):

$$C_H = 1 + 8.98 \times 10^{-3} \frac{1}{1} (2.5 - 1) = 1.013$$

We select $C_s = 1$.

$C_L = 2.466N_c^{-0.056} = 4.9404 \times 1.0^{-0.1045} = 4.9404$, $C_T = 1$, $C_R = 0.85$ (Table 13.3), $C_m = K_m = 1.55$ (Table 13.2), $K_a = C_a = 1.25$ (Table 13.1), $C_F = 1$, therefore:

$$K = \left(\frac{S_H C_s C_L C_H}{C_T C_R C_p} \right)^2 \left(\frac{\cos \varphi C_v}{C_0 C_m C_f} \right)$$

$$= \left(\frac{686 \times 10^6 \times 1 \times 2.466 \times 1.013}{1 \times 0.85 \times 1.9 \times 10^5} \right)^2 \left(\frac{\cos(20°) \times 0.7}{1.25 \times 1.55 \times 1} \right) = 38.2 \times 10^6$$

$$M_1 = \frac{60H}{2\pi n_1} = \frac{60 \times 25,000}{2\pi \times 1000} = 238.7 \text{ Nm}$$

Step 6: Select the (width/center distance) b/a ratio in the range 0.25–0.4.

We select $b/a = 0.35$.

Step 7: Use Equation (13.28a) to find the center distance a. Find the bearing diameters and width. Use Equation (13.27a) to find d_1 and d_2.

$$d_1^3 = \frac{4M_1}{(1 + R)(b/a)IK}$$

$$= \frac{4 \times 238.7}{(1 + 2.5)(0.35) \times 0.115 \times 38.2 \times 10^6}, \, d_1 = 0.056 \text{ m}$$

$$d_1 = 56 \text{ mm}, \, d_2 = Rd_1 = 2.53 \times 56 = 141.6 \text{ mm}$$

Step 8: Find geometry factor J.

From Figure 13.6, for $\phi = 20°$, $N_1 = 17$, regular gear teeth, $J = 0.24$.

Step 9: From Equation (13.28), find the module m. If the number of teeth d_1/m is greater than the minimum selected in step 4, go to step 4 and assume the new number of teeth N_1 and repeat steps 5–9.

$$
m = \frac{4M_1 K_0 K_m K_R K_T}{K_v(b/a)d_1^2(1 + R)JK_L S_n''}
$$

$$
= \frac{4 \times 238.7 \times 1.24 \times 1.55 \times 0.85 \times 1}{0.7 \times 0.35 \times 0.056^2 \times (1 + 2.5) \times 0.24 \times 4.9404 \times 170 \times 10^6} = 0.0029 \text{ m}
$$

We select $m = 3$ mm and then $N_1 = d_1/m = 56/3 = 18.6$, which is greater than the minimum allowed $N_1 = 17$. We select $N_1 = 19$.

Second iteration

The peripheral velocity $u = \pi n_1 d_1/60 = \pi \times 1500 \times 0.102/60 = 8.01$ m/s. From Figure 13.5, $K_v = 0.72$. Now:

$$
m = \frac{4M_1 K_0 K_m K_R K_T}{K_v(b/a)d_1^2(1 + R)JK_L S_n'}
$$

$$
= \frac{4 \times 238.7 \times 1.24 \times 1.55 \times 0.85 \times 1}{0.72 \times 0.35 \times 0.056^2 \times (1 + 2.5) \times 0.24 \times 2.466 \times 170 \times 10^6} = 0.0029 \text{ m}
$$

No further iteration is needed.

Step 10: Find the contact ratio from Equation (2.69).

$$
m_c = \frac{[(N_2 + 2)^2 - N_2^2 \cos^2 \varphi]^{1/2} - N_2 \sin \varphi + [(N_1 + 2)^2 - N_1^2 \cos^2 \varphi]^{1/2} - N_1 \sin \varphi}{2\pi}
$$

$$
= \frac{[(43 + 2)^2 - 43^2 \cos^2 20°]^{1/2} - 43 \sin 20° + [(17 + 2)^2 - 17^2 \cos^2 20°]^{1/2} - 17 \sin 20°}{2\pi}
$$

$$
= 1.523
$$

Design summary: Pair of spur gears, $N_1 = 19$, $N_2 = 48$, $d_1 = 56$ mm, $d_2 = 3 \times 48 = 144$ mm, center distance $a = (d_1 + d_2)/2 = (56 + 144)/2 = 100$ mm, gear width $b = a(b/a) = 100 \times 0.35 = 35$ mm.

Example 13.4. An electric motor of rated power 12 kW at 1785 rpm runs an air blower at 714 rpm with one stage of standard 20° involute spur gearing. Ninety percent reliability will suffice, and no shocks are expected in the motor, with light shocks in the blower. The gears are to last for at least 1 million revolutions at a temperature only slightly exceeding that one of the environment. The gears will be made of AISI 1015 steel surface-hardened to 220 BHN and will be machined but not ground. Design the proper gearing, obtaining the solution through a spreadsheet.

Solution From Chapter 6, AISI 1015 carbon steel has a fatigue strength of 170 MPa. For a 270 BHN, the contact fatigue strength is (Table 13.4):

TABLE E13.4

A	B	C	D	E	F	G	H
2	**PROCEDURE #13.1. DESIGN OF SPUR GEARS-Example 13.4**						
3	(Implementing procedure #13.1, modified for CAD implementation)					**UNITS: N, m, s**, unless otherwise indicated	
4	**Input:**						Step 1:
5	Speed of Rotation RPM			1785	Power, W	12000	
6	Fatigue Strength, Pa	Pinion:		1,70E+08	Gear:	1,70E+08	
7	Surface Strength, Pa			6,80E+08		6,80E+08	
8	Youngs Modulus, Pa			2,10E+11		2,10E+11	
9	BHN			2,70E+02		2,70E+02	
10	b/a ratio =	0,3	$R =$	2,5	Angle $\phi =$	2,00E+01	
11	Operating Temperature			70	degrees C		
12	**INITIAL GUESS:**		$d_{1\text{-in}} =$	0,088505	Solver changes cells on the left		Step 2:
13			modul m =	2,24E-03			

14	$U =$	8,271878					Step 3:
15	$K\_v =$	0,42444	$C\_L =$	1	$C\_R =$	1,1	
16	$K\_0 =$	1	$C\_H =$	1,00E+00	$C\_p =$	2,71E+05	
17	$K\_m =$	1,4	$C\_T =$	1,002915	$I =$	0,114784	
18			$C\_s =$	0,9			
19					$K =$	3,10E+06	
20							Step 4:
21	Computer-assumed diameter (must be 0)				$d\_l - d\_{lin} =$	0	must be 0
22	Computed N_1 must be greater than minimum:						Step 5:
23	$N_{1-min} =$	17	$N_{1comp} =$	39	$N_{1comp}\text{-}N_{1-min} =$	22	must be >0
24	$b =$	0,046465	$N_{2comp} =$	97	Volume $=$	0,045218	
25	$C\_s =$	0,85	$C\_f =$	0,85	$J =$	0,270645	Step 5:
26							Step 6:
27	Computed modul-assumed module (must be 0)				$m\text{-}m\_in$	0,00E+00	Must be 0
28	Use Solver to change cells E12 and E13 for the target cell G21 to be zero, with constraints G23>0 and G27=0. For optimization, set target cell G24 to minimum with constraints G21=0, G23>0 and G27=0						

$$S_H = 2.8 \times 270 - 70 = 680 \text{ MPa}$$

The solution follows the Design Procedure 13.1 and is shown in Table E13.4.

13.3. DESIGN OF HELICAL GEARS

13.3.1. Theory

In spur gears, each tooth is parallel to the axis of rotation. The involute geometry ensures a constant speed transmission ratio. However, inaccuracies in the tooth geometry, nonuniform rotation, or errors in misalignment and center distance force the gearing to deviate from the perfect involute geometry. This leads to nonconstant speed ratio. Abrupt variation of the speed means acceleration and dynamic forces. This reduces the life of gearing and introduces noise and vibration. To overcome this, the generatrix of the tooth is inclined in respect to the axis of rotation by an angle ψ, forming a helix on the cylindrical surface of the gear. In this way, many teeth are in contact at the same time, equalizing the errors and giving longer life and quiet operation. In addition, breakage of one tooth is not detrimental to the gearing operation because many teeth are engaged at the same time.

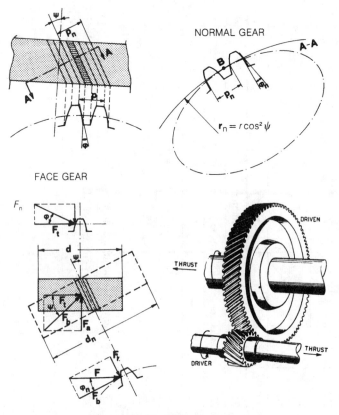

Figure 13.11. Helical gearing.

If we look at a helical gearing, a circular pitch p, module $m = p/\pi$, and pressure angle ϕ are defined at a section perpendicular to the axis of rotation in the gear plane. At a plane $A–A$ perpendicular to the generatrix of the tooth, another tooth profile appears, called "normal":

$$F_n = F_t \cos \varphi \qquad (13.29)$$

$$\tan \phi_n = \tan \varphi \cos \psi \qquad (13.30)$$

$$m_n = m \cos \psi \qquad (13.31)$$

The section of the plane $A–A$ with the cylindrical pitch surface of the gear is obviously an ellipse. At point B, however, there is a circle tangent to the ellipse, the closest circle, with a radius from the theory of conic sections:

$$d_n = \frac{d}{\cos^2 \psi} \qquad (13.32)$$

The operation of the gearing can be considered as equivalent with a gear in the plane $A–A$ having a diameter d_n, pitch p_n, module m_n, and pressure angle φ_n. The number of teeth of this hypothetical (virtual) gear is:

$$N_n = \frac{d_n}{m_n} = \frac{d}{m \cos^3 \psi} = \frac{N}{\cos^3 \psi} \qquad (13.33)$$

13.3.2. Force Analysis

The normal force F_r and a force F_D at a plane tangent to the gear cylinder surface at the point of contact (pitch point). In turn, force F_D can be analyzed into the tangential force F_t and axial force F_a. The tangential force is again

$$F_t = \frac{H}{(d/2)\omega} \qquad (13.34)$$

The analysis of the contact force F into the components F_r, F_t, and F_a yields (Figure 13.12):

$$\left.\begin{array}{l} F_r = F_t \tan \varphi = F_D \tan \phi_n \\[2mm] F_a = F_t \tan \psi \\[2mm] F_D = \dfrac{F_t}{\cos \psi} \\[2mm] F = F_D \cos \varphi_n = \dfrac{F_t}{\cos \psi \cos \varphi_n} \\[2mm] F_r = \dfrac{F_t \tan \varphi_n}{\cos \psi} \end{array}\right\} \qquad (13.35)$$

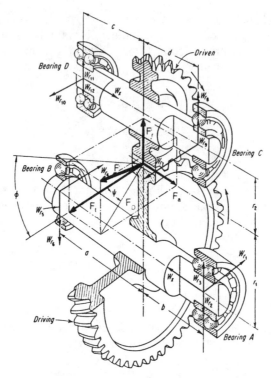

Figure 13.12. Forces in helical gear drives. (Adapted from Dudley 1962.)

13.3.3. Design

Standard pitch/module are usually specified at the normal gear because of the methods of cutting the teeth. Beam strength and surface strength are referred to the normal gear. There is one difference, however. In the spur gear, the whole width of the tooth is loaded, while in helical gears the situation is more complicated. The greater the helix angle ψ, the greater the number of helical teeth in contact. Therefore, the number of teeth in contact must be proportional to the contact ratio m_c and inversely proportional to cos ψ.

AGMA suggests a 7% reduction of the mounting factor K_m and 5% reduction of the contact ratio m_c owing to better performance of helical gearing in mounting and profile errors. They also issue standards with recommended geometry factors. The design procedure is similar to the spur gear design. In fact, one can use the same procedure with a new definition of the design parameters. Therefore, AGMA recommends the following design equations for helical gears:

Beam strength:

$$\frac{F_t K_a (0.93 K_m)}{K_v \cos \psi b J m} \leq \frac{S_n' K_L}{K_T K_R} \qquad (13.36)$$

Contact fatigue strength:

$$\sigma_H = C_P \left[\frac{F_t C_a (0.93 C_m)(C_y \cos \psi / 0.95 m_c)}{b d_1 I} \right]^{1/2} \le \frac{S_H C_L C_H}{C_T C_R} \quad (13.37)$$

where m_c is the contact ratio (see Chapter 2). To this end, the tangential force F_t must be such that it will equal F_t of the normal gear. Since the tangential force is proportional to the torque and inversely proportional to diameter:

$$F_D = \frac{F_t}{\cos \psi} = \frac{2M_1}{d_{1n} \cos^3 \psi} \quad (13.37a)$$

Therefore, if the power P, proportional to M_1, is divided by $\cos^3 \psi$, it will yield, with the dimensions of the normal gear, the right tangential force for the computation of the module and the center distance a_n. In turn, the resulting m_n and a_n have to be transformed into the helical gear quantities and $m = m_n / \cos \psi$ and $a = a_n \cos^2 \psi$. Also, according to the AGMA recommendations, the width ratio must be adjusted:

$$\left(\frac{b}{a} \right)_n = \frac{b}{a} \frac{0.95 m_c}{0.93 \cos \psi} \quad (13.37b)$$

The design equations for the spur gears are applicable here, if the quantities d, m, a, and b are replaced, according with what was said above, with the expressions for the normal quantities of the virtual gear, $d / \cos^2 \psi$, $m \cos \psi$, $a / \cos^2 \psi$, $b / \cos \psi$, respectively, with the adjustments discussed above. This yields

$$d^3 = \frac{3.91 M_1 \cos \psi}{m_c (1 + R)(b/a) I K} \quad (13.37c)$$

$$m = \frac{3.72 M_1 K_a K_m K_T K_R}{K_v (b/a) d_1^2 \cos \psi (1 + R) J K_L S_n'} \quad (13.37d)$$

Moreover, the contact of two teeth is not a line contact, as for spur gears, but a point (more precisely, a small area) contact. On the other hand, the greater the helix angle, the more teeth are in contact along the axis of the gearing. The geometry factors have to be different. AGMA suggests a different J factor for helical gears (see Figures 13.14 and 13.15) in the form

$$J = J_1 J_2 \quad (13.37e)$$

and for the geometry factor the expression (Shigley and Vicker 1980):

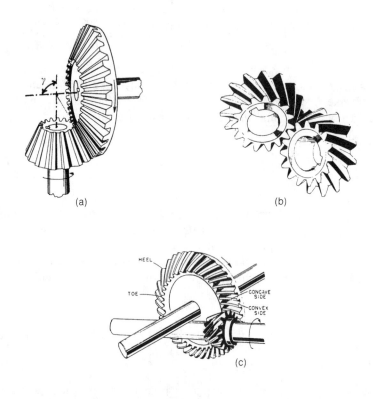

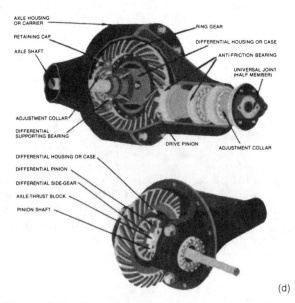

Figure 13.13. Helical gear drives. (Courtesy Mobil Oil Co.)

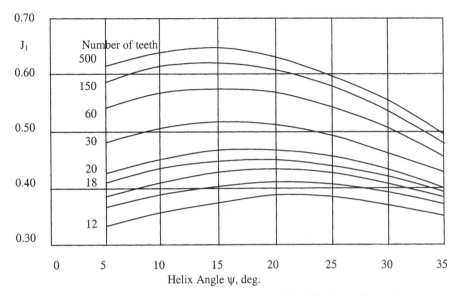

Figure 13.14. J_1 factor for helical gears cut with a full fillet hob. For other cases, see appropriate AGMA tables. (Adapted from AGMA note 218.01.)

$$I = \frac{(\sin 2\varphi_n)(N_2/N_1)}{4(1 + N_2/N_1)m_n} \tag{13.37f}$$

where m_N is the load sharing factor:

$$m_n = \frac{\pi m}{0.95 \cos \psi \sqrt{(d_1/2 + m) - (d_1 \cos \phi)^2} + \sqrt{(d_2/2 + m) - (d_2 \cos \phi)^2} \ \pi - a \sin \varphi} \tag{13.37g}$$

Because several parameters are functions of the number of teeth and dimensions, the computation should be iterative.

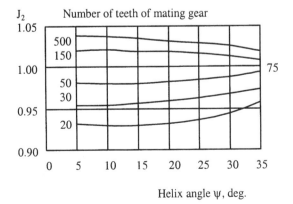

Figure 13.15. J_2 factor for helical gears. (Adapted from AGMA note 218.01.)

The module GEARDES of MELAB 2.0 performs design of helical gears. Alternatively, Design Procedure 13.2 can be used.

Design Procedure 13.2: Design of Helical Gear Drives

Input: Gear ratio R, helix angle ψ, transmitted power P, rotating speed n_1. **Output:** Gear design

Step 1: Sketch the gear drive and place on the sketch loads and motion data.

Step 2: Select bearing materials, find fatigue strength S_n' and S_H (Table 13.4) and Young's moduli for the materials E_1 and E_2 (1 = pinion).

Step 3: Determine factor K_v (Equation 13.7). Use initially $K_v = 1$.

Step 4: Select a pressure angle, normally 20°, from Table 13.5 and find the minimum pinion number of teeth N_1. From the gear ratio R find $N_2 = RN_1$.

Step 5: From Equations (13.17) and (13.18), find C_p and I. Find torque $M_1 = 60H/2\pi n_1$.

Step 6: Select the (width/center distance) b/a ratio in the range 0.25–0.4.

Step 7: Use Equation (13.36c) to find the pinion diameter d_1. Find the gear diameter and width.

Step 8: Find factors K_0 (Table 13.1), K_m (Table 13.2), K_v (Equation 13.7a), C_R, C_t, C_f, C_s from Chapter 5. Find J for helical gears.

Step 9: From Equation 13.36, find the module m. If the number of teeth d_1/m is greater than the minimum selected in step 4, go to step 4 and assume the new number of teeth N_1 and repeat steps 5–9.

Step 10: Find the contact ratio from Equation (2.69).

Example 13.5 An electric motor of rated power 12 kW at 1785 rpm drives an air blower at 714 rpm with one state of standard 20° involute gearing. Ninety percent reliability will suffice, and no shocks are expected in the motor, with light shocks in the blower. The gears are to last for at least 1 million revolutions at a temperature only slightly exceeding that of the environment. The gears will be made of AISI 1015 steel surface-hardened to 270 BHN and will be machined but not ground. Design a helical gearing of 30° helix angle. Compare the resulting gearing with the spur gearing of Example 13.4, which was designed for the same application.

Solution The gear is $R = 1785/714 = 2.5$. From Chapter 4, AISI 1015 carbon steel has a fatigue strength of 170 MPa. For a 270 BHN, the contact fatigue strength is (Table 13.4):

$$S_H = 2.8 \times 270 - 70 = 686 \text{ MPa}$$

The solution uses Design Procedure 13.2 and is shown in Table E13.5.

The 24 teeth of the pinion result in operation without interference. The contact ratio is 1.6, in the acceptable range.

The overall design is judged acceptable.

In comparison to the spur gearing of Example 13.4, the helical gearing has 28 teeth instead of 39 and center distance 146 instead of 155 mm, resulting in smaller gears. In fact, the volume is 0.040 m³, which is 10% smaller than the 0.045 m³ of the spur gearing.

TABLE E13.5

A	B	C	D	E	F	G	H
2	**PROCEDURE #13.2. DESIGN OF HELICAL GEARS-Example 13.5**						
3	(Implementing procedure #13.2, modified for CAD)					**UNITS: N, m, s**	
4	**Input:**						**Step 1:**
5	Speed of Rotation RPM				1785	Power	12000
6	Fatigue Strength	Pinion:			1,70E+08	Gear:	1,70E+08
7	Surface Strength				6,80E+08		6,80E+08
8	Youngs Modulus				2,10E+11		2,10E+11
9	BHN				2,20E+02		2,20E+02
10	b/a ratio =	0,3	Gear Ratio =		2,5	Angle ϕ =	2,00E+01	degrees
11	Operating Temperature				70	Angle ψ =	30	degrees
12	**INITIAL GUESS:**		$d_{1\text{-in}}$ =		0,083427	Solver changes cells on the left (d_1 and m)		**Step 2:**
13			modul =		2,92E-03			
14	U =	7,797331						**Step 3:**

TABLE E13.5. (*Continued*)

A	B	C		E	F	G	H
15	$K_v =$	0,438933	$C_L =$	1	$C_R =$	1,1	
16	$K_0 =$	1	$C_H =$	1,00E+00	$C_p =$	2,71E+05	
17	$K_m =$	1,4	$C_T =$	1,002915	$I =$	0,114784	
18	$C_s =$		$C_s =$	0,9			
19			$K =$	3,20E+06			
20							**Step 4:**
21							
22	Computer-assumed diameter (must be 0)				$d_l - d_{\text{lin}} =$	1,22E-11	must be 0
23							**Step 5:**
24	$N_{1\text{-min}} =$	17	$N_{1\text{comp}} =$	28	$N_{1\text{comp}} - N_{1\text{-min}} =$	11	must be >0
25	$b =$	0,043799	$N_{2\text{comp}} =$	70	Volume $=$	0,040179	
26	$C_s =$	0,85	$C_f =$	0,85	$J =$	0,260857	**Step 5:**
27							**Step 6:**
28	Computed modul-assumed modul (must be 0)				$m - m_{\text{in}}$	−2,38E-11	Must be 0
29	Use Solver to change cells E12 and E13 for the target cell G21 to be zero, with constraints G23>0 and G27=0. For optimization, set target cell G24 to minimum with constraints G21=0, G23>0 and G27=0						

13.4. DESIGN OF BEVEL GEARS

Bevel gears are used to transmit rotation between intersecting or nearly intersecting shafts. They may have straight, skew, spiral, or other curvilinear teeth (Figure 13.16).

Straight bevel gears are used for low peripheral velocities up to 2 or 3 m/s. For higher velocities, curved-tooth bevel gears are used because the gears mesh more smoothly and have a higher load-carrying capacity.

Bevel gears intersecting at an angle γ operate by rolling of two pitch cones with angles γ_1 and γ_2 for the pinion and gear respectively, with:

$$\gamma_1 + \gamma_2 = \gamma \tag{13.38}$$

Each of the two gears is formed by the pitch cone of angle γ_i, $i = 1$ or 2, and front and back cones intersecting with the pitch cone at right angles, having the same axis of symmetry and angle $\pi/2 - \gamma_i$.

Again, operation of the straight tooth can be modeled (Figure 13.17) by assigning a virtual spur gear having the same tooth center on the axis of symmetry.

The tooth of a bevel gear has variable thickness because of the geometry. The virtual spur gear has teeth of the same shape as the bevel gear teeth at some locations. Usually the shape of the tooth at the back cone (larger thickness) is standard and the strength calculations are performed with a virtual spur gear with tooth thickness equal to that of the bevel gear at the middle of the width of the tooth. The basic geometric relationships are, for the pinion gear 1:

	Bevel Gear	Virtual Gear
Maximum pitch diameter	d_1	$d_{1v} = d_1/\cos \gamma$
Pitch at maximum pitch diameter	p_1	p_1
Module at maximum pitch diameter	m_1	m_1
Mean diameter	$d_{1m} = d_1 - b \sin \gamma_1$	$d_{1v} = d_{1m}/\cos \gamma_1$
Pitch at mean diameter	p_{1m}	p_{1m}
Module at mean diameter	m_{1m}	m_{1m}
Width	b	b
Number of teeth	$N_1 = d_{1m}/m_{1m}$	$N_{1v} = N_1/\cos \gamma_1$
Pressure angle	φ	φ

The basic force relationships are, at the mean diameter:

	Bevel Gear	Virtual Gear
Tangential force	$F_t = H/(d_{1m}/2)\omega_1$	F_t
Transmitted force	$F = F_t/\cos \varphi$	F
Normal force	$F_n = F \sin \varphi = F_t \tan \varphi$	F_n
Axial force	$F_a = F_n \sin \gamma_1 = F_t \tan \varphi \sin \gamma_1$	0
Radial force	$F_r = F_n \cos \gamma_1 = F_t \tan \varphi \cos \gamma_1$	F_n

With respect to the variation in the size of the teeth along their length, bevel gears are classified into three forms:

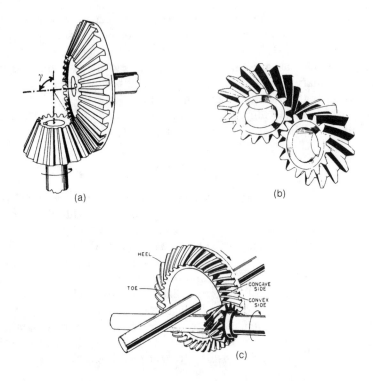

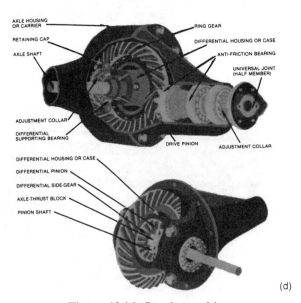

Figure 13.16. Bevel gear drives.

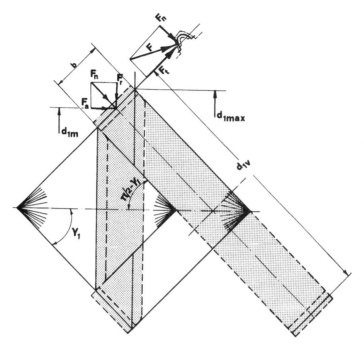

Figure 13.17. Geometry of bevel gearing.

Form I: Normally converging teeth; the apexes of the pitch and root (dedendum) cones coincide. This is the most widely used form for straight and skew teeth bevel gears. It is also applied for curved-tooth bevel gearing with $N_1 < 25$.

Form II: The apex of the root cone is located so that the width of the bottom land (width of the tooth space on the root cone) is constant and the tooth thickness along the pitch cone increases proportionally to the distance from the apex. This form enables both sides of the teeth to be machined simultaneously by a single tool. It is the principal form used for curved-tooth bevel gears.

Form III: The teeth are of equal whole depth; the elements of the pitch and root cones are parallel. This form is used for curved-tooth bevel gearing with $N_2 > 40$.

The spiral angle ψ is selected from the consideration that higher values yield smoother gear engagement but the axial thrust is also increased. Most commonly used for spiral bevel gears is $20° < \psi < 35°$.

The minimum allowable numbers of teeth are given in Table 13.6.

The load is nonuniformly distributed along the length of the teeth in bevel gearing. The elastic deflections at various sections of the teeth are not the same but are proportional to the distance from the axes of the gears or from the common apex of the pitch cones.

At the same time, the specific rigidity of geometrically similar teeth does not depend on their linear dimensions (module), since the effect of increasing the arm

TABLE 13.6. Minimum Number of Teeth for Helical Bevel Gears

	N_1 at $\psi =$		
N_2/N_1	0°–15°	20°–25°	30°–40°
1	17	17	17
1.6	15	15	14
2	13	12	11
>3.15	12	10	8

of the load is compensated for by the increase in the height of the tooth cross-section. It is usually assumed, therefore, that the unit load is distributed along the length of straight or skew teeth of bevel gears proportionally to the elastic deflections, that is, according to a trapezoid.

The point of application of the resultant load (the center of gravity of the trapezoidal load diagram) is displaced from the middle of the length of the teeth toward the large end of the teeth. However, to simplify calculations, they are carried out for the middle cross-section (located at the middle along the length of the teeth). This increases the factor of safety to some extent.

The design equation will be for surface strength and fatigue strength, respectively:

$$d^3 = \frac{4M_1}{(1 + R)(b/a)IK} \tag{13.39}$$

$$m = \frac{4M_1 K_0 K_m K_R K_T}{K_v(b/a)d_1^2(1 + R)JK_L S_n'} \tag{13.40}$$

Geometry factor J and all the derating factors must be selected on the basis of the large diameter of the pinion (location of the largest thickness of the teeth). Equations (13.39) and (13.40) ought to be applied for the virtual gear. AGMA, however, has issued standards for the values of the geometry factor for several pressure, shaft, and spiral angles that are obtained on the basis of the module and diameter of the gear at the location of the maximum tooth width. The geometry factors J and I for straight bevel gears are given in Figures 13.18 and 13.19, respectively.

For spiral bevel gears, the virtual gear is a helical one and the force relationships are:

$$F_a = \frac{F_t (\tan \varphi_n \sin \gamma_1 \pm \sin \psi \cos \gamma_1)}{\cos \psi} \tag{13.41a}$$

$$F_r = \frac{F_t (\tan \varphi_n \cos \gamma_1 + \sin \psi \sin \gamma_1)}{\cos \psi} \tag{13.41b}$$

where ψ is the helix angle. The + sign applies to a driving pinion with right-hand

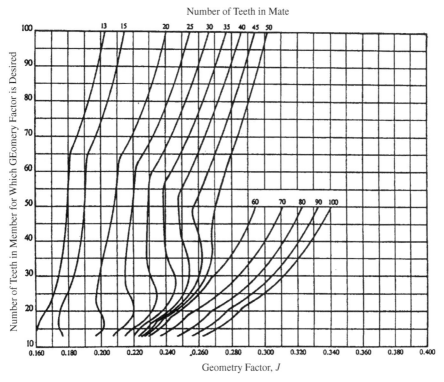

Figure 13.18. Geometry factor J for straight 20° bevel gears. (AGMA Standard 2003.)

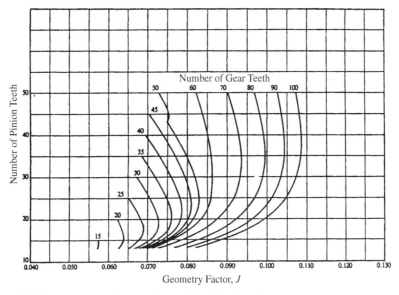

Figure 13.19. Geometry factor I for straight 20° bevel gears. (AGMA Standard 2003.)

spiral rotating clockwise or left-hand rotating counterclockwise as viewed from the end cone.

Again, $\tan \varphi_n = \tan \varphi \cos \psi$.

The two design equations are sufficient to yield the gear width b and the module m for an assumed diameter. AGMA values for the tooth geometry factors I and J can be approximated for straight 20° bevel gears, for preliminary design and CAD:

$$J_{\text{bevel}} = 240 J_{\text{spur}} (100 + N_2)$$

$$I_{\text{bevel}} = 1.04 I_{\text{spur}} \left(\frac{1 - 100}{N_1^2} \right) \tag{13.42}$$

Because the computation for spur, helical, and bevel gears is similar, these procedures have been coded in module GEARDES of MELAB 2.0.

Example 13.6   An electric motor of rated power 12 kW at 1785 rpm runs an air blower at 714 rpm with one stage of standard 20° involute bevel gearing at right angles. Ninety percent reliability will suffice, and no shocks are expected in the motor, with light shocks in the blower. The gears are to last for at least 1 million revolutions at a temperature only slightly exceeding that of the environment. The gears will be made of AISI 1015 steel surface-hardened to 270 BHN and will be machined but not ground. Design the proper bevel gearing with straight teeth.

Solution   The shaft angle is 90°. The gear ratio is $R = 1785/714 = 2.5$. From Chapter 4, AISI 1015 carbon steel has a fatigue strength of 170 MPa. For a 270 BHN, the contact fatigue strength is (Table 13.4):

$$S_H = 2.8 \times 270 - 70 = 686 \text{ MPa}$$

The calculations are performed by the program GEARDES of the MELAB 2.0 package. The interactive computer session is in Figure E13.6b.

The 22 teeth of the pinion result in operation without interference. The contact ratio is 1.57, in the acceptable range.

The overall design is judged acceptable.

13.5. DESIGN OF CROSSED GEARING

In gearing with intersecting shafts and bevel gears, one of the gears is mounted on an overhanging shaft. Because it is associated with high deflection and wear of the gear, this is not a sound design. This can be corrected with gearing with nonparallel and nonintersecting shafts because both shafts can extend beyond the gearing in both directions. Smoothness of operation is typical of such gearing, but the gears also can have higher sliding velocities, wear, and friction losses.

In crossed helical gearing (Figure 13.20), the teeth have initial point contact under conditions of considerable sliding velocities. Therefore, the load-carrying capacity of this gearing is limited.

The diameters of the pitch and rolling cylinders of uncorrected gears (Figure 13.20) are:

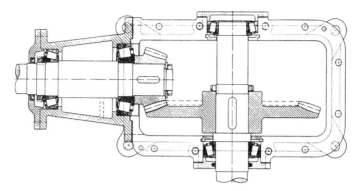

Figure E13.6a

GEAR DESIGN PROGRAM

Menu:

SPUR GEARS	1
HELICAL GEARS	2
BEVEL GEARS	3

Your selection ?3

Bevel gears

DATA		RESULTS		
			Pinion	Gear
Driving power, kW	12			
Pinion speed, RPM	1785	No of teeth	22	55
Gear ratio	2.5	Diameter	68.62	171.55
No of cycles	1E6	BHN	220	220
Pinion BHN, Young mod E	220, 2.1E5	Young mod	210000	210000
Gear BHN, Young mod E	220, 2.1E5	Fatigue strength	170	170
		Surface strength	680	680
Operating Temperature, deg C	70	SYSTEM PARAMETERS:		
Shaft angle, deg	90	Generatrix length	105.138741	
Involute angle (14.5, 20, 25)	20	Shaft angle	90	
Helix angle, deg	0	Modul	5	
Service Factor Ko	1.25	Pressure angle	20	
Width/a ratio	.25	Diametral pitch	5.08	
Reliability %	90	Contact ratio	1.57601579	
Precision:		Width	42.7271033	
High precision ⟨1⟩		b/a ratio	.25	
Ground ⟨2⟩		Tangential force	2999.36699	
Non-ground ⟨3⟩	3	Radial force	680.664891	
Fatigue strength SN1, SN2	170, 170	Axial force	852.73047	
Surface strength SH1, SH2	600, 600	Gearing volume/1E6	1.14511891	
Cycle asymmetry factor	1.4	WANT CHANGES?(Y/N)	N	

Figure E13.6b

$$d_1 = m_{t1}N_1 = \frac{m_n N_1}{\cos \psi_1} \quad \text{and} \quad d_2 = m_{t2}N_2 = \frac{m_n N_2}{\cos \psi_2} \quad (13.43)$$

where m_{t1} and m_{t2} are face modules of the pinion and gear and ψ_1 and ψ_2 are helix angles of the pinion and gear teeth.

The distance of the axes is $a = 0.5(d_1 + d_2)$. The speed ratio is:

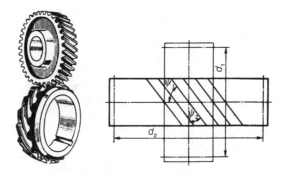

Figure 13.20. Crossed helical gearing.

$$R = \frac{n_1}{n_2} = \frac{\omega_1}{\omega_2} = \frac{N_2}{N_1} \qquad (13.43a)$$

Expressing N_1 and N_2 in terms of the pitch diameters, we obtain:

$$R = \frac{d_2}{d_1} \tan \psi_1 \qquad (13.44)$$

In power drives requiring a sufficiently high efficiency, the helix angles are selected equal or approximately the same. In certain auxiliary gearing, particularly in some instruments, it proves convenient to have a pinion and gear with the same pitch diameter. Then the required speed ratio is obtained by proper selection of the helix angles of the teeth.

Hypoid gearing consists of spiral bevel gears with crossed axes (Figure 13.16c). The speed ratio of such gears is usually in the range 2–10. In addition to the above-mentioned advantages of gearing with nonparallel, nonintersecting axes, hypoid gearing also has a high load-carrying capacity. This is primarily because in hypoid gearings, in contrast to crossed helical gearings, tooth contact close to linear contact is achieved with optimal size and shape of the tooth bearing contact pattern. In this respect, hypoid gearing is similar to curved-tooth bevel gearing. The sliding velocity in hypoid gearing is substantially less than in crossed helical gearing. For the same diameter of gear and speed ratio, the diameter of a hypoid pinion is larger than in bevel gearing. Besides, the mating teeth in hypoid gearing run-in well and are not subject to appreciable distortion because the sliding action along the working surfaces is sufficiently uniform. Because several pairs of teeth are in mesh simultaneously, hypoid gearing can be employed in high-precision mechanisms, such as in the indexing gear trains of precision gear-cutting machines.

Hypoid gearing has found wide application in automobiles and other transportation machinery.

Contact stress and beam strength calculations for hypoid gearing can be carried out as for bevel gearing having the same diameters, face widths, and large end modules.

13.6. DESIGN OF WORM GEAR DRIVES

13.6.1. Geometry and Kinematics of Worm Gear Drives

Worm gearing (Figures 13.21 and 13.22) is one of the oldest gear drives, invented by Archimedes. It consists of a worm, which is a screw with trapezoidal, or approximately trapezoidal, threads, and a worm gear, which is a toothed gear with teeth of special shape obtained by geometric interaction with the threads of the worm.

Worm gearing is one kind of crossed helical gearing, but it has initial contact along a curved line, giving better load distribution and higher strength. Other advantages are high speed reduction, smooth and silent operation, and small drive size, many times a tenth that of ordinary gear drives.

Disadvantages are low efficiency and expensive antifriction materials necessary for the worm gear.

The speed ratio R of worm gearing is determined from the condition that for each revolution of the worm the worm gear turns through a number of teeth equal to the number of starts, or threads, N_1 on the worm. Thus:

$$R = \frac{n_1}{n_2} = \frac{\omega_1}{\omega_2} = \frac{N_2}{N_1} \qquad (13.45)$$

It must be pointed out that now the speed ratio is no longer equal to the ratio of the diameters.

Due to low efficiency, worm gear drives are used for low and medium power transmission up to 50 kW and for torques up to 500 kNm. The speed ratio usually ranges from 8–100 and, in special cases, up to 1000.

Geometrical calculations for worm gearing are similar to those for ordinary toothed gearing.

The worm gears are cut with a tool (hob) having the shape and size of the worm, obtained automatically.

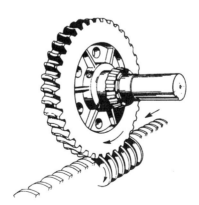

Figure 13.21. Worm gear system.

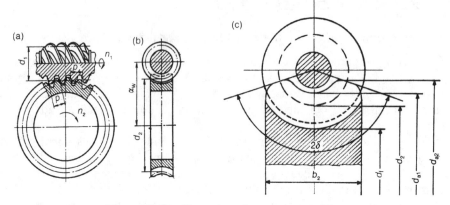

Figure 13.22. Geometry of a worm gear drive.

The types of worms used are Archimedean basic rack, involute helicoidal, and worms with a concave thread profile.

Archimedean worms (Figure 13.23a) are screws with threads that are straight-sided in an axial section (trapezoidal profile).

The number of starts, or threads, of the worm depends upon the speed ratio and is usually $z = 1$, 2, or 4.

The pitch helix angle of the worm threads is the screw angle:

$$\tan \psi = \frac{h}{\pi d_1} = \frac{N_1 \pi n}{\pi d_1} = \frac{N_1 m}{d_1} \tag{13.46}$$

The addendum h_a and dedendum h_f of the worm threads are:

$$h_{a1} = h_{a1}^* m \text{ and } h_{f1} = h_{f1}^* m \tag{13.47}$$

where the addendum factor $h_{a1}^* = 1$ and the dedendum factor $h_{f1}^* = 1.2$ for Archimedean worms and $h_{f1}^* = 2.2 \cos \psi - 1$ for involute helicoidal worms.

The outside d_{a1} and d_{f1} root diameters of the worm are:

$$d_{a1} = d_1 + 2h_{a1} \text{ and } d_{f1} = d_1 - 2h_{f1} \tag{13.48}$$

The length b_1 of the threaded portion of the worm depends on the number of teeth of the worm gear N_2. Thus:

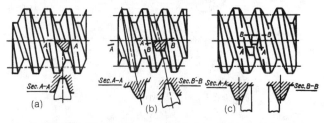

Figure 13.23. Geometry of the worm thread.

$$b_1 > (c_1 + c_2 N_2)m \tag{13.49}$$

where for $N_1 = 1$ or 2, $c_1 = 11$ and $c_2 = 0.06$; and for $N_1 = 4$, $c_1 = 12.5$ and $c_2 = 0.09$.

The minimum number of teeth N_2 of the worm gear is taken equal to 17 or 18 for auxiliary kinematic drives with a single-start worm. For power drives, the minimum number $N_2 = 26$–28 (only 17 if the worm is of the involute helicoidal type).

The pitch circle diameter is:

$$d_2 = mN_2 \tag{13.50}$$

The throat diameter d_{a2} and root diameter d_{f2} of the worm gear are:

$$d_{a2} = d_2 + 2h_{a1} \text{ and } d_{f2} = d_2 - 2h_{f1} \tag{13.51}$$

In multiple-start drives, the effective field of engagement is less than for single-start worm gearing and therefore the outside diameter and face width of the worm gear are taken less than for the corresponding single-start drive (with the same values of d_2, d_{a2}, and m). The maximum outside diameter should be:

$$d_{am2} < d_{a2} + \frac{6m}{N_1 + 2} \tag{13.52}$$

The face width of the worm gear is assigned in accordance with the outside diameter of the worm (Figure 13.22c):

$$\text{for } N_1 = 1 \text{ or } 2, \ b_2 < 0.75 d_{a1}; \text{ for } N_1 = 3 \text{ or } 4, \ b_2 < 0.67 d_{a1} \tag{13.52a}$$

The conditional worm gear face angle 2δ, used in strength calculations, is found from the points of intersection of a circle of diameter $d_{a1} = 0.5m$ with the end faces (contour) of the worm gear. Thus:

$$\sin \delta = \frac{b_2}{d_{a1} - 0.5m} \tag{13.53}$$

The center distance is equal to one half of the sum of pitch diameters of the worm and worm gear that is:

$$a = \frac{d_1 + d_2}{2} = \frac{(N_1 + N_2)m}{2} \tag{13.54}$$

In designing worm gearing, it is necessary to plan for axial adjustment of the worm gear in assembly so that it can be made to coincide with the axial plane of the worm.

In contrast to rolling of spur gear surfaces, worm gears work primarily on sliding contact. To increase the load-carrying capacity, a geometry providing an oil wedge, as in the bearings, must be selected.

Sliding cannot be totally avoided, and therefore the principal causes of worm gearing failure are surface damage, tooth wear, and seizing.

Seizing is especially dangerous if the worm gear is made of hard materials such as hard bronzes or cast iron. In this case, seizing is severe, with damage to the surfaces followed by rapid wear of the teeth by particles of the worm gear material welded onto the threads of the worm. With worm gears of softer materials, seizing of a less dangerous type is observed: the material of the worm gear (bronze) is "smeared" on the worm.

Fatigue pitting is observed mainly in gearing with a worm gear made of a seizure-resistant bronze. Such pitting, as a rule, occurs only on the worm gear.

Breakage can be observed mainly after severe wear, and usually only the teeth of the worm gear are broken.

In power drives, the worms are made, as a rule, of steel heat-treated to considerable hardness. The most durable gearing has the worm made of a case-hardening steel having a hardness of 56–63 R after heat treatment. Also extensively used are worms of medium-carbon steels surface or through-hardened to 45–55 R.

13.6.2. Force Transmission in Worm Gear Drives

Force calculations for the worm are based on the screw equations, while for the worm gear the situation is similar to that for helical gears.

Due to geometry if the worm and gear axes are perpendicular (Figure 13.25), the tangential force of the one equals the axial force of the other. Therefore (Figure 13.24):

$$F_{2a} = F_{1t} = \frac{H_1}{(d_1/2)\omega_1}$$

$$F_{2t} = F_{1a} = \frac{H_2}{(d_2/2)\omega_2}$$

(13.55)

To compute the other forces, we must take the friction force into account. It is proportional to the normal force F_n with a direction opposing the sliding direction. For coefficient of friction f:

$$F_{2t} = F_{1a} = F_n \cos \varphi_n \cos \psi - fF_n \sin \psi$$

$$F_{1t} = F_{2a} = F_n \cos \varphi_n \sin \psi + fF_n \cos \psi = 2H_1/(\omega_1 d_1)$$

$$F_n = \frac{2H_1}{\omega_1 d_1 (\cos \varphi_n \sin \psi + f \cos \psi)}$$

(13.56)

$$F_{1r} = F_{2r} = F_{1t} \frac{\cos \varphi_n \sin \psi + f \cos \psi}{\cos \varphi_n \cos \psi - f \sin \psi}$$

Therefore:

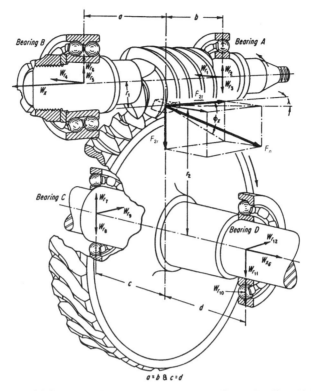

Figure 13.24. Forces in worm gear systems. (From Dudley 1962.)

$$\frac{F_{2t}}{F_{1t}} = \frac{\cos \varphi_n \cos \psi - f \sin \psi}{\cos \varphi_n \sin \psi + f \cos \psi} \tag{13.57}$$

From Figure 13.25, the sliding velocity u_s is analyzed into the worm and gear tangential velocities u_1 and u_2, respectively. It is:

$$\tan \psi = \frac{u_2}{u_1} \tag{13.57a}$$

The efficiency of the system is the ratio of the worm tangential force without friction to the one with friction. Therefore:

$$\eta = \frac{\cos \varphi_n \sin \psi - f \tan \psi \sin \psi}{\cos \varphi_n \sin \psi + f \cos \psi} \tag{13.58}$$

The friction coefficient f depends on sliding velocity. For usual applications AGMA recommends values

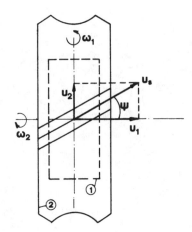

Figure 13.25. Kinematics of worm gears.

$$f = 0.123[1 - 0.23 \log_{10}(197u_s)] \tag{13.59}$$

where $u_s = u_1 \cos \psi$ (m/s). Therefore, f depends on the helix angle

$$f = 0.123[1 - 0.28\log_{10}(197u_1 \cos \psi)] \tag{13.60}$$

This suggests also that the helix angle plays an important role in efficiency. Optimization yields values for optimum ψ of about 40°, but the improvement in efficiency is small above $\psi = 25°$. Because lower ψ yields higher gear ratio, values of between 25° and 30° are usually selected.

If the gear is driving, the efficiency will be the ratio of F without and with friction:

$$\eta' = \frac{\cos \varphi_n \cos \psi - f \sin \psi}{\cos \varphi_n \cos \psi + f \cos^2 \psi / \sin \psi} \tag{13.61}$$

For $f > \cos \phi_n \cos \psi$ the efficiency is negative, that is, the system cannot move, it is self-locked. For a given friction coefficient, the helix angle must have small values if self-locking is desired. This is the case, for example, in drives where safety demands it, such as in elevators. For drives of reverse driving, conditions of self-locking should not exist, that is, ψ must be large enough.

13.6.3. Design for Strength of Worm Gear Drives

Strength of worm gearing follows the same lines as for helical gears, with emphasis on surface fatigue strength. The worm is usually much stronger than the gear, and calculations are performed on the worm gear.

Special demands are placed on the materials for worm gear drives. There is considerable sliding friction between worm and gear. The worm has high speed and needs to be made of a hard material, usually surface-hardened or heat-treated steel. The gear has low speed and the frictional heat raises its temperature substantially, unless it is made of material with high thermal conductivity, such as copper and aluminum alloys. Moreover, the combination of materials for worm and gear must yield low coefficient of friction. Materials for the worm and the gear are tabulated in Table 13.7 together with some useful design parameters.

In principle, the Lewis and Buckingham equations are applicable to worm gear drives. However, for the worm, tooth strength and surface durability are seldom an issue, and failure is usually by bending of the cylindrical body of the worm as a beam. Failure of the gear teeth is more common because the gear is made of a material usually much weaker than that of the worm. Moreover, the geometry of the worm gear tooth is much more complex than the spur, helical, or bevel gears and the conditions of friction and heat transmission dominate. Thus, the Lewis and Buckingham equations lead to their empirical adaptation for worm gear drives.

The design equation for beam strength of the gear teeth is the following modification (Thomas and Chatchut 1957) of the Lewis equation (*watch the units!*):

$$m = 12.45 \sqrt[3]{\frac{H_1 \eta}{K_N c(b_2/p)N_1 n_1}} \tag{13.62}$$

where H_1 is the power of the worm (W), η is the efficiency of the drive, c is a material strength factor (MPa) from Figure 13.26, N_1 is the number of the worm starts (leads), n_1 is the worm speed (rpm), b_2 is the face width of the gear, and $p = \pi m$ is the pitch of the gear. The (b_2/p) ratio is usually between 2 and 2.5. Furthermore, K_N is a factor that depends on the number of gear teeth:

$$K_N = 0.6776 Ln(N_2) - 1.3753 \tag{13.63}$$

AGMA recommends the following design equation for surface fatigue strength of the gear in worm gear drives (*watch the units!*):

TABLE 13.7. Materials for Worm Gear Drives[a]

Symbol	Material	Manufacturing	BHN	Allowable Pairs	Notes
		WORM			
A	Carbon steel C15	Case-hardened	600	1, 2, 3	Ground
C	Carbon steel C45	Heat-treated	180–208	1, 2	Machined
		GEAR			
1	Phosphor-bronze	Sand casting	60–95	A, B	
2	Cast iron	Sand casting	197–241	A, B	$v_s < 3$ m/s
3	Aluminum alloy	Chilled casting	95–105	A	

[a]Adapted from Thomas and Chatchut, 1957.

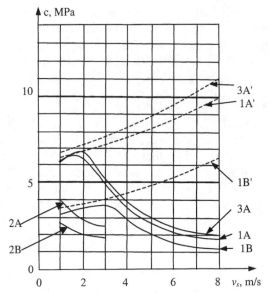

Figure 13.26. Fatigue strength of gears for worm gear drives. Solid lines are for splash lubrication and air cooling of the drive. Dotted lines are for pumped lubrication with external worm gear, as in Table 13.7.

$$F_{2t} = \frac{2H_2}{\omega_2 d_2/1000} = 13.2 \times 10^{-3} C_s d_2^{0.8} b_2 C_m C_v \qquad (13.64)$$

$$d_2^{1.8} = \frac{60\eta H_1}{13.2 \times 10^{-6}\pi n_2 C_s b_2 C_m C_v} \qquad (13.64a)$$

where F_{2t} is the tangential force of the gear (N), d_2 is the pitch diameter of the gear (mm), and b_2 the effective face width of the gear (mm) (Equation (13.52a)). The factors C are given below (AGMA Standard 6034-A87, Mott 1999):

Materials factor C_s (d_2 in mm)

> For *sand-cast bronzes*, $C_s = 520.2 - 476.5 \log_{10}(d_2)$ for $d_2 > 200$ mm, $= 1000$ otherwise
> For *chill-cast* or *forged bronzes*, $C_s = 771.3 - 455.8 \log_{10}(d_2)$ for $d_2 > 200$ mm, $= 1000$ otherwise
> For *centrifugally cast bronzes*, $C_s = 999.2 - 476.5 \log_{10}(d_2)$ for $d_2 > 625$ mm, $= 1000$ otherwise

Gear ratio correction factor C_m

> $C_m = 0.020(-R^2 + 40R - 76)^{0.5} + 0.46$, for gear ratio R between 6 and 20
> $C_m = 0.0107(-R^2 + 56R\ 5145)^{0.5}$, for R between 20 and 76
> $C_m = 1.148 - 0.00658R$, for $R > 76$.

Velocity factor C_v

$C_v = 0.659e^{-0.197v_s}$, for v_s between 0 and 3.5 m/s
$C_v = 0.651v_s^{-0.571}$, for v_s between 3.5 and 15 m/s
$C_v = 1.096v_s^{-0.774}$, for $v_s > 15$ m/s

The module is now $m = d_2/N_2$. While $N_2 = RN_1$, N_1 is the number of starts of the worm. Again, the normal module is adjusted to available sizes.

The diameter of the worm cannot be computed from the results available up to this point. Usually it is computed from considerations of strength of the carrying shaft, since the forces can be computed with the gear geometry already computed. For preliminary design calculations, a ratio d_G/d_W between 3 and 6 is suggested.

The stress diameter of the worm is $d_s = d_1 - 2.5\ m$. If the distance of the worm bearings is L, the tangential and radial forces of the worm are F_{1t} and F_{1r} and the bending force is:

$$F_b = \sqrt{F_{1t}^2 + F_{1r}^2} \tag{13.65}$$

The stress equation for the worm is then:

$$\sigma_{max} = \frac{MC}{I} = \frac{(F_bL/4)(d_s/2)}{\pi d_s^4/64} = \frac{8F_bL}{\pi d_s^3} \tag{13.66}$$

Therefore:

$$d_s = \sqrt[3]{\frac{8F_bLN}{\pi S_y}} \tag{13.67}$$

and the worm diameter is:

$$d_1 = d_s + 2.5\ m \tag{13.68}$$

Example 13.7 A worm gear drive has input speed $n_1 = 960$ rpm, gear ratio $R = 52$, number of worm starts (leads) $N_1 = 1$, module $m = 8$ mm, worm pitch diameter $d_1 = 80$ mm, $\phi_n = 20°$. The worm is machined and heat-treated steel and the gear is made of machined phosphor-bronze. The lubrication is pressurized with external oil cooling. Find the rated power of the drive.

Solution The spreadsheet is shown in Table E13.7.

Example 13.8 Design a worm gear drive for input speed $n_1 = 960$ rpm, rated power $H_1 = 25$ kW, gear ratio $R = 52$, number of worm starts (leads) $N_1 = 1$, worm pitch diameter $d_1 = 80$ mm, $\phi_n = 20°$. The worm should be machined and heat-treated steel and the gear will be made of machined phosphor-bronze. The lubrication is pressurized with external oil cooling.

TABLE E13.7

Data:

$n_1 =$	960	RPM	Involute angle $\phi_n = 20°$,	0.349066	rad
$R =$	52				
Worm starts $N_1 =$	1				
Module $m =$	0.008	m			
Worm Diameter $d_1 =$	0.08	m			
SOLUTION					
Gear number of teeth $= N_1 R =$	52				
Gear speed $n_2 = n_1/R = 960/52 =$	18.46154	RPM			
Gear Diameter $d_2 = mN_2 = 0.08 \times 52 =$	0.416	m			
The peripheral speed of the worm is					
$v_1 = (2\pi n_1/60)(d_1/2) = (2\pi \times 960/60)(0.080/2) =$			4.021239	m/sec	
The peripheral speed of the gear is					
$v_2 = (2\pi n2/60)(d2/2) = (2\pi \times 960/60)(0.080/2) =$			0.077332	m/sec	

The helix angle ψ is (equation 13.57a) $\psi = a\,\tan(v_2/v_1) = 0.077/4.02) =$			0.019228		
The friction coefficient is (eq. 13.60)					
$= 0.123[1-0.28\log 10(197\times4.02\cos(0.019228)]$	0.023172				
The efficiency, from equation 13.58					
$\eta=[\cos(0.349)\sin(0.0192)-0.023\tan(0.0192)]/[\cos(0.349)\times\sin(0.0192)+0.023\times\cos(0.0192)] =$				0.4273549	
A. Bending strength					
Coefficient K_m in equation 13.62, $K_N = 0.6776\mathrm{Ln}(52) - 1.3753 =$		1.302063			
From Figure 13.26, curve 1B', for $v_s = v_1 = 4$ m/s, $c =$	4.3 MPa	We select $b/p = 2.1.$			
Equation 13.62 is now solved for the power H_1: $= (8/12.45)^{\wedge}3\times1.3\times4.3\times2.1\times1\times960/0.427 =$				**7007.5**	Watts
B. Surface strength					
Equation 13.64a will be solved for the power $H_1 = 0.0132\times d_2^{1.8}\times\pi\times n_2\times C_s\times b_2\times C_m\times C_v/(60\eta)$					
$C_s = 771.3 - 455.8\log 10(406) =$	944.9173				
$C_m = 0.0107(-52^2 + 56\times52 + 5145)^{0.5} =$	1.109967				
$C_v = 0.651 v_s^{-0.571} =$	0.294098	The face width $b_2 = (b_2/p)p = 2.1\times\pi\times8 =$		52.778	mm
Therefore, $H_1 = 416^{1.8}\times0.0000132\times\pi\times18.46\times945\times52.78\times1.11\times0.294/(60\times0.427) =$				**25181.**	Watts
RESULT:					
The smallest of the two values is the rated power:		**7007.52**	**Watts**		

TABLE E13.8

Data:

$n1 =$	960	RPM	Rated power $H_1 =$	25000	Watts
$R =$	52			
Worm starts $N_1 =$	1		Involute angle $\phi_n =$	0.349066	rad
Module $m =$	0.012224	m	(we give initially any arbitrary value, say 0.020)
Worm Diameter $d1 =$	0.08	m	

SOLUTION

Gear number of teeth $= N_1 R =$	52		
Gear speed $n_2 = n_1/R =$	18.46154	RPM	
Gear Diameter $d_2 = m N_2 =$	0.635651	m	
The peripheral speed of the worm is			
$v_1 = (2\pi n_1/60)(d_1/2) =$		4.021239	m/sec
The peripheral speed of the gear is			
$v_2 = (2\pi n_2/60)(d_2/2) =$		0.077332	m/sec
The helix angle ψ is (equation 13.57a) $\psi = a\tan(v_2/v_1) =$			0.019228
The friction coefficient is (eq. 13.60)		$=$	0.023171803

888

The efficiency		0.427355		
A. Bending strength				
Coefficient K_m in equation 13.62 $K_m = 0.6776\text{Ln}(N_2) - 1.3753 =$		1.302063		
From Figure 13.26, curve 1B', for vs = v_i = 4 m/s, c =	4.3	MPa. We select b/p =	2.1	
Equation 13.62 is now solved for the power H1:				
=		**25000**		Watts
B. Surface strength				
Equation 13.64a will be solved for the power H1 = 0.0132×d2^1.8×p×n2×Cs×b2×Cm×Cv/(60h)				
Cs = 771.3 − 455.8log10(d.2)=	860.993			
Cm = 0.0107(−R^2 + 56*R + 5145)^0.5 =	1.109967			
Cv = 0.651vs −0.571 =	0.294098	The face width b2 = (b2/p)p =	80.64629923	mm
Therefore, H1 =		**75202.27**		Watts
RESULT:				
The smallest of the two values is the rated power:	**25000**	Watts		
The module m will be found with **Solver** to make the difference (computed rated power − given rated power) = 0				
		Target difference =	−6.8E-08	

Solution   The spreadsheet is shown in Table E13.8.

The result is a worm gear drive with module $m = 12.2$ mm, gear diameter $d_2 = 635$ mm.

13.6.4. Heat Capacity and Thermal Design of Worm Gear Drives

Worm gearing operates at relatively low efficiency and generates a substantial amount of heat. But the heating of the oil to a temperature exceeding the limiting value $T = 95°C$ leads to the danger of seizing and rapid wear.

The amount of heat by worm gearing operating continuously with an efficiency η and transmitting the power H is:

$$W_H = (1 - \eta)H \tag{13.69}$$

The heat removed by convection from the free surface of the housing and by conduction through the foundation plate or frame is:

$$W'_H = h_t(T - T_0)A(1 + \xi) \tag{13.69a}$$

where A is the free surface of the housing from which heat is removed to cool the drive (included is 50% of the surface of the fins), T and T_0 are the temperatures of the oil and of the surrounding air (°C), h_t is the heat transfer coefficient, and ξ is a factor taking into account heat transfer to the foundation plate or frame of the machine and amounting to 0.3 when the housing seating surface is large.

The heat balance, $W_H = W'_H$, can be employed to find the working temperature, or the power H that can be transmitted by the worm gearing, complying with the condition that the oil temperature does not exceed T_{max}. Thus:

$$T = \frac{T_0 + (1 - \eta)H}{h_t A(1 + \xi)}$$

$$H = \frac{h_t(T_{max} - T_0)A(1 + \xi)}{1 - \eta} \tag{13.70}$$

If the equilibrium temperature is high, additional cooling surface must be provided. This is done by addition of fins to the housing, mechanical ventilation, coils with cooling fluid in the oil bath, or cooling facilities.

The fins are designed for adequate air flow over them. Because heated air rises, fins should be arranged vertically for natural ventilation. For mechanical ventilation, the fins are arranged along the stream of air blown by the fan, usually horizontally.

Mechanical ventilation is accomplished by a fan mounted on the worm shaft. Air cooling is much simpler and less expensive than water cooling and is consequently more widely employed.

Example 13.9   A worm gear drive has input speed $n_1 = 960$ rpm, rated power $H_1 = 25$ kW, efficiency $\eta = 0.43$, outer surface area including cooling fins $A = 1.5$ m², and surface heat transfer coefficient with forced air cooling $h_t = 150$ W/°C m². From measurements in similar drives it is known that $\xi = 0.25$. Find the maximum temperature at rated load for ambient temperature $T_0 = 20°C$.

Solution   Equation (13.70) gives:

$$T = \frac{20 + (1 - 0.43) \times 25{,}000}{150 \times 1.5 \times (1 + 0.25)} = 50.74°C$$

13.7. CHAIN DRIVES

13.7.1. Field of Application

There are three methods of transmitting power mechanically from shaft to shaft: chain drives, gears, and flexible belts. The chief advantage of chains and belts over gears is that chains can be used with arbitrary shaft center distances. Also, chains are simpler and generally less costly than gears.

All drives can be engineered to provide a specific capacity and service life; however, within specific size constraints, chains generally do not have capacities or service lives equal to those of gears. Compared with belts, chains offer advantages in capacity and service life at temperature extremes. Chains also provide positive drive ratios.

Chains are available in a range of accuracies extending from classifications of "precision" to "nonprecision." There is no firm boundary between the two, but nonprecision chains are those that do not provide extremely close fit between sprockets and link and are not designed to articulate with exceptional smoothness. In this class are detachable, pintle, and welded-steel chain. These types are low-cost chains intended primarily for low-speed drives below 50 hp.

Precision chains are designed for smooth, free-running operation at high speeds and high power, ranging to over 1000 hp. The most common precision chain is the roller chain.

Horsepower ratings for chains have been established by the American National Standards Institute (ANSI) and are based on the criterion according to which the chain supposedly wears until it jumps the sprockets before failing from fracture.

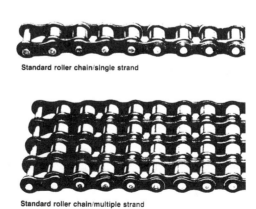

Standard roller chain/single strand

Standard roller chain/multiple strand

Figure 13.27. Roller chains.

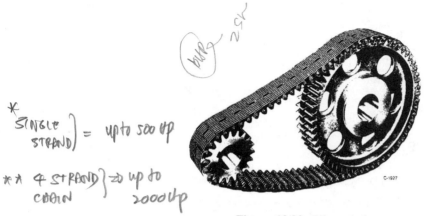

[handwritten margin notes:]

[top of figure:] bhp 2gr

SINGLE STRAND } = up to 500 hp

** A 4 STRAND } ⇒ up to
CHAIN   }   2000 hp

Figure 13.28. Silent chain.

Roller chains are the basic type of power chain used throughout industry (Figure 13.27). They are rated at up to about 500 hp for a single-strand chain. Multi-strand types are available at significantly higher horsepower. For example, ratings for four-strand chains exceed 2000 hp. Precision construction provides efficient, low-loss, quiet operation. Rated values are predicated upon adequate lubrication covered by ANSI B29.1. In the smaller sizes (below 0.5 in. pitch; actually "bushing" chains), speeds can exceed 9,000 fpm. Self-lubricating types, or stud-bushing chains, have oil-impregnated sleeve bearings. They are capable of running at almost the same loads as true roller chain, but at the lower end of the speed range. They provide a service life intermediate between nonlubricated and well-lubricated roller chain.

Double-pitch chains are designed for service less rigorous than that handled by standard roller chains. They provide the same precision construction and efficient operation as standard roller chains, except that pitch is twice as long, providing a lighter-weight construction, and ratings go only to 92.5 hp. They function well on drives with long center distances and are covered by ANSI B29.3.

Inverted silent tooths are a more expensive type of chain that operates smoothly, quietly, and dependably in rigorous applications (Figure 13.28). They are often used as a power takeoff from the prime mover in heavy equipment and some automobiles, and also used as a timing chain (camshaft drive) in engines. Power capabilities are generally equivalent to those of roller chains, except that silent chains reach maximum power at maximum rpm, whereas roller chains reach maximum power far below their maximum rpm.

Offset sidebars are the most costly of precision chains—more costly than detachable, pintle, or welded steel—but they are able to carry heavier loads, up to 425 hp, and run at speeds to 2250 fpm. Because of their "open" construction, offset sidebars tend to be tolerant of dirt and debris that might cause binding in a roller chain. Offset sidebars are also more tolerant of misalignment or twisting of the sprocket axes. This type is rugged and durable and often is used to drive construction machinery. They serve well as conveyor chains in high-temperature ovens. They generally do not provide the speed or power capability available with roller chains. They are well suited for use over cast sprockets at reduced speed. Larger sizes are covered by ANSI B29.10.

Bead chains are a light-duty type of chain generally used for manual-control systems such as those in television tuners and air-conditioner controls. They are also used in low-powered systems such as paper drives for business machines and laboratory recorders. They are available in four standard bead diameters: of 3/32, 1/8, 3/16, and 1/4 in., rated at 15, 25, 40, and 75 lb.

Detachable chains are the lightest, simplest, and least costly of all chains capable of transmitting significant power up to 25 hp at up to 350 fpm, but they are not as smooth running as precision chains. They do not require lubrication and thus are well suited for operation where lubricant might wash or bake away or where dust or granular materials would stick to a lubricated chain. They are commonly used in farm machinery. Detachable chains of steel and malleable iron are covered by ANSI B29.6 and B29.7.

Pintle and welded steel chains are used for slightly more rigorous service than detachable chains. Applications are basically similar to those of detachable chains, and they are often designed to run on the same sprockets. They are rated to operate at up to 40 hp and 450 fpm. They do not require lubrication. They are not as smooth running, strong, or durable as precision chains. Welded steel chains generally can carry heavier loads and offer greater wear resistance than pintle chains.

13.7.2. Kinematics of Chain Drives

A roller chain drive consists of a series of rollers connected with linear components, and the pitch of the chain conforms with the pitch of the geared wheel, the sprocket. The pitch angle (Figure 13.29a):

$$\beta = \frac{2\pi}{N_1} \tag{13.71}$$

where N_1 is the number of teeth of the driving sprocket.

If the sprocket rotates with a constant angular velocity ω, the peripheral velocity of the chain is $u = \omega R$, where R is the distance of the leaving branch of the chain from the center of rotation. In the location shown, $R = OB = OA \cos(\beta/2) = R_p \cos(\beta/2)$, but after rotation of the sprocket by angle $b/2$, $R = OA = R_p$, where R_p is the pitch radius of the sprocket. Thus, the chain speed varies between two values

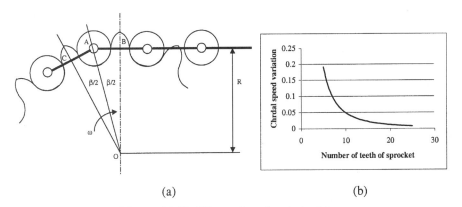

(a)   (b)

Figure 13.29. Kinematics of a chain drive.

$\omega R_p \cos(\beta/2)$ and ωR_p. Therefore, the relative variation of the speed during operation, the *cordal speed variation*, is defined as:

$$\frac{\Delta u}{u} = \frac{\omega R_p - \omega R_p \cos(\beta/2)}{\omega R_p} = 1 - \cos\frac{\beta}{2} = 1 - \cos\frac{\pi}{N_1} \qquad (13.72)$$

This variation is plotted in Figure 13.29b. It is evident that for number of teeth below 10, the speed variation exceeds 5%. Usual design practice is to require seventeen teeth or more for the driving sprocket, depending on the speed. For low speeds, a smaller number of teeth can be used.

Silent chains have involute teeth, like the gear teeth, that engage with a gear and keep the distance of the chain from the center of the gear constant (Figure 13.28). Thus, there is no speed variation during rotation.

13.7.3. Design of Chain Drives

The two common modes of failure for chain drives are by fracture of the chain links and by wear of the chain rollers and bearings. Usually, stretching the links to their limits results in rapid roller deterioration and wear. Thus, we will design the roller chain here for link strength.

Table 13.8 lists the roller chain designations by ANSI and ISO and some properties that are useful for design.

TABLE 13.8. Properties of Roller Chains[a]

ANSI No.	ISO No.	Pitch p, mm	Max speed, rpm[b]	Mass/unit length, kg/m (1 strand)	Static strength P, N No. of strands		
					1	2	3
	05B	8	2000	0.18	380.8101	662.2785	943.7468
35	06B	9.525	2000	0.14	753.3418	1432.177	2102.734
40	08B	12.7	1600	0.70	1506.684	2632.557	4520.051
50	10B	15.875	1250	0.95	1879.215	3758.43	5637.646
60	12B	19.05	1250	1.25	2442.152	4884.304	7326.456
80	16B	25.4	1000	2.70	3568.025	7136.051	10704.08
100	20B	31.75	800	3.6	5447.241	10894.48	16341.72
120	24B	38.1	630	6.7	8261.924	16523.85	24785.77
140	28B	44.45	400	8.3	10894.48	21788.96	32683.44
160	32B	50.8	400	10.5	14272.1	28544.2	42816.3
200	40B	63.5	250	16	22161.49	44322.99	66484.48
240	48B	76.2	160	25	33801.04	67602.08	101403.1

[a]Sources: ANSI B29.1-1975 and ISO-SIN 8187, 8188.
[b]For higher speeds, refer to specific manufacturers' recommendations.

The static chain load F_1 is:

$$F = \frac{M}{d_1/2} = \frac{60H}{\pi n_1 d_1} \tag{13.73}$$

where $d_1 = pN_1/\pi$ is the pitch diameter, p is the pitch, N_1 is the number of teeth of the driving sprocket, and n_1 the sprocket speed (rpm). Therefore:

$$F_1 = \frac{60H}{n_1 p N_1} \tag{13.74}$$

Due to the speed of rotation, there is a centrifugal force on the chain as in the belts (see Chapter 10):

$$F_c = qu^2 = q\left(\frac{\pi d_1 n_1}{60}\right)^2 = q\left(\frac{n_1 p N_1}{60}\right)^2 \tag{13.75}$$

where q is the mass per unit length of the chain (see Chapter 10). q is tabulated for single-strand chains in Table 13.8 For multiple strands, the values of q from Table 13.8 need to be multiplied by the number of strands z. Therefore, the design equation is:

$$F = F_1 + F_c = \frac{H}{n_1 p N_1 /60} + zq\left(\frac{n_1 p N_1}{60}\right)^2 \le \frac{P}{C} \tag{13.76}$$

$$P = \frac{CH}{n_1 p N_1 /60} + zq\left(\frac{n_1 p N_1}{60}\right)^2 \tag{13.76a}$$

where P is the design strength of the chain (demand), which should not exceed the values for the chain strength (capacity) given in Table 13.8, C is a service factor that depends on operating conditions of the driving and the driven machine (Table 13.9).

For certification, the rated power of a chain drive can be obtained by solving Equation 13.76 for the power H:

$$H = \frac{n_1 p N_1}{60}\left[\frac{P}{C} - zq\left(\frac{n_1 p N_1}{60}\right)^2\right] \tag{13.77}$$

Chains need to be lubricated. Grease lubrication is used for peripheral speeds up to 0.3 m/s. Frequent manual or drip lubrication is used for speeds up to 2 m/s. Splash or dipped lubrication is used for speeds up to 10 m/s, and oil-stream lubrication should be used for chain speeds above 10 m/s. Manufacturers' recommendations need to be closely followed.

An important note about the *safety* and the *service* factors: Recall from Chapter 4 that a safety factor N accounts for the uncertainty in the material properties (capacity) and the machine loads (demand). We introduced here the *service*

TABLE 13.9. Service Factors for Machinery $C = c_1 + c_2$

Driving machines	
Electric motors	0.5
Turbomachinery	0.75
Reciprocating engines:	1.5–2.8
4-cylinder	
6-cylinder	1–2
Driven machines	c_2
Machine tools	3
Rolling mills	1.6–2
Blowers	1.5
Textile machinery	1.6–2.2
Rotary pumps, compressors	1.5
Reciprocating pumps,	2.2–3.2
compressors	
Paper mills	1.8–3
Mixing machines	1.8–3
Wood processing machines	2–3
Electric generators	1.1–1.3
Hoists	1.5

factor C to account for the load uncertainty only. In use, the service factor C multiplies the rated power P to yield the maximum expected power, while the safety factor N divides the theoretical material strength (yield strength S_y, for example) to yield the *allowable* stress. In a design, use of both factors is allowed *if the safety factor* N *accounts only for the dispersion in the material strength*. If the safety factor N accounts also for the dispersion or uncertainty of the load, the service factor should not be used.

Example 13.10 Chain Design (Certification)   Find the rated power of a chain drive with seventeen teeth, 1200 rpm driving sprocket, two-strand ANSI 40 chain, $C = 2$.

Solution   From Table 13.8, for the ANSI 40 chain, $p = 12.7$ mm, $q = 0.70$ kg/m, $P = 2632$ N for two strands, $z = 2$. The pitch diameter $d_1 = pN_1/\pi = 0.0127 \times 17/\pi = 0.0687$ m. Therefore:

$$H = \frac{n_1 p N_1}{60} \left[\frac{P}{C} - zq \left(\frac{\pi d_1 n_1}{60} \right)^2 \right]$$

$$= \frac{1200 \times 0.0127 \times 17}{60} \left[\frac{2632}{2} - 2 \times 0.70 \times \left(\frac{\pi \times 0.0687 \times 1200}{60} \right)^2 \right]$$

$$= 5499 \text{ W or } 5.5 \text{ kW}$$

Example 13.11 Chain Design (Design)   Design a chain drive for an agitator, driving sprocket speed 1200 rpm, power $H = 12$ kW, output speed 600 rpm, $C = 2$.

Solution We start with $N_1 = 17$. From Table 13.8, for the ANSI 40 chain, $p = 19.05$ mm, $q = 1.25$ kg/m, $z = 1$. From Equation (13.76):

$$\frac{H}{n_1 p N_1/60} + zq \left(\frac{n_1 p N_1}{60}\right)^2 \leq \frac{P}{F_s}$$

Solving for P:

$$P = F_s \frac{H}{n_1 p N_1 /60} + zq \left(\frac{n_1 p N_1}{60}\right)^2$$

$$= 2 \frac{12,000}{1,200 \times 0.0127 \times 17/60} + 1 \times 0.70 \left(\frac{1200 \times 0.0127 \times 17}{60}\right)^2 = 3810 \text{ N}$$

which is greater than the single-strand strength of 2442 N (Table 13.8). We try a double strand, $z = 2$:

$$P = C \frac{H}{n_1 p N_1/60} + zq \left(\frac{n_1 p N_1}{60}\right)^2$$

$$= \frac{12,000}{1,200 \times 0.0127 \times 17/60} + 1 \times 0.70 \left(\frac{1200 \times 0.0127 \times 17}{60}\right)^2 = 3915 \text{ N}$$

From Table 13.8, the #60 chain has strength for double strand $P = 4884$ N, thus the ANSI No. 60–double-strand chain is adequate. The pitch diameters are:

$$d_1 = \frac{p N_1}{\pi} = \frac{0.01905 \times 17}{\pi} = 0.103 \text{ m} \qquad d_2 = d_1 \frac{n_1}{n_2} = 0.103 \frac{1200}{600} = 0.206 \text{ m}$$

The number of teeth of the driven sprocket is:

$$N_2 = N_1 \frac{n_1}{n_2} = 17 \frac{1200}{600} = 34$$

Example 13.12 Develop a spreadsheet for the design of roller chains. For application, design a chain drive for an agitator, driving sprocket speed 1200 rpm, power $H = 12$ kW, output speed 600 rpm, $C = 2$.

Solution We select $N_1 = 17$. In the following spreadsheet, we have computed the required strength and printed it on the right of Table E13.12 for comparison with the chain strength. We observe that there are two possible solutions: the ANSI #60, double-strand chain has required strength 3915 N and available strength 4884 N. Moreover, the ANSI #80, single-strand chain has required strength 3182 N and available strength 3569 N. The latter solution, however, violates the maximum rpm allowed in Table E13.12, which is 1000 rpm for the ANSI #80 chain.

TABLE E13.12

Example 13.12. Design of a roller chain

DATA:

Service Factor	$C =$	2
Power, W	$H =$	12000
Speed of driving sprocket, RPM	$n_1 =$	1200
Number of teeth of driving sprocket	$N_1 =$	17

									RESULTS	
					STRENGTH OF CHAIN (capacity)			REQUIRED STRENGTH (demand)		
					Number of strands			Number of strands		
ANSI No	ISO No	p, mm	max RPM	q, kg/m	1	2	3	1	2	3
	05B	8	2000	0.18	380.8101	662.2785	943.7468	8826.193	8828.856	8831.52
35	06B	9.525	2000	0.14	753.3418	1432.177	2102.734	7413.775	7416.712	7419.648
40	08B	12.7	1600	0.7	1506.684	2632.557	4520.051	5584.232	5610.335	5636.438
50	10B	15.88	1250	0.95	1879.215	3758.43	5637.646	4501.856	4557.208	4612.561
60	12B	19.05	1250	1.25	2442.152	4884.304	7326.456	3810.298	3915.177	4020.056
80	16B	25.4	1000	2.7	3568.025	7136.051	10704.08	3181.799	3584.534	3987.268
100	20B	31.75	800	3.6	5447.241	10894.48	16341.72	3062.282	3901.313	4740.343
120	24B	38.1	630	6.7	8261.924	16523.85	24785.77	4101.312	6349.913	8598.515
140	28B	44.45	400	8.3	10894.48	21788.96	32683.44	5379.523	9171.009	12962.49
160	32B	50.8	400	10.5	14272.1	28544.2	42816.3	7654.294	13919.06	20183.82
200	40B	63.5	250	16	22161.49	44322.99	66484.48	16027.72	30943.82	45859.92
240	48B	76.2	160	25	33801.04	67602.08	101403.1	34487.58	68048.8	101610

CASE STUDY 13.1: Design of a Two-Speed Coiler Gear Box[1]

Peter Sawchuck, a design engineer, was given the task of designing a two speed coiler drive gear box. The specifications are shown in the following work order:

WORK ORDER

Date: January 9th, 1972

Work Order No.:   936852

Customer: Colossal Consultants Inc.

Quantity: 2

Description: 100 H.P. Two Speed Coiler Drive Gear Box

DESIGN SPECIFICATIONS

Prime Mover: 100 H.P. DC Motor

Design Speed: 850 r.p.m.

Maximum Speed: 2000 r.p.m.

COILER REDUCER

Dual output speed gear reducer   38.7 & 12.9 output r.p.m.

Gear ratios: 66 to 1 and 22 to 1

Mechanical rating: 100 H.P. at AGMA S.F. 1.0

Thermal rating: 100 H.P.

Gear shifter: Manual

Length of low speed shaft: 58"

Shaft to support: 52" wide x 72" dia. coil maximum 30,000 lbs.

SPECIAL INSTRUCTIONS: —

These units are part of a cold rolling mill being supplied to Brazilian
Enterprises. It is expected that at a later date additional units will
be required. Gear set is to be designed to AGMA specifications.
Preliminary sketch of gear box is attached based on a previous design.
Foundation mounting will be in accordance with these dimensions.

Figure CS13.1

For steel and rolling mills, coilers are used to reel and unreel the steel coils. Their characteristic is low speed and consequently large size and high cost. Thus, low cost is a must in this design.

Helical gears are a logical choice for an efficient and compact drive. Further, single helical gears were selected instead of double or herringbone helical gears. They have the disadvantage of thrust loads but are simpler and more economical to manufacture.

---

[1] Based on Case Study ECL 255 of the Engineering Case Library, to which the reader is referred for further details. Originally prepared by G. Kardos of Carleton University. By permission of the trustees of the Rose-Hulman Institute of Technology. Names and places have been disguised to preserve anonymity.

A triple reduction gear reducer was decided upon. At maximum speed the total gear ratio is 22, and at lower speed it is 66. Thus, the first shaft will have design speed 850 rpm, the second 207 rpm, the third 57.2 rpm, and the fourth either 38.7 or 12.9 rpm. The speed of the fourth shaft will be controlled by engaging selectively a 1.48 or a 4.4 gear ratio pair of helical gears by way of a clutch.

Next, the designer sized the gears using the bending strength and the surface strength formulas. He selected 20° pressure angle. For the high-speed gears 1 and 2, he selected SAE 4140 steel material surface hardened to 300 BHN for the gear and SAE 43L40 chrome–moly–nickel steel, through-hardened to 310-34 BHN for the pinions. For the low-speed gears 3 and 4 he selected material SAE 41L40 steel. Due to their large size, they were fabricated and mild steel A36 was used for the hubs. After the welding, stress relieving was applied at 850–900°F.

The results are summarized in Table CS13.1 (dimensions in in.).

The shafts were designed using the AGMA recommendations (AGMA code 260.01), which imposed limitations on shaft deflection to avoid gear misalignment.

For the thrust, SKF 6314 deep-groove ball bearings were selected and the resulting bearing life was $L_{10} = 32,700$ hours.

The housing was fabricated out of ASTM A-36 mild steel.

The speed reducer would operate safely to a 200°F temperature. Assuming a 6% loss, the heat generated in the gear box is:

$$Q = 100 \times 0.06 \times 42.41 = 254.6 \text{ Btu/min or } 0.06 \times 100 = 6 \text{ hp}$$

The exposed area of the housing is 67.5 ft². Using a heat transfer coefficient of 0.0017 hp/(ft² °F):

$$T = T_0 + \Delta T = 85 + \frac{6}{0.0017 \times 67.5} = 137°F$$

which is acceptable.

A #4 EP (extreme pressure) oil was selected with a viscosity of 700–1000 SUV at 100°F.

A sketch of the gear box is shown in Figure CS13.2.   □

TABLE CS13.1

Description	Pinion #1	Gear #1	Pinion #2	Gear #2	Pinion #3	Gear #3	Pinion #4	Gear #4
Speed, rpm	850	207	207	57.2	57.2	38.7	57.2	12.9
No. of teeth	20	83	26	94	44	65	20	88
Pitch diameter	6.99	29.01	5.41	19.58	7.67	11.33	3.52	15.48
Face width	10.5		5.5		4.5			
Helix angle	17		16		7		18	

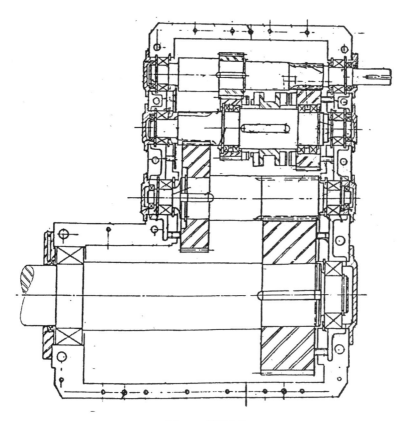

Figure CS13.2

CASE STUDY 13.2: The Case of the Runaway Lift[2]

An electrically powered traction elevator moved at 350 fpm speed and operated in an 18-story hotel. While at the 9th floor it suddenly accelerated upwards and the cabin hit the roof of the elevator well. The two passengers in the elevator suffered severe injuries.

Investigation

The elevator is powered by a dc motor that is rigidly coupled to a worm gear set and equipped with an externally contracting spring-operated brake between the motor and the worm gear set. The gear operates the sheave of the hoist cables. The cabin of the elevator is partially balanced; that is, there is a dead weight counterbalancing the cabin, usually weighing one-half the fully loaded cabin, to minimize the maximum power supplied by the motor. The upward acceleration was due to the fact that the cabin was very lightly loaded at the time of the accident and thus the dead weight was heavier than the cabin. The

---

[2] Based on Manning (1987), to which the reader is referred for more details. By permission of the American Society of Mechanical Engineers.

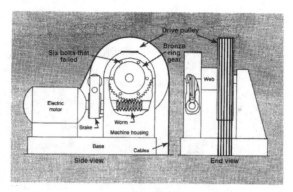

Figure CS13.3

worm is usually designed to be self-locking, providing additional safety. It is apparent that the failure was at the worm gear set so that the cabin would move out of control. Indeed, an examination of the worm gear set revealed that all six bolts attaching the gear rim to the gear web (Figures CS14.2 and CS14.3) were broken. Thus, the worm could not provide the motor braking or the self-locking action.

Examination of the broken surfaces of the bolts revealed fatigue failure in all of them and that failure occurred at the connection of the shank and head of the bolt.

The motor produced 25 hp at 1150 rpm. The worm was a two-lead, 1.0 in. pitch, 3.25 in. pitch diameter. The gear had 48 teeth.

The circumferential dynamic gear load was calculated to be 9300 lb and was assumed in the analysis to be equally shared among the six bolts, which of course is not true, as shown in Chapter 8. The separating gear load was 3200 lb, assumed to be carried by the gear fit. The axial load of the gear was 1940 lb, assumed to be carried by each bolt as it passed through the worm area. This latter load was the dynamic load on the bolt, in addition to the prestress load. The manufacturer specified prestress 75% of the bolt material yield stress.

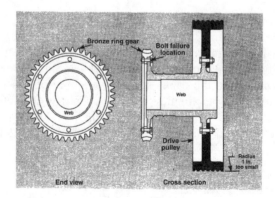

Figure CS13.4

Analysis of the bolt fatigue strength revealed a safety factor less than 0.6. Thus, a failure like this would be expected.

Redesign of the bolts for adequate fatigue strength would be straight-forward. □

REFERENCES

American Gear Manufacturers Association Standards.

Buckingham, E. 1949. *Analytical Mechanics of Gears.* New York: McGraw-Hill.

Decker, K.-H. 1973. *Maschinenelemente.* Munich: C. Hanser Verlag.

Dolan, T. J., and E. L. Broghamer. 1942. "A Photoelastic Study of Stresses in Gear Tooth Profiles." Bull. No 335, Eng. Exp. Sta., Univ. of Illinois, Urbana, Ill.

Dudley, D. W., ed. 1962. *Gear Handbook.* New York: McGraw-Hill.

Engel, L., and H. Klingele. 1981. *An Atlas of Metal Damage.* Munich: C. Hanser Verlag.

Lewis, W. 1892. *Investigation of the Strength of Gear Teeth.* Philadelphia: Engineer's Club of Philadelphia.

Manning, L. 1987. "The Case of the Runaway Lift." *Mechanical Engineering* 109, no. 5 (May): 62–67.

Mott, R. L. 1999. *Machine Elements in Machine Design.* Upper Saddle River, N.J.: Prentice-Hall.

Shigley, J. E., and J. J. Vicker. 1980. *Theory of Machines and Mechanisms.*

Thomas, A. K., and V. Chatchut. 1957. *Die Tragfähigkeit der Zahnräder* [The Strength of Gears]. 3rd ed. Munich: C. Hanser Verlag.

ADDITIONAL READING

British Petroleum. *Lubrication Theory and Application.* London: BP Trading Co., 1969.

Mechanical Drives: Reference Issue. *Machine Design.* Cleveland: Penton/IPC, 1978.

Niemann, G. *Maschinenelemente.* Berlin: Springer, 1965.

PTC Corp. *Link-belt, Roller and Silent Chains.* Atlanta: PTC, 1987.

Shigley, J. E., and C. R. Minschke. *Mechanical Engineering Design*, 5th ed. New York: McGraw-Hill, 1989.

Wills, J. G. *Lubrication Fundamentals.* New York: Marcel Dekker, 1980.

PROBLEMS[3]

13.1. [C] The standard 20° involute spur gears shown (Figure P13.1) transmit power $P = 16$ kW at pinion speed 3500 rpm. The pinion is on the top and the gear ratio is $R_{12} = 4$. Determine the gear interaction forces and the bearing loads. Draw force and bending moment diagrams for the shafts.

---

[3] [C] = certification, [D] = design, [N] = numerical, [T] = theoretical problem.

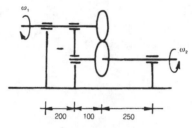

Figure P13.1

13.2. [C] In the system shown in Figure P13.2, the lower gear, rotating at 1750 rpm, transmits power $P = 20$ kW. The second gear rotates at 875 rpm and also transmits power 12kW to the third gear at 730 rpm while transmitting the balance 8 kW through its shaft to the right. Determine the gear forces and bearing rotations.

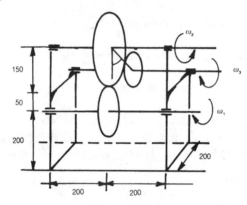

Figure P13.2

13.3. [C] A spur gear drive (Figure P13.3) has power input at the upper gear $H = 6$ kW at 2800 rpm and output speed 3600 rpm at the lower gear. The center distance is 100 mm. Determine the gear and bearing forces.

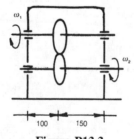

Figure P13.3

13.4. [C] A spur gear drive (Figure P13.4) has power input at the upper gear $H = 12$ kW at 3400 rpm and output speed 1700 rpm at the lower gear. The center distance is 120 mm. Determine the gear and bearing forces.

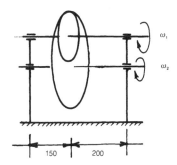

Figure P13.4

13.5. [C] A spur gear drive (Figure P13.5) has power input at the upper gear $H = 12$ kW at 3000 rpm and output speed 1500 rpm at the lower gear. The center distance is 80 mm. Determine the gear and bearing forces.

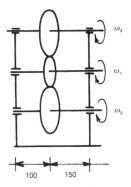

Figure P13.5

13.6. [C] Using the program GEARPLOT, plot a pair of spur gears with $N_1 = 28$, $N_2 = 56$, $m = 8$ mm. Also make a sketch of the gearing and draw the pressure line and contact limits.

13.7. [C] Using the program GEARPLOT, plot a pair of spur gears with $N_1 = 16$, $N_2 = 16$, $m = 8$ mm. Also make a sketch of the gearing and draw the pressure line and contact limits.

13.8. Using the program GEARPLOT, plot a pair of spur gears with $N_1 = 20$, $N_2 = 20$, $m = 10$ mm. Also make a sketch of the gearing and draw the pressure line and contact limits. Draw the teeth for a rotation by half pitch.

13.9. [C] Using the program GEARPLOT, plot a pair of spur gears with $N_1 = 16$, $N_2 = 1000$, $m = 10$ mm, to approximate a pinion–rack system. Also make a sketch of the gearing and draw the pressure line and contact limits.

13.10. [C] Using the program GEARPLOT, plot a pair of gears with $N_1 = 8$, $N_2 = 8$, $m = 10$ mm. Also make a sketch of the gearing and draw the pressure line and contact limits. Comment on interference.

13.11. [C] A spur gear drive has $d_1 = 80$ mm, $N_1 = 17$ teeth, $d_2 = 160$ mm, gear width $b = 50$ mm, pinion speed $n_1 = 1750$ rpm. Carbon steel is used SAE 1020 heat-treated and surface-hardened to 220 BHN. The operation is smooth at normal temperatures, and a reliability of 99% with 2 million cycles of rotation required life. The gears are ground, $Q_v = 10$, and the bearings are at both sides of the gears. Find the maximum transmitted power.

13.12. [C] A spur gear drive has $d_1 = 100$ mm, $N_1 = 17$ teeth, $N_2 = 51$ teeth, gear width $b = 60$ mm, pinion speed $n_1 = 3500$ rpm. The gears are made of carbon steel SAE 1020 heat-treated and surface-hardened to 170 BHN. The operation is smooth at normal temperatures, and a reliability of 99% with 2 million cycles of rotation required life. The gears are machined, $Q_v = 7$, and the bearings are at both sides of the gears. Find the maximum transmitted power.

13.13. [C] A spur gear drive has center distance $a = 160$ mm, $N_1 = 17$ teeth, $N_2 = 51$ teeth, gear width $b = 60$ mm, pinion speed $n_1 = 3000$ rpm. Carbon steel is used SAE 1020 heat-treated and surface-hardened to 220 BHN. The operation is smooth at normal temperatures, and a reliability of 99% with 2 million cycles of rotation required life. The gears are machined, $Q_v = 7$, and the bearings are at both sides of the gears. Find the maximum transmitted power.

13.14. [C] A 6 kW spur gear drive has $d_1 = 80$ mm, $N_1 = 17$ teeth, $d_2 = 160$ mm, gear width $b = 50$ mm, pinion speed $n_1 = 1750$ rpm. If the pinion speed is reduced to 900 RPM, find the maximum transmitted power, all other operating requirements remaining the same. The gears are machined with $Q_v = 7$.

13.15. [C] A spur gear drive has $d_1 = 80$ mm, $N_1 = 17$ teeth, $d_2 = 160$ mm, gear width $b = 50$ mm, pinion speed $n_1 = 1750$ rpm. Carbon steel is used SAE 1020 heat-treated and surface-hardened to 220 BHN. The operation is smooth at normal temperatures, and a reliability of 99% with 2 million cycles of rotation required life. The gears are machined to quality $Q_v = 6$, and the bearings are at both sides of the gears. To increase the power transmitted, it was decided to make the gears with precision grounding after machining to quality $Q_v = 10$. Find the increase in power of the drive that was achieved with the change.

13.16. [D] The spur drive shown in Figure P13.1 transmits 15 kW at 1450 rpm. Carbon steel is used SAE 1020 heat-treated and surface-hardened. The operation is smooth, at normal temperatures, and a reliability of 99% with 1000 hours operation is required. The gears will be ground, $Q_v = 10$. Design the proper gearing and draw a sketch of the system.

13.17. [D] The spur drive shown in Figure P13.2 transmits 15 kW at 1870 rpm. Carbon steel is used SAE 1020 heat-treated and surface-hardened. The operation is with medium shocks at normal temperatures, and a reliability of 99% with 500 hours operation is required. The gears will be ground, $Q_v = 10$. Design the proper gearing and draw a sketch of the system. Then redesign for 1000 hours operation and comment on the difference.

13.18. [D] The spur gear drive shown in Figure P13.3 transmits 15 kW at 1450 rpm. Carbon steel is used SAE 1020 heat-treated and surface-hardened. The operation is smooth, at normal temperatures, and a reliability of 99% with 500 hours operation is required. The gears will be ground, $Q_v = 10$. Design the proper gearing and draw a sketch of the system. Redesign for 99.99% reliability and comment on the difference.

13.19. [D] A spur gear drive is shown in Figure P13.4. Carbon steel is used SAE 1020 heat-treated and surface-hardened. The operation is smooth at normal temperatures, and a reliability of 99% with long operation is required. The gears will be ground, $Q_v = 10$. Design the proper gearing and draw a sketch of the system.

13.20. [D] The dual drive shown in Figure P13.5 will have spur gears made of carbon steel SAE 1020 heat-treated and surface-hardened. The operation is smooth at normal temperatures, and a reliability of 99% with long operation is required. The spur gears will be machined, $Q_v = 7$. Design the proper gearing and draw a sketch of the system.

13.21. [D] A spur gear pair is to be designed for the second gear of an automotive transmission with rear ratio $R = 1.5$ and input power 50 kW at 3500 rpm. It is required that the center distance be approximately $a = 90$ mm. For 1000 hours operation and 99.9% reliability, determine the required material properties. Select a proper material and specify the required heat treatment.

13.22. [D] A spur gear pair is to be designed for the power pack of an overhead crane with gear ratio $R = 3$ and input power 12 kW at 1200 rpm. It is required that the module be 5 mm and the gear width $b = 40$ mm. For 4000 hours operation and 99.9% reliability, determine the required material properties. Select a proper material and specify the required heat treatment.

13.23. [D] A spur gear pair is to be designed for the drive of a machine tool with gear ratio $R = 8$ and input power 20 kW at 3500 rpm for occasional operation, equally in forward and backward motion and totaling no more than 50 hours. Because it will mesh with forward and backward gears, the pinion diameter should be approximately $d = 80$ mm. For 99% reliability, determine the required material properties. Select a proper material and specify the required heat treatment.

13.24. [D] For the system of Problem 13.22, determine the width if the material for pinion and gear is SAE 1020 surface-hardened to BHN = 200.

13.25. [D] For the system of Problem 13.23, determine the width if the material for pinion and gear is SAE 1020 surface-hardened to BHN = 200.

13.26. [C] A helical gear drive has helix angle $\psi = 30°$, $d_1 = 80$ mm, $N_1 = 17$ teeth, $d_2 = 160$ mm, gear width $b = 50$ mm, pinion speed $n_1 = 1750$ rpm. Carbon steel is used SAE 1020 heat-treated and surface-hardened to 220 BHN. The operation is smooth at normal temperatures, and a reliability of 99% with 2 million cycles of rotation required life. The gears are ground, $Q_v = 10$, and the bearings are at both sides of the gears. Find the maximum transmitted power.

13.27. [C] A helical gear drive has helix angle $\psi = 35°$, $d_1 = 100$ mm, $N_1 = 17$ teeth, $N_2 = 51$ teeth, gear width $b = 60$ mm, pinion speed $n_1 = 3500$ rpm. The gears are made of carbon steel SAE 1020 heat-treated and surface-hardened to 170 BHN. The operation is smooth at normal temperatures, and a reliability of 99% with 2 million cycles of rotation required life. The gears are machined, $Q_v = 7$, and the bearings are at both sides of the gears. Find the maximum transmitted power.

13.28. [C] A helical gear drive has helix angle $\psi = 30°$, center distance $a = 160$ mm, $N_1 = 17$ teeth, $N_2 = 51$ teeth, gear width $b = 60$ mm, pinion speed $n_1 = 3000$ rpm. Carbon steel is used SAE 1020 heat-treated and surface-hardened to 220 BHN. The operation is smooth, at normal temperatures, and a reliability of 99% with 2 million cycles of rotation required life. The gears are machined, $Q_v = 7$, and the bearings are at both sides of the gears. Find the maximum transmitted power.

13.29. [C] A 6 kW helical gear drive has helix angle $\psi = 25°$, $d_1 = 80$ mm, $N_1 = 17$ teeth, $d_2 = 160$ mm, gear width $b = 50$ mm, pinion speed $n_1 = 1750$ rpm. If the pinion speed is reduced to 900 rpm, find the maximum transmitted power, all other operating requirements remaining the same. The gears are machined with $Q_v = 7$.

13.30. [C] A helical gear drive has helix angle $\psi = 20°$, $d_1 = 80$ mm, $N_1 = 17$ teeth, $d_2 = 160$ mm, gear width $b = 50$ mm, pinion speed $n_1 = 1750$ rpm. Carbon steel is used SAE 1020 heat-treated and surface-hardened to 220 BHN. The operation is smooth at normal temperatures, and a reliability of 99% with 2 million cycles of rotation required life. The gears are machined and the bearings are at both sides of the gears. To increase the power transmitted, it was decided to make the gears with precision grounding after machining to quality $Q_v = 10$. Find the increase in power of the drive that was achieved with the change.

13.31. [D] The helical drive shown in Figure P13.1 transmits 15 kW at 1450 rpm. Carbon steel is used SAE 1020 heat-treated and surface-hardened. The helix angle $\psi = 30°$. The operation is smooth at normal temperatures, and a reliability of 99% with 1000 hours operation is required. The gears will be ground, $Q_v = 10$. Design the proper gearing and draw a sketch of the system.

13.32. [D] The helical drive shown in Figure P13.2 transmits 15 kW at 1870 rpm. The helix angle $\psi = 20°$. Carbon steel is used SAE 1020 heat-treated and surface-hardened. The operation is with medium shocks at normal temperatures, and a reliability of 99% with 500 hours operation is required. The gears will be ground, $Q_v = 10$. Design the proper gearing and draw a sketch of the system. Then redesign for 1000 hours operation and comment on the difference.

13.33. [D] The helical gear drive shown in Figure P13.3 transmits 15 kW at 1450 rpm. Carbon steel is used SAE 1020 heat-treated and surface-hardened. The helix angle $\psi = 25°$. The operation is smooth at normal temperatures, and a reliability of 99% with 500 hours operation is required. The gears will be ground, $Q_v = 10$. Design the proper gearing and draw a sketch of the system. Redesign for 99.99% reliability and comment on the difference.

13.34. [D] A helical gear drive is shown in Figure P13.4. Carbon steel is used SAE 1020 heat-treated and surface-hardened. The helix angle $\psi = 30°$. The operation is smooth at normal temperatures, and a reliability of 99% with long operation is required. The gears will be ground, $Q_v = 10$. Design the proper gearing and draw a sketch of the system.

13.35. [D] The dual drive shown in Figure P13.5 will have helical gears made of carbon steel SAE 1020 heat-treated and surface-hardened. The helix angle $\psi = 28°$. The operation is smooth at normal temperatures, and a reliability of 99% with long operation is required. The helical gears will be machined, $Q_v = 7$. Design the proper gearing and draw a sketch of the system.

13.36. [D] A bevel gear drive must transmit at right angles 20 kW at $n = 600$ rpm, gear ratio $R = 3$, and minimum pinion teeth 20. Design the gears for medium shocks, 2000 hours operation at 99% reliability with cast iron material.

13.37. [D] A bevel gear drive must transmit at right angles 20 kW at $n = 2500$ rpm, gear ratio $R = 2.5$, and module $m = 8$. Design the gears for medium shocks, 2000 hours operation at 99% reliability with carbon steel SAE 1015 material.

13.38. [D] A bevel gear drive must transmit at right angles 20 kW at $n = 600$ rpm, gear ratio $R = 3$, and maximum pinion diameter $d = 90$ mm. Design the gears for medium shocks, 2000 hours operation at 99% reliability. Specify material with the required heat and surface treatment.

13.39. [D] A bevel gear drive must transmit at right angles 20 kW at $n = 600$ rpm, gear ratio $R = 3$. The tooth width must not exceed 40 mm. Design the gears for medium shocks, 2000 hours operation at 99% reliability with cast iron material.

13.40. A vertical drill must transmit at right angles torque 400 Nm at $n = 500$ rpm to the vertical gear from a horizontal pinion rotating at 1500 rpm. Design the gears for medium shocks, 2000 hours operation at 99% reliability with cast iron material.

13.41. [D] Design a worm gear drive for input power 15 kW at 2000 rpm, with center distance $a = 200$ mm, transmission ratio $R = 15$, number of worm starts (leads) $N_1 = 2$ and module $m = 5$. The worm must be made of carbon steel heat-treated and surface-hardened. A gear bronze must be selected for the gear. Determine the efficiency and the bearing forces making reasonable assumptions for the enclosure dimensions for a self-contained drive. Make a sketch for the general arrangement if the worm is on top of the gear.

13.42. [D] Design a worm gear drive for input power 22 hp at 3700 rpm with approximate diameters $d_1 = 50$, $d_2 = 150$ mm, transmission ratio $R = 14$ and number of worm starts (leads) $N_1 = 2$. The worm must be made of carbon steel heat-treated and surface-hardened. A gear bronze must be selected for the gear. Determine the efficiency and the bearing forces making reasonable assumptions for the enclosure dimensions for a self-contained drive. Make a sketch for the general arrangement if the worm is on top of the gear.

13.43. [D] Design a worm gear drive for input power 12 kW at 3000 rpm with worm diameter 50 mm, transmission ratio $R = 15$, number of worm starts (leads) $N_1 = 2$, and module $m = 10$. The worm must be made of carbon steel heat-treated and surface-hardened. A gear bronze must be selected for the gear. Determine the efficiency and the bearing forces making reasonable assumptions for the enclosure dimensions for a self-contained drive. Make a sketch for the general arrangement if the worm is on top of the gear.

13.44. [D] Design a worm gear drive for input power 20 kW at 3550 rpm with transmission $R = 30$, number of worm starts (leads) $N_1 = 2$, number of gear teeth $N_2 = 60$. The worm must be made of carbon steel heat-treated and surface-hardened. A gear bronze must be selected for the gear. Determine the efficiency of the bearing forces making reasonable assumptions for the enclosure dimensions for a self-contained drive. Make a sketch for the general arrangement if the worm is on top of the gear.

13.45. [D] Design a worm gear drive for output power 10 kW at 200 rpm with center distance $a = 200$ mm, transmission ratio $R = 15$, number of worm starts (leads) $N_1 = 2$, and gear teeth $N_2 = 30$. The worm must be made of carbon steel heat-treated and surface-hardened. A gear bronze must be selected for the gear. Determine the efficiency and the bearing forces making reasonable assumptions for the enclosure dimensions for a self-contained drive. Make a sketch for the general arrangement if the worm is on top of the gear.

13.46. [C] A roller chain drive has a double-strand, ANSI #30 chain. The driving sprocket has 15 teeth, speed of rotation 1300 rpm, and the driven sprocket has 30 teeth. Find the rated power for service factor $C = 1.5$.

13.47. [C] A roller chain drive has a single-strand, ANSI #120 chain. The driving sprocket has 19 teeth, speed of rotation 700 rpm, and the driven sprocket rotates at 300 RPM. Find the rated power for service factor $C = 2.5$.

13.48. [C] A roller chain drive has a triple-strand, ANSI #50 chain. The driving sprocket has 21 teeth, speed of rotation is 1300 rpm, and the driven sprocket has 30 teeth. Find the rated power for service factor C, the sprocket diameters, and the chain length and total weight.

13.49. [C] A roller chain drive has a triple-strand, ANSI #50 chain. The driving sprocket has 17 teeth, speed of rotation is 800 rpm, and the driven sprocket has 30 teeth. Find the sprocket diameters, the chain length and total weight, and the cordal speed variation.

13.50. [C] A roller chain drive has a single-strand, ANSI #120 chain. The driving sprocket has 13 teeth, speed of rotation is 800 rpm, and the driven sprocket has speed 400 rpm. Find the sprocket diameters, the chain length and total weight, and the cordal speed variation.

13.51. [D] Design a roller chain for a drive that connects an electric motor to a mixing machine with the following operational requirements: power $H = 15$ kW, input speed 900 rpm, output speed 400 rpm. Then find the sprocket diameters, the chain length and total weight, and the cordal speed variation.

13.52. [D] Design a roller chain for a drive driven by a 4-cylinder diesel engine, and driving a wood processing mill, and having the following operational requirements: power $H = 20$ kW, input speed 1100 rpm, output speed 600 rpm. Then find the sprocket diameters, the chain length and total weight, and the cordal speed variation.

13.53. [D] Design a roller chain for a drive driven by an electric motor, driving a machine tool, and having the following operational requirements: power $H = 10$ kW, input speed 1100 rpm, output speed 800 rpm. Then find the sprocket diameters, the chain length and total weight, and the cordal speed variation.

13.54. [D] Design a roller chain for a drive driven by a 6-cylinder diesel engine, driving a food processing machine, and having the following operational requirements: power $H = 5$ kW, input speed 1400 rpm, output speed 800 rpm. Then find the sprocket diameters, the chain length and total weight, and the cordal speed variation.

13.55. [D] Design a roller chain for a drive driven by an electric motor, driving a hoist, and having the following operational requirements: power $H = 3$ kW, input speed 1500 rpm, output speed 1000 rpm. Then find the sprocket diameters, the chain length and total weight, and the cordal speed variation.

CHAPTER 14

DESIGN FOR TORSION: SHAFTS, COUPLINGS, AND PINS

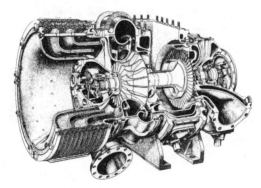

Exhaust gas turbocharger for a ship's diesel engine. One of the most critical applications of shafts is in turbomachinery.

OVERVIEW

In Chapters 8 and 9, we have developed design methods based mostly on material strength and flexibility, respectively. In Chapter 10, we studied design methods based on failures due to dry friction, with applications to clutches and brakes. In Chapter 11 we studied machine element design methods based on the prevention of failures related to hydrostatic and hydrodynamic lubrication, the use of a flowing fluid to prevent the rubbing surfaces from touching one another during sliding friction. In Chapter 12, we studied the design of lubricated elements with rolling friction. In Chapter 13, we studied gears where all the above design considerations apply.

There is a true connection among all these elements: they are usually placed on rotating shafts that transmit power from one element to another and from one machine to another one or more. Design of shafts will be the subject of this chapter.

14.1. ROTARY MOTION

Rotary motion is, by its nature, very advantageous for use in machines. Almost invariably, machines have members with rotary motion.

The main advantage of rotary motion over linear motion is the absence of speed variation or reversal associated with high acceleration (or deceleration), inertia loads, and finally inertia stresses. However, this is not always the case. Besides the centrifugal inertia loads and stresses, there are factors that change the speed of rotation and cause acceleration and dynamic loads, such as start-up and coasting-down, intermitted loads, and accidental or intended braking.

The static torque for a transmitted power P at rotating speed is n (see Chapter 3): *Uniform speed torque, or*

$$\text{Static torque} \Rightarrow T_0(\text{Nm}) = \frac{7121P(\text{hp})}{n(\text{rpm})} \tag{14.1}$$

If the transmitted power is W (watts) and the angular velocity is ω (rad/s), then:

$$T_0(\text{Nm}) = \frac{P(\text{W})}{\omega(\text{rad/s})} = \frac{9549P(\text{kW})}{n(\text{rpm})} \tag{14.2}$$

Dynamic loads can be computed numerically in most cases, by simulation techniques on models such as in Figure 14.1.

In the absence of the required information for analytical evaluation for dynamic loads, for preliminary design the static torque is increased to account for these loads with the empirical formula

Dynamic loads w/o starting torque

$$T_d = T_0(c_1 + c_2) \tag{14.3}$$

where c_1 depends on the driving machine and c_2 depends on the driven machine (Table 13.9).

Starting torques are not included in the factors of Table 13.9. They must be accounted for separately because sometimes they can reach very high values.

Factors determining the starting torque are the start-up time and the inertia of the system. The starting torque is also limited by the starting characteristics of the

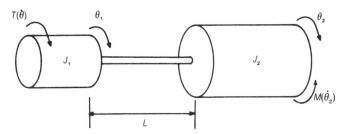

Figure 14.1. Rotors of the driving and driven machines connected with a shaft, a typical machine configuration.

driving machine. For most driving machines, the maximum torque is known. Because in many cases this torque is high, the design objective is to increase the start-up time in order to have lower starting torque. Then the starting torque, assumed constant, is, for constant acceleration motion:

$$T_{su} = \frac{\lambda J \omega}{t} + T_f \tag{14.4}$$

where J is the rotating system inertia, ω is the operating angular velocity, t is the start-up time, T_f is the constant friction torque, and λ is a factor expressing the deviation from the assumed constant acceleration. If no measurements are available, λ must be taken between 1.2 and 2.

This approach yields acceptable results for compact machines designed with high safety factors. In recent years, computer simulation has been used extensively, yielding very accurate results provided that the torque–speed characteristics for the driving machine and the resisting torque are known. In most situations, knowledge of this information can be assumed and we can then proceed with the system modeling and simulation, which gives detailed information for the dynamic loads for the system.

Torque–speed characteristics are usually available from the motor manufacturer and have the forms indicated in Figure 14.2. For typical induction motors, Figures 14.3 and 14.4 give the torque versus speed of rotation.

14.2. SHAFT DESIGN

14.2.1. Fundamentals

Shafts are solids of revolution intended for transmitting a torque along their axes and supporting rotating machine components. Because the transmission of torque is associated with the development of forces applied to the shafts, such as forces acting on the teeth of gears and belt tension, shafts are usually subjected to transverse forces and bending moments in addition to the torque. Shafts transmitting

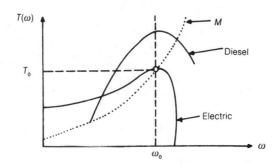

Figure 14.2. Torque characteristics of prime movers.

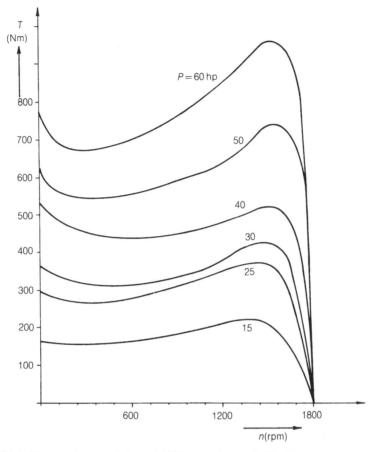

Figure 14.3. Torque characteristics of 1800 rpm, four-pole/60 Hz asynchronous motors.

forces and moments between their two ends without intermediate loads are known as *axles.*

Shafts can be classified with respect to their purpose; as transmission shafts carrying drive members, such as toothed gears, pulleys, chain sprockets, and clutches (Figure 14.5*a, b*), and as main shafts of machines and other special shafts that, in addition to drive members, carry the operating members of engines of machine tools. Such members may include turbine wheels and disks and cranks (Figure 14.5*c, d, e*).

With respect to the shape of their geometric axes, shafts are classified as *straight shafts* and *crankshafts.* Crankshafts (Figure 14.5*e*) are used to convert reciprocating motion into rotary motion or vice versa. They combine the functions of ordinary shafts with those of cranks in slider-crank mechanisms. In a separate group are shafts with a geometric axis of variable shape (Figure 14.5*f, g*).

The supporting sections of shafts are called *journals* (Figure 14.6).

As to their shape, straight shafts can be plain (constant-diameter) shafts or stepped shafts. The diameter of a shaft along its length is determined by the load

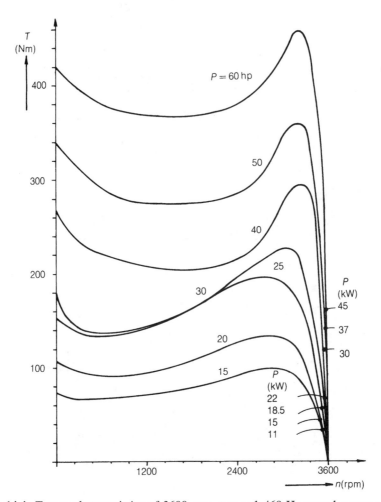

Figure 14.4. Torque characteristics of 3600 rpm, two-pole/60 Hz asynchronous motors.

distribution, bending moments and torques, axial loads, and conditions imposed by the manufacturing and assembly processes used.

Equal-strength shafts are preferable to shafts of constant cross-section. Usually they are of stepped design. Such a shape is convenient in manufacture and assembly. The shoulders of the shaft can carry large axial forces.

Shafts may be hollow in design. A hollow shaft with a hole-to-outside-diameter ratio of 0.75 is lighter by about 50% than a solid shaft of equal strength and rigidity.

The endurance of shafts is influenced by stress concentration. For this reason, special design and processing methods to raise the endurance of shafts are employed. Design methods for raising the endurance of shafts at the seating (fit) surfaces by reducing edge pressure are illustrated in Figure 14.7. The endurance limit of shafts can be increased by 80–100% by strengthening the surfaces for seating hubs by work (strain) hardening (using a roll or ball burnishing process). This is effective for shafts up to 500 or 600 mm in diameter. Such hardening procedures and shot peening are widely used.

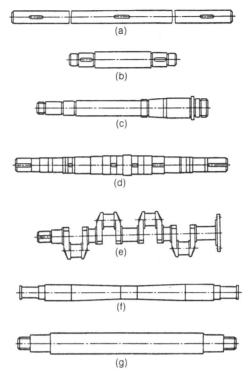

Figure 14.5. Different shaft designs.

Axial loads acting on shafts or on components mounted on the shafts are transmitted as follows (Figure 14.8):

1. *Heavy loads:* Transmission by having the loaded components bear against shoulders of the shaft, by mounting the components or locating rings with an interference fit (Figure 14.8*a, b*).
2. *Medium loads:* Transmission by means of nuts, pins passing directly through the loaded components or through locating rings, or by clamp joints (Figure 14.8*c, d, e*).
3. *Light loads:* Transmission by means of set screws holding either the component or a locating ring, by clamp joints, or by snap locating rings (Figure 14.8*e, f, g*).

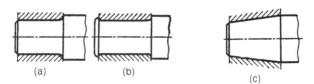

Figure 14.6. Mounting of pins.

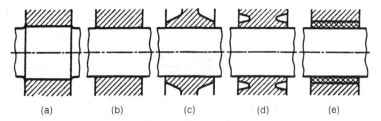

Figure 14.7. Mounting of hubs.

Two-way axial securing of a shaft is not compulsory if it is held in place by a constant force that prevents displacement (for instance, the weight of the components in the case of heavy vertical shafts).

The transition between two shaft steps of different diameters is usually designed with a fillet of a single radius or a transition surface (Figure 14.9). The radius of the fillet is to be taken as smaller than that of the edge round or radial dimension of the chamfer of the components to be mounted on the shaft step against the shoulder. It is desirable to have the fillet radius of heavily stressed shafts larger than or equal to $0.1d$.

Designing fillets of optimal shape that extend over a considerable length of the shaft makes it possible to practically eliminate stress concentration. In Figure 14.9, the optimal shapes for shafts subject to bending, torsion, and tension are shown. The transition zone extends over a length of the shaft equal to its diameter. For this reason it is possible to use such fillets only in rare cases, such as for torsion shafts (i.e., shafts serving as springs and operating in torsion) or for the free sections of heavily stressed shafts.

An effective method for increasing the strength of shafts in their transition zones is the removal of low-stressed material by providing load-relief grooves (Figure 14.9f) or by drilling an axial hole in the larger step (Figure 14.9g). This yields a more uniform stress distribution and reduces stress concentration.

Strain hardening of shaft fillets (by rolling or, for large shafts, shot peening) can raise the load-carrying capacity of shafts by 50–100%.

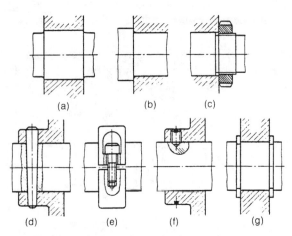

Figure 14.8. Securing hubs for axial motion.

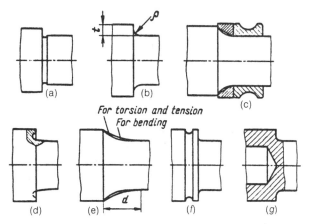

Figure 14.9. Shaft fillets for low stress concentration.

14.2.2. Shaft Materials

The selection of the material and heat treatment for shafts and axles depends upon their criteria of working capacity, including the criteria for the journals with their supports. The latter may be the determining factors in the case of sliding-friction supports.

Because of their high strength, high modulus of elasticity, and capacity to be strengthened by hardening, and the ease of obtaining the required bar stock or blanks by mill rolling, carbon and alloy steels are the chief materials for manufacturing shafts and axles.

Shafts and axles that are to satisfy the criterion of rigidity and are not to be heat-treated are made mainly of steels C5 and C6. The great majority of shafts are made of medium-carbon and alloy steels, grades 45 and 40X, which are capable of being heat-treated. Alloy steels, grades 40XH, 40XH2MA, 30X T, 30X CA, etc., are steels for highly stressed shafts of critical machinery. Shafts of these steels are usually subjected to structural improvement, hardening followed by high-temperature tempering, or induction surface hardening followed by low-temperature tempering (spline shafts).

High-speed shafts running in sleeve bearings require journals of especially high hardness. They are made of case hardening steels, grades 20X, 12XH3A, 18X T, or nitriding steels, type 38X2MIOA. Chromium-plated shafts have highest wear resistance. According to the data of automotive engineering, chromium plating of the crank-pins and main journals of crankshafts lengthens their service life up to regrinding three- to fivefold.

It should be borne in mind that high-strength heat-treated steels can be efficiently employed for shafts whose dimensions are determined by conditions of sufficient rigidity only when dictated by requirements of the durability of splines and other surfaces subject to wear.

Requirements made for the economy of metals and increased producibility lead to the manufacture of large-diameter shafts of pipes with welded or forced-on flanges, or welded roll-formed plate steel, also with welded flanges. The use of welded shafts for powerful hydraulic turbines effects a savings in metals of from 20–40%.

In addition to steels, high-strength cast iron (nodular cast iron) and inoculated cast iron are being employed to make shaped shafts: lower strength of the cast iron shafts is compensated for, to a considerable degree, by their more refined shapes (especially crankshafts), the lower susceptibility of cast iron to stress concentration, the lower sensitivity in multiple-support shafts to displacements of the supports (due to the lower modulus of elasticity), and the lower dynamic loads because of the higher damping capacity.

Steel shafts up to 150 mm in diameter are usually made of round rolled bar stock; larger shafts and shaped shafts are made of forgings.

Shafts are turned in lathes, and the searing (fit) surfaces are subsequently ground in cylindrical grinders. Highly stressed shafts are ground all over. Depending upon the class of accuracy of the bearing and its diameter, the surface finish for the seat of an antifriction bearing is assigned from the 6th through the 9th class of surface roughness, mainly the 7th and 8th class. The journals for sleeve bearings are finished to the 7th through 9th class of surface roughness and even higher, depending upon the operating conditions.

The ends of shafts should be made with chamfers to facilitate the forcing-on of mounted components, to avoid denting of the shaft surfaces and damage to the workers' hands.

14.2.3. Design Criteria

Shafts and rotating axles are usually designed as beams on hinged supports. This assumption corresponds with sufficient accuracy to the actual conditions for shafts running in antifriction bearings and having a single bearing in each support (Figure 14.10a). For shafts running in antifriction bearings and having two bearings in each support (Figure 14.10b), the support reaction is carried mainly by the bearing on

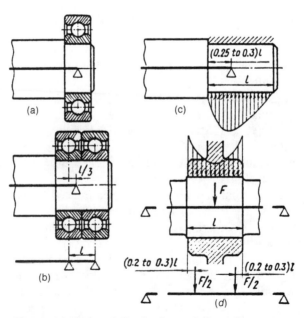

Figure 14.10. Load distribution at hubs and supports.

the side of the loaded span. The outer bearings are subject to much less load, and if they are not installed up against the inner bearings, they may be subject to a support reaction in the opposite direction. For this reason, it is more accurate to have the nominal hinged supports coincide with the middle of the inner bearings or located one-third of the distance between the bearings of each support, closer to the inner bearings. More precise calculations for such shafts should take into account the combined behavior of the shaft with the bearings as for multiple-support beams on elastic supports.

For shafts running in non-self-aligning sleeve bearings (Figure 14.10c), the pressure along the length of the bearings is nonsymmetrically distributed due to deformation of the shaft. The nominal hinged support should be located at the distance (0.25–0.3) l from the inner face of the bearing, but not over one-half of the shaft diameter from the bearing face on the side of the loaded span. More precise calculations for these shafts should take their combined behavior with the bearings into consideration.

Forces are transmitted to a shaft through the components mounted on it: toothed gears, chain sprockets, pulleys, couplings, and so on. For the simplest calculations, it is assumed that the components transmit forces and torques to the shaft at the middle of their width, and design is based on the corresponding cross-sections. Actually, forces of interaction between hubs and shafts are distributed along the length of the hubs, and the latter function in conjunction with the shafts (Figure 14.10d). It is more accurate to regard the moments as being in cross-sections at distance of (0.2–0.3) l from the faces of the hub, where l is the hub length, and to assume that the concentrated forces of interaction between the hub and shaft are in the same cross-sections. The smaller distances from the design cross-sections to the hub faces correspond to interference fits and rigid hubs; the larger distances, to clearance fits and less rigid hubs.

Of the various strength criteria for most shafts of modern high-speed machinery, endurance is of prime significance. Fatigue failures make up to 40 or 50% of the cases in which shafts become inoperative. Low-cycle fatigue may develop in operation with high overloads. For slow-speed shafts of normalized, structurally improved or hardened steels with high-temperature tempering, the limiting criterion may also turn out to be the static load capacity at peak loads (absence of impermissible permanent deformation). Finally, the criterion for the same kind of shafts, but of brittle or low-ductility materials (cast iron or low-temper steel), is the resistance to brittle failure.

Rigidity criteria of shafts are determined by the conditions of proper operation of toothed gears and bearings, as well as vibration-proof properties.

14.2.4. Design for Strength

The main loads acting on shafts are forces due to power transmission. Forces constant in magnitude and direction cause constant stresses in stationary axles, and stresses that vary in an alternating symmetrical cycle in rotating axles and in shafts. Constant loads rotating together with the axles and shafts, due, for instance, to unbalanced rotating components, cause constant stresses.

Tentative calculations to determine the diameter of a shaft, required to make a sketch of the shaft for the subsequent main calculations, are carried out by means of empirical formulas or by conditional calculations of the shaft in torsion. Thus, the end diameter of the input shaft of a reducing gear is taken from 0.8 to 1.2 of the diameter of the driving motor shaft; the diameter of the driven shaft of each stage of a spur or helical reducing gear is taken from 0.3 to 0.4 of the center-to-center distance of the stage. The diameter of the crank-pins of a crankshaft is determined by empirical formulas based on the diameter of the engine cylinder; the diameters of machine tool spindles depend empirically on the principal geometric dimension of the machine tool, etc.

If reliable formulas are not available for the given machine unit, the conditions for torsion calculations should be applied. This form of calculation is resorted to because the length dimensions of the shaft have not yet been determined, so it is impossible to compute the bending moments.

The torsion strength condition $S_{sy}/N = Tr_{max}I_p$ yields:

$$T = \frac{9549P}{n} = \frac{\pi d^3}{16} \frac{S_{sy}}{N} \tag{14.5}$$

from which an initial value of the shaft diameter can be obtained:

$$d_{in} = 36.5 \left(\frac{P}{n} \frac{N}{S_{sy}} \right)^{1/3} \tag{14.6}$$

where T is the torque (Nm), P is the power (kW), n is the speed of rotation (rpm), d is the shaft diameter (m), and S_{sy}/N is the allowable shear stress (Pa).

Because bending is not taken into account, a lower value of S_{sy}/N is used. It is often selected in the range 120–200 MPa.

For the main calculations of the strength of shafts and axles, it is necessary to determine the bending moment and torque in all the dangerous cross-sections. In calculations for shafts subject to complex loads, bending and torsional moment diagrams are constructed. If the shaft is subject to loads in different planes, the loads are usually projected onto two perpendicular planes. If the forces act in planes up to 15° from the coordinate planes, they can be assumed to coincide with the latter. To determine the resultant moment, bending moments M_x and M_y in perpendicular planes are added geometrically according to the vector addition equation:

$$M_b = (M_x^2 + M_y^2)^{1/2} \tag{14.7}$$

The several stresses are computed and the failure criterion is the applied to yield the design equation.

The dangerous cross-section is determined by the moment diagrams, size of the shaft cross-sections, and stress concentration. With some practice, the location of the dangerous cross-section can be readily determined without any calculations. In certain cases, calculations are carried out for two likely cross-sections.

The equivalent stress in the dangerous cross-section is:

$$\sigma_{eq} = (\sigma^2 + a\tau^2)^{1/2} \leq \frac{S_y}{N} \tag{14.8}$$

where σ and τ are normal and shear stresses.

Expressing the stresses in terms of the moments and assuming $a = 4$ (according to the maximum shear theory of failure), we can write:

$$\sigma_{eq} = \frac{32}{\pi d^3}(M_b^2 + T^2)^{1/2} \leq \frac{S_y}{N} \tag{14.9}$$

In calculations based on the static strength at overloads, M_b and T are to be regarded as the nominal moments multiplied by the overload factor.

Owing to their convenience, static strength calculations based on nominal stresses are efficiently applied in design to determine the diameters of axles and shafts with subsequent checking calculations based on endurance. Here, the calculations are usually carried out to the nominal moments and the allowable stresses are taken from Table 14.1. In precise calculations, the stresses are increased. Strengthening techniques enable the allowable stresses to be raised substantially.

The diameter of an axle subject to bending is determined from the preceding formula, taking $T = 0$. Thus:

$$d = \left(\frac{32M_b}{\pi S_y/N}\right)^{1/3} \tag{14.10}$$

The diameter of a shaft subject to both bending and torsion is:

$$d = \left[\frac{32(M_b^2 + T^2)^{1/2}}{\pi S_y/N}\right]^{1/3} \tag{14.11}$$

14.2.5. Fatigue Strength

Endurance calculations take into consideration the character of stress variation, static and fatigue characteristics of the materials, stress concentration, the scale factor, surface conditions, and surface hardening. These calculations are usually in the form of a check of the safety factor. Known values for calculations must include the constant (M_m and T_m) and variable (M_r and T_r) stress components.

The equivalent static stress from the Goodman diagram (see Chapter 5) is:

$$\sigma_{eq} = \sigma_m + \frac{S_u}{S_e}\sigma_r \leq \frac{S_u}{N} \tag{14.12}$$

Using Equation (14.9):

$$d = \left\{ \frac{32N}{\pi S_u} \left[(M_m^2 + T_m^2)^{1/2} + \frac{S_u}{S_e} (M_r^2 + T_r^2)^{1/2} \right] \right\}^{1/3} \qquad (14.13)$$

and for the yield line, respectively:

$$\sigma_{eq} = \sigma_m + \sigma_r \leq \frac{S_y}{N} \qquad (14.14)$$

$$d = \left\{ \frac{32N}{\pi S_y} [(M_m^2 + T_m^2)^{1/2} + (M_r^2 + T_r^2)^{1/2}] \right\}^{1/3} \qquad (14.15)$$

The design diameter is the largest from the ones obtained with Equations (14.13) and (14.15).

These design equations must be used iteratively because in the determination of the fatigue strength some factors depend on the diameter.

Stress concentration factors are used in conjunction with the nature of the material, as explained in Chapter 5. In short, fatigue stress concentration factors are always used if variable stresses are present. On the steady stresses, stress concentration factors are used only for brittle materials.

In Equation (14.15), the range of the torque T_r must be used with caution because torsion in the Goodman diagram appears as a horizontal line. Therefore, Equation (14.15) can be used for small values of the torque range only, which is almost always the case.

Under average conditions, the fatigue safety factor is $N_f = 1.5$–2.5. If the diameters of shafts are determined from conditions of sufficient rigidity, the value of N_f may be substantially greater. If the design loads have been determined with enough accuracy, and with accurate calculations and homogeneous materials, a lower fatigue safety factor can be chosen. The lowest advisable value is $N_f = 1.3$.

If a shaft operates under conditions of non-steady-state loading and it is necessary to completely utilize all strength margins, calculations are carried out on the basis of Miner's rule (Equation (5.77)) and the fatigue curve, Equation (5.70)). Thus:

$$\sigma_{eq} = \left[\left(\sum_{i=1}^{m} n_{ri} \right) / (N_0 a) \right]^{1/m} \qquad (14.16)$$

where N_0 is the number of cycles up to the inflection point of the fatigue curve, taken equal to (3 to 4) $\times$ 10^6 for shafts of small cross-section and 10^7 for large shafts, n_{ri} is the total number of loading cycles at stress σ_r, m is an exponent of the fatigue curve, usually taken equal to 9 (for shafts with press-fitted components, $m = 6$), and σ_{ri} is an alternating stress in the shaft at the maximum load applied during an appreciable length of time.

The Miner fraction a is shown in Figure 14.11. It is greater, in particular, for soft steels. In general, a is greater than 1 in the zone $\sigma_2 > \sigma_1 > S_e$, and reaches a maximum in the zone $\sigma_2/\sigma_1 = 1.1$–1.2. If experimental data are not available, it is advisable to take a equal to 1, which gives a slight increase in the factor of safety and more conservative design.

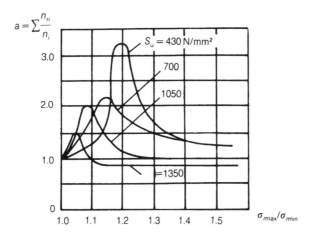

Figure 14.11. Miner's ratio for cumulative fatigue damage.

An important note about the *safety* and the *service* factors: Recall from Chapter 4 that a safety factor N accounts for the uncertainty in the material properties (capacity) and the machine loads (demand). In Chapters 10 and 13, we introduced the *service factor* C to account for the load uncertainty only. In use, the service factor C multiplies the rated power H to yield the maximum expected power, while the safety factor N divides the theoretical material strength (yield strength S_y, for example) to yield the *allowable* stress. In a design, use of both factors is allowed *if the safety factor N accounts only for the dispersion in the material strength*. If the safety factor N also accounts for the dispersion or uncertainty of the load, the service factor should not be used.

Example 14.1 An 11 kW, 1500 rpm rated motor drives a pump through a stub-shaft of length 2 m, connected through a flexible coupling. If the shaft is made of AISI 1015 carbon steel, find the diameter.

Solution The torque loading is nonreversing, and there is no bending to induce alternating stresses. From Chapter 4, for AISI 1015 steel the fatigue strength is 273 MPa and the yield strength 430 MPa. Because the loading is steady, the design will be based on the yield strength.

The service factor, from Table 13.9, is $C = 1.8$ and will be used instead of the safety factor N. No stress concentration will be used because the material is ductile. Therefore, for $S_{sy} = S_y/2 = 215$ MPa, Equation (14.6) yields:

$$d = 36.5 \left[\frac{(11 \times 10^3)/1462}{(215 \times 10^6)/1.80} \right]^{1/3} \text{ m or } 14 \text{ mm}$$

Example 14.2 The shaft shown rotates at 930 rpm. It carries a helical gear at B with helix angle $\psi = 35°$, standard involute teeth with $\phi = 20°$ and diameter 250 mm. On the right end, D carries a flywheel of mass $m = 200$ kg. Ball bearings at A and C support the shaft. The operation is intermittent and with heavy torsional impacts equalized by the flywheel.

The power of 30 kW is supplied at the left end. Determine the shaft loads and draw force and moment diagrams.

Solution The flywheel weighs $mg = 200 \times 9.81 = 1962$ N. The angular velocity is:

$$\frac{2\pi n}{60} = \frac{2\pi \times 930}{60} = 97.4 \text{ rad/s}$$

The torque between A and B is:

$$T = \frac{P}{\omega} = \frac{30,000}{97.4} = 308 \text{ Nm}$$

The flywheel supplies power to keep the shaft running, and it is assumed that the

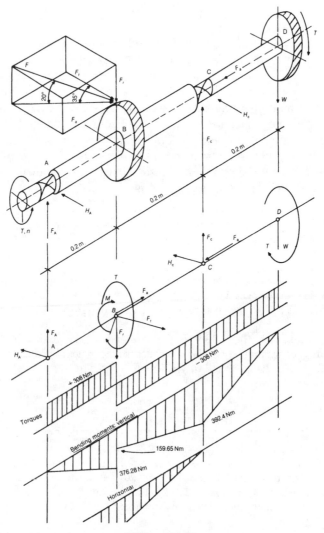

Figure E14.2

maximum power supplied equals the system power. Therefore, the torque of 308 Nm continues to point D.

The forces on the gear are:

$$F_t = \frac{2T}{d} = \frac{2 \times 308}{0.25} = 2464 \text{ N}$$

$$F_r = \frac{F_t \tan \varphi'}{\cos \psi} = 1095 \text{ N}$$

$$F_a = F_t \tan \psi = 1725 \text{ N}$$

The axial force is support at the bearing at C and gives bending moment

$$M = \frac{F_a d}{2} = 215.6 \text{ Nm}$$

In the vertical plane, the reactions F_A and F_C:

$$\Sigma F = 0: F_A - 1095 + F_C - 1962 = 0$$

$$\Sigma M_A = 0: 215.63 - 1095 \times 0.2 + 0.4 F_C - 0.6 \times 1962 = 0$$

$$F_A = 1876 \text{ N}, F_C = 1180.6 \text{ N}$$

In the horizontal plane, the force F_t is carried to point B with a torque T. The reactions are:

$$\Sigma F = 0: H_A + H_C + F_t = 0$$

$$\Sigma M_A = 0: 0.2 F_t + H_C \times 0.4 = 0$$

$$H_A = 1232N, H_C = 1232 \text{ N}$$

The force and moment diagrams are shown in Figure E14.2.

Example 14.3 A shaft transmits 20 kW at 800 rpm through a 20° involute spur gear at the mid-span of diameter 160 mm. The gear and bearing dimensions require that the span be at least 260 mm. The gear is secured on the shaft with a key in a 10 mm deep keyway. The shaft is made of hot-rolled AISI 1015 steel, machined, and a reliability of 90% is sufficient. Determine the shaft diameter for a service factor $N = 2$.

Solution The torque is:

$$T = \frac{9550P}{n} = \frac{9550 \times 20}{800} = 238.75 \text{ Nm}$$

The tangential force of the gear is:

$$F_t = \frac{2T}{d} = \frac{2 \times 238.75}{0.16} = 2984 \text{ N}$$

The bending moment due to the tangential force is then:

$$M_{b1} = \frac{F_t}{2} \frac{L}{2} = 194 \text{ Nm}$$

The radial force is:

$$F_r = F_t \tan 20° = 1086 \text{ N}$$

The bending moment is:

$$M_{b2} = \frac{F_r}{2} \frac{L}{2} = \frac{1086}{2} \times \frac{0.260}{2} = 70.6 \text{ Nm}$$

The two bending moments are in planes perpendicular to one another and they are added vertically, giving

$$M_b = (M_{b1}^2 + M_{b2}^2)^{1/2} = (194^2 + 70.6^2)^{1/2} = 206 \text{ Nm}$$

This gives reversing stresses. Then the mean and range values of bending moments and torques are:

$$T_m = T = 239 \text{ Nm}, \ T_r = 0, \ M_m = 0, \ M_r = M_b = 206 \text{ Nm}$$

For AISI steel $S_u = 527$, $S_y = 430$, and $S_n' = 270$ MPa. A first estimate of the diameter (Equation (14.6)) for $S_{sy} = S_y/2 = 430/2 = 215$ MPa gives, assuming $S_e = S_{n'}/3 = 90$ MPa,

$$d^3 = \frac{32N}{\pi S_u} \left(T_m + \frac{S_u}{S_e} M_r \right)$$

$$= 3.14 \times \frac{32 \times 2}{527 \times 10^6} \left(239 + \frac{527}{90} \times 206 \right) = 0.000056, \ d = 0.039 \text{ m or } 39 \text{ mm}$$

is an initial estimate of the diameter.

From Appendix A, the stress concentration factor for the keyway is $K_t = 2.15$. From Figure 5.27, the notch sensitivity factor for notch radius 2 mm is $q = 0.75$. Therefore, the fatigue stress concentration factor is (Equation (5.71)):

$$K_f = 1 + q(K_t - 1) = 1 + 0.75(2.15 - 1) = 1.86$$

The surface finish factor $C_f = 0.80$ and the size factor $C_s = 0.70$ (Figures 5.25, 5.26). For 90% reliability, Figure 4.29 gives $C_R = 1/1.1 = 0.90$. Therefore, from Equation (5.70):

$$S_e = \frac{0.8 \times 0.7 \times 0.9 \times 273}{1.86} = 74 \text{ MPa}$$

Equation (14.15) yields the shaft diameter:

$$d^3 = \frac{32N}{\pi S_u}\left(T_m + \frac{S_u}{S_e}M_r\right)$$

$$= 3.14 \times \frac{32 \times 2}{527 \times 10^6}\left(239 + \frac{527}{74} \times 206\right) = 0.00006595 \text{ and then } d = 41.7 \text{ mm}$$

No further iteration is needed.

14.2.6. Design for Rigidity

Strength of rigidity analysis of shafts of complicated geometry and loading, multiple supports, and so on can be performed with one of the structural programs with prismatic beam elements presented in Chapter 4.

Recall that the program performs static analysis of a multi-node beam in one plane. Therefore, all loads have to be vector-analyzed into the vertical and horizontal planes. At every position on the shaft, the resulting loads and displacements have to be vector-added to yield load and displacement at this position. This analysis has to be performed separately for the static and dynamic loading. Then equivalent loads must be computed and compared with the allowed stress of the chosen material.

Distributed loads are usually equally lumped at the nearest nodes. In most cases this is adequate, and it also simplifies stress analysis because with concentrated masses, the maxima of shear forces and bending moments exist at the nodes, and therefore analysis between nodes is not necessary for stress analysis.

For deflection analysis this is not true, but placing some nodes in the vicinity of the place where the maximum deflection is expected is usually adequate.

Using the program TRANSFEF MATRIX as an analysis subroutine, a program SHAFTDES incorporates the above module of the MELAB 2.0 package.

Applied torques are computed separately, something usually very simple. If substantial axial loads exist, the shaft has to be tested with the program TMSTABIL for buckling.

Finally, a dynamic analysis is necessary in most high-speed shafts, which will be discussed later.

Shafts with continuously variable cross-section can be analyzed by way of taking enough nodes and considering constant section properties between nodes. A constant strength along the shaft length is the aim of an optimum design.

Static elastic angular deformation of kinematic trains can affect the accuracy of machine performance, for instance that of precision engine lathes and gear-cutting machines or dividing machines. The elastic deformations of drives for low-speed machinery may cause stick–slip motion. For this reason, for example, the angles of twist of long feed shafts of heavy machine tools are limited to values on the order of 0.5° per meter length. The elastic deformations of divided drives powered from a single motor and used for traversing overhead cranes, portals, cross-

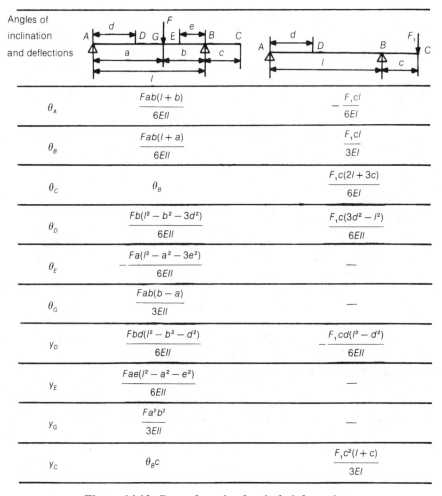

Angles of inclination and deflections		
θ_A	$\dfrac{Fab(l + b)}{6Ell}$	$-\dfrac{F_1cl}{6EI}$
θ_B	$\dfrac{Fab(l + a)}{6Ell}$	$\dfrac{F_1cl}{3EI}$
θ_C	θ_B	$\dfrac{F_1c(2l + 3c)}{6EI}$
θ_D	$\dfrac{Fb(l^2 - b^2 - 3d^2)}{6Ell}$	$\dfrac{F_1c(3d^2 - l^2)}{6Ell}$
θ_E	$-\dfrac{Fa(l^2 - a^2 - 3e^2)}{6Ell}$	—
θ_G	$\dfrac{Fab(b - a)}{3Ell}$	—
y_D	$\dfrac{Fbd(l^2 - b^2 - d^2)}{6Ell}$	$-\dfrac{F_1cd(l^2 - d^2)}{6Ell}$
y_E	$\dfrac{Fae(l^2 - a^2 - e^2)}{6Ell}$	—
y_G	$\dfrac{Fa^2b^2}{3Ell}$	—
y_C	θ_Bc	$\dfrac{F_1c^2(l + c)}{3EI}$

Figure 14.12. Beam formulas for shaft deformations.

members of heavy machine tools, and so on may result in jamming of the slideways. In the transmission shafts of the mechanisms for traversing traveling cranes, angles of twist range from 0.15–0.20° per meter of length.

Insufficient torsional rigidity of pinion gears cut integral with their shafts may lead to load concentration along the face width of the teeth.

Torsional rigidity is of special significance for shafts for which torsional vibrations are dangerous, for example in the drives of piston engines. It can prevent resonance vibrations and increase the service life of the gearing.

For most shafts, torsional rigidity is of no essential importance and there is no necessity to check the rigidity of such shafts. An allowable angle of twist equal to 1/4° per meter length, frequently given in the literature, has become obsolete and cannot be technically substantiated. In certain cases it is exceeded by many times. This is especially true for small-diameter shafts because the stress is inversely proportional to the cube of the shaft diameter and the angle of twist per unit length, to the fourth power. For instance, the angles of twist of the propeller shafts of automobiles (30–50 mm in diameter) may reach several degrees per meter of length. The angle of twist of the cylindrical portion of length of a shaft subject to the torque T is:

$$\Delta \varphi = \frac{TL}{GJ} = cT \tag{14.17}$$

where G is the shear modulus, J is the polar moment of inertia of the shaft cross-section, and c is the torsional compliance of the cylindrical portion of the shaft.

For a portion weakened by keyways, a factor of reduction in rigidity k is introduced into the right-hand member of the equations:

$$k = \left(1 - \frac{4nh}{d}\right)^{-1}, \ \Delta \varphi c k T \tag{14.18}$$

where h is depth of the keyway and n is a factor equal to 0.5 for one key, 1 for two keys at an angle of 90° and 1.2 at an angle of 180°, and 0.4 for two tangent keys at on angle of 120°.

Peterson (1974) gives much useful information on this subject.

Example 14.4   A high-pressure turbine rotor running at 3600 rpm has four stages sharing equally the production of 50,000 kW power. The shaft material is AISI 12 Cr steel with $S_u = 830$ MPa, $S_y = 540$ MPa, $S_e = 320$ MPa. Due to high temperatures and creep, a service factor of 2.5 must be used. Design the shaft for 99% reliability. Compute the maximum deflection owing to the shaft's own weight. The wheels weigh 3, 3.55, 4.1, and 4.1 kN from left to right, equally spaced between the two end bearings on a rotor of total length of 5 m. The power is transmitted to the right, and the thrust is negligible.

Solution   The power per wheel is:

$$\frac{50,000}{4} = 12,500 \text{ kW}$$

The torque on each wheel is then:

$$T = \frac{9550P}{n} = \frac{9550 \times 12,500}{3660} = 33,159 \text{ Nm}$$

The transition from one diameter to the other is assumed smooth without stress concentration.

For 99.9% reliability, Figure 4.29 gives safety factor 1.4 and $C_R = 1/1.4 = 0.714$. The surface finish factor for machined shaft $C_f = 0.72$ (Figure 5.25), and the size factor $C_s = 0.6$ (Figure 5.26). These are tentative values because the exact diameters are not yet known. Therefore:

$$S_e = 0.72 \times 0.6 \times 0.714 \times 320 = 98.7 \text{ MPa}$$

The program SHAFTDES (of the MELAB 2.0 package) is used to yield the diameters and deflections. The output is shown in Figure E14.4b.

```
ENTER DATA:
Number of elements                            ? 5
Node  1 : Enter FX,FY,MX,MY                    ? 0,0,0,0
          Support <S>, spring <P>, free <F>    ? s
Node  2 : Enter FX,FY,MX,MY                    ? 0,3000,0,0
          Support <S>, spring <P>, free <F>    ? f
Node  3 : Enter FX,FY,MX,MY                    ? 0,3550,0,0
          Support <S>, spring <P>, free <F>    ? f
Node  4 : Enter FX,FY,MX,MY                    ? 0,4100,0,0
          Support <S>, spring <P>, free <F>    ? f
Node  5 : Enter FX,FY,MX,MY                    ? 0,4500,0,0
          Support <S>, spring <P>, free <F>    ? f
Node  6 : Enter FX,FY,MX,MY                    ? 0,0,0,0
          Support <S>, spring <P>, free <F>    ? s
Element  1 :Length,QloadX,QloadY                ? 1,0,0
            Element torque,thrust               ? 0,0
Element  2 :Length,QloadX,QloadY                ? 1,0,0
            Element torque,thrust               ? 33000,0
Element  3 :Length,QloadX,QloadY                ? 1,0,0
            Element torque,thrust               ? 66000,0
Element  4 :Length,QloadX,QloadY                ? 1,0,0
            Element torque,thrust               ? 99000,0
```

Figure E14.4a

```
Element  5 :Length,QloadX,QloadY                ? 1,0,0
            Element torque,thrust               ? 132000,0
Enter Material Data:
                    Young Modulus      ? 2.1e11
                    Fatigue strength   ? 98.7e6
                    Yield Strength     ? 540e6
                    Specific weight    ? 7.8
                    Service Factor     ? 2.5


Hit ENTER to continue...
?
VERTICAL PASS...
DATA:  5  ELEMENTS,   6   NODES

  LENGTH    DIAMETER   FORCE      LOAD      SPRING
----------------------------------------------------------------
  100.E-02  232.E-03    0.E+00    0.E+00    0.E+00 SUPPORT
  100.E-02  232.E-03  300.E+01    0.E+00    0.E+00
  100.E-02  232.E-03  355.E+01    0.E+00    0.E+00
  100.E-02  232.E-03  410.E+01    0.E+00    0.E+00
  100.E-02  232.E-03  450.E+01    0.E+00    0.E+00
    0.E+00  232.E-03    0.E+00    0.E+00    0.E+00 SUPPORT
computing...
SOLUTION ...
STATION DEFLECTION   SLOPE      MOMENT       SHEAR       REACTION
----------------------------------------------------------------
  1   615.E-08  623.E-06    0.E+00     0.E+00   707.E+01
  2   589.E-06  504.E-06 -707.E+01  -707.E+01
  3   953.E-06  200.E-06 -111.E+02  -407.E+01
  4   964.E-06 -181.E-06 -117.E+02  -520.E+00
  5   608.E-06 -511.E-06 -808.E+01   358.E+01
  6   703.E-08 -646.E-06    0.E+00   123.E-03   808.E+01
Hit ENTER to continue?
RESULTS:
NODE   D(-)        D(+)
-------------------------------------------------
  1    232.E-03    116.E-03    ...bearing
  2    117.E-03    147.E-03
  3    147.E-03    168.E-03
  4    168.E-03    185.E-03
  5    184.E-03    727.E-04
  6      0.E+00    727.E-04    ...bearing
Hit ENTER to continue...?
Do you want bearing calculations (Y/N) ? n
```

Figure E14.4b

14.2.7. Vibration of Rotating Shafts

A considerable part of the system is involved in the vibrations observed in machines, including the main kinematic train and the basic parts. The individual vibrations of the separate transmission shafts, such as the shafts of the gearbox, play no essential role in the dynamics of a machine and are therefore not considered separately. In contrast, the vibrations of main shafts with their attached assemblies and supports (turbine rotors, crankshafts of piston engines, machine tool spindles with workpieces, etc.) may be the governing factor in design.

August Föppl (1854–1924) was born in Gross-Umstadt, in Hesse, Germany, in the family of a doctor. He studied structural engineering at the Polytechnic Institute of Darmstadt and was so impressed by the teachings of Mohr that when Mohr moved to the Polytechnic Institute of Karlsruhe, he followed him there, where he got his degree in 1874. He wanted to work as a bridge designer but was unable to find a job. He worked some years as a high school teacher. He did important research and published a book on Maxwell's theory in 1892. In 1893, he was elected professor of mechanics at the Polytechnic Institute of Munich, succeeding Bauschinger in both teaching and his famous research laboratory. An excellent lecturer, he taught classes of 500 students. His lectures were published in several volumes. Among his contributions was the explanation of the whirling of shafts. It was generally accepted that operation of shafts above the critical speed was impossible because the restoring forces were smaller than the inertia (centrifugal forces). But de Laval was operating his turbines well above the critical speed! Föppl solved the puzzle by proving that the shaft vibration changed phase angle above the critical speed. Thus, the restoring forces and the inertia forces were both directed towards the center of rotation and operation was possible. Even more, the vibration amplitude would tend to diminish above the critical speed, something that de Laval knew already from his experiments.

Of main practical importance in shaft design is the determination of the natural frequency of vibration to avoid resonance of vibrations, that is, an increase in the amplitude of vibrations when the frequency of the exciting forces coincides with the natural frequency of vibrations. Observed in shafts are transverse, or bending, vibrations, angular, or torsional, vibrations, and also combined bending and torsional vibrations. The natural frequency of simple shafts and axles is computed by design formulas given in Table 14.1.

The most widely employed are calculations of the fundamental frequencies of vibrations. These vibrations are usually dangerous because most machines operate near the lowest critical speed. Heavy or very high-speed machinery, such as turbomachinery, might operate near the second or the third.

The fundamental frequency of natural vibration of a shaft with a concentrated mass, taking into consideration the dead weight of the shaft, is most readily found if the reduced mass of the shaft is added to the concentrated mass. The reduction

TABLE 14.1. Natural Frequencies of Some Shafts[a]

Shaft Description	Lateral Natural Frequency ω_n^2
Simply supported, uniform diameter d, length L, mass m at mid-span,	$k = 48EI/L^3 = 48E(pd^4/64)/L^3$, $\omega_n^2 = k/m$
Uniform, diameter d, length L, density ρ	$\omega_n^2 = n\pi EI/(\rho AL) = n\pi Ed^2/(16\rho L)$

[a]From Dimarogonas 1996.

factor for transverse vibrations of an overhanging axle of constant cross-section with a mass at its end is 0.235. For a two-support shaft or axle with the mass at the middle, it is 0.48. For torsional vibrations of a shaft with one end fixed and with a disk at the other end, it is 0.33.

14.3. COUPLINGS

14.3.1. Field of Application

Joining two shafts is done with mechanical couplings. Couplings should be capable of transmitting the rated torque capacity of the shaft while accommodating any misalignment between the shafts.

Rigid couplings (Figure 14.13a) are simple and low-cost, but they demand almost perfect alignment of the mating shafts. The shafts must also have stable bearing supports. Misalignment, whether present initially or developed from wear, causes undue forces and accelerated wear on the shafts, coupling, shaft bearings, or machine housing.

In most coupling applications, misalignment is the rule rather than the exception. It comes from such sources as bearing wear, structural deflection, thermal expan-

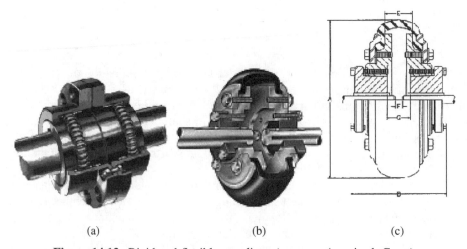

(a)	(b)	(c)

Figure 14.13. Rigid and flexible couplings (courtesy Ameritech Corp.).

sion, or settling machine foundations. When misalignment is expected, a flexible coupling must be used (Figure 14.13*b*, *c*).

Common selection factors include:

1. Amount of torque
2. Positive shaft engagement
3. Misalignment tolerance
4. Lubrication/maintenance
5. Ease of installation/removal
6. Operation under adverse conditions
7. Service life
8. Cost

Couplings are classified according to:

1. Relative shaft position (parallel, collinear, intersecting)
2. Means of torque transmission (fluid, mechanical, electrical)
3. Function (adjusting to misalignment, extending the shaft, etc.)
4. Level of performance (heavy-duty, high-speed, etc.)

Some of the relevant AGMA standards for flexible couplings are 510.02/1969 (nomenclature), 511.02/1969 (bore and keyway sizes), 512.03/1974 (keyways), 513.01/69 (taper bores), and 514.02/1971 (load classification and service factors).

14.3.2. Design of Mechanical Couplings

A standard rigid coupling (Figure 14.14*a*) consists of two flanges that attach to the two shafts that are to be connected, by way of a shrink fit or a key. The torque is

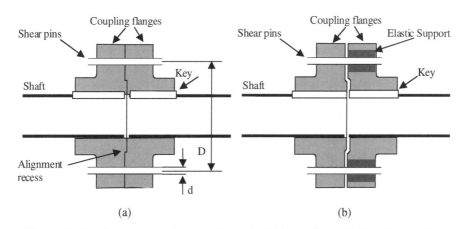

Figure 14.14. Components of a coupling: (*a*) rigid coupling; (*b*) flexible coupling.

transferred through a number of pins of diameter d symmetrically located over a circle of diameter D. Failure of the pin is by shear.

Because the pins are usually made of low-carbon steel, the pins and holes are carefully machined, and stress redistribution ensures uniform loading of the pins, the average shear stress in the pin cross-section is:

$$\tau_{av} = \frac{F}{A} = \frac{T/(D/2)}{z\pi d^2/4} \tag{14.19}$$

where T is the torque, z is the number of pins, d is the pin diameter, and D is the pin circle diameter. Because the torque $T = CH/\omega$, where H is the power transmitted, C is the service factor (Table 13.9), and ω is the angular velocity of the shaft, the design equation is:

$$\tau_{av} = \frac{2CH}{\omega D z \pi d^2} = \frac{S_{sy}}{N} \tag{14.20}$$

where S_{sy} is the shear yield strength ($= 0.577S_y$, if the equivalent distortion energy theory of failure is employed) and N is the safety factor. From Equation (14.20), the diameter of the pin can be determined, if the number of pins z and the pin circle diameter D are specified. Alternatively, if the pin diameter and the pin circle diameter D are specified, the number of pins z can be determined from Equation (14.20).

A standard flexible coupling (Figure 14.14b) also consists of two flanges that attach to the two shafts that are to be connected, by way of a shrink fit or a key. The torque is transferred through a number of pins of diameter d symmetrically located over a circle of diameter D. However, an elastic medium is placed between the pins and one of the two flanges. Sometimes this elastic medium is a rubber hollow cylinder, but in other designs it can have various forms. Failure of the pin is by bending and shear.

Flexible couplings have several advantages:

1. They can tolerate substantial misalignment.
2. They do not require accurate machining of the pin surface and flange holes.
3. They attenuate impacts so that the fatigue load of the shafts and other system components due to impact is drastically reduced.

The loading of the pin is shown in Figure 14.15. The elastic medium applies the peripheral force $P = T/(zD/2)$ uniformly over the width L of the elastic medium at each of the pins. Assuming that the resultant force P is applied in the middle of the loaded length of the pin, the maximum bending stress at the root of the pin is:

$$\sigma_{max} = \frac{Mc}{I} = \frac{[T/(zD/2)](L/2)(d/2)}{\pi d^3} = \frac{32TL}{zD\pi d^3} \tag{14.21}$$

and the design equation for bending strength of the pin is, for $T = CH/\omega$:

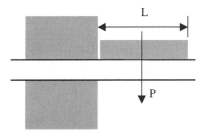

Figure 14.15. Loading of a pin in a flexible coupling.

$$\sigma_{max} = \frac{32CHL}{z\omega D\pi d^3} = \frac{S_y}{N} \qquad (14.22)$$

from which the diameter of the pin can be determined, if the number of pins z and the pin circle diameter D are specified. Alternatively, if the pin diameter and the pin circle diameter D are specified, the number of pins z can be determined from Equation (14.22).

Example 14.5 A coupling is to connect a 1500 rpm, 12 kW motor to a water pump. The two flanges are connected by two bolts located symmetrically at diameter $D = 80$ mm. If the yield strength of the bolt material is $S_y = 220$ MPa and the safety factor $N = 4$, determine the bolt diameter (a) for a rigid coupling, (b) for a flexible coupling if $L = 30$ mm.

Solution

1. The angular velocity $\omega = 2\pi n/60 = 2\pi \times 1500/60 = 157$ rad/s. From Table 13.9, $c_1 = 0.5$ (electric motor), $c_2 = 1.5$ (pump). Therefore, $C = c_1 + c_2 = 0.5 + 1.5 = 2$. Solving Equation (14.20) for the diameter, with $S_{sy} = 0.577 \, S_y$:

$$d^2 = \frac{2CHN}{\omega Dz\pi S_{sy}}$$

$$= \frac{2 \times 2 \times 12,000 \times 4}{157 \times 0.080 \times 2\pi \times 0.577 \times 220 \times 10^6}, \; d = 0.0044 \text{ m or 5 mm}$$

2. Solving Equation (14.22) for the diameter:

$$d^3 = \frac{32CHLN}{S_y z\omega D\pi} = \frac{32 \times 2 \times 12000 \times 0.030 \times 4}{220 \times 10^6 \times 2 \times 157 \times 0.080 \times \pi}, \; d = 0.0023 \text{ m or 3 mm}$$

14.4. KEYS

14.4.1. Field of Application

In critical applications, shafts are made integral with hubs of gears, sheaves, sprockets, couplings, and so on. This is an expensive solution, however, and in usual applications these components are made separate from the shaft and are connected to the shaft with various fits. In most cases, there must be a way to carry the torque to the associated part, and for this purpose we use keys, pins, and splines.

Keys are shear elements that are placed in slots of the shaft and the hub and provide positive engagement between them (Figure 14.16). In most applications, the hub is much longer than the width of the gear or coupling to accommodate a long key that will have similar strength to the shaft in torsion. The advantages of keys are low cost, simplicity, and easy assembly. Disadvantages are the required machining and the weakening of the shaft.

Keys have been standardized by AGMA 620.02 and ANSI B17.1, B17.2. Standard metric sizes of keys of square cross-section are given in Table 14.2. Dimensions of rectangular keys are given in Table C.2, Appendix C.

14.4.2. Design of Keys

Failure of keys is observed with two modes:

1. There can be shear failure on the horizontal plane, at the point where the hub meets the shaft. The average shear stress is:

$$\tau_{av} = \frac{F}{A} = \frac{T(d/2)}{bL} = \frac{2CH}{\omega dbL} \tag{14.23}$$

where d is the shaft diameter, T is the transmitted torque $= CH/\omega$, C is the service factor (Table 13.9), H is the power, ω is the angular velocity, b is the width of the key, and L is the length of the key. The design equation for shear strength is then:

$$\tau_{av} = \frac{2CH}{\omega dbL} = \frac{S_{sy}}{N} \tag{14.24}$$

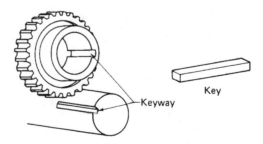

Figure 14.16. A key and a keyway in a shaft–gear assembly (Courtesy Deere & Co. Technical Services).

TABLE 14.2 Standard Metric Sizes of Squares Keys

Shaft Diameter D, mm	Key Thickness a, mm	Shaft Diameter D, mm	Key Thickness a, mm
12–15	3	50–60	14
15–20	4	60–70	16
20–30	6	70–80	18
30–40	8	80–90	20
40–50	10	90–100	24

2. There can be crushing failure on the side of the key due to bearing pressure. The average bearing stress is:

$$\sigma_{\text{av}} = \frac{F}{A} = \frac{T(d/2)}{(h/2)L} = \frac{4CH}{\omega dhL} \tag{14.25}$$

The design equation for bearing strength is then:

$$\sigma_{b-\text{av}} = \frac{4CH}{\omega dhL} = \frac{S_b}{N} \tag{14.26}$$

where S_b is the bearing strength of the *weakest material* (usually the hub) that can be taken equal to $1.1S_u$, as discussed in Chapter 5 or from AGMA standard 260.2.

In Equations (14.24) and (14.26), a safety factor of at least $N = 4$ is recommended for nonsliding hubs and $N = 6$ for sliding hubs.

Equations (14.24) and (14.26) may be used to find the key dimensions b, a, L. One of the three needs to be selected and the other two can then be found from Equations (14.24) and (14.26).

Example 14.6 A key of length $L = 60$ mm is to connect a 50 mm diameter shaft to a hub to transmit 25 kW power at 1200 rpm. The key will be made of carbon steel with yield

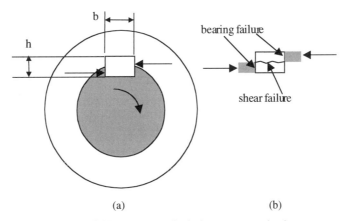

(a)   (b)

Figure 14.17. Stress analysis in a rectangular key.

strength S_y = 220 MPa and ultimate strength S_u = 450 MPa, while the hub is made of cast iron with ultimate strength S_u = 250 MPa. For a service factor C = 3 and a safety factor N = 4, find the dimensions of a key of rectangular cross-section.

Solution The angular velocity is $\omega = 2\pi n/60 = 2\pi \times 1200/60 = 126$ rad/s. For shear strength of the key we shall apply Equation (14.24) and solve for the key width b:

$$b = \frac{2CHN}{\omega dLS_{sy}} = \frac{2 \times 3 \times 25000 \times 4}{126 \times 0.050 \times 0.060 \times 0.577 \times 220 \times 10^6} = 0.0125 \text{ m or } 12.5 \text{ mm}$$

For bearing strength of the hub, the bearing strength of the weakest material (the hub) is $S_b = 1.1S_u = 1.1 \times 300 = 330$ MPa. We shall apply Equation (14.26) and solve for the key thickness h:

$$h = \frac{4CHN}{\omega dLS_b} = \frac{4 \times 3 \times 25000 \times 4}{126 \times 0.050 \times 0.060 \times 1.1 \times 300 \times 10^6} = 0.0096 \text{ m or } 10 \text{ mm}$$

From Table C2, Appendix C, we find that the nearest standard key has b = 14 mm, h = 10 mm.

CASE STUDY 14.1: The Go-Matic Accessory for Motorcycles[1]

Go-Power Corporation, a Palo Alto, California, company manufactured and marketed the Go-Matic, an accessory for lightweight motorcycles. Its function was to enable the user to change from a trail (speed range 5–20 mph) to a street version (speed range 20–50 mph). It is essentially a small gear box having two output sprockets, to either one of which the chain can be selectively engaged, providing the switch function.

After Go-Matic had been on the market for two months, there were several failures, mostly of the main shaft and some breakdowns of the main shaft bearings. Within a year, about 40% of Go-Matics were sent back for replacement.

Investigation

The investigation and redesign followed the typical small company attitude: Without any design analysis, engineers, managers, or consultants state opinions as to the cause of the problem, changes are suggested, decided upon, and implemented, and the redesigned product is marketed without much serious testing.

In the case of the Go-Matic, the shaft that failed was made of 1020 steel, and the decision was made to change to 8620 steel hardened to 40 Rockwell C. In the bearing area, the shaft was hardened by carbo-nitriding to 55 RC. Moreover, the shaft was redesigned to reduce stress concentration without altering the main stress concentration of concern, the sharp change in diameters at the point of failure (Figure CS14.1).

---

[1] Based on Case Study ECL 113 of the Engineering Case Library, to which the reader is referred to for further details. Originally prepared by P. Z. Bulkeley and P. C. Garg. By permission of the trustees of the Rose-Hulman Institute of Technology.

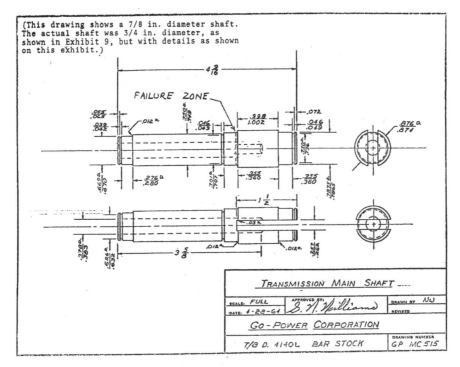

Figure CS14.1

The new design performed better, but due to the large number of failures already encountered, sales declined and the company decided not to continue marketing the product. A total of 2700 units were sold.   □

REFERENCES

Anon. *Shaft Couplings*. Farmingdale, N.Y.: Renold Crofts, 1972.

Dimarogonas, A. D. 1996. *Vibration for Engineers*, 2nd ed. Upper Saddle River, N.J.: Prentice Hall.

Peterson, R. E. 1974. *Stress Concentration Factors*. New York: John Wiley & Sons.

ADDITIONAL READING

Dimarogonas, A. D., and S. A. Paipetis. *Analytical Methods in Rotor Dynamics*. London: Applied Science, 1983.

Duggan, T. V., and J. Byrne. *Fatigue as a Design Criterion*. London: Macmillan, 1977.

Hänchen, R., and K.-H. Decker. *Neue Festigkeitsberechnung für den Maschinenbau*. Munich: C. Hanser Verlag, 1967.

Hindhede, U., et al. *Machine Design Fundamentals*. New York: John Wiley & Sons, 1983.
Niemann, G. *Maschinenelemente*. Berlin: Springer, 1965.

PROBLEMS[2]

14.1. [T] Classify the shafts on the basis of shape, rigidity, in-span conditions.

14.2. [T] Discuss modes of failure of shafts.

14.3. [T] Discuss advantages and disadvantages of solid and hollow shafts.

14.4. [T] Identify stress raisers in shafts.

14.5. [T] In which cases are heat treatment and surface-hardening applied to shafts?

14.6. [D] The stub-shaft connecting a 250 hp, 875 rpm motor to the gearbox of a steel-rolling machine is made of AISI 1040 carbon steel. A key is used at the coupling. Determine the shaft diameter for safety factor $N = 1.5$. Is reliability important?

14.7. [D] An automobile engine delivers maximum torque at 110 hp at 3200 rpm. At low gear, the transmission ratio is $R = 4.5$ and the efficiency $\eta = 87\%$. At high gear, $R = 1$ and $\eta = 97\%$. Determine the main axle diameter for safety factor $N = 2$ if it is made of carbon steel AISI 1035.

14.8. [D] A 25 hp, 1750 rpm electric motor drives a lathe with intermittent operation and light shocks. Determine the axle diameter for safety factor $N = 1.2$ if no appreciable bending moment is applied and a spline is used at each end. The material is AISI 1040 carbon steel.

14.9. [D] A 20 kW four-cylinder internal combustion engine is connected to an irrigation water pump through a shaft made of AISI 1030 carbon steel. Determine the shaft diameter for safety factor $N = 1.8$.

14.10–14.20. [D] The shaft of Figure P14.10–20 rotates with angular velocity ω (rad/s.). It is driven from the left with power P_0 (kW), which is transmitted partly through two pulleys 1 and 2, shown schematically, powers P_1 and P_2, respectively, and the remaining power is transmitted to a driven machine at the right end with a very flexible coupling. Assuming no shear forces and bending moments transmitted to left and right ends, draw the shear force, torque, and bending moment diagrams in the vertical plane. Then determine the shaft diameters, selecting proper materials. The reliability should be 99%, the surfaces ground, and the fillet radii should not exceed 3 mm. The service factor should be 2.1, and the safety factor $N = 1.5$.

The data are given in the table below. H_1 and H_2 are horizontal forces at pulleys 1 and 2, not shown in Figure P14.10–20. For all problems, $P_0 = 8$ kW.

---

[2][C] = certification, [D] = design, [N] = numerical, [T] = theoretical problem.

Problem	$\omega(\text{rad/s})$	$P_1(\text{kW})$	$P_2(\text{kW})$	$F_1(\text{N})$	$F_2(\text{N})$	$H_1(\text{N})$	$H_2(\text{N})$
14.16	200	4	4	1000	2000	—	—
14.17	200	4	2	1000	2000	2000	2000
14.18	—	—	—	2000	2000	2000	2000
14.19	200	2	4	—	—	—	—
14.20	200	2	4	2000	2000	—	—
14.21	200	2	4	—	—	2000	2000
14.22	200	6	—	3000	1000	—	—
14.23	400	4	4	1000	2000	—	—
14.24	400	4	2	1000	2000	2000	2000
14.25	400	—	—	2000	2000	2000	2000

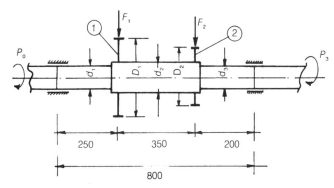

Figure P14.10–20

14.21. [D] The stub-shaft connecting a 250 hp, 875 rpm motor to the gearbox of a steel-rolling machine is made of AISI 1040 carbon steel. A key is used at the coupling. Design a rigid coupling for safety factor $N = 1.5$. Is reliability important?

14.22. [D] An automobile engine delivers maximum torque at 110 hp at 3200 rpm. At low gear, the transmission ratio is $R = 4.5$ and the efficiency $\eta = 87\%$. At high gear, $R = 1$ and $\eta = 97\%$. Design a rigid coupling for safety factor $N = 2$ if it is made of carbon steel AISI 1035.

14.23. [D] A 25 hp, 1750 rpm electric motor drives a lathe with intermittent operation and light shocks. Design an elastic coupling for safety factor $N = 1.2$ if no appreciable bending moment is applied and a spline is used at each end. The material is AISI 1040 carbon steel.

14.24. [D] A 20 kW four-cylinder internal combustion engine is connected to an irrigation water pump through a shaft made of AISI 1030 carbon steel. Design an elastic coupling for safety factor $N = 1.8$.

14.25. [D] A 30 kW six-cylinder internal combustion engine is connected to a chemicals-mixing machine through a shaft made of AISI 1020 carbon steel. Design an elastic coupling for safety factor $N = 2$.

14.26. [D] The stub-shaft connecting a 250 hp, 875 rpm motor to the gearbox of a steel-rolling machine is made of AISI 1040 carbon steel. A key of rectangular cross-section and length $L = 120$ mm is used to connect the 100 mm diameter shaft to the coupling. Determine the rectangular key dimensions if the shaft, key, and hub are made of the same material for safety factor $N = 1.5$.

14.27. [D] An automobile engine delivers maximum torque at 110 hp at 3200 rpm. At low gear, the transmission ratio is $R = 4.5$ and the efficiency $\eta = 87\%$. At high gear, R = 1 and $\eta = 97\%$. A key of rectangular cross-section and length $L = 120$ mm is used to connect the 80 mm diameter shaft to the coupling. Determine the rectangular key dimensions if the shaft, key, and hub are made of the same material for safety factor $N = 2$. The material is carbon steel AISI 1035.

14.28. [D] A 25 hp, 1750 rpm electric motor drives a lathe with intermittent operation and light shocks. A key of rectangular cross-section and length $L = 80$ mm is used to connect the 50 mm diameter shaft to the coupling. Determine the rectangular key dimensions if the shaft, key, and hub are made of the same material for safety factor $N = 1.2$ if no appreciable bending moment is applied. The material is AISI 1040 carbon steel.

14.29. [D] A 20 kW four-cylinder internal combustion engine is connected to an irrigation water pump through a shaft made of AISI 1030 carbon steel. A key of rectangular cross-section and length $L = 60$ mm is used to connect the 50 mm diameter shaft to the coupling. Determine the rectangular key dimensions if the shaft, key, and hub are made of the same material for safety factor $N = 1.8$.

14.30. [D] A 20 kW six-cylinder internal combustion engine is connected to a chemicals-mixing machine through a shaft made of AISI 1020 carbon steel. A key of rectangular cross-section and length $L = 50$ mm is used to connect the 50 mm diameter shaft to the coupling. Determine the rectangular key dimensions if the shaft, key, and hub are made of the same material for safety factor $N = 1.5$.

STRESS CONCENTRATION FACTORS

TABLE A.1. Theoretical Stress Concentration Factors

	σ_0	$K_t = \sigma_{max}/\sigma_0$
	$4F/\pi d^2$	$1 + \dfrac{(r/d)^{-0.36-0.2(D/d)}}{5 + 0.12/(D/d - 1)}$
	$16T/\pi d^3$	$1 + \dfrac{(r/d)^{-0.3-0.2(D/d)}}{13 + 03/(D/d - 1)}$
	$32M/\pi d^3$	$1 + \dfrac{(r/d)^{-0.73-0.42(D/d-1)}}{5 + 4.38/(D/d - 1)^{0.16}}$
	$T \left/ \left(\dfrac{\pi D^3}{16} - \dfrac{dD^2}{6} \right) \right.$	$1 + 1.47(d/D)^{-0.197}$
	$\dfrac{M}{\pi D^3/32 - dD^2/6}$	$1 + 0.65(d/D)^{-0.275}$
	$4F/\pi d^2$	$1 + \dfrac{(r/d)^{-0.511-(D/d - 1)0.34}}{3 + 0.507(D/d - 1)^{-0.42}}$

TABLE A.1. (*Continued*)

	σ_0	$K_t = \sigma_{max}/\sigma_0$
	$\dfrac{32M}{\pi d^3}$	$1 + \dfrac{(r/d)^{-0.59 - (D/d-1)0.184}}{5 + 0.0812/(D/d-1)}$
	$\dfrac{16T}{\pi d^3}$	$1 + \dfrac{(r/d)^{-0.609 - (D/d-1)\times 0.146}}{5 + 3.73(D/d-1)^{-0.252}}$
	$F/(w-d)t$	$1 + \dfrac{(d/w)^{-0.179}}{0.9}$
	$\dfrac{6M}{(w-d)h^2}$	$1 + \dfrac{(d/w)^{-0.21 - 0.09(d/h)^{0.3}}}{0.954 + 0.966(d/h)^{0.65}}$
	σ_0	$1 + 2(t/r)^{1/2}$
	$F/(w-d)t$	$(0.780 + 2.243(t/r)^{1/2})[0.993 + 1.80(2t/D)$ $- 1.060(2t/D)^2 + 1.710(2t/D)^3]$ $\times (1 - 2t/D)$
	$\dfrac{6M}{td^2}$	$1 + \dfrac{(r/D)^{-0.55 - 0.3(w/d-1)}}{4 + 0.31(w/d-1)^{1.35}}$
	F/dt	$1 + \dfrac{(r/d)^{-0.63 + (D/d-1)\times 0.1}}{4 + 0.22/(D/d-1)^{0.945}}$
	$\dfrac{6M}{td^2}$	$1 + \dfrac{(r/d)^{-0.64 - (D/d-1)\times 0.08}}{5 + 1.8/(D/d-1)^{0.376}}$
	$\dfrac{32M}{\pi d^3}$	$1 + \dfrac{(r/d)^{-0.66}}{11.14}$ $(b = d/4, \ t = d/8)$

TABLE A.2. Fatigue Stress Concentration Factors

| | | K_f | |
| | | Loading Mode | |
Heat Treatment		Bending	Torsion
Annealed (less than 200 BHN)		1.3	1.3
Quenched and drawn (over 200 BHN)		1.6	1.6

		K_f	
Annealed (less than 200 BHN)		1.6	1.3
Quenched and drawn (over 200 BHN)		2.0	1.6

ISO threads

		$K_f = k_1 k_2$			
Bolt	Cut	Rolled	Annealed	Surf. Hard.	Core Hard.
k_1	1	1.2	1.3	1.4	1.6
Nut	Steel, cut		Steel, rolled	A1	Bronze
k_2	1		1.05	1.1	1.15

APPENDIX B

STRESS INTENSITY FACTORS FOR CRACKED STRUCTURES

TABLE B.1. Stress Intensity Factors

Geometry and Loading	$K = f(a/b)\sigma(\pi a)^{1/2}$, $\lambda = a/b$
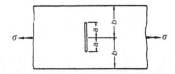	$f(a/b) = (\lambda) = (1 - 0.5\lambda + 0.37\lambda^2 - 0.044\lambda^3)/(1 - \lambda)^{1/2}$
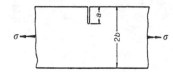	$f(\lambda) = \left(\dfrac{2}{\pi\lambda}\right)^{1/2} \dfrac{0.752 + 2.02\lambda + 0.37(1 - \sin(\pi\lambda/2))}{\cos(\pi\lambda/2)}$
	$f(\lambda) = (1 + 0.122\cos^4(\pi\lambda/2))[(2/\pi\lambda)\tan(\pi\lambda/2)]^{1/2}$
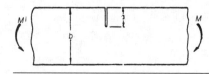	$f(\lambda) = [(2/\pi\lambda)\tan(\pi\lambda/2)]^{1/2} \dfrac{0.923 + 0.199(1 - \sin(\pi\lambda/2))^4}{\cos(\pi\lambda/2)}$

TABLE B.1. (*Continued*)

Geometry and Loading	$K = f(a/b)\sigma(\pi a)^{1/2}$, $\lambda = a/b$

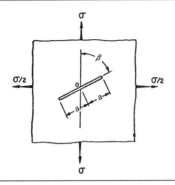

$$K_1 = \tfrac{1}{2}\sigma(\alpha)^{1/2}[1 + \sin^2 \beta]^{1/2}, \quad K_2 = \tfrac{1}{2}\sigma(\alpha)^{1/2} \sin \beta \cos \beta$$

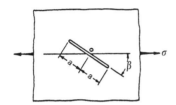

$$K_1 = \tfrac{1}{2}\sigma(\alpha)^{1/2} \sin^2 \beta, \quad K_2 = \sigma(\alpha)^{1/2} \sin \beta \cos \beta$$

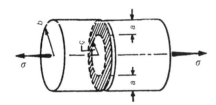

$$K = \frac{\sigma(\pi\alpha)^{1/2}}{2} \left(\frac{c}{b}\right)^{1/2} \left(1 + \frac{1c}{2b} + \frac{3c^2}{8b^2} - 0.363 \frac{c^3}{b^3} + 0.731 \frac{c^4}{b^4}\right)$$

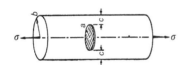

$$K = \frac{\sigma(\pi b)^{1/2}}{(\alpha)^{1/2}E(k)} (\alpha^2 \sin^2 \phi + b^2 \cos^2 \phi)^{1/4}$$

$$\xi = \frac{2}{\pi} \sigma(\pi a)^{1/2}, \quad E(\xi) = \text{elliptic integral}$$

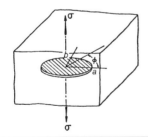

$$K = \frac{\sigma(\pi a)^{1/2}}{(1 - a^2/b^2)} \left(\frac{c}{b}\right)^{1/2} \left[\frac{2}{\pi} \left(1 + \frac{1}{2}\frac{a}{b} - \frac{5}{8}\frac{a^2}{b^2}\right) + 0.268 \frac{c^3}{b^3}\right]$$

APPENDIX C

STANDARDIZED MACHINE ELEMENTS

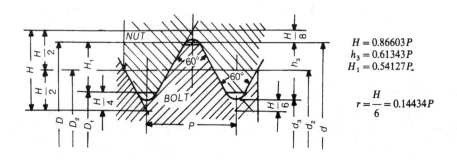

$$H = 0.86603P$$
$$h_3 = 0.61343P$$
$$H_1 = 0.54127P$$

$$r = \frac{H}{6} = 0.14434P$$

TABLE C.1. Bolt Dimensions

Size	Pitch (P)	Major Diameter $(d = D)$	Pitch Diameter $(d_2 = D_2)$	Minor Diameter Bolt (d_3)	Minor Diameter Nut (d_1)	Depth of Thread (h_3)	Max. Depth of Engagement (H_1)	Stress Area, A_s (mm²)
Coarse series								
M2.5	0.45	2.5	2.208	1.948	2.013	0.276	0.244	3.39
M3	0.5	3	2.675	2.387	2.459	0.307	0.271	5.03
M4	0.7	4	3.545	3.141	3.242	0.429	0.379	8.78
M5	0.8	5	4.480	4.019	4.134	0.491	0.433	14.2
M6	1	6	5.350	4.773	4.918	0.613	0.541	20.1
M8	1.25	8	7.188	6.466	6.647	0.767	0.677	36.6
M10	1.5	10	9.026	8.160	8.376	0.920	0.812	58.0
M12	1.75	12	10.863	9.853	10.106	1.074	0.947	84.3
M16	2	16	14.701	13.546	13.835	1.227	1.083	157
M20	2.5	20	18.376	16.933	17.294	1.534	1.353	245
M24	3	24	22.051	20.320	20.752	1.840	1.624	353
M30	3.5	30	27.727	25.706	26.211	2.147	1.894	561
M33	3.5	33	30.727	28.706	29.211	2.147	1.894	694
M36	4	36	33.402	31.093	31.67	2.454	2.165	817

TABLE C.1. (*Continued*)

Size	Pitch (P)	Major Diameter (d = D)	Pitch Diameter ($d_2 = D_2$)	Minor Diameter		Depth of Thread (h_3)	Max. Depth of Engagement (H_1)	Stress Area, A_s (mm²)
				Bolt (d_3)	Nut (d_1)			
Fine series								
M8 × 1	1	8	7.350	6.773	6.918	0.613	0.541	39.2
M10 × 1.25	1.25	10	9.188	8.466	8.647	0.767	0.677	61.2
M12 × 1.25	1.25	12	11.188	10.466	10.647	0.767	0.677	92.1
M16 × 1.5	1.5	16	15.026	14.16	14.376	0.920	0.812	167
M20 × 1.5	1.5	20	19.026	18.16	18.376	0.920	0.812	272
M24 × 2	2	24	22.701	21.546	21.835	1.227	1.083	384
M30 × 2	2	30	28.701	27.546	27.835	1.227	1.083	621
M36 × 3	3	36	34.051	32.32	35.752	1.840	1.624	865

Stress area $A_s = \dfrac{\pi}{4}\left(\dfrac{d_2 + d_3}{2}\right)^2$

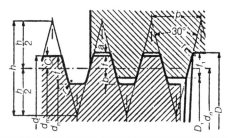

$h = 1.866p$
$f = 0.5p + a$
$f_1 = 0.5p + 2a - b$
$t = 0.5p + a - b$
$c = 0.25p$

d (mm)	d_n (mm)	A_s (mm²)	f (mm)	r (mm)	D (mm)	D_1 (mm)	f_1 (mm)	$r_1(1)$ (mm)	p (mm)	d_m (mm)	t (mm)	a (mm)	b (mm)
10	**6.5**	**33**	**1.75**	**0.25**	**10.5**	**7.5**	**1.5**	**0.20**	**3**	**8.5**	**1.25**	**0.25**	**0.5**
12	**8.5**	**57**	**1.75**	**0.25**	**12.5**	**9.5**	**1.5**	**0.20**	**3**	**10.5**	**1.25**	**0.25**	**0.5**
14	**9.5**	**71**	**2.25**	**0.25**	**14.5**	**10.5**	**2**	**0.20**	**4**	**12**	**1.75**	**0.25**	**0.5**
16	**11.5**	**104**	**2.25**	**0.25**	**16.5**	**12.5**	**2**	**0.20**	**4**	**14**	**1.75**	**0.25**	**0.5**
18	**13.5**	**143**	**2.25**	**0.25**	**18.5**	**14.5**	**2**	**0.20**	**4**	**16**	**1.75**	**0.25**	**0.5**
20	**15.5**	**189**	**2.25**	**0.25**	**20.5**	**16.5**	**2**	**0.20**	**4**	**18**	**1.75**	**0.25**	**0.5**
(22)	16.5	214	2.75	0.25	22.5	18	2.25	0.20	5	19.5	2	0.25	0.75
25	**19.5**	**299**	**2.75**	**0.25**	**25.5**	**21**	**2.25**	**0.20**	**5**	**22.5**	**2**	**0.25**	**0.75**
(28)	22.5	398	2.75	0.25	28.5	24	2.25	0.20	5	22.5	2	0.25	0.75
30	**23.5**	**434**	**3.25**	**0.25**	**30.5**	**25**	**2.75**	**0.20**	**6**	**27**	**2.5**	**0.25**	**0.75**
(32)	25.5	511	3.25	0.25	32.5	27	2.75	0.20	6	29	2.5	0.25	0.75
35	**28.5**	**638**	**3.25**	**0.25**	**35.5**	**30**	**2.75**	**0.20**	**6**	**32**	**2.6**	**0.25**	**0.75**
(38)	30.5	731	3.75	0.25	38.5	32	3.25	0.20	7	34.5	3	0.25	0.75
40	**32.5**	**830**	**3.75**	**0.25**	**40.5**	**34**	**3.25**	**0.20**	**7**	**36.5**	**3**	**0.25**	**0.75**
(42)	34.5	935	3.75	0.25	42.5	36	3.25	0.20	7	38.5	3	0.25	0.75
45	**36.5**	**1 046**	**4.25**	**0.25**	**45.5**	**38**	**3.75**	**0.20**	**8**	**41**	**3.5**	**0.25**	**0.75**
(48)	39.5	1 225	4.25	0.25	48.5	41	3.75	0.20	8	44	3.5	0.25	0.75
50	**41.5**	**1 353**	**4.25**	**0.25**	**50.5**	**43**	**3.75**	**0.20**	**8**	**46**	**3.5**	**0.25**	**0.75**
(55)	45.5	1 626	4.75	0.25	55.5	47	4.25	0.20	9	50.5	4	0.25	0.75
60	**50.5**	**2 003**	**4.75**	**0.25**	**60.5**	**52**	**4.25**	**0.20**	**9**	**55.5**	**4**	**0.25**	**0.75**
(65)	54.5	2 333	5.25	0.25	65.5	56	4.75	0.20	10	60	4.5	0.25	0.75
70	**59.5**	**2 781**	**5.25**	**0.25**	**70.5**	**61**	**4.75**	**0.20**	**10**	**65**	**4.5**	**0.25**	**0.75**
(75)	64.5	3 267	5.25	0.25	75.5	66	4.75	0.20	10	70	4.5	0.25	0.75
80	**69.5**	**3 794**	**5.25**	**0.25**	**80.6**	**71**	**4.75**	**0.20**	**10**	**75**	**4.5**	**0.25**	**0.75**

TABLE C.2. Key Dimensions

Rectangular keys

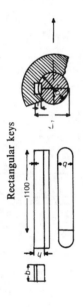

D (mm)	b (mm)	h (mm)	z (mm)	l (mm)	t (mm)	t_1 (mm)
10 to 12	4	4	0.3	10 12 15 18 20 25 30	2.5	$D + 1.5$
12 to 17	5	5	0.3	10 12 15 18 20 25 30 35 40	3	$D + 2$
17 to 22	6	6	0.3	12 15 18 20 25 30 35 40	3.5	$D + 2.5$
22 to 30	8	7	0.3	20 25 30 35 40 45 50	4	$D + 3$
30 to 38	10	8	0.3	20 25 30 35 40 45 50 60 70	4.5	$D + 3.5$
38 to 40	12	8	0.3	30 35 40 45 50 60 70 80 90 100 120	4.5	$D + 3.5$
44 to 50	14	9	0.4	35 40 45 50 60 70 80 90 100 120 140	5	$D + 4$
50 to 58	16	10	0.4	45 50 60 70 80 90 100 120 140	5	$D + 5$
58 to 68	18	11	0.4	50 60 70 80 90 100 120 140 160 180 200	6	$D + 5$
68 to 78	20	12	0.4	60 70 80 90 100 120 140 160 180 200 220	6	$D + 6$
78 to 92	24	14	0.4	70 80 90 100 120 140 160 180 200 220 250 280	7	$D + 7$
92 to 110	28	16	0.5	80 90 100 120 140 160 180 200 220 250 280 300	8	$D + 8$
110 to 130	32	18	0.5	90 100 120 140 160 180 200 220 250 280 300 350	9	$D + 9$
130 to 150	36	20	0.5	100 120 140 160 180 200 220 250 280 300 350 400	10	$D + 10$
150 to 170	40	22	0.5	100 120 140 160 180 200 220 250 280 300 350 400	11	$D + 11$
170 to 200	45	25	0.5	160 180 200 220 250 280 300 350 400	13	$D + 12$
200 to 230	50	28	0.5	160 180 200 220 250 280 300 350 400	14	$D + 14$

TABLE C.2. (*Continued*)

Woodroof keys

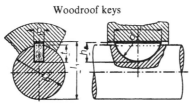

D (mm)	b × h (mm) × (mm)	t (mm)	t_1 (mm)	D (mm)	b × h (mm) × (mm)	t (mm)	t_1 (mm)
3 to 4	1 × 1.4	0.9	D + 0.6	22 to 28	6 × 9 6 × 10 6 × 11 6 × 13	7.4 8.4 9.4 11.4	D + 1.8
4 to 5	1.5 × 1.4 1.5 × 2.6	0.9 2.1	D + 0.6				
5 to 7	2 × 2.6 2 × 3.7	1.8 2.9	D + 0.9	28 to 38	8 × 11 8 × 13 8 × 15 8 × 16 8 × 17	9.5 11.5 13.5 14.5 15.5	D + 1.7
7 to 9	2.5 × 3.7	2.9	D + 0.9				
9 to 13	3 × 3.7 3 × 5 3 × 6.5	2.5 3.8 5.3	D + 1.3	38 to 48	10 × 16 10 × 17 10 × 19 10 × 24	14 15 17 22	D + 2.2
13 to 17	4 × 5 4 × 6.5 4 × 7.5	3.8 5.3 6.3	D + 1.4	48 to 58	12 × 19 12 × 24	16.5 21.5	D + 2.7
17 to 22	5 × 6.5 5 × 7.5 5 × 9 5 × 10	4.9 5.9 7.4 8.4	D + 1.8				

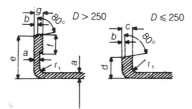

TABLE C.3. Slider Bearings

	a	b	c	d	e	f	g	r	r_1	r_3
25 to 50	0.5	0.5	1	7	—	—	—	2.5	1.5	1.5
50 to 75	1	1	1.5	9	—	—	—	4	2.5	4
75 to 100	1.8	1	2	10	—	—	—	4	2.5	4
100 to 150	2	1	2.5	12	—	—	—	6	4	6
150 to 200	3	1.5	3.5	15	—	—	—	10	4	6
200 to 250	3.5	1.5	4	17	—	—	—	10	6	10
250 to 300	4	1.5	—	—	22	7	6.5	15	6	10
300 to 400	5	2	—	—	28	9	7.5	18	10	15
400 to 500	5.5	2	—	—	35	10	8.5	22	15	20
500 to 600	6	2	—	—	40	12	9.5	25	15	22

TABLE C.3. (*Continued*)

D	a	b	c	d	e	f	g	h	i	k	l	m	n	o	p	q	s	t	u	v	r	r_1	r_3
25 to 50	2	1.5	8	6	8	8	12	1	4	5	7	—	—	—	—	—	—	—	—	18	2.5	1.5	1.5
50 to 75	2.5	2	9	8	10	10	14	1	4.5	5.5	9	—	—	—	—	—	—	—	—	20	4	2.5	4
75 to 100	3	2	10	9	12	12	16	1	5	6	10	—	—	—	—	—	—	—	—	25	4	2.5	4
100 to 150	3.5	2.5	12	10	14	14	19	1.5	6	7.5	12	—	—	—	—	—	—	—	—	30	6	4	6
150 to 200	4	2.5	14	12	17	19	22	1.5	7	8.5	17	—	—	—	—	—	—	—	—	36	10	4	6
200 to 250	4.5	3	16	14	19	23	25	1.5	8	10	20	—	—	—	—	—	—	—	—	40	12	6	10
250 to 300	5	3.5	18	16	22	28	28	2	—	—	—	7	10	10	2	22	20	7	12	45	15	6	10
300 to 400	6	3.5	20	18	24	32	32	2	—	—	—	8	11	12	2	28	24	8	14	50	18	10	15
400 to 500	7	4	22	20	27	37	36	2	—	—	—	9	12.5	14	2	35	28	9	16	55	22	15	20
500 to 600	8	4	24	22	30	40	40	2	—	—	—	10	14	16	2	40	32	10	18	60	25	15	22

954

TABLE C.4. Nominal Dimensions of Standard V-Belts

TYPE	Designation	Dimension	Size									
Heavy Duty	English					A	B	C	D	E		
	Metric		6C	8C	10C	13C	16C	22C	32C			
		Top Width				0.5	0.718	0.875	2.75	5.5	in.	
			6	8	10	13	18	22	70	140	mm	
		Thickness				0.343 75	0.437 5	0.562 5	0.75	1	in.	
			4	5	6	9	11	14	19	25	mm	
Narrow Section	English		3V	5V	8V							
	Metric		9N	15N	25N							
		Top width	0.375	0.625	1							
			10	16	25							
		Thickness	0.312	0.531	0.875							
			8	13	22							
Light Duty	English		2L	3L	4L	5L						
	Metric		6R	9R	12R	16R						
		Top width	0.281	0.375	0.5	0.656	in.					
			7	10	13	17	mm					
		Thickness	0.197	0.256	0.315	0.433	in.					
			5	6.5	8	11	mm					
Automotive	English		0.25	0.315	0.38	0.44	0.5	0.6	0.66	0.79	0.91	
	Metric		6A	8A	10A	11A	13A	15A	17A	20A	23A	
		Top width	0.25	0.315	0.38	0.44	0.5	0.6	0.66	0.79	0.91	in.
			6	8	10	11	13	15	17	20	23	mm
		Thickness	0.197	0.236	0.256	0.315	0.394	0.433	0.512	0.591	0.709	in.
			5	6	6.5	8	10	11	13	15	18	mm

APPENDIX D

HOT ROLLED SECTIONS

TABLE D.1. Double T Sections

x	h	b	$s=r_1$	t	r_2	F (cm^2)	G (kg/m)	U (m^2/m)	J_x (cm^4)	W_x (cm^3)	i_x (cm)	J_y (cm^4)	W_y (cm^3)	$i_y=i_i$ (min) (cm)	S_x (cm^3)	s_x (cm)
80	80	42	3.9	5.9	2.3	7.57	5.94	0.304	77.8	19.5	3.20	6.29	3.00	0.91	11.4	6.84
100	100	50	4.5	6.8	2.7	10.6	8.34	0.370	171	34.2	4.01	12.2	4.88	1.07	19.9	8.57
120	120	58	5.1	7.7	3.1	14.2	11.1	0.439	328	54.7	4.81	21.5	7.41	1.23	31.8	10.3
140	140	66	5.7	8.6	3.4	18.2	14.3	0.502	573	81.9	5.61	35.2	10.7	1.40	47.7	12.0
160	160	74	6.3	9.5	3.8	22.8	17.9	0.575	935	117	6.40	54.7	14.8	1.55	68.0	13.7
180	180	82	6.9	10.4	4.1	27.9	21.9	0.640	1 450	161	7.20	81.3	19.8	1.71	93.4	15.5
200	200	90	7.5	11.3	4.5	33.4	26.2	0.709	2 140	214	8.00	117	26.0	1.87	125	17.2
220	220	98	8.1	12.2	4.9	39.5	31.1	0.775	3 060	278	8.80	162	33.1	2.02	162	18.9
240	240	106	8.7	13.1	5.2	46.1	36.2	0.844	4 250	354	9.59	221	41.7	2.20	206	20.6
260	260	113	9.4	14.1	5.6	53.3	41.9	0.906	5 740	442	10.4	288	51.0	2.32	257	22.3
280	280	119	10.1	15.2	6.1	61.0	47.9	0.966	7 590	542	11.1	364	61.2	2.45	316	24.0
300	300	125	10.8	16.2	6.5	69.0	54.2	1.03	9 800	653	11.9	451	72.2	2.56	381	25.7
320	320	131	11.5	17.3	6.9	77.7	61.0	1.09	12 510	782	12.7	555	84.7	2.67	457	27.4
340	340	137	12.2	18.3	7.3	86.7	68.0	1.15	15 700	923	13.5	674	98.4	2.80	540	29.1
360	360	143	13.0	19.5	7.8	97.0	76.1	1.21	19 610	1 090	14.2	818	114	2.90	638	30.7
380	380	149	13.7	20.5	8.2	107	84.0	1.27	24 010	1 260	15.0	975	131	3.02	741	32.4
400	400	155	14.4	21.6	8.6	118	92.4	1.33	24 210	1 460	15.7	1 160	149	3.13	857	34.1
425	425	163	15.3	23.0	9.2	132	104	1.41	36 970	1 740	16.7	1 440	176	3.30	1 020	36.2
450	450	170	16.2	24.3	9.7	147	115	1.48	45 850	2 040	17.7	1 730	203	3.43	1 200	38.3
475	475	178	17.1	25.6	10.3	163	128	1.55	56 480	2 380	18.6	2 090	235	3.60	1 400	40.4
500	500	185	18.0	27.0	10.8	179	141	1.63	68 740	2 750	19.6	2 480	268	3.72	1 620	42.4
550	550	200	19.0	30.0	11.9	212	166	1.80	99 180	3 610	21.6	3 490	349	4.02	2 120	46.8
600	600	215	21.6	32.4	13.0	254	199	1.92	139 000	4 630	23.4	4 670	434	4.30	2 730	50.9

TABLE D.2. U Sections

C	h	b	s	t = *r₁	r₂	F (cm²)	G (kg/m)	U (m²/m)	J_x (cm⁴)	W_x (cm³)	i_x (cm)	J_y (cm⁴)	W_y (cm³)	i_y (cm)	S_x (cm³)	s_x (cm)	e_y (cm)	x_M (cm)
												$x - x$				**$y - y$**		
30 × 15	30	15	4	4.5	2	2.21	1.74	0.103	2.53	1.69	1.07	0.38	0.39	0.42	—	—	0.52	0.74
30	30	33	5	7	3.5	5.44	4.27	0.174	6.39	4.26	1.08	5.33	2.68	0.99	—	—	1.31	2.22
40 × 20	40	20	5	5.5*	2.5	3.66	2.87	0.142	7.58	3.79	1.44	1.14	0.86	0.56	—	—	0.67	1.01
40	40	35	5	7	3.5	6.21	4.87	0.199	14.1	7.05	1.50	6.68	3.08	1.04	—	—	1.33	2.32
50 × 25	50	25	5	6	3	4.92	3.86	0.181	16.8	6.73	1.85	2.49	1.48	0.71	—	—	0.81	1.34
50	50	38	5	7	3.5	7.12	5.59	0.232	26.4	10.6	1.92	9.12	3.75	1.13	—	—	1.37	2.47
60	60	30	6	6	3	6.46	5.07	0.215	31.6	10.5	2.21	4.51	2.16	0.84	—	—	0.91	1.50
65	65	42	5.5	7.5	4	9.03	7.09	0.273	57.5	17.7	2.52	14.1	5.07	1.25	—	—	1.42	2.60
80	80	45	6	8	4	11.0	8.64	0.312	106	26.5	3.10	19.4	6.36	1.33	15.9	6.65	1.45	2.67
100	100	50	6	8.5	4.5	13.5	10.6	0.372	206	41.2	3.91	29.3	8.49	1.47	24.5	8.42	1.55	2.93
120	120	55	7	9	4.5	17.0	13.4	0.434	364	60.7	4.62	43.2	11.1	1.59	36.3	10.0	1.60	3.03
140	140	60	7	10	5	20.4	16.0	0.489	605	86.4	5.45	62.7	14.8	1.75	51.4	11.8	1.75	3.37
160	160	65	7.5	10.5	5.5	24.0	18.8	0.546	925	116	6.21	85.3	18.3	1.89	68.8	13.3	1.84	3.56
180	180	70	8	11	5.5	28.0	22.0	0.611	1 350	150	6.95	114	22.4	2.02	89.6	15.1	1.92	3.75
200	200	75	8.5	11.5	6	32.2	25.3	0.661	1 910	191	7.70	148	27.0	2.14	114	16.8	2.01	3.94
220	220	80	9	12.5	6.5	37.4	29.4	0.718	2 690	245	8.48	197	33.6	2.30	146	18.5	2.14	4.20
240	240	85	9.5	13	6.5	42.3	33.2	0.775	3 600	300	9.22	248	39.6	2.42	179	20.1	2.23	4.39
260	260	90	10	14	7	48.3	37.9	0.834	4 820	371	9.99	317	47.7	2.56	221	21.8	2.36	4.66
280	280	95	10	15	7.5	53.3	41.8	0.890	6 280	448	10.9	399	57.2	2.74	266	23.6	2.53	5.02
300	300	100	10	16	8	58.8	46.2	0.950	8 030	535	11.7	495	67.8	2.90	316	25.4	2.70	5.41
320	320	100	14	17.5	8.75	75.8	59.5	0.982	10 870	629	12.1	597	80.6	2.81	413	26.3	2.60	4.82
350	350	100	14	16	8	77.3	60.6	1.047	12 840	734	12.9	570	75.0	2.72	459	28.6	2.40	4.45
380	380	102	13.5	16	8	80.4	63.1	1.110	15 750	829	14.0	615	78.7	2.77	507	31.1	2.38	4.58
400	400	110	14	18	9	91.5	71.8	1.182	20 350	1 020	14.9	846	102	3.04	618	32.9	2.65	5.11

TABLE D.3. T Sections

h	b	s=t =r₁	r₂	r₃	F (cm²)	G (kg/m)	U (m²/m)	e_x (cm)	J_x (cm⁴)	W_x (cm³)	I_x (cm)	J_y (cm⁴)	W_y (cm²)	i_y = i_i (cm)	d (mm)	w₁ (mm)	w₂ (mm)
									x − x			y − y					
T																	
20	20	3	1.5	1	1.12	0.88	0.075	0.58	0.38	0.27	0.58	0.20	0.20	0.42	3.2	—	—
25	25	3.5	2	1	1.64	1.29	0.094	0.73	0.87	0.49	0.73	0.43	0.34	0.51	3.2	15	14
30	30	4	2	1	2.26	1.77	0.114	0.85	1.72	0.80	0.87	0.87	0.58	0.62	4.3	17	17
35	35	4.5	2.5	1	2.97	2.33	0.133	0.99	3.10	1.23	1.04	1.57	0.90	0.73	4.3	19	19
40	40	5	2.5	1.5	3.77	2.96	0.153	1.12	5.28	1.84	1.18	2.58	1.29	0.83	6.4	21	22
45	45	5.5	3	1.5	4.67	3.67	0.171	1.26	8.13	2.51	1.32	4.01	1.78	0.93	6.4	24	25
50	50	6	3	1.5	5.66	4.44	0.191	1.39	12.1	3.36	1.46	6.06	2.42	1.03	6.4	30	30
60	60	7	3.5	2	7.94	6.23	0.229	1.66	23.8	5.48	1.73	12.2	4.07	1.24	8.4	34	35
70	70	8	4	2	10.6	8.32	0.268	1.94	44.5	8.79	2.05	22.1	6.32	1.44	11	38	40
80	80	9	4.5	2	13.6	10.7	0.307	2.22	73.7	12.8	2.33	37.0	9.25	1.65	11	45	45
90	90	10	5	2.5	17.1	13.4	0.345	2.48	119	18.2	2.64	58.5	13.0	1.85	13	50	50
100	100	11	5.5	3	20.9	16.4	0.383	2.74	179	24.6	2.92	88.3	17.7	2.05	13	60	60
120	120	13	6.5	3	29.6	25.2	0.459	3.82	366	42.0	3.51	178	29.7	2.45	17	70	70
140	140	15	7.5	4	39.9	21.3	0.537	3.80	660	46.7	4.07	330	47.2	2.88	21	80	75

TABLE D.3. (*Continued*)

T	h	b	s	t	F (cm²)	G (kg/m)	U (m²/m)
20	20	20	3	3	1.11	0.871	0.080
25	25	25	3.5	3.5	1.63	1.28	0.100
30	30	30	4	4	2.24	1.76	0.120
35	35	35	4.5	4.5	2.95	2.31	0.140
40	40	40	5	5	3.75	2.94	0.160

L	a	s	F (cm²)	G (kg/m)	U (m²/m)
20 × ³⁄₄	20	3 / 4	1.11 / 1.44	0.871 / 1.13	0.080
25 × ³⁄₄	25	3 / 4	1.44 / 1.84	1.11 / 1.44	0.100
30 × ³⁄₄	30	3 / 4	1.71 / 2.24	1.34 / 1.76	0.120
35 × 4	35	4	2.64	207	0.140
40 × ⁴⁄₅	40	4 / 5	3.04 / 3.75	2.38 / 2.94	0.160
45 × 5	45	5	4.25	3.34	0.180
50 × 5	50	5	4.75	3.73	0.200

TABLE D.4 Angle Sections

L	a	s	r_1	r_2	F (cm²)	G (kg/m)	U (m²/m)	$e_x = e_y$ (cm)	w (cm)	v_1 (cm)	v_2 (cm)	$J_x = J_y$ (cm⁴)	$W_x = W_y$ (cm³)	$i_x = i_y$ (cm)	J_ξ (cm⁴)	i_ξ (cm)	J_η (cm⁴)	W_η (cm)	i_η (cm)
														$x-x = y-y$		$\xi - \xi$		$\eta - \eta$		
20 ×	20	3	3.5	2	1.12	0.88	0.077	0.60	1.41	0.85	0.70	0.39	0.28	0.59	0.62	0.74	0.15	0.18	0.37
		4			1.45	1.14		0.64		0.90	0.71	0.48	0.35	0.58	0.77	0.73	0.19	0.21	0.36
25 ×	25	3	3.5	2	1.42	1.12	0.097	0.73	1.77	1.03	0.87	0.79	0.45	0.75	1.27	0.95	0.31	0.30	0.47
		4			1.85	1.45		0.76		1.08	0.89	1.01	0.58	0.74	1.61	0.93	0.40	0.37	0.47
		5			2.26	1.77		0.80		1.13	0.91	1.18	0.69	0.72	1.87	0.91	0.50	0.44	0.47
30 ×	30	3	5	2.5	1.74	1.36	0.116	0.84	2.12	1.18	1.04	1.41	0.65	0.90	2.24	1.14	0.57	0.48	0.57
		4			2.27	1.78		0.89		1.24	1.05	1.81	0.86	0.89	2.85	1.12	0.76	0.61	0.58
		5			2.78	2.18		0.92		1.30	1.07	2.16	1.04	0.88	3.41	1.11	0.91	0.70	0.57
35 ×	35	3	5	2.5	2.04	1.60	0.136	0.96	2.47	1.36	1.23	2.29	0.90	1.06	3.63	1.34	0.95	0.70	0.68
		4			2.67	2.10		1.00		1.41	1.24	2.96	1.18	1.05	4.68	1.33	1.24	0.88	0.68
		5			3.28	2.57		1.04		1.47	1.25	3.56	1.45	1.04	5.63	1.31	1.49	1.10	0.68
		6			3.87	3.04		1.08		1.53	1.27	4.14	1.71	1.04	6.50	1.30	1.77	1.16	0.68
40 ×	40	3	6	3	2.35	1.84	0.155	1.07	2.83	1.52	1.40	3.45	1.48	1.21	5.45	1.52	1.44	0.95	0.78
		4			3.08	2.42		1.12		1.58	1.40	4.48	1.56	1.21	7.09	1.52	1.86	1.18	0.78
		5			3.79	2.97		1.16		1.64	1.42	5.43	1.91	1.20	8.64	1.51	2.22	1.35	0.77
		6			4.48	3.52		1.20		1.70	1.43	6.33	2.26	1.19	9.98	1.49	2.67	1.57	0.77
45 ×	45	4	7	3.5	3.49	2.74	0.174	1.23	3.18	1.75	1.57	6.43	1.97	1.36	10.2	1.71	2.68	1.53	0.88
		5			4.30	3.38		1.28		1.81	1.58	7.83	2.43	1.35	12.4	1.70	3.25	1.80	0.87
		6			5.09	4.00		1.32		1.87	1.59	9.16	2.88	1.34	14.5	1.69	3.83	2.05	0.87
		7			5.86	4.60		1.36		1.92	1.61	10.4	3.31	1.33	16.4	1.67	4.39	2.29	0.87

TABLE D.4 (*Continued*)

L	a	s	r_1	r_2	F (cm²)	G (kg/m)	U (m²/m)	$e_x = e_y$ (cm)	w (cm)	v_1 (cm)	v_2 (cm)	$J_x = J_y$ (cm⁴)	$W_x = W_y$ (cm³)	$i_x = i_y$ (cm)	J_ξ (cm⁴)	i_ξ (cm)	J_η (cm⁴)	W_η (cm)	i_η (cm)
														x – x = y – y		*ξ – ξ*		*η – η*	
50 ×	50	4	7	3.5	3.89	3.06	0.194	1.36	3.54	1.92	1.75	8.97	2.46	1.52	14.2	1.91	3.73	1.94	0.98
		5			4.80	3.77		1.40		1.98	1.76	11.0	3.05	1.51	17.4	1.90	4.59	2.32	0.98
		6			5.69	4.47		1.45		2.04	1.77	12.8	3.61	1.50	20.4	1.89	5.24	2.57	0.98
		7			6.56	5.15		1.49		2.11	1.78	14.6	4.15	1.49	23.1	1.88	6.02	2.85	0.98
		8			7.41	5.82		1.52		2.16	1.80	16.3	4.68	1.48	25.7	1.86	6.87	3.19	0.98
		9			8.24	6.47		1.56		2.21	1.82	17.9	5.20	1.47	28.1	1.85	7.67	3.47	0.98
55 ×	55	5	8	4	5.32	4.18	0.213	1.56	3.89	2.15	1.93	14.7	3.70	1.66	23.3	2.09	6.11	2.84	1.07
		6			6.31	4.95		1.64		2.21	1.94	17.3	4.40	1.66	27.4	2.08	7.24	3.28	1.07
		8			8.23	6.46		1.64		2.32	1.97	22.1	5.72	1.64	34.8	2.06	9.35	4.03	1.07
		10			10.1	7.90		1.72		2.43	2.00	26.3	6.97	1.62	41.4	2.02	11.3	4.65	1.07
60 ×	60	5	8	4	5.82	4.57	0.223	1.64	4.24	2.32	2.11	19.4	4.45	1.82	30.7	2.30	8.03	3.46	1.17
		6			6.91	5.42		1.69		2.39	2.11	22.8	5.29	1.82	36.1	2.29	9.43	3.95	1.17
		8			9.03	7.09		1.77		2.50	2.14	29.1	6.88	1.80	46.1	2.26	12.1	4.84	1.17
		10			11.1	8.69		1.85		2.62	2.17	34.9	8.41	1.78	55.1	2.23	14.6	5.57	1.17
65 ×	65	6	9	4.5	7.53	5.91	0.252	1.80	4.60	2.55	2.28	29.2	6.21	1.97	46.3	2.48	12.1	4.74	1.27
		7			8.70	6.83		1.85		2.62	2.29	33.4	7.18	1.96	53.0	2.47	13.8	5.27	1.27
		8			9.85	7.73		1.89		2.67	2.31	37.5	8.13	1.95	59.4	2.46	15.6	5.84	1.27
		9			11.0	8.62		1.93		2.73	2.32	41.3	9.04	1.94	65.4	2.44	17.2	6.30	1.27
		11			13.2	10.3		2.00		2.83	2.36	48.8	10.8	1.91	76.8	2.42	20.7	7.31	1.27

APPENDIX E

PROPERTIES OF SOME ENGINEERING MATERIALS

TABLE E.1. Physical Properties of Engineering Materials

Material Number	Material Name	Specific Weight	Young's Modulus (GPa)	Shear Modulus (GPa)	Thermal Coeff. ($\times 10^6$)/°C	Sp. Heat (kJ/kg °C)	Thermal Cond. (kJ/mh °C)	Electrical Resistance $\mu(\Omega m)$	Poisson Ratio
1	Aluminum	2.70	62.1	23.3	22.2	0.921	775	0.027	0.34
2	Wrought Al alloys	2.72	74	28	22	0.921	500	0.045	0.34
3	Cast Al alloys	2.7	68	28	23	0.921	560	0.053	0.34
4	Structural steels	7.85	210	85	11.45	0.477	190	0.17	0.27
5	Alloy steels	7.85	210	84	11.4	0.510	120	0.7	0.27
6	Stainless steels	7.7	200	86	18	0.5	45	0.7	0.29
7	Heat res. steels	7.83	210	82	11.45	0.4	45	0.8	0.35
8	Copper	8.97	117	50	16	0.385	1400	0.017	0.295
9	Bronze	8.5	112	41	17	0.385	600	0.045	0.295
10	Brass	8.5	109	40	17	0.377	245	0.08	0.295
11	Cast iron	7.5	66–170	9.6–28	10	0.586	180	0.9	0.2
12	Cast steel	7.83	207	77	12.5	0.48	134	1	0.31
13	Mg alloys	1.8	45	16.6	26	1.05	300	0.14	0.3
14	Titanium	4.51	107	41	8.5	0.469	50	0.12	0.34

TABLE E.2. Mechanical Properties of Cast Iron

Material Number	Commercial Name	ISO Designation	USA Standard	USA Designation	DIN Designation	Ultimate Strength (MPa)	Yield Strength (MPa)	Fatigue Strength (MPa)	Elongation (%)	Notes
1	Cast iron	185 GR 10	ASTM A159	G1800	GG-10	100	100	70	0.37	Machine frames 3-30 mm thick
2	Cast iron	185 GR 15	ASTM A159	G2500	GG-15	550–700	150–200	70	0.37	Above
	Cast iron	185 GR 20	ASTM A159	G3000	GG-20	600–850	200–250	90	0.37	Above
4	Cast iron	185 GR 25	ASTM A159	G3500	GG-25	700–1000	250–300(T)	120	0.37	Above
5	Cast iron	185 GR 30	ASTM A159	G4000	GG-30	800–1200	300–350(T)	140	0.33	Above
6	Cast iron	185 GR 35	ASTM A48	CLASS 50	GG-35	950–1400	340–400(T)	150	0.33	Above
7	Cast iron	185 GR 40	ASTM A48	CLASS 55	GG-40	1100–1400	400–450(T)	160	0.33	Above
8	Cast iron/steel	1083 GR 38-17	ASTM A536	GR 60-40-18	GGG-38	370	230	160	17	Cast parts with impact loads
9	Cast iron/steel	1083 GR 42-12	ASTM A536	GR 60-45-12	GGG-40	410	270	170	12	Above
10	Cast iron/steel	1083 GR 50-7	ASTM A536	GR 80-55-06	GGG-50	490	340	200	20	Above
11	Cast iron/steel	1083 GR 60-2	NA	NA	GGG-60	590	390	230	23	Above
12	Cast iron/steel	1083 GR 70-2	ASTM A536	GR 100-70-03	GGG-70	690	440	260	26	Above
13	Cast iron/steel	NA	ASTM A536	GR 120-90-02	GGG-80	800	450	270	27	Above
14										

TABLE E.3. Mechanical Properties of Austenitic Cast Iron

Material Number	Commercial Name	ISO Designation	USA Standard	USA Designation	DIN Designation	Ultimate Strength (MPa)	Yield Strength (MPa)	Fatigue Strength (MPa)	Elongation (%)	Notes
1	Austenitic cast iron	S-NiMn13 7			GGG-NiMn13 7	390	210	NA	15	
2	Austenitic cast iron	S-NiCr20 2	ASTM A439	TYPE D-2	GGG-NiCr20 2	370	210	NA	7	
3	Austenitic cast iron	S-NiCr20 3	ASTM A439	TYPE D-2B	GGG-NiCr20 3	390	210	NA	7	
4	Austenitic cast iron	S-NiSiCr-2 0 5 2			GGG-NiSiCr 20 4 2	370	210	NA	10	
5	Austenitic cast iron	S-Ni22	ASTM A439	TYPE D-2C	GGG-Ni 22	370	170	NA	20	
6	Austenitic cast iron	S-NiMn23 4			GGG-NiMn23 4	440	210	NA	25	
7	Austenitic cast iron	s-NiCr30 1	ASTM A439	TYPE D-3A	GGG-NiCr30 1	370	210	NA	13	
8	Austenitic cast iron	s-NiCr30 3	ASTM A439	TYPE D-3	GGG-NiCr30 3	370	210	NA	7	
9	Austenitic cast iron	S-NiSiCr-3 0 5 5	ASTM A439	TYPE D-4	GGG-NiSiCr 30 5 5	390	240		7	

No.	Material	Designation	Standard	Type	German designation					Remark
10	Austenitic cast iron	s-Ni35	ASTM A439	TYPE D-5	GGG-Ni35	370	210	NA	20	
11	Austenitic cast iron	S-NiCr35 3	ASTM A439	TYPE D-5B	GGG-NiCr35 3	370	210		7	
12	Austenitic cast iron	L-NiCu-Cr1 5 6 2	ASTM A436	TYPE 1	GGT-NiCuCr 15 6 2	170				Laminated gray
13	Austenitic cast iron	L-NiCu-Cr1 5 6 3	ASTM A436	TYPE 1b	GGL-NiCuCr 15 6 3	190				Laminated gray
14	Austenitic cast iron	L-NiCr-20 2	ASTM A436	TYPE 2	GGL-NiCr20 2	170				
15	Austenitic cast iron	L-NiCr20 3	ASTM A436	TYPE 2b	GGL-NiCr20 3	190				
16	Austenitic cast iron	L-NiCr-30 3	ASTM A436	TYPE 3	GGNiCr30 3	190				
17	Austenitic cast iron	L-NiSi-Cr3 0 5 5	ASTM A436	TYPE 4	GGL-NiSiCr 30 5 5	170				
18	Austenitic cast iron	L-Ni35	ASTM A436	TYPE 5	GGL-Ni35	120				
19										

TABLE E.4. Mechanical Properties of Steel Castings

Material Number	Commercial Name	ISO Designation	USA Standard	USA Designation	DIN Designation	Ultimate Strength (MPa)	Yield Strength (MPa)	Fatigue Strength (MPa)	Elongation (%)	Notes
1	Mall iron/BL	943 GR C				290			6	Small parts/Light loads
2	Mall iron/BL	943 GR B	ASTM A47	GR 32510		310	190	NA	10	Above
3	Mall iron/BL	943 GR A	ASTM A47	GR 35018	GTS-35	340	210	180	12	Engine parts up to 100 kg thick
4	Mall iron/BL	942 GR B			GTW-35	310	NA	NA	4	Whitehart CI/Parts up to 50 kg
5	Mall iron/Perl		ASTM A220	GR 40010		410	270	NA	10	
6	Mall iron/Perl	944 GR E	ASTM A220	GR 45006	GTS-45	440	270	NA	7	As 4 above
7	Mall iron/Perl	944 GR D	ASTM A220	GR 50005		490	310	NA	5	
8	Mall iron/Perl	944 GR C	ASTM A220	GR 60004	GTS-55	540	350	NA	4	See 3 above
9	Mall iron/Perl	944 GR B	ASTM A220	GR 80002	GTS-65	640	420	NA	3	See 3 above
10	Mall iron/Perl	944 GR A 4444 GR 44-44	ASTM A220	GR90001 GR444-444	GTS-70	690 4444	540	NA	2	See 3 above

11	Steel castings	3755 GR 20-40	ASTM A27	GR 60-30	GS 38	400	200	NA	25
12	Steel castings	3755 GR 26-52	ASTM A27	GR 70-40	GS-52	520	260	18	18
13	Steel castings	3755 GR 26-52	ASTM A27	GR 70-40	GS-52	520	260	NA	18
14	Steel castings	3755 GR 30-57	ASTM A148	GR 80-50	GS-60	570	300	NA	15
15	Steel castings		ASTM A148	GR 90-60	GS-62	620	410	NA	20
16	Steel castings		ASTM A148	GR 105-85	GS-70	720	590	NA	17
17	Steel castings		ASTM A148	GR 120-95		830	650		14
18	Steel castings		ASTM A148	GR 150-125		1030	860	NA	9
19									
20	Steel castings	944 GR D	ASTM A220	GR 50005		490			
21	Steel castings	944 GR C	ASTM A220	GR 60004	GTS-55	540			
22	Steel castings	944 GR B	ASTM A220	GR 80002	GTS-65	640			
23	Steel castings	944 GR A	ASTM A220	GR 90001	GTS-70	690			
24	Steel castings	3755 GR 20-40	ASTM A27	GR 60-30	GS-38	400			
25	Steel castings	3755 GR 23-45	ASTM A27	GR 65-35	GS-45	450			
26	Steel castings	3755 GR 26-52	ASTM A27	GR 70-40	GS-52	520			
27	Steel castings	3755 GR 30-57	ASTM A148	GR 80-50	GS-60	570			
28	Steel castings		ASTM A148	GR 90-60	GS-62	620			
29	Steel castings		ASTM A148	GR 105-85	GS-70	720			
30	Steel castings		ASTM A148	GR 120-95		830			
31	Steel castings		ASTM A148	GR 150-125		1030			
32									
33									

TABLE E.5. Mechanical Properties of Structural Steel

Material Number	Commercial Name	ISO Designation	USA Standard	USA Designation	DIN Designation	Ultimate Strength (MPa)	Yield Strength (MPa)	Fatigue Strength (MPa)	Elongation (%)	Note
1	Structural steel	630 Fe 37-A	ASTM A284	GRADE D	USt37-1	360	230	170	26	General use
2	Structural steel	630 Fe 42-A	ASTM A570	GRADE D	USt42-1	410	250	190	23	Impact and dynamic loads
3	Structural steel	630 Fe 44-A	ASTM A470	GRADE E	RSt46-2	430	270	210	23	
4	Structural steel	630 Fe 52-B	ASTM A572	GRADE 50	St52-3	490	350	265	22	High strength crane components
5	Structural steel	1052 Fe 50-1	ASTM A572	GRADE 50	St52-3	690	290	265	20	
6	Structural steel	1052 Fe 60-1	ASTM A572	GRADE 50	St60-1	590	330	290	15	High strength power transmissions
7	Structural steel	1052 Fe 70-2	ASTM A572	GRADE 55	St70-2	690	360	320	11	High local stresses/dies /rolls

TABLE E.6. Mechanical Properties of Carbon Steels

Material Number	Commercial Name	ISO Designation	USA Standard	USA Designation	DIN Designation	S_u Ultimate Strength (MPa)	S_y Yield Strength (MPa)	S_n' Fatigue Strength (MPa)	Elongation (%)	Notes
1	Carbon steels		AISI	1010	C10	460	355	250	25	Cold drawn/General use in machinery
2	Carbon steels		AISI	1015	C15	527	430	273	25	As above
3	Carbon steels		AISI	1020	C22	500–650	300–420	190	20	As above
4	Carbon steels	C25	AISI	1025		540–690	360–	NA	18	As above
5	Carbon steels	C30	AISI	1030		580–730	390–	NA	17	As above
6	Carbon steels	C35	AISI	1035	C35	620–760	420–500	290	17	As above
7	Carbon steels	C40	AISI	1040		660–800	450–	NA	16	
8	Carbon steels	C45	AISI	1045	C45	700–840	480–500	330	14	
9	Carbon steels	C50	AISI	1050		720–880	510–	NA	13	
10	Carbon steels	C55	AISI	1055	C55	780–930	540–	NA	12	
11	Carbon steels	C60	AISI	1060	C60	830–980	570–600	400	11	
12										

TABLE E.7. Mechanical Properties of Stainless Steels, Ferritic/Martensitic

Material Number	Commercial Name	ISO Designation	USA Standard	USA Designation	DIN Designation	Ultimate Strength (MPa)	Yield Strength (MPa)	Fatigue Strength (MPa)	Elongation (%)	Notes
1	Stainless FERR/MART	683/13 GRADE 1	AISI	403	X7CR13	440–640	250	NA	20	
2	Stainless FERR/MART	GRADE 2	AISI	405	X7CRAL13	410–610	250	NA	20	
3	Stainless FERR/MART	GRADE 8	AISI	430	X8CR17	440–640	250	NA	18	
4	Stainless FERE/MART	GRADE 8A	AISI	430F	X8CR17	440–640	250	NA	15	
5	Stainless FERR/MART	GRADE 8B	AISI	—	X8CRTI17	440–640	250	NA	18	
6	Stainless FERR/MART	GRADE 9C	AISI	436	X6CRMO17	440–640	250	NA	18	
7	Stainless FERR/MART	GRADE 3	AISI	410	X10CR13	590–780	410	NA	16	
8	Stainless FERR/MART	GRADE 7	AISI	416	X12CRS13	640–830	440	NA	12	
9	Stainless FERR/MART	GRADE 4	AISI	420	X20CR13	690–880	490	NA	14	
10	Stainless FERR/MART	GRADE 9	AISI	431	—	830–1030	640	NA	10	
11	Stainless FERR/MART	GRADE 9B	AISI	—	X22CINI17	880–1130	690	NA	9	
12	Stainless FERR/MART	GRADE 5	AISI	420 FSE	X30CR13	780–980	590	380	11	
13										

TABLE E.8. Mechanical Properties of Alloy, Direct Hardening Steels

Material Number	Commercial Name	ISO Designation	USA Standard	USA Designation	DIN Designation	Ultimate Strength (MPa)	Yield Strength (MPa)	Fatigue Strength (MPa)	Elongation (%)	Notes
1	Alloy direct hardening steels	R683 PART 2 GR 1	AISI	4130	24CrMo4	880–1080	690–	NA	12	Large parts heavily stressed
2	Alloy direct hardening steels	R683 PART 2 GR 2	AISI	4135	34CrMo4	980–1080	690–850	450	11	As above
3	Alloy direct hardening steels	R683 PART 2 GR 3	AISI	4140	43CrMo4	1080–1270	880–1000	500	10	As above
4	Alloy direct hardening steels	R683 PART 5 GR 1	AISI	1527	28Mn6	780–930	880–1000	350	13	Low stresses/ automotive use
5	Alloy direct hardening steels	R683 PART 6 GR 1			32CrMo12	1080–1270	880–1050	630	10	Large parts highly stressed
6	Alloy direct hardening steels	R683 PART 7 GR 1	AISI	5132	34Cr4	880–1080	500–700	NA	12	As above
7	Alloy direct hardening steels	R683 PART 7 GR 2	AISI	5135	37Cr4	930–1130	740–	NA	11	As above
8	Alloy direct hardening steels	R683 PART 7 GR 3	AISI	5140	41Cr4	980–1240	780–	NA	11	As above

TABLE E.8. (*Continued*)

Material Number	Commercial Name	ISO Designation	USA Standard	USA Designation	DIN Designation	Ultimate Strength (MPa)	Yield Strength (MPa)	Fatigue Strength (MPa)	Elongation (%)	Notes
9	Alloy direct hardening steels	R683 PART 8 GR 1	AISI	8740		1030–1230	830–	NA	10	
10	Alloy direct hardening steels	R683 PART 8 GR 2	AISI	9840		1030–1230	830–	NA	10	
11	Alloy direct hardening steels	R683 PART 8 GR 3	AISI	4340	34CrNiMo6	1180–1370	980–	NA	9	
12	Alloy direct hardening steels	R683 PART 8 GR 5			30CrNiMo8	1230–1420	1030–	NA	9	
13	Alloy direct hardening steels	R683 PART 8 GR 6				1230–1420	1030–	NA	9	

TABLE E.9. Mechanical Properties of Free-Cutting Steels

Material Number	Commercial Name	ISO Designation	USA Standard	USA Designation	DIN Designation	Ultimate Strength (MPa)	Yield Strength (MPa)	Fatigue Strength (MPa)	Elongation (%)	Notes
1	Free cutting steels	R683/9 GR 1	AISI	1211	9S20	490–780	390	NA	8	Non hardened use in machines
2	Free cutting steels	R683/9 GR 2	AISI	1212	9SMn28	510–800	410	NA	7	Above
3	Free cutting steels	R683/9 GR 2Pb	AISI	12L13	9SMnPb28	510–800	410	NA	7	Above
4	Free cutting steels	R683/9 GR 3	AISI	1214	9SMn36	540–830	430	NA	7	Above
5	Free cutting steels	R683/9 GR 3Pb	AISI	12L14	9SMnPb36	540–830	430	NA	7	Above
6	Free cutting steels	R683/9 GR 4	AISI	1108	10S20	490–780	390	NA	8	Case hardened
7	Free cutting steels	R683/9 GR 4Pb	AISI	11L08	10SPb20	490–780	390	NA	8	Above
8	Free cutting steels	R683/9 GR 5	AISI	1117		510–800	410	NA	7	Above
9	Free cutting steels	R683/9 GR 6	AISI	1115		540–830	410	NA	7	Above
10	Free cutting steels	R683/9 GR 7	AISI	1138	35S20	570–760	390	NA	14	Hardened
11	Free cutting steels	R683/9 GR 8	AISI	1140		620–810	420	NA	14	Above
12	Free cutting steels	R683/9 GR 9	AISI	1137		740–930	510	NA	12	Above
13	Free cutting steels	R683/9 GR 10	AISI	1146	45S20	650–840	450	NA	11	Above

TABLE E.10. Mechanical Properties of Nitriding, Case Hardening Steels

Material Number	Commercial Name	ISO Designation	USA Standard	USA Designation	DIN Designation	Ultimate Strength (MPa)	Yield Strength (MPa)	Fatigue Strength (MPa)	Elongation (%)	Notes
1	Nitriding steel	R683 PART 10 GR 1			31CrMo12	1080–1270	880	NA	10	Components resistant to wear
2	Nitriding steel	R683 PART 10 GR 2			39CrMoV139	1270–1470	1080	NA	8	Above
3	Nitriding steel	R683 PART 10 GR 3	ASTM CLASS D	A355	34CrA1Mo5	780–930	590	NA	14	Above
4	Nitriding steel	R683 PART 10 GR 4	ASTM CLASS A	A355	41CrA1Mo7	930–1130	740	NA	12	Above
5	Case hardening steel	R683 PART 11 GR 1	AISI	1010	C10	490–830	290	250	13	
6	Case hardening steel	R683 PART 11 GR 2	AISI	1015	C15	590–930	340	170	12	
7	Case hardening steel	R683 PART 11 GR 3	AISI	1016		640–980	390	NA	10	
8	Case hardening steel	R683 PART 11 GR 4	AISI	5120	12Cr3	830–1180	540	320	10	

No.	Material	Standard		No.	Designation	Range			
9	Case hardening steel	R683 PART 11 GR 5	AISI	5115	16MnCr5	930–1270	640	440	9
10	Case hardening steel	R683 PART 11 GR 7	AISI			1030–1370	690	NA	8
11	Case hardening steel	R683 PART 11 GR 8	AISI	4118	20MoCr4	930–1270	640	NA	9
12	Case hardening steel	R683 PART 11 GR 9	AISI	4718		980–1320	640	NA	8
13	Case hardening steel	R683 PART 11 GR 10			15CrNi6	1030–1370	690	450	8
14	Case hardening steel	R683 PART 11 GR 11				980–1320	640	NA	8
15	Case hardening steel	R683 PART 11 GR 12	AISI	8620		980–1320	640	NA	8
16	Case hardening steel	R683 PART 11 GR 13	AISI	4320		1080–1420	740	NA	8
17	Case hardening steel	R683 PART 11 GR 14	AISI	9310		1130–1470	780	NA	8
18	Case hardening steel	R683 PART 11 GR 15				1270–1620	880	NA	7
19									

TABLE E.11. Mechanical Properties of Stainless Steels, Austenitic

Material Number	Commercial Name	ISO Designation	USA Standard	USA Designation	DIN Designation	Ultimate Strength (MPa)	Yield Strength (MPa)	Fatigue Strength (MPa)	Elongation (%)	Notes
1	Stainless steels-austenitic	683/13 GR 10	AISI	304L	X2CrNi18 9	440	180	NA	40	
2	Stainless steels-austenitic	683/13 GR 15	AISI	321	X12CrNiTi1 8 9	490	210	NA	35	
3	Stainlness steels-austenitic	683/13 GR 16	AISI	347	X10CrNiNb1 8 9	490	210	NA	35	
4	Stainless steels-austenitic	683/13 GR 11	AISI	304	X5CrNi18 9	490–690	200	NA	40	
5	Stainless steels-austenitic	683/13 GR 12	AISI	302	X12CrNi 18 8	490–690	210	40	35	S
6	Stainless steels-austenitic	683/13 GR 17	AISI	303	X12CrNiS 18 8	490–690	210	NA	35	
7	Stainless steels-austenitic	683/13 GR 13	AISI	305	X5CrNi19 1 1	490–690	180	NA	40	
8	Stainless steels-austenitic	683/13 GR 14	AISI	301	DIN 17224 X12CrNi 17 7	590–780	220	NA	NA	
9	Stainless steels-austenitic	683/13 GR 19	AISI	316L	X2CrNiMo 18 10	440–660	200	270	40	
10	Stainless steels-austenitic	683/13 GR 20	AISI	316	X5CrNiMo 18 10	490–690	210	270	40	
11	Stainless steels-austenitic	683/13 GR 21			X10CrNiMoT i18 10	490–690	220	270	35	
12	Stainless steel	ISO683/13 GR 23			X10CrNiMoN b18 10	490–690	220	270	35	
13	Stainless austenitic steel	GRADE 19A	AISI	316L	X2CrNiMo 18 12	440	200	270	40	

No.	Material	Specification	Standard	AISI No.	Designation				
14	Stainless austenitic steel	GRADE 20A	AISI	316	X5CrNiMo 18 12	490	210	270	40
15	Stainless austenitic steel				X10CrNiMoT i18 12	490–690	220	270	35
16	Stainless austenitic steel	GRADE 23A	AISI			490–690	220	NA	35
17	Stainless austenitic steel	GRADE 24	AISI	317	X2CrNiMo 18 16	490–690	200	NA	35
18	Stainless austenitic steel	GRADE A-3	AISI	202	X8CrMnNi 18 9	640–830	300	NA	40
19									
20	Stainless steels-austenitic	683/13 GR 15	AISI	321	X12CrNiTi1 8 9	490			
21	Stainless steels-austenitic	683/13 GR 16	AISI	347	X10CrNiNb1 8 9	490			
22	Stainless steels-austenitic	683/13 GR 11	AISI	304	X6CrNi18 9	490			
23	Stainless steels-austenic	683/13 GR 12	AISI	302	X12CrNi 18 8	490			
24	Stainless steels-austenitic	683/13 GR 17	AISI	303	X12CrNiS 18 8	490			
25	Stainless steels-austenitic	683/13 GR 13	AISI	305	X5CrNi19 1 1	490			
26	Stainless steels-austenitic	683/13 GR 14	AISI	301	DIN 17224 X12CrNi 17 7	590			
27	Stainless steels-austenitic	683/13 GR 19	AISI	316L	X2CrNiMo 18 10	440			
28	Stainless steels-austenitic	683/13 GR 20	AISI	316	X5CrNiMo 18 10	490			
29	Stainless steels-austenitic	683/13 GR 21			X10C2NiMoT i18 10	490			

TABLE E.11. (*Continued*)

Material Number	Commercial Name	ISO Designation	USA Standard	USA Designation	DIN Designation	Ultimate Strength (MPa)	Yield Strength (MPa)	Fatigue Strength (MPa)	Elongation (%)	Notes
30	Stainless steel	ISO683/13-GR 23			X10CrNiMoN b18 10	490–690	220	270	35	
31	Stainless austenitic steel	GRADE 19A	AISI	316L	X2CrNiMo 18 12	440	200	270	40	
32	Stainless austenitic steel	GRADE 20A	AISI	316	X5CrNiMo 18 12	490	210	270	40	
33	Stainless austenitic steel				X10CrNiMoT il8 12	490	220	270	35	
34	Stainless austenitic steel									
35										
36										

TABLE E.12. Mechanical Properties of Spring Steels

Material Number	Commercial Name	ISO Designation	USA Standard	USA Designation	DIN Designation	Ultimate Strength (MPa)	Yield Strength (MPa)	Fatigue Strength (MPa)	Elongation (%)	Notes
1	Spring steels	683 PART 1 4 GR 1	AISI	1074 1080	D75-2	1180	880	NA	6	
2	Spring steels	683 PART 1 4 GR 2				1180	880	NA	6	
3	Spring steels	683 PART 1 4 GR 3			46Si7	1270	1080	NA	6	
4	Spring steels	683 PART 1 4 GR 4			51Si7	1320	1130	550	6	
5	Spring steels	683 PART 1 4 GR 5	AISI	9255	55Si7	1320	1130	550	6	
6	Spring steels	683 PART 1 4 GR 6	AISI	9260		1370	1180	NA	5	
7	Spring steels	683 PART 1 4 GR 7			60Cr7	1370	1180	650	5	
8	Spring steels	683 PART 1 4 GR 8	AISI	5155	55Cr3	1370	1180	650	6	
9	Spring steels	683 PART 1 4 GR 9	AISI	5160		1370	1180	650	5	
10	Spring steels	GRADE 10	AISI	51B60		1370	1180	650	6	
11	Spring steels	GRADE 11	AISI	—		1370	1180	650	6	
12	Spring steels	GRADE 12	AISI	4161		1370	1180	650	6	
13	Spring steels	GRADE 13	AISI	6150	50CrV4	1370	1180	650	6	
14	Spring steels	GRADE 14	AISI	—	51CrMoV4	1370	1180	650	6	
15										

TABLE E.13. Mechanical Properties of Flame, Induction, Hardening Steels

Material Number	Commercial Name	ISO Designation	USA Standard	USA Designation	DIN Designation	Ultimate Strength (MPa)	Yield Strength (MPa)	Fatigue Strength (MPa)	Elongation (%)	Notes
1	Flame and induction hardening	683/12 GR 1	AISI	1035	Cf35	620–760	420	NA	17	
2	Flame and induction hardening	683/12 GR 2	AISI	1040		660–800	450	NA	16	
3	Flame and induction hardening	683/12 GR 3	AISI	1045	Cf45	700–840	400	NA	14	
4	Flame and induction hardening	683/12 GR 4	AISI	1050		740–880	510	NA	13	
5	Flame and induction hardening	683/12 GR 5	AISI	1055	Cf53	740–880	510	NA	12	
6	Flame and induction hardening	683/12 GR 6	AISI	5145	45Cr2	880–1080	640	NA	12	
7	Flame and induction hardening	683/12 GR 7	AISI	5135	38Cr4	930–1130	740	NA	11	
8	Flame and induction hardening	683/12 GR 8	AISI	5140	42Cr4	980–1180	780	NA	11	

9	Flame and induction hardening	683/12	AISI	4140	41CrMo4	1080–1270	880	NA	10
10	Flame and induction hardening	683/12 GR 10	AISI	8640 8740		1030–1230	830	NA	10
11	Flame and induction hardening	683/13 GR 11				1030–1230	830	NA	10
12									

APPENDIX F

PROPERTIES OF SECTIONS

TABLE F.1. Properties of Sections

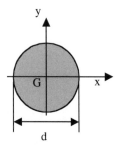

$A = \pi d^2/4$

$I_x = I_y = \pi d^4/64, \; c = d/2$

$I_p = \pi d^4/32, \; R_{max} = d/2$

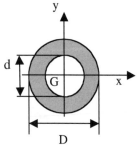

$A = \pi(D^2 - d^2)/4$

$I_x = I_y = \pi(D^4 - d^4)/64, \; c = D/2$

$I_p = \pi(D^4 - d^4)/32, \; R_{max} = D/2$

TABLE F.1. (*Continued*)

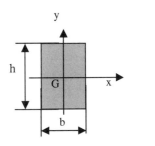

$$A = bh$$

$$I_x = bh^3/12, \ c = h/2$$

$$I_y = hb^3/12, \ c = b/2$$

$$I_p = I_x + I_y, \ R_{max} = (h^2 + b^2)/4$$

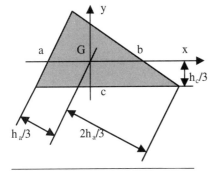

$$A = ch_c/2$$

$$I_x = ch_c^3/36$$

$$I_p = ch_c^3/12$$

[a]A = section area, I_x = section area moment of inertia about the x-axis, I_p = section polar area moment of inertia about G, G = section area center.

INDEX